Environmental Pollution

Volume 25

Series Editors
Brian J. Alloway
University of Reading Department of Soil Science, Reading, United Kingdom

Jack T. Trevors
Ontario Agricultural College, University of Guelph School of Environmental Sciences, Guelph, Ontario, Canada

The Environmental Pollution book series includes current, comprehensive texts on critical national and global environmental issues useful to scientists in academia, industry and government from diverse disciplines. These include water, air, and soil pollution, organic and inorganic pollution, risk assessment, human and environmental health, environmental biotechnology, global ecology, mathematics and computing as related to environmental pollution, environmental modelling, environmental chemistry and physics, biology, toxicology, conservation and biodiversity, agricultural sciences, pesticides, environmental engineering, bioremediation/biorestoration, and environmental economics. Environmental problems and solutions are complex and interrelated. Complex problems often require complex solutions. The linkage of many disciplines can result in new approaches to old and new environmental problems as well as pollution prevention. This knowledge will assist in understanding, maintaining and improving the biosphere in which we live.

Proposals for this book series can be sent to either of the Series' Editors:

Brian Alloway at B.J.Alloway@reading.ac.uk

Jack Trevors at jtrevors@uoguelph.ca

or the Publishing Editor, Paul Roos, at Paul.Roos@springer.com

More information about this series at http://www.springer.com/series/5929

Wim de Vries • Jean-Paul Hettelingh
Maximilian Posch
Editors

Critical Loads and Dynamic Risk Assessments

Nitrogen, Acidity and Metals in Terrestrial and Aquatic Ecosystems

Springer

Editors
Wim de Vries
Alterra Wageningen University
and Research Centre
Wageningen
The Netherlands

Jean-Paul Hettelingh
RIVM – Coordination Centre for Effects
Bilthoven
The Netherlands

Maximilian Posch
RIVM – Coordination Centre for Effects
Bilthoven
The Netherlands

ISSN 1566-0745
Environmental Pollution
ISBN 978-94-017-9507-4
ISBN 978-94-017-9508-1 (eBook)
DOI 10.1007/978-94-017-9508-1

Library of Congress Control Number: 2014960242

Springer Dordrecht Heidelberg New York London

Printed on acid-free paper

Springer is part of Springer Science+Business Media (www.springer.com)

Foreword

Initial concern on the negative effects of acidic deposition on natural ecosystems was voiced in Scandinavia during the 1970s, where fish death was attributed to surface water acidification caused by elevated sulphur deposition. In the beginning of the 1980s, extensive forest damage in Central Europe linked to high levels of acidic deposition, led to wide-spread public and political concern for the vitality of forests. In 1979 member states of the UN Economic Commission for Europe adopted the Convention on Long-range Transboundary Air Pollution. This Convention mobilised scientific research—including work by many authors in this book—for the support of air pollution abatement policies to mitigate impacts on human health and the environment.

During the 1980s critical loads were derived for acidification and nitrogen-induced eutrophication as early warning indicators of excessive atmospheric deposition of sulphur and nitrogen. Near the end of the 1990s, critical loads were also developed to assess the adverse effects of heavy metal deposition. In concert, various dynamic models were developed to evaluate the temporal impacts of atmospheric deposition of acidic and eutrophying compounds and heavy metals on ecosystems.

This book provides an overview of the development of critical load research for terrestrial and aquatic ecosystems, and applications in support of air pollution abatement policies during the past 30 years. It consists of 26 chapters, divided over five major themes:

1. Assessment of indicators and thresholds for air pollutant impacts;
2. Empirical and process-model based critical loads of nitrogen, acidity and metals for characteristic ecosystems;
3. Dynamic modelling approaches for the temporal assessment of abiotic changes owing to atmospheric deposition of nitrogen, sulphur and heavy metals at the site-specific scale;
4. Critical loads and dynamic model applications on a regional scale in Europe, Asia and North America;
5. Integrated assessment of changes in environmental quality and ecosystem services as affected by air pollution under a changing climate.

The book concludes with a review of the critical loads approach in the context of uncertainty and of possible future scientific and policy directions.

This book marks the successful collaboration over more than 30 years, between researchers in the field of environmental impacts of air pollution in general, and between RIVM and Alterra in particular. We congratulate the authors of this scientific record of achievements, and expect that this book will serve as a basis to stimulate effects-based research for the challenges ahead.

Prof.dr. A.N. van der Zande
Director General
National Institute for Public Health
and the Environment (RIVM)

Ir. K. Slingerland
Director General
Alterra Wageningen University
and Research Centre (WUR)

Acknowledgements

The methods and results presented in this book would be inexistent without the close collaboration with communities of scientists that work under the Effects Programme of the UNECE Convention on Long-range Transboundary Air Pollution (LRTAP Convention). Of similar importance are the research programmes of the European Union, North America and China, that enhanced the collaboration with, and between, institutions and individuals from various fields of environmental science that resulted in contributions to this book.

The stimulus of 25 years of bilateral collaboration between the Coordination Centre for Effects (CCE) at the National Institute for Public Health and the Environment (RIVM) and the Alterra Research Centre at Wageningen University and Research Centre (WUR), as well as the multi-lateral collaboration with the colleagues implied above, has been instrumental both to the idea, about five years ago, and the realisation of this work.

This book would not have materialized without the support of many. The editors wish to acknowledge in particular:

- The Directorate for Climate, Air and Noise of the Dutch Ministry of Infrastructure and the Environment for their continued support both of the CCE and Alterra activities enabling the design and compilation of knowledge on impacts of air pollution for European and national environmental policies,
- The Ministry of Economic Affairs that funded part of the work under the strategic research program "Sustainable spatial development of ecosystems, landscapes, seas and regions",
- The management of RIVM and Alterra for providing resources needed for the accomplishment of this effort,
- The Working Group on Effects and Task Force of the International Cooperative Programme on the Modelling and Mapping of Critical Levels and Loads and Air Pollution Effects, Risks and Trends (ICP M&M) under the LRTAP Convention for inspiration, collaboration, and relevant application options for the support of European air pollution policies,
- The EMEP Meteorological Synthesizing Centres East and West and the EMEP Centre for Integrated Assessment at the International Institute for Applied Systems Analysis for their collaboration in the field of atmospheric dispersion and integrated assessment modelling,

- The European Commission's LIFE III Preparatory Projects, for co-funding the participation of the CCE in the European Consortium for Modelling Air pollution and Climate Strategies (EC4MACS: *LIFE06 ENV/AT/PREP/06),*
- The co-funding of Alterra and CCE in the ECLAIRE project Effects of Climate Change on Air Pollution and Response Strategies for European Ecosystems (ECLAIRE: grant agreement no 282910) under the European Union's 7th Framework Programme for Research and Technological Development.

The editors are very grateful to have benefited from the skills of Jan Cees Voogd of Alterra in expertly handling a variety of tools for the editing of text, references and graphics.

Finally, authors and co-authors of the chapters are thanked for their patience with the editors of this book.

Contents

Contributors

Julian Aherne Environmental and Resource Studies, Trent University, Peterborough, ON, Canada

Didier Alard UMR BioGeco Université Bordeaux 1, Talence, France

Edith B. Allen Department of Botany and Plant Sciences and Center for Conservation Biology, University of California, Riverside, CA, USA

Michael Ashmore Environment Department, University of York, Heslington, York, UK

Dick Bal Ministry of Economic Affairs, Den Haag, The Netherlands

Jill S. Baron US Geological Survey, Fort Collins, CO, USA

Salim Belyazid Belyazid Consulting & Communication, Malmö, Sweden

Department of Chemical Engineering, Lund University, Lund, Sweden

Roland Bobbink B-Ware Research Centre, Radboud University Nijmegen, Nijmegen, The Netherlands

Luc T. C. Bonten Alterra, Wageningen University and Research Centre, Wageningen, The Netherlands

William D. Bowman University of Colorado, Boulder, CO, USA

Sabine Braun Institute for Applied Plant Biology, Schönenbuch, Switzerland

Adam Butler Biomathematics & Statistics Scotland, JCMB, Edinburgh, Scotland, UK

Christopher M. Clark US EPA, Washington DC, USA

Emmanuel Corket UMR BioGeco Université Bordeaux 1, Talence, France

B. Jack Cosby Centre for Ecology and Hydrology, Environment Centre Wales, Bangor, Gwynedd, United Kingdom

Department of Environmental Sciences, University of Virginia, Charlottesville, VA, USA

Chris J. Curtis School of Geography, Archaeology and Environmental Studies, University of the Witwatersrand, Johannesburg, South Africa

Wim de Vries Alterra, Wageningen University and Research Centre, Wageningen, The Netherlands

Heleen A. de Wit Norwegian Institute for Water Research, Oslo, Norway

Hugo Denier van der Gon Climate, Air and Sustainability, TNO, Utrecht, The Netherlands

Robin L. Dennis US EPA, Research Triangle Park, NC, USA

Thomas Dirnböck Environment Agency Austria, Vienna, Austria

Charles T. Driscoll Department of Civil and Environmental Engineering, Syracuse University, Syracuse, NY, USA

Lei Duan School of Environment, Tsinghua University, Beijing, P. R. China

Bridget Emmett Centre for Ecology and Hydrology, Environment Centre Wales, Bangor, UK

Chris D. Evans Centre for Ecology & Hydrology, Environment Centre Wales, Bangor, UK

Mark E. Fenn USDA Forest Service, Riverside, CA, USA

Jens Fölster Swedish University of Agricultural Sciences, Uppsala, Sweden

Martin Forsius Finnish Environment Institute (SYKE), Helsinki, Finland

Noémie Gaudio Université de Toulouse; INP, UPS; EcoLab (Laboratoire Ecologie Fonctionnelle et Environnement), ENSAT, Castanet Tolosan, France

CNRS, EcoLab, Castanet Tolosan, France

Thierry Gauquelin IMBE, Aix-Marseille University, Provence, France

Jean-Claude Gégout AgroParisTech, UMR 1092 INRA-AgroParistech, Laboratoire d'Etude des Ressources Forêt-Bois (LERFoB), Nancy, France

Linda H. Geiser USDA Forest Service, Corvallis, OR, USA

Frank S. Gilliam Marshall University, Huntington, WV, USA

Christine L. Goodale Cornell University, Ithaca, NY, USA

Tara L. Greaver US EPA, Research Triangle Park, NC, USA

Jan E. Groenenberg Alterra, Wageningen University and Research Centre, Wageningen, The Netherlands

Jane Hall Centre for Ecology and Hydrology, Environment Centre Wales, Bangor, UK

Sharon J. Hall Arizona State University, Tempe, AZ, USA

Jiming Hao School of Environment, Tsinghua University, Beijing, P. R. China

Peter A. Henrys Centre for Ecology and Hydrology, Lancaster Environment Centre, Lancaster, UK

Jean-Paul Hettelingh Coordination Centre for Effects (CCE), RIVM, Bilthoven, The Netherlands

Ilia Ilyin EMEP-Meteorological Synthesizing Centre East, Moscow, Russia

Dean Jeffries Environment Canada, Burlington, ON, Canada

Johannes Kros Alterra, Wageningen University and Research Centre, Wageningen, The Netherlands

Daniel Kurz EKG Geo-Science, Bern, Switzerland

Thorjørn Larssen Norwegian Institute for Water Research (NIVA), Oslo, Norway

Anne-Christine Le Gall LRTAP-Task Force Modelling and Mapping, INERIS, Verneuil-en-Halatte, France

Sophie Leguédois Université de Toulouse; INP, UPS; EcoLab (Laboratoire Ecologie Fonctionnelle et Environnement), ENSAT, Castanet Tolosan, France

CNRS, EcoLab, Castanet Tolosan, France

Erik A. Lilleskov USDA Forest Service, Houghton, MI, USA

Lingli Liu US EPA, Research Triangle Park, NC, USA

Stephen Lofts Centre for Ecology and Hydrology, Lancaster Environment Centre, Bailrigg, Lancaster, UK

Espen Lund Norwegian Institute for Water Research (NIVA), Oslo, Norway

Jason A. Lynch US EPA, Washington DC, USA

Rob Maas LRTAP-Task Force Integrated Assessment Modelling, RIVM, Bilthoven, The Netherlands

Arnaud Mansat Université de Toulouse; INP, UPS; EcoLab (Laboratoire Ecologie Fonctionnelle et Environnement), ENSAT, Castanet Tolosan, France

CNRS, EcoLab, Castanet Tolosan, France

Thomas Meixner Department of Hydrology and Water Resources, University of Arizona, Tucson, AZ, USA

Filip Moldan IVL Swedish Environmental Research Institute, Göteborg, Sweden

Janet P. Mol-Dijkstra Alterra, Wageningen University and Research Centre, Wageningen, The Netherlands

Knute J. Nadelhoffer University of Michigan, Ann Arbor, MI, USA

Bengt Nihlgård Department of Chemical Engineering, University of Lund, Lund, Sweden

Annika Nordin Umeå Plant Science Centre, Department of Forest Genetics and Plant Physiology, Swedish University of Agricultural Sciences, Umeå, Sweden

Carole Obeidy Université de Toulouse; INP, UPS; EcoLab (Laboratoire Ecologie Fonctionnelle et Environnement), ENSAT, Castanet Tolosan, France

CNRS, EcoLab, Castanet Tolosan, France

Linda H. Pardo USDA Forest Service, Burlington, VT, USA

Jean-Paul Party Sol-Conseil, Strasbourg, France

Tomasz Pecka Institute of Environmental Protection—National Research Institute, Warsaw, Poland

Steven J. Perakis US Geological Survey, Corvallis, OR, USA

Maximilian Posch Coordination Centre for Effects (CCE), RIVM, Bilthoven, The Netherlands

Anne Probst Université de Toulouse; INP, UPS; EcoLab (Laboratoire Ecologie Fonctionnelle et Environnement), ENSAT, Castanet Tolosan, France

CNRS, EcoLab, Castanet Tolosan, France

Leela E. Rao Department of Environmental Science and Center for Conservation Biology, University of California, Riverside, CA, USA

Gert Jan Reinds Alterra, Wageningen University and Research Centre, Wageningen, The Netherlands

Molly J. Robin-Abbott USDA Forest Service, Burlington, VT, USA

Ed C. Rowe Centre for Ecology and Hydrology, Environment Centre Wales, Bangor, UK

Thomas Scheuschner OEKO-DATA Strausberg, Strausberg, Germany

Angela Schlutow OEKO-DATA Strausberg, Strausberg, Germany

Kirsten Schütz Institute for Applied Plant Biology, Schönenbuch, Switzerland

Gudrun Schütze Umweltbundesamt, Dessau, Germany

Laura Shotbolt Geography Department, Queen Mary University of London, London, UK

Jaap Slootweg Coordination Centre for Effects (CCE), RIVM, Bilthoven, The Netherlands

Simon M. Smart Centre for Ecology and Hydrology, Lancaster Environment Centre, Lancaster, UK

Carly J. Stevens Lancaster Environment Centre, Lancaster University, Lancaster, UK

John L. Stoddard US EPA, Corvallis, OR, USA

Joachim Strengbom Department of Ecology, Swedish University of Agricultural Sciences, Uppsala, Sweden

Timothy J. Sullivan E&S Environmental Chemistry, Corvallis, OR, USA

Harald U. Sverdrup Department of Chemical Engineering, Lund University, Lund, Sweden

Industrial Engineering, University of Iceland, Reykjavik, Iceland

Edward Tipping Centre for Ecology and Hydrology, Lancaster Environment Centre, Bailrigg, Lancaster, UK

Hilde Tomassen B-Ware Research Centre, Radboud University Nijmegen, Nijmegen, The Netherlands

Oleg Travnikov EMEP-Meteorological Synthesizing Centre East, Moscow, Russia

Leon van den Berg B-Ware Research Centre, Radboud University Nijmegen, Nijmegen, The Netherlands

Aquatic Ecology and Environmental Biology, Radboud University Nijmegen, Nijmegen, The Netherlands

Han F. van Dobben Alterra, Wageningen University and Research Centre, Wageningen, The Netherlands

Arjen van Hinsberg Netherlands Environmental Assessment Agency (PBL), Bilthoven, The Netherlands

G. W. Wieger Wamelink Alterra, Wageningen University and Research Centre, Wageningen, The Netherlands

Kathleen C. Weathers Cary Institute of Ecosystem Studies, Millbrook, NY, USA

Maaike Weijters B-Ware Research Centre, Radboud University Nijmegen, Nijmegen, The Netherlands

Henricus J.J. Wieggers Alterra, Wageningen University and Research Centre, Wageningen, The Netherlands

Richard F. Wright Norwegian Institute for Water Research (NIVA), Oslo, Norway

Fengming Yuan Environmental Sciences Division, Oak Ridge National Laboratory, Oak Ridge, TN, USA

Yu Zhao School of the Environment, Nanjing University, Nanjing, P. R. China

Qingtao Zhou Department of Civil and Environmental Engineering, Syracuse University, Syracuse, NY, USA

List of Abbreviations

AAE	Average Accumulated Exceedance
ADI	Acceptable Daily Intake
AGH scenario	Above-Ground Harvesting scenario
ALARA	As Low As Reasonably Achievable
Alterra	Research Centre at Wageningen University (The Netherlands)
ALTM	Adirondack Long-Term Monitoring
AMF	Arbuscular Mycorrhizal Fungi
ANC	Acid Neutralizing Capacity (in surface waters)
AOT40	Accumulated hourly ozone concentration Over a Threshold of 40 ppb
BAU scenario	Business As Usual scenario
Bc	Base cations (Ca+Mg+K)
BC	Base cations (Bc+Na)
Bc/Al	Base cation to Aluminium ratio (in soil solution)
BERN	Bioindication for Ecosystem Regeneration towards Natural conditions model
BLM	Biotic Ligand Model
C/N	Carbon to Nitrogen ratio
Ca/Al	Calcium to Aluminium ratio
CBD	Convention on Biological Diversity
CC scenario	Maximum change in future climate (but no change in future forest management) scenario
CCE	Coordination Centre for Effects at RIVM (The Netherlands); Programme Centre of the ICP M&M
CCLU scenario	Maximum change in future climate, maximum utilisation of forests scenario
CEC	Cation Exchange Capacity (of a soil)
CEH	Centre for Ecology and Hydrology (United Kingdom)
CHUM-AM	CHemistry of the Uplands Model-Annual Metals model
CIS	Commonwealth of Independent States
CL	Critical Load (plural: CLs)
CL(M)	Critical Load of a (heavy) Metal

CLA	Critical Load of Acidity
$CL_{emp}N$	Empirical Critical Load of Nitrogen
CLE scenario	Current LEgislation scenario of emission (and related deposition) reductions
CLF	Critical Load Function (in the context of HM: 'Critical Limit Function')
$CL_{max}N$	Maximum Critical Load of Nitrogen
$CL_{max}S$	Maximum Critical Load of Sulphur
$CL_{min}N$	Minimum Critical Load of Nitrogen
$CL_{nut}N$	Critical Load of nutrient Nitrogen
CLRTAP	see LRTAP Convention
CMAQ	Community Multiscale Air Quality model
COD	Chemical Oxygen Demand
CORINE	CoORdination of INformation on the Environment of the EU
DayCent	Daily Century model
DDT	Damage Delay Time
DECOMP	Model of litter DECOMPosition (part of ForSAFE)
DFP	Distribution Function of the Possibility (of the occurrence of a plant species)
DM	Dynamic Model(ling)
DOC	Dissolved Organic Carbon
DOM	Dissolved Organic Matter
DPSIR	Driving forces, Pressures, State, Impact, Response integrated assessment
D-R	Dose-Response
EC	European Commission
ECHAM/OPYC3	A model of general circulation of atmosphere and ocean
ECLAIRE	Effects of CLimate change on Air pollution Impacts and Response strategies for European Ecosystems (EU Framework Programme 7)
EEA	European Environment Agency
EFI	European Forest Institute (Finland)
EFISCEN	European Forest Information SCENario
EIA	Environmental Impact Analysis
EMEP	The European Monitoring and Evaluation Programme (under the LRTAP Convention)
EMERGE	European Mountain lake Ecosystems: Regionalisation, diaGnostic & socio-economic Evaluation (EU Framework Programme 5)
EMF	EctoMycorrhizal Fungi
ESQUAD	European Soil & sea QUality due to Atmospheric Deposition project
ETD	Enhanced Trickle Down model
EU	European Union
EUgrow	EUropean forest growth model

EUNIS	EUropean Nature Information System
EXMAN	EXperimental MANipulation of forest ecosystems
FAB	First-order Acidity Balance model
FI scenario (FIHM)	Full Implementation (of the Aarhus HM protocol)
FIAM	Free Ion Activity Model
FOCUS	FOcal Center Utility Study (USA)
ForSAFE	Integrated process-oriented forest model for long-term sustainability assessments
GAINS	Greenhouse gas and Air pollution INteractions and Synergies model
GBMOVE	Dynamic model predicting plant species changes
GP scenario	Gothenburg Protocol emission reductions agreed under the LRTAP Convention
HadAM3	Hadley Centre Atmospheric Model 3
HM	Heavy Metal
IAM	Integrated Assessment Model
IAP	Institute for Applied Plant Physiology (Schönenbuch, Switzerland)
ICP	International Cooperative Programme (under the LRTAP Convention)
ICP-F	ICP on Assessment and Monitoring Air Pollution Effects on Forests
ICP-IM	ICP on Integrated Monitoring of Air Pollution Effects on Ecosystems
ICP-M&M	ICP on Modelling and Mapping Critical Levels and Loads and Air Pollution Effects, Risks and Trends
ICP-V	ICP on Assessment and Monitoring Effects of Air Pollution on Natural Vegetation and Crops
ICP-W	ICP on Assessment and Monitoring Effects of Air Pollution on Rivers and Lakes
ILWAS	Integrated Lake/Watershed Acidification Study (USA)
IMAGE	Integrated Model to Assess the Global Environment
IPCC	Intergovernmental Panel on Climate Change
IPCC-AR4	Fourth Assessment Report of the IPCC
IUCN	International Union for the Conservation of Nature
IVL	Swedish Environmental Research Institute (Sweden)
JRC	Joint Research Centre of the European Commission
Level II plots	Intensive monitoring sites of the ICP-F
LOEC	Lowest Observed Effects Concentration
LRTAP	Long-range Transboundary Air Pollution
LU scenario	No change in future climate, maximum utilisation of the forest
MADOC	Model of Acidity Dynamics and Organic Carbon
MAGIC	Model of Acidification of Groundwater In Catchments
MEA	Millennium Ecosystem Assessment

MELA	Management oriented large scale forestry model
MFR scenario	Maximum Feasible emission (or related deposition) Reduction scenario
MOVE	MOdel for terrestrial VEgetation
MPC	Maximum Permissible Concentration
MPLS	Mean Phreatic Level in Spring
MSC-E	Meteorological Synthesizing Centre-East of EMEP under the LRTAP Convention (Moscow)
MSC-W	Meteorological Synthesizing Centre-West of EMEP under the LRTAP Convention (Oslo)
MultiMOVE	A package of niche models for British vegetation
NAPAP	National Acid Precipitation Assessment Program (USA)
Natura 2000	Network of EU nature protection areas established under the 1992 habitats directive
NEC	National Emission Ceiling directive of the EU
NEG/ECP	New England Governors and Eastern Canadian Premiers
NFC	National Focal Centre (of the ICP-M&M)
NIJOS	Norwegian Institute for Soil and Forest Inventory
NITREX	NITRogen saturation EXperiment
NOEC	No Observed Effect Concentration
NRCS	Natural Resources Conservation Service (USA)
NTM	Natuur Technisch Model ('Nature Technical Model' *in Dutch*)
NUCSAM	NUtrient Cycling and Soil Acidification Model
OECD	Organisation for Economic Co-operation and Development
ORCHESTRA	Objects Representing CHEmical Speciation and TRAnsport: an object-oriented framework for implementing chemical equilibrium models
PELCOM	Pan-European Land COver Monitoring
PLA	Projected Leaf Area
PNEC	Predicted No Effect Concentration
PnET-BGC	Photosynthesis net and EvapoTranspiration-BioGeoChemistry
POC	Probability of OCcurrence
POD_Y	Phytotoxic Ozone Dose above a threshold of Y nmol m^{-2} s^{-1}
PROFILE	Steady-state soil chemistry model (including weathering)
PROPS	PRobability of Occurrence of Plant Species model
PULSE	A soil hydrology model
RAD	Reactive Airways Diseases
RAINS	Regional Acidification INformation and Simulation model
RDT	Recovery Delay Time
RENECOFOR	National network of forest health survey by the National Forest Office (France)
RESAM	REgional Soil Acidification Model
RIVM	National Institute for Public Health and the Environment (The Netherlands)
RMCC	Research and Monitoring Coordinating Committee (Canada)

SAFE	Soil Acidification in Forest Ecosystems model
SBH scenario	Stems plus Branches Harvesting scenario
SDMM	Simple Dynamic Model for Metals
SEI	Stockholm Environment Institute
SLU	Swedish University of Agricultural Sciences
SMART	Simulation Model for Acidification's Regional Trends
SMB	Simple Mass Balance model
SOH scenario	Stem Only Harvesting scenario
SOM	Soil Organic Matter
SOMO35	Sum Of Maximum daily 8-hour average O_3 levels above 35 ppb
SSD	Species Sensitivities Distribution
SSMB	Steady-State Mass Balance model
SSWC	Steady-State Water Chemistry model
SUMO	Model for vegetation forecast
SYKE	Finnish Environment Institute
TEEB	The Economics of Ecosystems and Biodiversity
TFH	Task Force on Health under the LRTAP Convention
TL	Target Load
TMDL	Total Maximum Daily Load
TOC	Total Organic Carbon
TSAP	Thematic Strategy on Air Pollution of the European Union
UBA	Umweltbundesamt (Federal Environmental Agency, Germany)
UNECE	United Nations Economic Commission for Europe
UNESCO	United Nations Educational, Scientific and Cultural Organization
USDA	United States Department of Agriculture
US-EPA	United States Environmental Protection Agency
USFS	United States Forest Service
VEG (or Veg)	Model for ground vegetation
VSD (VSD+)	Very Simple Dynamic model (plus)
WGE	Working Group on Effects under the LRTAP Convention
WHAM	Windermere Humic Aqueous Model
WHO	World Health Organisation
WTH scenario	Whole Tree Harvesting scenario

About the Editors

Wim de Vries (1959) is a senior research scientist at Alterra, part of Wageningen University and Research Centre. He is also professor at the Environmental Systems Analysis Group of Wageningen University, where he holds the chair "Integrated nitrogen impact modelling".

He holds both a Ph.D. (cum laude) and MSc. in Soil Chemistry at Wageningen University. From 1982–1983 he worked at the South Australian Department of Agriculture, Adelaide, South Australia, where he investigated the effects of land use changes on the salinity of a drinking water reservoir. He then moved to the Netherlands Soil Survey Institute (1983–1988) where he studied soil acidification in response to changed nitrogen and sulphur inputs. From 1989–1999 he worked at the DLO-Winand Staring Centre, mainly concentrating on critical loads of nitrogen, acidity and heavy metals in view of effects on terrestrial ecosystems. Since 2000, he works at Alterra. He has been project leader of numerous multi-partner national and international projects on integrated assessment of agricultural management on air, soil and water quality, on critical loads of nitrogen, acidity and metals on terrestrial ecosystems and on impacts of air quality and climate change on ecosystem services.

His research is currently organized around impacts of the changes in land management, air quality and climate on soil and water quality, biodiversity, forest growth and carbon sequestration. His specific expertise is related to the development, validation and application of soil models at various scales.

Alterra Wageningen University, Wageningen, The Netherlands

Jean-Paul Hettelingh (Amsterdam, 1954) directs the Coordination Centre for Effects (CCE) located at the Dutch National Institute for Public Health and the Environment (RIVM) since 1990. With his CCE colleagues he collaborates with a European network of scientific and policy institutions under the LRTAP Convention of the United Nations (UNECE) to support assessments of ecosystem effects of European air pollution policies.

His academic background is econometrician (MSc.) and he holds a Ph.D. in economics at the Free University (Amsterdam) with a focus on uncertainty in the modelling of regional environmental systems. He started with research positions in the field of integrated assessment modelling (1978–1985) at the Institute for Environmental Studies at the Free University. He then (1986–1989) moved to the International Institute for Systems Analysis (IIASA, Austria) to contribute to the development of the Regional Acidification, Information and Simulation (RAINS) model. He performed parts of his Ph.D. work at IIASA, and at the Environmental Sciences Division of Oak Ridge National Laboratory (Oak Ridge, Tennessee, USA). Since 1989 he joined RIVM from where he held a part time professorship in environmetrics (1997–2002) at the Institute of Environmental Sciences (CML) at Leiden University (Leiden, The Netherlands).

His research focus is on impact analysis in broad scale integrated assessment models. Results are published from work under the Asian Development Bank, the World Bank, the UNECE, the European Environment Agency and the European Union.

RIVM—Coordination Centre for Effects, Bilthoven, The Netherlands

Maximilian Posch (Vienna, 1953) is a senior researcher at the Coordination Centre for Effects (CCE) located at the Dutch National Institute for Public Health and the Environment (RIVM). The CCE is the data and modelling centre of the ICP Modelling & Mapping under the UNECE LRTAP Convention.

He holds a Ph.D. in Physics and a MSc. in Mathematics from the Technical University of Vienna. From 1981–1989 he worked at the International Institute for Applied Systems Analysis (IIASA) in Laxenburg (Austria), first on the minimisation of air pollution by district heating, funded by the City of Vienna, and then on the design and development of the integrated assessment model of acidification in Europe (the RAINS/GAINS model). During 1987–1989 he also worked for the Austrian Institute of Technology (formerly: Austrian Research Centers), finishing his Ph.D. in Theoretical Physics. From 1990–1994 he worked at the Finnish Water and Environment Research Institute (now: Finnish Environment Institute, SYKE) in Helsinki on the environmental impacts of acid deposition and agricultural practices. There he also participated in many collaborative projects funded by the Nordic Council of Ministers, developing the critical load methodology and mapping in the Nordic countries. Since 1995 he works at the RIVM.

His research focuses on developing and implementing tools and models dealing with the effects of air pollutants and climate change on terrestrial and aquatic ecosystems on a site and regional scale, and the transfer of that knowledge to policy making.

RIVM—Coordination Centre for Effects, Bilthoven, The Netherlands

Chapter 1
The History and Current State of Critical Loads and Dynamic Modelling Assessments

Wim de Vries, Jean-Paul Hettelingh and Maximilian Posch

1.1 Introduction

Concern of Elevated Air Pollutant Concentrations: Environmental conditions are rapidly changing on a global scale since the beginning of the industrial times. Since the late nineteenth century until its peak in the mid-1980s, sulphur (S) emissions, mostly in the form of sulphur dioxide (SO_2), increased in Europe by a factor of seven, while nitrogen (N) emissions, in the form of nitrogen oxides (NO_x) and ammonia (NH_3) increased by a factor near five, correlated with a similar increase in acid deposition (Schöpp et al. 2003). The concern for negative effects of acid deposition started in Scandinavia in the seventies (Odén 1976), where slow weathering bedrock and shallow soils render ecosystems sensitive, and widespread death of trout and salmon was attributed to the enhanced mobilisation and leaching of aluminium which damaged the fish (Baker and Schofield 1982).

Shortly afterwards, in the beginning of the eighties, the health and vitality of forest ecosystems became a subject of wide public and political concern due to the extensive forest damage observed in rural areas in Central Europe, which was connected to the high acid deposition (e.g Lammel 1984; Schütt et al. 1983). Forest condition in Europe has been monitored since the early 1980s because of concerns relating acid deposition to forest vitality (Fischer et al. 2007).

Initially the attention in the seventies focused on the adverse impacts of acid deposition induced by high S emissions. However, since the mid-1980s, attention changed towards the adverse impacts of elevated N deposition, due to declining S emissions and the high N emissions in Europe. Global emissions of NH_3 and NO_x strongly increased in the second half of the twenty-first century (Holland et al. 1999). Ammonia is volatilised from intensive agricultural systems, whereas

W. de Vries (✉)
Alterra Wageningen University and Research Centre, Wageningen, The Netherlands
e-mail: wim.devries@wur.nl

J.-P. Hettelingh · M. Posch
Coordination Centre for Effects (CCE), RIVM, Bilthoven, The Netherlands

W. de Vries et al. (eds.), *Critical Loads and Dynamic Risk Assessments*,
Environmental Pollution 25, DOI 10.1007/978-94-017-9508-1_1

nitrogen oxides originate mainly from burning of fossil fuel by traffic, industry and households (Truscott et al. 2005). Long- and short-range transport of these N compounds has resulted in increased atmospheric N deposition in many natural and semi-natural ecosystems across the world (Galloway et al. 2004). In addition to the different emission sources, oxidized and reduced nitrogen also differ from the point of view of atmospheric transport. NO_x deposition prevails in urban and industrial areas, whereas reduced NH_3 deposition clearly dominates in rural regions. Areas with high atmospheric N deposition (>20 kg N $ha^{-1}yr^{-1}$) are located mainly in central and western Europe, eastern United States and eastern Asia (e.g. Dentener et al. 2006; Galloway and Cowling 2002; Liu et al. 2013). In most regions with relatively high N depositions, a high proportion of the deposited N originates from NH_y (e.g. Asman et al. 1998; Fowler 2002; Sutton et al. 2008).

Critical Loads and Critical Levels of Air Pollutants: Concerning the effects of SO_2, NH_3 and NO_x on vegetation, two pathways can be distinguished: direct effects of the gas on the above-ground parts of the vegetation, and the indirect effects on the below-ground parts (roots), both in the form of eutrophication and acidification. For each of these effects, methods have been developed over the past decades to identify no-effect thresholds, called critical limits in view of direct effects and critical loads in view of indirect, soil-mediated effects.

Critical loads and critical levels of air pollutants have been used in support of effect-based air pollution abatement agreements (see e.g. Hettelingh et al. 2001, 2007). In Europe these include the National Emission Ceilings Directive (EC 2001; Hettelingh et al. 2013) and the revised protocol to abate acidification, eutrophication and ground-level ozone (Reis et al. 2012) under the Convention on Long-range Transboundary Air Pollution (LRTAP Convention; Sliggers and Kakebeeke 2004)

Critical loads, the focus of this book, are related to indirect, soil-mediated effects of elevated deposition and are defined as "a quantitative estimate of an exposure to one or more pollutants below which significant harmful effects on specified sensitive elements of the environment do not occur according to present knowledge" (Nilsson and Grennfelt 1988). Critical levels are defined as "the atmospheric concentrations of pollutants in the atmosphere above which adverse effects on receptors, such as human beings, plants, ecosystems or materials, may occur according to present knowledge" (see Mapping Manual, www.icpmapping.org). Note that the Mapping Manual has evolved from paper versions in 1996 (UBA 1996) and 2004 (UBA 2004) to an electronic version allowing its chapters to undergo regular updates, hence its reference to www.icpmapping.org.

The transboundary nature of atmospheric pollution and its widespread effects on the environment and public health is now well understood. Scientifically sound knowledge on the effects of deposition and ambient concentrations of these pollutants have become key to policy agreements and abatement strategies, in Europe in particular. Methods have been developed over the past decades to identify no-effect thresholds and understand the dynamics of impact mechanisms in soils and surface waters. This is how critical load and dynamic modelling assessments became scientifically sound tools in effect-based policy support.

Aim of this Book: The aim of this book is to describe the science behind steady-state critical load assessments and dynamic risk assessments of nitrogen, acidity and metals for terrestrial and aquatic ecosystems. It describes methods and reviews existing literature and data with respect to impacts, critical loads and dynamic models and their applications. The latter is key to gaining knowledge on time delays, of either ecosystem recovery or damage, as a result of policy-driven deposition changes with respect to these thresholds. Dynamic model applications include examples on a regional and national scale in Europe, in North America and in China.

The book is limited to the assessment of indirect, soil mediated effects of elevated deposition caused by emissions of SO_2, NO_x and NH_3 and metals. The impacts of N are related to eutrophication, those of N and S combined to acidification and those of metals to toxic impacts on terrestrial and aquatic organisms. While the health of ecosystems (vegetation diversity, soil and aquatic fauna, ecosystem functions) is the common endpoint in view of acidification, eutrophication and metal accumulation, the deposition and accumulation of metals in agricultural and natural soils and surface waters also affect human health. Endpoints of harmful effects from acidification include root functioning (biomass, length and uptake by fine roots) and fish, whereas geo-chemical indicators focus on dissolved concentrations of aluminium and acid neutralizing capacity. The endpoint of eutrophication focuses on vegetation changes, while the geochemical indicators are concentrated on dissolved N concentrations. The endpoints of toxic effects by metals focus on both the health of ecosystems (soil fauna and aquatic fauna) and of humans (food quality), whereas the indicators include metal concentrations in soil, soil solution and surface waters. This book attempts to comprehensively describe and substantiate indicators, and their (assumed) critical limits, that are key to computing critical loads and performing dynamic risk assessments.

This book focuses on the assessment of indirect, soil-mediated effects of elevated deposition of SO_2, NO_x, NH_3 and metals on terrestrial and aquatic ecosystems and related critical loads of those compounds. The treatment of direct effects is beyond the scope of this book. However, direct effects should be included when a complete understanding is required of all possible causes of negative impacts of air pollution. Therefore, an Annex has been included, summarizing direct effects of these pollutants on ecosystems and human health. Note that the reduction of exceedances of critical loads can also help to reduce direct effects. For example, NO_x does not only cause acidification and eutrophication, but is also one of the precursors of ozone (O_3). Ozone concentrations are a significant threat to both human health and the environment. Human health is also directly affected by these acidifying and eutrophying pollutants, as well as by metals as constituents of fine particulate matter. Except for the exception Annex, the concept of direct effects is not further discussed in this book.

In this chapter, we first discuss in more detail the concept of critical loads (Sect. 1.2). We then present an overview of the history and current state of critical load assessments (Sect. 1.3). Next, we present the use of dynamic modelling in effect-oriented research, which is aimed for policy support by deriving so-called target loads (Sect. 1.4). The chapter ends with a reading guideline to this book and the logic of its organization (Sect. 1.5).

1.2 The Concept of Critical Loads

Critical loads are used in the development of emission abatement strategies and policies as a target for atmospheric depositions. Reducing critical load exceedances in a cost-effective manner has become an important policy target in European and national air pollution abatement strategies. Following the definition of a critical load by Nilsson and Grennfelt (see above), the sustainability of the structure and function of an ecosystem is protected when a critical load is not exceeded by (atmospheric) deposition of pollutants, thus avoiding adverse effects and possibly irreversible damage in the future. Critical loads became widely used in the context of the acidification of soils and surface waters caused by 'acid rain'[1], i.e. the excessive atmospheric deposition of sulphur and nitrogen compounds. The use of critical loads to provide intelligence on the impacts of air pollution for the support of abatement strategies and policies is termed "the critical load approach".

Before the critical load approach became popular at the beginning of the 1990s, environmental policies consisted of reducing emissions and assessing technology and related costs of implementing these policies. Effects were then often regarded as a closing entry of a, often qualitative, environmental balance. Brydges (1991) made a comprehensive comparison between conventional environmental objectives and critical loads (Table 1.1). As can be concluded from Table 1.1, the critical load approach attempts to provide quantitative information on "damage levels" of ecosystems and (long-term) consequences of their exceedance[2], whereas conventional objectives were more focused on the level of organisms, with effects being based on laboratory experiments. Note, however, that laboratory experiments are not excluded in providing critical limits for key geochemical indicators used for critical load assessments, as also explained in this book.

Methodologies for deriving critical loads of sulphur and nitrogen have been developed since around 1990 (De Vries 1988; Hettelingh and De Vries 1992; Nilsson and Grennfelt 1988; Sverdrup et al. 1990). For metals the development of a methodology started in the late 1990s (De Vries and Bakker 1998; De Vries et al. 1998), but only recently European maps of critical loads have become available (Slootweg et al. 2007). Its use in international negotiations on the reduction of metal emissions is still under debate. The currently used methods are all described in a Mapping Manual (www.icpmapping.org).

---

[1] Robert Angus Smith is credited with coining the term 'acid rain' (Smith 1872). However, the term 'acid rain' (in French) had already been used in 1845 by Ducros in a scientific journal article (Ducros 1845).

[2] The word 'exceedance' is defined as "the amount by which something, especially a pollutant, exceeds a standard or permissible measurement" (The American Heritage Dictionary of the English Language, Fourth Edition 2000) and is a generally accepted term within the air pollution discipline. Nevertheless, the term 'critical load excess' or a different spelling ('exceedence') is preferred by some (English) speakers, Interestingly, the Oxford English Dictionary (OED) database has an example of 'exceedance' from 1836.

Table 1.1 Contrasts between (historic) conventional environmental objectives and critical loads. (Adapted from Brydges 1991)

Conventional objectives	Critical loads
Effects are generally experienced at the organism level	Effects usually manifest at the ecosystem level
Objectives are generally established on the basis of laboratory tests	Studies of geo-chemistry and biodiversity of terrestrial and aquatic ecosystems are required to establish values
Lethality of physiological effects are the usual response used in setting objectives	Ecosystem effects occur via direct (geo-chemical change) or indirect (food chain alteration) mechanisms
Compounds considered are not essential to life	Elements or compounds are essential building blocks for life
Environmental objectives are set well below known effects to provide some margin of safety	Objectives are set as close to damage levels as possible
No beneficial effects are likely to occur in the environment at any level	Changes may occur that are deemed beneficial (such as increased productivity) although such benefits may be very subjective
Environmental damage from exceedances is usually observed within a short time	Environmental damage from exceedances may occur over a long time (years-decades) and may be cumulative

1.3 The History and Current State of Critical Load Assessments

It was for work under the 1979 LRTAP Convention that the critical loads concept was adopted and further developed. While the earlier protocols to the Convention (1985 on sulphur, 1988 on nitrogen oxides, and 1991 on volatile organic compounds) were stand-still agreements or required flat-rate emission reductions, the 1994 Sulphur Protocol was the first to consider ecosystem vulnerability in terms of critical loads for the formulation of reduction requirements. The signing of this protocol was a first climax of the work of the Task Force on Mapping Critical Levels and Loads, which was established in 1989 under the Working Group on Effects of the LRTAP Convention. Under this Task Force, critical load data from individual countries are collected, collated and mapped by the Coordination Centre for Effects (see e.g. Hettelingh et al. 2007; Posch et al. 1999) and provided to the relevant bodies under the LRTAP Convention as well as the European Commission to formulate emission reduction strategies in Europe.

The scientific discussion on critical loads started at a workshop organised by the Nordic Council of Ministers (NMR) in 1986 in Sundvollen (Norway) and provided, for the first time, estimates of critical loads of acidity, induced by S and N deposition, for forest soils, groundwater, and surface waters (Nilsson 1986). The first workshop on critical loads held under the auspices of the United Nations

Economic Commission for Europe (UNECE), which provided the permanent secretariat for the LRTAP Convention, was organised in 1988 by the NMR at Skokloster (Sweden).

As the role of nitrogen in the acidification of soils and surface waters gained increasing attention at the end of the 1980s in both the scientific and policy arena, a workshop was organised by the NMR and the US-EPA on that topic in Copenhagen in 1988 (Malanchuk and Nilsson 1989). The purpose of that workshop was to review the state of science on the role of N in the acidification of the environment. Finally, the foundation for the mapping of critical loads in the ECE countries was laid at a UNECE workshop held in 1989 in Bad Harzburg (Germany), resulting in a manual for mapping critical levels and loads, which is updated periodically (www.icpmapping.org). Furthermore, in a workshop on critical loads of N organised by the NMR in Lökeberg (Sweden) in 1992, recommendations for deriving critical loads of N and their exceedances were elaborated (Grennfelt and Thörnelöf 1992). Remaining open questions were discussed at a UNECE workshop in Grange-over-Sands (England) in 1994 (Hornung et al. 1995).

Critical loads for heavy metals are a more recent development. After a pilot study carried out in the Netherlands (Van den Hout 1994), a draft manual for the calculation and mapping of metals (De Vries and Bakker 1996) was discussed and amended at a UNECE workshop in Bad Harzburg (Germany), followed by updated manuals in 1998 (De Vries and Bakker 1998; De Vries et al. 1998). An update of these manual has been made in 2005 (De Vries et al. 2005), and this approach has then been adopted under the LRTAP Convention for the calculation of critical loads for cadmium, lead and mercury (Mapping Manual, www.icpmapping.org). The approach has been applied to calculate and map critical loads for heavy metals (see e.g. Chap. 21), but has not been used in support of European pollution reduction policy. The 1998 protocol on metal emission abatement under the LRTAP Convention is based on flat rate reductions using best available abatement techniques, ignoring differences in susceptibility of receptors to metal input. The ultimate aim is to use a critical load approach as a next step in abating metal emissions under the LRTAP Convention, as it may lead to more cost-efficient emission reductions.

1.4 The Use of Dynamic Modelling

Critical loads are generally derived on the basis of a given chemical criterion in the soil solid phase, soil solution and/or surface water. Information on those chemical criteria for acidity, nitrogen and metals is given in Chap. 2, whereas the approach to derive critical loads on the basis of those criteria is given in Chap. 6 for acidity and nitrogen and in Chap. 7 for metals. Critical loads are, however, steady-state quantities, i.e. they do not have a time dimension. If depositions become equal to critical loads, one cannot say when the critical limit is attained. It may take a few years, or centuries, depending on the finite buffers in the soil, such as cation exchange capacity and N retention (these finite buffers do not influence steady state, i.e. critical loads). Adverse effects of excessive deposition does therefore not necessarily lead to immediate damages to ecosystems, e.g. violating geochemical balances or causing vegetation changes or forest dieback. To reconstruct and/or predict the temporal development of a soil (and vegetation) system, dynamic models that include the relevant time-dependent processes are required. Dynamic models can help inform us on time horizons to achieve a sustainable structure and function of ecosystems, i.e. of metrics for the geo-chemistry in soils and surface waters and of key parameters for biodiversity.

For almost three decades, dynamic geochemical soil and surface water models have been developed (Tiktak and van Grinsven 1995), initially for single site applications, but gradually—with increasing computing power and the availability of adequate databases—also for regional applications (e.g. Posch and Reinds 2009; Wallman et al. 2005). Although there have been many acidification models developed over time, such as ILWAS (Gherini et al. 1985), Trickle-Down (Nikolaidis et al. 1991) RESAM (De Vries et al. 1995), NUCSAM (Groenenberg et al. 1995), PnET-BGC (Gbondo-Tugbawa et al. 2001), there are only a limited number of dynamic models that are still used in regional assessments, and thus presented in this book, i.e. the models VSD/VSD+ (Posch and Reinds 2009; Reinds et al. 2001; Chaps. 8 and 13), SMART/SMARTml (Bonten et al. 2011; De Vries et al. 1989; Chaps. 8 and 22), MAGIC (Cosby et al. 1985, 2001; Chaps. 8 and 18), ForSAFE (Belyazid et al. 2006; Wallman et al. 2005; Chaps. 8, 12 and 13). These models dynamically predict chemical parameters related to acidity and N which can be compared to critical limits. These models can also be used to calculate so-called target loads and damage—or recovery delay times (Chap. 8). During the last decade, the regional application of those dynamic models has been introduced to support (European) emission reduction policies (Chap. 25).

More recently some of the dynamic geochemical models have been linked to vegetation models to assess the link between deposition and vegetation changes (see De Vries et al. 2010 and the Chaps. 11–14). However, the complexity and data requirements of these dynamic model chains still limit their large-scale applications. But the same ranges of (possibly interlinked) chemical criteria could be used to derive critical/optimal loads along the lines outlined above. This would allow for large-scale regional applications and an easier link of biodiversity issues with integrated assessment models.

1.5 Reading Guideline

Following its aim described above, we summarize in this book approaches to assess critical loads for nitrogen, acidity and metals in view of impacts on terrestrial and aquatic ecosystems. The book is divided into five major sections/themes:

1. Assessment of *critical limits* for *geochemical indicators* for air pollutant impacts on terrestrial and aquatic ecosystems (Chap. 2) and biological indicators for terrestrial ecosystems (Chap. 3).
2. Empirical and model based *critical loads* for nitrogen, acidity and metals for characteristic ecosystems (Chaps. 4–7).

Table 1.2 The place of Chaps. 2–25 in view of the problem area considered, the type of model applied, the indicators used or the scale of application

Type of approach	Indicator used	Scale of application	Problem area[a]		
			Acidification	Metal toxicity	Nitrogen eutrophication
Empirical	Geochemical and biological	Multiple sites	–	–	4,5
		Regional	–	–	23
Steady-state model	Geochemical	Multiple sites	2,6	2,7	2,6
		Regional	2, 15, 16,17	2, 20,21	2
Dynamic model	Geochemical	Multiple sites	2,8	2, 9	2,8
		Regional	2, 18,19	2	2,24,25
	Biological	Multiple sites	–	–	3,10,11,12,13,14
		Regional	–	–	3,22,24

[a] The type of model or scale of application does not relate to the description of indicators in Chaps. 2 and 3, as it refers to both model types and all scales

3. *Dynamic modelling* approaches for the assessment of critical limit violations and target loads for nitrogen, acidity and metals in time *at site scale* (Chaps. 8–14).
4. *Critical loads and dynamic model applications on a regional scale* in different world regions (Chaps. 15–23).
5. *Integrated assessment*, including combined critical loads and dynamic model applications in a changing climate and their use in policy (Chaps. 24, 25).

Each section deals with both terrestrial and aquatic ecosystems, with a focus on terrestrial ecosystems. The relationship between the themes is as follows. First air pollution impacts on terrestrial and aquatic ecosystems are described, including geochemical and biological indicators that describe the status of an environmental endpoint to be protected and related critical limits (Theme 1). These indicators with related critical limits are the basis for critical load assessments by either empirical approaches or steady-state models (Theme 2). They are also the metrics by which future impacts of critical load exceedances are evaluated using dynamic models (Theme 3). In the book a distinction is made between dynamic model applications on a site scale (Theme 2 and 3) and a regional scale (Theme 4). Most model applications focus on Europe, but empirical and model-based critical load assessments in United States, Canada and China are also included. The policy implications of the various approaches are discussed in Theme 5. The book concludes with a short summary of findings and a synthesis in the context of uncertainty and finally touches upon trade-offs between air quality, climate change and biodiversity (Chap. 26).

The scope of the various chapters with respect to the problem areas considered in this book (acidity, metals and nitrogen) is summarized in Table 1.2, including information on the type of model, its indicators and the scale of application. The assessment of empirical critical loads and the use of biological indicators in models are limited to the impacts of N only.

References

Asman, W. A. H., Sutton, M. A., & Schjorring, J. K. (1998). Ammonia: Emission, atmospheric transport and deposition. *New Phytologist, 139,* 27–48.

Baker, J. P., & Schofield, C. L. (1982). Aluminum toxicity to fish in acidic waters. *Water, Air, & Soil Pollution, 18,* 289–309.

Belyazid, S., Westling, O., & Sverdrup, H. (2006). Modelling changes in forest soil chemistry at 16 Swedish coniferous forest sites following deposition reduction. *Environmental Pollution, 144,* 596–609.

Bonten, L. T. C., Groenenberg, J. E., Meesenburg, H., & De Vries, W. (2011). Using advanced surface complexation models for modelling soil chemistry under forests: Solling forest, Germany. *Environmental Pollution, 159,* 2831–2839.

Brydges, T. G. (1991). *Critical loads, reversibility and irreversibility of damage to ecosystems.* Proceedings of a workshop "Electricity and the environment" held in Helsinki, Finland, International Atomic Energy Agency, Vienna, 13–17 May 1991.

Cosby, B. J., Hornberger, G. M., Galloway, J. N., & Wright, R. F. (1985). Modeling the effects of acid deposition: Assessment of a lumped parameter model of soil water and streamwater chemistry. *Water Resources Research, 21,* 51–63.

Cosby, B. J., Ferrier, R. C., Jenkins, A., & Wright, R. F. (2001). Modelling the effects of acid deposition: Refinements, adjustments and inclusion of nitrogen dynamics in the MAGIC model. *Hydrology and Earth System Sciences, 5,* 499–517.

De Vries, W. (1988). Critical deposition levels for nitrogen and sulphur on Dutch forest ecosystems. *Water, Air, & Soil Pollution, 42,* 221–239.

De Vries, W., & Bakker, D. J. (1996). *Manual for calculating critical loads of heavy metals for soils and surface waters. Preliminary guidelines for environmental quality criteria, calculation methods and input data.* (Report 114). Wageningen, the Netherlands: DLO Winand Staring Centre.

De Vries, W., & Bakker, D. J. (1998). *Manual for calculating critical loads of heavy metals for terrestrial ecosystems. Guidelines for critical limits, calculation methods and input data.* (Report 166). Wageningen, the Netherlands: DLO Winand Staring Centre.

De Vries, W., Posch, M., & Kämäri, J. (1989). Simulation of the long-term soil response to acid deposition in various buffer ranges. *Water, Air, & Soil Pollution, 48,* 349–390.

De Vries, W., Kros, J., & van der Salm, C. (1995). Modelling the impact of acid deposition and nutrient cycling on forest soils. *Ecological Modelling, 79,* 231–254.

De Vries, W., Bakker, D. J., & Sverdrup, H. U. (1998). *Manual for calculating critical loads of heavy metals for aquatic ecosystems. Guidelines for critical limits, calculation methods and input data.* (Report 165). Wageningen, the Netherlands: DLO Winand Staring Centre.

De Vries, W., Schütze, G., Lofts, S., Tipping, E., Meili, M., Römkens, P. F. A. M., & Groenenberg, J. E. (2005). *Calculation of critical loads for cadmium, lead and mercury. Background document to a Mapping Manual on Critical Loads of cadmium, lead and mercury.* (Report 1104). Wageningen, the Netherlands: Alterra.

De Vries, W., Wamelink, G. W. W., van Dobben, H., Kros, J., Reinds, G. J., Mol-Dijkstra, J. P., Smart, S. M., Evans, C. D., Rowe, E. C., Belyazid, S., Sverdrup, H. U., van Hinsberg, A., Posch, M., Hettelingh, J.-P., Spranger, T., & Bobbink, R. (2010). Use of dynamic soil–vegetation models to assess impacts of nitrogen deposition on plant species composition: An overview. *Ecological Application, 20,* 60–79.

Dentener, F., Drevet, J., Lamarque, J. F., Bey, I., Eickhout, B., Fiore, A. M., Hauglustaine, D., Horowitz, L. W., Krol, M., Kulshrestha, U. C., Lawrence, M., Galy-Lacaux, C., Rast, S., Shindell, D., Stevenson, D., Van Noije, T., Atherton, C., Bell, N., Bergman, D., Butler, T., Cofala, J., Collins, B., Doherty, R., Ellingsen, K., Galloway, J., Gauss, M., Montanaro, V., Müller, J. F., Pitari, G., Rodriguez, J., Sanderson, M., Solmon, F., Strahan, S., Schultz, M., Sudo, K., Szopa, S., & Wild, O. (2006). Nitrogen and sulfur deposition on regional and global scales: A multimodel evaluation. *Global Biogeochemical Cycles, 20,* GB4003.

Ducros, M. (1845). Observation d'une pluie acide. *Journal de Pharmacie et de Chimie, 3,* 273–277.

EC. (2001). *Directive 2001/81/EC of the European Parliament and of the Council of 23 October 2001 on national emission ceilings for certain atmospheric pollutants.* Brussel: European Commision.

Fischer, R., Mues, V., Ulrich, E., Becher, G., & Lorenz, M. (2007). Monitoring of atmospheric deposition in European forests and an overview of its implication on forest condition. *Applied Geochemistry, 22,* 1129–1139.

Fowler, D. (2002). Pollutant deposition and uptake by vegetation. In J. N. B. Bell & M. Treshow (Eds.), *Air pollution and plant life* (pp. 43–67). Chichester: Wiley.

Galloway, J. N., & Cowling, E. B. (2002). Reactive nitrogen and the world: 200 years of change. *Ambio, 31,* 64–71.

Galloway, J. N., Dentener, F. J., Capone, D. G., Boyer, E. W., Howarth, R. W., Seitzinger, S. P., Asner, G. P., Cleveland, C. C., Green, P. A., Holland, E. A., Karl, D. M., Michaels, A. F., Porter, J. H., Townsend, A. R., & Vörösmarty, C. J. (2004). Nitrogen cycles: Past, present and future. *Biogeochemistry, 70,* 153–226.

Gbondo-Tugbawa, S. S., Driscoll, C. T., Aber, J. D., & Likens, G. E. (2001). Evaluation of an integrated biogeochemical model (PnET-BGC) at a northern hardwood forest ecosystem. *Water Resources Research, 37,* 1057–1070.

Gherini, S. A., Mok, L., Hudson, R. J. M., Davis, G. F., Chen, C. W., & Goldstein, R. A. (1985). The ILWAS model: Formulation and application. *Water, Air, & Soil Pollution, 26,* 425–459.

Grennfelt, P., & Thörnelöf, E. (1992). *Critical loads for nitrogen.* Copenhagen, Denmark: Nord 1992:41 Nordic Council of Ministers.

Groenenberg, J. E., Kros, J., van der Salm, C., & De Vries, W. (1995). Application of the model NUCSAM to the solling spruce site. *Ecological Modelling, 83,* 97–107.

Hettelingh, J. P., & De Vries, W. (1992). *Mapping Vademecum.* (Report 259101002). Bilthoven, the Nederlands: National Institute of Public Health and Environmental Protection.

Hettelingh, J.-P., Posch, M., & de Smet, P. A. M. (2001). Multi-effect critical loads used in multi-pollutant reduction agreements in Europe. *Water, Air, & Soil Pollution, 130,* 1133–1138.

Hettelingh, J.-P., Posch, M., Slootweg, J., Reinds, G. J., Spranger, T., & Tarrasón, L. (2007). Critical loads and dynamic modelling to assess European areas at risk of acidification and eutrophication. *Water, Air, & Soil Pollution: Focus, 7,* 379–384.

Hettelingh, J.-P., Posch, M., Velders, J. M., Ruyssenaars, P., Adam, M., de Leeuw, F., Lükewille, A., Maas, R., Sliggers, J., & Slootweg, J. (2013). Assessing interim objectives for acidification, eutrophication and ground-level ozone of the EU National Emissions Ceilings Directive with 2001 and 2012 knowledge. *Atmospheric Environment, 75,* 129–140.

Holland, E. A., Dentener, F. J., Braswell, B. H., & Sulzman, J. M. (1999). Contemporary and pre-industrial global reactive nitrogen budgets. *Biogeochemistry, 46,* 7–43.

Hornung, M., Sutton, M. A., & Wilson, R. B. (1995). *Mapping and modelling of critical loads for nitrogen: A workshop report.* United Kingdom: Institute of Terrestrial Ecology.

Lammel, R. (1984). Endgültige Ergebnisse und bundesweite Kartierung der Waldschadenserhebung 1983. *AFZ-Der Wald, 39,* 340–344.

Liu, X., Zhang, Y., Han, W., Tang, A., Shen, J., Cui, Z., Vitousek, P., Erisman, J. W., Goulding, K., Christie, P., Fangmeier, A., & Zhang, F. (2013). Enhanced nitrogen deposition over China. *Nature, 494,* 459–462.

Malanchuk, J. L., & Nilsson, J. (1989). *The role of nitrogen in the acidification of soils and surface waters.* Nord 1989:92 Nordic Council of Ministers Copenhagen Denmark.

Nikolaidis, N. P., Muller, P. K., Schnoor, J. L., & Hu, H. L. (1991). Modeling the hydrogeochemical response of a stream to acid deposition using the enhanced trickle-down model. *Research Journal of Water Pollution Control Federation, 63,* 220–227.

Nilsson, J. (1986). *Critical loads for nitrogen and sulphur.* Copenhagen: Miljørapport 1986:11 Nordic Council of Ministers.

Nilsson, J., & Grennfelt, P. (1988). *Critical loads for sulphur and nitrogen*. Report from a Workshop held at Skokloster Sweden March 19–24 1988. Miljø rapport 1988: 15. Copenhagen Denmark Nordic Council of Ministers.

Odén, S. (1976). The acidity problem—An outline of concepts. *Water, Air, & Soil Pollution, 6,* 137–166.

Posch, M., & Reinds, G. J. (2009). A very simple dynamic soil acidification model for scenario analyses and target load calculations. *Environmental Modelling and Software, 24,* 329–340.

Posch, M., de Smedt, P. A. M., Hettelingh, J. P., & Downing, R. J. (1999). *Calculation and mapping of critical thresholds in Europe. CCE Status Report 1999.* (RIVM report 259101009). Bilthoven, The Netherlands: Coordination Center for Effects, National Institute of Public Health and the Environment.

Reinds, G. J., Posch, M., & De Vries, W. (2001). *A semi-empirical dynamic soil acidification model for use in spatially explicit integrated assessment models for Europe*. (Alterra rapport 84). Wageningen (Netherlands): Alterra.

Reis, S., Grennfelt, P., Klimont, Z., Amann, M., ApSimon, H., Hettelingh, J.-P., Holland, M., Le-Gall, A.-C., Maas, R., Posch, M., Spranger, T., Sutton, M. A., & Williams, M. (2012). From acid rain to climate change. *Science, 338,* 1153–1154.

Schöpp, W., Posch, M., Mylona, S., & Johansson, M. (2003). Long-term development of acid deposition (1880–2030) in sensitive freshwater regions in Europe. *Hydrological and Earth System Sciences, 7,* 436–446.

Schütt, P., Blaschke, H., Hoque, E., Koch, W., Lang, K. J., & Schuck, H. J. (1983). Erste Ergebnisse einer botanischen Inventur des 'Fichtensterbens'. *Forstwissenschaftliches Centralblatt, 102,* 177–186.

Sliggers, J., & Kakebeeke, W. (2004). *Clearing the air: 25 years of the convention on long-range transboundary air pollution*. Geneva: United Nations, Economic Commission for Europe.

Slootweg, J., Posch, M., & Hettelingh, J.-P. (Eds.). (2007). *Critical loads of nitrogen and dynamic modelling: CCE Progress Report 2007. Coordination Centre for Effects*. Bilthoven, Netherlands: MNP Report 500090001.

Smith, R. A. (1872). *Air and rain: The beginnings of a chemical climatology*. London: Longmans, Green & Co.

Sutton, M. A., Erisman, J. W., Dentener, F., & Miller, D. (2008). Ammonia in the environment: From ancient times to present. *Environmental Pollution, 156,* 583–604.

Sverdrup, H., De Vries, W., & Henriksen, A. (1990). *Mapping critical loads. A guidance manual to criteria calculation methods data collection and mapping*. Miljø rapport 1990: 14. Nordic Council of Ministers Copenhagen 1990.

Tiktak, A., & van Grinsven, J. J. M. (1995). Review of sixteen forest-soil-atmosphere models. *Ecological Modelling, 83,* 35–54.

Truscott, A. M., Palmer, S. C. F., McGowan, G. M., Cape, J. N., & Smart, S. (2005). Vegetation composition of roadside verges in Scotland: The effects of nitrogen deposition, disturbance and management. *Environmental Pollution, 136,* 109–118.

UBA. (1996). Manual on methodologies and criteria for mapping Critical Levels & Loads and geographical areas where they are exceeded. (UBA-Texte 71/96).

UBA. (2004). Mapping manual 2004. Manual on methodologies and criteria for modelling and mapping critical loads and levels and air pollution effects, risks and trends. www.icpmapping.org. Accessed 25 Nov 2014.

Van den Hout, K. D. (1994). *The impact of atmospheric deposition of non-acidifying pollutants on the quality of European forest soils and the North Sea*. (Report R93/329). Delft, the Netherlands: IMW-TNO.

Wallman, P., Svenssson, M., Sverdrup, H., & Belyazid, S. (2005). ForSAFE—An integrated process-oriented forest model for long-term sustainability assessments. *Forest Ecology and Management, 207,* 19–36.

Part I
Assessment of Indicators for Air Pollutant Impacts

Chapter 2
Geochemical Indicators for Use in the Computation of Critical Loads and Dynamic Risk Assessments

Wim de Vries, Maximilian Posch, Harald U. Sverdrup, Thorjørn Larssen, Heleen A. de Wit, Roland Bobbink and Jean-Paul Hettelingh

2.1 Introduction

To assess critical atmospheric deposition levels (critical loads) of acidity, nitrogen (N) and metals, an endpoint needs to be identified, such as tree growth, crop quality, plant species diversity and soil biodiversity. An endpoint can be defined as a 'specified sensitive element of the environment' that one wants to protect. To quantify impacts, a certain endpoint indicator is needed, which can be defined as 'a quantifiable measure describing the status of an endpoint'. In case of geochemical indicators, this is also called a chemical criterion. Finally, to assess (potential) risk, a critical limit is needed, which can be defined as a maximum or minimum value allowed for an endpoint indicator or chemical criterion. For example, an endpoint for N impacts is 'vegetation change', the 'endpoint indicator' or 'chemical criterion' is the dissolved N concentration and its critical limit is in the range 2.5–6.0 mg N l^{-1}.

With respect to the assessment of critical loads, a distinction can be made between an empirical approach and a model-based approach. In the empirical approach, critical loads are derived from observed relationships between atmospheric deposition and effects on 'specified sensitive elements' within an ecosystem (ecosystem status) by correlative or experimental research. Such an approach has, for example, been used for nitrogen in view of adverse impacts on biodiversity, since N has a dominat-

W. de Vries (✉)
Alterra Wageningen University and Research Centre, Wageningen, The Netherlands
e-mail: wim.devries@wur.nl

M. Posch · J.-P. Hettelingh
Coordination Centre for Effects (CCE), RIVM, Bilthoven, The Netherlands

H. U. Sverdrup
Department of Chemical Engineering, Lund University, Lund, Sweden

T. Larssen · H. A. de Wit
Norwegian Institute for Water Research, Oslo, Norway

R. Bobbink
B-Ware Research Centre, Radboud University Nijmegen, Nijmegen, The Netherlands

W. de Vries et al. (eds.), *Critical Loads and Dynamic Risk Assessments*,
Environmental Pollution 25, DOI 10.1007/978-94-017-9508-1_2

ing influence on species diversity of terrestrial vegetation. Details on this approach are given in Chaps. 4 and 5. Tables with empirical critical N load values and estimates of their reliability have been compiled since the beginning of the nineties (Bobbink et al. 1992), followed by an updated background paper (Bobbink et al. 1996), published as Annex III in the 1996 Mapping Manual (UBA 1996), followed by an extensive review and update in the period 2001–2002 (Achermann and Bobbink 2003) and recently updated (Bobbink and Hettelingh 2011), and now included in the Mapping Manual (www.icpmapping.org).

One of the drawbacks of empirical critical (nitrogen) loads is that the endpoints which they aim to protect vary between ecosystems. Moreover, they are generally based on comparatively short-term experiments and thus provide limited knowledge of the evolution of impacts over time. Finally, the variation in critical loads for a given ecosystem is not spatially explicit but only included by expressing the critical load as a range. An alternative approach is thus to use models that enable the assessment over time and space of the risk of adverse effects, indicated by the exceedance of critical limits for specifically chosen endpoint indicators.

The assessment of critical loads using a model-based approach includes the use of geochemical indicators and critical limits or threshold values in a steady-state model. The critical limits can also be used in a dynamic modelling approach to evaluate impacts over time. 'Risk' in this book refers to potential impacts caused by exceedances of critical loads (assessed by steady-state models) or critical limit exceedances (assessed by dynamic models). Risk is thus used more loosely than "the probability of occurrence of an (adverse) event, times its (negative) impact (damage)…" as defined by Helbing (2013). Risks refer to potential impacts which occur when exceedances of critical loads or limits are calculated. The use of models helps standardize (risk) assessments at temporal (see Chaps. 8 and 9) as well as regional scale (see Chaps. 21 and 25) for different environmental problems (e.g. Hettelingh et al. 2007).

In the model-based approach, critical loads are derived with steady-state models using critical limits for element concentrations or element ratios in terrestrial and aquatic ecosystems, based on dose-response relationships between these critical limits and the ecosystem status. Using this approach, a critical load of nitrogen, acidity or heavy metals equals the load causing a concentration nitrogen, acidity, aluminium or heavy metals in a compartment (e.g. soil, groundwater, plant, etc.) that does not exceed a critical limit, thus preventing 'significant harmful effects on specified sensitive elements of the environment'. Consequently, the selection of critical limits is a important step of major importance in deriving a critical load. In computing a critical load one aims at long-term protection of the ecosystem.

This chapter gives an overview of (i) the abiotic impacts of acid, metal and nitrogen deposition on soils and waters with their potential impacts on terrestrial and aquatic ecosystems and (ii) the related geochemical indicators with derived critical limits for acidity (Sect. 2.2), metals (Sect. 2.3) and nitrogen (Sect. 2.4), including a critical evaluation of the usefulness of those limits to assess effects in the field situation. Terrestrial ecosystems are all non-agricultural systems and include both soil and underlying ground water. Aquatic ecosystems include fresh waters, specifically lakes. The chapter concludes with an overview of derived or applied critical limits (Sect. 2.5).

2.2 Impacts of Acid Deposition on Terrestrial and Aquatic Ecosystems and Critical Limits for Geochemical Indicators

2.2.1 Terrestrial Ecosystems

2.2.1.1 Impacts of Acid Deposition

Atmospheric depositions of sulphur (S) and nitrogen compounds have an impact on terrestrial ecosystems through 'eutrophication by nitrogen', as discussed in Sect. 2.4, and soil acidification. Some forms of nitrogen (e.g. HNO_3) are directly acidic. In most cases however, N becomes acidifying when reduced forms of N are oxidized to NO_3^- via nitrification. Soil acidification is characterised by a wide variety of long-term effects. It is defined as the loss of Acid Neutralizing Capacity (ANC) and ultimately leads to a decrease in soil pH. Changes in soil pH are dependent on the buffering capacity of the soil (e.g. Ulrich 1981).

The link between acid deposition and changes in soil and soil solution chemistry is well documented. Acidifying compounds (N and S) deposited on calcareous soils (including substrates of young moraine regions) at first will not change soil acidity. In these soils protons (H^+) are exchanged on the soil complex for ions such as bicarbonate (HCO_3^-) and calcium (Ca^{2+}), thereby buffering the soil against acidifying processes. HCO_3^- and Ca^{2+} ions leach from the system, but the pH remains the same until almost all of the calcium carbonate has been depleted. In soils dominated by silicate minerals, buffering is taken over by cation exchange processes of the soil adsorption complexes. In these soils, protons are exchanged for calcium (Ca^{2+}), magnesium (Mg^{2+}) and potassium (K^+) and these cations are leached from the soil together with anions (mostly nitrate or sulphate). Subsequent leaching of Ca^{2+}, $Mg^{2+}$and K^+ leads to loss of the soil's buffering capacity by base cations and to nutrient imbalances for plant growth. Because of the restricted capacity of this buffering system, soil pH will decrease. Both soil solution chemistry and soil base cation pools are thus affected by acid deposition, both being interdependent. In many catchments, it has been shown that acid deposition has caused prolonged export of base cations, such as Ca^{2+} and Mg^{2+}, from forest soils, resulting in base cation nutrient depletion (Akselsson et al. 2007; Sverdrup et al. 2006; Watmough et al. 2005).

Buffering by cations and leaching will continue until all base cations are exchanged and continuing acidification will lead to a shift in the buffer range of the soil from base cation buffering to aluminium (Al^{3+}) buffering ($pH<4.5$). In mineral soils with a large cation exchange capacity and high base saturation, the base cation buffering may continue for several decades, even at relatively high acid inputs.

At low pH (<4.5), hydrous oxides of several metals start to dissolve. This causes a strong increase in dissolved concentrations of Al^{3+}(De Vries et al. 1989; Ulrich 1981) and other metals, such as manganese (Augustin et al. 2005b). Acid deposition on base-poor soils thus leads to mobilization of potentially toxic Al^{3+} (Mulder et al. 1989). Significant correlations between S deposition and enhanced concentrations of Al^{3+} in soil solutions have been demonstrated in acidic forest soils in Europe (De Vries et al. 2003).

Plant growth and species composition of vegetation can be seriously affected by soil acidification, associated with a decrease in pH and base saturation as well as an increase of the concentration of Al^{3+} in the soil solution. It may lead to dominance of acid-resistant plant species and the decline of characteristic species for the intermediate and higher soil pH range (Falkengren-Grerup 1986; Falkengren-Grerup and Tyler 1993). Furthermore, enhanced dissolution of Al by acid deposition has long been considered a probable threat to forest vitality (Ulrich et al. 1980; Ulrich 1984). Hypothesized mechanisms of Al toxicity include hampered root growth and inhibition of uptake of nutrients (Matzner and Murach 1995; Schulze 1989; Sverdrup et al. 1990; Sverdrup and Warfvinge 1993; Sverdrup et al. 1992; Warfvinge et al. 1993). Furthermore, several authors (e.g. Roelofs et al. 1985) showed that release of Al by soil acidification and imbalances of ammonium to base cations, due to excessive N inputs and reduced nitrification, may cause nutrient deficiencies, which may be aggravated by a loss of mycorrhiza or root damage. This coincided with field observations and foliage analyses, showing that deficiencies of Mg and K caused yellowing of needles of Norway spruce (Zöttl and Mies 1983). In the eighties, several authors (for example Hutchinson et al. 1986; Ulrich and Pankrath 1983) considered soil acidification, especially the increase of the concentration of Al^{3+} in soil solution, responsible for forest decline, since Al^{3+} is very likely to be toxic to plant roots (Cronan and Grigal 1995; Marschner 1990; Mengel 1991; Sverdrup and Warfvinge 1993). Currently, the risk of Al^{3+} for forest health in the field is considered lower (see Sect. 2.2.1.3), but the adverse impact of Al^{3+} on root functioning is an established fact, at least under laboratory conditions.

2.2.1.2 Critical Limits of Geochemical Indicators

Ulrich and co-workers (e.g. Ulrich and Matzner 1983) were among the first who postulated that increased Al concentrations, specifically inorganic Al, and elevated Al/Ca ratios in soil solution are a major cause of forest dieback, by damaging the root system of tree species. Effects of high concentrations of Al on trees were tested with seedlings, either grown in water cultures, pot trials or in a greenhouse, mainly during the 1980s. For overviews, we refer to Rengel (1992) and Kinraide (2003). The sensitivity of a tree to Al varies as a function of solution pH, Al speciation, Ca concentration, overall ionic strength, the form of inorganic N (NH_4 or NO_3), mycorrhiza interactions, soil moisture etc. Consequently, a wide range of Al toxicity thresholds for various tree species has been reported in the literature, varying between less than 1.5 and more than 30 mg l^{-1} (e.g. Cronan et al. 1989; Joslin and Wolfe 1988, 1989; Keltjens and van Loenen 1989; McCormick and Steiner 1978;

Ryan et al. 1986a, b; Smit et al. 1987; Steiner et al. 1980; Thornton et al. 1987). The sensitivity increases from red spruce, with significant biomass reductions starting to occur near 2 mg l^{-1} of inorganic Al, to Douglas fir, spruce and European beech, whereas Scots pine, oak and birch are relatively insensitive to Al (Cronan et al. 1989). Results in a variety of laboratory experiments listed above showed that the Ca/Al ratio was a better indicator for root impacts than inorganic Al (Cronan and Grigal 1995; Sverdrup and Warfvinge 1993; Sverdrup et al. 1992). As with Al, a wide range in toxicity thresholds for the Al/Ca ratio has been reported. Based on greenhouse experiments by Rost-Siebert (1983), a critical molar Al/Ca ratios of 1.0 for spruce and 10 for birch was proposed by Ulrich and Matzner (1983). A correlative field study between soil solution chemistry and forest vitality (Roelofs et al. 1985) yielded a critical molar Al/Ca ratio for black pine of 1.0, identical to that for spruce. Furthermore, laboratory experiments with two-year old black pine trees showed a strong decrease in Ca and Mg uptake at Al/Ca ratios above 1 (Boxman and van Dijk 1988). Based on the data given above, an inorganic Al concentration of 0.2 eq m^{-3} (about 2 mg l^{-1}) and a molar Al/Ca ratio of 1.0 has widely been used as geochemical risk indicators in the computation of critical acid loads (e.g. De Vries 1993).

Sverdrup and Warfvinge (1993) carried out a systematic review of impacts of Al on the growth of tree seedlings and plants in laboratory experiments, based on approximately 200 studies. The response in acid soils, as expressed by root growth, stem growth or plant growth in experiments, has been determined for different species of coniferous and deciduous trees. Studies showed that the plant response can be described better as a function of the base cation and Al concentration in soil solution than just as a function of Al alone or a Ca/Al ratio. The critical limit was expressed as a molar Bc/Al ratio, with Al referring to the total Al concentration and Bc denoting Ca + Mg + K. As an example, Fig. 2.1 shows the results of dose-response data collected for Norway spruce and the approach used to estimate a critical limit for Bc/Al based on a 20 % effect on biomass growth. Based on their study, the Ca/Al ratio was replaced by the Bc/Al ratio. Critical limits were derived for different tree species. A critical Bc/Al ratio is now the most commonly used critical limit (with biomass growth as endpoint) in the calculation of critical loads for forests (De Vries et al. 2003). In many calculations of critical loads of acid deposition to forest ecosystems, either a general limit value of 1 is used for Bc/Al, or a tree species specific value, ranging mostly between 0.5 and 2.0.

Critical acid loads have also been calculated with a critical limit that avoids depletion of secondary Al compounds (e.g. De Vries 1993; De Vries et al. 1994). This so-called Al depletion criterion is based on the idea that the acceptable rate of Al leaching should not be larger than the rate of Al mobilization by weathering of primary minerals. Otherwise, the remaining part of Al has to be supplied from secondary Al compounds. This causes depletion of these compounds, which might induce an increase in Fe buffering, which in turn leads to a decrease in the availability of phosphate (De Vries and Kros 1989). Negligible depletion of secondary Al compounds is achieved when Al leaching equals mobilization of Al from primary minerals. The critical inorganic Al concentration is thus calculated as:

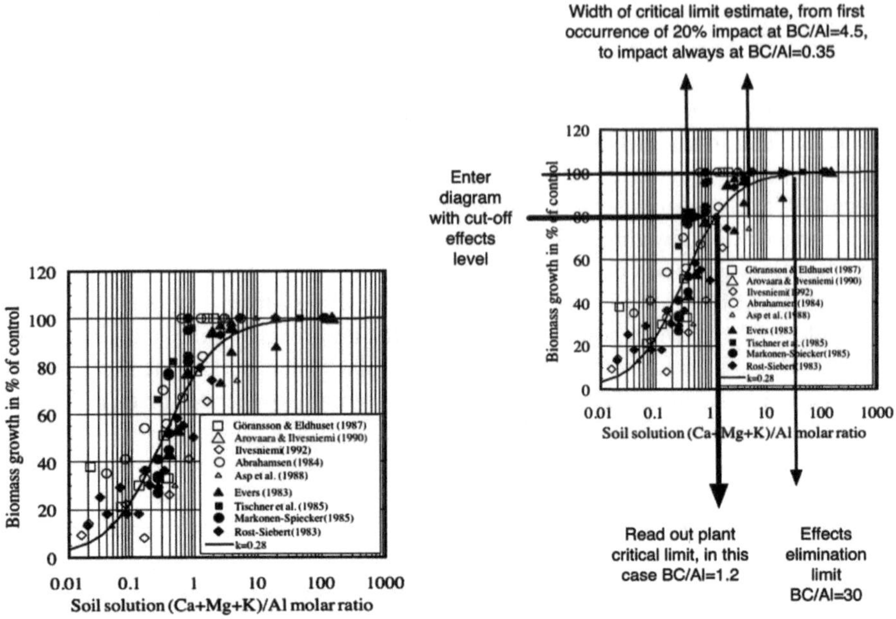

Fig. 2.1 Dose-response data collected for Norway spruce (*left*) and the approach that was used to estimate a critical limit for Bc/Al_{tot} based on a 20% effect on biomass growth (*right*). Using a fitted response, this yielded a critical limit for $Bc/Al_{tot}=1.2$, with 20% effect in all experiments at $Bc/Al_{tot}=0.35$, while the first incidence of 20% impact begins at $Bc/Al_{tot}=4.5$ (After Sverdrup and Warfvinge 1993)

$$[Al_i]_{crit} = r \cdot BC_{we} / Q \tag{2.1}$$

where $[Al_i]_{crit}$ is the critical inorganic Al concentration (eq m^{-3}), BC_{we} is base cation weathering (eq $ha^{-1}yr^{-1}$), r is the equivalent stoichiometric ratio of Al to BC in the congruent weathering of silicates (primary minerals) and Q is the precipitation excess (m^3 $ha^{-1}yr^{-1}$).

An aspect to be mentioned regarding Al is the fact that the free Al^{3+} is considered to be toxic, and not the organically complexed Al species. Unlike metals (see Sect. 2.3), critical limits have only been derived for total Al^{3+} concentration, but not for the free Al^{3+}. In Fig. 2.2 $[Al^{3+}]$, $[AlOH^{2+}]+[Al(OH)_2^+]+[Al(OH)_3^0]+[Al(OH)_4^-]$ and $[\Sigma Al_{org}]$ are shown as shares (in moles) of $[Al_{tot}]$ for a range of pH-values, computed for triprotic organic acids. For high values of *DOC* it shows that free Al dominates below pH=2.5, organic Al dominates between pH 4 and 6, whereas Al-hydroxides dominate above pH=7. The graph illustrates that the share of Al^{3+} is highly dependent on pH and DOC. The problem with re-analysis of laboratory data is that DOC values are generally not available, thus leaving Bc/Al_{tot} as the current risk indicator for acidification.

Apart from ecosystem impacts, Al may also cause health effects, especially by intake in drinking water. No health-based guideline value for aluminium was however recommended in the WHO Guidelines for drinking water quality (WHO

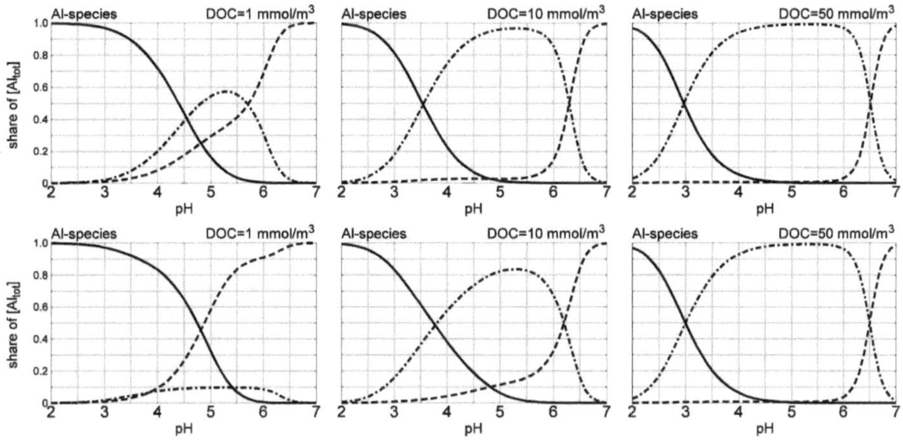

Fig. 2.2 $[Al^{3+}]$ (*solid line*), $[AlOH^{2+}]+[Al(OH)_2^+]+[Al(OH)_3^0]+[Al(OH)_4^-]$ (*dashed line*) and $[\Sigma Al_{org}]$ (*dashed-dotted line*) as shares of $[Al_{tot}]$ as a function of pH for $m \cdot DOC$= 1 (*left*), 10 (*centre*) and 50 mmol m^{-3} (*right*) for $[Al_{tot}]$= 1 (*top row*) and $[Al_{tot}]$= 10 µmol l^{-1} (*bottom row*). No complexation with other ions is considered

2004). It was concluded that epidemiological and physiological evidence did not support a causal role for aluminium in Alzheimer disease. An aluminium concentration of 0.2 mg l^{-1} in drinking-water is, however, provided a compromise between the practical use of aluminium salts in water treatment and discoloration of distributed water. This is used as an indicator parameter in the EU Drinking Water Directive (EC 1998).

2.2.1.3 Evaluation of Critical Limits of Geochemical Indicators in View of Field Effects

The relevance of laboratory experiments addressing Al toxicity under field conditions has been disputed (Binkley and Högberg 1997; De Wit et al. 2001b; Kreutzer 1995; Løkke et al. 1996). Indeed, healthy trees have been found at sites where high soil solution Al concentrations were measured (Huber et al. 2004), while nutrient deficiency symptoms in trees have been found at other sites with similar conditions (Alewell et al. 2000). In addition, whole-ecosystem experiments, designed to test effects of acid deposition on forests (Abrahamsen et al. 1993a; Beier et al. 1998; Huber et al. 2004; Kreutzer and Weiss 1998), have been inconclusive with respect to Al-toxicity effects on root growth and nutrient uptake. For example, field manipulations in forest ecosystems conducted to study effects of N and acid deposition on forest ecosystems, such as in Skogaby in Sweden (Nilsson and Wiklund 1995a, b), Nordmoen in Norway (Abrahamsen et al. 1993b), Höglwald in Germany (Huber et al. 2004; Kreutzer and Weiss 1998; Kreutzer et al. 1998; Nowotny et al. 1998; Rothe et al. 2002) and various experiments across Europe in the so-called NITREX and EXMAN sites (Beier et al. 1998; Boxman et al. 1995) are all difficult to interpret with respect to the specific effect of Al on forest health. Manipulations in the NITREX and the Skogaby experiments included the addition of N fertilizer to forest plots, potentially affecting both tree (Matson et al. 2002) and root growth (Persson and Majdi 1995), and are thus not suitable to specifically test the effects of Al toxicity. In EXMAN, roofs were constructed to avoid acid deposition from reaching the forest soil and enable clean water irrigation. The response of the forest was studied with the aim to understand recovery of forest under reduced acid deposition. Roof experiments also reduced N inputs compared to an open-sky situation. After 10 years of 'clean rain' in Solling (Germany) fine root growth increased, which was interpreted as an compensatory effect to reduced N inputs and not of reduced acidification (Lamersdorf and Borken 2004). In a previous acid manipulation experiment close to Nordmoen (Abrahamsen et al. 1993b), acid manipulation reduced the foliar content of Ca and Mg. This effect was, however, attributed to depleted soil pools of Ca and Mg. Impacts of Al toxicity could not be assessed, as soil solution chemistry was not investigated. In short, reasons for the inconclusiveness of field experiments include a lack of data collection describing below-ground processes and the addition or removal of N, which in itself may affect root and tree growth.

An important forest manipulation study explicitly designed to test the Al-toxicity hypothesis was a 7-year (1997–2003) experiment in a middle-aged Norway spruce stand at Nordmoen, Norway (De Wit et al. 2001a, b, 2010). The objective of the study was to quantify effects of chronically enhanced soil solution concentrations of inorganic Al on fine roots, on nutrient concentrations in needles, bark and fine roots, on tree growth and on crown condition. The site had low concentrations of Al and N in the control treatment. After three years of treatment, no effect was found on tree volume, annual increment, height increment, crown density or crown condition (De Wit et al. 2001b). Investigations of fine root growth using root ingrowth cores in the manipulated plots at Nordmoen also did not show any treatment responses in fine root biomass or necromass (Eldhuset et al. 2006; Nygaard and de Wit 2004). No other vegetation responses than changes in tree nutrient status were found. The most significant and consistent response to the experimental manipulation was the decline in Mg content of needles in the most extreme treatment, caused by reduced Mg uptake by the roots in response to Al addition. Strong competitive inhibition of Al^{3+} ions on Ca^{2+} and Mg^{2+} ion adsorption has been demonstrated in roots of *P. abies* (Schröder et al. 1988) and *P. rubens* (Cronan 1991). The presence of ectomycorrhiza associated with tree roots in the field, lacking in nutrient solution experiments, may have a protective effect against Al (Ahonen-Jonnarth et al. 2003). In summary, the field manipulation experiment at Nordmoen showed that the increased soil solution Al concentrations did not affect fine root growth and mortality, contrary to the hypothesized effects of toxic Al on fine roots, but reduced Mg uptake in extreme situations. One may criticize that 3 years (root impacts) to 7 years (foliar chemistry) of data is insufficient to test the hypothesis of reduced base cation uptake due to impacts of Al toxicity on root growth, because of a large base cation store in the soil, but it is at least an indication that results of (generally much shorter) laboratory experiments may deviate from those in field situations.

An often heard criticism of, mostly relatively short-term, field manipulation studies is the fact that the existence of a time delay between deposition changes and ecosystem effects may obscure rsults. This does not hold for studies comparing data on exposures and responses in long-term or large-scale field surveys. However, analyses of large data sets, obtained from European forest ecosystems monitoring programs, are unable to confirm the hypothesized relations between forest vitality and Al toxicity unambiguously due to confounding factors. Augustaitis and Bytnerowicz (2008) found that acidifying air compounds and their deposition are key factors explaining changes in Scots pine defoliation and stem growth in Lithuania. By applying a linear multiple regression technique, they found that 23–28 % of the variance of residual defoliation and 18.5 % of variability in stem increment residuals of pine trees was explained by acid deposition. However, the link to Al in soil solution is not clear from those analyses. Furthermore, on a European scale, clear relations have been demonstrated between acid deposition and soil acidification (De Vries et al. 2003) but not between forest health (crown condition) and acid deposition (Fischer et al. 2007; Klap et al. 2000). Rather, they point at additional confounding factors, such as site conditions, stand age, insect and fungi attacks, and weather conditions, which make it difficult to attribute observed patterns in crown condition to Al only. There is also a lack of clear relations between forest vitality and exceedances of critical loads (Løkke et al. 1996; Watmough and Dillon 2003). Akselsson et al. (2004) found no significant relation between forest damage in the most exposed part of the Czech Republic and the exceedance of critical loads, using the Bc/Al ratio as indicator. Nellemann and Frogner (1994) found a relationship between defoliation and modelled exceedance of critical loads based on about 100 forested plots throughout Europe, but Solberg et al. (2002), who used a stratum of about 100 old forest officers' plots, found no such relationship. Their soil solution Al/Ca data were, however, mostly below the critical limit of 1.0, demonstrating that modelled exceedance does not mean that unfavourable soil solution chemistry is present today, when critical loads are exceeded, but only that it might be reached sometime in the future. Hence, critical load exceedance may not be an appropriate variable in correlative studies searching for acid deposition effects.

Although current field data cannot show an unequivocal relationship between soil solution chemistry and forest vitality or forest growth, they do indicate that both Al concentrations and base cation stores should be considered in a risk evaluation of acid deposition. For example, long-term records from an acidified nutrient-poor spruce forest stand in Germany showed strong signs of nutrient deficiency with small pools of exchangeable Ca and Mg (Alewell et al. 2000). Furthermore, Lawrence et al. (2005) compared changes in soil chemistry, based on archived soil samples in 1926, 1964, and 2001, with patterns of tree growth in the same period near St. Petersburg, Russia. Results showed a depletion of exchangeable Ca and Mg in the period between 1926 and 1964 and an increase in exchangeable Al between 1964 and 2001. In accordance with soil buffering theories (De Vries et al. 1989; Ulrich 1981), inputs of acidity were first (1926–1964) neutralized by replacement of exchangeable Ca and Mg by H, while continued acid deposition (1964–2001) mobilized readily available Al, with related exchange of Ca, Mg and H by Al. The onset of Al mobilization coincided with decreased diameter growth and a suppression of climate-tree growth relationships in Norway spruce. This suggests that the trees were responding to both a decline in Ca and Mg availability and an increase in Al mobilization.

In summary, forest vitality variables, such as tree growth, fine root growth and defoliation, do not show a clear response to raised soil solution Al concentrations and decreased Bc/Al ratios in both field manipulation experiments and monitoring studies within the time horizon of these experiments and studies. This suggests that forests may be more resilient to (magnitude and duration of) acidification stress than previously thought. This is because the Bc/Al ratio, representing the acidification component of disturbance, is only one of many endpoint indicators affecting other damage parameters, such as water availability, N deposition, climate change including extreme events, ozone exposure, etc. (Sverdrup et al. 2007). Following the definition of critical loads, one should, however, keep in mind that critical limits represent minimum no-effect levels. Augustin et al. (2005a) thus argue that the critical limits for forest soil acidification aim to identify a starting point of an 'unfavourable acid stress' situation, that can ultimately lead to damage. The risk for damage includes, e.g., reducing the resilience of forests to extreme environmental conditions such as storms (Braun et al. 2003).

The limitation of the use of Bc/Al ratios and dissolved Al concentrations in critical load assessments has already been mentioned early in the nineties due to the lack of good correlations between those parameters and forest health. For example, in calculating critical load exceedances for European forests, De Vries et al. (1994) conclude their article with "For a correct interpretation of critical loads, it should be emphasized that an exceedance does not necessarily cause a visible effect on forests, due to the complex interactions in these ecosystems. They mainly have a signal function. However, an exceedance of the critical loads involves a certain risk to the health of forests which increases with the magnitude and the duration of the exceedance". It is in the light of this signal function that these parameters can be used.

2.2.2 Aquatic Ecosystems

2.2.2.1 Impacts of Acid Deposition

In acidic soils, acid deposition causes Al release from soils in the watershed that leaches into lakes and streams. There are well documented, large-scale effects of acid deposition on aquatic ecosystems. The earliest reports linking dead fish and declining fish populations to acid rain and surface water acidification appeared in the early 1970s in both Scandinavia and North-America (Beamish and Harvey 1972; Beamish 1974; Hultberg and Stenson 1970; Jensen and Snekvik 1972; Leivestad and Muniz 1976). Evidence for large scale die-back of fish populations in the Nordic countries became evident in the 1980s. For example, in Norway an estimated 8000 lakes lost their fish population (Hesthagen et al. 1999). Severe impacts, on fish as well as other organisms were also reported for Sweden, Finland, the UK, Canada and the USA (Gray et al. 2012; Monteith et al. 2005; Tammi et al. 2003).

The mechanisms inducing the fish kills and other biological effects are linked to low pH in combination with high Al concentrations. Both low pH (Schofield 1976) and increased Al levels (Baker and Schofield 1982) are directly toxic to fish. Effects include mortality of adult fish and recruitment failure, visible as different patterns in age classes of the fish population (Hesthagen et al. 2011). In addition, low pH and increased Al levels cause chronic stress that may not kill individual fish, but lead to a lower body weight and smaller size, either as a chronic effect on the fish or as indirect effects of changes in food and habitat (Lien et al. 1996). Some types of plants and animals are able to tolerate acidic waters. Others, however, are acid-sensitive and will be lost as the pH declines.

Time trend data have recently become available, not only showing clear evidence of chemical recovery from surface water acidification, but also biological recovery. One well-documented long-term dataset is from Lake Saudlandsvatn in southernmost Norway, where time trend data of brown trout (*Salmo trutta*), caddisfly (*Hydropsyche siltalai*) and zooplankton (*Daphnia longispina*) is available together with water chemistry data. The lake was highly acidified, with episodic pH below 5.0 in the 1970s and 1980s, and the critical load exceeded fivefold (Hesthagen et al. 2011). Chemical recovery following reduced S deposition is observed since the late 1990s and the S deposition dropped below the critical load around 2000. The brown trout population was stable until the early 1980s, then started to decline and nearly went extinct in the 1990s and started to recover ever since. Parallel with the trends in water chemistry it was also observed that the caddisfly disappeared from the tributaries in the 1980s, but reappeared in 1996 and increased in abundance from 2000 onwards. Recovery in all three organism groups coincided with an ANC of >20 meq m^{-3} and toxic inorganic Al concentration of <30 mg m^{-3} (Hesthagen et al. 2011).

2.2.2.2 Critical Limits of Geochemical Indicators

Critical limits for surface waters are generally derived for the protection of aquatic organisms, mostly fish species, but also invertebrates and aquatic plant species are used as indicators. The link between water chemistry and freshwater fauna has mostly been evaluated from survey data, used to establish empirical relationships between water chemistry parameters and presence-absence data of organisms. In general, low pH and high inorganic Al concentrations correlate with damaged and/or extinct organisms. In one of the earlier studies Schofield (1976) looked at the correlation between fish status and water pH, whereas Baker and Schofield (1982) report on the effects of inorganic Al on several fish species in their growth stages.

Jeffries and Lam (1993) used a critical $pH=6$ to determine critical loads (of wet sulphate deposition) for Canadian lakes with the aim of protecting fish and other aquatic organisms. Rask et al. (1995) derived critical values of pH and labile Al, using Finnish lake survey data. Critical annual average pH values, at which fish populations were affected, were 5.0 for perch and 5.8 for roach, while values at which these fish were lost were 4.8 and 5.5, respectively. Comparable values were derived by Hultberg (1988). Critical annual average concentrations of labile Al, at which fish populations were affected, varied from 90 mg m^{-3} for perch and 20 mg m^{-3} for roach. Hultberg (1988) gave a value of 50 mg m^{-3} for perch, but one has to keep in mind that water properties other than pH and labile Al also affect the toxicity of acidic water.

But it is mostly ANC that has been used to define critical chemical limits for surface waters. Of course, pH, [Al] and ANC are interrelated, as discussed in more detail in Chap. 6, but not one-to-one; their relationship depends, in addition to the partial pressure of CO_2, mostly on the amount of dissociated organic anions and complexed Al. The advantage of ANC is that it is a robust variable, e.g., it is not affected by the partial pressure of CO_2.

Lien et al. (1996) analysed the status of fish and invertebrate populations in the context of surface water acidification and loss of ANC in 1095 Norwegian lakes. The data for fish population status came from interview surveys among fishery managers, while lake chemistry data for the same lakes were mostly based on a regional lake survey carried out in 1986 (Henriksen et al. 1988; 1989). The critical level of ANC varied among fish species, with Atlantic salmon being the most sensitive, followed by brown trout. They concluded that Atlantic salmon appeared to be a good indicator of acidification of rivers, and trout seemed to be a useful indicator for the acidification of lakes. Based on an evaluation of fish and invertebrate populations, a lower limit of $[ANC]=20$ meq m^{-3} was suggested as the critical limit for Norwegian surface waters (Lien et al. 1996, see Fig. 2.3). This limit has been widely used (southern central Alps: Boggero et al. 1998; China: Duan et al. 2000; Kola, northern Russia: Moiseenko 1994; Nordic countries: Posch et al. 1997). Other values used were zero in the United Kingdom (CLAG 1995) and 40 meq m^{-3} in south-central Ontario, Canada (Henriksen et al. 2002).

Figure 2.3 indicates that in the ANC range 0–50 meq m^{-3} there is a decreasing probability from about 50 to 0% of damage to fish populations. This could mean that fish have responded to the same ANC differently in different lakes, indicating that a catchment-dependent ANC-limit would be more appropriate than a fixed value for all lakes. With this in mind, Henriksen et al. (1995) derived an ANC-limit that depends on catchment characteristics, such as deposition and runoff.

Lydersen et al. (2004) argued that the ANC-limit should be corrected for the amount of organic acids present in the lake. They revisited the database for water chemistry and fish status and showed that the fit between observed fish status and ANC can be (slightly) improved, if an 'organic acid adjusted' ANC, $[ANC]_{oaa}$, is used. This quantity they defined as:

$$[ANC]_{oaa} = [ANC] - \tfrac{1}{3} \cdot m \cdot TOC \tag{2.2}$$

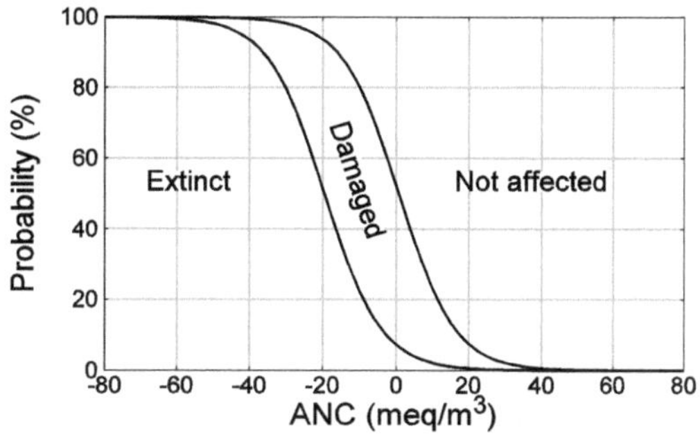

Fig. 2.3 Relationship between the ANC concentration in lake water and the probability for damage and extinction of fish (brown trout) populations in lakes, derived from Norwegian data (after Lien et al. 1996)

where $m \cdot TOC$ is the total organic carbon expressed in meq m^{-3} with m being the charge density, with $m = 10.2$ meq gC^{-1} chosen by Lydersen et al. (2004), and the factor 1/3 arising from the assumption that one third of organic acids should be counted as strong acids. Using a 95 % probability of no damage to fish populations—as was done for deriving the classical ANC-limit of 20 meq m^{-3} by Lien et al. (1996)—they obtained an $[ANC]_{oaa}$ limit of 8 meq m^{-3} for brown trout (*Salmo trutta*), –2 meq m^{-3} for perch (*Perca fluviatilis*) and 11 meq m^{-3} for arctic char (*Salvelinus alpinus*). From these the ANC-limit to be used in critical load calculations is obtained as:

$$[ANC]_{lim} = [ANC]_{oaa,lim} + \tfrac{1}{3} \cdot m \cdot TOC \tag{2.3}$$

i.e. the ANC-limit becomes dependent on the surface water considered. These lake- and fish-specific ANC-limits have been used in studies deriving critical loads for surface waters in Norway (Larssen et al. 2008) and Finland (Posch et al. 2012).

2.3 Impacts of Metal Deposition on Terrestrial and Aquatic Ecosystems and Critical Limits of Geochemical Indicators

2.3.1 Terrestrial Ecosystems

2.3.1.1 Impacts of Metal Inputs

Metals are naturally occurring constituents in the environment. Several metals, such as copper (Cu) and zinc (Zn) are essential for living organisms, because of their role in various physiological and biochemical processes. For other metals, such as cadmium (Cd), lead (Pb) and mercury (Hg), no biological functions are known (Clark 2001). A description of major pathways of metals in terrestrial ecosystems, including the link with aquatic ecosystems, is given in Fig. 2.4.

With respect to risks of heavy metal deposition for terrestrial ecosystems, a distinction can be made between effects on the health of (i) soil organisms/processes

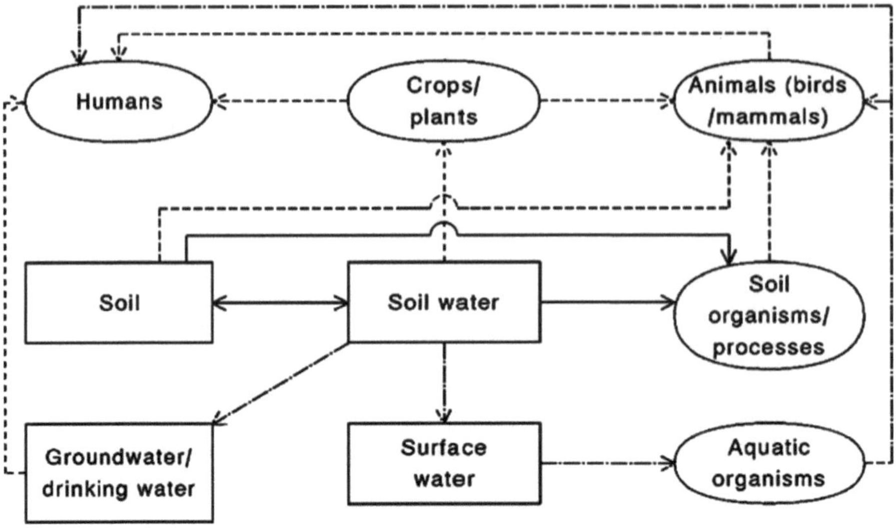

Fig. 2.4 Overview of the fluxes and impact pathways of metals from the soil to other compartments in terrestrial and aquatic ecosystems. *Boxes* are key 'pools' and *ovals* are key 'receptors'. (De Vries et al. 2007b)

and plants (Bringmark et al. 1998; Palmborg et al. 1998) (primary ecotoxicological risks), and (ii) animals, including both domestic and wild animals, and humans that use ground water as drinking water or that consume crops, meat or fish (Clark 2001) (secondary poisoning).

Effects on soil organisms, including micro-organisms/macro-fungi and soil fauna, such as nematodes and earthworms, include reduced species diversity, abundance and biomass, and changes in microbe-mediated processes (Bengtsson and Tranvik 1989; Giller et al. 1998; Vig et al. 2003). Effects on vascular plants include reduced development and growth of roots and shoots (toxicity symptoms), elevated concentrations of starch and total sugar, decreased nutrient contents in foliar tissues (physiological symptoms) and decreased enzymatic activity (biochemical symptoms) (Das et al. 1997; Prasad 1995). A review of these phytotoxic effects is given by Balsberg-Påhlsson (1989).

Effects on microorganisms, plants and, to a large extent, also on invertebrates occur through the soil solution (Ritchie and Sposito 2001). In accordance with the principles of the free ion activity model (FIAM) (Campbell 1995; Morel 1983) and the Biotic Ligand Model (BLM) (Di Toro et al. 2001; Thakali et al. 2006), metal uptake leading to toxicity can be considered to occur through interaction of the free metal ion (FMI) in soil solution (e.g. Cd^{2+}, Pb^{2+}, Cu^{2+}, Zn^{2+}) with the organism. The extent of this interaction, and thus the degree of toxic effect due to a given FMI concentration, also depends on the concentrations of solution cations (e.g. H^+, Na^+, Ca^{2+}) that in the BLM framework compete with the toxic metal for binding to the organism.

Next to effects on soil organisms, metals may be transferred via food chains affecting (i) humans through reduced food quality of crops and animal products and (ii) animal health through accumulation of metals in organs of cattle, birds and mammals. Heavy metal accumulation in food chains is important with respect to Cd and Hg, and to a lesser extent for Pb. In agriculture, it is specifically high Cd inputs that may lead to agricultural products violating food quality criteria (Alloway 1990; Fergusson 1990).

2.3.1.2 Critical Limits of Geochemical Indicators

Relevant receptors in terrestrial ecosystems are arable land, grassland and non-agricultural land (forests, heathlands, grasslands and peatlands). Possible effects on soil organisms and plants (phytotoxicity) and terrestrial fauna are of concern in all types of ecosystems. Furthermore, impacts on food quality are relevant for arable land and grassland (limits for animal food), whereas possible secondary poisoning effects on animals are relevant in grassland (cattle) and non-agricultural land (wild animals). Critical limits define the threshold of potential risk to a defined receptor, and have been defined in relation to ecotoxicological or human-toxicological risks, such as:

- Soil: critical limits related to effects on soil organisms (micro-organisms and soil invertebrates) and plants (mg kg^{-1}).
- Plants/terrestrial fauna: critical limits in plant tissue, animal products (meat) or target organs, such as kidney, related to effects on plants and/or animals and on humans by consumption (food quality criteria) (mg kg^{-1}).
- Ground water: critical limits in drinking water related to effects on humans by consumption (mg m^{-3} or μg l^{-1}).

Food quality criteria for metals in food crops and animal products, in view of human-toxicological risks, are relevant in agriculture. Critical limits for terrestrial ecosystems (often also denoted as Predicted No Effect Concentrations or PNECs) are limited to ecotoxicological effects on soil organisms and plants, while impacts on aquatic systems are related ecotoxicological effects on aquatic organisms. These limits are in principle derived from No Observed Effect Concentrations (NOECs) for metal contents in either soil, based on laboratory studies with plants and soil organisms, such as soil micro-biota and soil invertebrates (Tyler 1992) or surface water, based on laboratory studies with aquatic organisms.

In soils, the risks from heavy metals, however, depend on their availability, which is influenced by soil properties such as pH, clay and organic matter content (Boekhold et al. 1993; Groenenberg et al. 2010; Sauvé et al. 2000). The use of a single soil metal concentration as a critical limit for ecotoxicological effects upon soil organisms has therefore been criticized (Allen 1993) since it does not account for observed variations in the toxicity of cationic metals among soils of differing chemistry (Spurgeon and Hopkin 1996). The impacts of soil properties on the bioavailability and toxicity of metals thus has to be accounted for in critical limit derivations. Free metals in soil solution are the major pathway for metal impacts on plants, micro-organisms and soft-bodied invertebrates, such as earthworms (Saxe et al. 2001). Thus, an approach to set critical limits for the FMI is particularly appropriate to evaluate the risks of effects. In view of critical load assessments for terrestrial ecosystems, there is also a need for critical metal concentrations in soil solution since the critical metal leaching rate is the most important term in deriving critical loads.

Since total metal or free ion concentrations in soil solution are hardly ever measured, such concentrations need to be derived using models for solid-solution partitioning and speciation of metals in soil solution. In this context, empirical models (so-called transfer functions) for solid-solution partitioning can be used, which relate metal concentrations in solution to metal concentrations in the solid phase, accounting for the impact of soil properties.

2.3.1.3 Approach to Derive Critical Limits for Total Dissolved Metal Concentrations

Critical Dissolved Metal Concentrations Related to Impacts on Soil and Aquatic Organisms: The approach to derive critical limits for Cd, Pb, Hg, Cu and Zn for soil and surface water is based on the standard approach developed by the Organisation for Economic Co-operation and Development (OECD) calculating Maximum Permissible Concentrations (MPCs) of substances (OECD 1989). MPCs are calculated by extrapolation of No Observed Effects Concentrations (NOEC levels) or Lowest Observed Effects Concentration (LOECs) for metals in soils or surface water, based on chronic toxicity tests for single species, mostly assuming a log-logistic, or log-normal distribution of species sensitivities (Slooff 1992). From a range of NOEC data, the hazardous concentration at which p% of the species in an ecosystem are potentially affected (HC_p), implying that 100–p % are protected, can be derived from the 'species sensitivities distribution' (SSD). A critical limit for a certain compound is set at HC_5, meaning that theoretically 95% of the species within an ecosystem are protected. Using this method, the 95% protection level calculated with 50% confidence is regarded as the maximum permissible concentration ($MPC=HC_5$).

Critical Reactive and Total Metal Concentrations in the Soil Solid Phase: Critical reactive soil concentrations for Cd, Pb, Cu and Zn have been derived by Lofts et al. (2004) and De Vries et al. (2007b) on the basis of NOECs for the concentration of added (reactive) metal in laboratory experiments. The data set included (i) decomposers, comprising micro-organisms or microbe-mediated soil processes (e.g. enzymatic activity), (ii) consumers, such as invertebrates (earthworms and arthropods) and (iii) primary producers, specifically plants, drawn from several draft reports on EU Risk Assessment procedures for these metals. The impact of soil properties affecting the bioavailability and toxicity of metals has been accounted for by relating the critical value to soil solution pH and soil organic matter content, according to (log stands for $\log_{10}$):

$$\log ctM_{re,crit} = b_0 + b_1 \cdot pH + b_2 \cdot \log OM \tag{2.4}$$

where $ctM_{re,\,crit}$ is the critical reactive metal concentration in the soil (mg kg^{-1}); *pH* is the soil solution pH and *OM* is the soil organic matter content (%). Values derived for b_0, b_1 and b_2 are given in De Vries et al. (2007b). Critical total metal concentrations can be derived by a relationship with concentrations of reactive metal and soil properties, including pH, soil organic matter content and clay, according to (Römkens et al. 2004):

$$\log ctM_{tot,crit} = c_0 + c_1 \cdot \log ctM_{re,crit} + c_2 \cdot \log OM + c_3 \cdot \log clay \qquad (2.5)$$

where $ctM_{tot,\,crit}$ is the critical total metal concentration in the soil (mg kg^{-1}) and *clay* is the clay content (%). Values for c_0–c_3 are given in De Vries et al. (2007b).

Critical Limits for Free Metal Ion and Total Metal Concentrations in Soil Solution: Lofts et al. (2004) and De Vries et al. (2007b) modified the standard OECD approach for calculating MPCs to assess critical limits for metals in soil solution as a function of soil properties. Since NOEC data on free ionic Cd, Pb, Cu and Zn concentrations in soil solution are hardly available, critical limit functions for metals in soil and soil solution were derived from NOEC and EC_{10} (Effect Concentrations affecting 10 % of the organisms) toxicity data on (i) organisms which are exposed to the metal via the soil solution (plants, micro-organisms and soft bodied soil invertebrates), (ii) accompanied by data on soil properties (pH and organic matter content) to allow the calculation of dissolved concentrations by using Freundlich adsorption constants that depend on those properties (called transfer functions) and (iii) evaluated by a statistical approach deriving limits based on a 95 % protection level. The critical FMI concentrations were derived as a function of soil solution pH, according to (De Vries et al. 2007b, Lofts et al. 2004):

$$log[M]_{free,crit} = \alpha_{crit} \cdot pH + \gamma_{crit} \qquad (2.6)$$

where $[M]_{free,crit}$ is the critical FMI concentration (mol l^{-1}). The pH dependence of the critical FMI concentration is considered to result from a 'protective' effect of H^+ against toxicity, resulting from the competitive interactions of solution cations with the organism, as noted before. More recently, the same approach was applied for Ni and Hg (Tipping et al. 2010). Values used for the empirical coefficients α_{crit} and γ_{crit} for Cd, Cu, Pb, Zn, Ni and Hg are given in Table 2.1. More information on the approach and the data sets used is given in Lofts et al. (2004) and De Vries et al. (2007b) for Cd, Cu, Pb, Zn, in Ashmore et al. (2007) for Ni and in Tipping et al. (2010) for Hg. Results are based on the same data sets as those for which the critical reactive metal concentrations were derived. The critical limits for Hg are substantially lower than those of other divalent cationic metals (Table 2.1).

When the critical limit is defined in terms of a critical FMI concentration, the total metal concentration in soil solution, $[M]_{tot}$ (mol m^{-3}), has to be determined as the sum of the concentration of (i) the FMI, $[M]_{free}$ (mol m^{-3}), (ii) dissolved inorganic complexes, $[M]_{DIC}$ (mol m^{-3}) such as MOH^+, $MHCO_3^+$, MCl^+, and (iii) metals bound to dissolved organic matter, $[M]_{DOM}$ (mol kg^{-1} DOM), according to:

Table 2.1 Values for α_{crit} and γ_{crit} used to calculate critical limits for free metal ion ($[M]_{free}$) concentrations in soil solution

Metal[a]	α_{crit}			γ_{crit}		
	Lofts et al. (2004)	De Vries et al. (2007b)	Ashmore et al. (2007)[b]	Lofts et al. (2004)	De Vries et al. (2007b)	Ashmore et al. (2007)[b]
Cd	−0.43	−0.32	−0.31	−5.66	−6.34	−6.36
Pb	−0.83	−0.91	−0.93	−4.80	−3.80	−3.50
Cu	−1.21	−1.23	−1.26	−2.57	−2.05	−1.80
Zn	−0.34	−0.31	−0.25	−4.66	−4.63	−5.07
Ni	–	–	−0.42	–	–	−3.78
Hg	–	–	−2.15	–	–	−17.10

[a] Datasets used for the derivation include micro-organisms, invertebrates (earthworms and arthropods) and plants. The number of data points is 63 for Cd, 49 for Pb, 141 for Cu, 92 for Zn, 83 for Ni and 52 for Hg

[b] Data for Hg are based on results presented in Tipping et al. (2010), whereas the other results are taken from Ashmore et al. (2007)

$$[M]_{tot} = [M]_{free} + [M]_{DIC} + [M]_{DOM} \cdot [DOM] \tag{2.7}$$

where $[DOM]$ is the concentration of dissolved organic matter (kg m^{-3}). By assuming geochemical equilibrium, the partitioning and speciation of metals over the various fractions can be calculated, as described in more detail in Chap. 7.

Critical Dissolved Metal Concentrations Related to Impacts on Human and Animal Health: Human and animal health effects due to metal intake are caused by the uptake of heavy metals via drinking water and crops and in case of humans also livestock products and fish. To protect ground water that is used for drinking water, WHO (2004) uses limits for Cd, Pb and Hg concentrations of 3, 10 and 1 μg l^{-1}, respectively. These critical limits can directly be used in critical load calculations by applying them as limits for total concentrations in soil drainage water. Critical metal concentrations in plants or animal organs, in view of food quality or crop/animal health, can be combined with relationships for soil-plant transfer, soil-animal transfer and plant-animal transfer to assess related critical total metal concentrations in soil (e.g. De Vries et al. 2007b). Critical limits for the total metal concentration in soil solution can then be derived from a relationship with the total soil metal concentration, as discussed below.

Food quality criteria for metals in crops or target organs (liver and kidney) of grazing animals are available for Cd, Pb and Hg only (see e.g. De Vries et al. 2007b). The EU regulation (EG) No.466/2001 (EU 2001) used fresh weight critical limits for wheat of 0.10 mg kg^{-1} for Cd, 0.20 mg kg^{-1} for Pb and 0.03 mg kg^{-1} for Hg. Critical limits for vegetables were 0.20 mg kg^{-1} for Cd, 0.30 mg kg^{-1} for Pb and 0.03 mg kg^{-1} for Hg. The critical limit for Cd in wheat grains was later amended

Table 2.2 Values for the coefficients a_0–a_3 and m in the soil-plant relation (Eq. 2.8) for Cd in grass, maize, sugar beet, wheat and lettuce (after De Vries et al. 2008)

Crop	a_0	a_1	a_2	a_3	m	R^2	se-y_{est} [a]
Grass	1.45	−0.38	–	–	1.22	0.63	0.23
Maize	0.90	−0.21	–	−0.32	1.08	0.50	0.28
Sugar beet	1.33	−0.22	–	−0.13	0.62	0.83	0.15
Wheat	0.22	−0.12	−0.33	−0.04	0.62	0.64	0.20
Lettuce	2.55	−0.33	−0.39	−0.19	0.85	0.71	0.08

[a] Standard error of the y-estimate on a logarithmic basis

by (EC) No 629/2008 to 0.20 mg kg^{-1} fresh weight. This limit is, however, not based on effects but on the ALARA principle, i.e. "As Low As Reasonably Achievable". An effects-based critical limit for Cd in cereals (wheat) of 0.10 mg kg^{-1} (fresh weight) was recommended for use under the Long-range Transboundary Air Pollution (LRTAP) Convention (Schütze et al. 2003). Critical contents of Cd, Pb and Hg in meat of cows and sheep in view of food safety equal 0.05, 0.10 and 0.05 mg kg^{-1} (fresh weight), respectively (EU 2001). Data for the toxic impacts on animals can be derived for all metals based on an acceptable daily intake (ADI).

Critical total soil metal concentrations can be derived from critical metal contents in plants from a non-linear relationship between both metal concentrations, accounting for the impact of soil properties that control the (bio)availability of metals in soils (Adams et al. 2004; Brus et al. 2002), according to (e.g. De Vries et al. 2007b):

$$\log ctM_{plant,crit} = a_0 + a_1 \cdot pH_{KCl} + a_2 \cdot \log OM + a_3 \cdot \log clay + m \cdot \log ctM_{tot,crit} \quad (2.8)$$

where $ctM_{plant,\ crit}$ is the critical limit for metal concentration in plant (mg kg^{-1}) and m is a coefficient describing the non-linear relationship between the metal concentrations in plant and soil. Values for the various regression coefficients in Eq. (2.8) have been derived for Cd, Pb, Hg, Cu and Zn in grass, maize, sugar beet, wheat, potatoes, lettuce, endive and spinach (De Vries et al. 2008). The relationships obtained for Cd and Zn were much better than those for Pb, Hg and Cu. For the majority of crops, soil-plant relationships for Pb, Hg and Cu are not statistically significant and should therefore not be used. Furthermore, the critical limit of Zn related to phytotoxic effects by far exceeds the measured plant values in the database, which limits the application of the regression model (strong extrapolation). Critical soil concentrations related to food quality criteria are thus only relevant for Cd. Values for a_0–a_3 and m for Cd, derived for Dutch soils, are given in Table 2.2. Examples of limits depending on soil type and pH for various crops are given in De Vries et al. (2007b).

The derivation of critical soil metal concentrations from acceptable daily intakes (ADI) by animals is relevant for grazing animals (specifically cows). Acceptable

Table 2.3 Overview of transfer functions to calculate K_f according to Eq. (2.11) with ctM_{tot} in mmol kg^{-1} and $[M]_{tot}$ in mmol l^{-1} and both *OM* and *clay* in % (Römkens et al. 2004)

Metal	b_0	b_1	b_2	b_3	n	R^2	se-y_{est} [a]
Cd	−4.85	0.27	0.58	0.28	0.54	0.79	0.33
Cu	−3.55	0.16	0.48	0.18	0.47	0.62	0.35
Pb	−2.96	0.25	0.83	0.02	0.68	0.57	0.55
Zn	−4.51	0.45	0.39	0.35	0.74	0.82	0.40

[a] Standard error of the y-estimate on a logarithmic basis

daily intakes of metals can be related to critical metal contents in fodder (grass) and soil according to:

$$ADI = ctM_{plant,crit} \cdot I_{plant} + ctM_{tot,crit} \cdot I_{soil} \tag{2.9}$$

where *ADI* is acceptable daily intake of metals (mg d^{-1}); I_{plant} is intake of plants (fodder) (kg d^{-1}) and I_{soil} is intake of soil (kg d^{-1}). The critical soil metal concentration can be derived by combining Eq. (2.8) and (2.9) and solving for $ctM_{tot,\ crit}$ using a given ADI and given coefficients in the soil-plant relationship (see Table 2.2 for grass). As with food quality criteria for plants, the derivation of a critical soil metal concentration from ADI values is mainly relevant for Cd. Information on the derivation of ADI values for Cd, using food quality criteria for Cd in the kidney of cows, and data on the intake of plants and soil are given in De Vries et al. (2007b).

Critical concentrations for the total metal concentration in soil solution, $[M]_{tot}$, that are ultimately needed in a critical load calculation, can then be derived from a relationship between the total metal concentration in soil and in soil solution, according to (Römkens et al. 2004):

$$ctM_{tot} = K_f \cdot [M]_{tot}^n \tag{2.10}$$

where K_f is the Freundlich adsorption constant, a function of soil properties such as pH, organic matter content and clay content, and n is an exponent. An example of a transfer function describing the adsorption constant K_f is (Römkens et al. 2004):

$$\log K_f = b_0 + b_1 \cdot pH_{extract} + b_2 \cdot \log OM + b_3 \cdot \log clay \tag{2.11}$$

Values for b_0–b_3 and n for Cd, Cu, Pb and Zn, derived for Dutch soils, are given in Table 2.3. By using a critical limit for a metal in plants, a critical concentration in soil solution can thus be derived by a combination of Eq. (2.8, 2.10, and 2.11). A more involved way, often used in critical load calculations, is to use a relationship between: (i) the total metal content and reactive metal content, followed by (ii) a relationship between the reactive metal content and the free metal ion concentration (see e.g. Groenenberg et al. 2012) and then by (iii) a relationship between the free metal ion and total metal concentration in soil solution (e.g. De Vries and Groenenberg 2009).

2.3.2 *Aquatic Ecosystems*

2.3.2.1 Impacts of Metal Deposition

Concern about the input of metals to terrestrial ecosystems is also related to the ecotoxicological impact on aquatic organisms due to runoff to surface waters. Effects on aquatic organisms, including algae, crustacea and fish, include effects on gill function (Sola et al. 1995), nervous systems (Baatrup 1991), and growth and reproduction rates (Mance 1987). Human health effects of heavy metal uptake by consumption of aquatic organisms (fish) are specifically related to Hg (Meili 1997; Meili et al. 2003).

2.3.2.2 Critical Limits of Geochemical Indicators

Impacts of metals in aquatic ecosystems are related to contents in fish (this holds for Hg, for which a critical limit of 0.5 mg kg^{-1} fresh weight is used) and total dissolved metal concentrations (mg m^{-3} or μg l^{-1}) in surface water, being the relevant geochemical indicator for effects on aquatic organisms. Critical limits for metals in surface water are based on the OECD approach for calculating MPCs or critical limits of substances (OECD 1989) as presented in Sect. 2.3.1. Critical limits for total dissolved metal concentrations in surface water have been derived on the basis of chronic toxicity data for a variety of aquatic organisms, including the major taxonomic groups, i.e. algae, crustacea, macrophyta, molluscs and fish. A summary of effect-based critical limits, based on various EU Risk Assessment Reports for Cd, Pb, Cu, Zn and Hg, is presented in Table 2.4.

Values of the HC_5 are based on the 5th percentile cut-off value of various chronic toxicity data (Aldenberg and Slob 1993, Aldenberg and Jaworska 2000). For all metals, the HC_5 thus derived was divided by a so-called 'assessment factor' to assess the MPC. An assessment factor is a parameter that aims to calculate sufficiently 'safe' MPCs by accounting for e.g.: (i) limited endpoints covered, (ii) limited diversity and representativity of the taxonomic groups covered, (iii) statistical uncertainties around the 5th percentile estimate and (iv) limited validation of the HC_5 with multi-species mesocosm or field data. The necessity of such a factor, varying from 1 to 4 for the various metals, can be disputed. A comparison between the critical dissolved concentrations derived for soil solution at high pH, and surface water (Table 2.4) shows that the critical concentration is generally much lower in surface waters than in the soil solution for Cd, Zn and Pb and comparable for Cu and Hg (De Vries et al. 2007b). The differences between soil and surface water critical limits can be questioned. There is no theoretical reason why the sensitivities of soil and water organisms to metals should not be similar, assuming that the uptake of the free ion from the aqueous phase is the significant mechanism leading to toxicity. More research is needed to study the possibility to use similar limits for surface waters and soil solution. To avoid the risk of adverse effects, one may use the lowest values in deriving a critical load.

Table 2.4 Recommended critical limits for dissolved total Cd, Pb, Hg, Cu and Zn concentrations in surface waters (based on various EU risk assessment reports as summarized in De Vries et al. (2007b)

Metal	Data sources	HC5 concentration (mg m^{-3})	Assessment factor	Critical limit (mg m^{-3})
Cd	168 single species studies 9 multi species studies	0.38	2	0.19
Pb	19 freshwater NOECs/EC10s 11 saltwater NOECs/EC10s	5.0	3	1.6
Hg	30 freshwater and saltwater NOECs/EC10s	0.142	4	0.036
Cu	22 freshwater species specific NOECs/EC10s 4 multi species studies	8.2	1	8.2
Zn	18 freshwater NOECs/EC10s	15.6	2	7.8

2.4 Impacts of Nitrogen Deposition on Terrestrial and Aquatic Ecosystems and Geochemical Risk Indicators

2.4.1 Terrestrial Ecosystems

2.4.1.1 Impacts of Nitrogen Deposition

Theory on Ecosystem Nitrogen Saturation: Aber et al. (1989) launched the theory on ecosystem N saturation, focused on forest ecosystems, in which different stages can be identified in view of: (i) impacts on soil chemical processes such as mineralization, immobilization, nitrification, affecting N leaching, (ii) plant nutrition and forest growth and (iii) plant species diversity. Until a certain threshold level is reached, terrestrial ecosystems will react to additional N inputs by an increased biomass production until a physiological optimum (which is beyond the ecological optimum), but above that, production stays constant or even decreases. Below the threshold level for growth, however, changes in the ecosystem are already observed, especially the plant species diversity may gradually change towards more nitrophilic species (Bobbink et al. 1998; Bobbink and Hettelingh 2011; Ellenberg 1985). These impacts are discussed in more detail in Chaps. 4 and 5. In terrestrial (forest) ecosystems with a continuous elevated N input, the ecosystem may approach "N saturation" (Aber et al. 1989). In this stage, N leaching will increase above (nearly negligible) background levels, associated with soil acidification in terms of elevated leaching of base cations or aluminium, causing a decrease in ANC. Aber et al. (1998) hypothesize that the loss of mycorrhizal assimilation and exudation "could be a key process leading to increased nitrification and nitrate mobility". At the stage of "N saturation" or "N excess", the ecosystem may be destabilised by the interaction of a number of factors. It may cause nutrient imbalances, since the uptake of base cations (Ca, Mg, K) is reduced by increased levels of dissolved Al and NH_4 (Boxman et al. 1988). This effect may be aggravated in systems of low N status, where an elevated input of N will increase forest growth, thus causing an increased demand for base cations. Observations of increased tree growth of European forests (Spiecker et al. 1996) may be the effect of increased N inputs. An excess input of N can finally cause pollution of ground water due to high NO_3 leaching. An overview of possible effects on forests as a result of increased atmospheric acid and N deposition and/or exposure to air pollutants is presented by Erisman and De Vries (2000). An overview of impacts on ecosystem functions can also be found in Hettelingh et al. (2009) and De Vries et al. (2014) and in Chap. 24.

N deposition impacts on ecosystems can roughly be divided into the following categories: (i) direct toxicity of N gases and aerosols to individual species, (ii) eutrophication, (iii) acidification, (iv) differential effects of oxidised and reduced N, and (v) increased susceptibility to secondary stress and disturbance factors. An overview of the direct effect of nitrogenous gases and aerosols (NH_3, NO_2, NO, HNO_3 and NH_4^+) is summarized in the Annex, since direct effects and critical levels are not part of this book, while soil acidification effects have been described above. Below, we shortly elaborate on the other N-related effects.

Eutrophication: Increased N deposition results in an increase in the availability of inorganic N in the topsoil in the short term, except in bogs and fens. This gradually leads to an increase in plant productivity in N-limited vegetation, and thus to higher annual litter production and litter with high concentrations of N. Because of this, N mineralisation will also gradually increase, which, in turn, may increase plant productivity. This is a positive feedback, because higher N mineralisation leads to higher N uptake and its subsequent effects. Local plant species diversity increases with increasing resource availability when starting from originally very low levels. Above a certain level of primary productivity, however, local plant species diversity declines as production increases. Observational studies across N deposition gradients and many N-addition experiments demonstrate this effect in the long term. Competitive exclusion ('overshading') of characteristic species of oligotrophic or mesotrophic habitats occurs in the presence of relatively fast-growing nitrophilic species, with rare species at low abundances being especially at risk (e.g. Bobbink et al. 1998; Suding et al. 2005).

The rate of N cycling in the ecosystem clearly increases in case of high N inputs, although the response time to increased N inputs can be long in highly organic soils (with high C:N ratios), or, indeed in any soil with large potential N sinks. When N is no longer limiting in the ecosystem, plant growth becomes limited by other resources, such as phosphorus (P), K, Mg, or water. In this situation, the productivity of the vegetation does not increase further with continuing N input. However, N concentrations within the plants often tend to increase when N availability continues to increase and the increased N content may affect the palatability of the vegetation for herbivores or the sensitivity to pathogens. These processes will also affect the composition of microbial communities.

Differences in Effects of Oxidised Versus Reduced N: The response of sensitive plant species can be significantly affected by a change in N form. Species of calcareous or slightly acidic soils are able to use nitrate, or a combination of nitrate and ammonium, as their N source, whereas some studies showed that species in acidic habitats generally use ammonium (e.g. Gigon and Rorison 1972; Kinzel 1982). Laboratory and field studies demonstrate that the performance of most forest understory species of deciduous forests in southern Sweden improves when not only ammonium but also nitrate is available (Falkengren-Grerup et al. 1998; Olsson and Falkengren-Grerup 2000). One of the impacts of increased ammonium uptake is a reduced uptake of base cations (Ca, Mg and K) to the rhizosphere. Ultimately this can lead to severe nutritional imbalances (references in Bobbink et al. 2003, e.g. Nihlgård 1985; Van Dijk et al. 1990). High concentrations of ammonium in the soil solution are also toxic to many sensitive plant species, disrupting cell physiology, cell acidification, accumulation of N-rich amino acids, poor root development, and inhibition of shoot growth. Strong evidence exists that many endangered vascular plant species of grasslands, heathlands and soft-water lakes, and fen bryophytes, are very intolerant to increased concentrations of reduced N and to high NH_4^+:NO_3^- ratios (De Graaf et al. 1998; Kleijn et al. 2008; Paulissen et al. 2004; Van den Berg et al. 2008). Kleijn et al. (2008) for example, showed that growth of rare species typical to Dutch heaths, mat grass swards and fen meadows occurs only at low NH_4^+/NO_3^- ratios. Soil conditions (pH levels) were shown to severely affect toxicity levels of reduced N (Lucassen et al. 2003).

Increased Susceptibility to Secondary Stress and Disturbance Factors: Strong accumulation of N in foliage (e.g. as amino acids) may also affect frost hardiness and the intensity and frequency of insect and pathogenic pests. The sensitivity of plants to stress (defined here as external constraints, such as drought, frost, pathogens or herbivores, which limit dry-matter production), or disturbance factors (mechanisms which affect plant biomass by causing its partial or complete destruction), may be significantly affected by N deposition. With increasing N deposition, susceptibility to fungal pathogens and attacks by insects also increases. This is probably due to altered concentrations of phenolic compounds (leading to lower resistance) and soluble N compounds, such as free amino acids, in combination with a lower vitality of individual plants. Increased levels of pathogenic fungi have been found for several tree species in N-addition experiments and field surveys, but for most ecosystems data are lacking and the influence of such pathogens on diversity is still unclear (e.g. Bobbink et al. 2003; Flückiger et al. 2002).

In general, herbivory is affected by the palatability of plant material, which is strongly determined by its N content. Increased organic N content in plants, caused by N deposition, can thus result in increased insect herbivory (e.g. Throop and Lerdau 2004). Data on herbivory and N deposition are very scarce, but a link has been demonstrated in dry *Calluna* heathlands. The frequency and intensity of infestations of heather beetle (*Lochmaea suturalis*) are clearly related to atmospheric N inputs and N concentrations in the heather (e.g. Berdowski 1993; Brunsting and Heil 1985). N-related changes in plant physiology, phenology, biomass allocation (root:shoot ratios) and mycorrhizal infection can also differentially influence the sensitivity of plant species to drought or frost stress, leading to reduced growth in some species and possible changes in plant interactions.

Enhanced N Availability: Enhanced N availability may also cause water stress as a result of increased canopy size, increased shoot/root ratio and loss of mycorrhizal activity. This is due to: (i) water shortage, since a high N input favours growth of canopy biomass, whereas root growth is relatively unaffected (De Visser 1994), (ii) nutrient imbalances, since the increase in canopy biomass also causes an increased demand of base cation nutrients (Ca, Mg, K) whereas the uptake of these cations is reduced by increased levels of dissolved NH_4 (Boxman and van Dijk 1988) and (iii) an increased sensitivity to natural stress factors such as frost (Aronsson 1980; Bruck 1985) and attacks by fungi (Roelofs et al. 1985).

2.4.1.2 Critical Limits of Geochemical Nitrogen Indicators

Critical Soil C/N Ratios in View of Enhanced N Leaching: One possible indicator for N eutrophication impacts is the C/N ratio of either the organic layer or the mineral soil. There are indications that N retention is reduced with a decreasing soil C/N ratio, especially in the organic layer, as shown by Dise et al. (1998a, b) and Gundersen et al. (1998). Gundersen et al. (1998) presented a very limited C/N range in organic layers (25–30) to distinguish sites with high N retention, and thus low N leaching potential (>30), from those with low N retention and thus high low N leaching potential (<25). MacDonald et al. (2002), using a so-called IFEF dataset of published N budgets and C/N ratios of the organic layer, found the strongest relationships between N output and N input when the data were divided into 'N-rich' sites (C/N≤25) and 'C-rich' sites (C/N>25). Using an updated IFEF dataset, Dise et al. (2009) found the strongest relationships between N output and N input when the data were also divided into 'N-rich' sites and 'C-rich' sites, but with a C/N in the organic layer of 23 as threshold value. De Vries et al. (2007a), using a so-called ICP Forests level-II dataset of 121 input-output budgets compiled with comparable methods within this Pan-European Forest Monitoring Programme, found less clear relationships with C/N. Their results showed that N leaching fluxes were negligible at very high C/N ratios in the organic layer (>35), but at lower C/N ratios the scatter in N leaching rates was very high, although N leaching tended to increase with a decrease in C/N ratio. Using a subset of this ICP Forests level II database, subject to stringent quality control, Van der Salm et al. (2007) found that the highest explained variance in N leaching was found when including a positive relationship with temperature and an interaction between N throughfall and a critical C/N ratio of 23. The critical C/N ratio of 23, which is in line with results by Dise et al. (2009), was obtained by repetitive regression with different critical values. Considering the limited statistical relationship between soil C/N and N output one may argue that a C/N ratio in the organic layer between 20 and 25 is critical. From a dynamic point of view, one may argue that the C/N ratio should stay constant, as any lowering

Table 2.5 Acceptable N concentrations in soil solution given in the Mapping Manual (www.icpmapping.org), derived in De Vries et al. (2007c)

Impact	Critical N concentration (mg N l^{-1})
Vegetation changes in Northern Europe[a]	
Lichens to cranberry	0.2–0.4
Cranberry to blueberry	0.4–0.6
Blueberry to grass	1–2
Grass to herbs	3–5
Vegetation changes in Western Europe	
Coniferous forest	2.5–4
Deciduous forest	3.5–6.5
Grass lands	3
Heath lands	3–6
Other impacts on forests	
Nutrient imbalances	(0.2–0.4)[a]
Elevated nitrogen leaching/N saturation	1
Fine root biomass/root length	1–3
Sensitivity to frost and fungal diseases	3–5

[a] Note that these critical limits are not really substantiated by field observations (see main text)

leads potentially to an increase in N leaching. Since there is a very limited linkage between N output and C/N ratio without considering the N input, it is hardly used as a criterion in critical load calculations.

Critical Dissolved N Concentrations in View of Various Adverse Impacts: The main indicator for N impacts used in critical load assessments is the total inorganic N concentration in soil solution. This section includes a substantiation of reported critical limits summarized in Table 2.5, details of which can be found in De Vries et al. (2007c). The values have also been incorporated in the Mapping Manual (www.icpmapping.org).

Vegetation Changes: Critical limits for dissolved N concentrations in view of vegetation changes vary from 0.2–6.5 mg l^{-1}. The origin for the limits related to vegetation changes in Northern Europe (Table 2.5) is based on Warfvinge et al. (1992), who states that "limiting N soil solution concentrations have been suggested based on preliminary experiences from the Swedish Forest Survey program". As shown by De Vries et al. (2007c), the values are based on an inverse use of the SMB model, by using empirically derived critical N loads and deriving the critical N leaching rate by subtracting values for the related N uptake, N immobilization and denitrification and dividing this flux by the water flux. These authors noted that the numbers may lack sufficient experimental ground truth, since it is the increase in N availability through enhanced N cycling that triggers vegetation changes and not the leaching of N. The origin for the limits related to vegetation changes in Western Europe (Table 2.5) is based on application of the biodiversity impact model NTM in the Netherlands, indicating a range in median critical N concentration in soil solution of 0.1–5 mg l^{-1} for grass lands and heath lands and of 0.1–10 mg l^{-1} for forests, with a median value near 3 mg l^{-1} (De Vries et al. 2007c).

Note that an application of critical N concentrations in the given intervals may lead to critical loads that are outside the range of empirical data on vegetation changes. In this case it is relevant to limit the critical loads to either the lowest or highest value of the encountered range in empirical critical loads (see Chaps. 4 and 5).

Nutrient Imbalances: The Mapping Manual (www.icpmapping.org) gives in brackets critical limits in view of nutrient imbalances that vary from 0.2–0.4 mg l^{-1}. These values are based on the assumption that an imbalance occurs as soon as N leaching exceeds natural N leaching rates. There is, however, no substantiation for this assumption, neither from the literature nor by the concepts of plant physiology. Nutrient imbalance has been assumed to occur when the availability of base cations (Ca, Mg and K) instead of N becomes limiting for growth, but this is not clearly related to a critical N concentration in soil solution (De Vries et al. 2007c).

N Leaching and Ground Water Quality: De Vries et al. (2007c) suggest an upper limit of 1 mg N l^{-1} as differentiation between undisturbed and 'leaky' N saturated forest sites, based on Gundersen et al. (2006). These authors gave an overview of current water quality in forests by compiling a list of studies from the 1990s on nitrate concentration in seepage water from temperate forests, including >500 sites of seepage water from Europe. From the survey data it is difficult to conclude exactly at which level a forest ecosystem can be considered 'leaky', but they suggest an annual average N concentration level of 1 mg N l^{-1} for seepage water and 0.5 mg N l^{-1} for streams/catchments. Stoddard (1994) characterised four progressive stages of N saturation based on changes in seasonality and levels of nitrate leaching in steams and a value of 1 mg N l^{-1} coincides with his limit for the near final stage. It is clear that the value for a leaking site is still far below the critical limits for ground water. The EC critical limit for NO_3–N in drinking water is 11.3 mg N l^{-1} and the target value is 5.6 mg N l^{-1}.

Root Growth and Sensitivity to Frost and Diseases: Empirical data suggest that critical dissolved N concentrations in view of adverse impacts on fine root biomass/root length and an increased sensitivity to frost and fungal diseases vary between 1–3 mg N l^{-1} and 3–5 mg N l^{-1}, respectively (De Vries et al. 2007c). The critical values for impacts on fine root biomass and root length are based on Matzner and Murach (1995), who found that total fine root biomass of Norway spruce saplings decreased significantly when the dissolved N ($NO_3 + NH_4$) concentration was >2 mg N l^{-1}. Critical dissolved N concentrations in view of an increased sensitivity to frost and fungal diseases has been derived from a critical N concentration in the needles of 18 g kg^{-1}, above which this sensitivity increases. De Vries et al. (2007c) derived a relationship between foliar N contents and dissolved annual average N concentrations on the basis of the results for 120 Intensive Monitoring plots in Europe. Below 3 mg N l^{-1}, the N contents in foliage were always below 18 g kg^{-1}, while above 5 mg N l^{-1} values were nearly always above this value.

2.4.2 Aquatic Ecosystems

2.4.2.1 Impacts of Nitrogen Deposition

Eutrophication of Nutrient-Poor Aquatic Ecosystems: N deposition in natural ecosystems can lead to enhanced leaching of inorganic N from soils to surface waters, in cases where the supply of N is greater than plant and microbial demand, referred to as 'nitrogen saturation' (Aber et al. 1989; Stoddard 1994). In natural ecosystems, inorganic N species leached to surface waters are predominantly in the form of nitrate (NO_3). Effects of N leaching on surface water eutrophication were not considered likely until recently, because of P limitation of primary productivity (Stoddard 1994). Any additional N in a P-limited system could thus not be expected to stimulate algal growth. P limitation of primary productivity in surface waters has been the dominating view since the 1970s (Hecky and Kilham 1988; Schindler 1971, 1977). This view has been challenged recently in reviews (Elser et al. 2007; Lewis and Wurtsbaugh 2008; Sterner 2008), although earlier papers also stressed that N availability was important for algal production (Elser et al. 1990; Kratzer and Brezonik 1981).

An overview of literature concerning effects of N enrichment on the biology in nutrient-poor aquatic ecosystems in arctic, boreal and temperate regions is given in De Wit and Lindholm (2010). The choice for nutrient-poor ecosystems was guided by the focus under the LRTAP Convention on natural ecosystems, where local pollution sources (e.g. domestic sewage, industrial waste water, agriculture, etc.) are absent. Thus, any ecosystem responses to N will be most likely related to atmospheric N. Effects of N on freshwater systems were documented from various types of studies: historical records of lake sediments (paleolimnology), lake surveys and experimental studies focusing on algae (free-floating phytoplankton and sediment-dwelling benthic algae), water plants (macrophytes) and invertebrates (insects, snails).

Paleolimnological Evidence for N-Limitation: Lake sediment studies (paleolimnology) show shifts in algal communities and increases in algal growth related to higher N concentrations (Baron et al. 2009; Hobbs et al. 2010; Holmgren et al. 2009; Pla et al. 2009; Saros et al. 2003; Wolfe et al. 2001, 2006). Most paleolimnological papers on possible effects of N enrichment on phytoplankton communities in oligotrophic lakes are studies published from 2000 onwards. The studies have a wide geographical and climatic range and include temperate, alpine and arctic lakes from Europe and North-America, with N deposition levels between <1 and >10 kg N $ha^{-1}yr^{-1}$. The changes in community structures are in varying degrees attributed to N enrichment (Baron et al. 2000; Wolfe et al. 2001) or to interactions of climate and N deposition (Hobbs et al. 2010; Holmgren et al. 2009; Pla et al. 2009; Saros et al. 2003; Wolfe et al. 2006). There are also shifts in paleolimnological records that are attributed to climate only (Birks et al. 2004; Ruhland et al. 2008). The evidence presented support the hypothesis that N deposition increases algal productivity and can change algal community structure in oligotrophic arctic, alpine and temperate lakes, at levels of N deposition as low as <1 to 3 kg N $ha^{-1}yr^{-1}$, but especially in lakes receiving higher loads of atmospheric N (>10 kg N $ha^{-1}yr^{-1}$). However, not all studies show evidence of higher productivity related to increased N, and some changes in community structure are interpreted as climate effects rather than N enrichment effects. As geochemical risk indicator, depletion of stable ^{15}N isotopes was used to infer enrichment of sediments by airborne N pollution.

Experiments and Surveys Supplying Evidence for N Limitation: Whole-lake experiments, mesocosms and bioassays are accepted methods used to experimentally investigate nutrient controls on freshwater primary productivity, in declining order of ecological realism and costs. Lake surveys along various gradients (deposition, climate, vegetation cover) are also used to statistically relate water chemistry and indicators of algal growth. In many studies a combination of some or all of these methods is used to gain insights in nutrient controls on productivity.

Whole-lake manipulations have often been done by manipulating both N and P supply (Fee 1979; Schindler 1977), and conflicting interpretations exist with respect to whether N limitation has been detected (Fee 1979; Schindler 1977; Schindler et al. 2008; Scott and McCarthy 2010). However, increases in productivity of both phytoplankton and sediment-dwelling algae were observed (Axler and Reuter 1996; Jansson et al. 2001) where only N was added, and several authors interpret results from lakes receiving both N and P as showing evidence of N-limitation, in addition to P-limitation and co-limitation (see also Finlay et al. 2013).

Regional surveys in boreal lakes in Europe and North America with N deposition between <1 and >14 kg N $ha^{-1}yr^{-1}$ show higher chlorophyll concentrations per unit P in areas with higher N deposition, indicative of higher primary production (Bergström et al. 2005, 2008; Bergström and Jansson 2006). Experimental nutrient additions in lakes (mesocosm studies and bioassays) support the finding of the regional surveys by showing that co-limitation (N and P) or N limitation of algal growth is common, especially under conditions of low N availability (Andersson and Brunberg 2006; Lafrancois et al. 2004; Morris and Lewis 1988; Nydick et al. 2003, 2004; Saros et al. 2005). It is suggested that N deposition may have shifted primary productivity in oligotrophic lakes from N limitation to P limitation (Bergström and Jansson 2006; Elser et al. 2009).

2.4.2.2 Critical Limits of Geochemical Nitrogen Indicators

Critical limits of geochemical N indicators for eutrophication are not easy to quantify because effects do not depend on aqueous N concentrations only, but also on the availability of other nutrients. Lake chemistry, in combination with experimental evidence of impacts on phytoplankton, can be used to assess geochemical indicators for N limitation regarding phytoplankton. Total N to total P (TN:TP) mass ratios have been shown to correlate positively to N deposition in a large lake survey in Norway and Sweden (Elser et al. 2009). Nutrient enrichment studies in lakes in

Sweden, Norway and Colorado showed that N limitation occurred at TN:TP ratios below 20. However, some studies suggest that TN:TP ratios to evaluate nutrient deficiencies may underestimate N-deficiency, since total N in oligotrophic lakes is often partly unavailable (Axler et al. 1994; Maberly et al. 2002). Dissolved inorganic N (DIN) to TP ratios may be better predictors of responses to nutrient additions (Bergström 2010). Bergström (2010) evaluated the use of lake water concentration ratios as a geochemical indicator for N or P limitation of phytoplankton in oligotrophic lakes. The DIN:TP ratios correlated better with N limitation determined by bioassays than TN:TP ratios, and DIN:TP ratios were also better related to N deposition and catchment cover. For mass ratios of DIN:TP above 3.4, phytoplankton was usually P-limited, while N limitation dominated for a DIN:TP mass ratio below 1.5. For DIN:TP ratios between 1.5 and 3.4, co-limitation of N and P was common.

Camargo and Alonso (2006) performed a global assessment on the ecological and toxicological effects of inorganic N pollution in aquatic ecosystems. According to these authors, TN concentrations lower than 0.5–1.0 mg N l^{-1} might prevent aquatic ecosystems from developing eutrophication and acidification. Most oligotrophic freshwaters have TN concentrations below this threshold (Elser et al. 2009), while impacts on phytoplankton productivity and communities still have been documented. Thus, several assessments underline the importance of using both N and P concentrations for evaluation of N effects on phytoplankton productivity, and highlight especially the use of DIN rather than TN. However, no consensus appears to have emerged regarding the threshold values for these ratios. The mass ratio of DIN:TP of 1.5 suggested by Bergström (2010), below which N limitation was common, could be used as a first approximation. Using of this limit implies that models calculating a critical N load for nutrient-poor freshwater ecosystems have to account for P-availability. Ratios of N:P in stream waters vary strongly with water discharge and its controls remain poorly understood (Green and Finlay 2010). Catchment retention of N is furthermore very variable depending on land cover, climate and N deposition. The present state of insights in these complex interactions implies that model based critical N loads in view of eutrophication are difficult to derive. Instead, empirical critical N loads have been derived for nutrient-poor aquatic systems, as presented in Chap. 4.

2.5 Overview of Critical Limits of Geochemical Indicators and Evaluation

Table 2.6 provides an overview of the relevant geochemical indicators for N, acidity and metals in soil and water (soil solution, ground water and surface water) with critical limits in view of their impacts on terrestrial and aquatic ecosystems.

A critical evaluation of each geochemical endpoint has been given above. In summary, the critical values related to ANC in view of effects on aquatic organisms (fish) are most clearly related to visible impacts in the field. Critical limits for Al, Bc/Al ratios and metals in soil solution, related to effects on soil organisms

Table 2.6 Geochemical indicators and typical critical limits used to derive critical loads for nitrogen, acidity and heavy metals

Aspect	Soil	Soil water/drinking water	Surface water
Acidity: Ecosystem effects	Reactive Al pool: no decrease	$[Al^{3+}]_{free}=2$ mg l^{-1}	ANC = 20 (0–50) meq m^{-3} or ANC adjusted for TOC $[Al^{3+}]=20$ mg m^{-3} to avoid impacts on roach
	Exchangeable BC pool/base saturation: no decrease	Molar [Bc]/[Al] = 1	pH = 5.8 to avoid impacts on roach
Acidity: Human health effects		Indicator value of $[Al]_{tot}$ of 0.2 μg l^{-1} to protect drinking water	
Metals: Ecosystem effects (Cd, Pb, Hg, Cu and Zn)	Total and reactive metal concentration ($M_{tot,}$ M_{re}) = function of pH, clay and organic matter content	$[M]_{free}$ = function of pH $[M]_{tot}$ = function of pH, solution chemistry and DOM	$[M]_{tot}$ in mg $m^{-3}=0.19$ for Cd, 1.6 for Pb, 0.036 for Hg, 8.2 for Cu and 7.8 for Zn
Metals: Human health effects (Cd, Pb, Hg)	Total Cd concentration = function of pH, clay and organic matter content; based on food quality criterion (0.1 mg kg^{-1} fresh weight) and soil plant relationship	$[Cd]_{tot}$, $[Pb]_{tot}$ and$[Hg]_{tot}$ of 3, 10 and 1 μg l^{-1} to protect drinking water $[Cd]_{tot}$ = function of pH, clay and organic matter content; based on food quality criterion, soil plant relationship and soil solution relationship	$[Hg]_{tot}$ in fish = 0.5 mg kg^{-1} fresh weight
Nitrogen ecosystem effects	C/N ratio: no critical limit; Should not decrease to avoid N saturation	[N] = function of type of impact as shown in Table 2.5: varies from 0.2–6 mg N l^{-1}	[N] = 0.5–1.0 mg N l^{-1} DIN:TP = 1.5
Nitrogen: Human health effects		$[NO_3]=50$ mg l^{-1} implying $[NO_3-N]=11.3$ mg N l^{-1} to protect drinking water	

and plants, are based on a harmonized methodology, but mostly based on laboratory experiments. Consequently, the linkage between exceedances of critical limits and field effects is limited, and exceedances thus only serve as a risk indicator for terrestrial ecosystems in view of possible long-term chronic effects. Critical metals limits related to human health effects are based on acceptable daily intakes and models that assess related critical limits for food crops, drinking water and fish, using a precautionary principle. Critical limits for N in soil solution and surface water are mainly based on correlative studies under field conditions, but the impact is less clear than for ANC, amongst other due to interactions with phosphorus. Considering the above, field effects increase in probability going from acidification effects on terrestrial systems to eutrophication effects on terrestrial and aquatic systems to acidification effects on aquatic systems.

Uncertainties in the relation between critical chemical values and 'harmful effects' imply an associated uncertainty in the calculation of critical loads and their exceedances by atmospheric depositions. However, their influence on the robustness of 'risk' assessments, in the context described above, can be limited when moving from site-specific to regional model applications and from short to long time horizons (see, e.g., Chap. 25). All critical limits serve in critical load calculations and dynamic model calculations to enable standardized assessments on broad regional and temporal scales of the risks of atmospheric deposition to chosen endpoints for structure and function of sensitive ecosystems indicators. The violation of a critical limit points to a risk of field effects that may vary over temporal and spatial scales. In general, the 'risk' associated with an exceedance of a critical limit implies an increased probability of occurrence of adverse effects, either now or in the future.

References

Aber, J. D., Nadelhoffer, K. J., Steudler, P., & Melillo, J. M. (1989). Nitrogen saturation in northern forest ecosystems. *BioScience, 39,* 378–386.

Aber, J. D., McDowell, W., Nadelhoffer, K., Magill, A., Berntsen, G., Kamakea, M., et al. (1998). Nitrogen saturation in temperate forest ecosystems: Hypotheses revisited. *BioScience, 48,* 921–934.

Abrahamsen, G., Stuanes, A. O., & Tveite, B. (1993a). Introduction; study area; experimental design. In G. Abrahamsen, A. O. Stuanes, & B. Tveite (Eds.), *Long-term experiments with acid rain in Norwegian ecosystems* (pp. 3–33). New York: Springer-Verlag.

Abrahamsen, G., Stuanes, A. O., & Tveite, B. (1993b). Discussion and synthesis. In G. Abrahamsen, A. O. Stuanes, & B. Tveite (Eds.), *Long-term experiments with acid rain in Norwegian ecosystems* (pp. 297–331). New York: Springer-Verlag.

Achermann, B., & Bobbink, R. (Eds.). (2003). *Empirical critical loads for nitrogen: Expert workshop, Berne, 11–13 November 2002.* Berne: Swiss Agency for the Environment, Forests and Landscape.

Adams, M. L., Zhao, F. J., McGrath, S. P., Nicholson, F. A., & Chambers, B. J. (2004). Predicting cadmium concentrations in wheat and barley grain using soil properties. *Journal of Environmental Quality, 33,* 532–541.

Ahonen-Jonnarth, U., Göransson, A., & Finlay, R. D. (2003). Growth and nutrient uptake of ectomycorrhizal Pinus sylvestris seedlings in a natural substrate treated with elevated Al concentration. *Tree Physiology, 23,* 157–167.

Akselsson, C., Ardö, J., & Sverdrup, H. (2004). Critical loads of acidity for forest soils and relationship to forest decline in the Northern Czech Republic. *Environmental Monitoring and Assessment, 98,* 363–379.

Akselsson, C., Westling, O., Sverdrup, H., Holmqvist, J., Thelin, G., Uggla, E., et al. (2007). Impact of harvest intensity on long term base cation budgets in Swedish forest soils. *Water, Air and Soil Pollution: Focus, 7,* 201–210.

Aldenberg, T., & Jaworska, J. S. (2000). Uncertainty of the hazardous concentration and fraction affected for normal species sensitivity distributions. *Ecotoxicology and Environmental Safety, 46,* 1–18.

Aldenberg, T., & Slob, W. (1993). Confidence limits for hazardous concentrations based on logistically distributed NOEC toxicity data. *Ecotoxicology and Environmental Safety, 25,* 48–63.

Alewell, C., Manderscheid, B., Gerstberger, P., & Matzner, E. (2000). Effects of reduced atmospheric deposition on soil solution chemistry and elemental contents of spruce needles in NE-Bavaria, Germany. *Zeitschrift für Pflanzenernährung und Bodenkunde, 163,* 509–516.

Allen, H. E. (1993). The significance of trace metal speciation for water, sediment and soil quality standards. *Science of the Total Environment, 134,* 23–45.

Alloway, B. J. (Ed.). (1990). *Heavy metals in soils.* Glasgow: Blackie Academic & Professional, Chapman & Hall.

Andersson, E., & Brunberg, A. K. (2006). Inorganic nutrient acquisition in a shallow clearwater lake—Dominance of benthic microbiota. *Aquatic Sciences, 68,* 172–180.

Aronsson, A. (1980). Frost hardiness in Scots pine. II Hardiness during winter and spring in young trees of different mineral status. *Studia Forestalia Suecica, 155,* 1–27.

Ashmore, M., van den Berg, L., Terry, A., Tipping, E., Lawlor, A. J., Lofts, S., et al. (2007). *Development of an effects-based approach for toxic metals.* (Report to the UK Department for Environment, Food and Rural Affairs, the Scottish Executive, the National Assembly for Wales and the Department of the Environment in Northern Ireland. Contract CPEA 24). University of York.

Augustaitis, A., & Bytnerowicz, A. (2008). Contribution of ambient ozone to Scots pine defoliation and reduced growth in the Central European forests: A Lithuanian case study. *Environmental Pollution, 155,* 436–445.

Augustin, S., Bolte, A., Holzhausen, M., & Wolff, B. (2005a). Exceedance of critical loads of nitrogen and sulphur and its relation to forest conditions. *European Journal of Forest Research, 124,* 289–300.

Augustin, S., Stepanowitz, H., Wolff, B., Schröder, J., & Hoffmann, E. (2005b). Manganese in tree rings of Norway spruce as an indicator for soil chemical changes in the past. *European Journal of Forest Research, 124,* 313–318.

Axler, R. P., & Reuter, J. E. (1996). Nitrate uptake by phytoplankton and periphyton: Whole-lake enrichments and mesocosm-N-15 experiments in an oligotrophic lake. *Limnology and Oceanography, 41,* 659–671.

Axler, R. P., Rose, C., & Tikkanen, C. A. (1994). Phytoplankton nutrient deficiency as related to atmospheric nitrogen deposition in Northern Minnesota acid-sensitive lakes. *Canadian Journal of Fisheries and Aquatic Science, 51,* 1281–1296.

Baatrup, E. (1991). Structural and functional-effects of heavy-metals on the nervous-system, including sense-organs, of fish. *Comparative Biochemistry and Physiology Part C: Toxicology & Pharmacology, 199,* 253–257.

Baker, J. P., & Schofield, C. L. (1982). Aluminum toxicity to fish in acidic waters. *Water, Air, and Soil Pollution, 18,* 289–309.

Balsberg-Påhlsson, A. M. (1989). Toxicity of heavy metals (Zn, Cu, Cd, Pb) to vascular plants. *Water, Air, and Soil Pollution, 47,* 287–319.

Baron, J. S., Rueth, H. M., Wolfe, A. M., Nydick, K. R., Allstott, E. J., Minear, J. T., et al. (2000). Ecosystem responses to nitrogen deposition in the Colorado Front Range. *Ecosystems, 3,* 352–368.

Baron, J. S., Schmidt, T. M., & Hartman, M. D. (2009). Climate-induced changes in high elevation stream nitrate dynamics. *Global Change Biology, 15,* 1777–1789.

Beamish, R. J. (1974). Loss of fish populations from unexploited remote lakes in Ontario, Canada as a consequence of atmospheric fallout of acid. *Water Research, 8,* 85–95.

Beamish, R. J., & Harvey, H. H. (1972). Acidification of La Cloche Mountain lakes, Ontario, and resulting fish mortalities. *Journal of Fisheries Research Board of Canada, 29,* 1131–1143.

Beier, C., Blanck, K., Bredemeier, M., Lamersdorf, N., Rasmussen, L., & Xu, Y. J. (1998). Field-scale 'clean rain' treatments to two Norway spruce stands within the EXMAN project—Effects on soil solution chemistry, foliar nutrition and tree growth. *Forest Ecology and Management, 101,* 111–123.

Bengtsson, G., & Tranvik, L. (1989). Critical metal concentrations for forest soil invertebrates. *Water, Air, and Soil Pollution, 47,* 381–417.

Berdowski, J. J. M. (1993). The effect of external stress and disturbance factors on Calluna-dominated heathland vegetation. In R. Aerts, & G. W. Heil (Eds.), *Heathlands: Patterns and processes in a changing environment* (pp. 85–124). Dordrecht: Kluwer.

Bergström, A. K. (2010). The use of TN:TP and DIN:TP ratios as indicators for phytoplankton nutrient limitation in oligotrophic lakes affected by N deposition. *Aquatic Science, 72,* 277–281.

Bergström, A.-K., & Jansson, M. (2006). Atmospheric nitrogen deposition has caused nitrogen enrichment and eutrophication of lakes in the northern hemisphere. *Global Change Biology, 12,* 635–643.

Bergström, A.-K., Blomqvist, P., & Jansson, M. (2005). Effects of atmospheric nitrogen deposition on nutrient limitation and phytoplankton biomass in unproductive Swedish lakes. *Limnology and Oceanography, 50,* 987–994.

Bergström, A.-K., Jonsson, A., & Jansson, M. (2008). Phytoplankton responses to nitrogen and phosphorus enrichment in unproductive Swedish lakes along a gradient of atmospheric nitrogen deposition. *Aquatic Biology, 4,* 55–64.

Binkley, D., & Högberg, P. (1997). Does atmospheric deposition of nitrogen threaten Swedish forests? *Forest Ecology and Management, 92,* 119–152.

Birks, H. J. B., Jones, V. J., & Rose, N. L. (2004). Recent environmental change and atmospheric contamination on Svalbard as recorded in lake sediments—Synthesis and general conclusions. *Journal of Paleolimnology, 31,* 531–546.

Bobbink, R., & Hettelingh, J.-P. (2011). *Review and revision of empirical critical loads and dose-response relationships: Proceedings of an Expert Workshop, Noordwijkerhout, 23–25 June 2010.* (Report 680359002/2011). Bilthoven: Coordination Centre for Effects, National Institute for Public Health and the Environment.

Bobbink, R., Boxman, D., Fremstad, E., Heil, G., Houdijk, A., & Roelofs, J. (1992). Critical loads for nitrogen eutrophication of terrestrial and wetland ecosystems based upon changes in vegetation and fauna. In P. Grennfelt & E. Thörnelöv (Eds.), *Critical loads for nitrogen. Report from a workshop held at Lökeberg, Sweden, 6–10 April 1992* (pp. 111–161). Nordic Council of Ministers, Report 1992, 41.

Bobbink, R., Hornung, M., & Roelofs, J. G. M. (1996). Empirical nitrogen critical loads for natural and semi-natural ecosystems. In *Manual on methodologies and criteria for mapping critical loads/levels* (p. 54). UNECE Convention on Long-range Transboundary Air Pollution. Texte 71–96, III-1/III-54. Umweltbundesamt, Berlin.

Bobbink, R., Hornung, M., & Roelofs, J. G. M. (1998). The effects of air-borne nitrogen pollutants on species diversity in natural and semi-natural European vegetation. *Journal of Ecology, 86,* 717–738.

Bobbink, R., Ashmore, M., Braun, S., Flückiger, W., & van den Wyngaert, I. J. J. (2003). Empirical nitrogen critical loads for natural and semi-natural ecosystems: 2002 update. In B. Achermann, & R. Bobbink (Eds.), *Empirical critical loads for nitrogen* (pp. 43–170). Berne: Swiss Agency for Environment, Forest and Landscape SAEFL.

Boekhold, A. E., Temminghoff, E. J. M., & van der Zee, S. E. A. T. M. (1993). Influence of electrolyte composition and pH on cadmium sorption by an acid sandy soil. *Journal of Soil Science, 44,* 85–96.

Boggero, A., Barbieri, A., de Jong, J., Marchetto, A., & Mosello, R. (1998). Chemistry and critical loads of Alpine lakes in Canton Ticino (southern central Alps). *Aquatic Science, 60,* 300–315.

Boxman, A. W., & van Dijk, H. F. G. (1988). *Het effect van landbouw ammonium deposities op bos- en heidevegetaties*. Katholieke Universiteit Nijmegen.

Boxman, A. W., van Dijk, H. F. G., Houdijk, A. L. F. M., & Roelofs, J. G. M. (1988). Critical loads for nitrogen with special emphasis on ammonium. In J. Nilsson, & P. Grennfelt (Eds.), *Critical loads for sulphur and nitrogen. Report from a workshop held at Skokloster, Sweden, 19–24 March, 1988. Miljø rapport 1988 15.* (pp. 295–322). København: Nordic Council of Ministers.

Boxman, A. W., Vandam, D., Vandijk, H. F. G., Hogervorst, R. F., & Koopmans, C. J. (1995). Ecosystem responses to reduced nitrogen and sulfur inputs into 2 coniferous forest stands in the Netherlands. *Forest Ecology and Management, 71,* 7–29.
Braun, S., Schindler, C., Volz, R., & Flückiger, W. (2003). Forest damages by the storm 'lothar' in permanent observation plots in Switzerland: the significance of soil acidification and nitrogen deposition. *Water, Air, and Soil Pollution, 142,* 327–340.
Bringmark, L., Bringmark, E., & Samuelsson, B. (1998). Effects on mor layer respiration by small experimental additions of mercury and lead. *Science of the Total Environment, 213,* 115–119.
Bruck, R. I. (1985). Boreal montane ecosystem decline in the southern Appalachian Mountains: potential role of anthropogenic pollution. In H. S. Stubbs (Ed.), *Air pollution effects on forest ecosystems* (pp. 137–155). St. Paul, Minnesota: Acid Rain Foundation.
Brunsting, A. M. H., & Heil, G. W. (1985). The role of nutrients in the interactions between a herbivorous beetle and some competing plant species in heathlands. *Oikos, 44,* 23–26.
Brus, D. J., de Gruijter, J. J., Walvoort, D. J. J., De Vries, F., Bronswijk, J. J. B., Römkens, P. F. A. M., et al. (2002). Mapping the probability of exceeding critical thresholds for cadmium concentrations in soils in the Netherlands. *Journal of Environmental Quality, 31,* 1875–1884.
Camargo, J. A., & Alonso, A. (2006). Ecological and toxicological effects of inorganic nitrogen pollution in aquatic ecosystems: A global assessment. *Environment International, 32,* 831–849.
Campbell, P. G. C. (1995). Interactions between trace metals and aquatic organisms: A critique on the free-ion activity model. In A. Tessier, & D. R. Turner (Eds.), Metal speciation and bioavailability in aquatic systems (pp. 45–102). New York: Wiley.
CLAG, Critical Loads Advisory Group. (1995). *Critical loads of acid deposition for United Kingdom freshwaters*. London: ITE Edinburgh/Department of the Environment.
Clark, R. B. (2001). Metals. In Marine pollution (pp. 98–125). Oxford: Oxford Science Publishers.
Cronan, C. S. (1991). Differential adsorption of Al, Ca and Mg by roots of red spruce (*Picea rubens* Sarg.). *Tree Physiology, 8,* 227–237.
Cronan, C. S., & Grigal, D. F. (1995). Use of calcium/aluminum ratios as indicators of stress in forest ecosystems. *Journal of Environmental Quality, 24,* 209–226.
Cronan, C. S., April, R., Bartlett, R. J., Bloom, P. R., Driscoll, C. T., Gherini, S. A., et al. (1989). Aluminum toxicity in forests exposed to acidic deposition: The ALBIOS results. *Water, Air, and Soil Pollution, 48,* 181–192.
Das, P., Samantaray, S., & Rout, G. R. (1997). Studies on cadmium toxicity in plants: A review. *Environmental Pollution, 98,* 29–36.
De Graaf, M. C. C., Bobbink, R., Roelofs, J. G. M., & Verbeek, P. J. M. (1998). Differential effects of ammonium and nitrate on three heathland species. *Plant Ecology, 135,* 185–196.
De Visser, P. H. B. (1994). Growth and nutrition of Douglas-fir, Scots pine and pedunculate oak in relation to soil acidification. PhD Thesis, Wageningen Agricultural University, Wageningen, The Netherlands.
De Vries, W. (1993). Average critical loads for nitrogen and sulfur and its use in acidification abatement policy in the Netherlands. *Water, Air, and Soil Pollution, 68,* 399–434.
De Vries, W., & Groenenberg, J. E. (2009). Evaluation of approaches to calculate critical metal loads for forest ecosystems. *Environmental Pollution, 157,* 3422–3432.
De Vries, W., & Kros, J. (1989). The long-term impact of acid deposition on the aluminium chemistry of an acid forest soil. In J. Kämäri, D. F. Brakke, A. Jenkins, S. A. Norton, & R. F. Wright (Eds.). *Regional acidification models* (pp. 113–128). Berlin: Springer-Verlag.
De Vries, W., Posch, M., & Kämäri, J. (1989). Simulation of the long-term soil response to acid deposition in various buffer ranges. *Water, Air, and Soil Pollution, 48,* 349–390.
De Vries, W., Reinds, G. J., & Posch, M. (1994). Assessment of critical loads and their exceedance on European forests using a one-layer steady-state model. *Water, Air, and Soil Pollution, 72,* 357–394.
De Vries, W., Vel, E., Reinds, G. J., Deelstra, H., Klap, J. M., Leeters, E. E. J. M., et al. (2003). Intensive monitoring of forest ecosystems in Europe: 1. Objectives, set-up and evaluation strategy. *Forest Ecology and Management, 174,* 77–95.

De Vries, W., van der Salm, C., Reinds, G. J., & Erisman, J. W. (2007a). Element fluxes through European forest ecosystems and their relationships with stand and site characteristics. *Environmental Pollution, 148,* 501–513.

De Vries, W., Lofts, S., Tipping, E., Meili, M., Groenenberg, B. J., & Schütze, G. (2007b). Impact of soil properties on critical concentrations of cadmium, lead, copper, zinc and mercury in soil and soil solution in view of ecotoxicological effects. *Reviews of Environmental Contamination and Toxicology, 191,* 47–89.

De Vries, W., Kros, J., Reinds, G. J., Wamelink, G. W. W., Mol, J., van Dobben, H., et al. (2007c). *Developments in deriving critical limits and modelling critical loads of nitrogen for terrestrial ecosystems in Europe*. (Report 1382). Wageningen, The Netherlands, Alterra Wageningen UR.

De Vries, W., Römkens, P. F. A. M., Bonten, L., Rietra, R. P. J. J., Ma, W. C., & Faber, J. (2008). *De invloed van bodemeigenschappen op kritische concentraties voor zware metalen en organische microverontreinigingen in de bodem*. (Alterra-rapport 817). Wageningen: (in Dutch).

De Vries, W., Posch, M., Reinds, G. J., & Hettelingh, J.-P. (2014). Quantification of impacts of nitrogen deposition on forest ecosystem services in Europe. In M. A. Sutton, K. E. Mason, L. J. Sheppard, H. Sverdrup, R. Haeuber, & W. K. Hicks (Eds.), *Critical loads, nitrogen deposition and biodiversity. Chapter 43* (pp. 411–424). Springer.

De Wit, H. A., & Lindholm, M. (2010). *Nutrient enrichment effects of atmospheric N deposition on biology in oligotrophic surface waters—A review*. (NIVA report). Oslo: Norwegian Institute for Water Research.

De Wit, H. A., Mulder, J., Nygaard, P. H., & Aamlid, D. (2001a). Testing the aluminium toxicity hypothesis: A field manipulation experiment in mature spruce forest in Norway. *Water, Air, and Soil Pollution, 130,* 995–1000.

De Wit, H. A., Mulder, J., Nygaard, P. H., Aamlid, D., Huse, M., Kortnes, E., et al. (2001b). Aluminium: The need for a re-evaluation of its toxicity and solubility in mature spruce stands. *Water, Air, and Soil Pollution: Focus, 1,* 103–118.

De Wit, H. A., Eldhuset, T., & Mulder, J. (2010). Dissolved Al reduces Mg uptake in Norway spruce forest: Results from a long-term field manipulation experiment in Norway. *Forest Ecology and Management, 259,* 2072–2082.

Di Toro, D. M., Allen, H. E., Bergman, H. L., Meyer, J. S., Paquin, P. R., & Santore, R. C. (2001). Biotic ligand model of the acute toxicity of metals. 1. Technical basis. *Environmental Toxicology and Chemistry, 20,* 2383–2396.

Dise, N. B., Matzner, E., & Forsius, M. (1998a). Evaluation of organic horizon C:N ratio as an indicator of nitrate leaching in conifer forests across Europe. *Environmental Pollution, 102,* 453–456.

Dise, N. B., Matzner, E., & Gundersen, P. (1998b). Synthesis of nitrogen pools and fluxes from European forest ecosystems. *Water, Air, and Soil Pollution, 105,* 143–154.

Dise, N. B., Rothwell, J. J., Gauci, V., van der Salm, C., & De Vries, W. (2009). Predicting nitrate leaching in European forests using two independent databases. *Science of the Total Environment, 407,* 1798–1808.

Duan, L., Hao, J., Xie, S., & Du, K. (2000). Critical loads of acidity for surface waters in China. *Science of the Total Environment, 246,* 1–10.

EC. (1998). *Council Directive 98/83/EC of 3 November 1998 on the quality of water intended for human consumption*. Brussel: European Commision.

Eldhuset, T. D., Lange, H., & De Wit, H. A. (2006). Fine root biomass, necromass and chemistry during seven years of elevated aluminium concentrations in the soil solution of a middle-aged Picea abes stand. *Science of the Total Environment, 396,* 344–356.

Ellenberg, H. Jr. (1985). Veränderungen der Flora Mitteleuropas unter dem Einfluss von Düngung und Immissionen. *Schweizerische Zeitschrift für Forstwesen, 136,* 19–39.

Elser, J. J., Marzolf, E. R., & Goldman, C. R. (1990). Phosphorus and nitrogen limitation of phytoplankton growth in the fresh-waters of North-America—A review and critique of experimental enrichments. *Canadian Journal of Fisheries and Aquatic Science, 47,* 1468–1477.

Elser, J. J., Bracken, M. E. S., Cleland, E. E., Gruner, D. S., Harpole, W. S., Hillebrand, H., et al. (2007). Global analysis of nitrogen and phosphorus limitation of primary producers in freshwater, marine and terrestrial ecosystems. *Ecology Letters, 10,* 1135–1142.

Elser, J. J., Andersen, T., Baron, J. S., Bergström, A. K., Jansson, M., Kyle, M., et al. (2009). Shifts in lake N:P stoichiometry and nutrient limitation driven by atmospheric nitrogen deposition. *Science, 326,* 835.

Erisman, J. W., & De Vries, W. (2000). Nitrogen deposition and effects on European forests. *Environmental Reviews, 8,* 65–93.

EU. (2001). *Verordening nr. 466/2001 van 8 maart 2001, tot vaststelling van maximumgehalten aan bepaalde verontreinigingen in levensmiddelen.* Publicatieblad van de Europese Gemeenschappen L 77.

Falkengren-Grerup, U. (1986). Soil acidification and vegetation changes in deciduous forest in southern Sweden. *Oecologia, 70,* 339–347.

Falkengren-Grerup, U., & Tyler, G. (1993). Experimental evidence for the relative sensitivity of deciduous forest plants to high soil acidity. *Forest Ecology and Management, 60,* 311–326.

Falkengren-Grerup, U., van der Hoek, K. W., Erisman, J. W., Smeulders, S., & Wisniewski, J. R. (1998). Nitrogen response of herbs and graminoids in experiments with simulated acid soil solution. *Environmental Pollution, 102,* 93–99.

Fee, E. J. (1979). Relation between lake morphometry and primary productivity and its use in interpreting whole-lake eutrophication experiments. *Limnology and Oceanography, 24,* 401–416.

Fergusson, J. E. (1990). *The heavy elements: Chemistry environmental impact and health effects.* Oxford: Pergamon Press.

Finlay, J. C., Small, G. E., & Sterner, R. W. (2013). Human influences in nitrogen removal in lakes. *Science, 342,* 247–250.

Fischer, R., Mues, V., Ulrich, E., Becher, G., & Lorenz, M. (2007). Monitoring of atmospheric deposition in European forests and an overview of its implication on forest condition. *Applied Geochemistry, 22,* 1129–1139.

Flückiger, W., Braun, S., & Hiltbrunner, E. (2002). Effects of air pollutants on biotic stress. In J. N. B. Bell & M. Treshow (Eds.), *Air pollution and plant life* (2nd edn., pp. 379–406). Chichester: Wiley.

Gigon, A., & Rorison, I. H. (1972). The response of some ecologically distinct plant species to nitrate- and to ammonium-nitrogen. *Journal of Ecology, 60,* 93–102.

Giller, K. E., Witter, E., & McGrath, S. P. (1998). Toxicity of heavy metals to microorganisms and microbial processes in agricultural soils: A review. *Soil Biology and Biochemistry, 30,* 1389–1414.

Gray, D. K., Arnott, S. E., Shead, J. A., & Derry, A. M. (2012). The recovery of acid-damaged zooplankton communities in Canadian lakes: the relative importance of abiotic, biotic and spatial variables. *Freshwater Biology, 57,* 741–758.

Green, M. B., & Finlay, J. C. (2010). Patterns of hydrologic control over stream water total nitrogen to total phosphorus ratios. *Biogeochemistry, 99,* 15–30.

Groenenberg, J. E., Römkens, P. F. A. M., Comans, R. N. J., Luster, J., Pampura, T., Shotbolt, L., et al. (2010). Transfer functions for solid-solution partitioning of cadmium, copper, nickel, lead and zinc in soils: Derivation of relationships for free metal ion activities and validation with independent data. *European Journal of Soil Science, 61,* 58–73.

Groenenberg, J. E., Dijkstra, J. J., Bonten, L. T. C., De Vries, W., & Comans, R. N. J. (2012). Evaluation of the performance and limitations of empirical partition-relations and process based multisurface models to predict trace element solubility in soils. *Environmental Pollution, 166,* 98–107.

Gundersen, P., Callesen, I., & De Vries, W. (1998). Nitrate leaching in forest ecosystems is related to forest floor C/N ratios. *Environmental Pollution, 102,* 403–407.

Gundersen, P., Schmidt, I. K., & Raulund-Rasmussen, K. (2006). Leaching of nitrate from temperate forests—Effects of air pollution and forest management. *Environmental Reviews, 14,* 1–57.

Hecky, R. E., & Kilham, P. (1988). Nutrient limitation of phytoplankton in fresh-water and marine environments—A review of recent-evidence on the effects of enrichment. *Limnology and Oceanography, 33,* 796–822.
Helbing, D. (2013). Globally networked risk and how to respond. *Nature, 497,* 51–59.
Henriksen, A., Lien, L., Traaen, T. S., Sevaldrud, I. S., & Brakke, D. F. (1988). Lake acidification in Norway—Present and predicted chemical status. *Ambio, 17,* 259–266.
Henriksen, A., Lien, L., Rosseland, B. O., Traaen, T. S., & Sevaldrud, I. S. (1989). Lake acidification in Norway—Present and predicted fish status. *Ambio, 18,* 314–321.
Henriksen, A., Posch, M., Hultberg, H., & Lien, L. (1995). Critical loads of acidity for surface waters—Can the ANC_{limit} be considered variable? *Water, Air, and Soil Pollution, 85,* 2419–2424.
Henriksen, A., Dillon, P. J., & Aherne, J. (2002). Critical loads of acidity for surface waters in south-central Ontario, Canada: Regional application of the Steady-State Water Chemistry (SSWC) model. *Canadian Journal of Fisheries and Aquatic Science, 59,* 1287–1295.
Hesthagen, T., Sevaldrud, I. H., & Berger, H. M. (1999). Assessment of damage to fish population in Norwegian lakes due to acidification. *Ambio, 28,* 112–117.
Hesthagen, T., Fjellheim, A., Schartau, A. K., Wright, R. F., Saksgård, R., & Rosseland, B. O. (2011). Chemical and biological recovery of Lake Saudlandsvatn, a formerly highly acidified lake in southernmost Norway, in response to decreased acid deposition. *Science of the Total Environment, 409,* 2908–2916.
Hettelingh, J.-P., Posch, M., Slootweg, J., Reinds, G. J., Spranger, T., & Tarrasón, L. (2007). Critical loads and dynamic modelling to assess European areas at risk of acidification and eutrophication. *Water, Air, and Soil Pollution: Focus, 7,* 379–384.
Hettelingh, J.-P., Posch, M., & Slootweg, J. (2009). *Progress in the modelling of critical thresholds, impacts to plant species diversity and ecosystem services in Europe.* (CCE Status Report 2009). Bilthoven: RIVM, Coordination Centre for Effects.
Hobbs, W. O., Telford, R. J., Birks, H. J. B., Saros, J. E., Hazewinkel, R. R. O., Perren, B. B., et al. (2010). Quantifying recent ecological changes in remote lakes of North America and Greenland using sediment diatom assemblages. *PLoS One, 5,* e10026.
Holmgren, S. U., Bigler, C., Ingolfsson, O., & Wolfe, A. P. (2009). The Holocene–Anthropocene transition in lakes of western Spitsbergen, Svalbard (Norwegian High Arctic): Climate change and nitrogen deposition. *Journal of Paleolimnology, 43,* 393–412.
Huber, C., Kreutzer, K., Rohle, H., & Rothe, A. (2004). Response of artificial acid irrigation, liming, and N-fertilisation on elemental concentrations in needles, litter fluxes, volume increment, and crown transparency of a N saturated Norway spruce stand. *Forest Ecology and Management, 200,* 3–21.
Hultberg, H. (1988). Critical loads for sulphur to lakes and streams. In J. Nilsson, & P. Grennfelt (Eds.), *Critical loads for sulphur and nitrogen; Report from a workshop held at Skokloster, Sweden, 19–24 March, 1988. Miljø rapport 1988 15* (pp. 185–200). Copenhagen: Nordic Council of Ministers.
Hultberg, H., & Stenson, J. (1970). Försurningens effekter på fiskfaunan i två bohusländska småsjöar. *Fauna och Flora, 65,* 11–20.
Hutchinson, T. C., Bozic, L., & Munoz-Vega, G. (1986). Responses of five species of conifer seedlings to aluminum stress. *Water, Air, and Soil Pollution, 31,* 283–294.
Jansson, M., Bergstrom, A. K., Drakare, S., & Blomqvist, P. (2001). Nutrient limitation of bacterioplankton and phytoplankton in humic lakes in northern Sweden. *Freshwater Biology, 46,* 653–666.
Jeffries, D. S., & Lam, D. C. L. (1993). Assessment of the effect of acidic deposition on Canadian lakes: Determination of critical loads for sulphate deposition. *Water Science Technology, 28,* 183–187.
Jensen, K. W., & Snekvik, E. (1972). Low pH levels wipe out salmon and trout populations in southernmost Norway. *Ambio, 1,* 223–225.
Joslin, J. D., & Wolfe, M. H. (1988). Responses of red spruce seedlings to changes in soil aluminum in six amended forest soil horizons. *Canadian Journal of Forest Research, 18,* 1614–1623.

Joslin, J. D., & Wolfe, M. H. (1989). Aluminum effects on northern red oak seedling growth in six forest soil horizons. *Soil Science Society of America Journal, 53,* 274–281.

Keltjens, W. G., & van Loenen, E. (1989). Effects of aluminium and mineral nutrition on growth and chemical composition of hydroponically grown seedlings of five different forest tree species. *Plant and Soil, 119,* 39–50.

Kinraide, T. B. (2003). Toxicity factors in acidic forest soils: Attempts to evaluate separately the toxic effects of excessive Al^{3+} and H^{+} and insufficient Ca^{2+} and Mg^{2+} upon root elongation. *European Journal of Soil Science, 54,* 323–333.

Kinzel, S. (1982). *Pflanzenökologie und Mineralstoffwechsel.* Stuttgart: Ulmer.

Klap, J. M., Oude Voshaar, J. H., De Vries, W., & Erisman, J. W. (2000). Effects of environmental stress on forest crown condition in Europe. Part IV: Statistical analysis of relationships. *Water, Air, and Soil Pollution, 119,* 387–420.

Kleijn, D., Bekker, R. M., Bobbink, R., de Graaf, M. C. C., & Roelofs, J. G. M. (2008). In search for key biogeochemical factors affecting plant species persistence in heathland and acidic grasslands: A comparison of common and rare species. *Journal of Applied Ecology, 45,* 680–687.

Kratzer, C. R., & Brezonik, P. L. (1981). A Carlson-type trophic state index for Florida lakes. *Water Resources Bulletin, 17,* 713–715.

Kreutzer, K. (1995). Effects of forest liming on soil processes. *Plant and Soil, 168,* 447–470.

Kreutzer, K., & Weiss, T. (1998). The Höglwald field experiments—Aims, concept and basic data. *Plant and Soil, 199,* 1–10.

Kreutzer, K., Beier, C., Bredemeier, M., Blanck, K., Cummins, T., Farrell, E. P., et al. (1998). Atmospheric deposition and soil acidification in five coniferous forest ecosystems: A comparison of the control plots of the EXMAN sites. *Forest Ecology and Management, 101,* 125–142.

Lafrancois, B. M., Nydick, K. R., Johnson, B. M., & Baron, J. S. (2004). Cumulative effects of nutrients and pH on the plankton of two mountain lakes. *Canadian Journal of Fisheries and Aquatic Science, 61,* 1153–1165.

Lamersdorf, N. P., & Borken, W. (2004). Clean rain promotes fine root growth and soil respiration in a Norway spruce forest. *Global Change Biology, 10,* 1351–1362.

Larssen, T., Lund, E., & Høgåsen, T. (2008). *Exceedance of critical loads for aciddification and nitrogen for Norway—Update for the period 2002–2006.* (NIVA-report 5697–2008). Oslo: Norwegian Institute for Water Research.

Lawrence, G. B., Lapenis, A., Berggren, D., Aparin, B. F., Smith, K. T., Shortle, W. C., et al. (2005). Climate dependency of tree growth suppressed by acid deposition effects on soils in northwest Russia. *Environmental Science and Technology, 39,* 2004–2010.

Leivestad, H., & Muniz, I. P. (1976). Fish kill at low pH in a Norwegian river. *Nature, 259,* 391–392.

Lewis, W. M., & Wurtsbaugh, W. A. (2008). Control of lacustrine phytoplankton by nutrients: Erosion of the phosphorus paradigm. *Internationale Revue der Hydrobiologie, 93,* 446–465.

Lien, L., Raddum, G. G., Fjellheim, A., & Henriksen, A. (1996). A critical limit for acid neutralizing capacity in Norwegian surface waters, based on new analyses of fish and invertebrate responses. *Science of the Total Environment, 177,* 173–193.

Lofts, S., Spurgeon, D. J., Svendsen, C., & Tipping, E. (2004). Deriving soil critical limits for Cu, Zn, Cd and Pb: A method based on free ion concentrations. *Environmental Science and Technology, 38,* 3623–3631.

Løkke, H., Bak, J., Falkengren-Grerup, U., Finlay, R. D., Ilvesniemi, H., Nygaard, P. H., et al. (1996). Critical loads of acidic deposition for forest soils: Is the current approach adequate? *Ambio, 25,* 510–516.

Lucassen, E. C. H. E. T., Bobbink, R., Smolders, A. J. P., van der Ven, P. J. M., Lamers, L. P. M., & Roelofs, J. G. M. (2003). Interactive effects of low pH and high ammonium levels responsible for the decline of Cirsium dissectum (L.) Hill. *Plant Ecology, 165,* 45–52.

Lydersen, E., Larssen, T., & Fjeld, E. (2004). The influence of total organic carbon (TOC) on the relationship between acid neutralizing capacity (ANC) and fish status in Norwegian lakes. *Science of the Total Environment, 326,* 63–69.

Maberly, S. C., King, L., Dent, M. M., Jones, R. I., & Gibson, C. E. (2002). Nutrient limitation of phytoplankton and periphyton growth in upland lakes. *Freshwater Biology, 47,* 2136–2152.

MacDonald, J. A., Dise, N. B., Matzner, E., Armbruster, M., Gundersen, P., & Forsius, M. (2002). Nitrogen input together with ecosystem nitrogen enrichment predict nitrate leaching from European forests. *Global Change Biology, 8,* 1028–1033.

Mance, G. (1987). *Pollution threat of heavy metals in aquatic environments.* London: Elsevier Applied Science.

Marschner, H. (1990). *Mineral nutrition of higher plants.* London: Academic.

Matson, P., Lohse, K. A., & Hall, S. J. (2002). The globalization of nitrogen deposition: Consequences for terrestrial ecosystems. *Ambio, 31,* 113–119.

Matzner, E., & Murach, D. (1995). Soil changes induced by air pollutant deposition and their implication for forests in central Europe. *Water, Air, and Soil Pollution, 85,* 63–76.

McCormick, L. H., & Steiner, K. C. (1978). Variation in aluminum tolerance among six genera of trees. *Forest Science, 24,* 565–568.

Meili, M. (1997). Mercury in lakes and rivers. In A. Sigel & H. Sigel (Eds.), *Mercury and its effects on environment and biology* (pp. 21–51). New York: Marcel Dekker Inc.

Meili, M., Bishop, K., Bringmark, L., Johansson, K., Munthe, J., Sverdrup, H., et al. (2003). Critical levels of atmospheric pollution: Criteria and concepts for operational modelling of mercury in forest and lake ecosystems. *Science of the Total Environment, 304,* 83–106.

Mengel, K. (1991). *Ernährung und Stoffwechsel der Pflanze. 7th revised edition.* Jena.

Moiseenko, T. (1994). Acidification and critical load in surface waters: Kola, Northern Russia. *Ambio, 23,* 418–424.

Monteith, D. T., Hildrew, A. G., Flower, R. J., Raven, P. J., Beaumont, W. R. B., Collen, P., et al. (2005). Biological responses to the chemical recovery of acidified fresh waters in the UK. *Environmental Pollution, 137,* 83–101.

Morel, F. M. M. (1983). *Principles of aquatic chemistry.* New York: Wiley.

Morris, D. P., & Lewis, W. M. (1988). Phytoplankton nutrient limitation in Colorado mountain lakes. *Freshwater Biology, 20,* 315–327.

Mulder, J., Van Breemen, N., & Eijck, H. (1989). Depletion of soil aluminum by acid deposition and implications for acid neutralization. *Nature, 337,* 247–249.

Nelleman, C., & Frogner, T. (1994). Spatial patterns of spruce defoliation seen in relation to acid deposition, critical loads and natural growth conditions in Norway. *Ambio, 23,* 255–259.

Nihlgård, B. (1985). The ammonium hypothesis—An additional explanation for the forest dieback in Europe. *Ambio, 14,* 2–8.

Nilsson, L. O., & Wiklund, K. (1995a). Indirect effects of N and S deposition on a Norway spruce ecosystem. An update of findings within the Skogaby project. *Water, Air, and Soil Pollution, 85,* 1613–1622.

Nilsson, L. O., & Wiklund, K. (1995b). Nutrient balance and P, K, Ca, Mg, S and B accumulation in a Norway spruce stand following ammonium-sulfate application, fertigation, irrigation, drought and N-free-fertilization. *Plant and Soil, 168,* 437–446.

Nowotny, I., Dahne, J., Klingelhofer, D., & Rothe, G. M. (1998). Effect of artificial soil acidification and liming on growth and nutrient status of mycorrhizal roots of Norway spruce (Picea abies [L.] Karst.). *Plant and Soil, 199,* 29–40.

Nydick, K. R., Lafrancois, B. M., Baron, J. S., & Johnson, B. M. (2003). Lake-specific responses to elevated atmospheric nitrogen deposition in the Colorado Rocky Mountains, USA. *Hydrobiologia (incorporating JAQU), 510,* 103–114.

Nydick, K. R., Lafrancois, B. M., Baron, J. S., & Johnson, B. M. (2004). Nitrogen regulation of algal biomass, productivity, and composition in shallow mountain lakes, Snowy Range, Wyoming, USA. *Canadian Journal of Fisheries and Aquatic Sciences, 61,* 1256–1268.

Nygaard, P. H., & de Wit, H. A. (2004). Effects of elevated soil solution Al concentrations on fine roots in a middle-aged Norway spruce (*Picea abies* (L.) Karst.) stand. *Plant and Soil, 265,* 131–140.

OECD, (Organisation for Economic Cooperation and Development). (1989). *Report of the OECD workshop on ecological effects assessment.* Paris: OECD Environment Monographs No. 26.

Olsson, M. O., & Falkengren-Grerup, U. (2000). Potential nitrification as an indicator of preferential uptake of ammonium or nitrate by plants in an oak understory. *Annals of Botany, 85,* 299–305.

Palmborg, C., Bringmark, L., Bringmark, E., & Nordgren, A. (1998). Multivariate analysis of microbial activity and soil organic matter at a forest site subjected to low-level heavy metal pollution. *Ambio, 27,* 53–57.

Paulissen, M. P. C. P., van der Ven, P. J. M., Dees, A. J., & Bobbink, R. (2004). Differential effects of nitrate and ammonium on three fen bryophyte species in relation to pollutant nitrogen input. *New Phytologist, 164,* 551–458.

Persson, H., & Majdi, H. (1995). Effects of acid deposition on tree roots in Swedish forest stands. *Water, Air, and Soil Pollution, 85,* 1287–1292.

Pla, S., Monteith, D., Flower, R., & Rose, N. (2009). The recent palaeolimnology of a remote Scottish loch with special reference to the relative impacts of regional warming and atmospheric contamination. *Freshwater Biology, 54,* 505–523.

Posch, M., Kämäri, J., Forsius, M., Henriksen, A., & Wilander, A. (1997). Exceedance of critical loads for lakes in Finland, Norway and Sweden: Reduction requirements for acidifying nitrogen and sulfur deposition. *Environmental Management, 21,* 291–304.

Posch, M., Aherne, J., Forsius, M., & Rask, M. (2012). Past, present, and future exceedance of critical loads of acidity for surface waters in Finland. *Environmental Science and Technology, 46,* 4507–4514.

Prasad, M. N. V. (1995). Cadmium toxicity and tolerance in vascular plants. *Environmental and Experimental Botany, 35,* 525–545.

Rask, M., Mannio, J., Forsius, M., Posch, M., & Vuorinen, P. J. (1995). How many fish populations in Finland are affected by acid precipitation? *Environmental Biology of Fishes, 42,* 51–63.

Rengel, Z. (1992). Role of calcium in aluminum toxicity. *New Phytologist, 121,* 499–513.

Ritchie, G. S. P., & Sposito, G. (2001). Speciation in soils. In A. M. Ure & C. M. Davidson (Eds.), *Chemical speciation in the environment* (pp. 237–264). Oxford: Blackwell Science.

Roelofs, J. G. M., Kempers, A. J., Houdijk, A. L. F. M., & Jansen, J. (1985). The effect of airborne ammonium sulphate on Pinus nigra var. maritima in the Netherlands. *Plant and Soil, 84,* 45–56.

Römkens, P. F. A. M., Groenenberg, J. E., Bonten, L. T. C., De Vries, W., & Bril, J. (2004). *Derivation of partition relationships to calculate Cd, Cu, Ni, Pb and Zn solubility and activity in soil solutions.* (Alterra Rapport 305). Wageningen: Alterra.

Rost-Siebert, K. (1983). Aluminium-Toxizität und -Toleranz an Keimpflanzen von Fichte (Picea abies Karst.) und Buche (*Fagus silvatica L.*). *Allgemeine Forstzeitschrift, 26/27,* 686–689.

Rothe, A., Huber, C., Kreutzer, K., & Weis, W. (2002). Deposition and soil leaching in stands of Norway spruce and European Beech: Results from the Höglwald research in comparison with other European case studies. *Plant and Soil, 240,* 33–45.

Ruhland, K., Paterson, A. M., & Smol, J. P. (2008). Hemispheric-scale patterns of climate-related shifts in planktonic diatoms from North American and European lakes. *Global Change Biology, 14,* 2740–2754.

Ryan, P. J., Gessel, S. P., & Zasoski, R. J. (1986a). Acid tolerance of Pacific Northwest conifers in solution culture. I: Effect of high aluminium concentration and solution acidity. *Plant and Soil, 96,* 239–257.

Ryan, P. J., Gessel, S. P., & Zasoski, R. J. (1986b). Acid tolerance of Pacific Northwest conifers in solution culture. II: Effect of varying aluminium concentration at constant pH. *Plant and Soil, 96,* 259–272.

Saros, J. E., Interlandi, S. J., Wolfe, A. P., & Engstrom, D. R. (2003). Recent changes in the diatom community structure of lakes in the Beartooth Mountain Range, USA. *Arctic, Antarctic, and Alpine Research, 35,* 18–23.

Saros, J. E., Michel, T. J., Interlandi, S. J., & Wolfe, A. P. (2005). Resource requirements of Asterionella Formosa and Fragilaria crotonsis in oligotrophic alpine lakes: Implications for recent phytoplankton community reorganisations. *Canadian Journal of Fisheries and Aquatic Science, 62,* 1681–1689.

Sauvé, S., Norvell, W. A., McBride, M. B., & Hendershot, W. H. (2000). Speciation and complexation of cadmium in extracted soil solutions. *Environmental Science and Technology, 34,* 291–296.

Saxe, J. K., Impellitteri, C. A., Peijnenburg, W. J. G. M., & Allen, H. E. (2001). Novel model describing trace metal concentrations in the earthworm, Eisenia andrei. *Environmental Science and Technology, 35,* 4522–4529.

Schindler, D. W. (1971). Carbon, nitrogen, and phosphorus and eutrophication of freshwater lakes. *Journal of Phycology, 7,* 321–329.

Schindler, D. W. (1977). Evolution of phosphorus limitation in lakes. *Science, 195,* 260–262.

Schindler, D. W., Hecky, R. E., Findlay, D. L., Stainton, M. P., Parker, B. R., Paterson, M. J., et al. (2008). Eutrophication of lakes cannot be controlled by reducing nitrogen input: Results of a 37-year whole-ecosystem experiment. *Proceedings of the National Academy of Sciences of United States of America, 105,* 11254–11258.

Schofield, C. L. (1976). Acid precipitation: Effects on fish. *Ambio, 5,* 228–230.

Schröder, W. H., Bauch, J., & Endeward, R. (1988). Microbeam analysis of Ca exchange and uptake in the fine roots of spruce: Influence of pH and aluminum. *Trees, 2,* 96–103.

Schulze, E.-D. (1989). Air pollution and forest decline in a spruce (Picea Abies) forest. *Science, 244,* 776–783.

Schütze, G., Lorenz, U., & Spranger, U. (2003). *Expert meeting on critical limits for heavy metals and methods for their application.* Berlin: Proceedings, UBA Texte 47/2003. Berlin: Umweltbundesamt.

Scott, J. T., & McCarthy, M. J. (2010). Nitrogen fixation may not balance the nitrogen pool in lakes over timescales relevant to eutrophication management. *Limnology and Oceanography, 55,* 1265–1270.

Slooff, W. (1992). *RIVM guidance document: Ecotoxicological effect assessment. Deriving maximum tolerable concentrations from single-species toxicity data. (Report 719102018).* Bilthoven: National Institute of Public Health and the Environment (RIVM).

Smit, H. P., Keltjens, W. G., & van Breemen, N. (1987). Effects of soil acidity on Douglas fir seedlings. 2. The role of pH, aluminium concentration and nitrogen nutrition (pot experiment). *Netherlands Journal of Agricultural Science, 35,* 537–540.

Sola, F., Isaia, J., & Masoni, A. (1995). Effects of copper on gill structure and transport function in the rainbow trout, Oncorhynchus mykiss. *Journal of Applied Toxicology, 15,* 391–398.

Solberg, S., Kvindesland, S., Aamlid, D., & Venn, K. (2002). Crown condition and needle chemistry of Norway spruce in relation to critical loads of acidity in South-East Norway. *Water, Air, and Soil Pollution, 140,* 157–171.

Spiecker, H., Mielikäinen, K., Kölh, M., & Skovsgaard, J. P. (Eds.). (1996). *Growth trends in European forests. Studies from 12 Countries.* Berlin: Springer-Verlag.

Spurgeon, D. J., & Hopkin, S. P. (1996). Effects of variations of the organic matter content and pH of soils on the availability and toxicity of zinc to the earthworm *Eisenia fetida. Pedobiologia, 40,* 80–96.

Steiner, K. C., McCormick, L. H., & Canavera, D. S. (1980). Differential response of paper birch provenances to aluminium in solution culture. *Canadian Journal of Forest Research, 10,* 25–29.

Sterner, R. W. (2008). On the phosphorus limitation paradigm for lakes. *International Review of Hydrobiology, 93,* 433–445.

Stoddard, J. L. (1994). Long-term changes in watershed retention of nitrogen: Its causes and aquatic consequences. In L. A. Baker (Ed.). *Environmental chemistry of lakes and reservoirs* (pp. 223–284). Washington, DC: American Chemical Society.

Suding, K. N., Collins, S. L., Gough, L., Clark, C., Cleland, E. E., Gross, K. L., et al. (2005). Functional- and abundance-based mechanisms explain diversity loss due to N fertilization. *Proceedings of the National Academy of Sciences of the United States of America, 102,* 4387–4392.

Sverdrup, H., & Warfvinge, P. (1993). *The effect of soil acidification on the growth of trees, grass and herbs as expressed by the (Ca + Mg + K)/Al ratio.* Reports in ecology and environmental engineering 1993:2. Lund University, Department of Chemical Engineering II.

Sverdrup, H., De Vries, W., & Henriksen, A. (1990). *Mapping critical loads. A guidance manual to criteria calculation methods data collection and mapping*. Miljø rapport 1990:14. Nordic Council of Ministers Copenhagen 1990.

Sverdrup, H. U., Warfvinge, P., & Rosén, K. (1992). A model for the impact of soil solution Ca:Al ratio, soil moisture and temperature on tree base cation uptake. *Water, Air, and Soil Pollution, 61,* 365–383.

Sverdrup, H., Thelin, G., Robles, M., Stjernquist, I., & Sörensen, J. (2006). Assessing sustainability of different tree species considering Ca, Mg, K, N and P at Björnstorp Estate. *Biogeochemistry, 81,* 219–238.

Sverdrup, H., Belyazid, S., Nihlgård, B., & Ericson, L. (2007). Modelling change in ground vegetation response to acid and nitrogen pollution, climate change and forest management at in Sweden 1500–2100 A.D. *Water, Air, and Soil Pollution: Focus, 7,* 163–179.

Tammi, J., Appelberg, M., Beier, U., Hesthagen, T., Lappalainen, A., & Rask, M. (2003). Fish status survey of Nordic lakes: Effects of acidification, eutrophication and stocking activity on present fish species composition. *Ambio, 32,* 98–105.

Thakali, S., Allen, H. E., Di Toro, D. M., Ponizovsky, A. A., Rooney, C. P., Zhao, F.-J., et al. (2006). A terrestrial biotic ligand model. 1. Development and application to Cu and Ni toxicities to barley root elongation in soils. *Environmental Science and Technology, 40,* 7085–7093.

Thornton, F. C., Schaedle, M., & Raynal, D. J. (1987). Effects of aluminum on red spruce seedlings in solution culture. *Environmental and Experimental Botany, 27,* 489–498.

Throop, H. L., & Lerdau, M. T. (2004). Effects of nitrogen deposition on insect herbivory: Implications for community and ecosystem processes. *Ecosystems, 7,* 109–133.

Tipping, E., Lofts, S., Hooper, H., Frey, B., Spurgeon, D., & Svendsen, C. (2010). Critical Limits for Hg(II) in soils, derived from chronic toxicity data. *Environmental Pollution, 158,* 2465–2471.

Tyler, G. (1992). *Critical concentrations of heavy metals in the mor horizon of Swedish forests* (Report 4078). Solna: Swedish Environmental Protection Agency.

UBA. (1996). *Manual on methodologies and criteria for mapping critical levels & loads and geographical areas where they are exceeded.* (UBA-Texte 71/96). Berlin: Umweltbundesamt.

Ulrich, B. (1981). Ökologische Gruppierung von Böden nach ihrem chemischen Bodenzustand. *Zeitschrift für Pflanzenernährung und Bodenkunde, 144,* 289–305.

Ulrich, B. (1984). Effects of air pollution on forest ecosystems and waters: The principles demonstrated at a case study in Central Europe. *Atmospheric Environment, 18,* 621–628.

Ulrich, B., & Matzner, E. (1983). *Abiotische Folgewirkungen der weitraümigen Ausbreitung von Luftverunreinigung.* (Forschungsbericht 10402615). Umweltforschungsplan der Bundesminister des Inneren. BRD: (in German).

Ulrich, B., & Pankrath, J. (1983). *Effects of accumulation of air pollutants on forest ecosystems.* Dordrecht: D. Reidel Publishing Company.

Ulrich, B., Mayer, R., & Khanna, P. K. (1980). Chemical changes due to acid precipitation in a loess-derived soil in central Europe. *Soil Science, 130,* 193–199.

Van den Berg, L. J. L., Peters, C. J. H., Ashmore, M. R., & Roelofs, J. G. M. (2008). Reduced nitrogen has a greater effect than oxidised nitrogen on dry heathland vegetation. *Environmental Pollution, 154,* 359–369.

Van der Salm, C., De Vries, W., Reinds, G. J., & Dise, N. B. (2007). N leaching across European forests: Derivation and validation of empirical relationships using data from intensive monitoring plots. *Forest Ecology and Management, 238,* 81–91.

Van Dijk, H. F. G., de Louw, M. H. J., Roelofs, J. G. M., & Verburgh, J. J. (1990). Impact of artificial, ammonium-enriched rainwater on soils and young coniferous trees in a greenhouse. Part 2-Effects on the trees. *Environmental Pollution, 63,* 41–59.

Vig, K., Megharaj, M., Sethunathan, N., & Naidu, R. (2003). Bioavailability and toxicity of cadmium to microorganisms and their activities in soil: A review. *Advances in Environmental Research, 8,* 121–135.

Warfvinge, P., Sverdrup, H., & Rósen, K. (1992). Calculating critical loads for N to forest soils. In P. Grennfeld & E. Thörnelöf (Eds.), *Critical loads for nitrogen. NORD 1992:41* (pp. 403–418). Copenhagen: Nordic Council of Ministers.

Warfvinge, P., Falkengren-Grerup, U., Sverdrup, H., & Andersen, B. (1993). Modelling long-term cation supply in acidified forest stands. *Environmental Pollution, 80,* 209–221.

Watmough, S. A., & Dillon, P. J. (2003). Do critical load models adequately protect forests? A case study in south-central Ontario. *Canadian Journal of Forest Research, 33,* 1544–1556.

Watmough, S. A., Aherne, J., Alewell, C., Arp, P., Bailey, S., Clair, T., et al. (2005). Sulphate, nitrogen and base cation budgets at 21 forested catchments in Canada, the United States and Europe. *Environmental Monitoring and Assessment, 109,* 1–36.

WHO. (2004). *Guidelines for drinking-water quality. Volume 1: Recommendations*. Geneva.

Wolfe, A. P., Baron, J. S., & Cornett, R. J. (2001). Anthropogenic nitrogen deposition induces rapid ecological changes in alpine lakes of the Colorado Front Range (USA). *Journal of Paleolimnology, 25,* 1–7.

Wolfe, A. P., Cooke, C. A., & Hobbs, W. O. (2006). Are current rates of atmospheric Nitrogen deposition influencing lakes in the eastern Canadian Arctic? *Arctic, Antarctic and Alpine Research, 38,* 465–476.

Zöttl, H. W., & Mies, E. (1983). Nährelementversorgung und Schadstoffbelastung von Fichtenökosystemen im Südschwarzwald unter Immissionseinfluß. *Mitteilungen der Deutschen Botanischen Gesellschaft, 38,* 429–434.

Chapter 3
Plant Species Diversity Indicators for Use in the Computation of Critical Loads and Dynamic Risk Assessments

Han F. van Dobben, Maximilian Posch, G. W. Wieger Wamelink, Jean-Paul Hettelingh and Wim de Vries

3.1 Introduction

Ecological Effects of Atmospheric Nitrogen Deposition: Ecological effects of atmospheric deposition were first noticed in the 1960s (Odén 1967) and generated extensive public debate, especially after large-scale forest dieback had been predicted in the 1970s (Ulrich et al. 1979). Today the focus has shifted from deposition of acidity to deposition of nitrogen (N) compounds, and effects are now defined in terms of biodiversity loss rather than in terms of forest dieback, but there is still great concern about the ecological effects of deposition world-wide. There is ample evidence that increasing N availability causes overall declines in plant species diversity (Aerts and Berendse 1988; Bobbink et al. 2010; Galloway and Cowling 2002; Stevens et al. 2011; Tilman 1987). Sala et al. (2000) identified N deposition as the third most important driver of biodiversity change globally by the year 2100, after land use change and climate change, and as the most important driver overall for northern temperate forests.

The major impacts of N on terrestrial ecosystem diversity are through (1) eutrophication, (2) acidification, (3) direct foliar impacts, and (4) exacerbation of other stresses (see also Chap. 2). Of these processes the two first-mentioned have the most important impacts on plant species and communities, especially on those that are adapted to low nutrient levels or are poorly buffered against acidification. Two mechanisms play a role: (1) competition by species adapted to a high N availability and (2) decrease of the pH to values below the ecological range of sensitive species. In oligotrophic ecosystems, N is generally the most important growth-limiting element, and their species are adapted to a N-deficient environment. If the availability of N increases, a few species that use the available nitrogen most efficiently

H. van Dobben (✉) · W. Wamelink · W. de Vries
Alterra Wageningen University and Research Centre, Wageningen, The Netherlands
e-mail: han.vandobben@wur.nl

M. Posch · J.-P. Hettelingh
Coordination Centre for Effects (CCE), RIVM, Bilthoven, The Netherlands

W. de Vries et al. (eds.), *Critical Loads and Dynamic Risk Assessments*,
Environmental Pollution 25, DOI 10.1007/978-94-017-9508-1_3

will out-compete a large number of unproductive species adapted to N deficiency, resulting in a loss of species diversity. The composition of the vegetation can also be affected by acidification, leading to the decline of species that are characteristic for intermediate and higher soil pH, and to dominance of acid-resistant species. Details on the various soil chemical mechanisms behind the effects of N deposition are given in Chap. 2.

Effects of N deposition are now recognised in nearly all oligotrophic natural ecosystems; these include grasslands (including tundra and Mediterranean grasslands), heathland, coastal habitats, oligotrophic wetlands (mire, bog and fen), forests, aquatic habitats and marine habitats (Achermann and Bobbink 2003; Bobbink et al. 2003; Bobbink and Hettelingh 2011). Ecosystems of cold climates, including montane, boreal, tundra, subarctic and arctic habitats, are also vulnerable to N deposition (Bobbink et al. 2010; Dise et al. 2011). In Europe, many semi-natural grassland communities are dominated by species with low nutrient requirements and are sensitive to acidification, eutrophication, or both, and moreover occur in areas with elevated N deposition (see also Chaps. 4 and 22).

Approaches to Assess Impacts and Critical Loads of Atmospheric Nitrogen Deposition: There are three major types of evidence available to relate N deposition to biodiversity for terrestrial ecosystems:

- Manipulation experiments in which N deposition is increased, normally by application of NH_4^+ and/or NO_3^- in artificial rainwater;
- Spatial field surveys along a gradient of N deposition, generally with a regional, national or even continental focus;
- Re-surveys over time of previous vegetation studies.

Each of the above approaches has strengths and weaknesses and evidence from a variety of approaches thus provides the most convincing support for N-driven changes in biodiversity (Dise et al. 2011).

The above mentioned approaches, specifically N addition experiments, have been used to derive empirical critical loads, as elaborated in detail by Bobbink and Hettelingh (2011), and in this book by Bobbink et al. (Chap. 4) and Pardo et al. (Chap. 5). The great advantage of the use of empirical critical N loads is the direct field or experimental evidence for a relationship between N deposition and effects, but there are also disadvantages. First of all, the empirical critical load, being the lowest level of the N addition where effects occur, is influenced by the chosen interval in N addition levels, and by an uncertain background N deposition that has to be added to these levels. Furthermore, the lack of fixed criteria regarding significant effects for which a critical load is derived limits a strict comparability, and finally it is not possible to predict future impacts when N deposition and other environmental conditions change simultaneously (De Vries et al. 2010).

Dynamic soil—vegetation models, as elaborated in more detail in this book by Rowe et al. (Chap. 11), Belyazid et al. (Chap. 12), Probst et al. (Chap. 13) and Schlutow et al. (Chap. 14), do not have such drawbacks, although their results are hard to validate on the basis of field observations. A combination of both approaches is thus preferred (see also the contributions by Fenn et al. in Chap. 10 and Van Dobben et al. in Chap. 22). The principle of such model based approaches is that: (i) a dynamic soil model such as SMART2 (Mol-Dijkstra and Kros 2001), ForSAFE (Belyazid et al. 2011; Chap. 12), or VSD+ (Posch and Reinds 2009; Posch et al. 2010) predicts the changes in water and nutrient status (e.g. as nitrogen availability or C/N ratio) and soil acidity (e.g. as soil pH or base saturation) in response to atmospheric deposition, whereas (ii) a statistical model such as NTM (Wamelink et al. 2003), MOVE (Van Adrichem et al. 2010), GBMOVE (Rowe et al. 2009; Chap. 11) or a process based model such as VEG (Belyazid et al. 2011; Chap. 12) predicts changes in plant species composition in response to the changes in water, nutrient and acidity status, using plant species specific information on habitat preferences. Ecological models that focus on species composition are usually of the 'niche' type, that produce probabilities of occurrence per species at each given combination of environmental conditions. The impacts of atmospheric deposition scenarios on plant species diversity can be derived by linking these probabilities of occurrence, either by a per-species approach or at a higher integration level, to the abiotic conditions produced by a soil—atmosphere model. On the other hand such coupled models can be used in an inverse way to assess a critical load. In that case, a critical limit (i.e. a lowest acceptable value) for a biodiversity indicator has to be set, and translated to critical values for abiotic factors (e.g. N availability or soil pH) and subsequently used in the soil models to back calculate the critical N and acid loads.

Need for Biodiversity Indicators to Evaluate Impacts and Critical Loads of Atmospheric Nitrogen Deposition: To be useful in policy-driven effect studies, a strict quantitative relation between deposition (and other environmental conditions) and biodiversity has to be assessed. This requires a biodiversity indicator that is a single quantitative property of an ecosystem, related to: (i) its diversity, in terms of species number or 'evenness', or (ii) its conservation state in terms of the presence or absence of desired species. Many concepts exist to derive such an indicator. In this Chapter, we focus on plant species diversity and first evaluate available biodiversity indicators with a focus on (a) their utility to reflect diversity as appreciated from various viewpoints regarding ecosystem functions; and (b) their practical feasibility as output of coupled abiotic and biotic models to forecast diversity on the basis of abiotic conditions including atmospheric deposition. Next, we discuss a possible approach to quantify critical limits for plant species diversity indicators and derive critical values for abiotic factors (e.g. N availability or soil pH) based of these diversity indicators.

Section 3.2 gives an overview of plant species diversity indicators, starting with a general indicator description for presence/absence related indicators (3.2.1), followed by four possible weighting approaches to characterise plant species diversity as endpoints of model simulations (Sect. 3.2.2–3.2.5). A possible way to link plant species diversity indicators to critical values for abiotic factors is described in Sect. 3.3, while we end with a discussion in Sect. 3.4.

3.2 Indicators for Plant Species Diversity

3.2.1 Weighting Approaches

The term biodiversity is a container concept to summarize an ecosystem's 'richness', for which a precise definition should be agreed in each application. The present chapter limits itself to indicators for plant species diversity and their applicability in effect modelling and critical load assessment. Ideally, a biodiversity indicator has to meet the following characteristics (Van Dobben and Wamelink 2009):

- it should agree with conservationists' attitudes, i.e. an ecosystem's numerical value should correspond with conservationists' intuitive or generally accepted appreciation of that system;
- it should be ecologically meaningful and quantifiable on the basis of either model output or observed data that are already available or can be easily collected;
- it should be scale-independent and comparable over different regions;
- it should be politically useful i.e. there should be 'buttons to press' for politicians in order to influence it.

A study carried out for Europe enumerates a huge number of possible biodiversity indicators (EEA 2003), which can be grouped in 26 main groups (EEA 2007). Even these main groups contain a wide range of definitions, based on biotic, abiotic, administrative and societal indicators. In the present context of biological effects only definitions related to the presence or absence of species or species combinations are useful. Such a definition of a biodiversity indicator H can be given the following general form:

$$H = \sum_{i=1}^{n} weight_i \tag{3.1}$$

where $weight_i$ is the 'weight' assigned to species i, and n is the total number of species.

The functional form and the values of the weighting factors determine the behaviour of such an index, and the following approaches for weighting might be considered:

- no weighting, i.e. all species are equally important;
- weighting according to the abundances of the species. Some of the classical diversity measures, e.g. the Simpson and the Shannon-Weaver index (Sect. 3.2.3) use this form of weighting;
- weighting the intrinsic 'importance' of each species from a nature conservancy point of view (i.e. considering rareness, decline); the IUCN concept of 'Red Lists' (IUCN 2012), lists of endemics etc. may be the basis of such forms of weighting;
- weighting the desirability of certain species or species combinations (communities) to be present in a certain location; the Dutch concept of 'target species'

(which is derived from the Red List concept; Bal et al. 1995), or the European system of Natura 2000 with its protected habitat types and species (EC 1992) apply to this form of weighting.

Another option is weighting to functional groups; e.g. primary producers should be present anyway, an ecosystem is only 'complete' if large carnivores are present, or, in a forest, mycorrhiza-forming mushrooms should be present, etc. Such a criterion comes closest to the 'ecosystem service' concept, in the sense that certain species groups should be present to fulfil certain functions; such groups could consist of e.g. edible species, species with medical applications, highly productive species in terms of food, timber or fuel; etc. However, note that ecosystem services are not necessarily linked to species; other biotic or abiotic properties of the system can just as well be indicators.

Below, the four weighting approaches outlined above are elaborated and illustrated with examples of their use related to deposition impacts. The option of weighting to functional groups is not further elaborated here because its scope is far wider than only plant species diversity which is the focus of this chapter.

3.2.2 Simple Species Counting

Weighting approach (1) makes the biodiversity indicator equal to the number of species. Ecological theories that assume a system to become more stable if it contains more species, provide a rationale for species counting (Margalef 1963). However, according to modern insights such theories are probably not generally true (McCann 2000). For plant species, an additional drawback of simple species counting is that far more species occur on calcareous soils and in hot and dry climates than on acid soils and in cold en wet climates, independent of other environmental conditions. This causes a strong decrease in species richness from South to North over Europe, and makes a comparison over a larger geographical extent difficult (De Vries et al. 2002). For these reasons, some form of species weighting should be applied to arrive at a useful indicator.

3.2.3 Classical Diversity Measures

Classical diversity measures (weighting approach 2) have two components: (i) the number of species: an ecosystem with more species has a higher diversity and (ii) the relative abundances of the species: an ecosystem that is completely dominated by a single species is less diverse than a system with the same number of species in about equal proportions. Many diversity indices weight these two components to different degrees. It has been shown that all these indices can be expressed in a single formula (Baczkowski et al. 1997):

$$H(\alpha,\beta)=\sum_{i=1}^{n} p_i^{\alpha}(-\log p_i)^{\beta} \tag{3.2}$$

where n is the number of species, p_i is the quantity (abundance) of species i, and α and β are parameters defining (and determining the behaviour of) the index.

There are two popular special cases of the above formula:

- The Shannon-Weaver index $H(1,1)$ (Shannon 1948), which can be ecologically interpreted as one minus the chance to correctly guess the name of an individual taken randomly from the population;
- The Simpson index $H(2,0)$ (Simpson 1949), which can be ecologically interpreted as one minus the chance of two individuals taken randomly from the population belonging to the same species.

An example pertaining to the Simpson index is given by Rowe et al. (Chap. 11). The Shannon-Weaver index can be considered as a measure of information density and has many applications outside ecology (see e.g. Cover and Thomas 2012). However, in the present context these classical measures are less useful as their biological meaning is rather unclear despite their mathematical elegance (Hurlbert 1971). An obvious drawback of these measures is that they treat all species equally, i.e. independent of their known properties, and therefore attach equal weight to unwanted, invasive species and to highly desirable Red List species.

3.2.4 Conservancy Values Per Species

Weighting approach (3) probably comes nearest to biodiversity as it is intuitively appreciated by ecologists: if the Red List criterion is used, rare and declining species are given a high weight, and the indicator becomes a measure for a location's importance to prevent (local) extinction of species. Two elaborations of this criterion have been used to describe effects of deposition. One is the 'GBMOVE' approach that is extensively described by Rowe et al. (2009), and summarised in Chap. 11. Essentially, this approach is based on lists of wanted ('positive') and unwanted ('negative') indicator species for a given habitat, whose probabilities to be present are predicted by a niche model. These probabilities are used to calculate the local habitat quality, HQ, according to:

$$HQ = \frac{1}{p}\sum_{i=1}^{p}\frac{P_i}{P_{max,i}} - \frac{1}{n}\sum_{i=1}^{n}\frac{N_i}{N_{max,i}} \tag{3.3}$$

where P_i and N_i are the probabilities of occurrence of the p positive indicators and n negative indicators, and $P_{max,i}$ and $N_{max,i}$ are the maximum probabilities of occurrence for these species. Species not included in the indicator lists are not used in this calculation.

Another approach is used in NTM, described by e.g. Van Dobben and Wamelink (2009) and in this book by Rowe et al. (Chap. 11). Here, the basis is formed by the so-called Red Lists (IUCN 2012), officially approved documents containing lists of threatened species in various categories. These categories are combinations of rarity and decline, based on the assumption that a species that is both rare and declining has the highest probability to go extinct in the near future. Because the Red Lists only contain rare species they cannot be directly used to make an index, but their criteria rarity and decline, combined with a number of local merits, can be used to attribute a biodiversity index S to each species, irrespective of its ecological preferences. As an example, the elaboration of these criteria for The Netherlands is summarized in Table 3.1A, B, C and D. This index was originally designed to evaluate the diversity of vegetation relevés, which is calculated according to:

$$H = \sqrt{\frac{1}{\log \max\{5, n\}} \sum_{i=1}^{n} Q_i . S_i} \tag{3.4}$$

where H is the biodiversity score for the relevé, n is the number of species in the relevé, Q_i is the weighting factor for species i, and S_i is the index of species i, derived according to Table 3.1A, B, C and D. The weighting factor for each species is determined on the basis of its abundance, such that $Q=1$ for species present in very small quantities, $Q=2$ for species covering more than half of the plot, and logarithmically increasing with the species' abundance between >0 and 50 % cover.

In this way, the final score of the relevé is mostly determined by the summed indices per species, with corrections for species present in large quantities (weighted more heavily) and species-poor relevés (score somewhat increased). Finally, H is square-root transformed to achieve a normal distribution.

In the NTM model, a data set of 160,000 relevés was used to link H to abiotic conditions in terms of soil pH, water availability and N availability, and values for these three conditions determined by soil—atmosphere models can be directly translated into biodiversity estimates. Examples are given Rowe et al. (Chap. 11).

3.2.5 *Target Species and Communities*

In weighting approach (4) certain species are considered as prerequisites for a given location to have an ecological value anyway. These species may be given as such, or in the form of a plant community, i.e. a group of species that tend to co-occur, of which a certain proportion should be present. This approach was probably first elaborated for the Netherlands (Bal et al. 1995) and is now implemented in the European 'Habitats Directive' (EC 1992). Under this Directive, natural areas are set aside for the protection of given species or communities. The presence of such targets in their designated locations can be used to evaluate the ecological performance of each site. Many of the biodiversity indicators found in literature (e.g. Lamb et al. 2009) are based on this concept, i.e. comparing the actual species composition of a given site with a 'reference' species composition. This reference may be the species present in an 'intact' or 'pristine' community or under 'natural' conditions. An

Table 3.1 Determination of the biodiversity index S (Eq. 3.4) per species for the Netherlands used in the NTM3 model. The index is calculated per species as the sum of the species' scores determined by their rarity (A), decline (B), international rarity (C) and vulnerability (D). Note that (A) and (B) are the criteria of the Red Lists (IUCN 2012)

A: Scores for rarity. Figures in the left column are absolute numbers of 5×5 km^2 squares in which species has been found in the Netherlands (total number of squares in the Netherlands = 1677)

No. of grid squares	Score
0	50
1–3	45
4–10	40
11–29	35
30–79	30
80–189	25
190–410	20
411–710	15
711–1210	10
1211–1677	5

B: Scores for decline. Figures in the left column are changes in the class number as in Table 3.1A, between the periods 1900–1950 and 1950–1990, respectively

Change	Score
>+1	−10
+1	−5
0	0
−1	5
<−1	10

C: Scores for international rarity. Scores have been assigned for combinations of area size and relative rarity in the Netherlands. Area size classes: 'large' = both inside and outside Europe; 'intermediate' = exclusively European; 'small' = confined to western Europe; 'very small' ≤ 300,000 km^2

Frequency in The Netherlands compared to frequency elsewhere	Size of distribution area			
	'large'	'intermediate'	'small'	'very small'
Much less common	0	0	0	5
Less common	0	0	0	10
No difference	0	0	5	15
More common	0	5	10	20
Much more common	0	5	20	20

D: Scores for vulnerability (i.e., decline expected in the near future, based on expert judgement, taking account of ecological amplitude, seed longevity and dispersal capacity)

Vulnerability	Score
Decline expected	20
Decline possible	10
No change Expected	5
Increase possible	0
Increase expected	−5

obvious drawback of such methods is that many of the present-day ecosystems are man-made and it is often impossible to define a 'pristine' state, but this problem may be overcome by using a reference state that is defined a priori like in the Habitats Directive (Rowe et al. 2009). In that case the reference state is usually based on expert opinion of what should be present in that area. Buckland et al. (2005) and Jenssen (2009) emphasize the use of the first observation in a time series as the reference state, however time series are often unavailable or start in a state that is already considered as 'degraded' (Ellis et al. 2011). Alternatively, the reference state can simply be the model output under a 'clean' scenario; this is the approach used by Belyazid et al. (2011).

All reference-state based methods require a 'distance-to-target' measure to express the similarity between the actual (or predicted) state and the reference state. Like for biodiversity a great variety of such measures has been described, all of which can mathematically be considered as distance measures in an n-dimensional space (n = number of species). Sometimes the term similarity is used, where similarity = 1—distance (or dissimilarity). A comprehensive treatment of ecological distance measures is given by Legendre & Legendre (2012). All distance measures can be considered as elaborations of the Minkowski metric $M(r)$:

$$M(r)=\left(\sum_{I=1}^{n}\left|y_{1,i}-y_{2,i}\right|^{r}\right)^{1/r} \tag{3.5}$$

where $y_{1,i}$ and $y_{2,i}$ are the abundances of species i in samples (states) 1 and 2 (e.g. reference and actual), n is the number of species, and $r>0$ determines the behaviour of the measure.

As large values of r attach a very high importance to the largest differences in y, values above $r=2$ are seldom used in ecology. For $r=2$, the measure is the Euclidean distance, and for $r=1$ it is known as city-block distance or Manhattan metric. The $M(r)$ measure suffers from the 'double zero' problem, i.e. it has the somewhat counter-intuitive property that samples that do not share any species may have a shorter distance than samples that do share species. This is because that absence of a given species in both samples is weighted in the same way as presence in both samples, while ecologists tend to attach more weight to presence than to absence.

Many variants have been developed that apply different forms of weighting and/or standardization, the use of which depends upon the exact nature of the data (e.g. cover, presence/absence, counts etc.). A key to the optimal measure for various ecological applications is given by Legendre and Legendre (2012).

A popular measure is the Czekanowski index (mean character difference, 'durchschnittliche Differenz' in German):

$$C=\frac{1}{n}M(1)=\frac{1}{n}\sum_{i=1}^{n}\left|y_{1,i}-y_{2,i}\right| \tag{3.6}$$

Further elaborations apply a form of standardization (normalization) to the y's. A much-used form, which is well-suited for quantitative data, is the Bray-Curtis dissimilarity index, derived from $M(1)$:

$$BC = \frac{\sum_{i=1}^{n} |y_{1,i} - y_{2,i}|}{\sum_{i=1}^{n} (y_{1,i} + y_{2,i})} = 1 - \frac{2\sum_{i=1}^{n} min\{y_{1,i}, y_{2,i}\}}{\sum_{i=1}^{n} (y_{1,i} + y_{2,i})} = 1 - BC_s \quad (3.7)$$

The second equation holds because $x+y-|x-y|=2\min\{x, y\}$ for all x and y. BC_s is called the Bray-Curtis *similarity* index (Bray and Curtis 1957); and when applied to presence/absence data, this measure is identical to the well-known Sørensen similarity index. The 'Mondrian' index proposed by Belyazid et al. (2011) can be shown to be identical to the Bray-Curtis index due to the identity $\max\{x-y, 0\} = x - \min\{x, y\}$. Note that some authors call BC_s (also) the Czekanowski (similarity) index, which, however, seems to be a wrong attribution; see Somerfield (2008) for a discussion.

Also in Euclidean distances forms of weighting can be applied. A special case is the chi-square distance, in which the species abundances are normalized to 1, while the contribution of each species to the total sum of squares is weighted with the inverse of its standardized mean abundance over all samples (see, e.g. Legendre and Gallagher 2001):

$$\chi^2 = \left(\sum_{i=1}^{n} \frac{S_1 + S_2}{y_{1,i} + y_{2,i}} \left(\frac{y_{1,i}}{S_1} - \frac{y_{2,i}}{S_2} \right)^2 \right)^{1/2} \quad \text{with } S_k = \sum_{j=1}^{n} y_{k,j}, \ k = 1, 2 \quad (3.8)$$

The chi-square measure weighs the rarer species more heavily than other measures and is not sensitive to double zeros. Hence, it is more suitable to compare heterogeneous datasets that contain many rare species or that share only a few ones.

Minkowski metrics with $r=2$ are used in multivariate techniques that are popular in ecology: Euclidean distance in principal component analysis (PCA), and chi-square distance in correspondence analysis (CA). Both CA and PCA can be used to determine distances between ecosystem states; see Van Dobben and Slim (2012) for an example.

A special case is formed by distance measures derived from information theory, i.e. measures that estimate the amount of information (bits) needed to describe the difference between two states of a system. The Kullback or Kullback-Leibler divergence (Cover and Thomas 2012) is a measure that quantifies in bits how close a discrete probability distribution $q_1, \ldots, q_n$ is to a reference distribution $q_1^0, \ldots, q_n^0$:

$$K = \sum_{i=1}^{n} q_i^2 \log(q_i / q_i^0) \quad (3.9)$$

This measure is related to the Shannon-Weaver diversity measure (see above). Jenssen (2009) used this measure, in combination with the Bray-Curtis index and the number of protected species, to quantify the effect of deposition on forest ecosystem diversity.

3.3 Assessment of Critical Limits for Abiotic Conditions on the Basis of a Plant Species Diversity Indicator

As explained in 3.1 soil chemical models can be used to derive critical load values for ecosystems. To this end, such models need critical limits of soil chemical variables as inputs. These critical limits, in turn, have to be derived from ecological criteria. In the context of plant species diversity, an ecological model is needed that translates a 'maximum acceptable loss of plant species diversity' into soil chemical criteria. This implies that choices have to be made with respect to (i) the diversity index to be used and (ii) the critical value of this index. These aspects are further elaborated below.

3.3.1 Available Approaches

Many ecological indicators exist to quantify plant species diversity, as described in Sect. 3.2. However, there are only few examples of actual critical load estimates using such indicators. The studies by Schlütow et al. (Chap. 14) and Van Dobben et al. (Chap. 22) yield tables of critical load values per EU Habitat type. Basically, these authors use a niche model and impose lower limits (critical values) on the probabilities of occurrence per community. Next, the critical load is computed by a soil model as the deposition that leads to the abiotic conditions (in terms of e.g. base saturation, pH, C/N ratio) at the critical probabilities of occurrence. Also Belyazid et al. (2011, and Chap. 12) use a comparable method, however with a species-oriented approach. These authors use the modelled species composition at a low-deposition scenario as a reference state, and compute the distance between this state and a new state at a given deposition level in terms of the Bray-Curtis index (called 'Mondrian' index by them). A critical load can be derived by imposing a maximum distance between the reference state and the new state.

It may be argued that all above examples pertain to geographical regions where local circumstances strongly facilitate the rather sophisticated approaches used by their authors: Germany, UK and the Netherlands with a wealth of ecological data and therefore, rather well-known responses per community; or Scandinavia, with a relatively low number of species that makes a dynamic per-species approach feasible. For applications at a wider geographical extent, alternative methods should be sought for. A number of possible methods is outlined in Figs. 3.1, 3.2 and 3.3 and explained below. As niche models play a central role in these methods, the principle of these models is first explained in some detail.

3.3.2 Niche Models

In its simplest form, a niche model is a statistical description of the probability of occurrence (POC) of a community of plant species or of an individual plant species

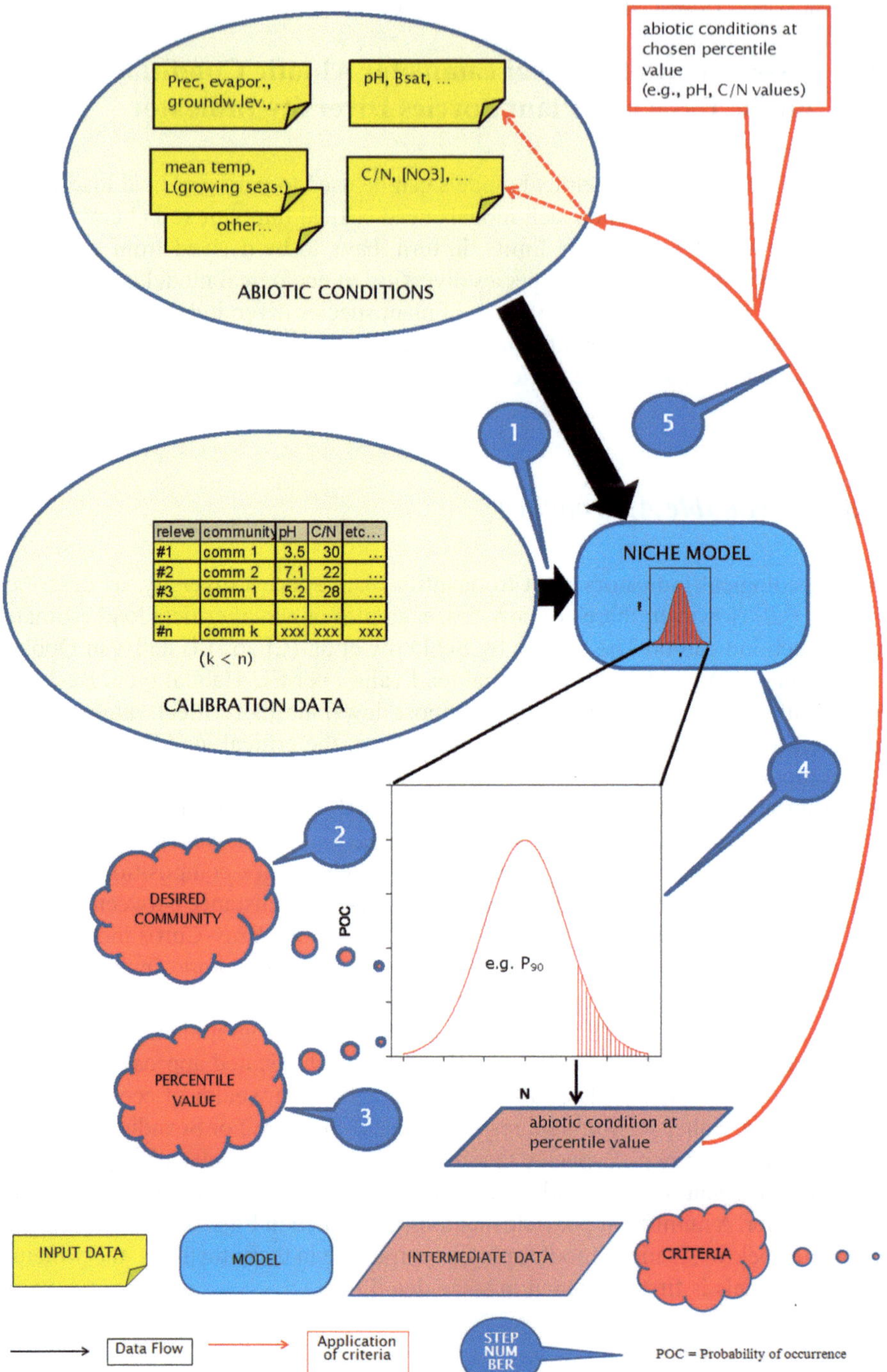

Fig. 3.1 Stepwise derivation of critical limits for abiotic conditions on the basis of a community-oriented niche model (see text for details): *Step 1*: Fit response curves of probability of occurrence (POCs) per community as a function of abiotic conditions (niche model calibration); *Step 2*: Assess desired community; *Step 3*: Set percentile value or minimum POC for the desired community's occurrence in relation to abiotic conditions; *Step 4*: Use response curve to determine the abiotic values that lead to the critical POC or percentile value from step 3; *Step 5*: The abiotic values determined in step 4 are the critical conditions that can be used as input for an inverted soil model to determine a critical load

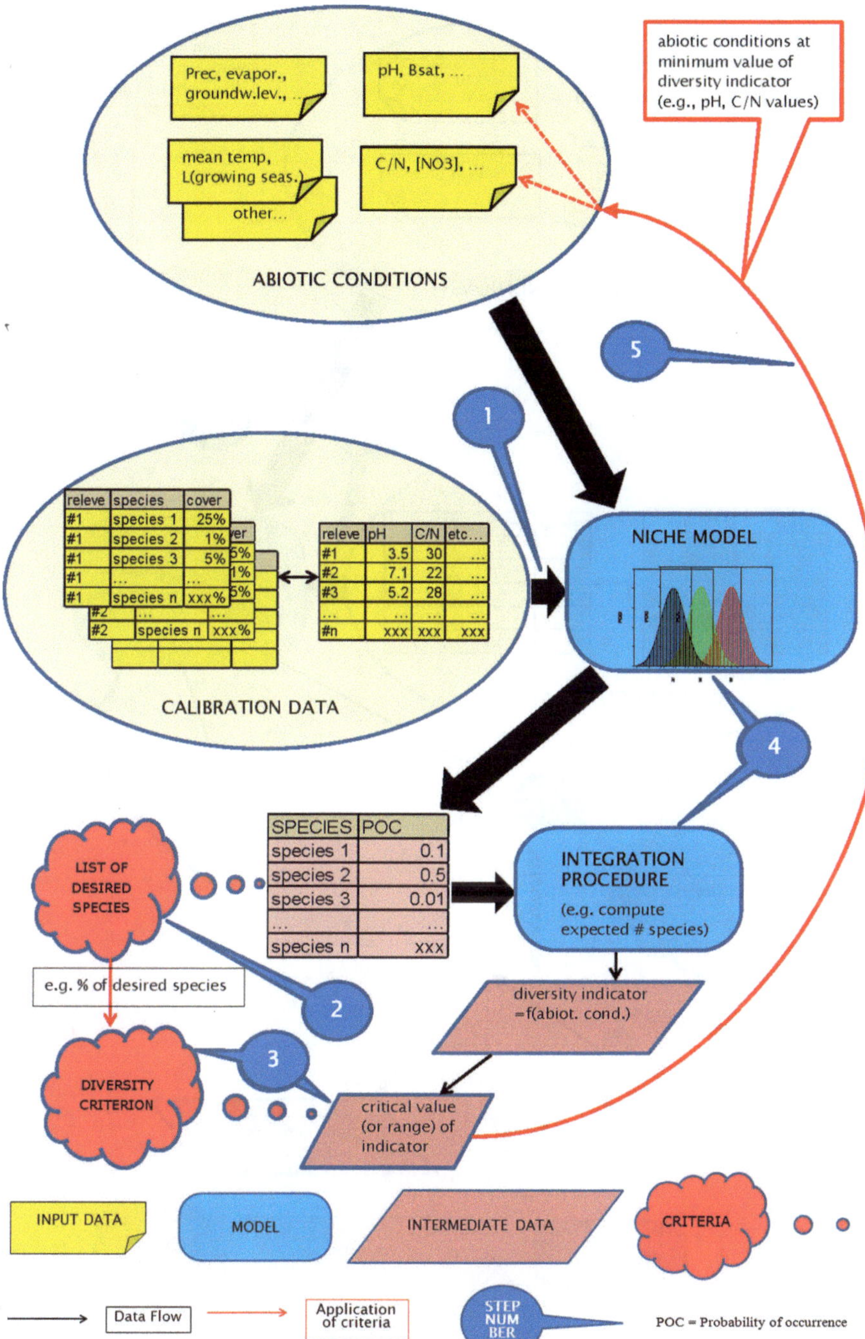

Fig. 3.2 Stepwise derivation of critical limits for abiotic conditions on the basis of a species-oriented niche model (see text for details): *Step 1*: Fit response curves of POCs per species as a function of abiotic conditions (niche model calibration); *Step 2*: Define diversity indicator; *Step 3*: Set critical limit (minimum acceptable value or maximum acceptable loss) for the diversity indicator; *Step 4*: Use response curve to determine the abiotic values that lead to the critical limit of the diversity indicator from step 3; *Step 5*: The abiotic values determined in step 4 are the critical conditions that can be used as input for an inverted soil model to determine a critical load

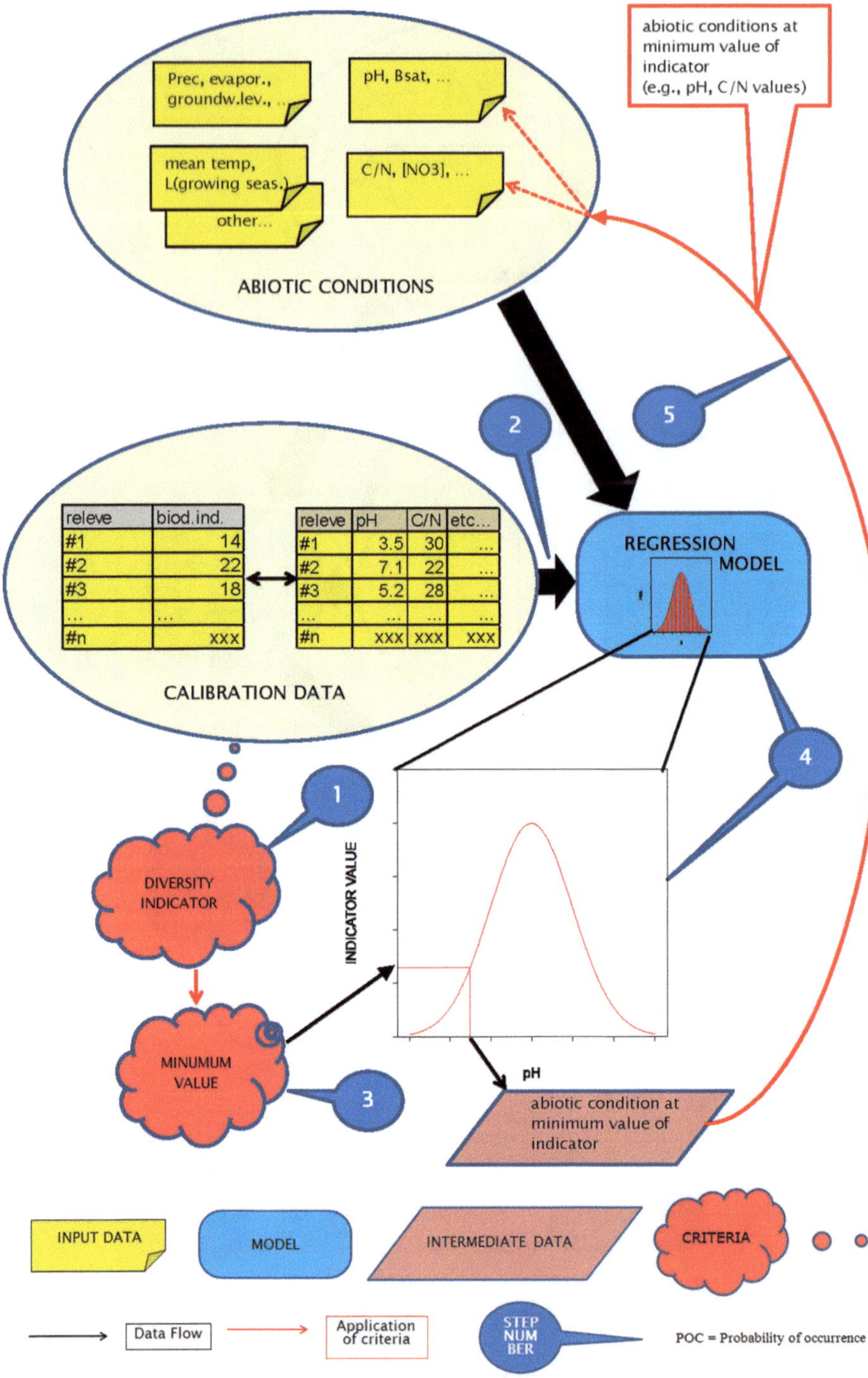

Fig. 3.3 Stepwise derivation of critical limits for abiotic conditions on the basis of a diversity indicator (see text for details): *Step 1*: Define indicator; *Step 2*: Fit response curve of indicator as a function of abiotic conditions (niche model calibration); *Step 3*: Set minimum acceptable value of indicator; *Step 4*: Use response curve to determine the abiotic values that lead to minimum value from step 3; *Step 5*: The abiotic values determined in step 4 are the critical conditions that can be used as input for an inverted soil model to determine a critical load

as a function of an abiotic variable or a combination of abiotic variables. If sufficient data are available such a model can be calibrated and used to predict the probability of occurrence at given abiotic conditions. In this context, calibration implies the fitting of regression parameters on 'training' data by minimizing the deviation between the prediction of the regression model and the data. The community's or species' response to the abiotic variables can be formulated in mathematical terms, e.g. as logistic or Gaussian curves, but this is not necessary. Also, spline functions that do not have a predefined shape can be used to derive response curves (Wamelink et al. 2005, 2011). More sophisticated methods use generalised linear models to describe the response to a combination of abiotic variables, possibly including interaction terms. A number of niche models (MOVE, MultiMOVE, PROPS, NTM4) are treated in detail in Chap. 11; the VEG model is essentially also a niche model although it has dynamic aspects.

Niche models can be used to predict the POC of both communities and species at given abiotic conditions as long as sufficient data for their calibration are available. Comparable statistical approaches may even be used to predict any other property of the community if calibration data are present. This is for example the case with the NTM3 model that directly predicts the value of a diversity indicator (H in Eq. 3.4) at given abiotic conditions. However, since niche models do not predict abundance, classical diversity measures (which are based on abundance) cannot be used as indicators to calculate a critical limit by this approach.

Although niche models are simple, straightforward and easy-to-explain methods to predict biotic properties on the basis of abiotic conditions, their use is strongly limited by the availability of calibration data. A sufficient number of observations has to be available to make reliable predictions, but it must also be kept in mind that the validity of these predictions is limited to the range of the abiotic data present in the calibration set. Usually the calibration data are in the form of vegetation relevés, i.e. descriptions of small plots (1–100 m^2) in terms of quantities per species. If the abiotic conditions in such a plot are available from local observations (e.g. soil analyses) or generic databases (e.g. soil type, hydrology, climate) or models (e.g. deposition), the combination of vegetation and abiotic conditions at that plot forms a single data point in the training set. The calibration data can be evaluated in two ways:

- community-oriented: the relevé is assigned to a plant community e.g. following the syntaxonomic classification of Braun-Blanquet (1964), the EUNIS classification (Davies et al. 2004; http://eunis.eea.europa.eu) or the Habitat classification (EC 1992); automated procedures for this assignment can be developed based on a reference set and a (combination of) distance measures (e.g. Van Tongeren et al. 2008). In this way the response of each community with sufficient data can be fitted to the abiotic conditions (step 1 in Fig. 3.1);
- species-oriented: the response of each species is directly fitted to the abiotic conditions on the basis of its presence or absence in each relevé (step 1 in Fig. 3.2).

If ecological targets are set in terms of communities the first approach is the most straightforward one; this approach is further explained in Fig. 3.1 and Sect. 3.3.3.

However, sufficiently large training sets, in which relevés have been assigned to communities in the desired typology, are often unavailable. In that case a work-around through species can be used, i.e. the desired communities are translated into lists of desired species and the POC is predicted for each of these species (Wamelink et al. 2011). In that case an integration procedure is necessary to combine and evaluate the POCs per species, as is further explained in Fig. 3.2 and Sect. 3.3.4.

In both approaches two choices have to be made that are arbitrary to a certain extent: (1) the diversity indicator to be used, and (2) its critical limit, i.e. the minimum acceptable value of the diversity indicator. These choices depend, e.g., on the nature of the diversity target, the available data, and the maximum loss of diversity (or: maximum distance to the target) that is judged acceptable. In practice it is highly improbable to find a state that is identical to the target state, so that a certain deviation from this state has to be accepted. Belyazid et al. (2011) accept a maximum modelled change in absolute cover percentage per species of 5 % which seems very strict. Van Dobben et al. (Chap. 22) accept a minimum of 20 % of the cumulative probability of occurrence per community which seems rather loose. In terms of the expected number of desired species one could set a limit of e.g. 10–20 % of the total number of desired species (compare Van Dobben 2011: in a modelling study in The Netherlands a maximum of 20 % of desired species was reported over a wide range of communities and abiotic conditions).

A disadvantage of the species approach that hitherto has received little attention is that it suffers from unsolved scale problems. This is because niche models are nearly always based on vegetation relevés that have a scale of about 1–100 m^2, whereas application is on the scale of forest stands, nature reserves or mapping units that usually have areas in the order of 1 ha or more. Note that species (in contrast to communities) have discrete individuals, and therefore—even in a perfectly homogeneous environment—their probability of occurrence depends in a non-linear way on the size of the examined area (Connor and McCoy 1979).

3.3.3 Community-Oriented Approach

The Habitats Directive (EC 1992) sets ecological targets in the form of habitat types (i.e., desired communities) for Europe's most important natural areas (the 'Natura 2000' network). In principle, the desired habitat types are known for each Natura 2000 site. These habitat types can be used as diversity targets (step 2 in Fig. 3.1). For the niche model a training set is needed of relevés that have known abiotic conditions and have been assigned to habitat types. If a relevé set is available in a typology that is different from the habitat typology, translation tables may be available, e.g. between syntaxonomic units (in the sense of Braun-Blanquet 1964) and habitat types (see Chap. 22 for an example).

The niche model produces a response curve in terms of the POC of each community as a function of the abiotic conditions under scrutiny. The critical value can be set in terms of probability of occurrence (POC) of the desired community but is more easily standardised by defining it in terms of a percentile value of its response curve (e.g. P_5 or P_{95}, depending on whether diversity is reduced by low or high values, respectively, of the abiotic condition considered) (step 3). The niche model can be used to determine the abiotic conditions at which the given percentile value occurs (step 4), which are the critical limits that can be fed into a soil model (step 5). This approach is used in Chap. 22.

3.3.4 Species-Oriented Approach

If a training set in the desired typology is unavailable, critical limits can be determined on the basis of POCs per species. In Europe, the EU requests each member state to produce a map of Habitat types, and syntaxonomic units can be localised through the map by Bohn et al. (2004). Using these data a list of desired species can be drawn up for each location in Natura 2000. Such desired species may be (1) species protected under the Habitats Directive, (2) 'typical species' for each Habitat type (each member state has to produce a table of such species), or (3) species that are characteristic of the Habitat type at a specific location, as derived from Bohn et al.'s (2004) map. Of course, the procedure described in this section can also be used when the target is defined in terms of species. Note that a list of species can be seen as a local ecological target or reference state (step 2 in Fig. 3.2). A limitation of this approach is that the desired species are often the rare ones for which insufficient data may be available for the calibration of the niche model.

The niche model produces response curves per species. A diversity indicator has to be defined on the basis of these response curves, and its critical limit has to be set (step 3). Plant species diversity indicators are treated in Sect. 3.2, and their choice should be mainly governed by the nature of the data in the training set (see 3.4). In making a choice it should be borne in mind that niche models do not produce 'hard' presences or abundances of species, but only probabilities of occurrence (POCs) under given abiotic conditions. The POC of (desired) species i, POC_i, is defined as:

$$POC_i = \frac{1}{m}\sum_{j=1}^{m} \pi_{i,j} \tag{3.10}$$

where $\pi_{i,j}$ is the presence/absence of species i at site j ($\pi_{i,j}$=0 or 1) and m is the number of sites.

The POCs of the desired species can be seen as a measure of target realisation, and what is needed is a simple metric to summarise the probabilities and make them comparable between communities or sites (step 4). Following Lamb et al. (2009), we suggest as a metric the fraction of desired species at site j, H_j, under given abiotic conditions:

$$H_j = \frac{1}{n}\sum_{i=1}^{n} \pi_{i,j} \tag{3.11}$$

where n is the number of desired species.

Note that the average of the POCs equals the average (i.e. expected) fraction of species at a site and equals the total number of occurrences of all species at all sites, divided by the product of number of sites with number of species:

$$\frac{1}{n}\sum_{i=1}^{n} POC_i = \frac{1}{nm}\sum_{i=1}^{n}\sum_{j=1}^{m}\pi_{i,j} = \frac{1}{m}\sum_{j=1}^{m} H_j. \tag{3.12}$$

It can be shown by simple algebra that for a site j the indicator H_j is equal to one minus the Czekanowski index C_j for that site (Eq. 3.6) if a hypothetical site, where all desired species are present, is used as the reference: Setting $y_{1,i}=1$ and $y_{2,i}=POC_i$ for all species i yields $|1-POC_i|=1-POC_i$ and thus one arrives at $C_j=1-H_j$.

It might be necessary to increase the comparability between communities by applying an extra standardisation. This is because the theoretical maximum value of the above indicators will only be reached if the predicted number of species equals the number of desired species. In practice, the predicted POC will seldom be equal to one even for a single species because this requires it to be present in all relevés at the optimal abiotic conditions in the training set. This is highly improbable or in other words, the maximum value of the response curve is nearly always lower than one and in practice, often very low (Van Dobben 2011). As the maximum value of the response curve is different per species (cf. Wamelink et al. 2005) the maximum value of the above indices will vary per community type. The most straightforward possibilities for standardisation are:

- standardise all response curves to maximum = 1 (cf. Van Adrichem et al. 2010). A consequence of this form of standardisation is that the sum of the POCs is no longer equal to the expected number of species;
- use the modelled POCs at a selected site with optimal abiotic conditions as a reference. This approach is comparable to the one of Belyazid et al. (2011);
- standardise the POC to values 0 or 1. This can be achieved by using the response curve to determine a range where a species is assumed to be present; this can be e.g. the range between the 5 and 95 percentile values (Wamelink et al. 2011).

Finally, critical limits for abiotic conditions can be determined as the values that lead to the critical value of the diversity indicator (step 5).

3.3.5 Alternative Approaches

As an alternative for the probability-of-occurrence approach one might also choose to directly back-predict a diversity indicator whose value can be determined for each site in the training set. This method is particularly useful for indicators that are based on intrinsic values per species. Such values can be aggregated to a value for each relevé, and these values can be regressed on the relevés abiotic conditions. This procedure is explained in detail in 3.2.4 and in Chap. 11. In this approach, the first step is the choice of the indicator (step 1 in Fig. 3.3) and its calibration, which again requires a sufficiently large data set although not as large as for the species- or community-oriented approach. The relation between the indicator and abiotic conditions can be assessed by (multiple) regression (step 2). A minimum acceptable value for the indicator has to be set, which again is arbitrary to a certain extent (step 3), and the response curve can be used to back-predict that abiotic conditions where this minimum value occurs (step 4), which are the critical conditions (step 5).

Although there are no examples of critical loads derived in this way, a derivation is possible by setting a critical value for the diversity indicator, e.g. H in Eq. 3.4 for the Netherlands: $H < 14$, low diversity, very low probability to find red-list species; $H > 16$, high diversity, moderate to high probability to find red-list species (Van Dobben 2011 and Table 3.1A, B, C and D). In this example, a critical value of $H = 16$ leads to critical load estimates that are in the order of magnitude of empirical values (Van Dobben 2011, p. 72 ff.).

3.4 Discussion

A wide variety of biological indicators for the effect of deposition (or any other threat to an ecosystem) exists, based on either intrinsic values per species or on similarity to predefined 'target' or 'reference' communities. Each of these methods has its specific advantages or disadvantages, and an overview of these is given in Table 3.2.

Table 3.2 may give the impression that many methods exist that are expressions of more or less conflicting strategies. However, in practice most strategies have the same aim, namely the conservation of threatened species. Therefore, target species that appear in political agreements are usually also the ones that are highly valued by ecologists. And such species often appear on the Red Lists and therefore also have a high intrinsic value. Therefore the choice of an indicator is not always a very fundamental one and can to a certain extent be governed by practical considerations.

For species-oriented targets a distance measure may be used to determine the similarity between a given (i.e., observed or modelled) state and the target. Many such similarity measures exist, the optimal one depending on the nature of the data under scrutiny. It seems that up to now the choice for a given measure has usually been rather ad-hoc; a comprehensive comparison of all available measures is lacking for most applications. A more stringent application of Table 7.4 in Legendre and Legendre (2012, p. 324 ff.) is therefore a recommendation for future work. However, distance measures are not absolutely necessary for the quantification of deposition effects. In the community-oriented approach the probability of occurrence of the target community is directly predicted on the basis of abiotic conditions, and a lower limit can be imposed on this probability. This seems a more straightforward approach than the one using lists of 'desired' or 'target' species. However its application may be limited by the limited availability of suitable training sets (i.e., consisting of relevés assigned to communities in the desired typology).

Table 3.2 Summary of advantages and disadvantages of biodiversity indicators (adapted from Table 5.1 in Van Dobben et al. 2010)

Method	Reference, target	Advantages	Disadvantages
Species-oriented targets	Any given list of species	Long history of model application	Validation not feasible, or bad performance in validation
		Conceptually simple, easy to explain	Unsolved scale problems, i.e. calibration and prediction on different scales
		Flexible, any list of species can be set as a target	Desired species are often the rare ones for which too few calibration data are available
	List of species present in 'pristine', 'natural' state	Ecologically meaningful, i.e. aiming at conservation of ecosystem's original state	Usually no data available on 'pristine', 'natural' state
	Predefined lists of 'target' species	Politically meaningful, i.e. yields metric of target realisation	Target may be ecologically rather arbitrary
Community-oriented targets	Diversity indices	Easy to compute, mathematically elegant, related to information theory	Ecological meaning unclear
	Habitat types	Immediately relevant for EU Habitat Directive	Targets ecologically rather arbitrary
	Other typology	May be relevant as indicator for local target realisation	Rigid, inflexible
Intrinsic value-oriented targets	List of intrinsic values per species	Can be directly related to conservancy targets e.g. protection of Red List species	List of intrinsic value per species must be available (e.g. Table 3.1A, 3.1B, 3.1C and 3.1D)
		Applicability is location-independent	Intrinsic values may be regionally different
		Ecologically meaningful, i.e. aiming at conservation of threatened species	No direct relevance for policy-set targets, difficult to explain, many alternative approaches possible

References

Achermann, B., & Bobbink, R. (Eds.). (2003). *Empirical critical loads for nitrogen: Expert workshop, Berne, 11–13 November 2002*. Swiss Agency for the Environment, Forests and Landscape.

Aerts, R., & Berendse, F. (1988). The effect of increased nutrient availability on vegetation dynamics in wet heathlands. *Vegetatio, 76,* 63–69.

Baczkowski, A. J., Joanes, D. N., & Shamia, G. M. (1997). Properties of a generalized diversity index. *Journal of Theoretical Biology, 188,* 207–213.

Bal, D., Beije, H. M., Hoogeveen, Y. R., Jansen, S. R. J., & van de Reest, P. J. (1995). *Handboek natuurdoeltypen in Nederland*. (Rapport IKC-N 11). Wageningen: IKC-Natuurbeheer.

Belyazid, S., Kurz, D., Braun, S., Sverdrup, H., Rihm, B., & Hettelingh, J. P. (2011). A dynamic modelling approach for estimating critical loads of nitrogen based on plant community changes under a changing climate. *Environmental Pollution, 159,* 789–801.

Bobbink, R., & Hettelingh, J.-P. (2011). Review and revision of empirical critical loads and dose-response relationships: Proceedings of an expert workshop, Noordwijkerhout, 23–25 June 2010. (Report 680359002/2011). Bilthoven: Coordination Centre for Effects, National Institute for Public Health and the Environment.

Bobbink, R., Ashmore, M., Braun, S., Flückiger, W., & van den Wyngaert, I. J. J. (2003). Empirical nitrogen critical loads for natural and semi-natural ecosystems: 2002 update. In B. Achermann, & R. Bobbink (Eds.), *Empirical critical loads for nitrogen* (pp. 43–170). Berne: Swiss Agency for Environment, Forest and Landscape SAEFL.

Bobbink, R., Hicks, K., Galloway, J., Spranger, T., Alkemade, R., Ashmore, M., Bustamante, M., Cinderby, S., Davidson, E., Dentener, F., Emmett, B., Erisman, J. W., Fenn, M., Gilliam, F., Nordin, A., Pardo, L., & De Vries, W. (2010). Global assessment of nitrogen deposition effects on terrestrial plant diversity: A synthesis. *Ecological Applications, 20,* 30–59.

Bohn, U., Neuhäusl, R., Gollub, G., Hettwer, C., Neuhäuslová, Z., Raus, T., Schlüter, H., & Weber, H. (2004). *Map of the Natural Vegetation of Europe scale 1: 2 500 000*. Münster: Landwirtschaftsverlag.

Braun-Blanquet, W. (1964). *Pflanzensoziologie. Grundzüge der Vegetationskunde*. (3. Aufl.). Wien: Springer.

Bray, R. J., & Curtis, J. T. (1957). An ordination of the upland forest communities of southern Wisconsin. *Ecological Monographs, 27,* 325–349.

Buckland, S. T., Magurran, A. E., Green, R. E., & Fewster, R. M. (2005). Monitoring change in biodiversity through composite indices. *Philsophical Transactions of the Royal Society Biological Science,* 360, 243–254 (London).

Connor, E. F., & McCoy, E. D. (1979). The statistics and biology of the species-area relationship. *The American Naturalist, 113,* 791–833.

Cover, T. M., & Thomas, J. A. (2012). *Elements of information theory*. Wiley Interscience.

Davies, C. E., Moss, D., & Hill, M. O. (2004). *EUNIS Habitat Classification revised 2004*. European Environment Agency, European topic centre on nature protection and biodiversity.

De Vries, W., Reinds, G. J., van Dobben, H., de Zwart, D., Posch, M., Voogd, J. C. H., Auee, J. & Vel, E. M. (2002). *Intensive monitoring of forest ecosystems in Europe. Technical Report 2002.* Geneva and Brussels: UN/ECE and EC, Forest Intensive Monitoring Coordinating Institute

De Vries, W., Wamelink, G. W. W., van Dobben, H., Kros, J., Reinds, G. J., Mol-Dijkstra, J. P., Smart, S. M., Evans, C. D., Rowe, E. C., Belyazid, S., Sverdrup, H. U., van Hinsberg, A., Posch, M., Hettelingh, J.-P., Spranger, T., & Bobbink, R. (2010). Use of dynamic soil–vegetation models to assess impacts of nitrogen deposition on plant species composition: An overview. *Ecological Applications, 20,* 60–79.

Dise, N. B., Ashmore, M., Belyazid, S., Bleeker, A., Bobbink, R., De Vries, W., Erisman, J. W., van den Berg, L., Spranger, T., & Stevens, C. (2011). Nitrogen as a threat to European terrestrial biodiversity Chap. 20. In M. A. Sutton, C. M. Howard, J. W. Erisman, G. Billen, A.

Bleeker, P. Grennfelt, H. van Grinsven, & B. Grizzetti (Eds.), *The European Nitrogen Assessment* (pp. 463–494). Cambridge: Cambridge University Press.

EC. (1992). *Council Directive 92/43/EEC of 21 May 1992 on the conservation of natural habitats and of wild fauna and flora*. Brussel: European Commision.

EEA. (2003). *An inventory of biodiversity indicators in Europe*. Luxembourg: Office for Official Publications of the European Communities.

EEA (European Environment Agency). (2007). *Halting the loss of biodiversity by 2010: Proposal for a first set of indicators to monitor progress in Europe*. (EEA Technical report 11/2007). Luxembourg: Office for Official Publications of the European Communities.

Ellis, C. J., Yahr, R., & Coppins, B. J. (2011). Archaeobotanical evidence for a massive loss of epiphyte species richness during industrialization in southern England. *Proceedings of the Royal Society B, 278,* 3482–3489.

Galloway, J. N., & Cowling, E. B. (2002). Reactive nitrogen and the world: 200 years of change. *Ambio, 31,* 64–71.

Hurlbert, S. H. (1971). The nonconcept of species diversity: A critique and alternative parameters. *Ecology, 52,* 577–586.

IUCN. (2012). *IUCN Red List Categories and Criteria: Version 3.1. Second edition*. Gland, Switzerland and Cambridge, UK: IUCN.

Jenssen, M. (2009). Assessment of the effects of top-soil changes on plant species diversity in forest, due to nitrogen deposition. In J. P. Hettelingh, M. Posch, & J. Slootweg (Eds.), *Progress in the modelling of critical thresholds, impacts to plant species diversity and ecosystem services in Europe* (pp. 83–100). Bilthoven: Netherlands Environmental Assessment Agency.

Lamb, E. G., Bayne, E., Holloway, G., Schieck, J., Boutin, S., Herbers, J., & Haughland, D. L. (2009). Indices for monitoring biodiversity change: Are some more effective than others? *Ecological Indicators, 9,* 432–444.

Legendre, P., & Gallagher, E. D. (2001). Ecologically meaningful transformations for ordination of species data. *Oecologia, 129,* 271–280.

Legendre, P., & Legendre, L. (2012). *Numerical ecology. Developments in environmental modelling Vol 24*. Amsterdam: Elsevier.

Margalef, R. (1963). On certain unifying principles in ecology. *The American Naturalist, 97,* 357–374.

McCann, K. S. (2000). The diversity—stability debate. *Nature, 405,* 228–233.

Mol-Dijkstra, J. P., & Kros, H. (2001). Modelling effects of acid deposition and climate change on soil and run-off chemistry at Risdalsheia, Norway. *Hydrology and Earth System Sciences, 5,* 487–498.

Odén, S. (1967). *Nederbördens försurning*. In Dagens Nyheter, October 24.

Posch, M., & Reinds, G. J. (2009). A very simple dynamic soil acidification model for scenario analyses and target load calculations. *Environmantal Modelling and Software, 24,* 329–340.

Posch, M., Belyazid, S., Kurz, D., & Reinds, G. J. (2010). The VSD-Veg Model: Progress and Prospects. In J. Slootweg, M. Posch, & J.-P. Hettelingh (Eds.), *Progress in the modelling of critical thresholds and dynamic modelling, including impacts on vegetation in Europe, CCE Status Report 2010* (pp. 49–54). Bilthoven: RIVM.

Rowe, E. C., Emmett, B. A., & Smart, S. M. (2009). A single metric for defining biodiversity damage using Habitats Directive criteria. In J. -P. Hettelingh, M. Posch, & J. Slootweg (Eds.), *Progress in the modelling of critical thresholds, impacts to plant species diversity and ecosystem services in Europe: CCE Status Report 2009* (pp. 101–107). Bilthoven: Coordination Centre for Effects.

Sala, O. E., Chapin, F. S., III, Armesto, J. J., Berlow, E., Bloomfield, J., Dirzo, R., Huber-Sanwald, E., Huenneke, L. F., Jackson, R. B., Kinzig, A., Leemans, R., Lodge, D. M., Mooney, H. A., Oesterheld, M., LeRoy Poff, N., Sykes, M. T., Walker, B. H., Walker, M., & Wall, D. H. (2000). Global Biodiversity Scenarios for the Year 2100. *Science, 287,* 1770–1774.

Shannon, C. E. (1948). A mathematical theory of communication. *The Bell System Technical Journal, 27,* 370–423, 623–656.

Simpson, E. (1949). Management of diversity. *Nature, 163,* 688.

Somerfield, P. J. (2008). Identification of the Bray-Curtis similarity index: Comment on Yoshioka (2008). *Marine Ecology-Progress Series, 372,* 303–306.

Stevens, C. J., Manning, P., van den Berg, L. J. L., de Graaf, M. C. C., Wamelink, G. W. W., Boxman, A. W., Bleeker, A., Vergeer, P., Arroniz-Crespo, M., Limpens, J., Lamers, L. P. M., Bobbink, R., & Dorland, E. (2011). Ecosystem responses to reduced and oxidised nitrogen inputs in European terrestrial habitats. *Environmental Pollution, 159,* 665–676.

Tilman, D. (1987). Secondary succession and the pattern of plant dominance along experimental nitrogen gradients. *Ecological Monographs, 57,* 189–214.

Ulrich, B., Mayer, R., & Khanna, P. K. (1979). *Die Deposition von Luftverunreinigungen und ihre Auswirkungen in Waldokosystemen im Solling.* Schriften aus der Forstl. Fak. D. Univ. Göttingen und der Nierders. Vers. Anst. Bd. 58.

Van Adrichem, M. H. C., Wortelboer, F. G., & Wamelink, G. W. W. (2010). *MOVE: MOdel for terrestrial VEgetation version 4.0.* (WOT werkdocument 153). Wageningen: Wettelijke Onderzoekstaken Natuur & Milieu.

Van Dobben, H. F. (2011). *Naar eenvoudige dosis-effectrelaties tussen natuur- en milieucondities; een toetsing van de mogelijkheden van de Natuurplanner.* (WOt-werkdocument 282). Wageningen: Wettelijke Onderzoekstaken Natuur & Milieu.

Van Dobben, H. F., & Slim, P. A. (2012). Past and future plant diversity of a coastal wetland driven by soil subsidence and climate change. *Climatic Change, 110,* 597–618.

Van Dobben, H., & Wamelink, W. (2009). A Red-List-based biodiversity indicator and its application in model studies in the Netherlands. In J.-P. Hettelingh, M. Posch, & J. Slootweg (Eds.), *Progress in the modelling of critical thresholds, impacts to plant species diversity and ecosystem services in Europe: CCE Status Report 2009* (pp. 77–81). Bilthoven: Coordination Centre for Effects.

Van Dobben, H., Hettelingh, J. P., De Vries, W., Slootweg, J. & Reinds, G. J. (2010). Plant species diversity indicators for impacts of nitrogen and acidity and methods for their simulation: An overview. In J. Slootweg, M. Posch, & J. P. Hettelingh (Eds.), *CCE Status Report 2010.*

Van Tongeren, O., Gremmen, N., & Hennekens, S. (2008). Assignment of relevés to pre-defined classes by supervised clustering of plant communities using a new composite index. *Journal of Vegetation Science, 19,* 525–536.

Wamelink, G. W. W., ter Braak, C. F. J., & van Dobben, H. F. (2003). Changes in large-scale patterns of plant biodiversity predicted from environmental economic scenarios. *Landscape Ecology, 18,* 513–527.

Wamelink, G. W. W., Goedhart, P. W., van Dobben, H. F., & Berendse, F. (2005). Plant species as predictors of soil pH: Replacing expert judgement with measurements. *Journal of Vegetation Science, 16,* 461–470.

Wamelink, G. W. W., Goedhart, P. W., Malinowska, A. H., Frissel, J. Y., Wegman, R. J. M., Slim, P. A., & van Dobben, H. F. (2011). Ecological ranges for the pH and NO_3 of syntaxa: A new basis for the estimation of critical loads for acid and nitrogen deposition. *Journal of Vegetation Science, 22,* 741–749.

Part II
Empirical and Model-Based Critical Loads and Target Loads

Chapter 4
Effects and Empirical Critical Loads of Nitrogen for Europe

Roland Bobbink, Hilde Tomassen, Maaike Weijters, Leon van den Berg, Joachim Strengbom, Sabine Braun, Annika Nordin, Kirsten Schütz and Jean-Paul Hettelingh

4.1 Introduction

4.1.1 Effects of Nitrogen Deposition on Ecosystems

The availability of nutrients is one of the most important abiotic factors that determine plant species composition in ecosystems. N is the primary limiting nutrient for plant growth in many natural and semi-natural ecosystems, especially for oligotrophic and mesotrophic habitats. Most of the plant species in such ecosystems are adapted to nutrient-poor conditions, and can only survive or compete successfully on soils with low N availability (e.g. Aerts and Chapin 2000; Tamm 1991). The series of events that occurs when N inputs increase in an area with originally low

R. Bobbink (✉) · H. Tomassen · M. Weijters · L. van den Berg
B-Ware Research Centre, Radboud University Nijmegen, P.O. Box 6558,
6503 GB Nijmegen, The Netherlands
e-mail: r.bobbink@b-ware.eu

L. van den Berg
Aquatic Ecology and Environmental Biology, Radboud University Nijmegen,
6525 ED Nijmegen, The Netherlands

J. Strengbom
Department of Ecology, Swedish University of Agricultural Sciences,
P.O. Box 7044, 75007, Uppsala, Sweden

S. Braun · K. Schütz
Institute for Applied Plant Biology, 4124 Schönenbuch, Switzerland

A. Nordin
Umeå Plant Science Centre, Department of Forest Genetics and Plant Physiology,
Swedish University of Agricultural Sciences, SE-901 87 Umeå, Sweden

J.-P. Hettelingh
Coordination Centre for Effects (CCE), RIVM, Bilthoven, The Netherlands

W. de Vries et al. (eds.), *Critical Loads and Dynamic Risk Assessments*,
Environmental Pollution 25, DOI 10.1007/978-94-017-9508-1_4

background deposition is highly complex, as many ecological processes interact and operate at different temporal and spatial scales. As a consequence, high variations in sensitivity to atmospheric N deposition have been observed between different natural and semi-natural ecosystems. In general, however, the most obvious effects of increased N deposition are significant changes in the N cycle, vegetation composition and biodiversity (see also Bobbink et al. 1998; Bobbink et al. 2010 and references herein).

The severity of the impacts of atmospheric N deposition depends on a number of factors, of which the most important are (numbers not being a ranking): (i) the duration and total amount of inputs, (ii) the chemical and physical form of the airborne N input, (iii) the intrinsic sensitivity of the plant and animal species present, (iv) the abiotic conditions, including climate, (v) the past and present land use or management, and (vi) the biogeochemical properties of the soil. With respect to the last; acid neutralising capacity (ANC), soil nutrient availability, and soil factors that influence the nitrification potential, N immobilisation and denitrification rates, are especially important. Despite the wide variety in factors that may determine N deposition impacts in different habitats, some generalisation of processes and their impacts is possible, including: (i) direct toxicity of N gases and aerosols to individual species, (ii) eutrophication, (iii) acidification, (iv) differences in effects of oxidised versus reduced N and (v) increased susceptibility to secondary stress and disturbance factors (Fig. 4.1). See Chap. 2 for a detailed description of those effects.

4.1.2 History of Empirical N Critical Loads and its Revisions

Empirical N critical loads for natural and semi-natural ecosystems were first presented in a background document for the 1992 workshop on N critical loads held under the UNECE Convention on Long-range Transboundary Air Pollution (LRTAP Convention) at Lökeberg, Sweden (Bobbink et al. 1992). After detailed discussions the proposed values were agreed at that meeting (Grennfelt and Thörnelöf 1992). Additional information from the 1992–1995 period was evaluated and summarised in an updated background paper and published as Annex III (Bobbink et al. 1996) of the Mapping Manual (www.icpmapping.org). The updated critical loads of N were discussed and set with full consensus at a 1995 expert meeting held under the LRTAP Convention in Geneva (Switzerland). They were also used for the development of the second edition of the Air Quality Guidelines for Europe by the World Health Organization's Regional Office for Europe (WHO 2001). Furthermore, the empirical critical loads for N were extensively reviewed and updated in 2001–2002 (Berne workshop; Achermann and Bobbink 2003). In that update, classification of the receptor ecosystems was brought in line with that of the European Nature Information System (EUNIS) (mostly level 3) (Davies and Moss 2002; Davies et al. 2004; Hall et al. 2003), in addition to the incorporation of results from new N-impact studies from the 1996–2002 period (Bobbink et al. 2003). In 2008 and 2009, it was recognised at CCE workshops and Task Force meetings of the International Co-

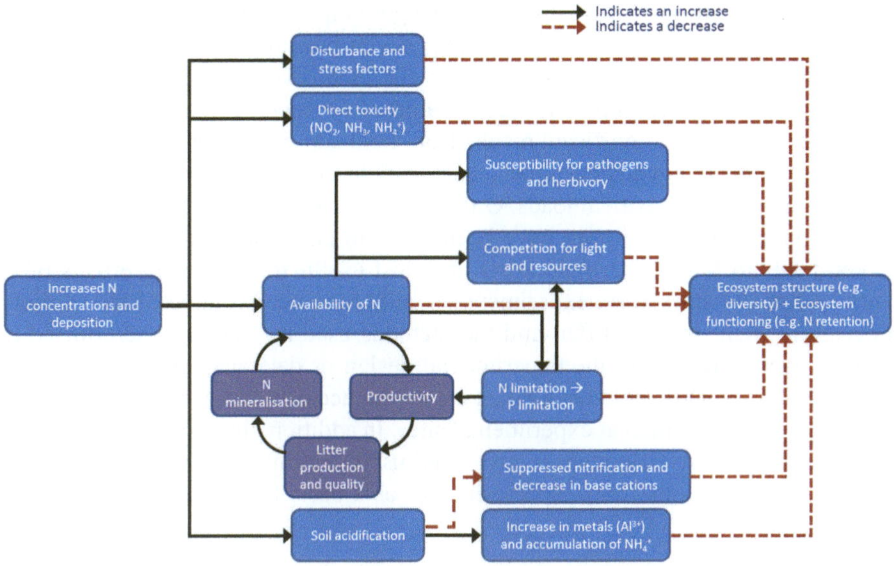

Fig. 4.1 Scheme of the main impacts of increased N deposition on ecosystems. Stress is considered to occur when external constraints limit the rate of dry matter production of the vegetation, whereas disturbance consists of mechanisms which affect soils and plant biomass by causing its partial or total destruction.

operative programme on Modelling and Mapping Critical Levels & Loads and Air Pollution Effects, Risks and Trends in Berne (ICP Modelling and Mapping 2008) and Stockholm (ICP Modelling and Mapping 2009) that considerable new insights into, and data on, the impacts of N deposition to natural and semi-natural vegetation have become available since the compilation of the 2003 document (Achermann and Bobbink 2003). A review and revision of the background material based on the availability of new scientific evidence for many N-sensitive ecosystems was adopted by the Working Group on Effects (WGE 2009) under the LRTAP Convention. Following this work, the empirical N critical loads for Europe were reviewed and revised in 2010 (Noordwijkerhout Workshop; Bobbink and Hettelingh 2011). Note, that in recent years empirical critical loads of N have also been established in the United States; they are described and discussed in Chap. 5 (see also Pardo et al. 2011).

4.1.3 Approach for Determining Empirical Critical Loads of Nitrogen for Europe

Overall Approach: Empirical critical loads of N are in almost all cases based on observations of negative effects of N input to an ecosystem, ecosystem functioning or processes in that ecosystem. Primarily changes in species abundance, composition and/or diversity ('structure'), N leaching, decomposition or mineralisation rate ('functioning') are recorded. For a more complete overview of indicators, see Løkke et al. (2000). Effects have been evaluated for specific ecosystems. Statistically and biologically significant results from field addition experiments and mesocosm experiments conducted under close-to-field conditions have been used for quantifying empirical critical loads. Only studies on independent N treatments with a duration of 2 years or more have been used. In particular data from long-term experiments in low-background areas are most useful for observing effects of N enrichment. However, since experimental studies have been conducted for a variety of reasons, their designs differ, and the methods used are carefully scrutinised to identify factors related to the experimental design or data analysis that may constrain their use. This includes evaluation of the accuracy of estimated values of background N deposition at experimental sites. In addition, the results from correlative or retrospective field studies have been used, but only as additional evidence to support conclusions from experiments, or as a basis for an 'expert judgement' rating.

Revision: Revision of empirical N critical loads from earlier versions is based on all available empirical data per considered EUNIS class (see 4.1.2) and has a strong emphasis on review by international experts in the field. Typically, in a first phase, all available data from peer-reviewed publications, PhD theses, book chapters, nationally published papers, and 'grey' reports by institutes or organisations are used to set, or revise and review, N critical loads for each habitat by a team of invited experts (*internal review*). After the internal review, the proposed critical loads are reviewed by a team of international experts on the impacts of N in natural and semi-natural ecosystems (*external review*). Each specific EUNIS class is in this phase evaluated by at least 2 external experts. In a final phase, the N critical loads are reviewed once more and finalised after adding the input of specialists from different scientific and applied fields that meet at an international workshop (*workshop-review*).

Types of Evidence Used: There are three major types of evidence available to relate N deposition to changes in ecosystems.

The first is from long-term field addition (or manipulation) experiments, in which N deposition is artificially increased, normally by application of increased concentrations of $NH_4^+$and/or NO_3^- in water and, less frequently via fumigation with NH_3 gas (e.g. Sheppard et al. 2009). Data is then used and, when significant impacts are detected, inferred with confidence that these are caused by N deposition. Experiments can provide information on the response time of different components of the system (structure and functioning) to N addition and can be designed to assess interactions, for example with other stresses or management. Experiments can identify thresholds for effects. However, since most experiments examine effects of increases in deposition, it is difficult to identify thresholds from experiments in areas with a relatively long history of elevated levels of N deposition, where there may already have been significant impacts of N deposition prior to the start of the experiments. Other limitations of experimental studies are that they typically assess relatively short-term responses (even the longest-running experiments seldom exceed 20 years) and that peculiarities of the experiment (e.g. very high concentrations of the applied pollutant compared with environmentally realistic burdens) or site-specific factors might also explain part of the observed response.

A second approach is through targeted field surveys of sites covering a gradient of N deposition. These may use short but intense gradients of N deposition (e.g. close to intensive animal husbandry units) or have a national or regional focus. Such field surveys may provide information on longer-term responses, cover a wider (and more realistic) range of N deposition than experiments, and avoid experimental artefacts. Since gradients of N deposition may be highly correlated with those of other potential drivers (e.g. S deposition, climate or management), these need to be measured and considered in statistical analyses and interpretation. For this reason, targeted surveys cannot prove causality, but can help to infer the role of N deposition as a possible driver of changes in, for instance, biodiversity. Targeted surveys and experiments allow effects on relatively common species and functional groups to be evaluated, but they have little statistical power to detect effects on rare or very scarce species.

The third approach consists of the use of data sets from a broader ecological surveillance which cover a wider range of communities, representative of the region of interest. These typically record the presence or absence of species in larger areas (e.g. 10 km × 10 km^2). Such data reflect the impact of land use and a range of climatic, edaphic and management factors, and hence attribution of any change to N deposition is difficult. However, they do provide important signals of change in biodiversity at regional and national scales. These data sets, potentially, are also able to detect effects on (very) rare and scarce species, but they are not directly useful to set critical loads.

Reliability: The reliability of empirical N critical loads is labelled qualitatively, distinguishing 3 levels of reliability: 'reliable', 'quite reliable' and 'expert judgement'. Current empirical N critical loads are determined on statistically and biologically significant outcomes of field addition experiments and mesocosm studies. Only studies which have independent N treatments and realistic N loads and durations (below 100 kg N $ha^{-1}yr^{-1}$ and more than 1 year) were used for the updating and refinement of critical load values. Studies with higher N additions or shorter experimental periods are only interpreted with respect to the understanding of mechanisms, possible N limitation or sensitivity of the ecosystem. The methods used in those studies have been carefully scrutinised to identify factors related to the experimental design or data analysis, which may constrain their use in assessing critical loads. This includes evaluation of the precision of the estimated values of background deposition at the experimental site, which is necessary to obtain insight into the total N load in both the N-treated and the control vegetation.

The results from targeted field studies are used as additional evidence with respect to the outcome of N-application studies, or to estimate a critical load value based on expert judgement, if experimental studies are lacking. When available, the outcomes from dynamic ecosystem models provide additional insight into underlying mechanisms, such as increased frequencies of pests and diseases. In general, pot or microcosm studies are not used for setting critical loads, except for studies on bryophytes. However, the outcomes of these studies, in some selected cases, are used as an indication of the N sensitivity of the most important or sensitive plant species of an ecosystem.

4.1.4 Contents of this Chapter

In this chapter, all groups of natural and semi-natural ecosystems have been classified according to the EUNIS habitat classification for Europe (Davies and Moss 2002; Davies et al. 2004) down to level 3 of the EUNIS hierarchy. Table 4.1 summarises the empirical critical loads of N for European habitats. In the table, EUNIS classification as well as European Habitat Codes that are used for Natura 2000 (Commission of the European Communities 2003) are provided. Table 4.1 shows the upper and lower boundaries of the empirical N critical loads for each of the EUNIS classes (and habitats), including the reliability of this range. The reliability of the empirical critical load ranges is expressed as 'reliable, 'quite reliable' and based on 'expert opinion'. In addition, the most significant biological changes in the ecosystem structure and/or functioning are shown when the N deposition inputs exceed the maximum of the critical load range. In the paragraphs below, background information for each of the habitats and the current N critical loads are discussed.

4.2 Marine Habitats (EUNIS Class A)

4.2.1 System Description

Marine habitats are distinguished from other ecosystems by their direct connection to the sea. Most of these ecosystems are either not covered with plants or fully aquatic, and therefore excluded here.

However, coastal salt marshes above the high spring tide in tidal regions are included in marine habitats (Class A). This habitat develops where fine sediments accumulate along sheltered coastlines in the temperate and high-latitude regions of the world. They are often associated with estuaries but also frequently occur in areas where the coastline is protected by islands and sandbars. They are typically intertidal and are periodically covered with salt water. The dominant plants are rooted macrophytes which are adapted to the environmental stresses associated with sea water inundation (Archibold 1995).

Table 4.1 Overview of empirical critical loads of N (kg N $ha^{-1}yr^{-1}$, column 4) to natural and semi-natural ecosystems (column 1), classified according to EUNIS (column 2), and classified as European Habitat (column 3).The reliability is qualitatively indicated by ## reliable; # quite reliable and (#) expert judgement (column 5). Column 6 provides a selection of effects that can occur when critical loads are exceeded. * indicates EU habitat that is strongly related to the EUNIS habitat. (Adapted from Bobbink and Hettelingh 2011)

Ecosystem type	EUNIS code	EU habitat	Empirical CL (kg N $ha^{-1}yr^{-1}$)	Reliability	Indication of exceedance
Marine habitats (A)					
Mid-upper salt marshes	A2.53	1330	20–30	(#)	Increase in dominance of graminoids
Pioneer salt marshes and low-mid salt marshes	A2.54 and A2.55	1310, 1320	20–30	(#)	Increase in late-successional species, increase in productivity
Coastal habitats (B)					
Shifting coastal dunes	B1.3	2110, 2120	10–20	(#)	Biomass increase, increased N leaching
Coastal stable dune grasslands (grey dunes)	B1.4[a]	2130	8–15	#	Increase in tall graminoids, decrease in prostrate plants, increased N leaching, soil acidification, loss of typical lichen species
Coastal dune heaths	B1.5	2150	10–20	(#)	Increase in plant production, increased N leaching, accelerated succession
Moist to wet dune slacks	B1.8[b]	2190	10–20	(#)	Increased biomass tall graminoids
Inland surface waters (C)					
Soft-water lakes (permanent oligotrophic waters)	C1.1[c]	3110, 3130	3–10	##	Change in the species composition of macrophyte communities, increased algal productivity and a shift in nutrient limitation of phytoplankton from N to P
Dune slack pools (permanent oligotrophic waters)	C1.16	3140*	10–20	(#)	Increased biomass and rate of succession
Permanent dystrophic lakes, ponds and pools	C1.4[d]	3160	3–10	(#)	Increased algal productivity and a shift in nutrient limitation of phytoplankton from N to P
Mires, bogs and fens (D)					

Table 4.1 (continued)

Ecosystem type	EUNIS code	EU habitat	Empirical CL (kg N $ha^{-1}yr^{-1}$)	Reliability	Indication of exceedance
Raised and blanket bogs	D1[e]	7110, 7130	5–10	##	Increase in vascular plants, altered growth and species composition of bryophytes, increased N in peat and peat water
Valley mires, poor fens and transition mires	D2[f]	7140	10–15	#	Increase in sedges and vascular plants, negative effects on bryophytes
Rich fens	D4.1[g]	7230	15–30	(#)	Increase in tall graminoids, decrease in bryophytes
Montane rich fens	D4.2[g]	7240	15–25	(#)	Increase in vascular plants, decrease in bryophytes
Grasslands and lands dominated by forbs, mosses or lichens (*E*)					
Sub-Atlantic semi-dry calcareous grasslands	E1.26	6210	15–25	##	Increase in tall grasses, decline in diversity, increased mineralisation, N leaching, surface acidification
Mediterranean xeric grasslands	E1.3	6220	15–25	(#)	Increased production, dominance by graminoids
Non-Mediterranean dry acidic and neutral closed grasslands	E1.7[b]	6230, 6270	10–15	##	Increase in graminoids, decline in typical species, decrease in total species richness
Inland dune pioneer grasslands	E1.94[b]	2330	8–15	(#)	Decrease in lichens, increase in biomass
Inland dune siliceous grasslands	E1.95[b]	2330	8–15	(#)	Decrease in lichens, increase in biomass, increased succession
Low- and medium-altitude hay meadows	E2.2	6510	20–30	(#)	Increase in tall grasses, decrease in diversity
Mountain hay meadows	E2.3	6520	10–20	(#)	Increase in nitrophilous graminoids, changes in diversity
Moist and wet oligotrophic grasslands					

Table 4.1 (continued)

Ecosystem type	EUNIS code	EU habitat	Empirical CL (kg N $ha^{-1}yr^{-1}$)	Reliability	Indication of exceedance
Molinia caerulea meadows	E3.51	6410	15–25	(#)	Increase in tall graminoids, decreased diversity, decrease in bryophytes
Heath (*Juncus*) meadows and humid (*Nardus stricta*) swards	E3.52		10–20	#	Increase in tall graminoids, decreased diversity, decrease in bryophytes
Moss- and lichen-dominated mountain summits	E4.2		5–10	#	Effects on bryophytes or lichens
Alpine and subalpine acidic grasslands	E4.3	6150	5–10	#	Changes in species composition; increase in plant production
Alpine and subalpine calcareous grasslands	E4.4	6170	5–10	#	Changes in species composition; increase in plant production
Heathland, scrub and tundra (F)					
Tundra	F1		3–5	#	Changes in biomass, physiological effects, changes in species composition in bryophyte layer, decrease in lichens
Arctic, alpine and subalpine scrub habitats	F2	4060, 4080	5–15	#	Decline in lichens, bryophytes and evergreen shrubs
Northern wet heath	F4.11				
'U' *Calluna*-dominated wet heath (Upland moorland)	F4.11[e, h]		10–20	#	Decreased heather dominance, decline in lichens and mosses, increased N leaching
'L' *Erica tetralix*-dominated wet heath (Lowland)	F4.11[e, h]	4010	10–20	(#)	Transition from heather to grass dominance
Dry heaths	F4.2[e, h]	4030	10–20	##	Transition from heather to grass dominance, decline in lichens, changes in plant biochemistry, increased sensitivity to abiotic stress
Mediterranean scrub	F5	5210*, 5230*, 5310*, 5330*	20–30	(#)	Change in plant species richness and community composition

Table 4.1 (continued)

Ecosystem type	EUNIS code	EU habitat	Empirical CL ($kg\ N\ ha^{-1}yr^{-1}$)	Reliability	Indication of exceedance
Woodland, forest and other wooded land (*G*)					
Fagus woodland	G1.6	9110, 9120, 9130, 9140, 9150	10–20	(#)	Changes in ground vegetation and mycorrhiza, nutrient imbalance, changes in soil fauna
Acidophilous *Quercus*-dominated woodland	G1.8	9190, 91A0	10–15	(#)	Decrease in mycorrhiza, loss of epiphytic lichens and bryophytes, changes in ground vegetation
Mesotrophic and eutrophic *Quercus* woodland	G1.A	9160, 9180, 9170, 9020	15–20	(#)	Changes in ground vegetation
Mediterranean evergreen (*Quercus*) woodland	G2.1	9330, 9340	10–20	(#)	Changes in epiphytic lichens
Abies and *Picea* woodland	G3.1	9410	10–15	(#)	Decreased biomass of fine roots, nutrient imbalance, decrease in mycorrhiza, changed soil fauna
Pinus sylvestris woodland south of the taiga	G3.4	91C0	5–15	#	Changes in ground vegetation and mycorrhiza, nutrient imbalances, increased N_2O and NO emissions
Pinus nigra woodland	G3.5	9530	15	(#)	Ammonium accumulation
Mediterranean *Pinus* woodland	G3.7	2270	3–15	(#)	Reduction in fine-root biomass, shift in lichen community
Spruce taiga woodland	G3.A [i]	9010, 9050	5–10	##	Changes in ground vegetation, decrease in mycorrhiza, increase in free-living algae
Pine taiga woodland	G3.B [i]	9010	5–10	#	Changes in ground vegetation and in mycorrhiza, increase occurrence of free-living algae
Mixed taiga woodland with *Betula*	G4.2		5–8	(#)	Increased algal cover

Table 4.1 (continued)

Ecosystem type	EUNIS code	EU habitat	Empirical CL ($kg\ N\ ha^{-1} yr^{-1}$)	Reliability	Indication of exceedance
Mixed *Abies-Picea Fagus* woodland	G4.6[j]		10–20	(#)	
Overall					
Broadleaved deciduous woodland	G1[k, l]		10–20	##	Changes in soil processes, nutrient imbalance, altered composition mycorrhiza and ground vegetation
Coniferous woodland	G3[k, l]		5–15	##	Changes in soil processes, nutrient imbalance, altered composition mycorrhiza and ground vegetation

[a] For acidic dunes, the 8–10 $kg\ N\ ha^{-1}\ yr^{-1}$ range should be applied, for calcareous dunes this range is 10–15 $kg\ N\ ha^{-1}\ yr^{-1}$

[b] Apply the lower end of the range to habitats with a low base availability; and the higher end of the range to those with high base availability

[c] This critical load should only be applied to oligotrophic waters with low alkalinity with no significant agricultural or other human inputs. Apply the lower end of the range to boreal and alpine lakes, and the higher end of the range to Atlantic soft waters

[d] This critical load should only be applied to waters with low alkalinity with no significant agricultural or other direct human inputs. Apply the lower end of the range to boreal and alpine dystrophic lakes

[e] Apply the high end of the range to areas with high levels of precipitation and the low end of the range to those with low precipitation levels; apply the low end of the range to systems with a low water table, and the high end of the range to those with a high water table. Note that water tables can be modified by management

[f] For EUNIS category D2.1 (valley mires): use the lower end of the range (#)

[g] For high-latitude systems, apply the lower end of the range

[h] Apply the high end of the range for areas where sod cutting has been practiced; use the lower end of the range for areas under low intensity management

[i] In Achermann and Bobbink (2003) presented as overall value for boreal forests

[j] Included in studies that were classified under EUNIS categories G1.6 and G3.1

[k] In Achermann and Bobbink (2003) presented as overall value for temperate forests

[l] For application at broad geographical scales

4.2.2 *Nitrogen Deposition Effects and Critical Loads of N*

Although these ecosystems receive large amounts of nutrients from surface water and export similarly large amounts of nutrients through surface water and denitrification (for N), it is generally accepted that salt-marsh vegetation is primarily N-limited (Mitsch and Gosselink 2000). During salt-marsh succession, N accumulates in organic material, and N mineralisation increases as marshes age, as shown by Olff et al. (1993) and Van Wijnen et al. (1999). This accumulation of N is considered a major driving force behind succession, as competition for nutrients is replaced by competition for light. Increased N deposition will accelerate this natural succession, but, because it does not affect the accretion rate of salt marshes, it will result in a net loss of salt marshes of a low successional age (EUNIS categories A2.55 and A2.54).

Unfortunately the number of studies on this ecosystem is limited. Based on the few studies available and the fact that there is a strong risk for N retention leading to succession, the empirical N critical load range for this EUNIS class was set between 20 and 30 kg N $ha^{-1}yr^{-1}$ as a conservative 'quite reliable' estimate.

4.3 Coastal Habitats (EUNIS Class B)

4.3.1 *System Description*

Coastal habitats are those ecosystems that are situated above the high spring tide limit (or above mean water level in non-tidal waters) with coastal features and characterised by their proximity to the sea. They include coastal dunes (dry grasslands, wet to moist dune slacks, scrub and wooded dunes), beaches and cliffs (Davies et al. 2004). Dune slack pools are discusses in EUNIS class C (inland surface waters), but consistency between critical loads for dune slacks and dune slack pools is maintained to facilitate practical applications of empirical N critical loads.

4.3.2 *Nitrogen Deposition Effects and Critical Loads of N*

Shifting coastal dunes are mobile sand habitats of the boreal, nemoral, steppe, Mediterranean and warm-temperate humid zones of Europe. They include embryonic shifting dunes and shifting dunes along the shoreline with *Ammophila arenaria* ('white dunes').

Unfortunately hardly any evidence from N addition experiments is available on this habitat type, except for a study in Iceland with rather high N addition rates (Greipsson and Davy 1997) and a correlative study on 11 sand dune sites in the UK (Jones et al. 2002, 2004). Both studies indicate N effects on productivity and support the idea that the sites with the higher N inputs have been impacted.

Based on the observed relationships and generally low nutrient status of these soils, the empirical critical load range for shifting coastal dunes (B1.3) was set at 10–20 kg N $ha^{-1}yr^{-1}$, but since it was based on a correlative study and a study with high N loads, the reliability was graded 'expert judgement'.

A large number of stable dune grasslands (B1.4) are located along the coasts of Europe, from the boreal to the Mediterranean and warm-temperate zones. They are found in fixed dunes, usually with herbs and graminoids as the dominant life form, although in certain areas in the northern and western systems, bryophytes make up a substantial component of the biomass—up to 70 %, particularly where grazed (e.g. Plassmann et al. 2009). In early corynephorus stages, mosses and lichens may dominate, both in abundance and in species richness. They are characterised by dry soil conditions and occur on calcareous to acidic sandy soils, thus from high to low base status (e.g. Davies et al. 2004). In general, these stable dune grasslands have a high species diversity and many characteristic plant and animal species. Studies on stable dune grasslands (e.g. Plassmann et al. 2009; Remke et al. 2009a; Remke et al. 2009b) and recent data suggested that the previous set critical load range for stable dune grasslands (grey dunes) of 10–20 kg N $ha^{-1}yr^{-1}$ (Bobbink et al. 2003) needed adjustment down to 8–15 kg N $ha^{-1}yr^{-1}$ in the most recent critical loads review (Bobbink and Hettelingh 2011). N input of rates >5–10 kg N $ha^{-1}yr^{-1}$ increased N content of lichens (*Cladonia sp.*) and caused a shift from lichen-rich short grass vegetation towards species-poor vegetation dominated by *Carex arenaria* (Fig. 4.2). In addition, soil acidification and accumulation of organic matter was observed in acidic (decalcified) grasslands. Based on these experiments, a critical loads range of 8–10 kg N $ha^{-1}yr^{-1}$ for acidic or decalcified dune grasslands, and 10–15 kg N $ha^{-1}yr^{-1}$ for calcareous stable dune grasslands was recommended.

Next to dry dune grasslands, heathland vegetation is also present in the coastal dunes in north-western Europe (Ellenberg 1988; Gimingham et al. 1979). These natural coastal dune heaths are mostly dominated by the typical dwarf shrub *Empetrum nigrum*, while *Calluna vulgaris* is less common. Coastal heaths (B1.5) are classified within EUNIS as a subcategory of coastal dune and sand habitats. These dune heaths are generally unmanaged, and thus have low N removal from the system. The empirical critical load range of N for coastal heath (B1.5) has been set at 10–20 kg $ha^{-1}yr^{-1}$ ('expert judgement').

Moist to wet dune slacks (EUNIS B1.8) of primary or secondary origin, are hot spots of plant diversity in the sandy dune regions of Europe. They are characterised by typical graminoids (sedges, rushes and grasses), together with many rare, basiphilous forb species. Groundwater level is at or above soil level in winter, whereas in the growing season groundwater level is considerably lower in these dune slacks. Nowadays, many dune slacks have been incorporated in nature reserves, and these are sometimes maintained through management, such as hay production and harvest or sod cutting (e.g. Davies et al. 2004; Lammerts and Grootjans 1997). Because of their isolation in the landscape and their successional position, they mostly receive nutrients via atmospheric inputs. In addition, they are very sensitive to desiccation from groundwater extraction.

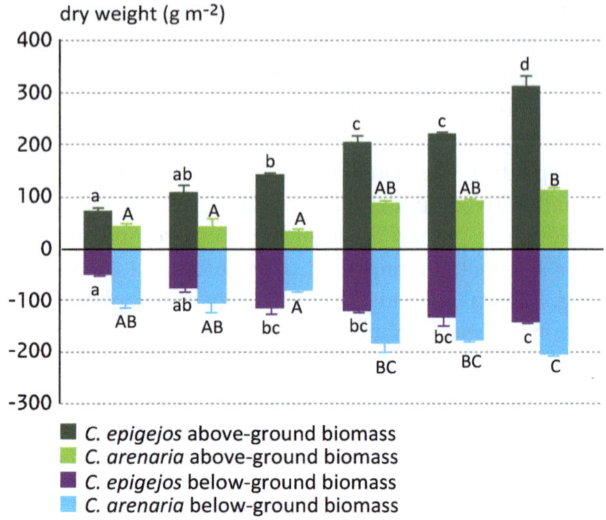

Fig. 4.2 Above- and below-ground biomass (g m^{-2}) of *Calamagrostis epigejos* and *Carex arenaria* after 2 years of N application in coastal stable dune grassland (B1.4) mesocosms. (Tomassen et al. 1999; Van den Berg et al. 2005). From *left* to *right*: 1, 10, 20, 40, 60 and 80 kg N ha^{-1}yr^{-1}

Recent values for empirical N critical loads are set to 10–20 kg N ha^{-1}yr^{-1} for moist to wet dune slacks, rather than 10–25 kg N ha^{-1}yr^{-1}as recommended in 2003. This is based on studies from the UK, where N input above 15 kg N ha^{-1}yr^{-1}resulted in changes in species composition (via growth of *Carex* species and germination of species from the seed bank) and changes of the N content of plants (Jones et al. 2004; Plassmann et al. 2008). Moreover, it turns out that there is a need to diminish accumulated nutrient rich material and maintain biodiversity by removing sludge and sods every 20 years in coastal areas of the Netherlands with a deposition above 20 kg ha^{-1}yr^{-1}.

4.4 Inland Surface Water Habitats (EUNIS Class C)

4.4.1 System Description

Inland surface water habitats are non-coastal, open, fresh or brackish water bodies (e.g., lakes and pools, rivers, streams, ditches and springs), including their littoral zones. This class includes constructed inland freshwater, brackish or saline water bodies (e.g., canals and ponds) which support a semi-natural community of both plants and animals, and seasonal water bodies which may dry out for part of the year (temporary or intermittent rivers and lakes and their littoral zones). Freshwater littoral zones include those parts of banks or shores that are sufficiently frequently inundated to prevent the formation of closed terrestrial vegetation. Permanent snow and ice are excluded from this EUNIS class (Davies et al. 2004). The main division of permanent standing waters containing fresh water is based on the trophic status of the waters, from oligotrophic (C1.1) via mesotrophic (C1.2) to eutrophic lakes, ponds and pools (C1.3). Dystrophic water bodies (C1.4) which are rich in humic substances and often with a brown colour, are also included in C1. Unfortunately, information on the gradient in alkalinity (very soft to hard water), which is an important factor for, and partly independent of, the nutrient status of the water (e.g. Brouwer et al. 2002; Moss 1988) is not separately distinguished in the EUNIS habitat hierarchy, making it difficult to classify these systems under EUNIS.

4.4.2 Nitrogen Deposition Effects and Critical Loads of N

In the lowlands of western Europe, many shallow soft waters are found on sandy sediments, which are poor in or almost devoid of calcium carbonate. These waters are fully mixed, poorly buffered and shallow. They have fluctuating water levels, and are mainly fed by rain water or water from acidic catchment soils, and therefore oligotrophic (C1.1). These soft-water ecosystems are characterised by plant communities from the phyto-sociological alliance *Littorelletea* (Arts 1990; Schoof-van Pelt 1973; Wittig and Pott 1982). The stands of these communities are characterised by the presence of rare and endangered isoetid plants, such as *Littorella uniflora, Lobelia dortmanna* and *Luronium natans*.

The effects of N pollutants on these soft-water systems have been intensively studied in the Netherlands, both in field surveys and experimental studies. From these studies it became clear that many macrophytes have been lost due to the effects of acidification and eutrophication as a result of increased N deposition. Evidence suggests that acidophilic and eutrophic species such as *Juncus bulbosus* or aquatic mosses, such as *Sphagnum cuspidatum* and *Drepanocladus fluitans* increase at the cost of more sensitive species in areas with atmospheric N loads of 10–13 kg N $ha^{-1}yr^{-1}$ (Arts 1990; Roelofs 1983; Schuurkes et al. 1987). More recently, however, it has or more become obvious that freshwater systems, especially in pristine areas of alpine, sub-Arctic, Arctic or boreal regions are N limited (rather than P limited in many of the lowland systems, e.g. Bergström and Jannson 2006; Saros et al. 2005; Wolfe et al. 2006). Experimental studies and surveys of boreal lakes (Bergström et al. 2005, 2008; Elser et al. 2009) suggest that atmospheric N deposition increases the stoichiometric N:P ratio in these lakes above N loads of 3 kg N $ha^{-1}yr^{-1}$, causing ecological nutrient limitation to shift.

For the permanent oligotrophic waters, a critical load range is therefore set at 3–10 kg N $ha^{-1}yr^{-1}$, where the lower boundary of the range applies to boreal, sub-Arctic and alpine lakes and the upper boundary applies to Atlantic soft-water lakes.

Evidence has been found that dystrophic lakes (C1.4) may respond to N deposition in a similar way as oligotrophic lakes (C1.1). Based on experiments and on expert judgement, the critical load range of 3–10 kg $ha^{-1}yr^{-1}$is therefore also applied to dystrophic lakes (C1.4), with the note that the lower end of the range applies to boreal, sub-Arctic and alpine dystrophic lakes. The N critical loads for these inland waters can be applied in particular to waters with low alkalinity and free of significant agricultural or other direct anthropogenic inputs that are often high, possibly causing severe adverse effects to biodiversity regardless of atmospheric inputs.

Empirical evidence is limited that supports the N critical loads for the relatively small but permanent clear-water pools in the European coastal dune areas (EUNIS class C1.16). These relatively small but permanent clear-water pools are found in the European coastal dune areas. The water is nutrient-poor and these dune slack pools are mostly insensitive to acidification, because of their high alkalinity (>1000 μeq l^{-1}). They are also characterised by clear water and a diverse, submerged macrophyte vegetation (e.g. *Potamogeton* and *Chara* species, and littoral isoetids). Eutrophication in these dune slack pools is more likely to be caused by atmospheric inputs or by high densities of waterfowl than by the inflow of enriched surface water, because of the hydrological isolation of these habitats.

Only a few studies have investigated the sensitivity of these dune slack pools with respect to critical load setting (Brouwer et al. 1996; Lammerts and Grootjans 1997) and showed changes at N inputs exceeding 20 kg $ha^{-1}yr^{-1}$. The critical load range for this ecosystem was therefore set at 10–20 kg N $ha^{-1}yr^{-1}$.

There is only limited data available that provide insights in the relation between N deposition and N leaching (Aber et al. 1989; Aber et al. 1998; Dise and Wright 1995; Goodale et al. 2005; Wright et al. 2001). Consequently, future N leaching, especially given prolonged N enrichment and climate change, is difficult to predict (Stoddard 1994; Wright et al. 2006) and partly hampers reliable estimation of N critical loads for freshwater ecosystems.

4.5 Mire, Bog and Fen Habitats (EUNIS Class D)

4.5.1 System Description

Class D of EUNIS includes a wide range of wetland systems that have their water table at or above soil or sediment level for at least half of the year, dominated by either herbaceous or ericoid vegetation (Davies and Moss 2002; Davies et al. 2004). Nutrient budgets in wetland ecosystems are characterised by inputs and outputs via groundwater and surface water, and are tightly linked with local hydrology. The extent to which these systems receive and lose nutrients with in- and out-flowing water largely determines their sensitivity to excess N. Several wetland types (e.g., those in categories D1, D2 and D4) are characterised by a partly or incomplete decomposition of the plant litter, resulting in peat formation (Malmer et al. 2003; Mitsch and Gosselink 2000). The capacity for peat formation is a major component in the nutrient balance of these wetland systems, and in many situations is strongly linked to the presence of peat mosses (*Sphagnum* species) (Van Breemen 1995).

4.5.2 Nitrogen Deposition Effects and Critical Loads of N

Ombrotrophic (raised) bogs, which receive all their nutrients from the atmosphere, are particularly sensitive to airborne N loads. These bogs are systems of acidic wet areas and are very common in the boreal and temperate regions of Europe. Because abiotic conditions prevail, decomposition rates are low, favouring the development of peat. In western Europe and far northern latitudes, typical plant species include bog mosses (*Sphagnum* species), sedges (*Carex, Eriophorum*) and heathers (*Andromeda, Calluna* and *Erica*).

Even though there has been only a limited number of long-term (>3–4 years) N manipulation experiments in bog ecosystems (D1), a clear picture is emerging from the potential impact of elevated N deposition on bog habitats. Responses to N additions as low as 10 kg N $ha^{-1}yr^{-1}$ have been observed in a number of bog species (in terms of survivorship, flowering, and density) (e.g. Gunnarsson et al. 2002; Fig. 4.3).

Bryophyte species, in particular *Sphagnum* species, appear to be susceptible to the increase of anthropogenic N pollution, showing a decline in favour of grass and other competitive species, changes in competition between *Sphagnum* species, and changes in physiological and biochemical characteristics (Gerdol et al. 2007; Limpens et al. 2004; Tomassen et al. 2003). The concept of a limited capacity for the retention of N in the moss layer, above which N availability in the rhizosphere increases

Fig. 4.3 Numbers of individuals of *Drosera rotundifolia* in an ombrotrophic raised bog (D1) near Stockholm (Sweden) during 4 years of N additions. (Redbo-Torstensson 1994)

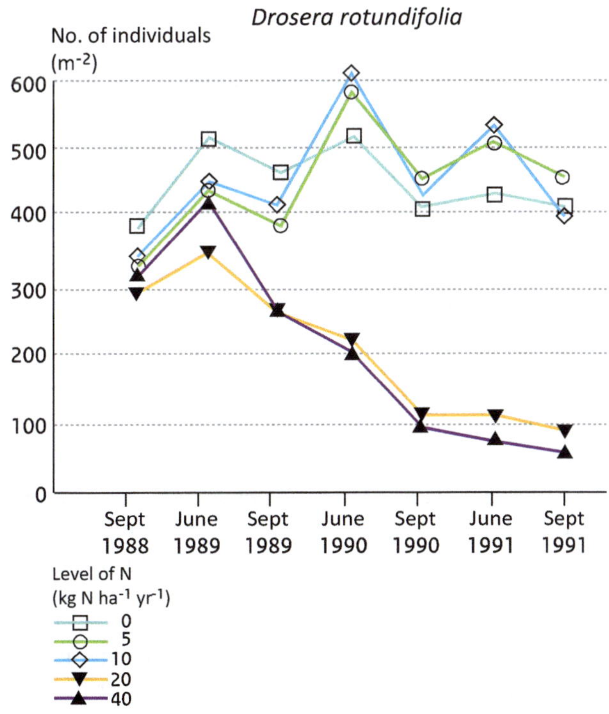

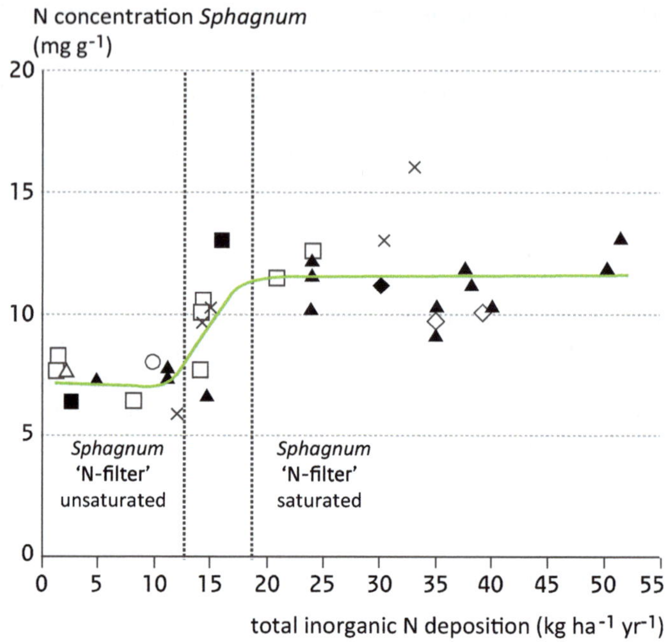

Fig. 4.4 The N concentration (mg g^{-1} dry weight) in raised bog (D1) *Sphagnum* species (apical parts) in Europe and the United States, in relation to total atmospheric N inputs (estimated at twice the wet deposition). (Lamers et al. 2000)

(Fig. 4.4; Bragazza et al. 2004), offers a tool for the assessment of factors which may modify the critical loads for these systems. Initial estimates using this method are consistent with a long-term response threshold of 10 kg N $ha^{-1}yr^{-1}$. The empirical N critical load range for these ecosystems is therefore set to 5–10 kg N $ha^{-1}yr^{-1}$, based on a considerable body of field evidence and experiments and are therefore judged to be 'reliable'. Expert judgement, based on observations of responses to N that are smaller in wetter bog areas, has established that bogs receiving high effective precipitation (e.g. in northern and western United Kingdom, Norway) are less sensitive to N than those in drier regions (e.g. in the Netherlands, Sweden). Therefore, precipitation is recommended as a modifying factor to be taken into account when assigning critical loads to individual sites.

Valley mires, poor fens and transition mires are weakly to strongly acidic peatlands, flushes and vegetated rafts formed in situations where they receive water from the surrounding landscape or are intermediate between land and water (Davies et al. 2004). All systems have permanently waterlogged soils, with groundwater just below or at the soil surface. This water supply is rather poor in base cations, leading to an acidic system, where peat mosses, but also small sedges and some brown moss species, dominate the vegetation. The distinction between valley mires, poor fens and transition mires is made on the basis of water level and water origin, which may have some implications for the level of their critical loads. Compared to poor fens, based on the generalisation from Morris (1991) on the link between N sensitivity and hydrology, valley mires are expected to be slightly less sensitive, and quaking bogs and transition mires to be more sensitive to excess N. Unfortunately, the low number of studies does not allow a further distinction, and the limited information that is available to date comes mainly from poor fen systems (D2.2). The critical load range previously recommended for poor fens (D2.2) of 10–15 kg N $ha^{-1}yr^{-1}$ is based on a number of field experiments and, hence, was judged to be 'quite reliable'. A study by Malmer et al. (2003) showed significant effects from additions of 20 kg N $ha^{-1}yr^{-1}$ (total input 25 kg N $ha^{-1}yr^{-1}$) on poor fen vegetation within a 2-year period. In the long term, effects are predicted to be even more pronounced. Since *Sphagnum* mosses function as N filters in poor fens, similar to those in ombrotrophic bogs, it is likely that 20 kg N $ha^{-1}yr^{-1}$ as an upper range of the critical load is too high.

Similar to poor fens, base-rich fens have developed on permanently waterlogged soils, but in these systems there is a base-rich, nutrient-poor and often calcareous water supply buffering the system. They are largely occupied by calcicolous small sedges and brown moss communities (Davies et al. 2004). Despite the fact that rich fens are the habitat of a range of specialised and rare species, very few field experiments have been conducted with enrichments of ecologically relevant doses of N to determine the effects of increased N deposition (e.g. Beltman et al. 1996; Boeye et al. 1997; Wassen et al. 1998). A study on Irish fens (Dorland et al. 2008) showed that additions of 35 kg N $ha^{-1}yr^{-1}$(under low ambient N depositions of 6–8 kg N $ha^{-1}yr^{-1}$) resulted in severe adverse effects on the bryophyte layer within 5 years. Considering these strong effects within a relatively short period, the N critical load range for rich fens was set at 15–30 kg N $ha^{-1}yr^{-1}$and is considered an 'expert judgement'. The higher end of the range is applied to managed or non-N-limited systems, while the lower end of the range is applied to N-limited systems. High latitude systems and rich fens in the mountains (D4.2) are considered slightly more sensitive and their critical loads range is set at 15–25 kg N $ha^{-1}yr^{-1}$.

4.6 Grasslands and Land Dominated by Forbs, Mosses and Lichens (EUNIS Class E)

4.6.1 System Description

A large number of grassland ecosystems are found across Europe: from very dry to wet habitats, acidic to alkaline conditions, inland saline soils, those adapted to high concentrations of heavy metals and very different climatic regimes (e.g. Davies et al. 2004; Ellenberg 1988). Only a small proportion of these grasslands is of natural origin (e.g. dry steppe grasslands, alpine grasslands), while most of these habitats are covered by semi-natural vegetation. Traditional agricultural use or management is thus an important factor influencing the structure and function of these grassland systems. These grasslands have long been an important part of the European landscape and contain many rare and endangered plant and animal species; a number of them have been set aside as nature reserves in several European countries. Semi-natural grasslands of conservation importance are generally nutrient-poor, because of the low levels of manure addition and the removal of plant material by grazing or hay making. The vegetation is characterised by many species of low stature and nutrient-poor soils. Some semi-natural meadow communities of high nature conservation value, particularly those on deep alluvial soils in river flood plains subject to periodic inundation or inputs of farmyard manure, can be moderately fertile. These ecosystems are likely to have a higher proportion of relatively fast-growing species. It is expected that several of these species-rich grasslands, especially those of oligotrophic or mesotrophic soils, will be sensitive to increased atmospheric N inputs. Moreover, species-rich grasslands that are found on weakly buffered or almost neutral soils, are vulnerable to soil acidification and very sensitive to negative impacts of ammonium accumulation in the case of high deposition of reduced N.

Grasslands and lands dominated by forbs, mosses or lichens (EUNIS class E) occur on dry or only seasonally wet soils covered by more than 30 % with vegetation. An important division criterion of class E is based on soil water availability (dry (E1), mesic (E2) and wet (E3) grasslands). Most of the studies on N effects in grassland habitats have been carried out for ecosystems which are classified as dry grasslands (E1). Furthermore, the impacts of N inputs have only been studied in parts of the other major EUNIS categories (E2–E7).

4.6.2 Nitrogen Deposition Effects and Critical Loads of N

Calcareous grasslands (E1.2) are communities on limestone, which are widespread in the hilly and mountainous regions of western and central Europe. Sub-soils consist of different kinds of limestone with high contents of calcium carbonate (>90 %), covered by shallow well-buffered rendzina soils low in phosphorus (P) and N (pH of the top soil: 7–8 with a calcium carbonate content of approx.10 %). Here, plant productivity is low and calcareous grasslands are among the most species-rich plant communities in Europe, including a large number of rare and endangered species.

Increased N availability is probably of major importance in a number of European calcareous grasslands (E1.2). In N-limited calcareous grasslands, increased availability of N is indicated by increased growth of some 'tall' grasses, especially of stress-tolerant species which have a slightly higher potential growth rate and more efficient N utilisation (Bobbink et al. 1988). However, under P-limited conditions, vegetation responses are slow and loss of species is associated with acidification and decreased base saturation. N mineralisation and nitrification are increased under P limitation, and in N-saturated systems with shallow soils. This results in N leaching, although a large proportion of the N inputs are retained in the system. The available studies that are performed at realistically low N deposition levels together with the conclusions from a modelling study on the species composition changes as a result of enhanced N inputs have confirmed the empirical N critical load range of 15–25 kg N $ha^{-1}yr^{-1}$.

Unfortunately, only a limited number of studies is available addressing the response to N deposition of xeric Mediterranean grasslands (E1.3). These dry grasslands are thought to behave similar to calcareous grasslands and a first estimate of the critical load range for the EUNIS category E1.3, based on expert judgement and a single Italian study (Bonanomi et al. 2006) is 15–25 kg N $ha^{-1}yr^{-1}$.

The non-Mediterranean dry acid and neutral closed grassland (E1.7) groups all dry, base-deficient grasslands on acidic and neutral soils with closed vegetation in Atlantic or sub-Atlantic lowland and montane regions of northern and middle Europe and the western part of the Iberian Peninsula. Typical phyto-sociological units are *Violion caninae*, *Nardetalia strictae* and *Agrostion curtisii* (Davies et al. 2004).

Rare species, such as *Arnica montana, Antennaria dioica, Thymus vulgaris* and *Dactylorhhiza maculata,* have been observed to disappear from these grasslands before high and dense-growing grasses started to dominate the vegetation in the Netherlands (e.g. Bobbink et al. 1996). The soils of these grasslands are weakly buffed and enhanced N input was linked to decreased acid neutralising capacity (ANC), and subsequently soil pH, and increased ammonium accumulation (due to hampered nitrification at $pH<4.5$) affecting many of these rare and endangered species (e.g. De Graaf et al. 1998, 2009; Van den Berg et al. 2005). Studies in the UK on these grassland communities have confirmed that these ecosystems are vulnerable to N inputs and, coupled with soil acidification, show a significant increase in grasses. This leads to increased competition and exclusion of smaller species such as rosette forbs, at additions of 15–20 kg N $ha^{-1}yr^{-1}$ or more (Arróniz-Crespo et al. 2008; Horswill et al. 2008; Maskell et al. 2010; Morecroft et al. 1994; Stevens et al. 2004; Stevens et al. 2010). In P-limited systems, the response in the vascular vegetation is much slower, but there is a clear and rapid effect on bryophytes (Carroll et al. 2003). In addition to these effects, N and P economy of the soil is significantly affected, which, among other effects, is likely to result in significant leaching of N (Carroll et al. 2003; Phoenix et al. 2003). Furthermore, the effects of N deposition became very prominent in a survey in the UK, showing lower species richness at above 10–15 kg N $ha^{-1}yr^{-1}$(Maskell et al. 2010) and a European survey in 2010 (Dorland et al. 2011; Fig. 4.5). Based on these studies, the N critical load range for dry acid grasslands (E1.7) is set at 10–15 kg N $ha^{-1}yr^{-1}$ and qualified as 'reliable'.

There is a lack of studies that help determine empirical N critical loads for the non-Mediterranean dry acid and neutral open grasslands, including inland dune grasslands (E1.9). However, these inland dune grasslands have a species composition and ecological functioning comparable to that of coastal (grey) dune grasslands. Because of this similarity, the critical load range for inland dune pioneer grasslands (E1.94) and inland dune siliceous grasslands (E1.95) is set equal to that of coastal grey dune grasslands (8–15 kg N $ha^{-1}yr^{-1}$). This estimation is based on 'expert judgement'.

Grasslands situated at low or medium altitudes and that are managed for making hay (E2.2) are the only class of mesic grasslands for which field experiments have been conducted with N-only treatments in realistic doses, showing loss in diversity

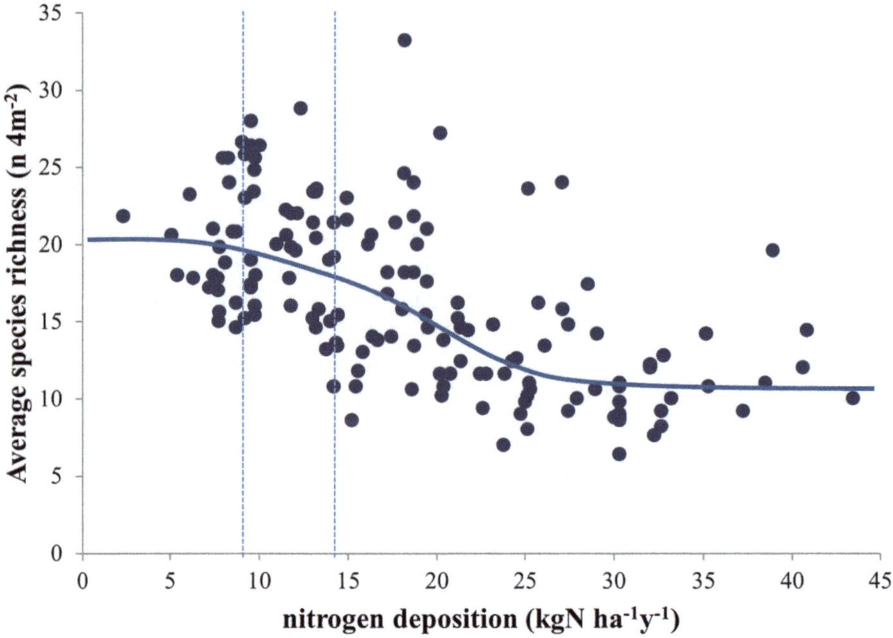

Fig. 4.5 Average species richness (n 4m^{-2}) of acidic grasslands (E1.7) across a gradient of N deposition in the Atlantic region of Europe. Dashed lines represent the lower and upper critical load for nitrogen deposition for this EUNIS class. (Adopted from Dorland et al. 2011)

and increased grass encroachment. Because hay management removes N from the system annually, the critical load range for low- and medium-altitude hay meadows in this category is set at 20–30 kg N ha^{-1}yr^{-1}, based on 'expert judgement'.

Many semi-natural grassland types occur in montane regions across Europe, containing many rare and endangered plant and animal species (e.g. Ellenberg 1996). It is important to emphasise the effects of N loads also on these montane grasslands (E2.3), because N deposition has certainly increased in mountainous regions in central Europe. However, the lack of information on effects of N on montane grasslands has been identified as a major gap in knowledge (Bobbink et al. 2003). Therefore the critical load range (10–20 kg N ha^{-1}yr^{-1}) for mountain hay meadows (E2.3) is based on 'expert judgement' only.

Moist or wet oligotrophic grassland habitats (E3.5) are characterised by oligotrophic and moist to wet peaty soil conditions. This category consists mostly of hay meadows under original agricultural management and they are especially rich in typical plant and animal species. Two subcategories have been distinguished in EUNIS, namely (i) *Molinia caerulea* meadows (E3.51; 'litter meadows' or 'fen meadows') and, (ii) heath meadows and humid *Nardus stricta* swards (E3.52). Because of their long history of traditional land use with low additional inputs of nutrients, these grassland communities are likely to be sensitive to extra nutrient inputs. Several fertilisation experiments in these wet oligotrophic grasslands have demonstrated

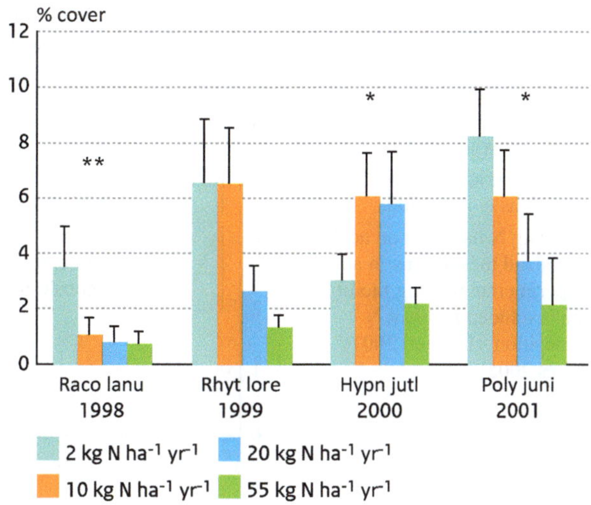

Fig. 4.6 Cover of moss species in *Nardus stricta* grassland (E3.52) mesocosms, exposed (since 1997) to four N treatments (2, 10, 20, and 55 kg N $ha^{-1}yr^{-1}$; from *left* to *right*). The mosses are *Racomitrium lanuginosum*, *Rhytidiadelphus loreus*, *Hypnum jutlandicum* and *Polytrichum juniperinum*. The year denotes when the optimum N level for each species became apparent. (Jones et al. 2002)

limitation by either N or P or even co-limitation by these elements (e.g. Egloff 1987; Olde Venterink et al. 2001; Spink et al. 1998; Van Duren et al. 1998; Vermeer 1986). In the case of N limitation, grass productivity, especially of the dominant *Molinia caerulea*, increased and species diversity declined (e.g. Vermeer 1986). The impacts of N additions on species richness have been quantified in flower-rich, oligotrophic wet hay meadows (E3.51) in Somerset, UK (Kirkham et al. 1996; Mountford et al. 1994; Tallowin and Smith 1994). N additions of 25 kg N $ha^{-1}yr^{-1}$ and more (with an estimated background load of 15–25 kg N $ha^{-1}yr^{-1}$) for 6 years significantly reduced the number of species, while several grasses increased in dominance (*Lolium perenne*, *Holcus lanatus* and *Bromus hordeaceus*). As a result, forbs characteristic of these meadows, declined sharply in number, and some, for example, *Cirsium dissectum*, *Lychnis flos-cuculi* and *Lotus pedunculatus* disappeared from N-treated plots altogether.

Several moist or wet oligotrophic grasslands (E3.5) of high conservational value have been shown to be sensitive to N eutrophication. Increases in dominant grasses and decreases in diversity have been observed following increasing levels of N inputs. A study by Jones et al. (2002) showed strong effects of N treatments of below 20 kg N $ha^{-1}yr^{-1}$ and the emergence of different optima for bryophyte species (Fig. 4.6), indicating strong shifts in species composition as a result of N input. Furthermore, interactions with P limitation are also apparent. In view of these studies, the critical load range for moist to wet oligotrophic grasslands is set at 10–20 kg N $ha^{-1}yr^{-1}$ specifically for *Nardus stricta* swards (E3.52), and considered ‘quite reliable’. The critical load range for *Molinia caerulea* meadows (E3.51), however, is based only on expert judgement, and estimated to be somewhat higher (15–25 kg N $ha^{-1}yr^{-1}$), because, to date, this habitat has barely been studied. Base status has been identified as a significant modifying factor for the application of critical loads; systems with low base status are likely to be more sensitive, while higher base saturation causes habitats in the E3.52 category to be less sensitive.

Fig. 4.7 Change in cover, detected by the number of hits using the pinpoint method, of the moss *Racomitrium lanuginosum*, following N additions (>20 kg N $ha^{-1}yr^{-1}$) as well as N reductions (<20 kg N $ha^{-1}yr^{-1}$) on acidic grassland mesocosms in an experimental misting facility that excluded ambient N deposition. (Emmett 2007; Jones 2005)

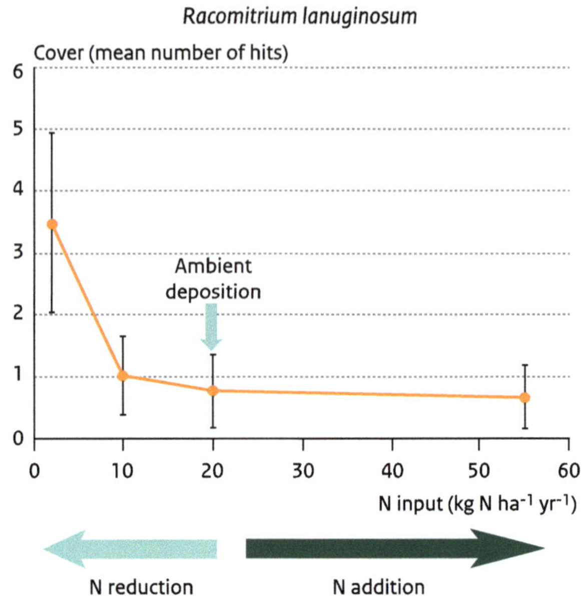

An important sub-category of alpine and subalpine grasslands (E4) are communities without extensive snow cover which are dominated by moss and lichen species (E4.2). Since these communities are nutrient limited and many moss and lichen species are highly responsive to increased N deposition, it is likely that they are sensitive and should be assigned a low critical load. However, the only substantive evidence to support a specific critical load is based on the N sensitivity of *Racomitrium* heath, which is found on mountain summits in Britain and in montane areas of Arctic and sub-Arctic zones (Thompson and Baddeley 1991). Although experimental evidence is limited to three studies with somewhat contrasting results, the fact that large and rapid changes in growth or cover have been observed in two of these studies at deposition rates of 10 kg N $ha^{-1}yr^{-1}$ or lower, suggests that a critical load range for moss and lichen dominated mountain summits, ridges and exposed slopes (E4.2) may be set at 5–10 kg N $ha^{-1}yr^{-1}$ and be qualified as 'quite reliable'.

The many natural grassland types that are found in the alpine and subalpine regions of Europe, on both acidic (E4.3) and calcareous (E4.4) soils are known for their high biodiversity (Körner 2003) and also include many bryophytes that are particularly vulnerable to enhanced N input (see Fig. 4.7). Although these habitats may have similar soil types as compared to the lowland habitats, the atmospheric pollution levels of these habitats have been historically low, but start to increase in these areas in the more recent past. Due to the extreme climatic conditions in combination with adaptation to low levels of nutrient availability, responses to enhanced N inputs are expected to be high in these alpine regions.

In a long-term N-addition experiment in a species-rich subalpine acid grassland area (*Geo-montani Nardetum*), at 2000 m above sea level, in the central Swiss Alps (Bassin et al. 2007, 2009) inputs of 5, 10, 25 and 50 kg N $ha^{-1}yr^{-1}$ have been applied, with an ambient background of about 4 kg N $ha^{-1}yr^{-1}$. From this experiment it became clear that the above-ground biomass of the vegetation significantly increased with the added N levels and that plant species composition changed in favour of sedges. Based on this study, alpine and subalpine acid grasslands (E4.3) are considered to be very sensitive to N loads and the N critical loads for these habitats were set at 5–10 kg N $ha^{-1}yr^{-1}$. Studies on calcareous soils have shown no changes in total plant cover following the application of N, both compared with the control area and with the cover at the start of the experiment. However, significant shifts in species composition were found, especially for the sedge species (Bassin et al. 2007, E. Hiltbrunner pers. comm.). Several sedges increased significantly in cover, following treatments of more than 5 kg N $ha^{-1}yr^{-1}$. Based on this study, alpine and subalpine calcareous grasslands are also considered to be very sensitive to N loads and the N critical load range for alpine and subalpine calcareous grasslands (E4.4) was set at 5–10 kg N $ha^{-1}yr^{-1}$.

4.7 Heathland, Scrub and Tundra Habitats (EUNIS Class F)

4.7.1 System Description

Historically, heathlands have played an important role in the western European landscape. Heathlands are classified together with scrub and tundra habitats in EUNIS class F. This class includes all dry and seasonally wet inland vegetation (cover >30 %) that is dominated by shrubs, dwarf shrubs or scrubs (Davies and Moss 2002; Davies et al. 2004). In a subcategory of these systems, the vegetation is determined by climate, and succession towards woodland is inhibited by drought, low temperature or length of the frost period (e.g. categories F1 and F2). In contrast, the extensive inland, lowland dwarf-shrub heathlands in Altantic and sub-Atlantic Europe (F4) are man-made and development towards woodland has been prevented by mowing, burning, sheep grazing or sod removal. They are widely dominated by some *Ericaceae*, especially *Calluna vulgaris* in the dry heathlands and *Erica tetralix* in the wet heathlands, or *Erica cinerea* in the western Atlantic heathlands (e.g. Gimingham et al. 1979). Heathland communities are found on nutrient-poor mineral soils with a low pH (3.5–4.5), which makes them sensitive to the effects of both eutrophication and acidification caused by increased N deposition. Because of their high nature conservation value, many heathlands have become nature reserves, in recent years.

Alpine and Arctic habitats have many ecological characteristics in common. The growing season is short, temperatures are low, winds are frequent and strong, and the distribution of plant communities depends on the distribution of snow during winter and spring. In classifying these communities under the EUNIS system, it is necessary to distinguish between tundra (F1) and Arctic, alpine and subalpine scrub habitats (F2). Tundra is defined as vegetated land with graminoids, shrubs, mosses or macro-lichens, overlying permafrost (Davies et al. 2004). The presence of permafrost prevents root penetration and often keeps the ground waterlogged in summer. European tundra is limited to Spitsbergen and northern Russia. Vegetation of the same species also occurs on boreal mountains and in the low Arctic region, far away from the main permafrost region, notably in Fennoscandia and Iceland. These habitats are listed under alpine and subalpine grasslands (E4), or Arctic, alpine and subalpine scrub habitats (F2) (Davies et al. 2004). The latter occur north of or above the climatic tree line, but outside the permafrost zone, or scrub occurring close to but below the climatic tree line, where trees are suppressed either by late-season snow, wind, or repeated browsing (Davies et al. 2004).

4.7.2 Nitrogen Deposition Effects and Critical Loads of N

Despite the limited number of long-term experiments, a clear picture is emerging of the potential impact of long-term N deposition on tundra (F1) ecosystems. Ecosystem response to N has been observed at deposition rates of as low as 5 kg N $ha^{-1}yr^{-1}$ (Gordon et al. 2001). To a large extent, however, the response to atmospheric N inputs of tundra ecosystems may well depend on other factors, such as P status and temperature. Experiments by Arens et al. (2008) confirmed, however, that tundra ecosystems are very sensitive to additional loads of N, since significant effects were already seen at additions of 5 kg N $ha^{-1}yr^{-1}$ supporting an empirical N critical load range of 3–5 kg $ha^{-1}yr^{-1}$.

There are a few published studies that are relevant for critical load evaluation of Arctic, alpine and subalpine scrubs (F2). It turns out that shoot extension of *Calluna vulgaris* was stimulated after 5 years by N additions of 10 kg $ha^{-1}yr^{-1}$ indicating that alpine heathlands are very sensitive to low levels of N deposition (Britton and Fisher 2008). Biodiversity was significantly reduced at additions of more than 10 kg N $ha^{-1}yr^{-1}$, primarily through reductions in lichen diversity (Fig. 4.8). Data from this 5-year fertilisation experiment (Britton and Fisher 2007, 2008) supports a N critical load range of 10–15 kg $ha^{-1}yr^{-1}$ for the F2 habitats.

In temperate regions, wet and dry heathlands have been placed within EUNIS class F4, because they occur in the Atlantic climate region and are dominated by ericoid shrubs. This EUNIS class has been divided into subcategories of wet heaths (F4.1), which are damp and characterised by peat soils, and dry heaths (F4.2). Both upland *Calluna* moorlands and lowland wet heaths dominated by *Erica tetralix* fall within the category of 'northern' wet heaths (F4.11). However, since these two communities ecologically clearly differ, it is important that this habitat distinction is retained when N critical loads are assigned.

The number of experiments with N addition within the range of the proposed critical load is limited. A study by Pilkington et al. (2007), was found useful for critical load estimates and found that lichen cover virtually disappeared within 4 years in plots receiving a total N input of approximately 26 kg N $ha^{-1}yr^{-1}$. The empirical critical load is therefore set at 10–20 kg $ha^{-1}yr^{-1}$, as proposed by Bobbink et al. (1996, 2003), for northern wet heath dominated by *Calluna vulgaris* (F4.11). This range is chosen partly because there is no clear justification for a higher criti-

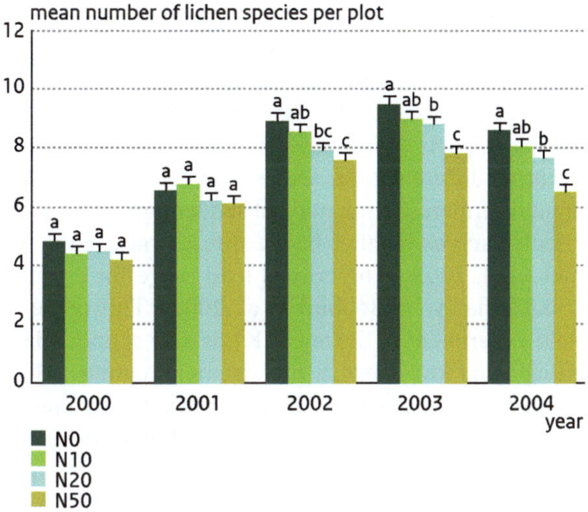

Fig. 4.8 Effect on mean lichen species richness of N addition treatments of 10, 20 and 50 kg $ha^{-1}yr^{-1}$. Means for years not sharing the same letter are significantly different ($P<0.05$); error bars show the standard error of the mean. (Source: Britton and Fisher 2007)

cal load other than for dry heaths (F4.2, for which there is a reliable estimate), since the dominant species, the indicators of exceedance and the modifying factors are all comparable. Furthermore, the range is mostly based on extrapolation, as the available experimental data are based on total N inputs (background plus treatments) that are significantly above the range of the proposed critical loads. This critical load range is dependent on management practices, with the high end of the range applying to wet *Calluna*-dominated heath with high-intensity management, and the low end of the range to wet *Calluna*-dominated heath with low-intensity management.

The wet habitats in western European lowland heathlands are dominated by the dwarf-shrub *Erica tetralix* (Ellenberg 1988) and classified within EUNIS as northern wet heath (F4.11). The lowland wet heathland communities are generally richer in plant species than dry heaths. In recent decades a drastic change in species composition has been observed in Dutch wet heathlands. Nowadays, many wet heathlands, which were originally dominated by *Erica tetralix*, have become monospecific stands of the grass *Molinia caerulea*. Alongside *Erica tetralix,* almost all of the rare plant species have disappeared from that system. It has been hypothesised that this change was caused by increased competitive vigour of *Molinia caerulea* as a direct result of atmospheric N eutrophication. Based on overlap in species and habîtat characteristics with both upland wet heaths (species composition, including N-sensitive bryophytes) and lowland dry heaths (oligotrophic mineral soils) and a simulation (empirical) model by Berendse (1988), the critical load range was set at 10–20 kg N $ha^{-1}yr^{-1}$.

A shift in species composition towards (monospecific) grasses-dominated habitats is also observed in many lowland heaths (F4.2), despite the conservation and management efforts in nature reserves. An evaluation using aerial photographs, has demonstrated that more than 35 % of Dutch heaths have developed into grasslands, during the 1980s (Van Kootwijk and van der Voet 1989). It has been suggested that a strong increase in atmospheric N deposition might have been a significant factor in the observed transition towards grassland. Comparable patterns were found in the

United Kingdom over the past 50 years (Pitcairn et al. 1991). Although the reduction of traditional management practices, such as grazing, burning or sod cutting, was partly responsible for observed changes, the decline in species in heathlands was clearly linked to increasing levels of N deposition.

The impacts of increased N inputs to dry inland heaths (F4.2) are complex and occur at different time scales. They involve N effects on productivity, enhancement of mineralisation and subsequent growth of competitive plants and secondary effects such as heather beetle infestations, winter/drought injuries (see also Bobbink et al. 2003). The critical load range for dry inland heaths has been set at 10–20 kg N $ha^{-1}yr^{-1}$ (Bobbink et al. 2003). This range, based primarily on a long-term field experiment in Surrey (UK), is also supported by results from simulation modelling using low intensity management regimes. The low end of the range should be applied to lichen-rich dry heaths, as shown by the study in Surrey (UK).

Maquis, arborescent, matorral and thermo-Mediterranean brushes (F5; in short Mediterranean scrub) are important habitats (in terms of diversity and cover area) in the Mediterranean area. This dry habitat is often a tall shrub vegetation, dominated by species such as *Quercus sp.*, *Juniperus communis* and *Rhamnus sp.*

The first N manipulation field study on these habitats, however, was started not until 2007, in southern Portugal, investigating the effects of N doses (40 and 80 kg N $ha^{-1}yr^{-1}$) on maquis vegetation at the Natura 2000 Arrábida/Espichel site. In contrast to most N-addition studies (Bobbink et al. 2010) the first results suggested that 1 year of N enrichment had already influenced the soil bacterial community, AMF sporal community and plant composition (Dias et al. 2011). In addition, changes in plant community were observed (increase in biodiversity) and related to soil N availability. The observed changes over such a short period of N addition corresponded with the coexistence of N 'savers' and the appearance of 'spender' species. This coexistence has been reported for epiphytic lichens (Mitchell et al. 2005; Pinho et al. 2008), but has probably not been observed in many studies on northern temperate ecosystems, since these systems may already have had a higher N status before the start of such a study. The results from this study can be considered as a first indication of the possible sensitivity of these Mediterranean scrub communities. Given the fast response of this ecosystem, a first estimate of the critical load range, based on 'expert judgement', for Maquis, arborescent matorral and thermo-Mediterranean brushes (F5) was set at 20–30 kg N $ha^{-1}yr^{-1}$.

4.8 Woodland, Forest and Other Wooded Land (EUNIS Class G)

4.8.1 System Description

EUNIS class G is defined as woodland or recently cleared or burnt land where the dominant vegetation is, or was until very recently, trees with a canopy cover of at least 10%. Trees are defined as woody plants, typically single-stemmed, that can reach a height of at least 5 m at maturity, unless stunted by poor climate or soil. This definition includes lines of trees, coppices, regularly tilled tree nurseries, and tree-crop plantations, but excludes stands of climatically-limited dwarf trees (*krummholz*) <3 m high, such as occur at the Arctic or alpine tree limit. According to the EUNIS classification, woodland and forest habitats (EUNIS level 2; G1, G2, G3, and G4) are separated from other wooded habitats (G5), such as lines of trees, small anthropogenic woodlands (<0.5 ha), recently felled woodlands, early-stage woodlands and coppice. Forests are characterised in EUNIS by the dominant tree types, which may be mixtures of species within the categories *broadleaved deciduous* (G1), *broadleaved evergreen* (G2), *coniferous* (G3), and *mixed broadleaved and coniferous* (G4). EUNIS class G emphasises dominant tree species, soil hydrology and management practices, and is less specific about soil chemistry (see Davies et al. 2004 and the EUNIS website).

4.8.2 Nitrogen Deposition Effects and Critical Loads of N

As a result of enhanced N input, soil processes may be severely affected. A surplus of N, originating from deposition or enhanced nitrification due to accumulated N in the soil, will lead to eutrophication. Field-based ^{15}N studies demonstrate that a large proportion of incoming N (11–56%) was retained in the forest floor through biotic and abiotic processes within the first 2 years of N enrichment (Emmett et al. 1998; Tietema et al. 1998). Accumulation of ammonium was shown in areas with high deposition of reduced N (Boxman et al. 1991; Roelofs et al. 1985; Schulze et al. 1989; Van Dijk and Roelofs 1988). In addition, nitrate that was not taken up by the plants or incorporated into organic matter leached from soils. This nitrate leaching provides an indicator of ecosystem N status, but strongly depends on the C:N ratio in the humus (Augustin et al. 2005; Gundersen et al. 1998). Below a C:N ratio of approximately 25 and above an annual N deposition of 10 kg N $ha^{-1}yr^{-1}$ the level of nitrate leaching drastically increases and endangers groundwater quality (Borken and Matzner 2004; Lorenz et al. 2005). Forest soils also suffer from acidification leading to accelerated leaching of base cations and, in poorly buffered soils, to increased dissolution of aluminium, which can damage fine root development and mycorrhiza, and thereby reduce nutrient and water uptake (Ritter 1990; Ulrich 1983).

Next to effects on soil processes, enhanced N input was shown to affect tree growth. In southern Norway, an analysis of increment cores from over 31,000 spruce forest plots showed a stem growth increase between 1960 and 1970, followed by a decline between 1980 and 1990. A growth decrease started in plots with a modelled wet N deposition of 7–15 kg N $ha^{-1}yr^{-1}$ in the 1990s (Nellemann and Thomsen 2001). This pattern of initial growth stimulation, followed by a subsequent growth decline, was also observed in an N-addition experiment (10, 20, 40, 80 and 160 kg N $ha^{-1}yr^{-1}$) in a young beech stand in Switzerland, on calcareous soil (Flückiger and Braun 2011) and at the NITREX experimental plot in the Netherlands, where ambient N deposition was reduced from 56 to 4 kg N $ha^{-1}yr^{-1}$. Trees

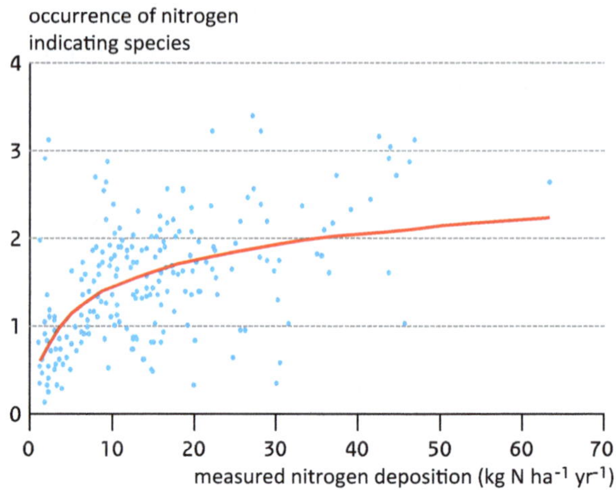

Fig. 4.9 Relationship between the occurrence of N-indicating species and N deposition. (Redrawn from UNECE 2006)

in a roofed environment with low N grew better than in the roofed control environment with high N (Boxman et al. 1998). Depressed growth was correlated with significant nutrient disharmonies within needles (in particular Mg and K), a significant shift in productivity from tree layer to ground vegetation, and massive water stress mainly induced by increased evaporation of dominating nitrophilous grasses, such as *Calamagrostis epigejos* (Hofmann et al. 1990). Enhanced atmospheric N input resulted in changes in tree physiology with notably higher N content in leaves but a decline in total P and base cations such as Mg, K and Ca (Sauter 1991; Van den Burg 1990). Such changes in nutrient status are suggested to influence frost hardiness, drought tolerance and tolerance to pests and fungi by affecting carbon production, respiration and allocation, as well as via changes in membrane properties and osmotic potential (Bigras et al. 2001). An unbalanced ratio between N and P has also been shown to affect forest growth (Braun et al. 2010).

It is well known that N fertilisation of forest stands can reduce mycorrhizal development (Wallenda and Kottke 1998). In addition, many correlative studies have suggested that N inputs above 10 kg N $ha^{-1}yr^{-1}$ affect forest vegetation and shift towards grass dominated vegetation with more nitrophilous species such as *Deschampsia flexuosa* (Dirkse and Van Dobben 1989; Falkengren-Grerup 1986, 1995; Tyler 1987). Many of the endangered species that are present in forests are plants with a low N indicator value (Ellenberg 1988), and the number of nitrophilous species has been shown to increase drastically at deposition loads of as low as 10 kg N $ha^{-1}yr^{-1}$ (UNECE 2006; Fig. 4.9).

In both coniferous forests (G3) as well as in broadleaved forests (G1, G2) processes such as the above occur and determine sensitivity to N input. Since ecosystem responses in broadleaved forests have been recorded above 10 kg N $ha^{-1}yr^{-1}$, the empirical N critical load for this habitat was set at 10–20 kg N $ha^{-1}yr^{-1}$, and is considered 'reliable'. Coniferous forest (G3) are thought to be slightly more sensitive, based on studies by Strengbom et al. (2003) and Poikolainen et al. (1998), where significant shifts (a reduced *Vaccinium vitis-idaea* cover and increased algal

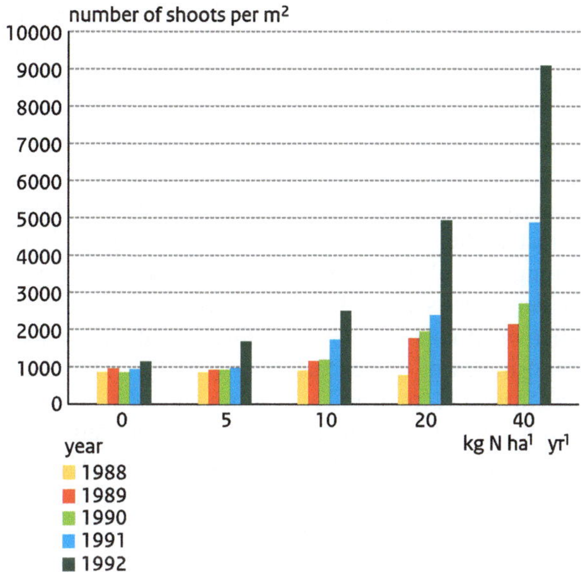

Fig. 4.10 Number of shoots of *Deschampsia flexuosa* vs. nitrogen supply (kg N ha^{-1}yr^{-1}) between 1988 and 1992 in a boreal forest (G3.A) located outside Söderhamn, central Sweden. (Kellner and Redbo-Torstensson 1995)

occurrences, respectively) occurred at low (5–10 kg N ha^{-1}yr^{-1}) levels of N input. The critical load range for this habitat is therefore set at 5–15 kg N ha^{-1}yr^{-1}.

Together, spruce (G3.A) and pine (G3.B) taiga woodlands constitute the westernmost part of the continuous Eurasian northern taiga belt (Davies et al. 2004). Spruce taiga woodlands (G3.A) include the boreal spruce and spruce-pine forests of Fennoscandia, north-eastern Poland, the Baltic States, Belarus and European Russia. Pine taiga woodlands (G3.B) include the boreal pine forests in the regions mentioned above. Both spruce and pine taiga woodlands are mostly very poor in N, and any increase in N input may influence major soil processes. Soil responses to increased N input are likely to be similar for both habitat classes.

In Sweden additions of 6 kg N ha^{-1}yr^{-1} over a 4 year period (background deposition around 2 kg N ha^{-1}yr^{-1}) were found to increase the abundance of *Deschampsia flexuosa* by around 50% (UNECE 2007), whereas abundance and occurrence of *Vaccinium myrtillus* was lower when N deposition exceeded this level (Strengbom et al. 2003). Additions of 5 kg N ha^{-1}yr^{-1} over 4 years resulted in 70% higher shoot densities of *Deschampsia flexuosa* in N-treated plots, compared to control plots (Kellner and Redbo-Torstensson 1995; Fig. 4.10). In addition, increased occurrence of *Deschampsia flexuosa* in Norway, between 1988 and 1993, was also attributed to N deposition (Økland 1995). Large effects on species composition and increased sensitivity to leaf pathogens have been reported from N additions of 12.5 kg N ha^{-1}yr^{-1} within a decade (Nordin et al. 2005, 2006; Strengbom et al. 2002). The low reversibility of N-induced effects (Strengbom et al. 2001) support the suggestion that spruce taiga woodlands is a habitat class that is sensitive to N deposition. Based on these results the empirical N critical load range for spruce taiga woodlands (G3.A) was set at 5–10 kg N ha^{-1}yr^{-1}, and is considered 'reliable'.

Compared to spruce taiga woodlands, there are indications that for pine forests negative effects on biodiversity from increased N input start to occur at higher N input rates. However, data that would support this finding is currently not sufficient to make such a distinction between pine and spruce forests. There are few studies showing effects on biodiversity from N inputs below 10 kg N $ha^{-1}yr^{-1}$ in pine forests; with two exceptions: a study by Poikolainen et al. (1998), who found increased occurrence of green algae on tree branches (no distinction between spruce and pine forests) when N inputs exceeded 3 kg N $ha^{-1}yr^{-1}$, and a study by Bråkenhielm and Quinghong (1995), who found decreased occurrence of N-sensitive lichens when N depositions exceeded 5–8 kg N $ha^{-1}yr^{-1}$. Strengbom et al. (2003) found that *Vaccinium vitis-idaea* in pine forests had a significantly lower occurrence when N deposition was higher than 6 kg N $ha^{-1}yr^{-1}$, but they also found that N depositions had to be > 12 kg N $ha^{-1}yr^{-1}$ before *Vaccinium myrtillus* would occur at significantly lower rates. Experiments using higher N loads (over 20–30 kg N $ha^{-1}yr^{-1}$) often show large effects on the composition of ground vegetation (Dirkse and Martakis 1992; Strengbom et al. 2001; Van Dobben et al. 1999), while effects of N on plant community composition (Strengbom et al. 2001) and N mineralisation (Chen and Högberg 2006) also appear to be reversed only slowly after external N inputs are stopped. These results suggest that pine taiga woodlands are sensitive to increased N deposition. With the support of gradient studies that revealed effects at N loads of less than 10 kg N $ha^{-1}yr^{-1}$, the N critical load range for Pine taiga woodlands (G3.B) was set at 5–10 kg N $ha^{-1}yr^{-1}$, and considered 'quite reliable'.

4.9 Discussion and Conclusions

4.9.1 Modifying Factors

A debate is ongoing about the use of so-called modifying factors to help establish whether the lower, middle or upper part of an empirical critical load range should be applied in situations with insufficient data on specific ecosystems. Modifying factors are general relationships between abiotic factors and critical loads for N as given in Table 4.2 (adapted after Bobbink et al. 2003). More information is needed to apply modifying factors to the various forest EUNIS categories.

Furthermore, strong responses to N in situations where P was also applied are an indication of N and P co-limitation, identifying P as an important modifier of the critical load. This is observed in studies on grasslands and heathlands. For these ecosystems, higher critical loads should be applied to systems which are limited by P, and lower critical loads to systems which are not.

In conclusion, additional qualitative information had been assigned to a number of modifying factors, in comparison to recommendations provided in 2003 (Bobbink et al. 2003) on how to interpret the agreed ranges of critical loads in specific situations for the various ecosystems. In the critical loads workshop in Noordwijkerhout (Bobbink and Hettelingh 2011) consensus was reached to use the minimum value of

Table 4.2 Suggestions when to apply lower, middle or upper parts of the critical load ranges for terrestrial ecosystems (excluding wetlands), if national data are insufficient

Action	Temperature/ frost period	Soil wetness	Base cation availability	Management intensity
Move to lower part	Cold/long	Dry	Low	Low
Use middle part	Intermediate	Normal	Intermediate	Usual
Move to higher part	Hot/none	Wet	High	High

the ranges of empirical critical loads in every EUNIS class, especially when these are used in the assessment of air quality policies in view of resulting N effects on plant species diversity. Furthermore, in the United Kingdom and the Netherlands, several new national approaches have been developed to estimate or calculate the critical loads for specific ecosystems that are part of one larger EUNIS class with an agreed range of empirical critical loads (Hall and Wadsworth 2010; Van Dobben and van Hinsberg 2008). This is an important development to create a more complete list of N critical load values for the large number of ecosystem types that are present.

4.9.2 Gaps in Knowledge and Research Needs

Although considerable progress has been made for several habitat groups between 2003 and 2010, there are still serious gaps in knowledge on the effects of increased N deposition on semi-natural and natural ecosystems. The following gaps in knowledge have been recognised as most important:

- More research and data are required to establish a critical load for the following ecosystems: steppe grasslands, all Mediterranean vegetation types, wet swamp forests, many mires and fens and several coastal habitats;
- More research is needed to increase the robustness of critical load estimates that are based on an 'expert' judgement rating, or for which only few studies on the effects of N impacts are available;
- Impacts of N enrichment in (sensitive) freshwater and shallow marine ecosystems (including coastal waters) need further research;
- Additional effort is needed to allocated observed N effects to the appropriate EUNIS forest habitat sub-types (level 3);
- An increasing number of gradient (survey) studies with respect to atmospheric N deposition have been either reported or recently initiated. More rigorous guidelines should be identified for the evaluation of these studies, covering estimation of deposition rates, quantification of confounding factors and application of methods for statistical analysis.
- Possible effects that are specific to the form (NO_y or NH_x) by which nitrogen is deposited are not sufficiently well known for the setting of critical loads for oxidised and reduced nitrogen separately;

- In order to refine current critical loads, long-term (5–10 years) N-addition experiments using high frequency of N-treatments, between 5 and 50 kg N $ha^{-1}yr^{-1}$, in regions with low background depositions, or in mesocosms, are useful. It would increase the reliability of the derived critical loads when the lowest treatment level does not exceed the critical load.

It is crucial to understand the long-term effects of both increasing and decreasing N deposition on ecosystem processes for a representative range of ecosystems. It is thus important to quantify the effects of N loads by the manipulation of N inputs in long-term ecosystem studies in both unaffected and affected areas. These data are essential to validate the set of critical loads and to develop robust dynamic ecosystem models and/or multiple correlative species models that are reliable enough to use in the calculation of critical loads of N deposition for natural and semi-natural ecosystems. Such models are also of increasing importance to help assess natural recovery rates for N-affected systems. It is clear that with on-going atmospheric N inputs, management of ecosystems needs to be adjusted to maintain species diversity and habitat status. There are however complicated interactions between management and N depositions that may affect estimation of N critical loads. Nitrogen deposition is globally increasing, especially in newly emerging industrial and agricultural areas, e.g. in Asia and South America. It is of importance to estimate the effects of enhanced N input to the different ecosystems that may be affected. For this, scientific work on these habitats is urgently needed. Since 2010, new data have emerged that will help in determining empirical critical N loads for habitats that are, due to the lack of experimental data, poorly covered in the most recent critical load review (Bobbink and Hettelingh 2011) (e.g. Mediterranean and (sub-)Arctic habitats). In addition, progress in statistical approaches and the availability of large vegetation data sets may help in cross-calibration of empirical critical loads and may be included in future revisions. A regular revision and review of empirical N critical loads is recommended to meet requirements for effective environmental policies that address the interaction of impacts of air pollution, climate change and biodiversity.

4.9.3 Empirical Critical Loads of Nitrogen in the Future

Most of the Earth's biodiversity is present in semi-natural and natural ecosystems. It is thus crucial to control atmospheric N loads, in order to prevent negative effects on these ecosystems. Fine-resolution maps of the sensitive ecosystems of high conservation value are needed per country, to map critical loads of N for these systems. It is advised to use both the mass balance and empirically derived N critical loads to improve the robustness on broad regional scales of effect-based assessments of natural areas at risk of nitrogen deposition scenarios (see Hettelingh et al. 2007). To assess negative impacts of exceedances on broad, regional scales in Europe specific dose-response relationships between N load and relevant indicators can be considered, provided that results are needed to compare the environmental risk of emission reduction scenarios in relative terms (see Chap. 25).

Countries are advised to identify those highly sensitive receptor ecosystems within the EUNIS classification relating to their individual interest. Efforts should be directed to produce fine-resolution maps of sensitive ecosystems of high conservation value. The current empirical critical loads of N have been set in values of total atmospheric N. More information is needed on the relative effects of oxidised and reduced N deposition. It was emphasised during the last two UNECE expert meetings on empirical critical loads held in Berne (Achermann and Bobbink 2003) and in Noordwijkerhout (Bobbink and Hettelingh 2011) that there is increasing evidence of significant more adverse effects of reduced N in comparison to those caused by oxidised N. Particularly, lichens and sometimes bryophytes in a number of ecosystems, and several, mostly weakly buffered ecosystems in EUNIS classes F, E, C and B are thought to be more sensitive to deposition of reduced N. However, unlike critical thresholds for ambient concentrations of NH_3 that have, for example, been established for vegetation with a high richness of lichens and bryophytes (Cape et al. 2009; Sutton et al. 2009), it is not yet possible to set critical loads for the deposition of reduced and oxidized N separately.

References

Aber, J. D., Nadelhoffer, K. J., Steudler, P., & Melillo, J. M. (1989). Nitrogen saturation in northern forest ecosystems. *Bioscience, 39,* 378–386.

Aber, J. D., McDowell, W., Nadelhoffer, K., Magill, A., Berntsen, G., Kamakea, M., McNulty, S., Currie, W., Rustad, L., & Fernandez, I. (1998). Nitrogen saturation in temperate forest ecosystems: Hypotheses revisited. *Bioscience, 48,* 921–934.

Achermann, B., & Bobbink, R. (Eds.). (2003). *Empirical critical loads for nitrogen: Expert workshop (11–13 November 2002)*. Berne: Swiss Agency for the Environment, Forests and Landscape.

Aerts, R., & Chapin, F. S. (2000). The mineral nutrition of wild plants revisited: A re-evaluation of processes and patterns. *Advances in Ecological Research, 30,* 1–67.

Archibold, O. W. (1995). *Ecology of world vegetation*. London: Chapman & Hall.

Arens, S. J. T., Sullivan, P. F., & Welker, J. M. (2008). Nonlinear responses to nitrogen and strong interactions with nitrogen and phosphorus additions drastically alter the structure and function of a high Arctic ecosystem. *Journal of Geophysical Research, 113*(G3), 1–10.

Arróniz-Crespo, M., Leake, J. R., Horton, P., & Phoenix, G. K. (2008). Bryophyte physiological responses to, and recovery from, long-term nitrogen deposition and phosphorus fertilisation in acidic grassland. *New Phytologist, 180,* 864–874.

Arts, G. H. P. (1990). *Deterioration of Atlantic soft-water systems and their flora*. Ph. D. Thesis, the Netherlands, Catholic University of Nijmegen.

Augustin, S., Bolte, A., Holzhausen, M., & Wolff, B. (2005). Exceedance of critical loads of nitrogen and sulphur and its relation to forest conditions. *European Journal of Forest Research, 124,* 289–300.

Bassin, S., Volk, M., Suter, M., Buchmann, N., & Fuhrer, J. (2007). Nitrogen deposition but not ozone affects productivity and community composition of subalpine grassland after 3 years of treatment. *New Phytologist, 175,* 523–534.

Bassin, S., Werner, R. A., Sörgel, K., Volk, M., Buchmann, N., & Fuhrer, J. (2009). Effects of combined ozone and nitrogen deposition on the in situ properties of eleven key plant species of a subalpine pasture. *Oecologia, 58,* 474–756.

Beltman, B., Kooijman, A. M., Rouwenhorst, G., & van Kerkhoven, M. (1996). Nutrient availability and plant growth limitation in blanket mires in Ireland. *Proceedings of the Royal Irish Academy, 96B,* 77–87.

Berendse, F. (1988). *De nutriëntenbalans van droge zandgrondvegetaties in verband met de eutrofiëring via de lucht. Deel 1 Een simulatiemodel als hulpmiddel bij het beheer van vochtige heidevelden*. Wageningen: Centrum voor Agrobiologisch Onderzoek.

Bergström, A.-K., & Jannson, M. (2006). Atmospheric nitrogen deposition has caused nitrogen enrichment and eutrophication of lakes in the northern hemisphere. *Global Change Biology, 12,* 635–643.

Bergström, A.-K., Blomqvist, P. & Jannson, M. (2005). Effects of atmospheric nitrogen deposition on nutrient limitation and phytoplankton biomass in unproductive Swedish lakes. *Limnology and Oceanography, 50,* 987–994.

Bergström, A.-K., Jonsson, A. & Jansson, M. (2008). Phytoplankton responses to nitrogen and phosphorus enrichment in unproductive Swedish lakes along a gradient of atmospheric nitrogen deposition. *Aquatic Biology, 4,* 55–64.

Bigras, F. J., Ryyppö, A., Lindström, A., & Stattin, E. (2001). Cold acclimation and deacclimation of shoots and roots of conifer seedlings. In F. J. Bigras & S. J. Colombo (Eds.), *Conifer cold hardiness*. Dordrecht: Kluwer Academic Publishers.

Bobbink, R., & Hettelingh, J.-P. (2011). *Review and revision of empirical critical loads and dose-response relationships.* Proceedings of an expert workshop, Noordwijkerhout (23–25 June 2010. (Report 680359002/2011)). Bilthoven: Coordination Centre for Effects, National Institute for Public Health and the Environment.

Bobbink, R., Bik, L., & Willems, J. H. (1988). Effects of nitrogen fertilization on vegetation structure and dominance of *Brachypodium pinnatum* (L.) Beauv. in chalk grasslands. *Acta Botanica Neerlandica, 37,* 231–242.

Bobbink, R., Boxman, D., Fremstad, E., Heil, G., Houdijk, A., & Roelofs, J. (1992). Critical loads for nitrogen eutrophication of terrestrial and wetland ecosystems based upon changes in vegetation and fauna. In P. Grennfelt & E. Thörnelöv (Eds.), *Critical loads for nitrogen* (pp. 111-161). Sweden: Nordic Council of Ministers (Report 1992, 41. Report from a workshop held at Lökeberg (6–10 April 1992)).

Bobbink, R., Hornung, M., & Roelofs, J. G. M. (1996). Empirical nitrogen critical loads for natural and semi-natural ecosystems. In UNECE convention on long-range transboundary air pollution. *Manual on methodologies and criteria for mapping critical loads/levels and geographical areas where they are exceeded.* (1996). Umweltbundesambt, 71–96, Annex III. Berlin. 54 pp.

Bobbink, R., Hornung, M., & Roelofs, J. G. M. (1998). The effects of air-borne nitrogen pollutants on species diversity in natural and semi-natural European vegetation. *Journal of Ecology, 86,* 717–738.

Bobbink, R., Ashmore, M., Braun, S., Flückiger, W., & van den Wyngaert, I. J. J. (2003). Empirical nitrogen critical loads for natural and semi-natural ecosystems: 2002 update. In B. Achermann & R. Bobbink (Eds.), *Empirical critical loads for nitrogen* (pp. 43–170). Berne: Swiss Agency for Environment, Forest and Landscape SAEFL.

Bobbink, R., Hicks, K., Galloway, J., Spranger, T., Alkemade, R., Ashmore, M., Bustamante, M., Cinderby, S., Davidson, E., Dentener, F., Emmett, B., Erisman, J. W., Fenn, M., Gilliam, F., Nordin, A., Pardo, L., & De Vries, W. (2010). Global assessment of nitrogen deposition effects on terrestrial plant diversity: A synthesis. *Ecological Applications, 20,* 30–59.

Boeye, D., Verhagen, B., van Haesebroeck, V., & Verheyen, R. F. (1997). Nutrient limitation in species-rich lowland fens. *Journal of Vegetation Science, 8,* 415–424.

Bonanomi, G., Caporaso, S., & Allegrezza, M. (2006). Short-term effects of nitrogen enrichment, litter removal and cutting on a Mediterranean grassland. *Acta Oecologica, 30,* 419–425.

Borken, W., & Matzner, E. (2004). Nitrate leaching in forest soils: An analysis of long-term monitoring sites in Germany. *Zeitschrift fur Pflanzenernährung und Bodenkunde, 167,* 277–283.

Boxman, A. W., Krabbendam, H., Bellemakers, M. J. S., & Roelofs, J. G. M. (1991). Effects of ammonium and aluminium on the development and nutrition of *Pinus nigra* in hydroculture. *Environmental Pollution, 73,* 119–136.

Boxman, A. W., van der Ven, P. J. M., & Roelofs, J. G. M. (1998). Ecosystem recovery after a decrease in nitrogen input to a Scots pine stand at Ysselsteyn, The Netherlands. *Forest Ecology and Management, 101,* 155–163.

Bragazza, L., Tahvanainen, T., Kutnar, L., Rydin, H., Limpens, J., Hàjek, M., Grosvernier, P., Hansen, I., Iacumin, P., & Gerdol, R. (2004). Nutritional constraints in ombrotrophic Sphagnum subject to increasing levels of atmospheric nitrogen deposition in Europe. *New Phytologist, 163,* 609–616.

Bråkenhielm, S., & Quinghong, L. (1995). Spatial and temporal variability of algal and lichen epiphytes on trees in relation to pollutant deposition in Sweden. *Water, Air & Soil Pollution, 79,* 61–74.

Braun, S., Thomas, V. F. D., Quiring, R., & Flückiger, W. (2010). Does nitrogen deposition increase forest production? The role of phosphorus. *Environmental Pollution, 158,* 2043–2052.

Britton, A. J., & Fisher, J. M. (2007). Interactive effects of nitrogen deposition, fire, grazing on diversity and composition of low-alpine prostrate *Calluna vulgaris* heathland. *Journal of Applied Ecology, 44,* 125–135.

Britton, A. J., & Fisher, J. M. (2008). Growth responses of low-alpine dwarf-shrub heath species to nitrogen deposition and management. *Environmental Pollution, 153,* 564–573.

Brouwer, E., Bobbink, R., Roelofs, J. G. M., & Verheggen, G. M. (1996). *Effectgerichte maatregelen tegen verzuring en eutrofiëring van oppervlaktewateren. Eindrapport monitoring tweede fase.* Katholieke Universiteit Nijmegen: Vakgroep Oecologie.

Brouwer, E., Bobbink, R., & Roelofs, J. G. M. (2002). Restoration of aquatic macrophyte vegetation in acidified and eutrophied softwater lakes: An overview. *Aquatic Botany, 73,* 405–431.

Cape, J. N., Van der Eerden, L. J., Sheppard, L. J., Leith, I. D., & Sutton, M. A. (2009). Evidence for changing the critical level for ammonia. *Environmental Pollution, 157,* 1033–1037.

Carroll, J. A., Caporn, S. J. M., Johnson, D., Morecroft, M. D., & Lee, J. A. (2003). The interactions between plant growth, vegetation structure and soil processes in semi-natural acidic and calcareous grasslands, receiving long-term inputs of stimulated pollutant nitrogen deposition. *Environmental Pollution, 121,* 363–376.

Chen, Y., & Högberg, P. (2006). Gross nitrogen mineralization rates still high 14 years after suspension of N input to a N-saturated forest. *Soil Biology and Biochemistry, 38,* 2001–2003.

Commission of the European Communities. (2003). *Interpretation manual of European Union habitats—EUR 25.* DG Environment—Nature and Biodiversity.

Davies, C. E., & Moss, D. (2002). *EUNIS habitat classification.* 2001 Work programme, final report to the european environment agency European topic centre on nature protection and biodiversity. Centre for Ecology and Hydrology.

Davies, C. E., Moss, D., & Hill, M. O. (2004). *EUNIS habitat classification revised 2004.* European Environment Agency, European topic centre on nature protection and biodiversity.

De Graaf, M. C. C., Bobbink, R., Roelofs, J. G. M., & Verbeek, P. J. M. (1998). Differential effects of ammonium and nitrate on three heathland species. *Plant Ecology, 135,* 185–196.

De Graaf, M. C. C., Bobbink, R., Smits, N. A. C., van Diggelen, R., & Roelofs, J. G. M. (2009). Biodiversity, vegetation gradients and key biogeochemical processes in the heathland landscape. *Biological Conservation, 10,* 2191–2201.

Dias, T., Malveiro, S., Chaves, S., Tenreiro, R., Branquinho, C., Martins-Loução, M. A., Sheppard, L., & Cruz, C. (2011). Effects of increased N availability on biodiversity of Mediterranean-type ecosystems: A case study in a Natura 2000 site in Portugal. In W. K. Hicks, C.P. Whitfield, W.J. Bealey, & M.A. Sutton (Eds.), *Nitrogen deposition and natura 2000: Science & practice in determining environmental impacts* (pp. 171–180). Brussels. COST729/Nine/ESF/CCW/JNCC/SEI Workshop Proceedings. Cost Office 2011.

Dirkse, G. M., & Van Dobben, H. F. (1989). Het effect van bemesting op de samenstelling van de kruidlaag van dennenbossen. *Natura, 9,* 208–212.

Dirkse, G. M., & Martakis, G. F. P. (1992). Effects of fertilizer on bryophytes in Swedish experiments on forest fertilization. *Biological Conservation, 59,* 155–161.

Dise, N. B., & Wright, R. F. (1995). Nitrogen leaching from European forests in relation to nitrogen deposition. *Forest Ecology Management, 71,* 153–161.

Dorland, E., Bobbink, R. & Robat, S. (2008). *Impacts of changing ratios of reduced and oxidized nitrogen deposition: Case studies in acid grasslands and fen ecosystems*. Proceedings 6th European Conference on Ecological Restoration, Ghent, Belgium. (8–12 September 2008).

Dorland, E., Bobbink, R., Soons, M., & Rotthier, S. (2011). Dalende stikstofdepositie is nog niet afdoende voor herstel van droge heischrale graslanden. *De Levende Natuur, 112,* 220–224.

Egloff, D. A. (1987). Food and growth relations of the marine microzooplankter, *Synchaeta cecilia* (Rotifera). *Hydrobiologia, 157,* 129–141.

Ellenberg, H. (1988). *Vegetation ecology of Central Europe*. Cambridge: Cambridge University Press.

Ellenberg, H. (1996). *Vegetation Mitteleuropas mit den Alpen* (5th ed.). Stuttgart: Ulmer.

Elser, J. J., Andersen, T., Baron, J. S., Bergström, A. K., Jansson, M., Kyle, M., Nydick, K. R., Steger, L., & Hessen, D. O. (2009). Shifts in lake N:P stoichiometry and nutrient limitation driven by atmospheric nitrogen deposition. *Science, 326,* 835.

Emmett, B. A. (2007). Nitrogen saturation of terrestrial ecosystems: Some recent findings and their implications for our conceptual framework. *Water, Air, & Soil Pollution: Focus, 7,* 99–109.

Emmett, B. A., Boxman, D., Bredemeier, M., Gundersen, P., Kjønaas, O. J., Moldan, F., Schleppi, P., Tietema, A., & Wright, R. F. (1998). Predicting the effects of atmospheric nitrogen deposition on conifer stand: Evidence from the NITREX ecosystem-scale experiments. *Ecosystems, 1,* 352–360.

Falkengren-Grerup, U. (1986). Soil acidification and vegetation changes in deciduous forest in southern Sweden. *Oecologia, 70,* 339–347.

Falkengren-Grerup, U. (1995). Long-term changes in flora and vegetation in deciduous forests of southern Sweden. *Ecological Bulletins, 44,* 215–226.

Flückiger, W., & Braun, S. (2011). *Auswirkung erhöhter Stickstoffbelastung auf die Stabilität des Waldes*. (*Synthesebericht*). Schönenbuch: Institut für Angewandte Pflanzenbiologie.

Gerdol, R., Petraglia, A., Bragazza, L., Iacumin, P., & Brancaleoni, L. (2007). Nitrogen deposition interacts with climate in affecting production and decomposition rate in *Sphagnum mosses*. *Global Change Biology, 13,* 1810–1821.

Gimingham, C. H., Chapman, S. B., & Webb, N. R. (1979). European heathlands. In R. L. Specht (Ed.), *Ecosystems of the world, 9A* (pp. 365–386). Amsterdam: Elsevier.

Goodale, C. L., Aber, J. D., Vitousek, P. M., & McDowell, W. H. (2005). Long-term decreases in stream nitrate: Successional causes unlikely; possible links to DOC? *Ecosystems, 8,* 334–337.

Gordon, C., Wynn, J. M., & Woodin, S. J. (2001). Impacts of increased nitrogen supply on high Arctic heath: The importance of bryophytes and phosphorus availability. *New Phytologist, 149,* 461–471.

Greipsson, S., & Davy, A. J. (1997). Responses of *Leymus arenarius* to nutrients: Improvement of seed production and seedling establishment for land reclamation. *Journal of Applied Ecology, 34,* 1165–1176.

Grennfelt, P., & Thörnelöf, E. (1992). *Critical loads for nitrogen*. Copenhagen: Nord 1992:41 Nordic Council of Ministers.

Gundersen, P., Callesen, I., & De Vries, W. (1998). Nitrate leaching in forest ecosystems is related to forest floor C/N ratios. *Environmental Pollution, 102,* 403–407.

Gunnarsson, U., Malmer, N., & Rydin, H. (2002). Dynamics or constancy in Sphagnum dominated mire ecosystems? A 40-year study. *Ecography, 25,* 685–704.

Hall, J., & Wadsworth, R. (2010). Estimating the effect of abiotic factors on modifying the sensitivity of vegetation to nitrogen deposition: An application of endorsement theory. *Water, Air & Soil Pollution, 212,* 441–459.

Hall, J., Davies, C., & Moss, D. (2003). Harmonisation of ecosystem definitions using the EUNIS habitat classification. In B. Achermann & R. Bobbink (Eds.), *Emperical critical loads for nitrogen—Proceedings expert workshop* (pp. 171–195). Berne: Swiss Agency for Environment, Forest and Landscape SAEFL.

Hettelingh, J.-P., Posch, M., Slootweg, J., Reinds, G. J., Spranger, T., & Tarrasón, L. (2007). Critical loads and dynamic modelling to assess European areas at risk of acidification and eutrophication. *Water, Air & Soil Pollution, 7,* 379–384.

Hofmann, G., Heinzdorf, D., & Krauß, H. H. (1990). Wirkung atmogener Stickstoffeinträge auf Produktivität und Stabilität von Kiefern-Forstökosystemen. *Journal Beiträge für die Forstwirtschaft, 24,* 59–73.

Horswill, P., O'Sullivan, O., Phoenix, G. K., Lee, J. A., & Leake, J. R. (2008). Base cation depletion, eutrophication and acidification of species-rich grasslands in response to long-term simulated nitrogen deposition. *Environmental Pollution, 155,* 336–349.

ICP Modelling and Mapping. (2008). MINUTES of the 24th meeting of ICP M & M task force. http://www.rivm.nl/thema/images/icpmm_24tfm_minutes_final_tcm61-48579.pdf. Accessed 1 Nov 2013.

ICP Modelling and Mapping. (2009). MINUTES of the 25th meeting of ICP M & M task force. http://www.rivm.nl/thema/images/icpmm_25tfm_minutes_ultimatedraft_tcm61-48601.pdf. Accessed 1 Nov 2013.

Jones, M. L. M. (2005). *Nitrogen deposition in upland grasslands: Critical loads, management and recovery.* PhD Thesis, UK, University of Sheffield.

Jones, M. L. M., Hayes, F., Brittain, S. A., Haria, S., Williams, P. D., Ashenden, T. W., Norris, D. A., & Reynolds, B. (2002). *Changing nutrient budget of sand dunes: Consequences for the nature conservation interest and dune management.* (CCW contract No. FC 73-01-347. Field survey). Bangor: Centre for Ecology and Hydrology.

Jones, M. L. M., Wallace, H. L., Norris, D., Brittain, S. A., Haria, S., Jones, R. E., Rhind, P. M., Reynolds, B. R., & Emmett, B. A. (2004). Changes in vegetation and soil characteristics in coastal sand dunes along a gradient of atmospheric nitrogen deposition. *Plant Biology, 6,* 598–605.

Kellner, O., & Redbo-Torstensson, P. (1995). Effects of elevated nitrogen deposition on the field layer vegetation in coniferous forests. In H. Staaf & G. Tyler (Eds.), *Effects of acid deposition and troposhperic ozone on forest ecosystems in Sweden. Ecological Bulletins 44* (pp. 227–237).

Kirkham, F. W., Mountford, J. O., & Wilkins, R. J. (1996). The effects of nitrogen, potassium and phosphorus addition on the vegetation of a Somerset peat moor under cutting management. *Journal of Applied Ecology, 33,* 1013–1029.

Körner, C. (2003). *Alpine plant life: Functional plant ecology of high mountain ecosystems.* Heidelberg: Springer.

Lamers, L. P. M., Bobbink, R., & Roelofs, J. G. M. (2000). Natural nitrogen filter fails in polluted raised bogs. *Global Change Biology, 6,* 583–586.

Lammerts, E. J., & Grootjans, A. P. (1997). Nutrient deficiency in dune slack pioneer vegetation: A review. *Journal of Coastal Conservation, 3,* 87–94.

Limpens, J., Berendse, F., & Klees, H. (2004). How phosphorus availability affects the impact of nitrogen deposition on Sphagnum and vascular plants in bogs. *Ecosystems, 7,* 793–804.

Løkke, H., Bak, J., Bobbink, R., Bull, K., Curtis, C., Falkengren-Grerup, U., Forsius, M., Gundersen, P., Hornung, M., Skjelkvåle, B. L., Starr, M., & Tybirk, K. (2000). *Critical Loads Copenhagen 1999.* Conference report prepared by members of the conference's secretariat, the scientific committee and chairmen and rapporteurs of its workshops in consultation with the UNECE secretariat, Critical loads. Denmark: National Environmental Research Institute (21–25 November 1999).

Lorenz, M., Becher, G., Mues, V., Fischer, R., Becker, R., Calatayud, V., Dise, N., Krause, G. H. M., Sanz, M., & Ulrich, E. (2005). *Forest condition in Europe.* (2005 technical report). Geneva: UNECE/EC.

Malmer, N., Albinsson, C., Svensson, B., & Wallén, B. (2003). Interferences between Sphagnum and vascular plants: Effects on plant community structure and peat formation. *Oikos, 100,* 469–482.

Maskell, L. C., Smart, S. M., Bullock, J. M., Thompson, K., & Stevens, C. J. (2010). Nitrogen deposition causes widespread loss of species richness in British habitats. *Global Change Biology, 16,* 671–679.

Mitchell, R. J., Truscot, A. M., Leith, I. D., Cape, J. N., van Dijk, N., Tang, Y. S., Fowler, D., & Sutton, M. A. (2005). A study of the epiphytic communities of Atlantic oak woods along an atmospheric nitrogen deposition gradient. *Journal of Ecology, 93,* 482–492.

Mitsch, W. J., & Gosselink, J. P. (2000). *Wetlands* (3rd ed.). New York: Wiley.

Morecroft, M. D., Sellers, E. K., & Lee, J. A. (1994). An experimental investigation into the effects of atmospheric nitrogen deposition on two semi-natural grasslands. *Journal of Ecology, 82,* 475–483.

Morris, J. T. (1991). Effects of nitrogen loading on wetland ecosystems with particular reference to atmospheric deposition. *Annual Review of Ecology and Systematics, 22,* 257–279.

Moss, B. (1988). *Ecology of fresh waters: Man and medium.* Oxford: Blackwell.

Mountford, M. O., Lakhani, K. H., & Holland, R. J. (1994). *The effects of nitrogen on species diversity and agricultural production on the Somerset Moors, Phase II: (a) After seven years of fertilizer application; (b) after cessation of fertilizer input for three years.* (Report to the institute for grassland and environmental research). Abbots Ripton. UK: Institute of Terrestrial Ecology.

Nellemann, C., & Thomsen, M. G. (2001). Long-term changes in forest growth: Potential effects of nitrogen deposition and acidification. *Water, Air, & Soil Pollution, 128,* 197–205.

Nordin, A., Strengbom, J., Witzell, J., Näsholm, T., & Ericson, L. (2005). Nitrogen deposition and the biodiversity of boreal forests: Implications for the nitrogen critical load. *Ambio: A Journal of Human Environment, 34,* 20–24.

Nordin, A., Strengbom, J., & Ericson, L. (2006). Responses to ammonium and nitrate additions by boreal plants and their natural enemies. *Environmental Pollution, 141,* 167–174.

Økland, R. H. (1995). Changes in the occurrence and abundance of plan species in a Norwegian boreal coniferous forest, 1988–1993. *Nordic Journal of Botany, 15,* 415–438.

Olde Venterink, H., van der Vliet, R. E. & Wassen, M. J. (2001). Nutrient limitation along a productivity gradient in wet meadows. *Plant and Soil, 234,* 171–179.

Olff, H., Huisman, J., & Van Tooren, B. F. (1993). Species dynamics and nutrient accumulation during early primary succession in coastal sand dunes. *Journal of Ecology, 81,* 693–706.

Pardo, L. H., Fenn, M., Goodale, C. L., Geiser, L. H., Driscoll, C. T., Allen, E., Baron, J., Bobbink, R., Bowman, W. D., Clark, C., Emmett, B., Gilliam, F. S., Greaver, T., Hall, S. J., Lilleskov, E. A., Liu, L., Lynch, J., Nadelhoffer, K., Perakis, S., Robin-Abbott, M. J., Stoddard, J., Weathers, K., & Dennis, R. L. (2011). Effects of nitrogen deposition and empirical nitrogen critical loads for ecoregions of the United States. *Ecological Applications, 21,* 3049–3082.

Phoenix, G. K., Booth, R. E., Leake, J. R., Read, D. J., Grime, J. P., & Lee, J. A. (2003). Effects of enhanced nitrogen deposition and phosphorus limitation on nitrogen budgets of semi-natural grasslands. *Global Change Biology, 9,* 1309–1321.

Pilkington, M. G., Caporn, S. J. M., Carroll, J. A., Cresswell, N., Lee, J. A., Emmett, B. A., & Bagchi, R. (2007). Phosphorus supply influences heathland responses to atmospheric nitrogen deposition. *Environmental Pollution, 148,* 191–200.

Pinho, P., August, S., Martins-Loução, M. A., Pereira, M. J., Soares, A., Máguas, C., & Branquinho, C. (2008). Causes of change in nitrophytic and oligotrophic lichen species in a Mediterranean climate: Impact of land cover and atmospheric pollutants. *Environmental Pollution, 154,* 380–389.

Pitcairn, C. E. R., Fowler, D., & Grace, J. (1991). *Changes in species composition of semi-natural vegetation associated with the increase in atmospheric inputs of nitrogen.* Penicuik: Institute of Terrestrial Ecology/NERC.

Plassmann, K., Brown, N., Jones, M. L. M., & Edwards-Jones, G. (2008). Can atmospheric input of nitrogen affect seed bank dynamics in habitats of conservation interest? The case of dune slacks. *Applied Vegetation Science, 11,* 413–420.

Plassmann, K., Edward-Jones, G., & Jones, M. L. M. (2009). The effects of low levels of nitrogen deposition and grazing on dune grassland. *Science of Total Environent, 407,* 1391–1404.

Poikolainen, J., Lippo, H., Hongisto, M., Kubin, E., Mikkola, K., Lindgren, M., van der Hoek, K. W., Erisman, J. W., Smeulders, S. & Wisniewski, J. R. (1998). On the abundance of epiphytic green algae in relation to the nitrogen concentrations of biomonitors and nitrogen deposition in Finland. *Environmental Pollution, 102,* 85–92.

Redbo-Torstensson, P. (1994). The demographic consequences of nitrogen fertilization of a population of sundew, *Drosera rotundifolia. Acta Botanica Neerlandica, 43,* 175–188.

Remke, E., Brouwer, E., Kooijman, A., Blindow, I. & Roelofs, J. G. M. (2009a). Low atmospheric nitrogen loads lead to grass encroachment in coastal dunes, but only on acid soils. *Ecosystems, 12,* 1173–1188.

Remke, E., Brouwer, E., Kooijman, A., Blindow, I., Esselink, H. & Roelofs, J. G. M. (2009b). Even low to medium nitrogen deposition impacts vegetation of dry, coastal dunes around the Baltic Sea. *Environmental Pollution, 157,* 792–800.

Ritter, G. (1990). Zur Wirkung van Stickstoffeinträgen auf Feinwurzelsystem und Mykorrhizabildung in Kieferbeständen. *Beitr fiir die Forstwirtsch, 24,* 100–104.

Roelofs, J. G. M. (1983). Impact of acidification and eutrophication on macrophyte communities in soft waters in the Netherlands. I. Field observations. *Aquatic Botany, 17,* 139–155.

Roelofs, J. G. M., Kempers, A. J., Houdijk, A. L. F. M. & Jansen, J. (1985). The effect of air-borne ammonium sulphate on *Pinus nigra* var. maritima in the Netherlands. *Plant and Soil, 84,* 45–56.

Saros, J. E., Michel, T. J., Interlandi, S. J., & Wolfe, A. P. (2005). Resource requirements of Asterionella Formosa and *Fragilaria crotonsis* in oligotrophic alpine lakes: Implications for recent phytoplankton community reorganisations. *Canadian Journal of Fisheries and Aquatic Sciences, 62,* 1681–1689.

Sauter, U. (1991). Zeitliche Variationen des Ernährungszustands nordbayerischer Kiefernbestände. *Forstwissenschaftliches Centralblatt, 110,* 13–33.

Schoof-van Pelt, M. M. (1973). *Littorelletea, a study of the vegetation of some amphiphytic communities of western Europe*. PhD thesis, Catholic University of Nijmegen.

Schulze, E. D., Lange, O. L., & Oren, R. (Eds.). (1989). *Forest decline and air pollution.* New York: Springer Berlin Heidelberg.

Schuurkes, J. A. A. R., Elbers, M. A., Gudden, J. J. F. & Roelofs, J. G. M. (1987). Effects of simulated ammonium sulphate and sulphuric acid rain on acidification, water quality and flora of small-scale soft water systems. *Aquatic Botany, 28,* 199–226.

Sheppard, L. J., Leith, I. D., Crossley, A., van Dijk, N., Cape, J. N., Fowler, D., & Sutton, M. A. (2009). Long term cumulative exposure exacerbates the effects of atmospheric ammonia on an ombrothrophic bog: Implications for critical levels. In M. A. Sutton, S. Reis, & S. M. H. Baker (Eds.), *Atmospheric ammonia—Detecting emission changes and environmental impacts.* Penicuik: Springer Science.

Spink, A., Sparks, R. E., van Oorschot, M. & Verhoeven, J. T. A. (1998). Nutrient dynamics of large river floodplains. *Regulated Rivers: Research & Management, 14,* 203–216.

Stevens, C. J., Dise, N. B., Mountford, J. O. & Gowing, D. J. (2004). Impact of nitrogen deposition on the species richness of grasslands. *Science, 303,* 1876–1879.

Stevens, C. J., Duprè, C., Dorland, E., Gaudnik, C., Gowing, D. J. G., Bleeker, A., Diekmann, M., Alard, D., Bobbink, R., Fowler, D., Corcket, E., Mountford, J. O., Vandvik, V., Aarrestad, P. E., Muller, S., & Dise, N. B. (2010). Nitrogen deposition threatens species richness of grasslands across Europe. *Environmental Pollution, 158,* 2940–2945.

Stoddard, J. L. (1994). Long-term changes in watershed retention of nitrogen: Its causes and aquatic consequences. In L. A. Baker (Ed.), *Environmental chemistry of lakes and reservoirs* (pp. 223–284). Washington DC: American Chemical Society.

Strengbom, J., Nordin, A., Näsholm, T., & Ericson, L. (2001). Slow recovery of boreal forest ecosystem following decreased nitrogen input. *Functional Ecology, 15,* 451–457.

Strengbom, J., Nordin, A., Näsholm, T., & Ericson, L. (2002). Parasitic fungus mediates change in nitrogen-exposed boreal forest vegetation. *Journal of Ecology, 90,* 61–67.

Strengbom, J., Walheim, M., Näsholm, T., & Ericson, L. (2003). Regional differences in the occurrence of understorey species reflect nitrogen deposition in Swedish forests. *Ambio: A Journal of Human Environment, 32,* 91–97.

Sutton, M. A., Reis, S., & Baker, S. M. H. (Eds.). (2009). *Atmospheric ammonia—Detecting emission changes and environmental impacts*. Penicuik: Springer Science.

Tallowin, J. R., & Smith, R. E. N. (1994). *The effects of inorganic fertilisers in flower-rich hay meadows on the Somerset Levels*. English Nature Research Report 87. Peterborough: English Nature.

Tamm, C. O. (1991). *Nitrogen in terrestrial ecosystems. Questions of productivity, vegetational changes and ecosystem stability*. Berlin: Springer.

Thompson, D. B. A., & Baddeley, J. A. (1991). Some effects of acidic deposition on montane *Racomitrium lanuginosum* heaths. In S. J. Woodin & A.M. Farmer (Eds.), *The effects of acid deposition on nature conservation in Great Britain* (pp. 17–28). Cambridgeshire: NCC Peterborough.

Tietema, A., Boxman, A. W., Bredemeier, M., Emmett, B. A., Moldan, F., Gundersen, P., Schleppi, P., Wright, R. F., van der Hoek, K. W., Erisman, J. W., Smeulders, S., & Wisniewski, J. R. (1998). Nitrogen saturation experiments (NITREX) in coniferous forest ecosystems in Europe: A summary of results. *Environmental Pollution, 102,* 433–437.

Tomassen, H., Bobbink, R., Peters, R., van der Ven, P., & Roelofs, J. (1999). *Kritische stikstofdepositie in heischrale graslanden, droge duingraslanden en hoogvenen: op weg naar meer zekerheid. Eindrapport in het kader van het Stikstof Onderzoek Programma (STOP), 1997–1999.* Katholieke Universiteit Nijmegen en Universiteit Utrecht: Nijmegen & Utrecht.

Tomassen, H. B. M., Smolders, A. J. P., Lamers, L. P. M., & Roelofs, J. G. M. (2003). Stimulated growth of *Betula pubescens* and *Molinia caerulea* on ombrotrophic bogs: Role of high levels of atmospheric nitrogen deposition. *Journal of Ecology, 91,* 357–370.

Tyler, G. (1987). Probable effects of soil acidification and nitrogen deposition on the floristic composition of oak (*Quercus robur* L.) forest. *Flora, 179,* 165–170.

Ulrich, B. (1983). Soil acidity and its relations to acid deposition. In B. Ulrich & J. Pankrath (Eds.), *Effects of accumulation of air pollutants in forest ecosystems* (pp. 127–146). Dordrecht: Reidel Publ. Co.

UNECE. (2006). *The condition of forests in Europe. 2006 executive report.* Hamburg: Federal Research Centre for Forests and Forest Products (BFH).

UNECE. (2007). *Recent results and updating of scientific and technical knowledge. Workshop on effects of low-level nitrogen deposition.* (UN report ECE/EB.AIR/WG.1/2007/15).

Van Breemen, N. (1995). How Sphagnum bogs down other plants. *Trends in Ecology and Evolution, 10,* 270–275.

Van den Berg, L. J. L., Tomassen, H. B. M., Roelofs, J. G. M. & Bobbink, R. (2005). Effects of nitrogen enrichment on coastal dune grassland: A mesocosm study. *Environmental Pollution, 138,* 77–85.

Van den Burg, J. (1990). Stickstoff- und Säuredeposition und die Nährstoffversorgung niederländischer Wälder auf pleistozänen Sandböden. *Forst und Holzwirt, 45,* 597–605.

Van Dijk, H. F. G., & Roelofs, J. G. M. (1988). Effects of excessive ammonium deposition on the nutritional status and condition of pine needles. *Physiologia Plantarum, 73,* 494–501.

Van Dobben, H. F., & van Hinsberg, A. (2008). *Overzicht van kritische depositiewaarden voor stikstof, toegepast op habitattypen en Natura 2000 gebieden.* (*Alterra Report 1654*). Wageningen: Alterra.

Van Dobben, H. F., ter Braak, C. J. F., & Dirkse, G. M. (1999). Undergrowth as a biomonitor for deposition of nitrogen and acidity in pine forest. *Forest Ecology and Management, 114,* 83–95.

Van Duren, I. C., Strykstra, R. J., Grootjans, A. P., ter Heerdt, G. N. J., & Pegtel, D. M. (1998). A multidisciplinary evaluation of restoration measures in a degraded Cirsio-Molinietum fen meadow. *Applied Vegetation Science, 1,* 115–130.

Van Kootwijk, E. J., & van der Voet, H. (1989). *De kartering van heidevergrassing in Nederland met de Landsat Thematic Mapper sattelietbeelden. (Report RIN 89/2).* Dutch: Arnhem.

Van Wijnen, H. J., van der Wal, R., & Bakker, J. P. (1999). The impact of herbivores on nitrogen mineralization rate: Consequences for salt-marsh succession. *Oecologia, 118,* 225–231.

Vermeer, J. G. (1986). The effects of nutrients on shoot biomass and species composition of wetland and hayfield communities. *Acta Oecologica/Oecologia Plantarum, 7,* 31–41.

Wallenda, T., & Kottke, I. (1998). Nitrogen deposition and ectomycorrhizas. *New Phytologist, 139,* 169–187.

Wassen, M. J., van der Vliet, R. E. & Verhoeven, J. T. A. (1998). Nutrient limitation in the *Biebrza fens* and floodplain (Poland). *Acta Botanica Neerlandica, 47,* 241–253.

WGE. (2009). *Report of the working group on effects on its twenty-eighth session*. http://www.unece.org/env/documents/2009/EB/wge/ece.eb.air.wg.1.2009.2.e.pdf. Accessed 1 Nov 2013.

WHO. (2001). *Air Quality guidelines for Europe*. Copenhagen: WHO Regional Publications (European Series, No. 91).

Wittig, R., & Pott, R. (1982). Verbreitung der Littorelletea-Arten in der Westfälischen Bucht. *Decheniana, 135,* 14–21.

Wolfe, A. P., Cooke, C. A., & Hobbs, W. O. (2006). Are current rates of atmospheric Nitrogen deposition influencing lakes in the eastern Canadian Arctic? *Arctic Antarctic, and Alpine Research, 38,* 465–476.

Wright, R. F., Alewell, C., Cullen, J. M., Evans, C. D., Marchetto, A., Moldan, F., Prechtel, A., & Rogora, M. (2001). Trends in nitrogen deposition and leaching in acid-sensitive streams in Europe. *Hydrological and Earth Systems Science, 5,* 299–310.

Wright, R. F., Aherne, J., Bishop, K., Camarero, L., Cosby, B. J., Erlandsson, M., Evans, C. D., Forsius, M., Hardekopf, D. W., Helliwell, R., Hruska, J., Jenkins, A., Kopacek, J., Moldan, F., Posch, M., & Rogora, M. (2006). Modelling the effect of climate change on recovery of acidified freshwaters: Relative sensitivity of individual processes in the MAGIC model. *Science of Total Environment, 365,* 154–166.

Chapter 5
Effects and Empirical Critical Loads of Nitrogen for Ecoregions of the United States

Linda H. Pardo, Molly J. Robin-Abbott, Mark E. Fenn, Christine L. Goodale, Linda H. Geiser, Charles T. Driscoll, Edith B. Allen, Jill S. Baron, Roland Bobbink, William D. Bowman, Christopher M. Clark, Bridget Emmett, Frank S. Gilliam, Tara L. Greaver, Sharon J. Hall, Erik A. Lilleskov, Lingli Liu, Jason A. Lynch, Knute J. Nadelhoffer, Steven J. Perakis, John L. Stoddard, Kathleen C. Weathers and Robin L. Dennis

5.1 Introduction

5.1.1 Effects of Nitrogen Deposition on Ecosystems

Human activity in the last century has led to a significant increase in nitrogen (N) emissions and deposition (Galloway et al. 2004). Total N emissions in the United States have increased significantly since the 1950s (Galloway 1998, Galloway et al. 2003). As S deposition has declined in response to regulation, the rate of N deposition relative to S deposition has increased since the 1980s (Driscoll et al. 2001, 2003) followed by a general decrease in NO_x emissions from electric utilities since the early 2000s. More recently, the relative proportion of NH_x ($NH_4^+ + NH_3$) to NO_x ($NO + NO_2$) emissions has also increased for many areas of the United States (Kelly et al. 2005; Lehmann et al. 2005).

L. H. Pardo (✉) · M. J. Robin-Abbott
USDA Forest Service, Burlington, VT, USA
e-mail: lpardo@fs.fed.us

M. E. Fenn
USDA Forest Service, Riverside, CA, USA

C. L. Goodale
Cornell University, Ithaca, NY, USA

L. H. Geiser
USDA Forest Service, Corvallis, OR, USA

C. T. Driscoll
Syracuse University, Syracuse, NY, USA

E. B. Allen
University of California, Riverside, CA, USA

W. de Vries et al. (eds.), *Critical Loads and Dynamic Risk Assessments*, Environmental Pollution 25, DOI 10.1007/978-94-017-9508-1_5

Because of past, and, in some regions, continuing increases in emissions (Lehmann et al. 2005; Nilles and Conley 2001), N deposition has reached a level that has caused or is likely to cause alterations in many United States ecosystems. In some ecoregions, the impact of N deposition has been severe, altering N cycling and biodiversity. Indicators of altered N cycling include increased N mineralization, nitrification, and nitrate (NO_3^-) leaching rates, as well as elevated plant tissue N concentration. The eventual outcome of increases in these processes can be N saturation, the series of ecosystem changes that occur as available N exceeds plant and microbial demand (Aber et al. 1989, 1998).

As N availability increases, there are progressive changes in biotic community structure and composition, including changes in diatom, lichen, mycorrhizal fungal and terrestrial plant communities. For example, in the Mediterranean California ecoregion, native plant species in some ecosystems have been replaced by invasive species that are more productive under elevated N deposition (Fenn et al. 2010;

J. S. Baron
US Geological Survey, Fort Collins, CO, USA

R. Bobbink
B-WARE Research Center, Nijmegen, Netherlands

W. D. Bowman
University of Colorado, Boulder, CO, USA

C. M. Clark · J. A. Lynch
US EPA, Washington DC, USA

B. Emmett
Centre for Ecology and Hydrology, Environment Centre Wales, Bangor, UK

F. S. Gilliam
Marshall University, Huntington, WV, USA

T. L. Greaver · L. Liu · R. L. Dennis
US EPA, Research Triangle Park, NC, USA

S. J. Hall
Arizona State University, Tempe, AZ, USA

E. A. Lilleskov
USDA Forest Service, Houghton, MI, USA

K. J. Nadelhoffer
University of Michigan, Ann Arbor, MI, USA

S. J. Perakis
US Geological Survey, Corvallis, OR, USA

J. L. Stoddard
US EPA, Corvallis, OR, USA

K. C. Weathers
Cary Institute of Ecosystem Studies, Millbrook, NY, USA

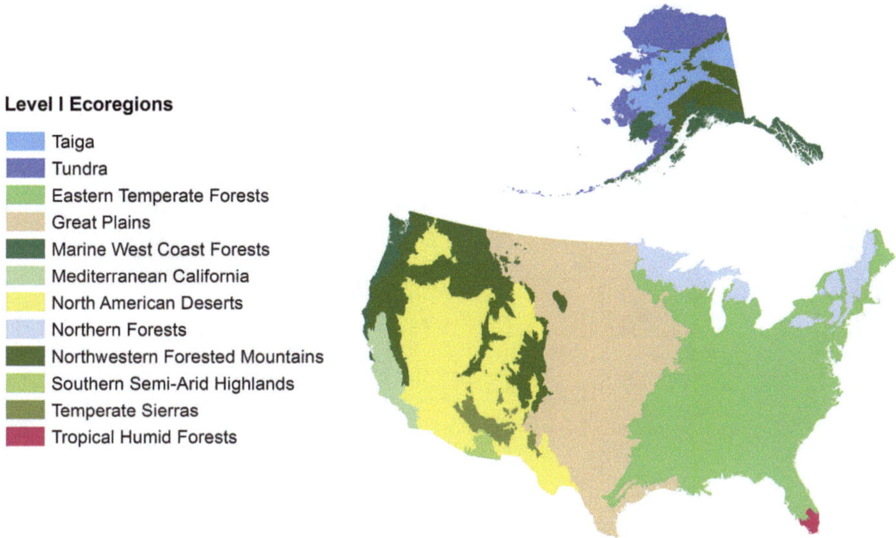

Fig. 5.1 Level I ecological regions in the United States. (Commission for Environmental Cooperation 1997)

Rao and Allen 2010; Rao et al. 2010; Weiss 1999; Yoshida and Allen 2004). Such shifts in plant community composition and species richness can lead to overall losses in biodiversity and further impair particular threatened or endangered species (Stevens et al. 2004), as has occurred for the checkerspot butterfly (Weiss 1999).

5.1.2 Approach for Determining Empirical Critical Loads of Nitrogen

Recently, Pardo et al. (2011a–d) synthesized research relating atmospheric N deposition to effects on terrestrial and freshwater ecosystems in the United States and quantified empirical critical loads of atmospheric N deposition, with one chapter devoted to each of 12 major ecoregions. This chapter summarizes those findings and includes a brief discussion of the approach used to set critical N loads.

For this synthesis, we reviewed studies of responses to N inputs for U.S. ecoregions as defined by the Commission for Environmental Cooperation (CEC) Level I ecoregions map for North America (CEC 1997; Fig. 5.1). We estimated critical loads based on data from > 3200 sites (Fig. 5.2). We identified the receptor of concern (organism or ecosystem compartment), the response of concern, the critical threshold value for that response, and the criteria for setting the critical load and extrapolating the critical load to other sites or regions. These methods are described in detail in Pardo et al. (2011a, b, d).

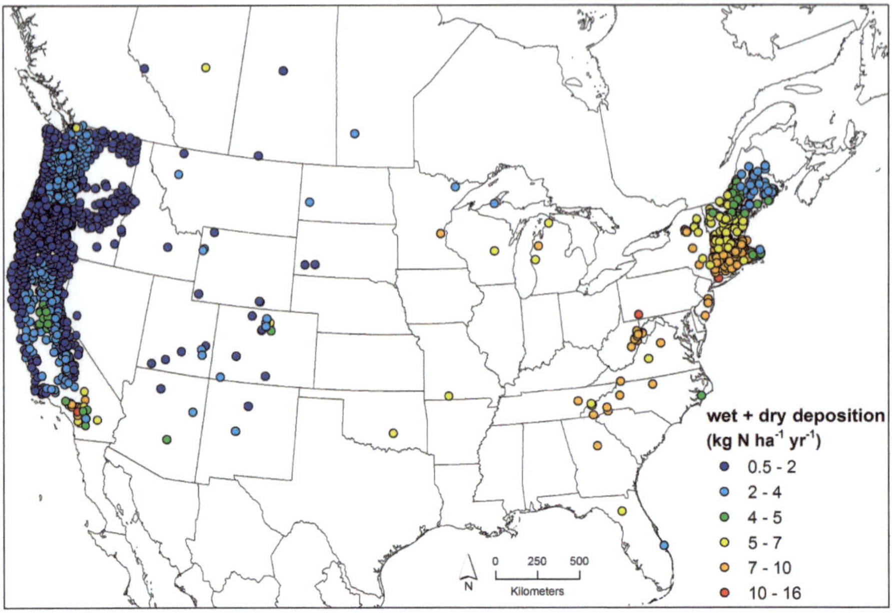

Fig. 5.2 Locations of > 3200 sites in the United States with modelled N deposition for which ecological responses are reported

The receptors evaluated included freshwater diatoms, mycorrhizal fungi, lichenized fungi (henceforth lichens), bryophytes, herbaceous plants, shrubs, and trees. Ecosystem impacts included: (1) biogeochemical responses and (2) individual species, population, and community responses. We considered N addition (fertilization) experiments, N deposition gradient studies and long-term monitoring studies in order to evaluate ecosystem response to N deposition inputs. Nitrogen deposition at sites included in this analysis (Weathers and Lynch 2011) was either based on the deposition reported in the publication or, when that was not available, we used modelled deposition quantified by the Community Multiscale Air Quality (CMAQ) model v.4.3 simulations of wet + dry deposition of oxidized (NO_y) and reduced (NH_x) N species (Fig. 5.2). Hereafter, this model is referred to as CMAQ 2001, as it uses 2001 reported data (Byun and Schere 2006; Byun and Ching 1999). In some areas of elevated N deposition, CMAQ at this grid scale (36 km) likely underestimates total N deposition. This is the case, for example, over much of California (Fenn et al. 2010). For more detail on deposition, see Weathers and Lynch (2011). We afforded greater weight to long-term fertilization studies (5–10 years) than to short-term studies. Single-dose forest fertilization studies exceeding 50 kg N ha^{-1} were generally not considered, but lower dose short-term studies were considered when other observations were limited.

We rarely had data to distinguish biotic or ecosystem response to reduced forms versus oxidized forms of N. There is some evidence that for some species, reduced forms of N may have more substantial impacts than oxidized forms (Bobbink et al. 2003; Cape et al. 2009; Kleijn et al. 2008; Sutton et al. 2009). Differences in uptake rates and preference for NH_4^+ versus NO_3^- across different plant taxa (Falkengren-Grerup 1995; McKane et al. 2002; Miller and Bowman 2002; Nordin et al. 2006) lead to differences in sensitivity to NH_x (Krupa 2003) and NO_y. However, not all species are more sensitive to NH_x than NO_y (Jovan et al. 2012); these responses vary by species and functional type. Some species are more sensitive to increases in NO_y, as was demonstrated for boreal forests (Nordin et al. 2006).

In general, we determined the critical load based on the observed response pattern to N inputs. In some cases, there was a clear dose-response relationship where the response changed above a certain threshold. In other cases, when response to increasing N was more linear, we estimated the "pristine" state of N deposition and the deposition that corresponded to a departure from that state. The criteria for setting critical loads are discussed in detail in Pardo et al. (2011a, b, d).

5.1.3 Contents of this Chapter

In this chapter we synthesize empirical critical loads of N reported for all the ecoregions of the United States, compare critical loads by life form or ecosystem compartment across all ecoregions, discuss the abiotic and biotic factors that affect the critical loads, present the significance of these findings and, finally, compare critical loads in the United States to those for similar ecoregions/ecosystems in Europe. For each receptor, we present maps of critical loads by ecoregion.

The range of critical loads of nutrient N reported for the United States ecoregions, inland surface waters, and freshwater wetlands is 1–39 kg N $ha^{-1}yr^{-1}$ (Table 5.1). This broad range spans the range of N deposition observed over most of the country (see Weathers and Lynch 2011). The number of locations for which ecosystem response data were available (Fig. 5.2) for an ecoregion is variable, which impacts the level of certainty of the empirical critical loads estimates.

5.2 Mycorrhizal Fungi

Mycorrhizal fungi reside at the interface between host plants and soils, exchanging soil resources, especially nutrients, with host plants in exchange for photosynthates (carbon compound). Due to this important and unique ecological niche, mycorrhizal fungi are at particular risk due to changes in either the soil environment or host carbon allocation.

Table 5.1 Summary of critical loads of nutrient N for United States ecoregions including their reliability

Ecoregion	Ecosystem component	CL for N deposition kg N $ha^{-1}yr^{-1}$	Reliability	Response	Comments	Study
Tundra	Prostrate dwarf shrubs	1–3	##	Changes in CO_2 exchange, cover, foliar N, and community composition of vascular plants	N addition study, Greenland high arctic, P enhanced N effects	(Arens et al. 2008)[a]
Tundra	Lichens	1–3	(#)	Changes in lichen pigment production and ultrastructure, changes in lichen and bryophyte cover	N addition studies, high and low arctic, P enhanced or moderated N effects	(Hyvärinen et al. 2003)[b]; (Makkonen et al. 2007)[b]; (Arens et al. 2008)[a],
Taiga	Lichen, moss, and algae in forests and woodlands	1–3	#	Changes in alga, bryophyte, and lichen community composition, cover, tissue N or growth rates		(Poikolainen et al. 1998)[b]; (Strengbom et al. 2003)[d]; (Vitt et al. 2003)[c], (Berryman et al. 2004)[c]; (Moore et al. 2004)[c]; (Berryman and Straker 2008)[c]; (Geiser et al. 2010)
Taiga	Spruce forests	5–7	(#)	Change in ectomycorrhizal fungal community structure	Expert judgment, extrapolated from Marine West coast spruce and northern spruce-fir forest	(Lilleskov 1999); (Lilleskov et al. 2001, 2002, 2008)
Taiga	Shrublands	6	##	Change in shrub and grass cover, increased parasitism of shrubs	Long term, low N addition study: shrub cover decreased, grass cover increased	(Strengbom et al. 2003)[d]; (Nordin et al. 2005)[d]
Northern Forests	Hardwood and Coniferous Forests	>3	#	Decreased growth of red pine, and decreased survivorship of yellow birch, scarlet and chestnut oak, quaking aspen, and basswood		(Thomas et al. 2010)

Table 5.1 (continued)

Ecoregion	Ecosystem component	CL for N deposition kg N $ha^{-1}yr^{-1}$	Reliability	Response	Comments	Study
Northern Forests	Lichens	4–6	(#)	Epiphytic lichen community change	Loss of oligotrophic species. Synergistic and/or confounding effects of acidic deposition not considered; assumes response threshold similar to Marine West Coast Forest	(Geiser et al. 2010)
Northern Forests	Ectomycorrhizal fungi	5–7	#	Change in fungal community structure		(Lilleskov et al. 2008)
Northern Forests	Herbaceous cover species	>7 and <21	#	Loss of prominent species	Response observed in low-level fertilization experiment	(Hurd et al. 1998)
Northern Forests	Hardwood and Coniferous Forests	8	##	Increased surface water NO_3^- leaching		(Aber et al. 2003)
Northern Forests	Old-growth montane red spruce	>10 and <26	#	Decreased growth and/or induced mortality	Response observed in low-level fertilization experiment	(McNulty et al. 2005)
Northern Forests	Arbuscular mycorrhizal fungi	< 12	(#)	Biomass decline and community composition change		(Van Diepen et al. 2007); (Van Diepen 2008)
Northwest Forested Mountains	Alpine lakes	1.5	##	Changes in diatom assemblages	As wet deposition only	(Baron 2006)
Northwest Forested Mountains	Lichens	1.2–3.7	(#)	Epiphytic lichen community change in mixed-conifer forests, Alaska	Application of western Oregon and Washington model	(Geiser et al. 2010)

Table 5.1 (continued)

Ecoregion	Ecosystem component	CL for N deposition kg N $ha^{-1}yr^{-1}$	Reliability	Response	Comments	Study
Northwest Forested Mountains	Lichens	2.5–7.1	##	Epiphytic lichen community change, thallus N enrichment in mixed-conifer forests, non-Alaska		(Fenn et al. 2008); (Geiser et al. 2010)
Northwest Forested Mountains	Sub-alpine forest	4	##	Increase in organic horizon N, foliar N, potential net N mineralization, and soil solution N, initial increases in N leaching below the organic layer		(Baron et al. 1994); (Rueth and Baron 2002)
Northwest Forested Mountains	Alpine lakes	4.0	#	Episodic freshwater acidification		(Williams and Tonnessen 2000)
Northwest Forested Mountains	Alpine grassland	4–10	##	Changes in plant species composition		(Bowman et al. 2006)
Northwest Forested Mountains	Ectomycorrhizal fungi	5–10	(#)	Changes in ectomycorrhizal fungal community structure in white, black, and Engelmann spruce forests	Expert judgment, extrapolated from Marine West Coast spruce and northern spruce-fir forest	(Lilleskov 1999); (Lilleskov et al. 2001, 2002, 2008)
Northwest Forested Mountains	Mixed conifer forest	17	## #	NO_3^- leaching reduced fine root biomass		(Fenn et al. 2008)
Marine West Coast Forests	Western OR and WA forests	2.7–9.2	##	Epiphytic lichen community change	Loss of oligotrophic species, enhancement of eutrophic species. CL increases with regional range in mean annual precipitation from 45–450 cm	(Geiser et al. 2010)

Table 5.1 (continued)

Ecoregion	Ecosystem component	CL for N deposition kg N $ha^{-1}yr^{-1}$	Reliability	Response	Comments	Study
Marine West Coast Forests	SE Alaska forests	5	(#)	Fungal community change; declines in ectomycorrhizal fungal diversity		(Whytemare et al. 1997); (Lilleskov 1999), (Lilleskov et al. 2001, 2002)
Eastern Temperate Forest	Eastern Hardwood Forest	>3	#	Decreased growth of red pine, and decreased survivorship of yellow birch, scarlet and chestnut oak, quaking aspen, and basswood		(Thomas et al. 2010)
Eastern Temperate Forest	Lichens	4–8	(#)	Epiphytic lichen community change	Loss of oligotrophic species. Synergistic/ confounding effects of acidic deposition not considered; based on application of model and estimated response threshold	(Geiser et al. 2010)
Eastern Temperate Forest	Southeast Coastal Plain	5–10	(#)	Ectomycorrhizal fungi community change		(Lilleskov et al. 2001, 2002, 2008); (Dighton et al. 2004)
Eastern Temperate Forest	Eastern Hardwood Forests	8	##	Increased surface water NO_3^- leaching		(Aber et al. 2003)
Eastern Temperate Forest	Michigan deposition gradient	<12	(#)	AMF biomass decline and community composition change		(Van Diepen et al. 2007); (Van Diepen 2008)
Eastern Temperate Forest	Herbaceous species	<17.5	(#)	Increases in nitrophilic species, declines in species-rich genera (e.g., *Viola*)		(Gilliam 2006, 2007); (Gilliam et al. 2006)

Table 5.1 (continued)

Ecoregion	Ecosystem component	CL for N deposition kg N $ha^{-1}yr^{-1}$	Reliability	Response	Comments	Study
Great Plains	Tall-grass prairie	5–15	#	Biogeochemical N cycling, plant and insect community shifts		(Tilman 1987, 1993); (Wedin and Tilman 1996); (Clark and Tilman 2008); (Clark et al. 2009)
Great Plains	Mixed-grass prairie	10–25	#	Soil NO_3^- pools, leaching, plant community shifts		(Clark et al. 2003, 2005); (Jorgensen et al. 2005)
Great Plains	Short-grass prairie	10–25	(#)		Inferred from mixed grass	(Epstein et al. 2001); (Barret and Burke 2002)
Great Plains	Mycorrhizal fungi	12	(#)	Decline in arbuscular mycorrhizal fungal activity		Egerton-Warburton, unpub. data
North American Desert	Lichens	3	(#)	Lichen community shifts, thallus N concentration	Uncertainty regarding modelled deposition estimates	(Porter 2007); (Geiser et al. 2008)
North American Desert	Shrubland, woodland, desert grassland	3–8.4	#	Vegetation response, vascular plant community change		(Inouye 2006); (Baez et al. 2007); (Allen et al. 2009); (Rao et al. 2010)
Mediterranean California	Coastal Sage Scrub	7.8–10	#	Invasive grass cover, native forb richness, arbuscular mycorrhizal fungi richness	Modelled and inferential N deposition estimates and published data for mycorrhizae, unpublished data for vegetation survey	[1]Allen unpublished data; (Egerton-Warburton and Allen 2000); (Tonnesen et al. 2007); (Fenn et al. 2010)
Mediterranean California	Chaparral; Lichens	3–6	#	Epiphytic lichen community change	Lichen critical loadis from modelled N deposition data and published data for lichens	(Jovan and McCune 2005); (Jovan 2008); (Fenn et al. 2010); (Geiser et al. 2010)

Table 5.1 (continued)

Ecoregion	Ecosystem component	CL for N deposition kg N $ha^{-1}yr^{-1}$	Reliability	Response	Comments	Study
Mediterranean California	Chaparral, Oak Woodlands, Central Valley	10–14	#	NO_3^- leaching; stimulated N cycling	Critical load for NO_3^- leaching of 10 kg N $ha^{-1}yr^{-1}$ is based on 1 year of throughfall data in Chamise Creek and an additional year of throughfall data from adjacent Ash Mountain, both in Sequoia National Park	(Fenn and Poth 1999); (Fenn et al. 2003a, 2003b, 2003c, 2009, 2011); (Meixner and Fenn 2004)
Mediterranean California	Mixed conifer forest; Lichens	3.1–5.2	##	Lichen chemistry and community changes	The lowest critical load is based on lichen tissue chemistry above the clean site threshold	(Fenn et al. 2008, 2010)
Mediterranean California	Mixed conifer forest	17	#	Reduced fine root biomass		(Grulke et al. 1998); (Fenn et al. 2008, 2010)
Mediterranean California	Mixed conifer forest	17–25.9	#	NO_3^- leaching; soil acidification		(Breiner et al. 2007); (Fenn et al. 2008, 2010)
Mediterranean California	Mixed conifer forest	24–39	(#)	Understory biodiversity; forest sustainability	N deposition from Fenn et al. (2008)	(Grulke et al. 1998, 2009); (Grulke and Balduman 1999); (Jones et al. 2004); (Allen et al. 2007)
Mediterranean California	Serpentine grassland	6	##	Annual grass invasion, replacing native herbs	Critical load based on a local roadside gradient; Serpentine grassland site is actually west of the Central Valley	(Weiss 1999); (Fenn et al. 2010)

Table 5.1 (continued)

Ecoregion	Ecosystem component	CL for N deposition kg N $ha^{-1}yr^{-1}$	Reliability	Response	Comments	Study
Temperate Sierras	Lichens	4–7	(#)	Epiphytic lichen community change	Increase in proportion of eutrophic species. Estimated from Marine West Coast Forests model, response threshold allows ~60 % eutrophs due to dry, hot climate, hardwood influence	(Geiser et al. 2010)
Temperate Sierras	*Pinus* forest	15	#	Elevated NO_3^- in stream and spring waters	Data are from *Pinus hartwegii* sites in the Desierto de los Leones National Park and Ajusco, Mexico	(Fenn et al. 1999, 2002); (Fenn and Geiser 2011)
Tropical and Subtropical humid Forests	N-rich forests	<5–10	(#)	NO_3^- leaching, N trace gas emissions	CL for N-rich forests should be lower than for N-poor forests based on possibility of N losses	ND
Tropical and Subtropical Humid Forests	N-poor forests	5–10	(#)	Changes in community composition; NO_3^- leaching, N trace gas emissions	CL for N-poor forests based on estimates for Southeastern Coastal Plain forests	ND
Wetlands	Freshwater wetlands	2.7–13	#	Peat accumulation and NPP change	CL for wetlands in the northeastern U.S. and southeastern Canada	(Rochefort et al. 1990)[c]; (Aldous 2002)[c]; (Vitt et al. 2003)[c]; (Moore et al. 2004)[c]
Wetlands	Freshwater wetlands	6.8–14	(#)	Sarracenia purpurea community change	CL based on northeastern populations	(Gotelli and Ellison 2002, 2006)

Table 5.1 (continued)

Ecoregion	Ecosystem component	CL for N deposition kg N $ha^{-1}yr^{-1}$	Reliability	Response	Comments	Study
Wetlands	Intertidal wetlands	50–100	##	Loss of eelgrass		(Latimer and Rego 2010)
Wetlands	Intertidal salt marshes	63–400	(#)	Changes in salt marsh community structure, microbial activity and biogeochemistry		(Wigand et al. 2003); (Caffrey et al. 2007)
Aquatic	Western Lakes	2	##	Freshwater eutrophication		(Baron 2006)
Aquatic	Eastern Lakes	8	#	NO_3^- leaching		(Aber et al. 2003)

reliable, # fairly reliable, (#) expert judgment
[a] based on data from Greenland
[b] based on data from Finland
[c] based on data from Canada
[d] based on data from Sweden
[1] Allen, E.B. unpublished data. Professor and Natural Resource Extension Specialist, Department of Botany and Plant Sciences and Center for Conservation Biology, University of California, Riverside, CA, 92521.

5.2.1 Effects of Nitrogen Deposition

Nitrogen deposition adversely affects mycorrhizal fungi (1) by causing decreased belowground C allocation by hosts and increased N uptake and associated metabolic costs (Wallander 1995) and (2) via soil chemical changes associated with eutrophication and acidification. There are two major groups of mycorrhizal fungi that are evolutionarily and ecologically distinct: arbuscular mycorrhizal fungi (AMF) and ectomycorrhizal fungi (EMF). Under sufficiently high N inputs, the progressive effect of elevated N is an early decline of sporocarp (reproductive structure) production for EMF and spore production for AMF, and subsequent decline in biological diversity and loss of taxa adapted to N-poor environments or sensitive to acidification (Lilleskov 2005). Sporocarp and spore production appears to be especially sensitive to N deposition, often declining before the communities on root tips have been substantially altered, presumably because sporocarps and spores are at the end of the carbon flux pathway from hosts.

Of the two plant-fungal symbioses examined here, mycorrhizal fungi appear to be less sensitive to N deposition than lichens, presumably because the soil environment buffers these soil fungi from some of the immediate impacts of N deposition to which lichens are directly exposed. Lichens have an advantage as indicators when compared with mycorrhizal fungi because they can be relatively easily inventoried. However, the critical role of mycorrhizal fungi as (i) root symbionts, central to plant nutrition and belowground production, (ii) repositories of a large part of the eukaryote diversity in forests, (iii) major components of food webs, and (iv) non-timber forest products of high economic value (edible sporocarps or mushrooms) (Amaranthus 1998), provides sufficient impetus to improve our understanding of their response to N deposition.

5.2.2 Critical Loads of Nitrogen

We reviewed empirical studies on mycorrhizal fungal response to N inputs to determine empirical critical loads for different ecoregions the United States (Table 5.1; Fig. 5.3). Nitrogen deposition sufficient to elevate inorganic N, especially NO_3^-, availability in soils can have measurable effects on mycorrhizal fungi. The data for EMF indicate that N deposition to N-limited conifer forests in the range of 5–10 kg $ha^{-1}yr^{-1}$ can significantly alter community structure and composition and decrease species richness (Dighton et al. 2004; Lilleskov 1999; Lilleskov et al. 2001, 2002, 2008). Similarly, the data for AMF suggest that N deposition levels of 7.8–12 kg $ha^{-1}yr^{-1}$ can lead to community changes, declines in spore abundance and root colonization, and changes in community function. This range is based on re-analysis of data from Egerton-Warburton et al. (2001) combined with N deposition data, decreases in fungal abundance (Van Diepen et al. 2007, Van Diepen 2008), and

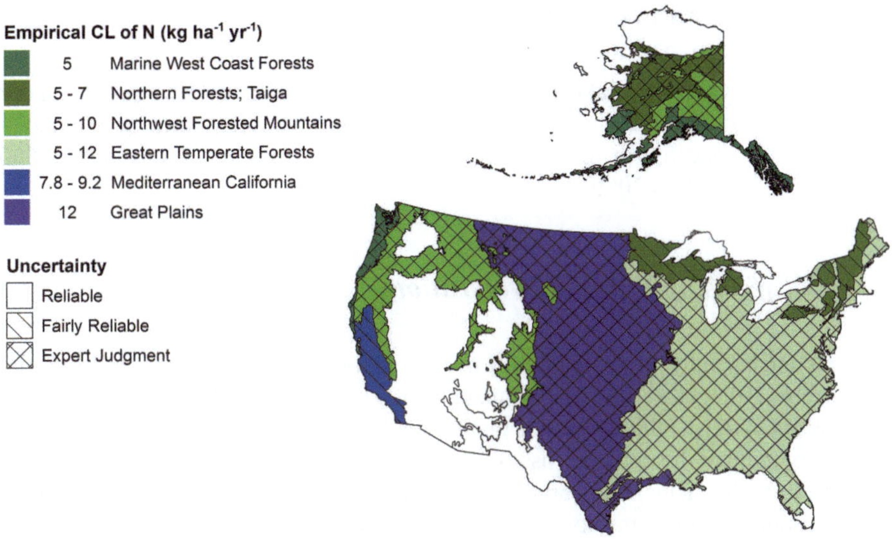

Fig. 5.3 Map of critical loads for mycorrhizal fungi by ecoregion in the United States (The *hatch marks* indicate increasing level of uncertainty: *no hatch marks* for the most certain "reliable" category, *single hatching* for the "fairly reliable" category, and *double hatching* for the "expert judgment" category. The colour sequence moves from red toward *blue* and *violet* as the critical load increases. As the range of the critical load gets broader, the saturation of the colour decreases)

declines in fungal activity[2]. The actual threshold for N effects on AMF could be even lower, because high background deposition precludes consideration of sites receiving deposition at or near pre-industrial levels. Therefore, the provisional expert judgment is that critical loads for mycorrhizal diversity for sensitive ecosystem types are 5–10 kg $ha^{-1}yr^{-1}$. The uncertainty of this estimate is high, because few studies have been conducted at low N deposition to further refine the critical load. Variation across ecoregions is associated with differences in EMF and AMF responses. Critical load values are lower in Marine West Coast Forests, Northern Forests, Taiga, and Northwestern Forested Mountains, with EMF as receptors. Eastern Forests, which include both EMF and AMF as receptors, have the greatest range in critical loads values. Mediterranean California and the Great Plains, with only values for AMF reported, have the highest critical loads.

[2] Egerton-Warburton, L.M. Unpublished data. Chicago Botanic Garden, 1000 Lake Cook Road, Glencoe, IL, 60022.

5.3 Lichens and Bryophytes

Lichens and bryophytes make substantial contributions to biodiversity. About 4100 lichens and 2300 bryophytes are known from North America north of Mexico—approximately one fourth of the number of vascular plant species (about 26,600 species; USDA NRCS 2009).

5.3.1 Effects of Nitrogen Deposition

Lichens and bryophytes are among the most sensitive bioindicators of N in terrestrial ecosystems (Blett et al. 2003; Bobbink et al. 2003; Fenn et al. 2003b, 2010; Glavich and Geiser 2008). Unlike vascular plants, lichens and bryophytes lack specialized tissues to mediate the entry or loss of water and gases (e.g., waxy epidermis, guard cells, root steele). Thus, they rapidly hydrate and absorb gases, water, and dissolved nutrients during high humidity or precipitation events. However, they dehydrate to a metabolically inactive state quickly as well, making them slow growing and vulnerable to contaminant accumulation. Consequently, the implementation of lichen or bryophyte-derived critical loads may prevent undesired impacts, such as declines in biological diversity, to much of the broader forest ecosystem (McCune et al. 2007).

Species of epiphytic lichens in wet and mesic forests that are most sensitive to N (i.e., the large pendant and foliose species) play important ecological roles that are not duplicated by the nitrophytic (i.e., N tolerant) species that may replace them. Dominant regional oligotrophs (e.g. *Alectoria, Bryoria, Lobaria, Ramalina, Usnea*) comprise the bulk of lichen biomass in old-growth forests, contribute to nutrient cycling through N_2 fixation, and are used for nesting material, essential winter forage for rodents and ungulates, and invertebrate habitat (McCune and Geiser 2009). Storage of water and atmospheric nutrients by these lichen genera and epiphytic bryophytes moderates humidity and provides a slow release system of essential plant nutrients to the soil (Boonpragob et al. 1989; Cornelissen et al. 2007; Knops et al. 1991; Pypker 2004). In the tundra, lichens and bryophytes represent a significant portion of the biomass, and reindeer lichens are a vital link in the short arctic food chain (Kytöviita and Crittenden 2007). Mosses comprise the bulk of the biomass of the extensive boreal peatlands. In the desert, together with other microbiota, lichens and bryophytes form cryptogamic mats important to soil stabilization and fertility.

5.3.2 Critical Loads of Nitrogen

The critical loads estimated (Pardo et al. 2011c) for lichens range from 1–9 kg N $ha^{-1}yr^{-1}$ (Fig. 5.4; Table 5.1). The certainty associated with these estimates for li-

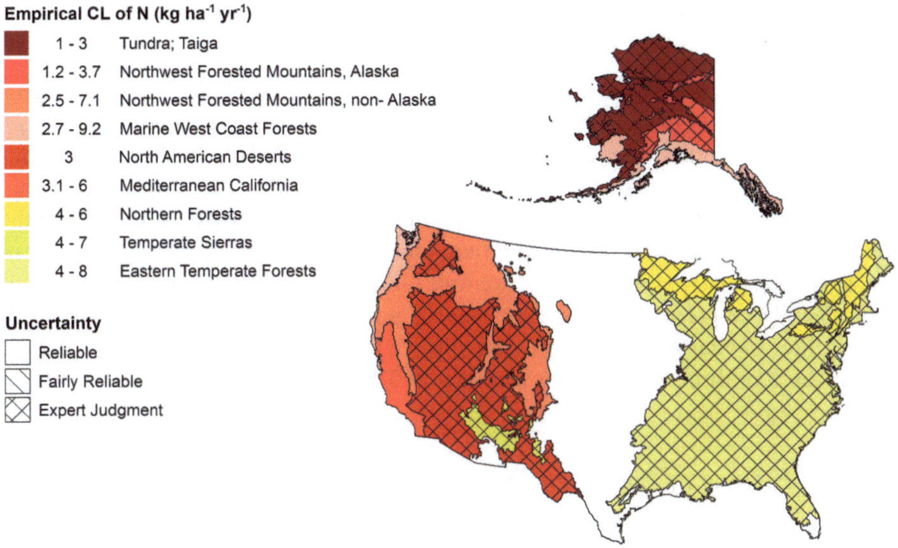

Fig. 5.4 Map of critical loads for lichens by ecoregion in the United States (see Fig. 5.3 for legend explanations)

chens varies considerably by ecoregion. This is partially because of differences in sampling scheme and intensity. For example, in the Pacific Northwest lichen communities were assessed intensively across wide environmental gradients spanning low to high N deposition on a fine grid over time, yielding highly reliable critical N load estimates (Geiser and Neitlich 2007; Jovan 2008), whereas assessments in the eastern United States are more problematic due to historical and contemporary S and N deposition. It is more difficult to determine the critical load where historical information necessary to identify a "pristine" or "clean" state is lacking, and the resulting confidence associated with the critical load is low.

The intensive studies in the Pacific Northwest facilitated the development of simple regressions to relate N deposition with shifts in community composition (Geiser and Neitlich 2007; Geiser et al. 2010; Jovan 2008) and thus to set critical loads. If such simple models could be tested and confirmed in other regions of the country, the confidence in the critical loads in those regions would improve.

The variation in critical loads for lichens across ecoregions (Fig. 5.4) is among others due to differences in ecosystem type, pre-existing lichen communities, and background N deposition. Marine West Coast Forests, with its broad range in environmental gradients, has the greatest range in critical loads. The low end of the critical load range in eastern ecoregions is higher than the low end of the critical load range in western ecoregions, likely as a result of higher historical S and N deposition in the eastern United States, which makes it difficult to establish critical loads for sensitive species.

5.4 Herbaceous Plants and Shrubs

Herbaceous species and shrubs are found in grasslands, shrublands, forests, deserts, and wetlands and comprise the majority of the roughly 26,600 vascular plant species found in North America north of Mexico (USDA NRCS 2009). Herbaceous plants play an important role in those ecosystems in which they are the dominant primary producers (e.g., grasslands, shrublands). In forests, however, the role of the herbaceous community in ecosystem function also has a significance disproportionate to its low relative biomass. For example, although they represent only ~0.2% of standing above-ground biomass, herbaceous understory species produce >15% of forest litter biomass and comprise up to 90% of forest plant biodiversity, including endangered or threatened species (Gilliam 2007).

5.4.1 Effects of Nitrogen Deposition

Herbaceous plants and some shrubs appear intermediate between cryptogam and tree species in their sensitivity to N deposition, due to specialized tissues that mediate the entry or loss of water and gases compared with cryptogams, and rapid growth rates, shallow rooting systems, and often shorter lifespan compared with trees. Thus, herbaceous species in a forest understory will likely respond more rapidly to changes in N deposition and to a greater degree than the trees with which they coexist. Herbaceous plants in alpine or tundra environments will respond later and to a lesser degree than the cryptogams with which they coexist.

5.4.2 Critical Loads of Nitrogen

The range of critical loads of N for herbaceous plants and shrubs across all ecoregions is 3–33 kg N $ha^{-1}yr^{-1}$ (Fig. 5.5; Table 5.1). Although this range is broader than those for lichens or mycorrhizal fungi, many of the critical loads for herbaceous plants fall into the range of 5–15 kg N $ha^{-1}yr^{-1}$. The uncertainty of these estimates is moderate. The shorter lifespan of some herbaceous plants can result in a more rapid response to N addition. This is especially relevant for annuals or perennials with little N storage. In grasslands, for example, elevated N deposition often leads to a rapid (1–10 years) increase in herbaceous production and a shift in biomass allocation towards more above-ground tissue. This often decreases light levels at ground surface and decreases the numbers of plant species, primarily of perennials, legumes, and natives (Clark and Tilman 2008; Suding et al. 2004; Tilman 1993).

As a result of this relatively rapid response, experimental studies of moderate to long duration (3–10 years) allow determination of the critical load with reasonable certainty. Longer studies (>10 years) would decrease the uncertainty further. In some cases, it can be difficult to determine whether the condition in reference plots or at the low end of a deposition gradient represents a "pristine" condition or

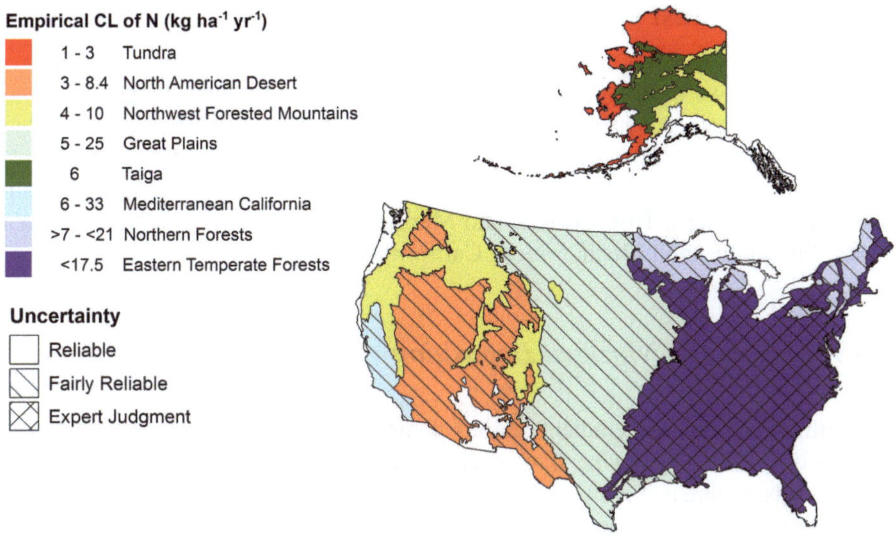

Fig. 5.5 Map of critical loads for herbaceous plants and shrubs by ecoregion in the United States (see Fig. 5.3 for legend explanations)

whether a site has already been altered by N deposition prior to or at the time of the study. For example, the Watershed Acidification Study at Fernow Experimental Forest, West Virginia, added 35 kg N $ha^{-1}yr^{-1}$ via aerial application in addition to ambient deposition of 15–20 kg N $ha^{-1}yr^{-1}$, which has led to changes in understory species composition (Adams et al. 2006). Recently, similar changes in understory species composition have occurred on the adjacent reference watershed receiving only ambient atmospheric deposition[3] (Gilliam et al. 1996) suggesting that the deposition to the reference watershed currently exceeds the critical load. Where deposition rates exceed the critical load, measurement of the rate of change of an ecological metric (e.g. plant abundance, diversity, or community composition) over a range of N inputs provides an estimate of the N level at which increased ecological change occurs (Bowman et al. 2006), but it is difficult to determine the critical load.

The large variation across ecoregions for herbaceous critical loads (Fig. 5.5) is caused, in part, by the differences in receptor species and ecosystems, the paucity of data in some ecoregions and historic N status. Where few studies are available, the range reported for the critical load is broad and is considered less reliable. Additional studies could narrow the range of the critical load and increase the reliability. N-poor sites and sites with relatively low productivity (e.g., Tundra, North American Deserts) have lower critical loads for herbaceous species than sites with more fertile soil and higher productivity (e.g., Great Plains). High levels of historical N deposition and lack of low-level N fertilization experiments mean that the critical loads for some ecoregions may be lower than currently reported.

[3] Gilliam, F.S. Unpublished data. Professor, Department of Biological Sciences, Marshall University, Huntington, WV 25755–2510.

5.5 Trees/Forest Ecosystems

In this section we discuss the responses of trees and the overall biogeochemical responses of forest ecosystems to N inputs, excluding the specific responses of mycorrhizal fungi, lichens, or understory herbaceous plants. Forest ecosystems represent about a third of landcover in the United States (USFS 2001) and are significant in Northern, Eastern, Tropical Wet, and Marine West Coast Forests, Northwestern Forest Mountains, and Mediterranean California ecoregions.

5.5.1 Effects of Nitrogen Deposition

In northeastern forests, gradient studies demonstrate that N deposition enhances growth in some fast-growing tree species, including many hardwoods with AMF associations, whereas it slows growth in some EMF species (red spruce, red pine), and has no detectable effect on still other species (Thomas et al. 2010). Similarly, N deposition enhances survivorship in a few species capable of forming AMF associations (black cherry, red maple, paper birch) and decreases survivorship in others, all ectomycorrhizal (Thomas et al. 2010). Survivorship under chronic N deposition, and possibly other co-occurring pollutants such as ozone, is often dependent on interactions with other stressors such as pests, pathogens, climate change, or drought (Grulke et al. 2009; McNulty and Boggs 2010). Over the long-term, these differential effects of N deposition on tree growth and survivorship are likely to shift species composition, possibly to more nitrophilic species, similar to patterns seen for organisms with shorter lifespans.

We have few data that show a major structural or functional shift in forest ecosystems, due to the long response times of trees and forest soils to changes in N inputs and N availability. This is caused by the relatively large pools of organic N in the forest floor, mineral soil, tree biomass, and detritus. Because of the long lag-time in response to N treatments, it can be difficult to determine the actual critical load of N for forest ecosystems based on short-term fertilization studies. If a response is observed over a relatively short period of time (i.e. years), it is nearly certain that the critical load is below the total N input at the treatment site and it can be difficult to further constrain the critical load. It is expected that the more complex and interconnected processes in forests will result in a higher critical load than other ecosystem types, in part because large N storage pools give forest ecosystems a greater capacity to buffer N inputs.

5.5.2 Critical Loads of Nitrogen

The range of critical loads reported for forest ecosystems is 4–39 kg N $ha^{-1}yr^{-1}$ (Fig. 5.6; Table 5.1). The threshold N deposition value which caused increased NO_3^- leaching from forest ecosystems into surface water was 8–17 kg N $ha^{-1}yr^{-1}$;

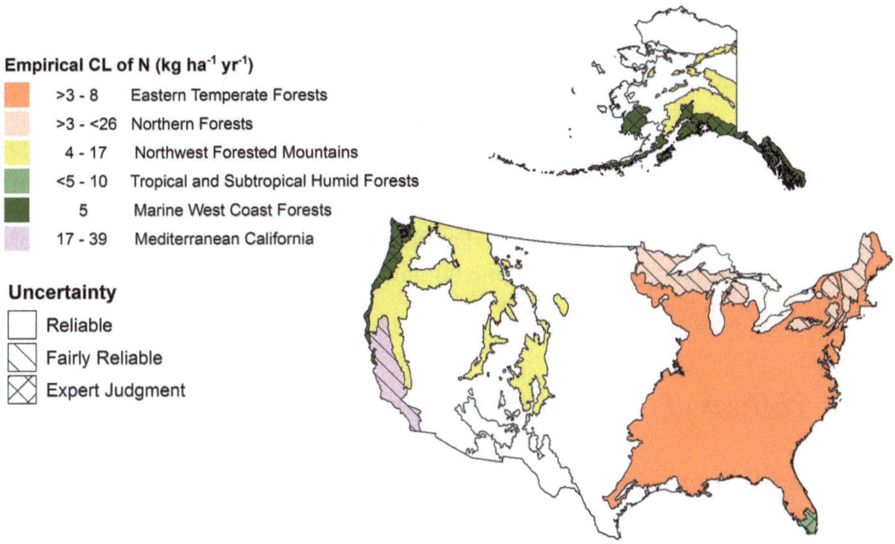

Fig. 5.6 Map of critical loads for forest ecosystems by ecoregion in the United States (This map does not include the responses of mycorrhizal fungi, lichens, or understory herbaceous plants already represented; see Fig. 5.3 for legend explanations)

the lower end of the range representing Northern and Eastern Forests, the upper end representing Mediterranean California mixed conifers (Fig. 5.6). At 4 kg N $ha^{-1}yr^{-1}$ in the Colorado Rockies, increasing NO_3^- concentration was reported in the organic horizon, which suggests incipient N saturation (Rueth and Baron 2002). The highest critical loads were reported for Mediterranean California mixed conifer forests for forest sustainability and for soil acidification caused by increased N deposition. These sites experience some of the highest N deposition reported in the United States, up to approximately 70 kg N $ha^{-1}yr^{-1}$ (Fenn et al. 2008).

Critical loads for forests vary across ecoregions due in part to reported receptors, site and soil characteristics, and background N deposition status. Critical loads values were lower for ecoregions where sensitive forest receptors, such as mycorrhizal fungi (Marine West Coast Forests) were used to set critical loads. Use of forest health and species composition resulted in a large range in critical loads in Northern Forests and Mediterranean California. In the Northwestern Forested Mountains, the critical load based on NO_3^- leaching ranged from a low value of 4 kg $ha^{-1}yr^{-1}$ in subalpine forests to 17 kg $ha^{-1}yr^{-1}$ in mixed conifer forests.

5.6 Freshwater and Wetland Ecosystems

Freshwater lakes and streams, and wetlands (freshwater and estuarine intertidal) are ecosystem types that occur in most ecoregions in North America. In freshwater lakes and streams, phytoplankton, algae that live in the water column, are sensitive to the chemical environment in which they reside, and many species can be used as indicators of the levels of nutrients or acidity because of individual species' preference for specific chemical conditions. Diatoms are used in this discussion because there has been more work published on these algae than others, but other types of algae also respond to N deposition (Lafrancois et al. 2004; Michel et al. 2006). Of the wetlands which occur in the conterminous United States, 95 % are freshwater and 5 % are estuarine or marine (USDI FWS 2005). The species composition differs between freshwater and intertidal wetlands, although together they support more than 4200 native plant species. Despite the high biodiversity, the effects of N loading are studied in just a few plant species.

5.6.1 Effects of Nitrogen Deposition

For the analysis of nutrient N effects to freshwater lakes and streams, we relied on papers and studies that linked aquatic biological and ecological response to atmospheric deposition, but the results are consistent with laboratory or *in situ* dose-response studies and even land use change studies. The productivity of minimally disturbed aquatic ecosystems is often limited by the availability of N, and slight increases in available N trigger a rapid biological response that increases productivity and rearranges algal species assemblages (Nydick et al. 2004; Saros et al. 2005). The mechanism for change is alteration of N:P ratios, which can increase productivity of some species at the expense of others (Elser et al. 2009). As with the terrestrial systems described above, the nutrient responses of lakes and streams are most evident where land use change and acidic deposition have been limited, thus most evidence of exceedance of critical loads comes from high elevations of the western United States (Baron et al. 2011). As with terrestrial plants, some diatoms respond rapidly to an increase in available N. An example that has been observed from a number of different lakes of the Rocky Mountains is dominance of two diatoms (*Asterionella formosa* and *Fragilaria crotonensis*) in lakes with higher N, in contrast to the flora of lakes with lower N deposition where there is a more even distribution, thus high biodiversity, of diatoms. Higher trophic levels (zooplankton, macroinvertebrates) may be secondarily affected by N, but further increases in primary, or autotrophic, production will be limited by other nutrients such as P or silica (Si).

Both freshwater and estuarine intertidal wetlands tend to be N-limited ecosystems (LeBauer and Treseder 2008; U.S. EPA 1993). Known responses to N enrichment are generally derived from nutrient-addition studies in the field and observations along gradients of N deposition. A variety of ecological endpoints are evaluated, such as altered soil biogeochemistry, increased peat accumulation, elevated primary production, changes in plant morphology, changes in plant population dynamics, and altered plant species composition (U.S. EPA 2008). In general, the sensitivity of wetland ecosystems to N is related to the fraction of rainfall (a proxy for atmospheric N deposition) in the total water budget. Most freshwater wetlands, such as bogs, fens, marshes and swamps, have relatively closed water and N cycles, thus are more sensitive to N deposition than estuarine intertidal wetlands, such as salt marshes and eelgrass beds (Greaver et al. 2011).

5.6.2 Critical Loads of Nitrogen

In general, critical loads for freshwater lakes and streams tend to be low, because the target organisms are unicellular algae that respond rapidly to changes in their chemical environment. The range of critical loads for eutrophication and acidity in freshwaters is 2–9 kg N $ha^{-1}yr^{-1}$ (Baron et al. 2011); the range reported for terrestrial ecosystems is much broader (Table 5.1; Fig. 5.7). Critical loads for NO_3^- leaching from terrestrial ecosystems ranged from 4–17 kg N $ha^{-1}yr^{-1}$, but many sensitive freshwaters at high altitudes are found above the tree-line where few watershed buffering mechanisms exist, due to sparse vegetation, poorly developed soils, short hydraulic residence time, and steep topography. These factors influence how rapidly a system exhibits elevated N leaching in response to increased N deposition, and how this increased N availability subsequently influences biota. In general, lakes have relatively rapid N turnover times compared to soil N pools and are at least seasonally well-mixed. They would thus be expected to have lower critical loads. Thus responses of terrestrial plants would not be expected to be as rapid as those of freshwater organisms.

Generally, freshwater wetlands are more sensitive to N deposition than estuarine intertidal wetlands, with critical loads for freshwater wetlands that range

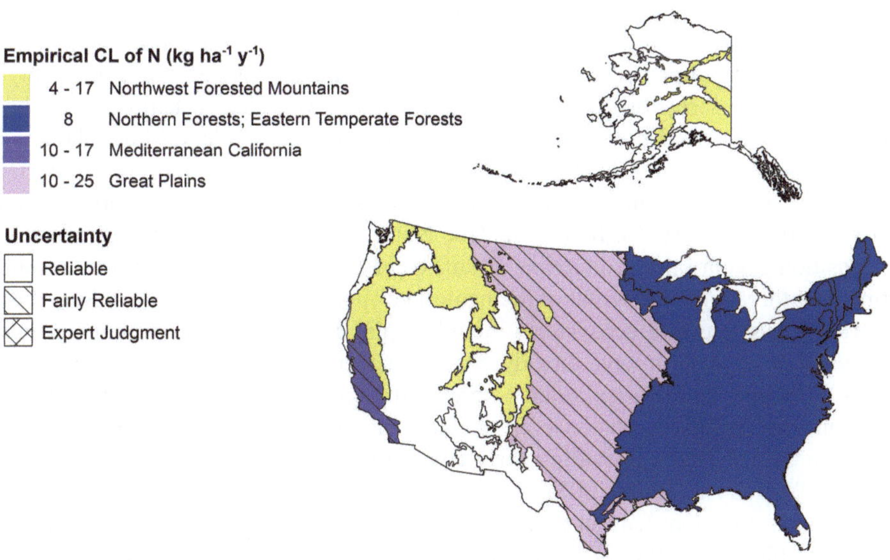

Fig. 5.7 Map of critical loads for freshwater and wetland ecosystems based on increased nitrate leaching by ecoregion in the United States (see Fig. 5.3 for legend explanations)

from 2.7–14 kg N $ha^{-1}yr^{-1}$ (Table 5.1; Greaver et al. 2011). The bryophyte genus *Sphagnum* and the carnivorous pitcher plant are the two taxa most commonly studied. The critical loads reported for freshwater wetlands (Greaver et al. 2011) fall between those reported for inland surface waters (Baron et al. 2011) and those reported for terrestrial ecosystems (Pardo et al. 2011c). This pattern may be related to the rate of N released by soils/sediment to the ecosystem. The critical load tends to be higher for estuarine intertidal wetlands than other types of ecosystems because they have open nutrient cycles which are often strongly affected by N loading sources other than atmospheric deposition. Based on field observations of N loading effects on plant growth and species composition on salt marsh and eel grass habitat, the critical load for estuarine intertidal wetlands ranges between 50–400 kg N $ha^{-1}yr^{-1}$ (Table 5.1).

5.7 Discussion and Conclusions

5.7.1 Effects of Nitrogen Deposition

The most significant changes that we are currently observing in the United States in response to elevated N deposition are changes in species composition: losses of N-sensitive species, shifts in dominance, and losses of native species in favour of exotic, invasive species. Shifts in diatom and lichen community composition away from N-intolerant (oligotrophic) species are observed across the country. Alterations in herbaceous species are broadly observed, but are not always clearly documentable because of the long-term pollution inputs and other disturbances (including land-use change) that caused changes prior to the initiation of careful observations.

Numerous examples illustrate the significance of these species- and community-level effects. In serpentine grasslands in California, it was clearly demonstrated that unless N inputs are decreased or N is removed in biomass, a larval host plant and numerous nectar source plants utilized by a threatened and endangered butterfly will decrease to levels unable to sustain the checkerspot butterfly population (Fenn et al. 2010; Weiss 1999). In Joshua Tree National Park in southern California, N deposition favours the production of sufficient invasive grass biomass to sustain fires that threaten the survival of the namesake species (Fenn et al. 2010; Rao et al. 2010). Other sensitive ecosystems include alpine meadows, where relatively low levels of N deposition have already changed species composition (Bowman et al. 2006). Changes in historical diatom community composition from N-limited to N-tolerant species have been observed in lake sediment cores at many locations in the western United States, providing early evidence of freshwater ecosystem eutrophication (Wolfe et al. 2001; 2003).

Changes in ecosystem structure are linked to changes in ecosystem function. For example, extirpation of lichens can alter food webs by reducing the availability of nesting material for birds, invertebrate habitat, and critical winter forage for mammals, and can also affect nutrient cycling (Cornelissen et al. 2007). In some arid low-biomass California ecosystems, N-enhanced growth of invasive species results in increased fire risk, even in areas where fire is normally infrequent (Allen et al. 2009; Fenn et al. 2010; Rao et al. 2010).

There is also evidence of N deposition contributing to multiple stress complexes, resulting in reduced forest sustainability (Grulke et al. 2009; McNulty and Boggs 2010). In North Carolina, elevated N deposition predisposed a pine ecosystem to a pest outbreak following a drought (McNulty and Boggs 2010). These types of complex interactions may be difficult to predict, but may intensify the impact of elevated N deposition in concert with other stressors, including climate change (Wu and Driscoll 2010). Further examples of changes in ecosystem structure and function are observed in coastal areas, where increased N export has led to toxic algal blooms (Rabalais 2002). As an example of N deposition effects on trace gas chemistry and climate change, N loading to ecosystems results in increased emissions of N trace gases, such as NO (nitric oxide, an ozone precursor), N_2O (nitrous oxide, a long-lived and powerful greenhouse gas); as well as declines in soil uptake of CH_4 (methane, another long lived and powerful greenhouse gas) (e.g. Liu and Greaver 2009).

5.7.2 Relative Sensitivities of Different Receptors, Ecosystem Types, and Regions

This synthesis demonstrates that empirical critical loads of N differ among life forms, tending to increase in the following sequence: diatoms < lichens and bryophytes < mycorrhizal fungi < herbaceous plants and shrubs < trees. Nitrogen deposition more rapidly affects those species that experience the most direct exposure to elevated N levels in the atmosphere (lichens and bryophytes) or receiving waters (diatoms), especially for those organisms that lack protective structures. By contrast, the capacity of soil organic matter to accumulate large quantities of N may delay adverse impacts on many herbs, shrubs, and trees. Altered N availability often appears to shift species composition most rapidly within those communities dominated by species with short lifespans (diatoms) compared to those with long lifespans (trees).

Critical loads vary more by receptor and response type than by region. For the same response of a given receptor, the western U.S. has generally similar critical load values to the eastern U.S., with the apparent exception that the critical load for NO_3^- leaching is approximately twice as high in Mediterranean California mixed conifers compared to northeastern forests (Fig. 5.7). In contrast, the critical load for NO_3^- leaching in high elevation catchments in the Colorado Front Range are lowest in the U. S., likely attributable to low biological N retention and storage capacity in these steep, rocky catchments (Baron et al. 2000; Fenn et al. 2003a, b; Sickman et al. 2002; Williams and Tonnessen 2000).

5.7.3 Factors Affecting the Critical Load

Multiple abiotic and biotic factors affect the critical load (Table 5.2). Abiotic influences include a range of climatic, hydrologic, and soil factors that can affect the timing and magnitude of N delivery to sensitive receptors. Climatic factors include temperature, precipitation amount and distribution, and the extent and rate of climate change. Hydrologic factors include catchment size, topographic relief, and flow path. Soil factors include soil type, age, depth, coverage, and parent material. Disturbance—forest fires or cutting—and past agricultural use can also affect soil N and thus the critical load.

Biological factors likely to contribute to lower critical loads of N include particularly sensitive species (diatoms, lichens, mycorrhizal fungi, certain plants), single species versus community responses, low biomass and low productivity ecosystems, short lifespan of receptor of concern, presence of invasive species, and presence of ozone-sensitive species (Fenn et al. 2008; Grulke et al. 1998, 2009; Grulke and Balduman 1999). For example, low-biomass ecosystems (e.g., grasslands, coastal sage scrub, desert) are more sensitive to N-enhanced growth of invasive species, if invasive pressure occurs. These low-biomass ecosystem types sometimes occur because of warm and dry climatic conditions. Because warmer temperatures often correspond to greater metabolic rates, longer periods of biological activity, greater biomass, and more rapid N cycling, one might expect that the critical load would increase with increasing temperature as has been suggested in Europe (Bobbink et al. 2003). We do not observe such a pattern across U.S. ecoregions in the critical loads reported in this synthesis, but Europe does not have warm and dry deserts with low critical loads as does the U.S. Note, however, that the uncertainty of the critical load estimates varies and is often fairly high, which may make it difficult to discern patterns in critical load values across regions. Moreover, a temperature pattern may be confounded by gradients in deposition form and quantity, moisture and elevation.

5.7.4 Comparison to Critical Loads in Europe

The range of critical loads of N deposition in U.S. ecoregions for terrestrial ecosystems is 1–39 kg N $ha^{-1}yr^{-1}$, which is close to the range for the most recently reported critical loads values for similar ecosystems in Europe (Bobbink and Hettelingh 2011). However, the low end of the critical loads range is nearly always lower in the U.S. than in Europe (Fig. 5.8; Table 5.3). There are several potential reasons why critical loads for the U.S. remain lower than European critical loads. These includes greater availability of pristine baselines in the U.S., more intensive land use in Europe; greater dominance of N deposition by reduced forms of N in Europe, and different threshold criteria.

Table 5.2 Assessment and interpretation of empirical critical loads of nutrient N for North American ecoregions

Ecoregion	Factors affecting the range of CL[a]	Comparison within ecoregion[b]
Tundra	1) Moisture 2) Competition between vascular plants and cryptogams 3) P-limitation 4) Temperature 5) pH	Critical loads are higher in wet and P-limited tundra; acidic tundra may be more sensitive to N deposition than non-acidic tundra. Increased N deposition may be more detrimental to lichens in the presence of graminoids and shrubs in the low and mid arctic than to lichens with less competition in the high arctic. Response time increases with latitude due to colder temperatures, less light, and poorer N and P mobilization
Taiga	1) Soil depth 2) Vegetation type and species composition 3) latitude	Morphological damage to lichens has been observed at a lower deposition in forests and woodlands than in shrub lands, bogs or fens; cryptogam dominated mats on thin soils become N saturated faster than forest islands
Northern Forest	1) Receptor 2) Tree species 3) Stand age 4) Site history 5) Pre-existing N status	CLs for lichen are generally lowest, followed by CLs for ectomycorrhizal fungi and NO_3^- leaching. CLs for herbaceous species and forests are generally higher than for other responses
Northwest Forested Mountains	1) Biotic receptor 2) Accumulated load of N 3) Ecosystem 4) Region	In alpine regions, diatom changes in lakes are seen at the lowest CL. Changes in individual plants are seen next, followed by vegetation community change, then soil responses. In subalpine forests, the CL of 4 kg N $ha^{-1}yr^{-1}$ for foliar and soil chemistry changes is similar to the lichen CL of 3.1–5.2 kg N $ha^{-1}yr^{-1}$ for lichen community change
Marine West Coast Forests	1) Background N status 2) Soil type 3) Species composition 4) Fire history 5) Climate	The midrange of responses reported for lichens (2.7–9.2 kg N $ha^{-1}yr^{-1}$) is broadly comparable to that for plant, soil, and mycorrhizal responses (5 kg N $ha^{-1}yr^{-1}$), despite limited studies for non-lichen responses.
Eastern Forests	1) precipitation 2) soil cation fertility and weathering 3) Biotic receptors	CLs for NO_3^- leaching, lichen community change, and ectomycorrhizal fungal response are within the same range. Arbuscular mycorrhizal fungal and herbaceous CLs are higher

Table 5.2 (continued)

Ecoregion	Factors affecting the range of CL[a]	Comparison within ecoregion[b]
Great Plains	1) N status 2) Receptor 3) Precipitation	CLs are lower in the tall grass prairie than in the mixed- and short-grass prairies. CLs in tall- and mixed-grass prairie are lower on N poor sites and sites with very N responsive plant species. CL in the short-grass prairie is likely lower in wet years than in dry years
North American Deserts	1) Receptor 2) Interaction of annual grasses with native forb cover 3) Precipitation	The lichen CL is lowest, at 3 kg N $ha^{-1}yr^{-1}$; vegetation CL varies from 3–8.4 kg N $ha^{-1}yr^{-1}$
Mediterranean California	1) Presence of invasive exotic annual grasses interacting with a highly diverse native forb community 2) N-sensitivity of mycorrhizal fungi 3) N-sensitivity of lichens 4) N retention capacity of catchments, catchment size 5) Co-occurrence of ozone and ozone-sensitive tree species	The lowest CLs in Mediterranean California are for sensitive lichen in chaparral and oak woodlands and mixed conifer forests. The CL for plant and mycorrhizal fungal community change in coastal sage scrub is higher, at 7.8–10 kg $ha^{-1}yr^{-1}$. CL for NO_3^- leaching is lower in chaparral and oak woodlands (10–14 kg $ha^{-1}yr^{-1}$) than in mixed conifer forests (17 kg $ha^{-1}yr^{-1}$). CLs are highest for mixed conifer forest plant community change and sustainability. Fine root biomass in ponderosa pine is reduced by both ozone and elevated soil N
Wetlands	1) Vegetation species 2) The fraction of rainfall in the total water budget 3) The degree of openness of N cycling	CL is much higher for intertidal wetlands (50–400 kg N $ha^{-1}yr^{-1}$) than for freshwater wetlands (2.7–14 kg N$ha^{-1}yr^{-1}$), which have relatively closed water and N cycles
Freshwaters	1) Extent of upstream vegetation development 2) Topographic relief 3) Land use/deposition history	CLs are lower in western mountain lakes/streams with poorly vegetated watersheds and steep catchments. CLs are greater in eastern lakes with prior land use and decades of acidic deposition

[a] Factors causing the critical load (CL) to be at the low or high end of the range reported
[b] Comparison of values and causes for differences if multiple critical loads are reported for an ecoregion

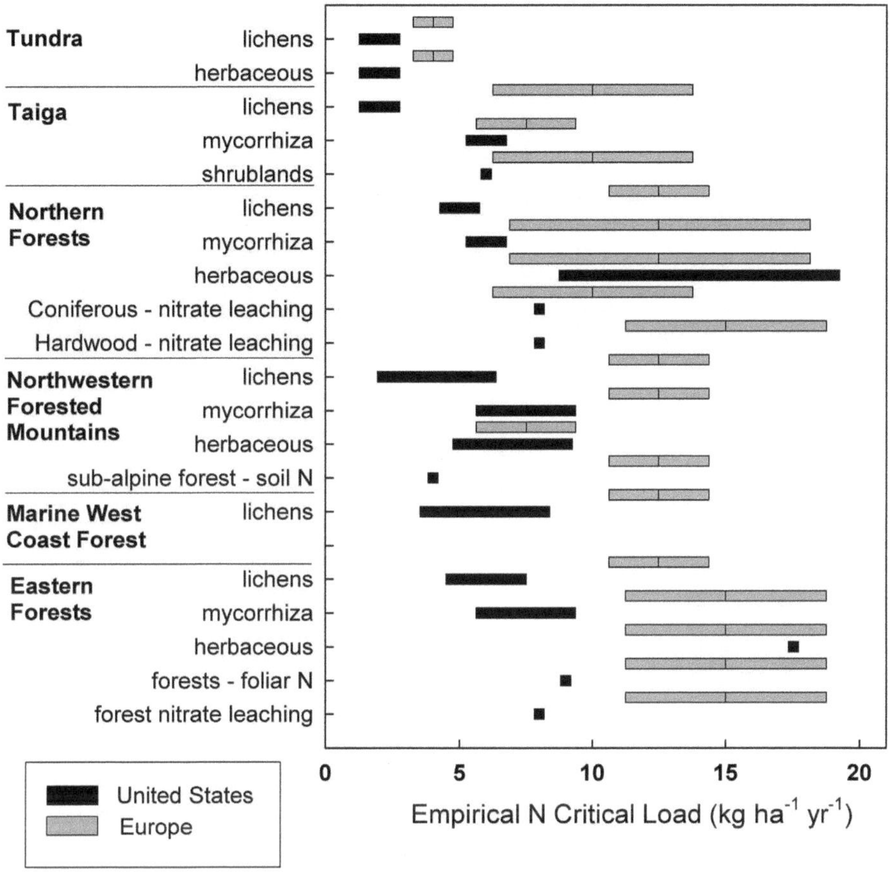

Fig. 5.8 Comparison of empirical critical loads of nutrient nitrogen for Europe. (based on Bobbink and Hettelingh 2011) and the United States

Availability of Pristine Baselines and Studies at Low Deposition: Because of high historic deposition levels, many European systems lack pristine baseline ecosystems as a reference to compare to those experiencing elevated N deposition. For example, past European critical loads for lichens were much higher than those in the U.S. (Bobbink et al. 2003). These loads were influenced by study sites in Scotland experiencing a deposition gradient from 10–22 kg N $ha^{-1}yr^{-1}$ from which critical loads were set at 11–18 kg N $ha^{-1}yr^{-1}$ (Mitchell et al. 2005). However, no oligotrophic species were observed, presumably because they were eliminated prior to the initial studies. The more recently reported European critical loads (Bobbink and Hettelingh 2011), used in our comparison, were set at 5–10 kg N $ha^{-1}yr^{-1}$. In some European ecosystems, such as dry grass lands, there is, however, still a need for more low N addition and deposition experiments (Bobbink and Hettelingh 2011).

Table 5.3 Critical loads for European ecosystems compared to critical loads for U.S. Ecoregions

European Ecosystem type (EUNIS code)	Critical load	Reliability	Indication of exceedance	U.S. Ecoregion; Ecosystem component	Critical load	Reliability	Indication of exceedance
Permanent oligotrophic lakes, ponds, and pools (C1.1)	3–10	##	Change in the species composition of macrophyte communities, increased algal productivity and a shift in nutrient limitation of phytoplankton from N to P	Aquatic; eastern and western high elevation lakes	2–8	#	Increased productivity, eutrophication, altered algal species assemblages
Raised and blanket bogs (D1)	5–10	##	Increase in vascular plants, altered growth and species composition of bryophytes, increased N in peat and peat water	Wetlands; bogs, fens, and swamps	2.7–14	#	Alterations in sphagnum accumulation and net primary productivity; alteration in threatened *Sarracenia purpurea* community
Sub-Atlantic semi-dry calcareous grasslands (E1.26)	15–25	##	Increase in tall grasses, decline in diversity, increased mineralization, N leaching; surface acidification	Great Plains; mixed-grass prairie	10–25	#	Change in soil NO_3^- pools, increased leaching, plant community shifts
Non-Mediterranean dry acidic and neutral closed grasslands (E1.7)	10–15	##	Increase in graminoids, decline in typical species, decrease in total species richness	Great Plains; tall-grass prairie	5–15	#	Change in biogeochemical N cycling, plant and insect community shifts
Tundra (F1)	3–5	#	Changes in biomass, physiological effects, changes in species composition in bryophyte layer, decrease in lichens	Tundra; prostrate dwarf shrub, lichens	1–3	#	Changes in lichen, bryophyte, and vascular plant cover. Changes in vascular plant CO_2 exchange, foliar N, and community composition; change in lichen pigment production and ultrastructure
Arctic, alpine, and subalpine scrub habitats (F2)	5–15	#	Decline in lichens, bryophytes and evergreen shrubs	Taiga; shrublands—lichen, moss, and algae	1–6	#	Change in shrub and grass cover, increased parasitism of shrubs; changes in alga, bryophyte, and lichen community composition, cover, tissue N, or growth rates

Table 5.3 (continued)

European Ecosystem type (EUNIS code)	Critical load	Reliability	Indication of exceedance	U.S. Ecoregion; Ecosystem component	Critical load	Reliability	Indication of exceedance
Broadleaved deciduous woodland (G1)	10–20	##	Changes in soil processes, nutrient imbalance, altered composition mycorrhiza and ground vegetation	Eastern and Northern Forests; hardwood forest, southeast coastal plain	3–21	#	Change in mycorrhizal fungal community structure and biomass, change in herb layer and loss of prominent species, increased surface water NO_3^- leaching, increased foliar N, change in tree growth and mortality
Coniferous woodland (G3)	5–15	##	Changes in soil processes, nutrient imbalance, altered composition mycorrhiza and ground vegetation	Northern Forests; coniferous forest	3–26	#	Change in mycorrhizal fungal community structure, biomass decline in arbuscular mycorrhizal fungi, loss of prominent herbaceous species, increased surface water NO_3^- leaching, change in tree growth and mortality
Abies and Picea woodland (G3.1)	10–15	(#)	Decreased biomass of fine roots, nutrient imbalance, decrease in mycorrhiza, changed soil fauna	Northwestern Forested Mountains; subalpine forest	4	##	Increase in organic horizon N; higher net N mineralization rates
Spruce taiga woodland (G3.A)	5–10	##	Changes in ground vegetation, decrease in mycorrhiza, increase in free living algae	Taiga; spruce-fir forests	5–7	(#)	Change in ectomycorrhizal fungal community structure
Temperate and boreal forest; lichen and algae (G)	5–10	#	Decline in lichens, increase in free-living algae	Eastern and Northern Forests, Northwestern Forested Mountains, Marine West Coast Forest; lichens	1–9	#	Lichen community change

reliable, # fairly reliable, (#) expert judgment

Land Use: A larger fraction of the forested landscape in Europe is heavily managed (harvested and planted) relative to the U.S. High rates of harvest removals of N in biomass, creating greater N demand and storage during re-establishment of the forest stand could contribute to higher critical loads in Europe than the U.S.

Forms and Mode of Measurement of N inputs: NH_4^+ inputs tend to be higher and represent a greater proportion of total N inputs in Europe, particularly in past decades; this is changing in the U.S. Some receptor species can be more sensitive to reduced than to oxidized forms of N inputs, and nitrification of NH_4^+ inputs can accelerate ecosystem acidification relative to inputs of NO_3^-.

Threshold Criteria: Another possible explanation for the higher critical loads is that the response thresholds utilized in Europe are sometimes higher. For example, choosing a threshold of a shift in lichen community composition will produce a much lower critical load than a threshold of near extirpation of lichen species as used in earlier European work (Bobbink et al. 2003). As a second example, choosing a threshold of initial changes in N biogeochemistry in the Colorado Front Range, interpreted as incipient responses of N saturation, led to a critical load <4 kg N $ha^{-1}yr^{-1}$ (Rueth et al. 2003). This is a subtle initial N enrichment response when compared to the magnitude of change (a later stage of N saturation) for the critical loads thresholds in Europe (10–15 kg $ha^{-1}yr^{-1}$). Finally, much of the same research was used to set critical loads for both European and U.S. tundra and taiga ecosystems (Bobbink and Hettelingh 2011; Pardo et al. 2011c). The difference in the critical loads for these ecosystems is primarily due to different threshold criteria.

Acknowledgements This project was funded, in part, by the US-EPA Clean Air Markets Division, DW-12-92196101. Initial work on this project was funded by the US Forest Service Air Resource Management Program. Funding was provided by the UCR Center for Conservation Biology and NSF grant DEB 04-21530.We would also like to thank Amanda Elliot Lindsey, Robert Johnson, and the University of Vermont Spatial Analysis Lab for their assistance in creating deposition and critical loads maps. We thank Sabine Braun, Doug Burns, and Jim Sickman for their reviews of an earlier version of the manuscript and Sabine Braun for the concept for Fig. 5.8.

References

Aber, J. D., Nadelhoffer, K. J., Steudler, P., & Melillo, J. M. (1989). Nitrogen saturation in northern forest ecosystems. *BioScience, 39,* 378–386.

Aber, J. D., McDowell, W., Nadelhoffer, K., Magill, A., Berntsen, G., Kamakea, M., McNulty, S., Currie, W., Rustad, L., & Fernandez, I. (1998). Nitrogen saturation in temperate forest ecosystems: Hypotheses revisited. *BioScience, 48,* 921–934.

Aber, J. D., Goodale, C. L., Ollinger, S. V., Smith, M.-L., Magill, A. H., Martin, M. E., Hallett, R. A., & Stoddard, J. L. (2003). Is nitrogen deposition altering the nitrogen status of Northeastern forests? *BioScience, 53,* 375–389.

Adams, M. B., DeWalle, D. R., Peterjohn, W. T., Gilliam, F. S., Sharpe, W. E., & Williard, K. W. J. (2006). Soil chemical responses to experimental acidification treatments. In M. B. Adams, D. R. DeWalle, & J. L. Hom (Eds.), *The Fernow watershed acidification Study* (pp. 41–69). Dordrecht: Springer.

Aldous, A. R. (2002). Nitrogen retention by Sphagnum mosses: Responses to atmospheric nitrogen deposition and drought. *Canadian Journal of Botany, 80,* 721–731.
Allen, E. B., Temple, P. J., Bytnerowicz, A., Arbaugh, M. J., Sirulnik, A. G., & Rao, L. E. (2007). Patterns of understory diversity in mixed coniferous forests of southern California Impacted by air pollution. *The Scientific World Journal, 7,* 247–263.
Allen, E. B., Rao, L. E., Steers, R. J., Bytnerowicz, A., & Fenn, M. E. (2009). Impacts of atmospheric nitrogen deposition on vegetation and soils in Joshua Tree National Park. In R. H. Webb, L. F. Fenstermaker, J. S. Heaton, D. L. Hughson, E. V. McDonald, & D. M. Miller (Eds.), *The Mojave Desert: Ecosystem processes and sustainability* (pp. 78–100). Las Vegas: University of Nevada Press.
Amaranthus, M. P. (1998). The importance and conservation of ectomycorrhizal fungal diversity in forest ecosystems: Lessons from Europe and the Pacific Northwest. General Technical Report PNW QTR-431. Portland: U.S. Department of Agriculture, Forest Service.
Arens, S. J. T., Sullivan, P. F., & Welker, J. M. (2008). Nonlinear responses to nitrogen and strong interactions with nitrogen and phosphorus additions drastically alter the structure and function of a high Arctic ecosystem. *Journal of Geophysical Research, 113,* 1–10.
Baez, S., Fargione, J., Moore, D. I., Collins, S. L., & Gosz, J. R. (2007). Atmospheric nitrogen deposition in the northern Chihuahuan desert: Temporal trends and potential consequences. *Journal of Arid Environments, 68,* 640–651.
Baron, J. S. (2006). Hindcasting nitrogen deposition to determine ecological critical load. *Ecological Applications, 16,* 433–439.
Baron, J. S., Ojima, D. S., Holland, E. A., & Parton, W. J. (1994). Analysis of nitrogen saturation potential in Rocky Mountain tundra and forest: Implications for aquatic systems. *Biogeochemistry, 27,* 61–82.
Baron, J. S., Rueth, H. M., Wolfe, A. M., Nydick, K. R., Allstott, E. J., Minear, J. T., & Moraska, B. (2000). Ecosystem responses to nitrogen deposition in the Colorado Front Range. *Ecosystems, 3,* 352–368.
Baron, J. S., Driscoll, C. T., & Stoddard, J. L. (2011). Inland surface waters. In L. H. Pardo, M. J. Robin-Abbott, & C. T. Driscoll (Eds.), *Assessment of N deposition effects and empirical critical loads of nitrogen for ecoregions of the United States. General Technical Report NRS-80* (pp. 209–228). Newtown Square: U.S. Department of Agriculture, Forest Service, Northern Research Station.
Barret, J. E., & Burke, I. C. (2002). Nitrogen retention in semiarid ecosystems across a soil organic-matter gradient. *Ecological Applications, 12,* 878–890.
Berryman, S., & Straker, J. (2008). Nitrogen loading and terrestrial vegetation—Assessment of existing regional monitoring and recommendations. (Report prepared for the Cumulative Environmental Management Association NOx-SO2 Management Working Group and Eutrophication Task Group). Sidney: Submitted by CE Jones and Associates.
Berryman, S., Geiser, L., & Brenner, G. (2004). *Depositional gradients of atmospheric pollutants in the Athabasca Oil Sands region, Alberta, Canada: An analysis of lichen tissue and lichen communities*. (Final Report Submitted to the Terrestrial Environmental Effects Monitoring (TEEM) Science Sub-committee of the Wood Buffalo Environmental Association (WBEA)). Ft McMurray: Lichen Indicator Pilot Program 2002–2003.
Blett, T., Geiser, L., & Porter, E. (2003). Air pollution-related lichen monitoring in national parks, forests, and refuges: Guidelines of studies intended for regulatory and management purposes. (Technical Report NPS D2292). Denver, CO: USDI National Park Service Air Resources Division.
Bobbink, R., & Hettelingh, J.-P. (2011). *Review and revision of empirical critical loads and dose-response relationships: Proceedings of an expert workshop, Noordwijkerhout, 23–25 June 2010*. (Report 680359002/2011). Bilthoven: Coordination Centre for Effects, National Institute for Public Health and the Environment.
Bobbink, R., Ashmore, M., Braun, S., Flückiger, W., & van den Wyngaert, I. J. J. (2003). Empirical nitrogen critical loads for natural and semi-natural ecosystems: 2002 update. In B. Achermann

& R. Bobbink (Eds.), *Empirical critical loads for nitrogen* (pp. 43–170). Berne: Swiss Agency for Environment, Forest and Landscape SAEFL.

Boonpragob, K., Nash, T. H. III, & Fox, C. A. (1989). Seasonal deposition patterns of acidic ions and ammonium to the lichen Ramalina menziesii Tayl. in Southern California. *Environmental and Experimental Botany, 29,* 187–197.

Bowman, W. D., Gatner, J. R., Holland, K., & Wiedermann, M. (2006). Nitrogen critical loads for alpine vegetation and terrestrial ecosystem response: Are we there yet? *Ecological Applications, 16,* 1183–1193.

Breiner, J., Gimeno, B. S., & Fenn, M. (2007). Calculation of theoretical and empirical nutrient N critical loads in the mixed-conifer ecosystems of southern California. *The Scientific World Journal, 7,* 198–205.

Byun, D. W., & Ching, J. K. S. (1999). *Science algorithms of the EPA models-3 community multiscale air quality model (CMAQ) modeling system.* (EPA/600/R-99/030). Washington, DC: U.S. Environmental Protection Agency, Office of Research and Development.

Byun, D., & Schere, K. L. (2006). Review of the governing equations, computational algorithms, and other components of the Models-3 Community Multiscale Air Quality (CMAQ) modeling system. *Applied Mechanics Reviews, 59,* 51–77.

Caffrey, J. M., Murrell, M. C., Wigand, C., & McKinney, R. (2007). Effect of nutrient loading on biogeochemical and microbial processes in a New England salt marsh. *Biogeochemistry, 82,* 251–264.

Cape, J. N., van der Eerden, L. J., Sheppard, L. J., Leith, I. D., & Sutton, M. A. (2009). Reassessment of critical levels for atmospheric Ammonia. In M. Sutton, S. Reis, & S. M. H. Baker (Eds.), *Atmospheric Ammonia* (pp. 15–40). Dordrecht: Springer Science.

CEC, (Commission for Environmental Cooperation). (1997). *Ecological regions of North America: Toward a common perspective.* Montreal: Commission for Environmental Cooperation.

Clark, C. M., & Tilman, D. (2008). Loss of plant species after chronic low-level nitrogen deposition to prairie grasslands. *Nature, 451,* 712–715.

Clark, J. E., Hellgren, E. C., Jorgensen, E. E., Tunnell, S. J., Engle, D. M., & Leslie, D. M. (2003). Population dynamics of hispid cotton rats (Sigmodon hispidus) across a nitrogen-amended landscape. *Canadian Journal of Zoology, 81,* 994–1003.

Clark, J. E., Hellgren, E. C., Jorgensen, E. E., & Leslie, D. M. (2005). Population dynamics of harvest mice (Reithrodontomys fulvescens and R. montanus) across a nitrogen-amended old field. *American Midland Naturalist, 154,* 240–252.

Clark, C. M., Hobbie, S., Venterea, R., & Tilman, D. (2009). Long-lasting effects on N cycling 12 years after treatments cease despite minimal N retention. *Global Change Biology, 15,* 1755–1766.

Cornelissen, J. H. C., Lang, S. I., Soudzilovskaia, N. A., & During, H. J. (2007). Comparative cryptogam ecology: A review of bryophyte and lichen traits that drive biogeochemistry. *Annals of Botany, 99,* 987–1001.

Dighton, J., Tuininga, A. R., Gray, D. M., Huskins, R. E., & Belton, T. (2004). Impacts of atmospheric deposition on New Jersey pine barrens forest soils and communities of ectomycorrhizae. *Forest Ecology and Management, 201,* 131–144.

Driscoll, C. T., Lawrence, G. B., Bulger, A. J., Butler, T. J., Cronan, C. S., Eagar, C., Lambert, K. F., Likens, G. E., Stoddard, J. L., & Weathers, K. C. (2001). Acidic deposition in the northeastern United States: Sources and inputs, ecosystem effects, and management strategies. *BioScience, 51,* 180–198.

Driscoll, C. T., Whitall, D., Aber, J., Boyer, E., Castro, M., Cronan, C., Goodale, C. L., Groffman, P., Hopkinson, C., Lambert, K., Lawrence, G., & Ollinger, S. (2003). Nitrogen pollution in the northeastern United States: Sources, effects, and management options. *BioScience, 53,* 357–374.

Egerton-Warburton, L. M., & Allen, E. B. (2000). Shifts in arbuscular mycorrhizal communities along an anthropogenic nitrogen deposition gradient. *Ecological Applications, 10,* 484–496.

Egerton-Warburton, L. M., Graham, R. C., Allen, E. B., & Allen, M. F. (2001). Reconstruction of the historical changes in mycorrhizal fungal communities under anthropogenic nitrogen deposition. *Proceedings of the Royal Society of London, Series B, 268,* 2479–2484.

Elser, J. J., Andersen, T., Baron, J. S., Bergström, A. K., Jansson, M., Kyle, M., Nydick, K. R., Steger, L., & Hessen, D. O. (2009). Shifts in lake N:P stoichiometry and nutrient limitation driven by atmospheric nitrogen deposition. *Science, 326,* 835.

Epstein, H. E., Burke, I. C., & Mosier, A. R. (2001). Plant effects on nitrogen retention in shortgrass steppe 2 years after ^{15}N addition. *Oecologia, 128,* 422–430.

Falkengren-Grerup, U. (1995). Interspecies differences in the preference of ammonium and nitrate in vascular plants. *Oecologia, 102,* 305–311.

Fenn, M. E., & Geiser, L. H. (2011). Temperate Sierra. In L. H. Pardo, M. J. Robin-Abbott, & C. T. Driscoll (Eds.), *Assessment of nitrogen deposition effects and empirical critical loads of nitrogen for ecoregions of the United States. General Technical Report NRS-80* (pp. 175–180). Newtown Square: U.S. Department of Agriculture, Forest Service, Northern Research Station.

Fenn, M. E., & Poth, M. A. (1999). Temporal and spatial trends in streamwater nitrate concentrations in the San Bernardino Mountains, southern California. *Journal of Environmental Quality, 28,* 822–836.

Fenn, M. E., de Bauer, L. I., Quevedo-Nolasco, A., & Rodriguez-Frausto, C. (1999). Nitrogen and sulfur deposition and forest nutrient status in the Valley of Mexico. *Water, Air, & Soil Pollution, 113,* 155–174.

Fenn, M. E., de Bauer, L. I., Zeller, K., Quevedo, A., Rodríguez, C., & Hernández-Tejeda, T. (2002). Nitrogen and sulfur deposition in the Mexico City Air Basin: Impacts on forest nutrient status and nitrate levels in drainage waters. In M. E. Fenn, L. I. de & T. Hernández-Tejeda (Eds.), *Urban air pollution and forests: Resources at risk in the Mexico City air basin. Ecological studies series,* Vol. 156 (pp. 298–319). New York: Springer-Verlag.

Fenn, M. E., Haeuber, R., Tonnesen, G. S., Baron, J. S., Grossman-Clarke, S., Hope, D., Jaffe, D. A., Copeland, S., Geiser, L., Rueth, H. M., & Sickman, J. O. (2003a). Nitrogen emissions, deposition, and monitoring in the western United States. *BioScience, 53,* 391–403.

Fenn, M. E., Baron, J. S., Allen, E. B., Rueth, H. M., Nydick, K. R., Geiser, L., Bowman, W. D., Sickman, J. O., Meixner, T., Johnson, D. W., & Neitlich, P. (2003b). Ecological effects of nitrogen deposition in the western United States. *BioScience, 53,* 404–420.

Fenn, M. E., Poth, M. A., Bytnerowicz, A., Sickman, J. O., & Takemoto, B. K. (2003c). Effects of ozone, nitrogen deposition, and other stressors on montane ecosystems in the Sierra Nevada. In A. Bytnerowicz, M. J. Arbaugh, & R. Alonso (Eds.), *Developments in environmental science, volume 2: Ozone air pollution in the Sierra Nevada: Distribution and effects on forests* (pp. 111–155). Amsterdam: Elsevier.

Fenn, M. E., Jovan, S., Yuan, F., Geiser, L., Meixner, T., & Gimeno, B. S. (2008). Empirical and simulated critical loads for nitrogen deposition in California mixed conifer forests. *Environmental Pollution, 155,* 492–511.

Fenn, M. E., Sickman, J. O., Bytnerowicz, A., Clow, D. W., Molotch, N. P., Pleim, J. E., Tonnesen, G. S., Weathers, K. C., Padgett, P. E., & Campbell, D. H. (2009). Methods for measuring atmospheric nitrogen deposition inputs in arid and montane ecosystems of western North America. In A. H. Legge (Ed.), *Developments in environmental science, Vol. 9: Air quality and ecological impacts: Relating sources to effects* (pp. 179–228). Amsterdam: Elsevier.

Fenn, M. E., Allen, E. B., Weiss, S. B., Jovan, S., Geiser, L., Tonnesen, G. S., Johnson, R. F., Rao, L. E., Gimeno, B. S., Yuan, F., Meixner, T., & Bytnerowicz, A. (2010). Nitrogen critical loads and management alternatives for N-impacted ecosystems in California. *Journal of Environmental Management, 91,* 2404–2423.

Fenn, M. E., Allen, E. B., & Geiser, L. H. (2011). Mediterranean California. In L. H. Pardo, M. J. Robin-Abbott, & C. T. Driscoll (Eds.), *Assessment of N deposition effects and empirical critical loads of nitrogen for ecoregions of the United States.* General Technical Report NRS-80 (pp. 143–170). Newtown Square: U.S. Department of Agriculture, Forest Service, Northern Research Station.

Galloway, J. N. (1998). The global nitrogen cycle: Changes and consequences. *Environmental Pollution, 102,* 15–24.

Galloway, J. N., Aber, J. D., Erisman, J. W., Seitzinger, S. P., Howarth, R. W., Cowling, E. B., & Cosby, B. J. (2003). The Nitrogen Cascade. *BioScience, 53,* 341–356.

Galloway, J. N., Dentener, F. J., Capone, D. G., Boyer, E. W., Howarth, R. W., Seitzinger, S. P., Asner, G. P., Cleveland, C. C., Green, P. A., Holland, E. A., Karl, D. M., Michaels, A. F., Porter, J. H., Townsend, A. R., & Vörösmarty, C. J. (2004). Nitrogen cycles: Past, present and future. *Biogeochemistry, 70,* 153–226.

Geiser, L. H., & Neitlich, P. N. (2007). Air pollution and climate gradients in western Oregon and Washington indicated by epiphytic macrolichens. *Environmental Pollution, 145,* 203–218.

Geiser, L. H., Ingersoll, A. R., Bytnerowicz, A., & Copeland, S. A. (2008). Evidence of enhanced atmospheric ammoniacal nitrogen in Hells Canyon NRA: Implications for natural and cultural resources. *Journal of the Air & Waste Management Association, 58,* 1223–1234.

Geiser, L. H., Jovan, S. E., Glavich, D. A., & Porter, M. (2010). Lichen-based critical loads for nitrogen deposition in western Oregon and Washington Forests, USA. *Environmental Pollution, 158,* 2412–2421.

Gilliam, F. S. (2006). Response of the herbaceous layer of forest ecosystems to excess nitrogen deposition. *Journal of Ecology, 94,* 1176–1191.

Gilliam, F. S. (2007). The ecological significance of the herbaceous layer in forest ecosystems. *BioScience, 57,* 845–858.

Gilliam, F. S., Adams, M. B., & Yurish, B. M. (1996). Ecosystem nutrient responses to chronic nitrogen inputs at Fernow Experimental Forest, West Virginia. *Canadian Journal of Forest Research, 26,* 196–205.

Gilliam, F. S., Hockenberry, A. W., & Adams, M. B. (2006). Effects of atmospheric nitrogen deposition on the herbaceous layer of a central Appalachian hardwood forest. *The Journal of the Torrey Botanical Society, 133,* 240–254.

Glavich, D. A., & Geiser, L. H. (2008). Potential approaches to developing lichen-based critical loads and levels for nitrogen, sulfur and metal-containing atmospheric pollutants in North America. *The Bryologist, 111,* 638–649.

Gotelli, N. J., & Ellison, A. M. (2002). Nitrogen deposition and extinction risk in the northern pitcher plant, Sarracenia purpurea. *Ecology, 83,* 2758–2765.

Gotelli, N. J., & Ellison, A. M. (2006). Forecasting extinction risk with nonstationary matrix models. *Ecological Applications, 16,* 51–61.

Greaver, T., Liu, L., & Bobbink, R. (2011). Wetlands. In L. H. Pardo, M. J. Robin-Abbott, & C. T. Driscoll (Eds.), *Assessment of N deposition effects and empirical critical loads of nitrogen for ecoregions of the United States*. General Technical Report NRS-80 (pp. 193–208). Newtown Square: U.S. Department of Agriculture, Forest Service, Northern Research Station.

Grulke, N. E., & Balduman, L. (1999). Deciduous conifers: High N deposition and O_3 exposure effects on growth and biomass allocation in ponderosa pine. *Water, Air, & Soil Pollution, 116,* 235–248.

Grulke, N. E., Andersen, C. P., Fenn, M. E., & Miller, P. R. (1998). Ozone exposure and nitrogen deposition lowers root biomass of ponderosa pine in the San Bernardino Mountains, California. *Environmental Pollution, 103,* 63–73.

Grulke, N. E., Minnich, R. A., Paine, T. D., Seybold, S. J., Chavez, D. J., Fenn, M. E., Riggan, P. J., & Dunn, A. (2009). Air pollution increases forest susceptibility to wildfires: A case study in the San Bernardino Mountains in southern California. In A. Bytnerowicz, M. J. Arbaugh, A. R. Riebau, & C. Andersen (Eds.), *Wildland fires and air pollution. Developments in environmental science*, Vol. 8 (pp. 365–403). Amsterdam: Elsevier.

Hurd, T. M., Brach, A. R., & Raynal, D. J. (1998). Response of understory vegetation of Adirondack forests to nitrogen additions. *Canadian Journal of Forest Research, 28,* 799–807.

Hyvärinen, M., Walter, B., & Koopmann, R. (2003). Impact of fertilisation on phenol content and growth rate of Cladina stellaris: A test of the carbon-nutrient balance hypothesis. *Oecologia, 134,* 176–181.

Inouye, R. S. (2006). Effects of shrub removal and nitrogen addition on soil moisture in sagebrush steppe. *Journal of Arid Environments, 65,* 604–618.

Jones, M. E., Paine, T. D., Fenn, M. E., & Poth, M. A. (2004). Influence of ozone and nitrogen deposition on bark beetle activity under drought conditions. *Forest Ecology and Management, 200,* 67–76.

Jorgensen, E. E., Holub, S. M., Mayer, P. M., Gonsoulin, M. E., Silva, R. G., West, A. E., Tunnell, S. J., Clark, J. E., Parsons, J. L., Engle, D. M., Hellgren, E. C., Spears, J. D. H., Butler, C. E., & Leslie, D. M. Jr. (2005). *Ecosystem stress from chronic exposure to low levels of nitrate.* (EPA/600/R-05/087). Cincinnati: U.S. Environmental Protection Agency, National Risk Management Research Laboratory.

Jovan, S. (2008). *Lichen bioindication of biodiversity, air quality, and climate: Baseline results from monitoring in Washington, Oregon, and California. (Gen. Tech. Rep. PNW-GTR-737).* Portland: U.S. Department of Agriculture, Forest Service, Pacific Northwest Research Station.

Jovan, S., & McCune, B. (2005). Air quality bioindication in the greater Central Valley of California with epiphytic macrolichen communities. *Ecological Applications, 15,* 1712–1726.

Jovan, S., Riddell, J., Padgett, P., & Nash, T. (2012). Eutrophic lichens respond to multiple forms of N: Implications for critical levels and critical loads research. *Ecological Applications, 22,* 1910–1922.

Kelly, V. R., Lovett, G. M., Weathers, K. C., & Likens, G. E. (2005). Trends in atmospheric ammonium concentrations in relation to atmospheric sulfate and local agriculture. *Environmental Pollution, 135,* 363–369.

Kleijn, D., Bekker, R. M., Bobbink, R., de Graaf, M. C. C., & Roelofs, J. G. M. (2008). In search for key biogeochemical factors affecting plant species persistence in heathland and acidic grasslands: A comparison of common and rare species. *Journal of Applied Ecology, 45,* 680–687.

Knops, J. M. H., Nash, T. H. III, Boucher, V. L., & Schlesinger, W. H. (1991). Mineral cycling and Epiphytic Lichens—Implications at the ecosystem level. *The Lichenologist, 23,* 309–321.

Krupa, S. V. (2003). Effects of atmospheric ammonia (NH_3) on terrestrial vegetation: A review. *Environmental Pollution, 124,* 179–221.

Kytöviita, M.-M., & Crittenden, P. D. (2007). Growth and nitrogen relations in the mat-forming lichens Stereocaulon paschale and Cladonia stellaris. *Annals of Botany, 100,* 1537–1545.

Lafrancois, B. M., Nydick, K. R., Johnson, B. M., & Baron, J. S. (2004). Cumulative effects of nutrients and pH on the plankton of two mountain lakes. *Canadian Journal of Fisheries and Aquatic Sciences, 61,* 1153–1165.

Latimer, J. S., & Rego, S. A. (2010). Empirical relationship between eelgrass extent and predicted watershed-derived nitrogen loading for shallow New England estuaries. *Estuarine Coastal and Shelf Science, 90,* 231–240.

LeBauer, D. S., & Treseder, K. K. (2008). Nitrogen limitation of net primary productivity in terrestrial ecosystems is globally distributed. *Ecology, 89,* 371–379.

Lehmann, C. M. B., Bowersox, V. C., & Larson, S. M. (2005). Spatial and temporal trends of precipitation chemistry in the United States, 1985–2002. *Environmental Pollution, 135,* 347–361.

Lilleskov, E. A. (1999). *Decline of above- and belowground ectomycorrhizal fungal diversity over an atmospheric nitrogen deposition gradient near Kenai, Alaska.* Ithaca: Cornell University.

Lilleskov, E. A. (2005). How do composition, structure, and function of mycorrhizal fungal communities respond to nitrogen deposition and ozone exposure? In J. Dighton, J. F. White, & P. Oudemans (Eds.), *The fungal community: Its organization and role in the ecosystem* (pp. 769–801). Boca Raton: Taylor & Francis.

Lilleskov, E. A., Fahey, T. J., & Lovett, G. M. (2001). Ectomycorrhizal fungal aboveground community change over an atmospheric nitrogen deposition gradient. *Ecological Applications, 11,* 397–410.

Lilleskov, E. A., Fahey, T. J., Horton, T. R., & Lovett, G. M. (2002). Belowground ectomycorrhizal fungal community change over a nitrogen deposition gradient in Alaska. *Ecology, 83,* 104–115.
Lilleskov, E. A., Wargo, P. M., Vogt, K. A., & Vogt, D. J. (2008). Mycorrhizal fungal community relationship to root nitrogen concentration over a regional atmospheric nitrogen deposition gradient in the northeastern US. *Canadian Journal of Forest Research, 38,* 1260–1266.
Liu, L., & Greaver, T. L. (2009). A review of nitrogen enrichment effects on three biogenic GHGs: The CO_2 sink may be largely offset by stimulated N_2O and CH_4 emission. *Ecology Letters, 12,* 1103–1117.
Makkonen, S., Hurri, R. S. K., & Hyvarinen, M. (2007). Differential responses of lichen symbionts to enhanced nitrogen and phosphorus availability: An experiment with Cladina stellaris. *Annals of Botany, 99,* 877–884.
McCune, B. M., & Geiser, L. H. (2009). *Macrolichens of the Pacific Northwest. 2nd Edition*. Corvallis: Oregon State University Press.
McCune, B., Grenon, J., Mutch, L. S., & Martin, E. P. (2007). Lichens in relation to management issues in the Sierra Nevada national parks. *Pacific Northwest Fungi, 2,* 1–39.
McKane, R. B., Johnson, L. C., Shaver, G. R., Nadelhoffer, K. J., Rastetter, E. B., Fry, B., Giblin, A. E., Kielland, K., Kwiatkowski, B. L., Laundre, J. A., & Murray, G. (2002). Resource-based niches provide a basis for plant species diversity and dominance in arctic tundra. *Nature, 415,* 68–71.
McNulty, S. G., & Boggs, J. L. (2010). A conceptual framework: Redefining forest soil's critical acid loads under a changing climate. *Environmental Pollution, 158,* 2053–2058.
McNulty, S., Boggs, J., Aber, J. D., Rustad, L., & Magill, A. (2005). Red spruce ecosystem level changes following 14 years of chronic N fertilization. *Forest Ecology and Management, 219,* 279–291.
Meixner, T., & Fenn, M. (2004). Biogeochemical budgets in a Mediterranean catchment with high rates of atmospheric N deposition-importance of scale and temporal asynchrony. *Biogeochemistry, 70,* 331–356.
Michel, T. J., Saros, J. E., Interlandi, S. J., & Wolfe, A. P. (2006). Resource requirements of four freshwater diatom taxa determined by in situ growth bioassays using natural populations from alpine lakes. *Hydrobiologia, 568,* 235–243.
Miller, A. E., & Bowman, W. D. (2002). Variation in nitrogen-15 natural abundance and nitrogen uptake traits among co-occurring alpine species: Do species partition nitrogen form? *Oecologia, 130,* 609–616.
Mitchell, R. J., Truscot, A. M., Leith, I. D., Cape, J. N., van Dijk, N., Tang, Y. S., Fowler, D., & Sutton, M. A. (2005). A study of the epiphytic communities of Atlantic oak woods along an atmospheric nitrogen deposition gradient. *Journal of Ecology, 93,* 482–492.
Moore, T., Blodau, C., Turunen, J., Roulet, N., & Richard, P. J. H. (2004). Patterns of nitrogen and sulfur accumulation and retention in ombrotrophic bogs, eastern Canada. *Global Change Biology, 11,* 256–367.
Nilles, M. A., & Conley, B. E. (2001). Changes in the chemistry of precipitation in the United States, 1981–1998. *Water, Air, & Soil Pollution, 130,* 409–414.
Nordin, A., Strengbom, J., Witzell, J., Näsholm, T., & Ericson, L. (2005). Nitrogen deposition and the biodiversity of boreal forests: Implications for the nitrogen critical load. *AMBIO A Journal of the Human Environment, 34,* 20–24.
Nordin, A., Strengbom, J., & Ericson, L. (2006). Responses to ammonium and nitrate additions by boreal plants and their natural enemies. *Environmental Pollution, 141,* 167–174.
Nydick, K. R., Lafrancois, B. M., Baron, J. S., & Johnson, B. M. (2004). Nitrogen regulation of algal biomass, productivity, and composition in shallow mountain lakes, Snowy Range, Wyoming, USA. *Canadian Journal of Fisheries and Aquatic Sciences, 61,* 1256–1268.
Pardo, L. H., Robin-Abbott, M. J., & Driscoll, C. T. (2011a). *Assessment of N deposition effects and empirical critical loads of nitrogen for ecoregions of the United States*. (General Technical Report NRS-80). Newtown Square: U.S. Department of Agriculture, Forest Service, Northern Research Station.

Pardo, L. H., Lilleskov, E. A., Geiser, L. H., & Robin-Abbott, M. J. (2011b). Detailed description of methods for estimating empirical critical loads of nitrogen. *Ecological Archives*, A021–137–A1. http://esapubs.org/archive/appl/A021/137/appendix-A.htm. Accessed 26 Nov 2014.

Pardo, L. H., Fenn, M., Goodale, C. L., Geiser, L. H., Driscoll, C. T., Allen, E., Baron, J., Bobbink, R., Bowman, W. D., Clark, C., Emmett, B., Gilliam, F. S., Greaver, T., Hall, S. J., Lilleskov, E. A., Liu, L., Lynch, J., Nadelhoffer, K., Perakis, S., Robin-Abbott, M. J., Stoddard, J., Weathers, K., & Dennis, R. L. (2011c). Effects of nitrogen deposition and empirical nitrogen critical loads for ecoregions of the United States. *Ecological Applications, 21,* 3049–3082.

Pardo, L. H., Lilleskov, E. A., Geiser, L. H., & Robin-Abbott, M. J. (2011d). Methods. In L. H. Pardo, M. J. Robin-Abbott, & C. T. Driscoll (Eds.), *Assessment of N deposition effects and empirical critical loads of nitrogen for ecoregions of the United States. General Technical Report NRS-80* (pp. 25–36). Newtown Square: U.S. Department of Agriculture, Forest Service, Northern Research Station.

Poikolainen, J., Lippo, H., Hongisto, M., Kubin, E., Mikkola, K., Lindgren, M., van der Hoek, K. W., Erisman, J. W., Smeulders, S., & Wisniewski, J. R. (1998). On the abundance of epiphytic green algae in relation to the nitrogen concentrations of biomonitors and nitrogen deposition in Finland. *Environmental Pollution, 102,* 85–92.

Porter, M. K. (2007). Regional modeling of nitrogen, sulfur and mercury atmospheric deposition in the Pacific Northwest. Pullman, WA, USA, Washington State University.

Pypker, T. G. (2004). Influence of canopy structure and epiphytes on the hydrology of Douglas-fir forests. Corvallis: Oregon State University, Ph. D. Dissertation, 180 pp.

Rabalais, N. N. (2002). Nitrogen in aquatic ecosystems. *AMBIO A Journal of the Human Environment, 31,* 102–112.

Rao, L. E., & Allen, E. B. (2010). Combined effects of precipitation and nitrogen deposition on native and invasive winter annual production in California deserts. *Oecologia, 62,* 1035–1046.

Rao, L. E., Allen, E. B., & Meixner, T. (2010). Risk-based determination of critical nitrogen deposition loads for fire spread in southern California deserts. *Ecological Applications, 20,* 1320–1335.

Rochefort, L., Vitt, D. H., & Bayley, S. E. (1990). Growth, production and decomposition dynamics of Sphagnum under natural and experimentally acidified conditions. *Ecology, 71,* 1986–2000.

Rueth, H. M., & Baron, J. S. (2002). Differences in Englemann spruce forest biogeochemistry east and west of the Continental Divide in Colorado, USA. *Ecosystems, 5,* 45–57.

Rueth, H. M., Baron, J. S., & Allstott, E. J. (2003). Responses of Engelmann spruce forests to nitrogen fertilization in the Colorado Rocky Mountains. *Ecological Applications, 13,* 664–673.

Saros, J. E., Michel, T. J., Interlandi, S. J., & Wolfe, A. P. (2005). Resource requirements of Asterionella Formosa and Fragilaria crotonsis in oligotrophic alpine lakes: Implications for recent phytoplankton community reorganisations. *Canadian Journal of Fisheries and Aquatic Sciences, 62,* 1681–1689.

Sickman, J. O., Melack, J. M., & Stoddard, J. L. (2002). Regional analysis of inorganic nitrogen yield and retention in high-elevation ecosystems of the Sierra Nevada and Rocky Mountains. *Biogeochemistry, 57,* 341–374.

Stevens, C. J., Dise, N. B., Mountford, J. O., & Gowing, D. J. (2004). Impact of nitrogen deposition on the species richness of grasslands. *Science, 303,* 1876–1879.

Strengbom, J., Walheim, M., Näsholm, T., & Ericson, L. (2003). Regional differences in the occurrence of understorey species reflect nitrogen deposition in Swedish forests. *AMBIO A Journal of the Human Environment, 32,* 91–97.

Suding, K. N., Gross, K. L., & Houseman, G. R. (2004). Alternative states and positive feedbacks in restoration ecology. *Trends in Ecology and Evolution, 19,* 46–53.

Sutton, M. A., Reis, S. & Baker, S. M. H. (Eds.). (2009). *Atmospheric ammonia—Detecting emission changes and environmental impacts*. Springer Science + Business Media B.V. Dordrecht, Netherlands.

Thomas, R. Q., Canham, C. D., Weathers, K. C., & Goodale, C. L. (2010). Increased tree carbon storage in response to nitrogen deposition in the US. *Nature Geoscience, 3,* 13–17.

Tilman, D. (1987). Secondary succession and the pattern of plant dominance along experimental nitrogen gradients. *Ecological Monographs, 57,* 189–214.

Tilman, D. (1993). Species richness of experimental productivity gradients: How important is colonization limitation. *Ecology, 74,* 2179–2191.

Tonnesen, G., Wang, Z., Omary, M., & Chien, C. J. (2007). *Assessment of Nitrogen Deposition: Modeling and Habitat Assessment.* (CEC-500–2005-032. http://www.energy.ca.gov/2006publications/CEC-500–2006-032/CEC-500–2006-032.PDF). California Energy Commission, PIER Energy-Related Environmental Research.

U.S. EPA, (Environmental Protection Agency). (1993). *Air quality criteria for oxides of nitrogen, volume II.* (Rep. No. EPA/600/8-91/049bf). Research Triangle Park, NC: Office of Health and Environmental Assessment, Environmental Criteria and Assessment Office.

U.S. EPA, (Environmental Protection Agency). (2008). *Integrated Science Assessment (ISA) for Oxides of Nitrogen and Sulfur-Ecological Criteria (Final Report).* (EPA/600/R- 08/082F). Research Triangle Park, NC: U.S. Environmental Protection Agency, National Center for Environmental Assessment–RTP Division, Office of Research and Development. http://cfpub.epa.gov/ncea/cfm/recordisplay.cfm?deid=201485. Accessed 18 Jan 2010.

USDA NRCS, (U.S. Department of Agriculture, Natural Resources Conservation Service). (2009). *The PLANTS database (http://plants.usda.gov, 11 December 2009).* Baton Rouge, LA 70874-4490: National Plant Data Center.

USDI FWS, (U.S. Department of the Interior, Fish and Wildlife Service). (2005). *Status and trends of wetlands in the conterminous United States 1998–2004.* Washington, DC: U.S. Department of Interior, Fish and Wildlife Service.

USFS, (U.S. Forest Service). (2001). *U.S. forest facts and historical trends.* (USDA Forest Service FS-696). Washington, DC: U.S. Department of Agriculture, Forest Service.

Van Diepen, L. T. A. (2008). The role and diversity of arbuscular mycorrhizal fungi in Acer saccharum dominated forest ecosystems under natural and N-amended conditions. Houghton: Michigan Technological University, Ph. D. Dissertation, 99 pp.

Van Diepen, L. T. A., Lilleskov, E. A., Pregitzer, K. S., & Miller, R. M. (2007). Decline of arbuscular mycorrhizal fungi in northern hardwood forests exposed to chronic nitrogen additions. *The New Phytologist, 176,* 175–183.

Vitt, D. H., Wieder, K., Halsey, L. A., & Turetsky, M. (2003). Response of Sphagnum fuscum to Nitrogen Deposition: A case study of ombrogenous peatlands in Alberta, Canada. *The Bryologist, 1062,* 235–245.

Wallander, H. (1995). A new hypothesis to explain allocation of dry matter between mycorrhizal fungi and pine seedlings in relation to nutrient supply. *Plant and Soil, 169,* 243–248.

Weathers, K., & Lynch, J. A. (2011). Deposition. In L. H. Pardo, M. J. Robin-Abbott, & C. T. Driscoll (Eds.), *Assessment of N deposition effects and empirical critical loads of nitrogen for ecoregions of the United States. General Technical Report NRS-80* (pp. 15–24). Newtown Square: U.S. Department of Agriculture, Forest Service, Northern Research Station.

Wedin, D. A., & Tilman, D. (1996). Influence of nitrogen loading and species composition on the carbon balance of grasslands. *Science, 274,* 1720–1723.

Weiss, S. B. (1999). Cars, cows, and checkerspot butterflies: Nitrogen deposition and management of nutrient-poor grasslands for a threatened species. *Conservation Biology, 13,* 1476–1486.

Whytemare, A. B., Edmonds, R. L., Aber, J. D., & Lajtha, K. (1997). Influence of excess nitrogen deposition on a white spruce (Picea glauca) stand in southern Alaska. *Biogeochemistry, 38,* 173–187.

Wigand, C., McKinney, R. A., Charpentier, M. A., Chintala, M. M., & Thursby, G. B. (2003). Relationships of nitrogen loadings, residential development, and physical characteristics with plant structure in New England salt marshes. *Estuaries, 26,* 1494–1504.

Williams, M. W., & Tonnessen, K. A. (2000). Critical loads for inorganic nitrogen deposition in the Colorado Front Range, USA. *Ecological Applications, 10,* 1648–1665.

Wolfe, A. P., Baron, J. S., & Cornett, R. J. (2001). Anthropogenic nitrogen deposition induces rapid ecological changes in alpine lakes of the Colorado Front Range (USA). *Journal of Paleolimnology, 25,* 1–7.

Wolfe, A. P., van Gorp, A. C., & Baron, J. S. (2003). Recent ecological and biogeochemical changes in alpine lakes of Rocky Mountain National Park (Colorado, USA): A response to anthropogenic nitrogen deposition. *Geobiology, 1,* 153–168.
Wu, W., & Driscoll, C. T. (2010). Impact of climate change on three-dimensional dynamic critical load functions. *Environmental Science and Technology, 44,* 720–726.
Yoshida, L. C., & Allen, E. B. (2004). ^{15}N uptake by mycorrhizal Artemisia californica and the invasive Bromus madritensis of a N-eutrophied shrubland. *Biology and Fertility of Soil, 39,* 243–248.

Chapter 6
Mass Balance Models to Derive Critical Loads of Nitrogen and Acidity for Terrestrial and Aquatic Ecosystems

Maximilian Posch, Wim de Vries and Harald U. Sverdrup

6.1 Introduction

The purpose of a model-based approach to calculating critical loads is to link, via mathematical equations, a chemical criterion (critical limit) with the maximum deposition(s) 'below which significant harmful effects on specified sensitive elements of the environment do not occur', i.e. for which the criterion is not violated. The quotations are from the definition of a critical load (Nilsson and Grennfelt 1988; see also Chap. 1). In most cases the 'sensitive element of the environment' will be of a biological nature (e.g., the vitality of a tree, a fish species in a lake), and thus the criterion should be a biological one. However, there is a dearth of simple yet reliable models that adequately describe the whole chain from deposition to biological impact. Therefore, chemical criteria are used instead, and physico-chemical relationships are used to derive critical loads. This simplifies the modelling process, but shifts the burden to find, or derive, appropriate chemical criteria (and critical limits) with established relationships to biological effects (see Chap. 2). The choice of the critical limit is an important step in deriving a critical load, and much of the uncertainty in critical load calculations stems from the uncertainty in the link between (soil or water) chemistry and biological impact.

The definition of a critical load does not include any time aspect. Obviously, the 'harmful effects' should not only be avoided in a particular year, but 'forever'. Thus, a critical load is a time-independent quantity, and this simplifies its modelling, since only steady-state models need to be considered. On the other hand—in contrast to

M. Posch (✉)
Coordination Centre for Effects (CCE), RIVM, Bilthoven, The Netherlands
e-mail: max.posch@rivm.nl

W. de Vries
Alterra Wageningen University and Research Centre, Wageningen, The Netherlands

H. U. Sverdrup
Department of Chemical Engineering, Lund University, Lund, Sweden

W. de Vries et al. (eds.), *Critical Loads and Dynamic Risk Assessments*,
Environmental Pollution 25, DOI 10.1007/978-94-017-9508-1_6

'classical' cause-effect modelling, where 'effect' is the explained variable—critical load models require the 'inverse', i.e. 'cause' becomes the output. A critical load model computes the deposition(s), called critical load(s), which lead to a prescribed critical limit at a given site. This becomes more involved if the processes considered in the critical load model depend themselves on the deposition.

Here we concentrate on the so-called Simple Mass Balance (SMB) model for soils (Sverdrup et al. 1990; Sverdrup and De Vries 1994) and the First-order Acidity Balance (FAB) model for surface waters (Henriksen and Posch 2001; Posch et al. 2012). Both models are single-layer models, i.e. the soil (and the water body) is treated as a single homogeneous compartment. Furthermore, it is assumed that the soil depth is (at least) the depth of the rooting zone, which allows neglecting the nutrient cycle and to deal with net growth uptake (harvested biomass) only. Additional simplifying assumptions include:

- all evapotranspiration occurs on the top of the soil profile,
- percolation is constant through the soil profile and occurs only vertically,
- physico-chemical constants are assumed uniform throughout the soil profile.

Since the critical load models describe steady-state conditions, they require long-term averages for input fluxes. Short-term variations—e.g., episodic, seasonal, inter-annual, due to harvest and as a result of short-term natural perturbations—are not considered, but are assumed to be included in the calculation of long-term means. In this context 'long-term' is defined as about 100 years, i.e. at least one rotation period for forests. Ecosystem interactions and processes like competition, pests, herbivore influences etc. are not considered in the models. Although the models are formulated for undisturbed (semi-natural) ecosystems, the effects of extensive management, such as grazing and fires, could be included.

6.2 Critical Loads of Nutrient Nitrogen for Soils

6.2.1 The SMB Model for Critical Loads of Nutrient Nitrogen

Elevated inputs of N can lead to undesirable changes in vegetation and/or to increased amounts of N in the leachate that can cause problems to ground and surface waters (see Chap. 2). The starting point for calculating critical loads of nutrient N with the SMB model is the mass balance of total N for the soil compartment under consideration (inputs = sinks + outputs):

$$N_{dep} + N_{fix} = N_{ad} + N_i + N_u + N_{de} + N_{eros} + N_{fire} + N_{vol} + N_{le} \qquad (6.1)$$

where:

N_{dep}	total N deposition
N_{fix}	N 'input' by biological fixation
N_{ad}	N adsorption
N_i	long-term net immobilisation of N in soil organic matter
N_u	net removal of N in harvested vegetation and animals
N_{de}	flux of N to the atmosphere due to denitrification
N_{eros}	N losses through erosion
N_{fire}	N losses in smoke due to (wild or controlled) fires
N_{vol}	N losses to the atmosphere via NH_3 volatilisation
N_{le}	leaching of N below the root zone

The units used are eq $ha^{-1}yr^{-1}$. The following assumptions are frequently made:

- N adsorption, e.g., the adsorption of NH_4 by clay minerals, can temporarily lead to an accumulation of N in the soil; however it is stored/released only when the deposition *changes*, and can thus be neglected under steady-state considerations;
- N fixation is negligible in most (forest) ecosystems, except for N-fixing species;
- The loss of N due to fire, erosion and volatilisation is small for many ecosystems and therefore neglected in the following discussion;
- The leaching of ammonium (NH_4) can be neglected in forest ecosystems due to its (preferential) uptake and complete nitrification within the root zone (i.e. $NH4_{,le}=0$, $N_{le}=NO3_{,le}$).

Under these simplifying assumptions Eq. 6.1 becomes:

$$N_{dep} = N_i + N_u + N_{de} + N_{le} \tag{6.2}$$

From this equation a critical load is obtained by defining a critical or acceptable limit to the leaching of N, $N_{le,acc}$, the choice of this limit depending on the 'sensitive element of the environment' to be protected. Inserting an acceptable leaching into Eq. 6.2, the deposition of N becomes the critical load of nutrient nitrogen, $CL_{nut}N$:

$$CL_{nut}N = N_i + N_u + N_{de} + N_{le,acc} \tag{6.3}$$

In this derivation of a critical load it is assumed that the sources and sinks of N do not depend on the deposition of N. This is unlikely to be the case, and thus all quantities should be taken 'at critical load'. However, to compute, e.g., 'denitrification at critical load' one needs to know the critical load, the very quantity one wants to compute. The only clean way to avoid this circular reasoning is to use a functional relationship between deposition and the sink of N, insert this function into Eq. 6.2 and solve for the deposition (to obtain the critical load). This has been done for denitrification: In the simplest case, denitrification is linearly related to the net input of N (De Vries et al. 1994):

$$N_{de} = f_{de} \cdot \left(N_{dep} - N_i - N_u\right)_+ \tag{6.4}$$

where $(x)_+$ or x_+ is a short-hand notation for max{x,0}, i.e. $x_+=x$ for $x>0$ and $x_+=0$ for $x\leq0$; and the denitrification fraction f_{de} $(0\leq f_{de}<1)$ is a site-specific quantity. Inserting Eq. 6.4 into Eq. 6.2 and solving for the deposition leads to the following expression for the critical load of nutrient N:

$$CL_{nut}N = N_i + N_u + \frac{N_{le,acc}}{1-f_{de}} \tag{6.5}$$

Based on data in Solberg et al. (2009), Posch et al. (2011) derived critical load expressions with net growth uptake as the sum of a constant term and a term proportional to N deposition. Ideally, all N fluxes depending on N deposition should be modelled as such and used in deriving a critical load. However, these might be so involved that no (simple) explicit expression for $CL_{nut}N$ can be obtained. Although this does not matter in principle, it would reduce the appeal and widespread use of critical loads. Therefore, when calculating critical loads from Eq. 6.3 or 6.5, the N fluxes are generally estimated as long-term averages derived from conditions not influenced by elevated anthropogenic N inputs.

6.2.2 Critical/Acceptable N Leaching

The value for the acceptable N leaching, $N_{le,acc}$, depends on the 'harmful effect' that should be avoided. In general, it is not the N leaching flux itself that is 'harmful', but the concentration of N in the leaching flux or a concentration in another compartment (e.g. tree foliage) that is related to the dissolved N concentration. This is related to the acceptable N leaching (in eq $ha^{-1}yr^{-1}$) by:

$$N_{le,acc} = Q \cdot [N]_{acc} \tag{6.6}$$

where $[N]_{acc}$ is the acceptable dissolved N concentration (eq m^{-3}) and Q is the precipitation surplus (in m^3 $ha^{-1}yr^{-1}$). Values for acceptable N concentrations are discussed in Chap. 2.

6.2.3 Sources and Derivation of Input Data

The obvious sources of input data for calculating critical loads are measurements at the site under consideration. However, in many cases these will not be available. A discussion on N sources and sinks can be found in De Vries (1993), Hornung et al. (1995) and UNECE (1995). Some data sources, default values and procedures to derive critical loads of N are summarised below.

Nitrogen Immobilisation: N_i refers to the long-term net immobilisation (accumulation) of N in the root zone, i.e., the continuous build-up of stable C-N-compounds in (forest) soils. In other words, this immobilisation of N should not lead to significant changes in the prevailing C/N ratio. This has to be distinguished from the high amounts of N accumulated in the soils over many years (decades) due to the increased deposition of N, leading to a decrease in the C/N ratio in the (top)soil.

Using data from Swedish forest soil plots, Rosén et al. (1992) estimated the average annual N immobilisation since the last glaciation at 0.2–0.5 kgN ha^{-1}yr^{-1}. Considering that the immobilisation of N is probably higher in warmer climates, values of up to 1 kgN ha^{-1}yr^{-1} could be used for N_i, without causing unsustainable accumulation of N in the soil. It should be pointed out, however, that even higher values (closer to present-day immobilisation rates) have been used in critical load calculations. Although studies on the capacity of forests to absorb N have been carried out (see, e.g. Sogn et al. 1999), there is no consensus on long-term sustainable immobilisation rates.

Nitrogen Uptake: The uptake flux N_u equals the long-term average removal of N from the ecosystem. For unmanaged ecosystems (e.g., national parks) the long-term (steady-state) net uptake is basically zero, whereas for managed forests it is the long-term net growth uptake. The harvesting practice is of crucial importance, i.e., whether stems only, stems plus (parts of) branches or stems plus branches plus leaves/needles (whole-tree harvesting) are removed. The uptake of N is then calculated as:

$$N_u = \frac{\text{N removed in harvested biomass (eq/ha)}}{\text{interval between harvests (rotation period) (yr)}} \qquad (6.7)$$

The amount of N in the harvested biomass (stems and branches) can be calculated as:

$$N_u = k_{gr} \cdot \rho_{st} \cdot \left(ctN_{st} + f_{br,st} \cdot ctN_{br} + f_{lv,st} \cdot ctN_{lv} \right) \qquad (6.8)$$

where k_{gr} is the average annual growth rate (m^3 ha^{-1}yr^{-1}), ρ_{st} is the density of stem wood (kg m^{-3}) and *ctN* is the N content in stems, branches and foliage (subscripts *st*, *br* and *lv*, respectively) (eq kg^{-1}). Furthermore, $f_{br,st}$ and $f_{lv,st}$ denote the (average) branch-to-stem and foliage-to-stem ratios (biomass expansion factors; kg kg^{-1}). Depending on the harvesting practice, the contribution of foliage and/or branches should be neglected.

Values for the quantities should be taken from local sources; alternatively they can be found in overview reports such as Kimmins et al. (1985) and Jacobsen et al. (2003). Growth rates used should be long-term average values, typical for the site. It has to be noted that recent growth rates are higher due to increased N input. Therefore it is recommended to use older investigations (yield tables), preferably from before the 1970s. Net uptake of N in non-forest (semi-)natural ecosystems is insignificant, unless they are used for extensive grazing.

Denitrification: Dutch and Ineson (1990) reviewed data on denitrification rates, which can be used directly for the computation of critical loads (Eq. 6.3) or enable the verification of computations made with Eq. 6.4. Typical values of N_{de} for boreal and temperate ecosystems are in the range of 0.1–3.0 kgN ha^{-1}yr^{-1} (=7.14–214.3 eq ha^{-1}yr^{-1}), where the higher values apply to wet(ter) soils; rates for well-drained soils are generally below 0.5 kgN ha^{-1}yr^{-1}. Concerning deposition-dependent denitrification, values for the denitrification fraction used by De Vries et al. (1994) were f_{de}=0.95 for peat soils, 0.8 for clay soils, 0.5 for sandy soils with gleyic features and 0.1 for sandy soils without gleyic features. Reinds et al. (2001) related the denitrification fraction to the drainage status of the soil, ranging from f_{de}=0.1 for well drained to 0.8 for (very) poorly drained soils.

Precipitation Surplus: The precipitation surplus Q is the amount of water percolating from the root zone. It is conveniently calculated as the difference between precipitation and actual evapotranspiration and it should be the long-term annual mean value. In many cases evapotranspiration will have to be calculated by a model using basic meteorological input data (precipitation, temperature, radiation etc.). For the basics of modelling evapotranspiration see Monteith and Unsworth (1990), and for an extensive collection of models see Burman and Pochop (1994).

6.3 Critical Loads of Acidity (N and S) for Soils

6.3.1 The SMB Model for Critical Loads of Acidity

Starting point for deriving critical loads of acidifying N and S for soils is the charge balance of the ions in the soil leaching flux:

$$H_{le} + Al_{le} + BC_{le} + NH_{4,le} = SO_{4,le} + NO_{3,le} + Cl_{le} + HCO_{3,le} + Org_{le} \quad (6.9)$$

where the subscript *le* stands for leaching, *Al* stands for the sum of all positively charged aluminium species, *BC* is the sum of base cations ($BC=Ca+Mg+K+Na$) and *Org* denotes the sum of organic anions. A leaching term is given by $X_{le}=Q\cdot[X]$, where $[X]$ is the soil solution concentration of ion X and Q is the precipitation surplus; again, all fluxes are expressed in eq ha^{-1}yr^{-1}. The concentrations of OH and CO_3 can be neglected in the pH-ranges considered. Defining the (leaching of) Acid Neutralising Capacity (ANC) as:

$$ANC_{le} = -H_{le} - Al_{le} + HCO_{3,le} + Org_{le} \quad (6.10)$$

Equation 6.9 becomes:

$$BC_{le} + NH_{4,le} - SO_{4,le} - NO_{3,le} - Cl_{le} = ANC_{le} \quad (6.11)$$

For more detailed discussions on the processes and concepts of (soil) chemistry encountered in the context of acidification see, e.g., the books by Reuss and Johnson (1986) or Ulrich and Sumner (1991).

Chloride is assumed to be a tracer, i.e. there are no sources or sinks of Cl within the soil compartment, and Cl leaching is therefore equal to Cl deposition (subscript *dep*):

$$Cl_{le} = Cl_{dep} \tag{6.12}$$

At steady-state the leaching of base cations has to be balanced by their net inputs:

$$BC_{le} = BC_{dep} + BC_{w} - Bc_{u} \tag{6.13}$$

where the subscripts w and u stand for weathering and net growth uptake, i.e. the net uptake by vegetation that is needed for long-term average growth. Bc stands for $Ca+Mg+K$, reflecting the fact that Na is not taken up by vegetation. The finite pool of base cations at exchange sites (cation exchange capacity) might buffer incoming acidity for decades, but will reach equilibrium with the soil solution under steady-state conditions, and thus need not be considered here.

The leaching of sulphate and nitrate can be linked to the deposition of these compounds by means of mass balances for S and N. For S this reads (De Vries 1991):

$$S_{le} = S_{dep} - S_{ad} - S_{i} - S_{u} - S_{re} - S_{pr} \tag{6.14}$$

where the subscripts *ad*, *i*, *re* and *pr* refer to adsorption, immobilisation, reduction and precipitation, respectively. An overview of sulphur cycling in forests by Johnson (1984) suggests that uptake, immobilisation and reduction of S are generally insignificant. Adsorption (and in some cases precipitation with Al complexes) can temporarily lead to a strong accumulation of sulphate (e.g. Johnson et al. 1982); however, it is only stored at or released from the adsorption complex when the input (deposition) *changes*. Consequently, as with base cations at exchange sites, S adsorption will reach equilibrium with the soil solution under steady-state conditions, and thus need not be considered in critical load calculations. Since sulphur is completely oxidised in the soil profile, $SO_{4,le}$ equals S_{le}, and thus:

$$SO_{4,le} = S_{dep} \tag{6.15}$$

For nitrogen, the mass balance in the soil has been derived above (see Eq. 6.2) assuming complete nitrification (i.e. $NH_{4,le}=0$):

$$NO_{3,le} = N_{dep} - N_{i} - N_{u} - N_{de} \tag{6.16}$$

Inserting all the mass balances into Eq. 6.11 leads to the following simplified charge balance for the soil compartment:

$$S_{dep} + N_{dep} = BC_{dep} - Cl_{dep} + BC_w - Bc_u + N_i + N_u + N_{de} - ANC_{le} \quad (6.17)$$

Strictly speaking, we should replace $NO_{3,le}$ in the charge balance not by the right-hand side of Eq. 6.16, but by $\max\{N_{dep}-N_i-N_u-N_{de},0\}$, since leaching cannot be negative; and the same holds true for base cations. However, this would lead to unwieldy critical load expressions; therefore we go ahead with Eq. 6.17, keeping this constraint in mind.

For given values for the sources and sinks of S, N and BC, Eq. 6.17 allows the calculation of the leaching of ANC, and thus assessment of the acidification status of the soil. Conversely, critical loads of S, *CLS*, and N, *CLN*, can be computed by defining a critical ANC leaching, $ANC_{le,crit}$:

$$CLS + CLN = BC_{dep} - Cl_{dep} + BC_w - Bc_u + N_i + N_u + N_{de} - ANC_{le,crit} \quad (6.18)$$

Note that Eq. 6.18 does not give a unique critical load of S or N. However, nitrogen sinks cannot compensate incoming sulphur acidity, and therefore the maximum critical load of sulphur is given by:

$$CL_{max}S = BC_{dep} - Cl_{dep} + BC_w - Bc_u - ANC_{le,crit} \quad (6.19)$$

Equation 6.19 gives the critical load as long as the N deposition is lower than the sum of the N sinks, termed the minimum critical load of N:

$$CL_{min}N = N_i + N_u + N_{de} \quad (6.20)$$

Finally, the maximum critical load of nitrogen (in the case of zero S deposition) is given by:

$$CL_{max}N = CL_{min}N + CL_{max}S \quad (6.21)$$

The three quantities $CL_{max}S$, $CL_{min}N$ and $CL_{max}N$ define the so-called *critical load function* (CLF; Fig. 6.1); and every deposition pair (N_{dep},S_{dep}) lying on the CLF are critical loads of acidifying N and S.

Deriving critical loads as above assumes that the sources and sinks of N do not depend on the N deposition. However, if we consider denitrification to be linearly related to the net input of N (see Eq. 6.4), we get the following expressions for $CL_{min}N$ and $CL_{max}N$ ($CL_{max}S$ remains unchanged):

$$CL_{min}N = N_i + N_u \quad \text{and} \quad CL_{max}N = CL_{min}N + \frac{CL_{max}S}{1-f_{de}} \quad (6.22)$$

where f_{de} ($0 \le f_{de} < 1$) is the denitrification fraction (see Eq. 6.4).

Since the main use of critical loads is to guide reductions in *anthropogenic* emissions (and thus depositions) of S and N, sea-salt derived sulphate should not be

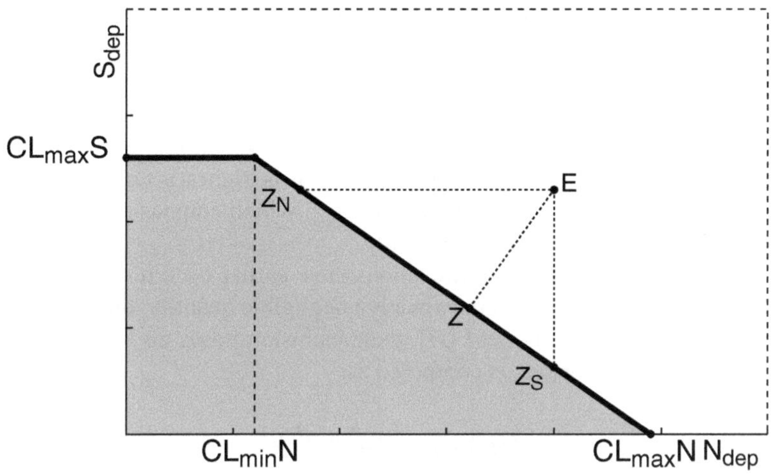

Fig. 6.1 Critical load function (CLF) of acidifying N and S, defined by the three quantities $CL_{max}S$, $CL_{min}N$ and $CL_{max}N$ in the (N_{dep}, S_{dep})-plane. Reducing emissions from point E one can reach the CLF, e.g., at points Z_N, Z, or Z_S, The *grey* area *below* the CLF denotes deposition pairs resulting in an ANC leaching greater than $ANC_{le,crit}$ (non-exceedance of critical loads)

considered in the balance. To retain charge balance, this is achieved by applying a sea-salt correction to sulphate, chloride and base cations, which is often done by using either Cl or Na as a tracer, whichever can be (safer) assumed to derive from sea-salts only. Denoting sea-salt corrected depositions with an asterisk, one has either $Cl^*_{dep}=0$ or $Na^*_{dep}=0$ (and $BC^*_{dep}=Bc^*_{dep}$), respectively. For procedures to compute sea-salt corrections see the Annex to this chapter.

6.3.2 Chemical Criteria and the Critical Leaching of ANC

To compute a critical load, the critical ANC leaching, $ANC_{le,crit}$, has to be defined. Critical chemical limits for the derivation of acidity critical loads are discussed in Chap. 2. For surface waters, for example, a critical limit for ANC is most often used. For soils, however, critical limits are mostly expressed as critical concentrations of Al ($[Al]_{crit}$) or H ($[H]_{crit}$), sometimes in conjunction with base cations; and in the following we show how to derive $ANC_{le,crit}$ from them.

The terms in the (leaching of) ANC, as given in Eq. 6.10, can all be expressed in terms of $[H]$. The relationship between $[H]$ and $[Al]$ is modelled as:

$$[Al] = K_{Alox} \cdot [H]^a \tag{6.23}$$

with an equilibrium constant K_{Alox} and an exponent a. For $a=3$ this describes the dissolution of gibbsite, but for $a<3$ this is an empirical relationship accounting also for the complexation of Al with organic matter (Cronan et al. 1986; Mulder and

Stein 1994), thus making it strongly dependent on soil type and soil depth. Values for K_{Alox} and a derived by Van der Salm and De Vries (2001) from several 100 Dutch forest soil solution samples show that a standard gibbsite constant ($K_{Alox}=K_{gibb}$) and $a=3$ are reasonable for sandy soils. Very different values, however, are obtained for peat soils and, to a lesser extent, also for loess and clay soils (especially for shallow parts of the soil, where the organic matter content is highest). Data also show that there is a strong correlation between a and K_{Alox}, which emphasises that these two parameters cannot be chosen independently.

Bicarbonates (HCO_3) and organic anions have earlier been neglected in critical load calculations. However, it is not always a negligible quantity; and since they can both be expressed as functions of $[H]$, their inclusion poses no real problem. The concentration of bicarbonates is computed as:

$$[HCO_3]=\frac{K_1 \cdot K_H \cdot p_{CO2}}{[H]} \tag{6.24}$$

where K_1 is the first dissociation constant, K_H is Henry's constant and p_{CO2} is the partial pressure of CO_2 in the soil solution. The simplest model describing organic acids assumes that only mono-protic organic anions are produced by the dissociation of dissolved organic carbon:

$$[Org]=\frac{m \cdot DOC \cdot K_{org}}{K_{org}+[H]} \tag{6.25}$$

where DOC is the concentration of dissolved organic carbon (in molC m^{-3}), m is the concentration of functional groups (the 'charge density', in mol $molC^{-1}$) and K_{org} the dissociation constant. Other models for the dissociation of organic acids have been suggested and are in use in dynamic models, such as di- and tri-protic models (see, e.g. Driscoll et al. 1994). In fact, any model for the dissociation of DOC can be used here as long as it depends only on $[H]$ and/or $[Al]$.

The combination of Eq. 6.10 with Eqs. 6.23–6.25 shows that $ANC_{le,crit}$ can be calculated if $[H]_{crit}$ is known; leaching fluxes are obtained by multiplying $[H]$ as well as $[Al]$, $[HCO_3]$ and $[Org]$ (which all depend on $[H]$) with Q, the precipitation surplus in $m^3\ ha^{-1}yr^{-1}$ (see above). If $[Al]_{crit}$ is known, $[H]_{crit}$ can be obtained by inverting Eq. 6.23. In the following we discuss different criteria in turn.

Aluminium Criteria: Aluminium criteria are considered most appropriate for mineral soils with low organic matter content. Three commonly used criteria are:

a. *Critical aluminium concentration:* Critical limits for Al have been suggested for forest soils, not only for effects on roots, but especially for drinking water (ground water) protection (see Chap. 2).
b. *Critical base cation to aluminium ratio:* Widely used for soils is the connection between soil chemical status and plant response (damage to fine roots) via a critical molar ratio of the concentrations of base cations (Bc = Ca+Mg+K) and Al in soil solution, denoted as $(Bc/Al)_{crit}$ (see Chap. 2). The critical Al leaching is calculated from the leaching of Bc:

$$Al_{le,crit} = 1.5 \cdot \frac{Bc_{le}}{(Bc/Al)_{crit}} = 1.5 \cdot \frac{Bc_{dep} + Bc_w - Bc_u}{(Bc/Al)_{crit}} \tag{6.26}$$

The factor 1.5 arises from the conversion of moles to equivalents (assuming that K is divalent). Note that the expression $Bc_{le}=Bc_{dep}+Bc_w-Bc_u$ has to be non-negative. In fact, it has been suggested that it should be above a minimum leaching or, more precisely, there is a minimum concentration of base cations in the leachate, below which they cannot be taken up by vegetation, i.e., Bc_{le} is set equal to $\max\{0, Bc_{dep}+Bc_w-Bc_u-Q\cdot[Bc]_{min}\}$, with $[Bc]_{min}$ in the order of 0.01 eq m^{-3}. If considered more appropriate, a critical molar ratio of calcium to aluminium in soil solution can be used, by replacing all the *Bc*-terms in Eq. 6.26 with *Ca*-terms.

c. *Critical Al mobilisation rate:* Critical ANC leaching can also be calculated using a criterion to prevent the depletion of secondary Al phases and complexes, which may cause structural changes in soils and a further pH decline. Aluminium depletion occurs when the acid deposition leads to an Al leaching in excess of the Al produced by the weathering of primary minerals. Thus the critical leaching of Al is given by:

$$Al_{le,crit} = Al_w = q \cdot BC_w \tag{6.27}$$

where Al_w is the weathering of Al from primary minerals, and q is the stoichiometric ratio of Al to BC weathering in primary minerals (eq eq^{-1}), with a default value of $q=2$ for typical mineralogy of northern European soils (range: 1.5–3.0).

Hydrogen Ion Criteria: A hydrogen ion (proton) criterion is generally recommended for soils with high organic matter content. Two such criteria are:

a. *Critical pH:* Critical pH limits have been suggested for forest soils, for example, $pH_{crit}=4.0$, corresponding to $[H]_{crit}=0.1$ eq m^{-3}.
b. *Critical base cation to proton ratio:* For organic soils that do not contain Al-(hydr)oxides (such as peat lands), it is suggested to use a critical molar base cation to proton ratio $(Bc/H)_{crit}$. In this case there is no Al leaching, and $H_{le,crit}$ is given by:

$$H_{le,crit} = 0.5 \cdot \frac{Bc_{dep} + Bc_w - Bc_u}{(Bc/H)_{crit}} \tag{6.28}$$

where the factor 0.5 comes from converting moles to equivalents. For organic soils the weathering in Eq. 6.28 will probably be negligible ($Bc_w=0$). Values suggested for $(Bc/H)_{crit}$ are expressed as multiples of $(Bc/Al)_{crit}$.

Critical Base Saturation: Base saturation, i.e. the fraction of base cations at the cation exchange complex, is an indicator of the acidity status of a soil, and one may want to keep this pool above a certain level, e.g., to avoid nutrient deficiencies.

Thus a base saturation could be chosen as a criterion for calculating critical loads of acidity (UNECE 2001). Base saturation is also used as criterion in the New England Governors/Eastern Canadian Premiers 'Acid Rain Action Plan' for calculating sustainable S and N depositions to upland forests (NEG/ECP 2001).

To relate base saturation to $[H]$ requires a model of the exchange of cations between the exchange complex and the soil solution. In general, this leads to a non-linear equation for $[H]_{crit}$ that has to be solved numerically (see UBA 2004). However, using the Gapon model for the exchange between H, Al and Bc = Ca+Mg+K—as implemented in the SAFE (Warfvinge et al. 1993) and the VSD (Posch and Reinds 2009) dynamic soil models—and gibbsite dissolution (i.e. $a=3$ in Eq. 6.23) the critical concentration $[H]_{crit}$ can be found as:

$$[H]_{crit} = K_{Gap} \cdot \sqrt{[Bc]} \cdot \left(\frac{1}{E_{Bc,crit}} - 1 \right) \quad \text{with} \quad K_{Gap} = \frac{1}{k_{HBc} + k_{AlBc} \cdot K_{gibb}^{1/3}} \tag{6.29}$$

where $E_{Bc,crit}$ is the selected critical base saturation, k_{HBc} and k_{AlBc} are the two (site-specific) selectivity coefficients describing the Gapon exchange, and $[Bc] = Bc_{le}/Q$. Assuming only Al and Bc exchange, Ouimet et al. (2006) used base saturation to select a chemical criterion for calculating acidity critical loads for eastern Canada. Values of selectivity coefficients for a range of (Dutch) soil types can be found in De Vries and Posch (2003).

6.3.3 Sources and Derivation of Input Data

The obvious sources of input data for calculating acidity critical loads are measurements at the site under consideration. However, in many cases these will not be available. For data on the different N quantities see Sect. 6.2.3. Some data sources and default values for the other variables, and procedures to derive them, are summarised here.

Base Cation and Chloride Deposition: The base cation and chloride depositions entering the critical load calculations should ideally be the (non-anthropogenic) deposition at critical load. Observations are available from the EMEP (www.emep.int) on a European scale, or from national sources. In a one-time study, EMEP has also modelled the deposition of base cations (for the year 2000) on a European scale (Van Loon et al. 2005).

Base Cation Weathering: Weathering here refers to the release of base cations from minerals in the soil matrix due to chemical dissolution, and the neutralisation and production of alkalinity connected to this process. This has to be distinguished from the denudation of base cations from ion exchange complexes (cation exchange) and the degradation of soil organic matter. Many methods for determining weathering rates have been suggested, and here we list those with the highest potential for regional applications (in order of increasing complexity).

Table 6.1 Weathering rate classes as function of texture and parent material classes

Parent material	Texture class				
	1	2	3	4	5
Acidic	1	3	3	6	6
Intermediate	2	4	4	6	6
Basic	2	5	5	6	6
Organic	Class 6 for Oe and class 1 for other organic soils				

a. *The Skokloster assignment:* This is a (semi-)empirical method devised in the early days of the critical loads development (UNECE Skokloster workshop, documented in Nilsson and Grennfelt 1988, p. 40, Table 1).
b. *The soil type—texture approximation*: Since mineralogy controls weathering rates, weathering rate classes were assigned to European (forest) soils by De Vries et al. (1993), see also Posch et al. (2003), based on texture class and parent material class (Table 6.1). Texture class, as a function of clay and sand content, is defined according to Eurosoil (1999) and parent material class has been defined for each FAO soil type (for details see UBA 2004).

The weathering rate (in eq $ha^{-1}yr^{-1}$) for a non-calcareous soil of depth z (in m) is then computed as:

$$BC_w = z \cdot 500 \cdot (WRc - 0.5) \cdot \exp\left(\frac{A}{281} - \frac{A}{273+T}\right) \tag{6.30}$$

where WRc is the weathering rate class (Table 6.1), T (°C) is the average annual (soil) temperature and $A=3600$ K (Sverdrup 1990). For calcareous soil, for which critical loads are not really of interest, one could, e.g., set $WRc=20$ in Eq. 6.30. The above equation gives weathering rates for BC=Ca+Mg+K+Na. However, in Eqs. 6.26 and 6.28 the weathering rate for Bc=Ca+Mg+K is needed; it can be approximated by multiplying BC_w with a factor between 0.70 for poor sandy soils and 0.85 for rich (sandy) soils.

c. *The total base cation content correlation*: Using the 'zirconium method', Olsson et al. (1993) derived from Swedish sites a correlation between historical average weathering rates of base cations and the total content of the respective element in the undisturbed bottom soil, with an additional temperature correction. For Ca, Mg and K the equations are (in eq $ha^{-1}yr^{-1}$):

$$\begin{aligned} Ca_w &= 0.13 \cdot (Ca)_{tot} \cdot ETS - 55.5 \\ Mg_w &= 0.23 \cdot (Mg)_{tot} \cdot ETS - 24.1 \\ K_w &= 0.05 \cdot (K)_{tot} \cdot ETS - 79.8 \end{aligned} \tag{6.31}$$

where $(X)_{tot}$ is the total content of element X (Ca, Mg or K in dry weight %) in the coarse fraction (<2 mm) of the undisturbed C-horizon soil and ETS is the annual sum of daily temperatures above a threshold of +5 °C. Care has to be taken when

Table 6.2 Ranges for K_{gibb} as a function of soil organic matter content

Soil type; layer	Organic matter (%)	K_{gibb} (m^6/eq^2)	$-pK_{gibb}$
Mineral soils; C-layer	<5	950–9500	8.5–9.5
Soils with low organic matter; B/C layers	5–15	300–3000	8–9
Soils with some organic material; A/E layers	15–30	100	7.6
Peaty and organic soils; organic layers	>70	9.5	6.5

applying these formulae, since they are based on Nordic geological history and they cover a limited number of soil types (mostly podzols). Using part of the Swedish data, this method was adapted in Finland for estimating weathering rates on a national scale (Johansson and Tarvainen 1997).

d. *The calculation of weathering rates:* Weathering rates can be computed with the kinetic Sverdrup-Warfvinge weathering model (Sverdrup and Warfvinge 1988) for (groups of) minerals in the soil. It is also integrated into the multi-layer steady-state model PROFILE (Warfvinge and Sverdrup 1992, 1995). Basic input data are the mineralogy of the site or a total element analysis, from which the mineralogy can be derived by a normative procedure or more general methods, such as the A2M model (Posch and Kurz 2007). The generic weathering rate of each mineral is modified by the pH, base cations, aluminium and organic anions as well as the partial pressure of CO_2 and temperature. The total weathering rate is proportional to soil depth and the wetted surface area of the minerals present. For the theoretical foundations of the weathering rate model, see Sverdrup (1990).
e. *Other methods*: Weathering rates can also be estimated from budget studies of small catchments (see, e.g. Paces 1983). However, budget studies can easily overestimate soil weathering rates where there is significant cation release due to weathering of the bedrock. More methods are listed and described in Sverdrup et al. (1990).

Base Cation Uptake: The uptake flux of base cations entering the critical load calculations, Bc_u, is the long-term average removal of base cations from the ecosystem. The data sources and calculations are the same as for the uptake of N (see Sect. 6.2.3). The (long-term) net uptake of base cations is limited by their availability through deposition and weathering (no depletion of exchangeable base cations at steady state). Furthermore, base cations might not be taken up below a certain concentration in soil solution, or due to other limiting factors, such a temperature. Thus the values entering critical load calculations should be constrained by:

$$Y_u \leq Y_{dep} + Y_w - Q \cdot [Y]_{min} \quad \text{for} \quad Y = Ca, Mg, K \tag{6.32}$$

Suggested values are 5 meq m^{-3} for $[Ca]_{min}$ and $[Mg]_{min}$, and zero for $[K]_{min}$ (Warfvinge and Sverdrup 1992). It should also be taken into account that vegetation takes up nutrients in fairly constant (vegetation-specific) ratios. Thus, when adjusting the uptake value for one element, the values for the other elements (including N) should be adjusted proportionally.

Other Parameters: The equilibrium constant relating the Al concentration to pH (Eq. 6.23) depends on the soil type, i.e. the type of Al (hydr)oxide. Table 6.2 presents ranges of K_{gibb} (and $pK_{gibb}=-\log_{10}(K_{gibb}$ in $(mol/l)^{-2}$) as a function of soil organic matter content. An often used default value is $K_{gibb}=10^8\ (mol/l)^{-2}=300\ m^6\ eq^{-2}$.

If sufficient empirical data are available to derive the relationship between $[H]$ and $[Al]$, these should be used in preference to the gibbsite equilibrium.

The two constants in the bicarbonate equilibrium (Eq. 6.24) are weakly temperature-dependent, and the value for their product at 8 °C is $K_1 \cdot K_H=10^{-1.7}=0.02\ eq^2\ m^{-6} atm^{-1}$. For systems open to the atmosphere, p_{CO2} is about 390 ppm or $3.9 \cdot 10^{-4}$ atm (in the year 2010). However, in soils p_{CO2} is generally higher (ranging from 10^{-2} to 10^{-1} atm; Bolt and Bruggenwert 1976), due to respiration and oxidation of below-ground organic matter. Respiratory production of CO_2 is temperature-dependent, and Gunn and Trudgill (1982) derived the following relationship:

$$\log_{10} p_{CO2} = -2.38 + 0.031 \cdot T \tag{6.33}$$

where T is the (soil) temperature (°C). Brook et al. (1983) present a similar regression equation based on data for 19 regions of the world.

Concerning organic acids, both DOC and m are site-specific quantities. While DOC estimates are often available, data for m are less easy to obtain. For example, Santore et al. (1995) report values of m between 0.014 for topsoil samples and 0.044 mol $molC^{-1}$ for a B-horizon in the Hubbard Brook experimental forest in New Hampshire. Since a single value of K_{org} does not always model the dissociation of organic acids satisfactorily, Oliver et al. (1983) derived an empirical relationship between K_{org} and pH:

$$pK_{org} = -\log_{10} K_{org} = a + b \cdot pH - c \cdot (pH)^2 \tag{6.34}$$

with $a=0.96$, $b=0.90$ and $c=0.039$ (and $m=0.12$ mol $molC^{-1}$). Note that Eq. 6.34 gives K_{org} in mol l^{-1}. In Fig. 6.2 the fraction of $m \cdot DOC$ dissociated is displayed as a function of pH for the Oliver model and a mono-protic acid with a 'widely-used'

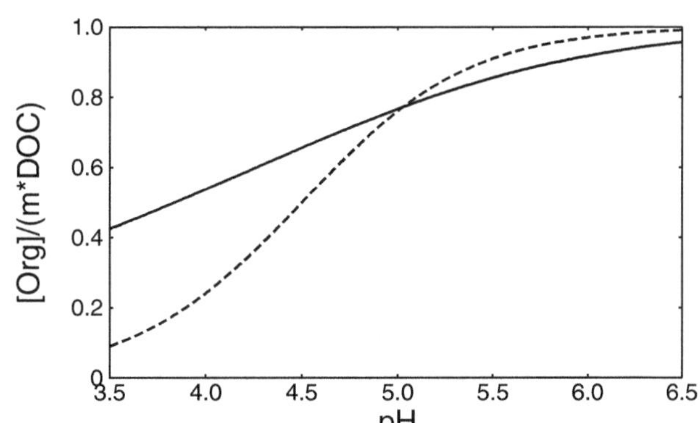

Fig. 6.2 Fraction of organic acids dissociated as a function of pH for the Oliver model (*solid line*) and the mono-protic model (Eq. 6.25) with $pK_1=4.5$ (*dashed line*)

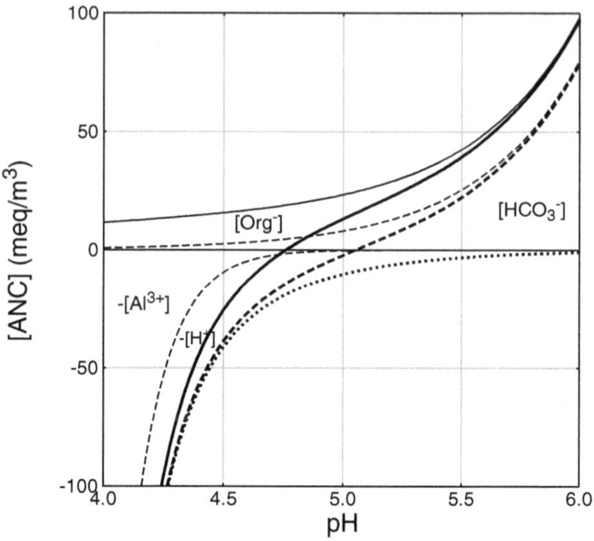

Fig. 6.3 ANC concentration (in meq m^{-3}) as a function of pH (*thick solid line*) with $[ANC]=-[H]-[Al]+[HCO_3]+[Org]$ (see Eq. 6.10). Also shown is the ANC in the absence of (or neglecting) organic anions (*thick dashed line*) and the ANC solely defined as $-[H]-[Al]$ (*thick dotted line*). The *thin dashed lines* show $[HCO_3]$ and $-[Al]$, and the *thin solid line* is $[HCO_3]+[Org]$, all as a function of pH (computed with $a=3$, $\log_{10}K_{gibb}=8$ and $m{\cdot}DOC=20$ mol m^{-3}).

value of $pK_{org}=4.5$. It shows that the share of dissociated organic acids can be substantial, even at low pH.

In Fig. 6.3 the contribution of the individual ions to the ANC is illustrated. It shows the ANC concentration as a function of pH (thick solid line), with the ANC defined as in Eq. 6.10. Also shown is the ANC in the absence of (or neglecting) organic anions (thick dashed line), and the ANC solely defined as $-[H]-[Al]$ (thick dotted line) as done in earlier works on critical loads (Sverdrup et al. 1990; Sverdrup and De Vries 1994) and which can never attain positive values.

6.4 Critical Loads of Acidity for Surface Waters

Critical loads for aquatic ecosystems estimate the maximum deposition(s) onto a catchment below which 'significant harmful effects' on biological species in the water body do not occur. Similar to terrestrial ecosystems, the links between water chemistry and biological impacts cannot be modelled adequately at present and thus (empirically derived) water quality criteria are used to determine critical loads for aquatic ecosystems. The models presented here are restricted to freshwater systems, since models for marine ecosystems do not seem to be documented in the open literature. Furthermore, only acidity critical loads are considered, since critical loads of nutrient N for surface waters have not been formulated/used to date; but see Smith et al. (1999), De Wit and Lindholm (2010) and Chap. 2 for a summary of effects and potential critical limits for N as a nutrient in oligotrophic surface waters.

In the following the most widely used models, the SSWC and the FAB model, are described. More details, especially also earlier work and references, can be found in the review by Henriksen and Posch (2001). As models of critical loads for surface waters also include processes taking place in their terrestrial catchment, it is advised to consult the sections above for some of the terminology and variables used in the context of critical loads for soils.

6.4.1 The Steady-State Water Chemistry (SSWC) Model

In the SSWC model (Henriksen et al. 1992; Henriksen and Posch 2001; Sverdrup et al. 1990) the critical load for a lake or stream is derived from annual mean values of present-day water chemistry. A critical load of acidity, *CLA*, is calculated from the same principles as in the SMB model for soils, i.e. it should not exceed the non-anthropogenic net base cation input minus a buffer to protect selected biota from being damaged:

$$CLA = BC_0^* - ANC_{limit} = Q\cdot\left([BC^*]_0 - [ANC]_{limit}\right) \tag{6.35}$$

where the second expression gives the critical load in terms of the catchment runoff Q and concentrations ($[X]=X/Q$). BC^*_0 is the pre-acidification sea salt corrected catchment-average net base cation flux (equal to $BC^*_{dep}+BC_{we}-Bc_{upt}$ when everything was in equilibrium; compare Eq. 6.13) and ANC_{limit} is the lowest ANC-flux that does not damage the selected biota. For estimating the sea salt correction, see the Annex to this chapter.

Although one might be able to estimate a pre-industrial base cation deposition, and one can assume that the pre-industrial average net removal of base cations by harvest is negligible, it is the catchment-average weathering rate that poses the problem, especially on a regional scale. Thus Henriksen (1984) argued that the change over time t in base cation concentration is proportional to the change in sulphate (plus nitrate) concentration:

$$[BC^*]_t - [BC^*]_0 = F\cdot\left([SO_4^*]_t - [SO_4^*]_0 + [NO_3]_t - [NO_3]_0\right) \tag{6.36}$$

with F denoting the proportionality factor (the 'F-factor') and where the subscripts t and 0 refer to present and pre-acidification concentrations, respectively. If $F=1$, all incoming protons are neutralised in the catchment (only soil acidification), at $F=0$ none of the incoming protons are neutralised in the catchment (only water acidification). The F-factor was estimated empirically to be in the range 0.2–0.4, based on the analysis of historical data from Norway, Sweden, the U.S.A. and Canada (Henriksen 1984). Brakke et al. (1990) later suggested that the F-factor should be a function of the base cation concentration:

$$F = \sin\left(\tfrac{\pi}{2}[BC^*]_t/[S]\right) \tag{6.37}$$

where $[S]$ is the base cation concentration at which $F=1$; and for $[BC^*]_t > [S]$ F is set to 1. For Norway $[S]$ has been set to 0.4 eq m^{-3} (Brakke et al. 1990). In Eq. 6.37 the present base cation concentration is used for practical reasons. However, to make the F-factor independent of the present base cation concentration (and to simplify the functional form), Posch et al. (1993) suggested the following relationship between F and the pre-acidification base cation concentration $[BC^*]_0$:

$$F = 1 - \exp\left(-[BC^*]_0 / [B]\right) \tag{6.38}$$

where $[B]$ is a scaling concentration estimated to be 0.131 eq m^{-3} from paleolimnological data from Finland (Posch et al. 1993). Inserting this expression into Eq. 6.36 gives a non-linear equation for $[BC^*]_0$ which has to be solved by an iterative procedure (see the Supporting Information in Posch et al. 2012 for details). The two expressions for the F-factor give similar results when used to calculate critical loads for surface waters in Norway (Henriksen and Posch 2001). It has also been suggested to replace the concentrations in the F-factor expressions by the corresponding fluxes to better capture differences in weathering characteristics instead of differences in runoff (Henriksen and Posch 2001).

Using data from lakes in the United Kingdom, Battarbee et al. (1996) developed the 'empirical diatom-based paleolimnological model'. The basic equation for the critical load of acidity in the empirical diatom model is:

$$CLA = \frac{[Ca^*]_0}{CR} \tag{6.39}$$

where CLA is in keq ha^{-1}yr^{-1}, $[Ca^*]_0$ in meq m^{-3}, and CR is the so-called critical ratio (e.g. 94:1). The pre-acidification Ca concentration is calculated with an F-factor approach according to Eqs. 6.36 and 6.37 (with BC replaced by Ca). For more details see Battarbee et al. (1996) and UBA (2004).

The pre-acidification sulphate concentration in lakes, $[SO_4^*]_0$, is assumed to consist of a constant atmospheric contribution and a geological contribution proportional to the concentration of base cations (Brakke et al. 1989):

$$[SO_4^*]_0 = a + b \cdot [BC^*]_t \tag{6.40}$$

The coefficients in this equation, estimated for different regions and by different authors, are summarised in Table 6.3. Larssen and Høgåsen (2003) suggested that the atmospheric contribution in Eq. 6.40 be derived from background S deposition (e.g. estimated by atmospheric transport models), i.e. $a = S_{dep,0}/Q$. Posch et al. (2012) also used background deposition to estimate a (with b set to zero).

The pre-acidification nitrate concentration $[NO_3]_0$ is generally assumed zero, since the background deposition of N is low and will most likely be taken up in the soil and the water.

The SSWC model has been developed for, and is particularly applicable to, dilute oligotrophic waters located on granitic and gneissic bedrock with thin soils,

Table 6.3 Constants to estimate the pre-acidification sulphate concentration (Eq. 6.40), derived from empirical data (*N* is the number of samples and *r* is the correlation coefficient)

a (meq m^{-3})	*b*	*N*	*r*	Region and reference
15	0.16	143	0.38	Lakes, Norway (Brakke et al. 1989)
8	0.17	289	0.78	Lakes, Norway (Henriksen and Posch 2001)
5	0.05	n.g.	n.g.	Groundwater, Sweden (Wilander 1994)
14	0.10	61	0.29	Lakes, Finland (Posch et al. 1993)
19	0.08	251	0.66	Lakes, N. Norway+Finland+Sweden (Posch et al. 1997)
9.5	0.08	60	0.66	Lakes, Ireland (Aherne et al. 2002)

such as in large parts of Fennoscandia, Scotland, Canada and Ireland. In such areas, surface waters are generally more sensitive to acid inputs than soils. In the model it is assumed that all sulphate in runoff originates from deposition alone, except for a small geologic contribution. In areas where geological conditions lead to more alkaline waters, the SSWC model has to be modified, since, e.g., significant amounts of sulphate from geological sources could be present in the runoff water.

The link between water chemistry and the impact on aquatic organisms is represented by the critical ANC-limit, $[ANC]_{limit}$ (see Eq. 6.35). Mostly fish species have been considered as the aquatic organisms that should be protected, and it has been mostly ANC itself that has been correlated with fish status, leading to values of $[ANC]_{limit}$ between zero and 50 meq m^{-3}, depending on location and study. An overview of critical limits for surface waters can be found in Chap. 2.

6.4.2 The First-order Acidity Balance (FAB) Model

The FAB model for calculating critical loads of S and N for a lake/catchment is the equivalent of the SMB model for a soil site, as described above. The original version of the FAB model was developed and applied to Finland, Norway and Sweden by Henriksen et al. (1993) and was also described by Posch et al. (1997). A modified version was first reported by Hindar et al. (2000) and fully described by Henriksen and Posch (2001), as well as in the Mapping Manual (UBA 2004). The description here follows closely that given in the Supporting Information of Posch et al. (2012).

The system we consider consists of the catchment soils and the lake water body, but vegetation is not part of the system. Thus N and base cation uptake are treated as sinks, but at steady state only the net uptake, i.e. the removal by harvest (on an annual average basis) is considered. Other N processes considered are denitrification and net N immobilization (see the discussion under the SMB model above). The lake and its catchment are assumed to be in a (semi-)natural state, unaffected by direct pollution (e.g. from agriculture), with all precipitation falling on the catchment entering the lake, i.e. no groundwater recharge bypassing the lake. The total catchment area (lake + terrestrial catchment) A consists of the lake area $A_l = A_0$ and m different sub-areas A_j ($j=1,\ldots,m$), comprising the terrestrial catchment:

$$A = A_l + \sum_{j=1}^{m} A_j = \sum_{j=0}^{m} A_j \tag{6.41}$$

where A_1 could be forested area, A_2 covered with grass, A_3 the area of bare rocks, etc. The starting point for the derivation of the FAB model is the charge balance in the surface water running off the catchment:

$$S_{runoff} + N_{runoff} = \sum\nolimits_Y Y_{runoff} - ANC_{runoff} \tag{6.42}$$

where Σ_Y stands for the sum of base cations minus chloride (Ca+Mg+K+Na–Cl). In the above equation we assume that the quantities are total amounts per time (e.g. eq yr^{-1}). To derive critical loads we have to link the ions in the lake water to their depositions, taking into account their steady-state sources and sinks in the terrestrial catchment and the lake. Steady-state mass balances in the lake are given by:

$$X_{runoff} = X_{in} - X_{ret}, \qquad X = S, N, Ca, Mg, K, Na, Cl \tag{6.43}$$

where X_{in} is the total amount of ion X entering the lake and X_{ret} the amount of X retained in the lake. The in-lake retention is assumed to be proportional to the input of the respective ion into the lake:

$$X_{ret} = \rho_X \cdot X_{in}, \qquad X = S, N, Ca, Mg, K, Na, Cl \tag{6.44}$$

where $0 \le \rho_X \le 1$ is a dimensionless retention factor. The mass balances then become:

$$X_{runoff} = (1 - \rho_X) \cdot X_{in}, \qquad X = S, N, Ca, Mg, K, Na, Cl \tag{6.45}$$

The total amount of S entering the lake is given by:

$$S_{in} = \sum_{j=0}^{m} A_j \cdot S_{dep,j} \tag{6.46}$$

where $S_{dep,j}$ is the total deposition of S per unit area onto sub-area *j*. It is assumed that all S deposited onto the catchment enters the lake, i.e. sources (e.g. weathering) or sinks (e.g. immobilization, reduction and uptake) are negligible in the terrestrial catchment. For N we assume that net uptake, net immobilization and denitrification can occur on all sub-areas, possibly at different rates. Thus the amount of N entering the lake is:

$$N_{in} = \sum_{j=0}^{m} A_j \cdot \left(N_{dep,j} - N_{i,j} - N_{u,j} - N_{de,j}\right)_+ \tag{6.47}$$

where $N_{dep,j}$ is the total N deposition, $N_{i,j}$ is the long-term net immobilization of N, $N_{u,j}$ the net growth uptake of N, $N_{de,j}$ is N lost by denitrification, all per unit area for sub-area j, and $(x)_+=\max\{x,0\}$. The effects of nutrient cycling are ignored in this model and the leaching of ammonium is considered negligible, i.e. it is assumed to be completely taken up and/or nitrified in the terrestrial catchment. Note that some of the terms in Eq. 6.47 can be zero for certain indices. Especially for $j=0$, i.e. the lake itself, one mostly assumes $N_{i,0}+N_{u,0}=0$.

While we assume immobilization and uptake to be independent of N deposition, denitrification is modelled, as in the SMB model, as a fraction of the available N:

$$N_{de,j} = f_{de,j} \cdot \left(N_{dep,j} - N_{i,j} - N_{u,j}\right)_+ \quad \text{for} \quad j = 0,\ldots,m \tag{6.48}$$

where $0 \leq f_{de,j} < 1$ is the denitrification fraction for sub-area j. Inserting Eq. 6.48 into Eq. 6.47 one obtains:

$$N_{in} = \sum_{j=0}^{m} A_j \cdot \left(1 - f_{de,j}\right) \cdot \left(N_{dep,j} - N_{i,j} - N_{u,j}\right)_+ \tag{6.49}$$

For base cations and chloride the amounts entering the lake are given by:

$$Y_{in} = \sum_{j=0}^{m} A_j \cdot \left(Y_{dep,j} + Y_{w,j} - Y_{u,j}\right)_+, \quad Y = Ca, Mg, K, Na, Cl \tag{6.50}$$

where $Y_{w,j}$ is the weathering rate and $Y_{u,j}$ the net uptake of ion Y for area j.

To obtain critical loads, a link has to be established between a chemical variable and effects on aquatic biota. The most commonly used criterion is a critical ANC-limit, i.e. a minimum concentration of ANC derived to avoid 'harmful effects' on aquatic organisms: $ANC_{runoff,crit}=A \cdot Q \cdot [ANC]_{limit}$, where Q is the catchment runoff. Other criteria, e.g., a critical pH or Al concentration can be employed, and the critical ANC concentration calculated from it.

Inserting Eqs. 6.46, 6.49 and 6.50 into Eqs. 6.45 and 6.42 and dividing by the total catchment area A yields the following equation to be fulfilled by critical depositions (loads) of S and N:

$$(1-\rho_S) \cdot \sum_{j=0}^{m} c_j \cdot S_{dep,j} + (1-\rho_N) \cdot \sum_{j=0}^{m} c_j \cdot \left(1 - f_{de,j}\right) \cdot \left(N_{dep,j} - N_{i,j} - N_{u,j}\right)_+ = L_{crit} \tag{6.51}$$

where we have defined the sub-area fractions $c_j=A_j/A$ and

$$L_{crit} = \sum_{Y} (1-\rho_Y) \cdot \sum_{j=0}^{m} c_j \cdot \left(Y_{dep,j} + Y_{w,j} - Y_{u,j}\right)_+ - Q \cdot [ANC]_{limit} \tag{6.52}$$

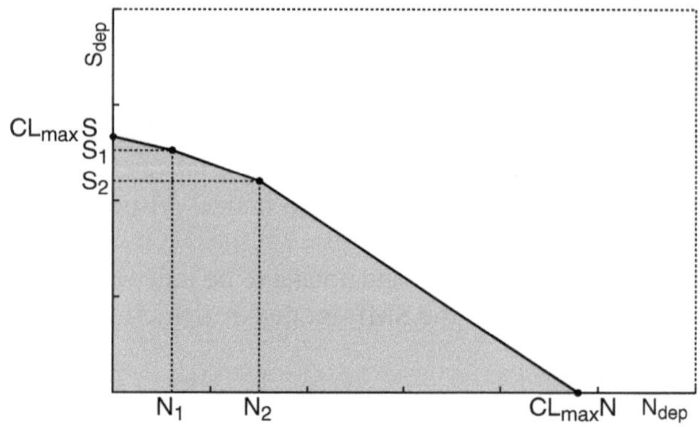

Fig. 6.4 Piece-wise linear critical load function (CLF) of S and acidifying N for a lake as defined by catchment properties, here shown for two land use classes characterized by $N_0=0$, (N_1,S_1) and (N_2,S_2) (see Eqs. 6.56 and 6.64). The *grey* area *below* the CLF denotes deposition pairs resulting in an ANC leaching greater than $Q\cdot[ANC]_{limit}$ (non-exceedance of critical loads; see Sect. 6.5 for the calculation of exceedances)

Denoting catchment average depositions with S_{dep} and N_{dep}, the depositions to the various sub-areas can be written as:

$$S_{dep,j} = s_j \cdot S_{dep} \quad \text{and} \quad N_{dep,j} = n_j \cdot N_{dep}, \quad j = 0,\ldots,m \tag{6.53}$$

where s_j and n_j are dimensionless factors describing the enhanced (or reduced) deposition onto sub-area j. Inserting them into Eq. 6.51 yields:

$$a_S \cdot S_{dep} + (1-\rho_N)\cdot \sum_{j=0}^{m} c_j \cdot (1-f_{de,j})\cdot n_j \cdot \left(N_{dep} - N_j\right)_+ = L_{crit} \tag{6.54}$$

where we define the dimensionless parameter:

$$a_S = (1-\rho_S)\cdot \sum_{j=0}^{m} c_j \cdot s_j \tag{6.55}$$

and the quantities:

$$N_j := \frac{N_{i,j} + N_{u,j}}{n_j}, \quad j = 0,\ldots,m \tag{6.56}$$

Equation 6.54 defines the critical load function (CLF, see Fig. 6.4) in the (N_{dep}, S_{dep})-plane, and in the following we look at this function in more detail. We assume that the sub-areas of the catchment are indexed in such a way that

$$N_{j-1} \le N_j \quad \text{for} \quad j = 1, \ldots, m \tag{6.57}$$

Between two successive values of N_j the CLF is linear, but at N_j it changes the slope (another of the brackets $(\ldots)_+$ in Eq. 6.54 becomes non-zero). The resulting piecewise linear function has (at most) $m+2$ segments, and every segment is of the form:

$$a_S \cdot S_{dep} + a_{N,k} \cdot N_{dep} = L_{N,k} + L_{crit} \;\; \text{for} \;\; N_{k-1} \le N_{dep} \le N_k, \quad k = 0, \ldots, m+1 \tag{6.58}$$

with the settings $N_{-1}=0$ and $N_m+1=\infty$. In Eq. 6.58 we introduced the dimensionless parameters:

$$a_{N,0} = 0, \quad a_{N,k} = (1-\rho_N) \cdot \sum_{j=0}^{k-1} c_j \cdot \left(1 - f_{de,j}\right) \cdot n_j, \quad k = 1, \ldots, m+1 \tag{6.59}$$

and the terms:

$$L_{N,0} = 0, \quad L_{N,k} = (1-\rho_N) \cdot \sum_{j=0}^{k-1} c_j \cdot \left(1 - f_{de,j}\right) \cdot \left(N_{i,j} + N_{u,j}\right), \quad k = 1, \ldots, m+1 \tag{6.60}$$

The maximum critical load of S is obtained by setting $N_{dep}=0$ in Eq. 6.54:

$$CL_{max}S = L_{crit} \,/\, a_S \tag{6.61}$$

To compute the maximum critical load of N one has to find the segment of the CLF that intersects the horizontal axis. The first segment is horizontal (since $a_{N,0} = 0$), and this segment extends till $N_{dep}=N_0$ (see Eqs. 6.56 and 6.57). Each of the following (at most) $m+1$ straight lines defined in Eq. 6.58 intersects the horizontal axis at (setting $S_{dep}=0$):

$$N_{0,k} = (L_{N,k} + L_{crit}) \,/\, a_{N,k}, \quad k = 1, \ldots, m+1 \tag{6.62}$$

And the $N_{0,k}$ that lies between the limits defined in Eq. 6.57 gives the maximum critical load of N. Denoting this specific index with K ($1 \le K \le m+1$), we have:

$$CL_{max}N = N_{0,K} = (L_{N,K} + L_{crit}) \,/\, a_{N,K} \quad \text{where} \quad N_{K-1} < N_{0,K} \le N_K \tag{6.63}$$

The first node of the CLF is $(0, CL_{max}S)$, the second one $(N_0, CL_{max}S)$. Note that in most applications N uptake and immobilization in the lake is assumed zero, i.e. $N_0=0$, and thus the second node coincides with the first. The next (maximum) $K-1$ nodes of this piecewise linear function are given by (N_k, S_k), where N_k is defined in Eq. 6.56 and the S_k are obtained as:

$$S_k = a_{N,k} \cdot (N_{0,k} - N_k) / a_S, \quad k = 1,\ldots,K-1 \tag{6.64}$$

And the last, at most (K+2)-nd, node is given by ($CL_{max}N$,0).

The base cation fluxes (Eq. 6.50) require, inter alia, estimates for the catchment weathering rates. Rapp and Bishop (2003) used the PROFILE model (Warfvinge and Sverdrup 1992, 1995) to estimate soil weathering rates in five Swedish catchments. In another study, Posch et al. (2007) used a geological weathering model (Lichtner 1992) to estimate base cation weathering in 100 high-altitude catchments in the southern Swiss Alps. Furthermore, the MAGIC model (e.g. Cosby et al. 2001) is routinely used to estimate average weathering rates for catchments, e.g. for the critical load calculations in Posch et al. (2012). The original—and by far most common—approach, however, is to estimate the net base cation flux with the SSWC model, using 'only' water chemistry data. This results in (compare Eq. 6.35):

$$L_{crit} = Q \cdot \left([BC^*]_0 - [ANC]_{lim} \right) \tag{6.65}$$

In addition to the input data required for the SSWC model, the FAB model needs also information on (a) the lake and catchment area, and land cover information, (b) terrestrial nitrogen sinks, and (c) parameters for in-lake retention of N and S.

While lake and catchment characteristics can be derived from (digital or paper) maps, data on the terrestrial N sinks are obtained in the same way as for the SMB model (see above), but expressed as catchment averages. In earlier FAB applications the denitrification fraction f_{de} has been estimated from the fraction of peatlands, f_{peat}, in the catchment as $f_{de} = 0.1 + 0.7 \cdot f_{peat}$ (Posch et al. 1997).

Concerning in-lake processes, the retention factor for nitrogen ρ_N is modelled by a kinetic equation (Kelly et al. 1987):

$$\rho_N = \frac{s_N}{s_N + z/\tau} = \frac{s_N}{s_N + Q/r} \tag{6.66}$$

where z is the mean lake depth, τ is the lake's residence time, $r = c_0$ is the lake:catchment ratio and s_N is the net mass transfer coefficient. Dillon and Molot (1990) give a range of 2–8 m yr^{-1} for s_N. Values for Canadian and Norwegian catchments are given in Kaste and Dillon (2003). An equation analogous Eq. 6.66 for ρ_S—with a mass transfer coefficient s_S—is used to model the in-lake retention of S. Baker and Brezonik (1988) give a range of 0.2–0.8 m yr^{-1} for s_S.

6.5 Calculation of Exceedances

Critical loads are derived to characterise the vulnerability (of parts/components) of ecosystems with respect to the deposition of one or more pollutants. In general, if the deposition of a pollutant at a given location is higher than the critical load of that pollutant at that location, it is said that the critical load is exceeded, and the difference is called *exceedance* (see Chap. 1 on the terminology). For a single pollutant, e.g., the critical load of nutrient N (see Eq. 6.5), the exceedance is thus defined as:

$$Ex_{nut}N = N_{dep} - CL_{nut}N \tag{6.67}$$

where N_{dep} is the (total) deposition of N. If the critical load is greater than or equal to the deposition, one says that it is not exceeded or there is non-exceedance of the critical load. An exceedance defined by Eq. 6.67 can obtain positive, negative or zero value. Since it is in most cases sufficient to know that there is non-exceedance (without being interested in the magnitude of non-exceedance), the exceedance of a single pollutant is also defined as:

$$Ex_{nut}N = \left(N_{dep} - CL_{nut}N\right)_{+} = \begin{cases} N_{dep} - CL_{nut}N & N_{dep} > CL_{nut}N \\ 0 & N_{dep} \leq CL_{nut}N \end{cases} \tag{6.68}$$

It should be noted that exceedances differ fundamentally from critical loads, as they are time-dependent. One can speak of *the* critical load (e.g. of nutrient N) for an ecosystem, but not of *the* exceedance of it. For exceedances the time, for which they have been calculated, has to be reported, since it is exceedances due to (past or future) *anthropogenic* depositions that are of interest.

Of course, the time-invariance of critical loads has its limitations, certainly when considering a geological time frame. But also during shorter time periods, such as decades or centuries, one can anticipate changes in the magnitude of critical loads due to global (climate) change, which influences the processes from which critical loads are derived. An example of a study of the (first-order) influence of temperature and precipitation changes on critical loads of acidity and nutrient N in Europe can be found in Posch (2002).

As shown (see Eqs. 6.19–6.21), there is no unique critical load of S and N acidity, and all deposition pairs (N_{dep},S_{dep}) lying on the critical load function (Fig. 6.1) lead to the critical leaching of ANC. Similarly, there is no unique exceedance of acidity critical loads, as illustrated in Fig. 6.5: Let the point E denote the (current) deposition of N and S. By reducing N_{dep} substantially, one reaches the point Z_N and thus non-exceedance without reducing S_{dep}; on the other hand one can reach non-exceedance by only reducing S_{dep} (by a smaller amount) until reaching Z_S; finally, with a reduction of both N_{dep} and S_{dep}, one can reach non-exceedance as well (e.g. point Z).

Intuitively, the reduction required in N and S deposition to reach point Z (see Fig. 6.5), i.e., the shortest distance to the critical load function, seems a good measure for exceedance. Thus we *define* the exceedance for a given pair of depositions (N_{dep},S_{dep}) as the sum of the N and S deposition reductions required to reach the critical load function by the 'shortest' path. Figure 6.6 depicts the five cases that can arise:

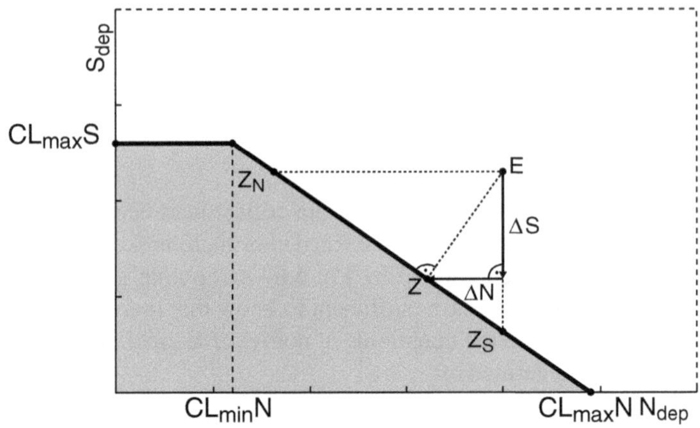

Fig. 6.5 Critical load function (CLF) of acidifying N and S in the (N_{dep}, S_{dep})-plane (see also Fig. 6.1). The *grey* area *below* the CLF denotes deposition pairs for which there is non-exceedance. Reducing emissions from point E one can reach, e.g., points Z_N, Z, or Z_S, showing that there is no unique exceedance. The critical load exceedance is calculated by adding the N and S deposition reductions needed to reach the CLF via the shortest path (E→Z): Ex=ΔS+ΔN

a. The deposition falls on or below the critical load function (Region 0). In this case the exceedance is defined as zero (non-exceedance);
b. The deposition falls into Region 1 (e.g. point E1). In this case the line perpendicular to the critical load function would yield a negative S_{dep}, and thus every exceedance in this region is defined as the sum of N and S deposition reduction needed to reach point Z1;
c. The deposition falls into Region 2 (e.g. point E2): this is the 'regular' case, the exceedance is given by the sum of N and S deposition reduction, *ExN+ExS*, required to reach the point Z2, such that the line E2-Z2 is perpendicular to the critical load function;
d. Region 3: every exceedance is defined as the sum of N and S deposition reduction needed to reach point Z3;
e. Region 4: the exceedance is simply defined as $S_{dep}-CL_{max}(S)$.

The exceedance function can be described by the following equation, in which the coordinates of point Z2 (see Fig. 6.6) are denoted by (N_0, S_0):

$$Ex(N_{dep}, S_{dep}) = \begin{cases} 0 & \text{if } (N_{dep}, S_{dep}) \in \text{Region 0} \\ N_{dep} - CL_{max}(N) + S_{dep} & \text{if } (N_{dep}, S_{dep}) \in \text{Region 1} \\ N_{dep} - N_0 + S_{dep} - S_0 & \text{if } (N_{dep}, S_{dep}) \in \text{Region 2} \\ N_{dep} - CL_{min}(N) + S_{dep} - CL_{max}(S) & \text{if } (N_{dep}, S_{dep}) \in \text{Region 3} \\ S_{dep} - CL_{max}(S) & \text{if } (N_{dep}, S_{dep}) \in \text{Region 4} \end{cases} \tag{6.69}$$

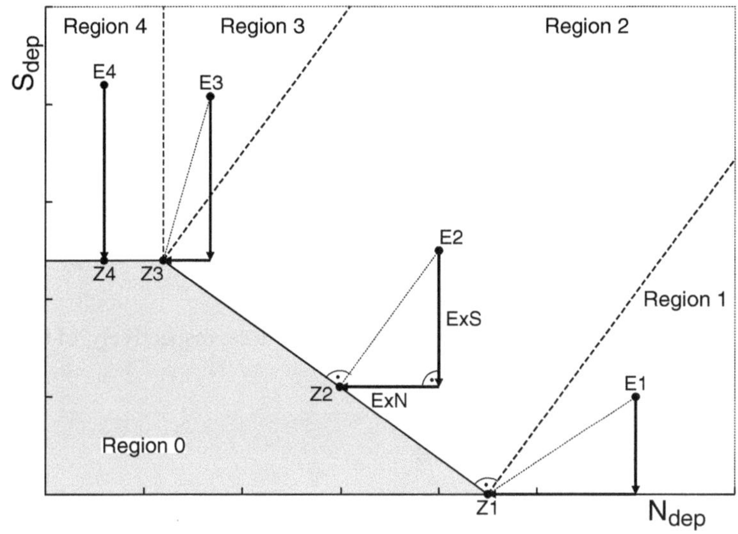

Fig. 6.6 The different cases for calculating the exceedance for a given critical load function

The function thus defined fulfils the criteria of a meaningful exceedance function: it is zero, if there is no exceedance of critical loads, positive when there is exceedance, and increasing in value when the point (N_{dep},S_{dep}) moves away from the critical load function.

Concerning critical loads for surface waters, exceedances of critical loads computed with the FAB model are defined in an analogous manner as exceedances for soils computed with the SMB model. Depending on the number of land cover classes within the catchment (and thus the number of nodes of the CLF; see Fig. 6.4), there are more 'cases' (and more 'regions' as defined in Fig. 6.6) of exceedances than for SMB critical loads, but the principles of calculation are the same.

Matters are different for critical loads according to the SSWC model. In that model, sulphate is assumed to be a mobile anion (i.e. leaching equals deposition), whereas N is assumed to a large extent to be retained in the catchment by various processes. Therefore, only the so-called present-day exceedance can be calculated from the leaching of N, N_{le}, determined as the sum of the measured concentrations of nitrate and ammonia in runoff. This 'present exceedance' is defined as (Henriksen and Posch 2001):

$$ExA = S_{dep} + N_{le} - CLA \tag{6.70}$$

where *CLA* is the critical load of acidity as computed with Eq. 6.35. No N deposition data are required for this exceedance calculation; however, *ExA* quantifies only the exceedance at present rates of retention of N in the catchment. Nitrogen processes are modelled explicitly in the FAB model (see above), and thus only that model can be used for comparing the effects of different N deposition scenarios on surface waters.

To compute a single exceedance value for a grid cell or region—needed, e.g., for mapping and scenario assessments (see Chap. 25)—the so-called average accumulated exceedance (AAE) has been introduced (UBA 2004; Posch et al. 2001). For *n* ecosystems within a grid cell or region it is defined as the area-weighted mean of the individual exceedances, i.e.:

$$AAE = \sum_{i=1}^{n} A_i Ex_i \,/\, \sum_{i=1}^{n} A_i \tag{6.71}$$

where Ex_i and A_i are the exceedance and ecosystem area, respectively, of ecosystem *i*.

6.6 Discussion and Outlook

The equations for computing critical loads of nutrient N as well as S and N acidity presented above are the standard forms widely applied in Europe and elsewhere (Forsius et al. 2010; Hettelingh et al. 1995, 2001, see also Chaps. 15–17) Needless to say, generalisations in many directions are possible. For example, the many N processes listed in Sect. 6.2 (e.g. N fixation), neglected in the current formulation of critical loads, could play a role at certain locations and could thus be included in the critical load equations. Furthermore, not only denitrification can be (modelled as) a function of deposition (see Eq. 6.4), other N processes will, to some extent, also be functions of the N deposition, e.g. N growth uptake. Also, these dependences on the N deposition need not be linear, leading to non-linear equations for the critical load, which might not be explicitly solvable. The lack of explicit equations may reduce the 'appeal' of the concept, but does not otherwise impede the calculation of critical loads. Also base cation fluxes may depend on the inputs of S and N (e.g., weathering depends on the pH and thus the input of acidifying N and S), and this could be incorporated as well.

The critical loads derived in this chapter are for total nitrogen, owing to the assumption of complete nitrification of all deposited ammonium. This is largely justified, since significant quantities of ammonium are rarely measure below the rooting zone. However, if reduced N is a major contributor and/or the critical load refers to the top layer of the soil, separate critical loads for reduced and oxidised N could be defined; and an example of separate oxidized and reduced nutrient N critical loads can be found in De Vries (1988) and Bonten et al. (2011). In the case of acidity the distinction between oxidised and reduced nitrogen leads to a 'critical load surface' in the $(NO_{x,dep}, NH_{y,dep}, S_{dep})$-space, with its shape depending on the nitrification model. In the acidity critical loads presented here base cation deposition has been taken as a constant at a given location (assuming that it is largely non-anthropogenic, as is the case in Europe). If, however, alkaline dust is part of the pollution problem, it makes sense to define critical load functions for which one of the axes is base cation deposition, thus providing a guide on how to 'optimally'

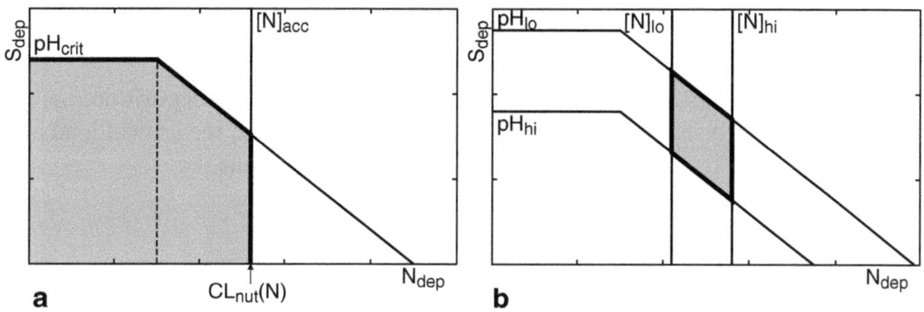

Fig. 6.7 **a** Critical load function (CLF) defined by an acceptable [N] *and* a critical pH. **b** CLF defined by *upper* and *lower* bounds for the chemical criteria, i.e. $[N]_{lo} < [N] < [N]_{hi}$ and $pH_{lo} < pH < pH_{hi}$. The *grey* areas denote those deposition pairs (N_{dep},S_{dep})—the 'optimal loads'—for which there is non-violation of the criteria

combine sulphur and base cation deposition reductions. For an example of such an approach in China, see Zhao et al. (2007).

Traditionally, critical loads of acidity and nutrient N are treated separately. However, if both the nutrient and the acidity status of an ecosystem are an issue, the chemical criteria for both effects can be considered simultaneously, and a single critical load function can be derived (see Fig. 6.7a). By definition, a critical load represents an upper limit for deposition(s): every deposition below the critical load is acceptable: the chemical criterion (or criteria) is (are) not violated and thus no 'harmful effects' are to be expected. In contrast, when correlating the occurrence/abundance of plant species with abiotic factors this often results in finite ranges for the abiotic factors under consideration. In the simplest case these intervals of 'good' values are independent, e.g. $[N]_{lo} < [N] < [N]_{hi}$ and $pH_{lo} < pH < pH_{hi}$. Using the critical load equations one can determine deposition pairs (N_{dep},S_{dep}) for which both of these pH and [N] conditions are fulfilled (see Fig. 6.7b). Every such deposition pair—which is also limited from below—could be called an 'optimal deposition' or 'optimal load'. Also in the case of correlated chemical limits, the critical load equations allow the determination of an 'optimal load region'; and for examples and a further discussion see Posch et al. (2011).

The appeal of critical loads lies in their simplicity, which makes them also applicable on a (large) regional scale and thus useful in the integrated assessment of emission reduction policies. As noted, critical loads are steady-state quantities, i.e. they do not have a time dimension. If depositions become equal to critical loads, one cannot say *when* the critical chemical value is attained, i.e. the danger of 'harmful effects' is averted. It may take a few years, or centuries, depending on the finite buffers in the soil, such as cation exchange capacity and N retention (these finite buffers do not influence steady state, i.e. critical loads). The time delays involved before a critical chemical value is attained under a given deposition (scenario) can only be estimated by dynamic models, i.e. models that take into account the finite buffers causing those delays; and some (simple) dynamic models and their applications are presented in Chap. 8.

6.7 Annex: Correcting for Sea Salts

Acidity critical loads are often compared with *anthropogenic* S deposition, i.e. the contribution due to sea spray is not included. If this is the case, the critical load of S has to be reduced by the S deposition originating from sea salts, i.e.

$$CL_{max}(S^*) = CL_{max}(S) - SO_{4,dep,ss} \tag{A.1}$$

where the asterisk indicates a sea salt corrected quantity and the subscript '*ss*' stands for sea salt derived. Ignoring ions such as Br, F, Sr, boric acid and bicarbonate, which occur only in traces in seawater, the charge balance of sea salt derived deposition reads:

$$SO_{4,dep,ss} = BC_{dep,ss} - Cl_{dep,ss} \tag{A.2}$$

Subtracting this from the critical load equation (Eq. 6.19) yields for the sea salt corrected critical load:

$$CL_{max}(S^*) = BC^*_{dep} - Cl^*_{dep} + BC_w - Bc_u - ANC_{le,crit} \tag{A.3}$$

The amounts of those ions in ocean water (unaffected by land drainage) are remarkably constant. This has been established by Dittmar (1884); and Dittmar's results were so consistent that later investigations introduced only minor changes, mostly with respect to more accurate atomic weights. Here we report the values given in Sverdrup et al. (1946), which are in turn based on data by Lyman and Fleming (1940). Table A.1 lists the amounts of the six major ions in seawater, their atomic weights and the calculated equivalents.

Assuming that either the Na or the Cl deposition at a given location derives only from sea salts, and using their globally constant ratio in sea water, the depositions

Table A.1 Major ions in seawater and their abundance

Ion	Amount in seawater[a] (g kg^{-1})	Molecular weight of ion[b] (mol g^{-1})	Equivalents in seawater (eq kg^{-1})
Ca^{2+}	0.4001	40.078	0.01997
Mg^{2+}	1.2720	24.305	0.10467
K^+	0.3800	39.098	0.00972
Na^+	10.5561	22.990	0.45916
Cl^-	18.9799	35.453	0.53545
SO_4^{2-}	2.6486	96.064	0.05514

[a] Sverdrup et al. (Sverdrup et al. 1946; p. 173)
[b] Weast et al. (1989)

Table A.2 Ion ratios $r_{XY}=[X]/[Y]$ (in eq/eq) in seawater. (Computed from Table A.1)

	X					
Y	Ca	Mg	K	Na	Cl	SO_4
Na	0.043	0.228	0.021	1	1.166	0.120
Cl	0.037	0.195	0.018	0.858	1	0.103

of base cations, sulphur and chloride (given in equivalents) are corrected for sea salts according to:

$$X^*_{dep} = X_{dep} - r_{XY} \cdot Y_{dep} \tag{A.4}$$

where $X=Ca,Mg,K,Na,Cl$ or SO_4, $Y=Na$ or Cl and r_{XY} is the ratio of ions X to Y in seawater. Ratios r_{XY} can by computed from the last column of Table A.1 and are shown in Table A.2 with 3-decimal accuracy. Note that if Na (Cl) is chosen to correct for sea salts, $Na^*_{dep}=0$ ($Cl^*_{dep}=0$).

It should be noted that the above procedure assumes that all quantities involved disperse in the atmosphere in the same way, which is not entirely true, especially for chloride. Nevertheless, given the dearth of dispersion modelling results for sea salts, the above procedure is widely used for locations not too far from the sea.

References

Aherne, J., Kelly-Quinn, M., & Farrell, E. P. (2002). A survey of lakes in the Republic of Ireland: Hydrochemical characteristics and acid sensitivity. *Ambio: A Journal of the Human Environment, 31,* 452–459.

Baker, L. A., & Brezonik, P. L. (1988). Dynamic model of in-lake alkalinity generation. *Water Resources Research, 24,* 65–74.

Battarbee, R. W., Allott, T. E. H., Juggins, S., Kreiser, A. M., Curtis, C., & Harriman, R. (1996). Critical loads of acidity to surface waters: An empirical diatom-based paleolimnological model. *Ambio: A Journal of the Human Environment, 25,* 366–369.

Bolt, G. H., & Bruggenwert, M. G. M. (1976). Composition of the soil. In G. H. Bolt & M. G. M. Bruggenwert (Eds.), *Soil chemistry A. Basic elements* (pp. 1–12). Amsterdam: Elsevier.

Bonten, L., Mol-Dijkstra, J., & Reinds, G. J. (2011). Validation of VSD+ and critical loads for nutrient N. In M. Posch, J. Slootweg, & J.-P. Hettelingh (Eds.), *Modelling critical thresholds and temporal changes of geochemistry and vegetation diversity: CCE status report 2011* (pp. 49–52). Bilthoven, The Netherlands: RIVM Report 680359003.

Brakke, D. F., Henriksen, A., & Norton, S. A. (1989). Estimated background concentrations of sulfate in dilute lakes. *Water Resources Bulletin, 25,* 247–253.

Brakke, D. F., Henriksen, A., & Norton, S. A. (1990). A variable F-factor to explain changes in base cation concentrations as a function of strong acid deposition. *Verh. Internat. Verein. Limnol., 24,* 146–149.

Brook, G. A., Folkoff, M. E., & Box, E. O. (1983). A world model of carbon dioxide. *Earth Surface Processes and Landforms, 8,* 79–88.

Burman, R., & Pochop, L. O. (1994). *Evaporation, evapotranspiration and climatic data*. Amsterdam: Elsevier.
Cosby, B. J., Ferrier, R. C., Jenkins, A., & Wright, R. F. (2001). Modelling the effects of acid deposition: Refinements, adjustments and inclusion of nitrogen dynamics in the MAGIC model. *Hydrology and Earth System Sciences, 5,* 499–517.
Cronan, C. S., Walker, W. J., & Bloom, P. R. (1986). Predicting aqueous aluminium concentrations in natural waters. *Nature, 324,* 140–143.
De Vries, W. (1988). Critical deposition levels for nitrogen and sulphur on Dutch forest ecosystems. *Water, Air, & Soil Pollution, 42,* 221–239.
De Vries, W. (1991). *Methodologies for the assessment and mapping of critical loads and the impact of abatement strategies on forest soils*. (Report 46). Wageningen, The Netherlands: DLO Winand Staring Center for Integrated Land, Soil and Water Research.
De Vries, W. (1993). Average critical loads for nitrogen and sulfur and its use in acidification abatement policy in the Netherlands. *Water, Air, & Soil Pollution, 68,* 399–434.
De Vries, W., & Posch, M. (2003). *Derivation of cation exchange constants for sand, loess, clay and peat soils on the basis of field measurements in the Netherlands* (Alterra Report 701). Wageningen: Alterra.
De Vries, W., Posch, M., Reinds, G. J., & Kämäri, J. (1993). *Critical loads and their exceedance on forest soils in Europe*. (Report 58 (revised version)). Wageningen, The Netherlands: DLO Winand Staring Centre.
De Vries, W., Reinds, G. J., & Posch, M. (1994). Assessment of critical loads and their exceedance on European forests using a one-layer steady-state model. *Water, Air, & Soil Pollution, 72,* 357–394.
De Wit, H. A., & Lindholm, M. (2010). *Nutrient enrichment effects of atmospheric N deposition on biology in oligotrophic surface waters—A review*. (NIVA-Report 6007). Oslo, Norway.
Dillon, P. J., & Molot, L. A. (1990). The role of ammonium and nitrate retention in the acidification of lakes and forested catchments. *Biogeochemistry, 11,* 23–43.
Dittmar, W. (1884). Report on researches into the composition of ocean water, collected by H.M.S. Challenger, during the years 1873–1876. *Physics and Chemistry, 1,* 1–251.
Driscoll, C. T., Lehtinen, M. D., & Sullivan, T. J. (1994). Modeling the acid-base chemistry of organic solutes in Adirondack, New York, lakes. *Water Resources Research, 30,* 297–306.
Dutch, J., & Ineson, P. (1990). Denitrification of an upland forest site. *Forestry, 63,* 363–377.
Eurosoil. (1999). *Metadata: soil geographical data base of Europe v. 3.2.8.0*. Ispra: Eurosoil.
Forsius, M., Posch, M., Aherne, J., Reinds, G. J., Christensen, J., & Hole, L. (2010). Assessing the impacts of long-range sulfur and nitrogen deposition on arctic and sub-arctic ecosystems. *Ambio: A Journal of the Human Environment, 39,* 136–147.
Gunn, J., & Trudgill, S. T. (1982). Carbon dioxide production and concentrations in the soil atmosphere: A case study from New Zealand volcanic ash soils. *Catena, 9,* 81–94.
Henriksen, A. (1984). Changes in base cation concentrations due to freshwater acidification. *Verh. Internat. Verein. Limnol., 22,* 692–698.
Henriksen, A., & Posch, M. (2001). Steady-state models for calculating critical loads of acidity for surface waters. *Water, Air, & Soil Pollution: Focus, 7,* 375–398.
Henriksen, A., Kämäri, J., Posch, M., & Wilander, A. (1992). Critical loads of acidity: Nordic surface waters. *Ambio: A Journal of the Human Environment, 21,* 356–363.
Henriksen, A., Forsius, M., Kämäri, J., Posch, M., & Wilander, A. (1993). *Exceedance of critical loads for lakes in Finland, Norway and Sweden: Reduction requirements for nitrogen and sulfur deposition*. (Report 2841). Oslo, Norway: Norwegian Institute for Water Research.
Hettelingh, J. P., Sverdrup, H., & Zhao, D. (1995). Deriving critical loads for Asia. *Water, Air, & Soil Pollution, 85,* 2565–2570.
Hettelingh, J.-P., Posch, M., & de Smet, P. A. M. (2001). Multi-effect critical loads used in multi-pollutant reduction agreements in Europe. *Water, Air, & Soil Pollution, 130,* 1133–1138.
Hindar, A., Posch, M., Henriksen, A., Gunn, J., & Snucins, E. (2000). *Development and application of the FAB model to calculate critical loads of S and N for lakes in the Killarney Provin-*

cial Park (Ontario, Canada). (Report 4202). Oslo, Norway: Norwegian Institute for Water Research.

Hornung, M., Sutton, M. A., & Wilson, R. B. (1995). *Mapping and modelling of critical loads for nitrogen: A workshop report*. United Kingdom: Institute of Terrestrial Ecology.

Jacobsen, C., Rademacher, P., Meesenburg, H., & Meiwes, K. J. (2003). *Element contents in tree compartments—Literature study and data collection* (*Berichte des Forschungszentrums Waldökosysteme*, Series B, Vol. 69). Göttingen: Niedersächsische Forstliche Versuchsanstalt (in German).

Johansson, M., & Tarvainen, T. (1997). Estimation of weathering rates for critical load calculations in Finland. *Environmental Geology, 29,* 158–164.

Johnson, D. W. (1984). Sulfur cycling in forests. *Biogeochemistry, 1,* 29–43.

Johnson, D. W., Henderson, G. S., Huff, D. D., Lindberg, S. E., Richter, D. D., Shriner, D. S., Todd, P. E., & Turner, J. (1982). Cycling of organic and inorganic sulphur in a chestnut oak forest. *Oecologia, 54,* 141–148.

Kaste Ø., & Dillon P. J. (2003). Inorganic nitrogen retention in acid-sensitive lakes in southern Norway and southern Ontario, Canada—A comparison of mass balance data with and empirical N retention model. *Hydrological Processes, 17,* 2393–2407.

Kelly, C. A., Rudd, J. W. M., Hesslein, R. H., Schindler, D. W., Dillon, P. J., Driscoll, C. T., Gherini, S. A., & Hecky, R. E. (1987). Prediction of biological acid neutralization in acid-sensitive lakes. *Biogeochemistry, 3,* 129–140.

Kimmins, J. P., Binkley, D., Chatarpaul, L., & de Catanzaro, J. (1985). *Biogeochemistry of temperate forest ecosystems literature on inventories and dynamics of biomass and nutrients* (Information Report PI-X-47E/F). Canada: Petawawa National Forestry Institute.

Larssen, T., & Høgåsen, T. (2003). *Critical loads and critical load exceedances in Norway [in Norwegian, with English summary and appendix]*. (Report 4722). Oslo, Norway: Norwegian Institute for Water Research.

Lichtner, P. C. (1992). Time-space continuum description of fluid/rock interaction in permeable media. *Water Resources Research, 28,* 3135–3155.

Lyman, J., & Fleming, R. H. (1940). Composition of sea water. *Journal of Marine Research, 3,* 134–146.

Monteith, J. L., & Unsworth, M. (1990). *Principles of environmental physics* (2nd ed.). London: Arnold.

Mulder, J., & Stein, A. (1994). The solubility of aluminum in acidic forest soils: Long-term changes due to acid deposition. *Geochimica et Cosmochimica Acta, 58,* 85–94.

NEG/ECP. (2001). *Protocol for assessment and mapping of forest sensitivity to atmospheric S and N deposition*. prepared by the NEG/ECP Forest Mapping Group, New England Governors/Eastern Canadian Premiers.

Nilsson, J., & Grennfelt, P. (1988). *Critical loads for sulphur and nitrogen*. Report from a Workshop held at Skokloster Sweden March 19–24 1988. Miljø rapport 1988: 15. Copenhagen Denmark Nordic Council of Ministers.

Oliver, B. G., Thurman, E. M., & Malcolm, R. L. (1983). The contribution of humic substances to the acidity of colored natural waters. *Geochimica et Cosmochimica Acta, 47,* 2031–2035.

Olsson, M., Rosén, K., & Melkerud, P. A. (1993). Regional modelling of base cation losses from Swedish forest soils due to whole tree harvesting. *Applied Geochemistry, 8,* 189–194.

Ouimet, R., Arp, P. A., Watmough, S. A., Aherne, J., & Demarchant, I. (2006). Determination and mapping critical loads of acidity and exceedances for upland forest soils in eastern Canada. *Water, Air, & Soil Pollution, 172,* 57–66.

Paces, T. (1983). Rate constants of dissolution derived from the measurements of mass balance in hydrological catchments. *Geochimica et Cosmochimica Acta, 47,* 1855–1863.

Posch, M. (2002). Impacts of climate change on critical loads and their exceedances in Europe. *Environmental Science Policy, 5,* 307–317.

Posch, M., & Kurz, D. (2007). A2M—A program to compute all possible mineral modes from geochemical analyses. *Computers and Geosciences, 33,* 563–572.

Posch, M., & Reinds, G. J. (2009). A very simple dynamic soil acidification model for scenario analyses and target load calculations. *Environmental Modelling and Software, 24,* 329–340.

Posch, M., Forsius, M., & Kämäri, J. (1993). Critical loads of sulfur and nitrogen for lakes 1: Model description and estimation of uncertainty. *Water, Air, & Soil Pollution, 66,* 173–192.

Posch, M., Kämäri, J., Forsius, M., Henriksen, A., & Wilander, A. (1997). Exceedance of critical loads for lakes in Finland, Norway and Sweden: Reduction requirements for acidifying nitrogen and sulfur deposition. *Environmental Management, 21,* 291–304.

Posch, M., Hettelingh, J.-P., & De Smet, P. A. M. (2001). Characterization of critical load exceedances in Europe. *Water, Air, & Soil Pollution, 130,* 1139–1144.

Posch, M., Hettelingh, J.-P., Slootweg, J., & Downing, R. J. (2003). *Modelling and mapping of critical thresholds in Europe—CCE status report 2003.* (RIVM Report 259101013). Bilthoven, The Netherlands.

Posch, M., Eggenberger, U., Kurz, D., & Rihm, B. (2007). *Critical loads of acidity for alpine lakes. A weathering rate calculation model and the generalized First-order Acidity Balance (FAB) model applied to alpine lake catchments.* (Environmental Studies 0709). Berne, Switzerland: Federal Office for the Environment.

Posch, M., Aherne, J., & Hettelingh, J.-P. (2011). Nitrogen critical loads using biodiversity-related critical limits. *Environmental Pollution, 159,* 2223–2227.

Posch, M., Aherne, J., Forsius, M., & Rask, M. (2012). Past, present, and future exceedance of critical loads of acidity for surface waters in Finland. *Environmental Science & Technology, 46,* 4507–4514.

Rapp, L., & Bishop, K. (2003). Modeling surface water critical loads with PROFILE: Possibilities and challenges. *Journal of Environmental Quality, 32,* 2290–2300.

Reinds, G. J., Posch, M., & De Vries, W. (2001). *A semi-empirical dynamic soil acidification model for use in spatially explicit integrated assessment models for Europe.* (Alterra rapport 84). Wageningen (Netherlands): Alterra.

Reuss, J. O., & Johnson, D. W. (1986). *Acid deposition and the acidification of soils and waters.* Berlin: Springer.

Rosén, K., Gundersen, P., Tegnhammar, L., Johansson, M., & Frogner, T. (1992). Nitrogen enrichment of Nordic forest ecosystems—The concept of critical loads. *Ambio: A Journal of Human Environment, 21,* 364–368.

Santore, R. C., Driscoll, C. T., & Aloi, M. (1995). A model of soil organic matter and its function in temperate forest soil development. In W. W. McFee & J. M. Kelly (Eds.), *Carbon forms and functions in forest soils* (pp. 275–298). Madison: Soil Science Society of America.

Smith, V. H., Tilman, G. D., & Nekola, J. C. (1999). Eutrophication: Impacts of excess nutrient inputs on freshwater, marine, and terrestrial ecosystems. *Environmental Pollution, 100,* 179–196.

Sogn, T. A., Stuanes, A. O., & Abrahamsen, G. (1999). The capacity of forest soil to absorb anthropogenic N. *Ambio: A Journal of Human Environment, 28,* 346–349.

Solberg, S., Dobbertin, M., Reinds, G. J., Lange, H., Andreassen, K., Fernandez, P. G., Hildingsson, A., & De Vries, W. (2009). Analyses of the impact of changes in atmospheric deposition and climate on forest growth in European monitoring plots: A stand growth approach. *Forest Ecology and Management, 258,* 1735–1750.

Sverdrup, H. U. (1990). *The kinetics of base cation release due to chemical weathering.* Sweden: Lund University Press.

Sverdrup, H., & De Vries, W. (1994). Calculating critical loads for acidity with the simple mass balance method. *Water, Air, & Soil Pollution, 72,* 143–162.

Sverdrup, H. U., & Warfvinge, P. G. (1988). Chemical weathering of minerals in the Gardsjön catchment in relation to a model based on laboratory rate coefficients. In J. Nilsson & P. Grennfelt (Eds.), *Critical loads for sulphur and nitrogen; Miljø rapport 1988 15* (pp. 131–149). Copenhagen: Nordic Council of Ministers.

Sverdrup, H. U., Johnson, M. W., & Fleming, R. H. (1946). *The oceans—Their physics, chemistry and general biology.* New York: Prentice-Hall.

Sverdrup, H., De Vries, W., & Henriksen, A. (1990). *Mapping critical loads. A guidance manual to criteria calculation methods data collection and mapping*. Miljø rapport 1990: 14. Nordic Council of Ministers Copenhagen 1990.

UBA. (2004). *Mapping manual 2004. Manual on methodologies and criteria for modelling and mapping critical loads and levels and air pollution effects, risks and trends*. Berlin, Germany: Texte 52/04, Federal Environmental Agency.

Ulrich, B., & Sumner, M. E. (Eds.). (1991). *Soil acidity*. Berlin: Springer.

UNECE. (1995). *Calculation of critical loads of nitrogen as a nutrient. Summary report on the development of a library of default values* (Document EB.AIR/WG.1/R.108). Geneva, Switzerland: United Nations Economic Commission for Europe.

UNECE. (2001). *Workshop on chemical criteria and critical limits* (Document EB.AIR/WG.1/2001/13). Geneva, Switzerland: United Nations Economic Commission for Europe.

Van der Salm, C., & De Vries, W. (2001). A review of the calculation procedure for critical acid loads for terrestrial ecosystems. *Science of the Total Environment, 271,* 11–25.

Van Loon, M., Tarrasón, L., & Posch, M. (2005). *Modelling base cations in Europe*. (EMEP Technical Report MSC-W 2/2005). Oslo, Norway.

Warfvinge, P., & Sverdrup, H. (1992). Calculating critical loads of acid deposition with PROFILE. A steady-state soil chemistry model. *Water, Air, & Soil Pollution, 63,* 119–143.

Warfvinge, P., & Sverdrup, H. (1995). *Critical loads of acidity to Swedish forest soils: Methods, data and results*. Lund: Dept. of Chemical Engineering, Lund University.

Warfvinge, P., Falkengren-Grerup, U., Sverdrup, H., & Andersen, B. (1993). Modelling long-term cation supply in acidified forest stands. *Environmental Pollution, 80,* 209–221.

Weast, R. C., Lide, D. R., Astle, M. J., & Beyer, W. H. (Eds.). (1989). *CRC handbook of chemistry and physics* (70th ed.). Boca Raton: CRC.

Wilander, A. (1994). Estimation of background sulphate concentrations in natural surface waters in Sweden. *Water, Air, & Soil Pollution, 75,* 371–387.

Zhao, Y., Duan, L., Larssen, T., Mulder, J., Hu, L., & Hao, J. (2007). Calculating critical loads for acidification for five forested catchments in China using an extended steady state function. *Science of the Total Environment, 387,* 54–67.

Chapter 7
Mass Balance Approaches to Assess Critical Loads and Target Loads of Metals for Terrestrial and Aquatic Ecosystems

Wim de Vries, Jan E. Groenenberg and Maximilian Posch

7.1 Introduction

Ecotoxicological and Human Health Impacts of Elevated Metal Inputs: Concern about of elevated concentrations of metals in terrestrial and aquatic ecosystems is related to: (i) ecosystem functions, including impacts on soil organisms/processes and plants (Bringmark et al. 1998; Palmborg et al. 1998) and on aquatic organisms, such as algae, crustacea and fish (Sola et al. 1995), (Baatrup 1991), and on (ii) human or animal health due to bioaccumulation in animal products (organs) and fish (Clark 2001). The latter effect is related to the phenomenon that certain metals accumulate through different trophic levels in a food chain (also denoted as secondary poisoning). Heavy metal accumulation in the food chain is considered important with respect to cadmium (Cd), mercury (Hg) and, to a lesser extent, lead (Pb). Accumulation ultimately causes toxic effects on: (i) humans by affecting food quality of crops (e.g. Kawada and Suzuki 1998) and animal products as well as drinking water quality and (ii) animal health by affecting fodder quality and due to direct intake of contaminated soil (Adriano 2001). For both humans and animals (e.g. cattle, birds and mammals), health effects arise mainly through accumulation in target organs like kidney and liver (Satarug et al. 2000). Apart from direct health effects related to intake of food and soil, elevated metal levels in soil also lead to an increase in leaching losses of metals to groundwater and surface water. This will, potentially after a considerable delay, affect both drinking water quality and aquatic organisms (Crommentuijn et al. 1997).

Metals are released naturally to soils and surface waters by chemical weathering, but it is the elevated inputs of metals from anthropogenic sources that are of concern. For example, Cd inputs by fertilizers and atmospheric deposition

W. de Vries (✉) · J. E. Groenenberg
Alterra Wageningen University and Research Centre, Wageningen, The Netherlands
e-mail: wim.devries@wur.nl

M. Posch
Coordination Centre for Effects (CCE), RIVM, Bilthoven, The Netherlands

W. de Vries et al. (eds.), *Critical Loads and Dynamic Risk Assessments*,
Environmental Pollution 25, DOI 10.1007/978-94-017-9508-1_7

increases its concentrations in agricultural soils and is associated with increasing Cd levels in crops (Anderson and Bingefors 1985; Jones et al. 1992), thus causing concern for the dietary intake of Cd. Atmospheric deposition is mostly the dominant input of Pb and Hg and this may cause effects on micro-organisms due to bioaccumulation in the organic layer of forests (e.g. Bringmark et al. 1998) or cause Hg contents in fish that affect food quality (e.g. Meili 1997; Meili et al. 2003).

Risk Assessment Approaches for Metal Contamination and the Critical Load Approach: Common practice in risk assessment of metal contamination is to compare present concentrations with critical concentrations at which adverse effects are to be expected. Critical limit values are set for food for human consumption (for an overview, see De Vries et al. 2007a) and for soil and surface waters to avoid ecotoxicological effects (for an overview, see De Vries et al. 2007b). The latter limits are also summarized in Chap. 2 as they are relevant for all non-agricultural systems, where atmospheric deposition is the only source of metal inputs. Such limit values, however, do not give information on the metal inputs for which those limit values are not exceeded in the long run. This shortcoming is overcome with the critical load concept. According to its definition (see Chap. 2), a critical metal load equals the metal input rate that will in the long run—at steady state—lead to a critical metal concentration (critical limit) for a defined receptor (e.g. soil solid phase, soil solution, groundwater, plant, sediment, fish). A critical load thus links—either empirically or through modelling—the limit value (critical limit) to the input.

A critical load does not consider finite buffers for the metal and thus neglects the delays between the (non-)exceedance of the critical load and the (non-)violation of the critical limit. To gain insight into this temporal behaviour, dynamic models are needed. The various possibilities that can occur are graphically illustrated in Fig. 7.1, showing the change in metal concentration as a function of time for various combinations of initial concentrations, the critical concentration and steady-state concentrations for a constant input. In the left figure, the initial metal concentration is below the critical limit. A metal input below the critical load results in a steady-state concentration below the critical limit. An input above the critical load results in a steady state above the critical limit, and it takes a finite time to exceed this critical limit, called the damage delay time (DDT). In the right figure, the concentration is initially above the critical limit (i.e., the system is already potentially at risk). An input equal to the critical load results (technically, after infinite time) in a steady state, with the concentration equalling the critical limit. Again, inputs above/below the critical load eventually result in steady states above/below the critical limit. And the time it takes to reach the critical limit from above is called the recovery delay time (RDT). The deposition level at which a critical limit is attained within a predefined time horizon is a so-called target load. Note that a target load is only meaningfully defined when the current metal concentration exceeds a critical limit. An in-depth discussion and illustrative examples of these concepts can be found in Posch and De Vries

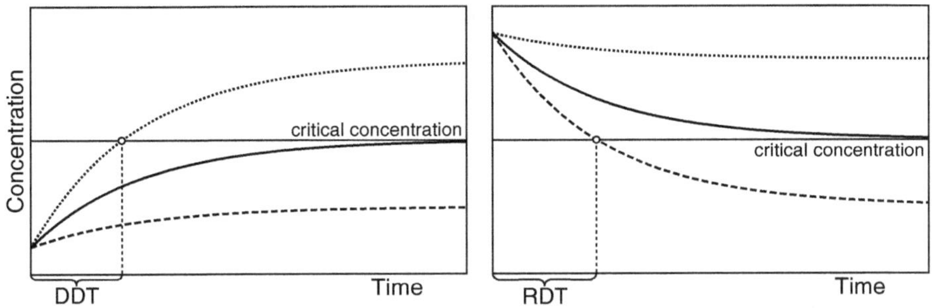

Fig. 7.1 Schematic diagram illustrating the change in metal concentration at constant input as a function of time with the initial concentration below (*left*) and above (*right*) the critical concentration (*DDT* damage delay time, *RDT* recovery delay time). The *dotted/solid/dashed lines* are for metal inputs above/at/below the critical load

(2009). More details on delay times and target loads in view of acidity can be found in Chap. 8.

The method to calculate critical loads of metals is based on the balance of all relevant metal fluxes in and out of a considered ecosystem in a steady-state situation. First approaches were described in the 'Manual for Calculating Critical Loads of Heavy Metals for Terrestrial Ecosystems' (De Vries and Bakker 1998) and '… for Aquatic Ecosystems' (De Vries et al. 1998). A state-of-the-art methodology manual, developed under the LRTAP Convention, has been adopted for the "priority metals" Cd, Pb and Hg, which are the subject of the 1998 Heavy Metals Protocol to the LRTAP Convention, as described further in the Mapping Manual (www.icpmapping.org) and its background document (De Vries et al. 2005). Until now, the approach has mainly been applied within Europe (e.g. Slootweg et al. 2007) and Canada (Doyle et al. 2003).

Contents of this Chapter: We first describe the methodology to compute critical loads and target loads based on critical limits for metal concentrations in soil solution and surface waters related to ecotoxicological impacts on terrestrial and aquatic ecosystems and human health impacts (Sect. 7.2; see also Chap. 2). We then present examples of applications on a local scale for critical loads of Cd in view of food quality, Cd, Pb, Cu and Zn in view of soil biodiversity and for Hg in view of impacts on fish (Sect. 7.3). Finally, limitations of the approach are discussed in view of the applicability of the critical load concept for heavy metals in future negotiations on the reduction in metal emissions (Sect. 7.4). Various aspects are described similarly in De Vries et al. (2012), but these authors limited their description to impacts on ecosystem functions, whereas the current chapter also considers critical metal loads in view of human health effects.

7.2 Methods to Compute Critical Loads and Target Loads

7.2.1 Steady-State Model to Compute Critical Loads

Simple Mass Balance Model to Compute Critical Metal Loads: Critical loads can be calculated from a simple mass balance of all relevant metal fluxes into and out of the relevant soil compartment at steady state according to:

$$M_{in} = M_{tu} + M_{le} - M_{we} - M_{lf} \tag{7.1}$$

where all terms are fluxes of metal M (in mg $m^{-2}yr^{-1}$). The subscripts *in* refer to the total input by deposition and other loads (e.g. fertilisers), *tu* to total uptake, *le* to leaching/runoff, *we* to weathering and *lf* to litterfall. In view of the uncertainties in metal adsorption and related leaching, it is considered appropriate to neglect the cycling of metals and simply include a net uptake of metals, as explained further in the Mapping Manual (www.icpmapping.org) and its background document (De Vries et al. 2005). Furthermore, neglecting metal weathering and describing the metal leaching/runoff as the product of a water percolation flux and a dissolved metal concentration, Eq. 7.1 can be simplified to:

$$M_{in} = M_u + Q \cdot [M]_{tot} \tag{7.2}$$

where M_u is the net metal uptake by the harvested parts of plants (mg ha^{-1} $year^{-1}$), Q is the drainage water flux (m^3 ha^{-1} $year^{-1}$) and $[M]_{tot}$ is the total metal concentration in soil solution (mg m^{-3}). The assumptions made for this steady-state model are: (i) the soil is homogeneously mixed, (ii) the soil is in an oxidised state and metal partitioning can be described with equilibrium sorption and complexation with DOC, (iii) the transport of water and metals only occurs vertically (no metal gain or loss by seepage flow, surface runoff or bypass flow), (iv) metal weathering and erosion are not considered, (v) the cycling of metals by litterfall, mineralization and uptake is neglected, and (vi) the metal input is chemically reactive and 100 % available for exchange with the solution phase. The limitations due to these assumptions are discussed in De Vries et al. (2005) and in De Vries and Groenenberg (2009).

The critical load (CL) of metal M for a given effect, *CL(M)*, is the input flux at which the metal concentration is at its critical value:

$$CL(M) = M_u + Q \cdot [M]_{tot,crit} \tag{7.3}$$

where $[M]_{tot,crit}$ is the critical total metal concentration in soil solution (mg m^{-3}).

Computing Critical Total Metal Concentrations in Soil Solution: A critical limit depends on the endpoint that has to be protected, such as crop quality, soil organisms or surface water organisms. In case of protection of surface water organisms, critical limits for dissolved total Cd, Pb, Hg, Cu and Zn concentrations in surface water are directly available (see Chap. 2; Table 2.3) and can be inserted to calculate a critical metal load. However, when using critical limits for metals in crops (food quality criteria), a critical limit in soil solution has to be back-calculated by using soil-plant relationships and a relation between dissolved and solid phase metal concentrations, all being dependent on soil properties, as described in Chap. 2 (see Eqs. 2.8–2.10).

Similarly, when aiming to protect soil organisms, a critical metal concentration in the soil or a critical free metal ion (FMI) concentration is used, as described in Chap. 2, requiring a procedure to derive a critical total metal concentration in soil solution. For example, the calculation of critical total metal concentration in soil solution from a critical FMI uses a geochemical speciation model and requires inputs of soil solution pH, dissolved organic carbon, % soil organic matter, and partial pressure of CO_2 in the soil. A full description of methodologies for calculating $[M]_{tot,crit}$ for the different indicators can be found in the Mapping Manual (www.icpmapping.org) and its background document (De Vries et al. 2005). Below, we summarize the method applied when using a critical FMI.

When the critical limit is defined in terms of a critical FMI concentration, the total metal concentration in soil solution, $[M]_{tot}$, has to be determined as the sum of the concentration of (i) the FMI, $[M]_{free}$, (ii) dissolved inorganic complexes, $[M]_{DIC}$, such as MOH^+, $MHCO_3^+$, MCl^+, and (iii) metals bound to dissolved organic matter, according to:

$$[M]_{tot} = [M]_{free} + [M]_{DIC} + [M]_{DOM} \cdot [DOM] \tag{7.4}$$

where $[M]_{DOM}$ is the metal concentration bound to dissolved organic matter (mol kg^{-1} DOM) and $[DOM]$ is the concentration of dissolved organic matter (kg m^{-3}). By assuming geochemical equilibrium, the partitioning and speciation of metals over the various fractions can be calculated. Given the free metal ion activity, the concentrations of the other metal species can be estimated with an equilibrium speciation model, such as WHAMVI (Tipping 1994, 1998). The calculation takes into account the dependence of metal binding to DOM on pH and competitive effects due to major cationic species of Mg, Al, Ca and Fe (Tipping et al. 2002; Tipping 2005). A customised program (W6-MTC), based on WHAMVI is often used in critical load calculations; for a detailed description of the calculation procedure see De Vries et al. (2005). DOM is calculated as 2·DOC, i.e. assuming a carbon content of 50 %. Binding of metals to DOM is calculated assuming that 65 % of DOM act as fulvic acid and 35 % of DOM are not involved in metal binding. This fraction is based on application of WHAMVI to field and laboratory data for waters and soils involving Al (Tipping et al. 1991, 2002), Cu (Bryan et al. 2002; Dwane and Tipping 1998; Vulkan et al. 2000), and Cd (Tipping 2002).

7.2.2 Simple Dynamic Models to Compute Target Loads

Critical loads are by definition a tool for quantifying the potential risk 'in the long term', where 'long term' means the length of time to (approximately) reach steady state. Dynamic modelling of metal cycling in terrestrial systems (De Vries and Groenenberg 2009; Paces et al. 2002; Posch and De Vries 2009) indicates that, starting from the current situation, it may take decades to millennia to reach a steady state. Consequently, in a soil in which the critical limit is currently exceeded, an input of metals equal to the critical load would cause the system to remain at a potential risk for a (very) long time. This time could be reduced by requiring a load *below* the critical load, termed a 'target load'. A target load is defined as 'the deposition for which a pre-defined chemical or biological status is reached in the specified target year and maintained (or improved) thereafter' (Posch et al. 2003). Target loads are currently quantified under the LRTAP Convention for acidifying pollutants (see Bonten et al. Chap. 8); and an approach for metals has been described in Posch and De Vries (2009). Target loads might also find application in maintaining concentrations of essential trace elements, such as copper and zinc, where this is important for agriculture and livestock rearing.

The computation of target loads requires that the changes in the solid and liquid metal pools are included in the model, which thus becomes:

$$\rho z \frac{d}{dt} ctM_{tot} + \theta z \frac{d}{dt}[M]_{tot} = M_{in} - M_u - Q \cdot [M]_{tot} \tag{7.5}$$

Where ctM_{tot} is the total content of heavy metal in the soil solid phase (mg kg^{-1}), ρ is the bulk density of the soil (kg m^{-3}), θ is the water content (m^3 m^{-3}) and z the thickness of the soil layer (m). Setting the time-derivatives in Eq. 7.5 to zero yields the steady-state equation (Eq. 7.2). To solve Eq. 7.5, a relationship between the total content and the total concentration in solution is required. In general, this is assumed to be in the form of a Freundlich isotherm:

$$ctM_{tot} = K \cdot [M]_{tot}^n \tag{7.6}$$

with a metal-dependent exponent n and a site-specific Freundlich adsorption constant K. In the case of a linear relation between the metal content in the solid phase and the concentration in solution ($n=1$), Eq. 7.5 can be solved analytically in terms of elementary functions (exponentials) and the concentration as a function of time t is obtained as:

$$[M]_{tot}(t) = [M]_{tot,ss} + ([M]_{tot,0} - [M]_{tot,ss})e^{-t/t_c} \tag{7.7}$$

where $[M]_{tot,ss}=(M_{in}-M_u)/Q$ is the steady-state concentration for the constant input M_{in}, $[M]_{tot}$,0 is the initial concentration in soil solution, and t_c is defined as:

$$t_c = \frac{\rho \cdot K \cdot z + \theta \cdot z}{Q} \tag{7.8}$$

The quantity t_c has the dimension of a time and is characteristic for the system under consideration; t_c is also called e-folding time, i.e. the time required to decrease a quantity to 100/e≈37% of the original amount. It equals the total metal pool in the soil at steady state divided by the steady-state net metal input, $Q \cdot [M]_{tot,ss}$. The characteristic time t_c is thus the time needed to completely fill the soil metal pool with (steady-state) net inputs. From Eq. 7.7 also the time to reach a pre-scribed concentration can be easily derived. For example, for $[M]_{tot,ss} > [M]_{tot,crit} > [M]_{tot,0}$ (see left panel in Fig. 7.1) the damage delay time, DDT, is obtained as:

$$DDT = t_c \cdot \log \frac{[M]_{tot,ss} - [M]_{tot,0}}{[M]_{tot,ss} - [M]_{tot,crit}} \tag{7.9}$$

The same formula holds for the recovery delay time, *RDT*; in this case $[M]_{tot,ss} < [M]_{tot,crit} < [M]_{tot}, 0$ (see right panel in Fig. 7.1). In the linear case also an explicit expression for the target load, *TL*(*M*), can be derived for any given target time (year) t_{tar}:

$$TL(M) = M_u + \frac{Q}{1 - e^{-t_{tar}/t_c}} \left([M]_{tot,crit} - [M]_{tot,0} \cdot e^{-t_{tar}/t_c} \right) \tag{7.10}$$

This equation has been used to compute target loads of Cd, Cu, Pb and Zn for a generic deciduous forest on five major soils in the Netherlands differing in texture and organic matter content (De Vries and Groenenberg 2009).

The dependence of the target load on the target year and initial condition can be investigated independently of site conditions by introducing the following scaled variables:

$$z_{tar} = \frac{TL(M) - M_u}{CL(M) - M_u}, \quad z_0 = \frac{[M]_{tot,0}}{[M]_{tot,crit}}, \quad T_{tar} = \frac{t_{tar}}{t_c} \tag{7.11}$$

In these dimensionless variables Eq. 7.10 reads:

$$z_{tar} = \frac{1 - z_0 e^{-T_{tar}}}{1 - e^{-T_{tar}}} \tag{7.12}$$

The target load as a function of the target time is shown in Fig. 7.2a. For very large target times the target load z_{tar} converges to one, irrespective of the initial condition, This reflects the fact that for $t \rightarrow \infty$ the target load approaches the critical load (Eq. 7.12)—in other words, the critical load is the target load for an infinite time horizon. For $z_0 > 1$, i.e. initial concentrations greater than the critical concentration, $z_{tar} < 1$, the target load is smaller than the critical load, i.e. one has to reduce inputs below the critical load to obtain the critical concentration in the prescribed time T_{tar}.

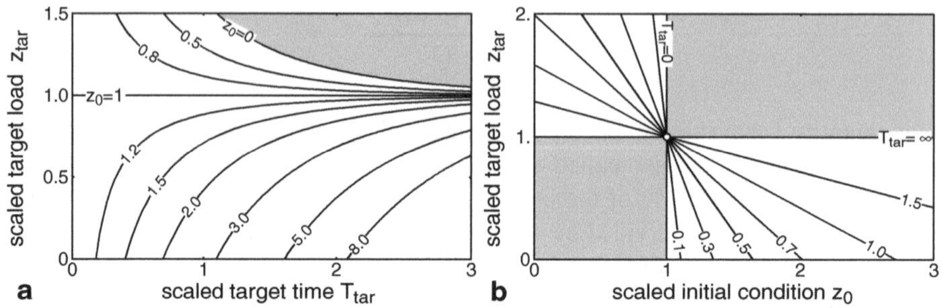

Fig. 7.2 Scaled target load z_{tar} (**a**) as a function of the scaled target time T_{tar} for selected initial conditions z_0; and (**b**) as a function of the scaled initial condition z_0 for selected target times T_{tar}

The dependence of the target load on the initial condition is illustrated in Fig. 7.2b; and in the special case of $n=1$ (Eq. 7.6) this dependence is linear. The grey areas in Fig. 7.2 cover the impossible combinations of target times and target loads. In both panels, also cases are displayed that are theoretically possible, but have little bearing on (political) reality: target loads greater than the critical load for initial conditions below the critical concentration (above $z_0=1$ in Fig. 7.2a and left of $z_0=1$ in Fig. 7.2b). The graphs in Fig. 7.2 are universal, i.e. independent of site properties; but for a concrete site any information derived from those graphs can be translated into 'real' numbers with the aid of Eq. 7.11.

The amount of metals in soil solution is in most cases small compared to the total amount of metals in the solid phase. Therefore the change in the liquid phase is often neglected in simple dynamic models (this can be done by setting $\theta=0$ in Eq. 7.5). For example, Posch and De Vries (2009) investigated the solutions of Eq. 7.5 for $\theta=0$ for the general non-linear case of metal partitioning (Eq. 7.6). Also in this case a characteristic time can be defined which, however, also depends on the constant net metal input:

$$t_c = \frac{\rho \cdot z \cdot K}{Q^n \cdot (M_{in} - M_u)^{1-n}} \tag{7.13}$$

A comparable approach is the CEH Simple Dynamic Model for Metals (SDMM) (Ashmore et al. 2004).

7.3 Examples of Critical Load and Target Load Calculations

To date, critical loads for surface waters have hardly been assessed. Surface water metal concentrations have been calculated using the dynamic CHUM-AM model (including WHAMVI), simulating impacts of different scenarios on both acidification and metal behaviour for catchments in Cumbria in north-west England (Tipping et al. 2006a, b), the Lochnagar catchment in Scotland (Tipping et al. 2007) and a number of other catchments in England (Ashmore et al. 2004). Model results suggested that the weakly sorbing metals, such as Cd and Zn, will respond on timescales of decades to centuries to changes in metal inputs or acidification status, while time scales for strongly sorbing metals, such as Cu and Pb, will be centuries to millennia. Unfortunately, the simulations have not yet been used to derive critical loads. Our site-scale examples are thus limited to available critical load calculations for soils, distinguishing between agricultural soils in view of food quality and non-agricultural soils in view of impacts on soil organisms. In both cases, the critical loads and target loads are calculated for typical land use soil combinations.

Critical Loads of Cadmium in View of Food Quality Impacts: De Vries and McLaughlin (2013) presented a model analysis, predicting Cd concentrations in top soils, soil solution and crops of four key Australian agricultural systems in response to changes in Cd inputs for the period 1900–2100, to assess potential risks. In addition they estimated critical Cd loads in view of food quality criteria to assess what quality of phosphate (P) fertilizer is needed to safeguard food quality in the long term. Results for three of those systems are given in Table 7.1. The agricultural systems represent extremes in terms of Cd inputs by fertilizer and hydrology, i.e.: (i) dryland cereals (wheat, canola, barley and wheat) typical for Southern and Western Australia in areas with very low rainfall, (ii) sugarcane and rotational crops (peanuts or soybeans) typical for coastal Queensland in areas with very high rainfall, and (iii) intensive rotational horticulture (potatoes, carrots, onions and broccoli) typical for areas nearby larger Australian cities with average rainfall. Critical Cd limits for the food crops included in the

Table 7.1 Calculated critical loads for Cd for typical Australian agricultural systems. (After De Vries and McLaughlin 2013)

Land use	Soil type	Cd_u [a] ($g\ ha^{-1}\ year^{-1}$)	Q[b] ($m^3\ ha^{-1}\ year^{-1}$)	$[Cd]_{tot}(crit)$[b] ($mg\ m^{-3}$)	CL of Cd[b] ($g\ ha^{-1}\ year^{-1}$)
Dry land cereals	Calcareous clay	0.30	300	0.24	0.38
	Sand	0.30	600	4.17	2.85
Sugarcane rotation	Heavy clay	0.37	9300	0.60	5.91
	Loam	0.37	7200	2.46	18.13
Intensive annual horticulture	Heavy clay	0.40	8300	0.38	7.16
	Loam	0.40	4700	3.68	21.20

[a] The net Cd uptake is derived by multiplying the harvested product ($t\ ha^{-1}\ year^{-1}$) of the most sensitive crop in the rotation, i.e. wheat in the dryland cereals, peanuts in the sugarcane rotation and potatoes in intensive annual horticulture, with a maximum level of 0.1 mg Cd kg^{-1} (g ton^{-1})

[b] The critical load of Cd is calculated as the sum of the net Cd uptake and the critical Cd leaching rate, being the product of the water flux leaving the depth of cultivation (Q) and the critical Cd concentration in soil solution $[Cd]_{tot}(crit)$

simulations (wheat, barley, peanuts, soybean, potatoes and vegetables) were are all set at 0.1 mg kg^{-1} fresh weight. For sugarcane and canola, there are no limits in Australia and therefore a limit of 0.2 mg kg^{-1} was used as applied in Europe.

A critical Cd load is determined by the net Cd uptake, the drainage water flux and a critical total Cd concentration in soil solution (see Eq. 7.3). The critical concentration, determined by the applied food quality criteria and the soil-plant and soil-solution relationships used (see also Chap. 2), is influenced by soil properties and is generally higher for sand and loamy soils than for clay soils (see Table 7.1). This concentration, in combination with the drainage flux, largely determined the critical Cd load, which varied between 0.4 and 21.2 g ha^{-1} $year^{-1}$. The calculated critical loads were (much) lower than the proposed annual application limit for Cd suggested by the National Cadmium Minimisation Strategy in Australia, being equal to 30 g ha^{-1} $year^{-1}$ averaged over 5 years. Calculated limits for Cd in P fertilizer for the three agricultural systems varied between 19 (dry land cereals on calcareous clay) and 363 mg Cd kg P^{-1} (sugarcane rotation on loam). The current average quality of P fertilizers actually in use in Australia (~60 mg Cd kg P^{-1}) only exceeded the critical limit of 19 for dryland cereal production on calcareous clay. In these systems the time to reach the critical soil Cd concentration was, however, over 600 years (De Vries and McLaughlin 2013).

Critical Loads and Target Loads of Cadmium, Lead, Copper and Zinc in View of Impacts on Soil Organisms: De Vries and Groenenberg (2009) illustrated the impacts of differences in various model approaches by examples for Cd, Cu, Pb and Zn for a deciduous forest on major soils differing in texture and organic matter content, including a non-calcareous sandy soil and a loess, clay and peat soil. The influence of soil type on the critical loads for Cd, Pb, Cu and Zn in the mineral layer, using Eq. 7.3, is shown in Table 7.2, using the approach described in this chapter. The small differences that occur are the result of differences in hydrology, forest growth, pH and DOC concentrations. Critical loads for Cd and Zn do not differ too much for the different soils, but for Cu and Pb a larger variation is found. This is due to the fact that the total concentration of these metals (including metal bound to DOC) depends much more strongly on pH and DOC concentrations than the Cd and Zn concentrations. The highest CLs are obtained for peat soils because of the high DOC concentration.

Table 7.2 Critical loads for Cd, Pb, Cu and Zn for different soil types (after De Vries and Groenenberg 2009)

Soil type	Critical load (mg m^{-2} $year^{-1}$)			
	Cd	Pb	Cu	Zn
Sand	1.5	9.6	12	47
Loess	1.0	5.0	5.0	39
Clay	2.7	5.6	14	53
Peat	1.9	20	28	62

Target loads (Eq. 7.10) were also calculated but in nearly all cases, the used current metal concentrations were below the critical limits, implying that "target loads" were significantly larger than the critical loads (De Vries and Groenenberg 2009). Note, however, that a target load is not meaningful here, since there are no policy targets to increase metal concentrations in soils up to a limit within a prescribed time horizon.

The examples by De Vries and Groenenberg (2009) assume a linear relationship between soil and dissolved metal. Posch and De Vries (2009) computed characteristic times for sand, clay and peat soils using average Dutch data for grasslands and a uniform soil depth $z=0.1$ m, while accounting for non-linear speciation (Table 7.3). As can be seen, Pb shows by far the slowest response for all soil classes, whereas Zn shows the widest variability among different soils. In general, the characteristic times for all metals and soils listed show that steady state is reached very slowly, and that they are far beyond the time horizons usually considered in (environmental) policies.

Critical Limit of Mercury in Precipitation in View of Impacts on Fish and Soil Organisms: Mercury is regarded as a metal of environmental concern in northern Europe because of its accumulation to toxic levels in freshwater fish, thus rendering them not edible, and to a lesser extent because of potential toxic effects on soil biota. Meili et al. (2003) applied a process- and scale-independent model formulation to assess a critical Hg concentration in precipitation. They used concentration ratios and transfer factors of Hg based on Hg concentrations in precipitation, large lacustrine fish, forest soils and lake sediments. They calculated a critical Hg concentration in precipitation of 2 ng l^{-1} to keep Hg in Swedish soils below 0.5 mg kg^{-1} in soil organic matter and Hg in fish (pike) in the most sensitive watersheds below 0.5 mg kg^{-1} fresh weight. This concentration is about half of the present level in the most remote areas of the northern hemisphere, and similar to the pre-industrial background, implying that virtually all regional as well as global anthropogenic emissions need to be eliminated to achieve the goal (Meili et al. 2003).

Table 7.3 Characteristic times of Cd, Pb, Cu and Zn for sand, clay and peat soils computed from the average data for grasslands in the Netherlands. (Posch and De Vries 2009)

Soil type	Characteristic time t_c (year)			
	Cd ($n=0.50$)	Pb ($n=0.67$)	Cu ($n=0.51$)	Zn ($n=0.61$)
Sand	297	2002	225	148
Clay	1179	5435	643	1296
Peat	648	4234	491	342

7.4 Discussion and Conclusions

7.4.1 Reliability of Critical Loads and Target Loads

Parameters Influencing the Reliability: The reliability of critical loads and target loads mainly depends on the reliability of the various parameters determining these loads. Both loads are influenced by metal uptake (yield and metal content in stem wood) and the water flux leaving the root zone. The uncertainty is further influenced by the critical dissolved metal concentration, while target loads are also influenced by the present metal content in soil and the Freundlich adsorption constant, *K*, which in turn is influenced by the uncertainty in the transfer function coefficients and soil properties, specifically pH and soil organic content (see Chap. 2). Ashmore et al. (2004) concluded from an uncertainty analysis that the uncertainty in critical limit function dominates the uncertainty in critical loads. As mentioned by these authors, this reflects the limited database available for the derivation of the critical limit function. The uncertainty in the critical total metal concentration in soil solution is further determined by the uncertainty in the chemical speciation model, calculating total metal concentrations from free metal ion concentrations. With respect to target loads, the uncertainty in *K* is most relevant. Both aspects are discussed below.

Speciation Modelling in Soil Solution: The uncertainty of the calculation of total metal concentrations from free metal concentrations with a chemical speciation model, such as WHAMVI (W6-MTC), is due to uncertainties in the model approach, the model inputs (DOC, Al and Fe concentrations in solution) and the parameterisation of the model. Especially the variable fraction and binding properties of fulvic acids in dissolved organic matter contribute to substantial uncertainties in predicted metal speciation (Groenenberg et al. 2010a). To get an idea of the adequacy of such an approach, De Vries and Groenenberg (2009) used the W6-MTC model to calculate total metal concentrations from FMI activities, similar to the approach in the CL calculations, for data sets in which both total and FMI concentrations in solution were measured (Pampura et al. 2006, 2007; Weng et al. 2001). Input data to W6-MTC were the FMI activity, the organic matter content, the pH and DOC. Because DOC is usually not available at regional scale, De Vries and Groenenberg (2009) also made calculations in which DOC was set to standard values (De Vries et al. 2005). The Mean Absolute Error (MAE) of the log-concentration (Table 7.4) varied between 0.11 for Zn and 0.45 for Pb when using the measured DOC concentrations. This means that the concentrations on a linear scale vary by a factor 1.3–2.8. The error is largest for the strongly complexing metals Cu and Pb, for which calculations are also most sensitive to variation in estimated DOC concentrations (compare DOC measured and DOC default in Table 7.4).

Uncertainties in Transfer Functions Between Soil and Soil Solution: Unlike critical loads, target loads are strongly influenced by the transfer functions for solid solution partitioning of metals. Sorption and ion exchange are reasonably well understood

Table 7.4 Mean Absolute Errors of dissolved total Cd, Pb, Cu and Zn concentrations (mg m^{-3}) calculated with W6-MTC compared to measured concentrations, using measured and default values for DOC. (De Vries and Groenenberg 2009)

Metal	Mean absolute error (log-concentration)	
	DOC measured	DOC default
Cd	0.32	0.38
Pb	0.45	0.55
Cu	0.40	0.57
Zn	0.11	0.13

Table 7.5 Calculated and measured dissolved total Cd, Pb, Cu and Zn concentrations in the mineral topsoil of 56 soil samples in 5 land-use types in the UK. Values are medians, with the range between the 5th and 95th percentile in parentheses. (De Vries and Groenenberg 2009)

Metal	Dissolved total metal concentration (mg m^{-3})	
	Calculated	Measured
Cd	0.6 (0.1–4.3)	0.6 (0.07–5.7)
Pb	53 (6.7–599)	12 (1.2–125)
Cu	5.5 (0.9–71)	8.2 (2.5–28)
Zn	216 (33–723)	44 (12–195)

under laboratory conditions, but measurements of the partition coefficient in the laboratory may differ from those needed for robust applications to field modelling. To get an indication of the adequacy of the combined use of all relationships used in modelling metal speciation in and between soil and soil solution, De Vries and Groenenberg (2009) compared results for calculated and measured total metal concentrations in soil solution. The measurements were obtained in the field from 56 soil samples of five land-use types (acid grassland, heathland, upland and lowland coniferous forest and deciduous woodland) collected from 'background' sites in the UK (Ashmore et al. 2004). Table 7.5 shows that the ranges (5–95 % percentile) of calculated total metal concentrations correspond well with measured ranges for Cd and Cu. Median values were, however, overestimated by a factor 4–5 for Pb and Zn. The MAE of the log-concentration is mostly between 0.3 and 0.5 (a factor 2–4 on a linear scale), which is within the error ranges calculated for the transfer function relating metal activities in solution to the reactive soil metal content (Groenenberg et al. 2010b).

7.4.2 Relevance of the Approach for Policy Making in View of Time Scales

As with their use for nitrogen and acidity, critical loads and targets loads for metals are a potential policy tool to optimise the protection of the environment for a given investment in pollution control. By definition, a critical metal load equals the load resulting at steady state in a metal concentration in a compartment (e.g. soil solution, plant, fish) that equals the critical limit for that compartment. However, unlike N and S, the time to reach steady state can be hundreds to thousands of years and this makes the approach less useful, unless used in addition to a dynamic modelling approach that provides insight into the past and future build-up of anthropogenic inputs of metals in soils. The reason for this is that variations in metal transformation and transport, moderated by climate change and soil acidification over time, may influence exposure and effects more than the magnitude of the metal input. Climate change influences the soil water balance, changes vegetation types and drive land use change, thus affecting the carbon cycle, and thereby the concentration of dissolved organic carbon. Soil acidification, in terms of a change in pH, will influence complexation, sorption, precipitation and ion exchange processes. Such changes are not included in the assessment of target loads. Dynamic modelling and target loads are therefore important to assess the future risk of metal inputs, in addition to the use of critical loads.

References

Adriano, D. C. (2001). *Trace elements in terrestrial environments: Biogeochemistry, bioavailability, and risks of metals*. New York: Springer-Verlag.

Anderson, A., & Bingefors, S. (1985). Trends and annual variations in Cd concentrations in grain of winter wheat. *Acta Agriculturae Scandinavica, 35,* 339–344.

Ashmore, M., Shotbolt, L., Hill, M., Hall, J., Spurgeon, D., Svendsen, C., Fawehinimi, J., Heywood, E., Tipping, E., Lofts, S., Lawlor, A., & Jordan, C. (2004). Further development of an effects (critical loads) based approach for cadmium, copper, lead and zinc. Final Report, EPG 1/3/188.

Baatrup, E. (1991). Structural and functional-effects of heavy-metals on the nervous-system, including sense-organs, of fish. *Comparative Biochemistry and Physiology Part C: Toxicology & Pharmacology, 199,* 253–257.

Bringmark, L., Bringmark, E., & Samuelsson, B. (1998). Effects on mor layer respiration by small experimental additions of mercury and lead. *Science of the Total Environment, 213,* 115–119.

Bryan, S. E., Tipping, E., & Hamilton-Taylor, J. (2002). Comparison of measured and modelled copper binding by natural organic matter in freshwaters. *Comparative Biochemistry and Physiology Part C: Toxicology & Pharmacology, 133,* 37–49.

Clark, R. B. (2001). Marine pollution (Metals, pp. 98–125). Oxford: Oxford Science Publishers.

Crommentuijn, T., Polder, M. D., & van de Plassche, E. J. (1997). *Maximum permissible concentrations and negligible concentrations for metals, taking background concentrations into account*. Report no. 601501 001. Bilthoven, The Netherlands: National Institute of Public Health and the Environment.

De Vries, W., & Bakker, D. J. (1998). *Manual for calculating critical loads of heavy metals for terrestrial ecosystems. Guidelines for critical limits, calculation methods and input data*. (Report 166). Wageningen, The Netherlands: DLO Winand Staring Centre.

De Vries, W., & Groenenberg, J. E. (2009). Evaluation of approaches to calculate critical metal loads for forest ecosystems. *Environmental Pollution, 157,* 3422–3432.

De Vries, W., & McLaughlin, M. J. (2013). Modeling the cadmium balance in Australian agricultural systems in view of possible impacts on food quality. *Science of the Total Environment, 461–462,* 240–257.

De Vries, W., Bakker, D. J., & Sverdrup, H. U. (1998). *Manual for calculating critical loads of heavy metals for aquatic ecosystems. Guidelines for critical limits, calculation methods and input data.* (Report 165). Wageningen, The Netherlands: DLO Winand Staring Centre.

De Vries, W., Schütze, G., Lofts, S., Tipping, E., Meili, M., Römkens, P. F. A. M., & Groenenberg, J. E. (2005). *Calculation of critical loads for cadmium, lead and mercury. Background document to a mapping manual on critical loads of cadmium, lead and mercury.* (Report 1104). Wageningen, The Netherlands: Alterra.

De Vries, W., Römkens, P. F. A. M., & Schütze, G. (2007a). Critical soil concentrations of cadmium, lead and mercury in view of health effects on humans and animals. *Reviews of Environmental Contamination and Toxicology, 191,* 91–130.

De Vries, W., Lofts, S., Tipping, E., Meili, M., Groenenberg, B. J., & Schütze, G. (2007b). Impact of soil properties on critical concentrations of cadmium, lead, copper, zinc and mercury in soil and soil solution in view of ecotoxicological effects. *Reviews of Environmental Contamination and Toxicology, 191,* 47–89.

De Vries, W., Groenenberg, J. E., Lofts, S., Tipping, E., & Posch, M. (2012). Critical loads of heavy metals for soils. In B. J. Alloway (Ed.), *Heavy metals in soils. Trace metals and metalloids in soils and their bioavailability* (Chap. 8, pp. 211–237). Dordrecht: Springer.

Doyle, P. J., Gutzman, D. W., Sheppard, M. I., Sheppard, S. C., Bird, G. A., & Hrebenyk, D. (2003). An ecological risk assessment of air emissions of trace metals from copper and zinc production facilities. *Human and Ecological Risk Assessment, 9,* 607–636.

Dwane, G. C., & Tipping, E. (1998). Testing a humic speciation model by titration of copper-amended natural waters. *Environment International, 24,* 609–616.

Groenenberg, J. E., Koopmans, G. F., & Comans, R. N. J. (2010a). Uncertainty analysis of the nonideal competitive adsorption-donnan model: Effects of dissolved organic matter variability on predicted metal speciation in soil solution. *Environmental Science and Technology, 44,* 1340–1346.

Groenenberg, J. E., Römkens, P. F. A. M., Comans, R. N. J., Luster, J., Pampura, T., Shotbolt, L., Tipping, E., & De Vries, W. (2010b). Transfer functions for solid-solution partitioning of cadmium, copper, nickel, lead and zinc in soils: Derivation of relationships for free metal ion activities and validation with independent data. *European Journal of Soil Science, 61,* 58–73.

Jones, K. C., Jackson, A., & Johnston, A. E. (1992). Evidence of an increase in cadmium content of herbage since the 1860s. *European Journal of Soil Science, 26,* 834–836.

Kawada, T., & Suzuki, S. (1998). A review on the cadmium content of rice, daily cadmium intake, and accumulation in the kidneys. *Journal of Occupational Health, 40,* 264–269.

Meili, M. (1997). Mercury in lakes and rivers. In A. Sigel & H. Sigel (Eds.), *Mercury and its effects on environment and biology* (pp. 21–51). New York: Marcel Dekker.

Meili, M., Bishop, K., Bringmark, L., Johansson, K., Munthe, J., Sverdrup, H., & De Vries, W. (2003). Critical levels of atmospheric pollution: Criteria and concepts for operational modelling of mercury in forest and lake ecosystems. *Science of the Total Environment, 304,* 83–106.

Paces, T., Corcimaru, S., Emmanuel, S., Erel, Y., Novak, M., Plyusnin, A., Veron, A., & Wickham, S. (2002). Critical loads of hazardous trace elements in soil-water system. *Journal of Field Science, 1,* 15–22.

Palmborg, C., Bringmark, L., Bringmark, E., & Nordgren, A. (1998). Multivariate analysis of microbial activity and soil organic matter at a forest site subjected to low-level heavy metal pollution. *Ambio, 27,* 53–57.

Pampura, T., Groenenberg, J. E., & Rietra, R. P. T. M. (2006). Comparison of methods for copper free ion activity determination in soil solutions of contaminated and background soils. *Forest Snow Landscape Research, 80,* 305–322.

Pampura, T., Groenenberg, J. E., Lofts, S., & Priputina, I. (2007). Validation of transfer functions predicting Cd and Pb free metal ion activity in soil solution as a function of soil characteristics and reactive metal content. *Water Air and Soil Pollution, 184,* 217–234.

Posch, M., & De Vries, W. (2009). Dynamic modelling of metals—Time scales and target loads. *Environmental Modelling & Software, 24,* 86–95.

Posch, M., Hettelingh, J.-P., & Slootweg, J. (2003). *Manual for dynamic modelling of soil response to atmospheric deposition*. (RIVM report 259101 012). Bilthoven, The Netherlands: National Institute for Public Health and the Environment.

Satarug, S., Haswell-Elkins, M. R., & Moore, M. R. (2000). Safe levels of cadmium intake to prevent renal toxicity in human subjects. *British Journal of Nutrition, 84,* 791–802.

Slootweg, J., Hettelingh, J.-P., Posch, M., Schütze, G., Spranger, T., De Vries, W., Reinds, G. J., van 't Zelfde, M., Dutchak, S., & Ilyin, I. (2007). European critical loads of cadmium, lead and mercury and their exceedances. *Water Air & Soil Pollution: Focus, 7,* 371–377.

Sola, F., Isaia, J., & Masoni, A. (1995). Effects of copper on gill structure and transport function in the rainbow trout, *Oncorhynchus mykiss*. *Journal of Applied Toxicology, 15,* 391–398.

Tipping, E. (1994). WHAM—A chemical equilibrium model and computer code for waters, sediments, and soils incorporating a discrete site/electrostatic model of ion-binding by humic substances. *Computers & Geosciences, 20,* 973–1023.

Tipping, E. (1998). Humic ion-binding Model VI: An improved description of the interactions of protons and metal ions with humic substances. *Aquatic Geochemistry, 4,* 3–47.

Tipping, E. (2002). *Cation binding by humic substances*. Cambridge: Cambridge University Press.

Tipping, E. (2005). Modelling Al competition for heavy metal binding by dissolved organic matter in soil and surface waters of acid and neutral pH. *Geoderma, 127,* 293–304.

Tipping, E., Woof, C., & Hurley, M. A. (1991). Humic substances in acid surface waters; modelling aluminium binding, contribution to ionic charge-balance, and control of pH. *Water Research, 25,* 425–435.

Tipping, E., Rey-Castro, C., Bryan, S. E., & Hamilton-Taylor, J. (2002). Al(III) and Fe(III) binding by humic substances in freshwaters, and implications for trace metal speciation. *Geochimica et Cosmochimica Acta, 66,* 3211–3224.

Tipping, E., Lawlor, A. J., & Lofts, S. (2006a). Simulating the long-term chemistry of an upland UK catchment: Major solutes and acidification. *Environmental Pollution, 141,* 151–166.

Tipping, E., Lawlor, A. J., Lofts, S., & Shotbolt, L. (2006b). Simulating the long-term chemistry of an upland UK catchment: Heavy metals. *Environmental Pollution, 141,* 139–150.

Tipping, E., Yang, H., Lawlor, A. J., Rose, N. L., & Shotbolt, L. (2007). Trace metals in the catchment, loch and sediments of Lochnagar: Measurements and modelling. In N. L. Rose (Ed.), *Lochnagar: The natural history of a mountain lake* (pp. 345–373). Dordrecht: Springer.

Vulkan, R., Zhao, F.-J., Barbosa-Jefferson, V., Preston, S., Paton, G. I., Tipping, E., & McGrath, S. P. (2000). Copper speciation and impacts on bacterial biosensors in the pore water of copper-contaminated soils. *Environmental Science and Technology, 34,* 5115–5121.

Weng, L. P., Temminghoff, E. J. M., & van Riemsdijk, W. H. (2001). Determination of the free ion concentration of trace metals in soil solution using a soil column Donnan membrane technique. *European Journal of Soil Science, 52,* 629–637.

Part III
Dynamic Modelling for the Assessment of Air Pollution Impacts at Site Scale

Chapter 8
Dynamic Geochemical Models to Assess Deposition Impacts and Target Loads of Acidity for Soils and Surface Waters

Luc T. C. Bonten, Gert Jan Reinds, Jan E. Groenenberg, Wim de Vries, Maximilian Posch, Chris D. Evans, Salim Belyazid, Sabine Braun, Filip Moldan, Harald U. Sverdrup and Daniel Kurz

8.1 Introduction

In the early days of the identification of adverse effects of 'acid rain' it gradually became clear that there was a time delay between acid deposition and the actual acidification of the soil and/or surface water. The systematic understanding of the soil and aquatic mechanisms involved in the time delays of (the recovery from) adverse effects led to the design of dynamic models, capable of simulating the temporal evolution of geo-chemical processes involved in the buffering of air pollution inputs.

A critical load (see Chap. 6) tells if current inputs to an ecosystem are sustainable or not. A critical load does *not* tell—if it is exceeded by current depositions—*when* adverse effects on the ecosystem can be expected; and vice versa, if an ecosystem has been affected, how fast a reduction of deposition below the critical load leads

L. T. C. Bonten (✉) · G. J. Reinds · J. E. Groenenberg · W. de Vries
Alterra Wageningen University and Research Centre, Wageningen, The Netherlands
e-mail: luc.bonten@wur.nl

M. Posch
Coordination Centre for Effects (CCE), RIVM, Bilthoven, The Netherlands

C. D. Evans
Centre for Ecology & Hydrology, Environment Centre Wales, Bangor, UK

S. Belyazid
Belyazid Consulting & Communication, Malmö, Sweden

S. Braun
IAP, Swiss Institute for Applied Plant Physiology, Schönenbuch, Switzerland

F. Moldan
IVL Swedish Environmental Research Institute, Göteborg, Sweden

H. U. Sverdrup
Department of Chemical Engineering, Lund University, Lund, Sweden

D. Kurz
EKG Geo-Science, Bern, Switzerland

W. de Vries et al. (eds.), *Critical Loads and Dynamic Risk Assessments*,
Environmental Pollution 25, DOI 10.1007/978-94-017-9508-1_8

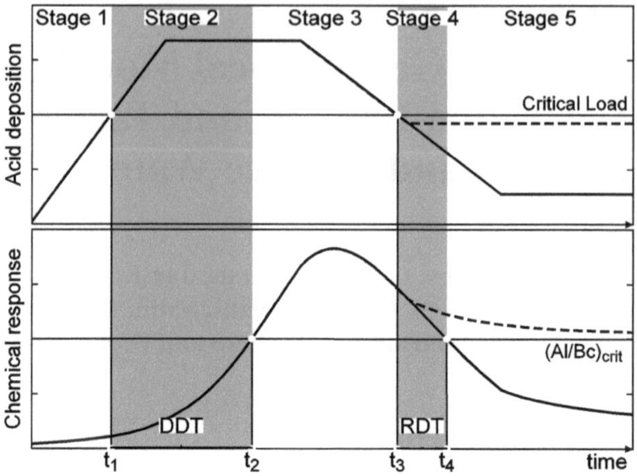

Fig. 8.1 'Typical' past and future development of the deposition effects on a soil chemical variable (Al/Bc-ratio) in comparison to the critical values of those variables and the critical load derived from them. The delay between the (non)exceedance of the critical load, the (non) violation of the critical chemical criterion is indicated in *grey* shades, highlighting the Damage Delay Time (*DDT*) and the Recovery Delay Time (*RDT*) of the system

to a restoration of good ecological conditions. With critical loads, i.e. in a steady-state situation (see Chap. 6), only two cases can occur when comparing them to deposition: (i) the deposition is below (or at) critical load; or (ii) the deposition is greater than the critical load (critical load exceedance). In the first case there is no (apparent) problem, whereas in the second case there is, by definition, an increased risk of damage to the ecosystem. It is often assumed that reducing deposition to (or below) critical loads immediately removes the risk of 'harmful effects', i.e. the chemical criterion immediately attains a 'safe' value. But the reaction of soils, especially their solid phase, to changes in deposition is delayed by (finite) buffers, the most important being the cation exchange capacity (CEC). Such buffers can delay the attainment of the chemical criterion; in fact, it might take decades or even centuries, before an equilibrium (steady state) is reached. Dynamic models are needed to estimate the times involved in attaining a certain chemical state in response to deposition scenarios.

Dynamic models are useful for assessing future effects of scenarios of emission (reduction) policies, climate change or land use changes, separate or in combination. Such 'scenario analyses' are the classical application of dynamic models, i.e. inputs and drivers (especially deposition) are specified, and the temporal development of selected output variables (e.g. soil solution concentrations) is simulated. In addition to scenario analyses, dynamic models can also be used to answer the following two questions: (A) For a given deposition scenario, how long does it take before a selected chemical criterion is attained? (B) For a given value of a selected output variable, the so-called 'chemical criterion' (e.g. pH or Al concentration), what are the deposition level(s), so-called target loads, at which the criterion is attained within a predefined time horizon?

Figure 8.1 summarises the possible development of a (soil) chemical variable in response to a 'typical' temporal deposition pattern. Five stages can be distinguished: *Stage 1*: Deposition was and is below the critical load (CL) and the chemical criterion is not violated. *Stage 2*: Deposition is above the CL, but the chemical criterion is not violated because there is a time delay due to buffer mechanisms (e.g. base cation supply from the CEC). No damage is likely to occur at this stage, despite exceedance of the CL. The time between the first exceedance of the CL and the first violation of the criterion is termed the *Damage Delay Time* (DDT=t_2–t_1). *Stage 3*: The deposition is above the CL and the chemical criterion is violated. *Stage 4*: Deposition is below the CL, but the chemical criterion is still violated and thus recovery cannot yet occur. The time between the first non-exceedance of the CL and the subsequent non-violation of the criterion is termed the *Recovery Delay Time* (RDT=t_4–t_3). *Stage 5*: Deposition is below the CL and the criterion is no longer violated. This stage is similar to Stage 1 and only at this stage can the ecosystem be considered to have recovered. Note that in addition to the delay in chemical recovery, there is likely to be a further delay before the desired biological state is reached, i.e. even if the chemical criterion is met, it will take time before biological recovery is achieved. But this can be included in the delay times defined above, if desired; for details see Posch et al. (2003) and Jenkins et al. (2003).

Reducing deposition(s) to (or below) the critical load ensures that the criterion will be met at some (unknown) time in the future. Dynamic models can be used to calculate these delay times (DDT or RDT) and thus answer question (A) above. Conversely, one may want to achieve recovery (i.e. attainment of the criterion) by a given year (question (B) above); and the deposition(s) that ensure that is called a *target load*, and the year in which the chemical criterion is met (for the first time) is called the target year. Target loads will, in general, be lower than the critical load of the system, especially for target years in the immediate future. Note that a critical load is an ecosystem property, whereas target loads depend on the target years. For more on target loads see Warfvinge et al. (1992), Posch et al. (2003) and Jenkins et al. (2003). Dynamic models are in general not designed to compute target loads, i.e. they are not designed to derive the driving force (deposition) for a given soil chemical variable. Thus trial and error (iteration) has to be used to find the deposition (path) which produces a prescribed chemical status. For the VSD model (see further), however, such an inverse procedure for target load calculations is part of the overall model design (Posch and Reinds 2009).

A large number of dynamic models have developed over the years, ranging from relatively simple to quite complex. In the United States, starting in the 1980s, modelling focussed on (individual, small) watersheds, and the most widely used models were the MAGIC model (see below), the Enhanced Trickle Down (ETD) model (Nikolaidis et al. 1989) and the most detailed of all: the Integrated Lake-Watershed Acidification Study (ILWAS) model (Gherini et al. 1985). Around the same time, in the context of developing the RAINS integrated assessment model for Europe (see Alcamo et al. 1990), a regional model of forest soil acidification in Europe was developed (Kauppi et al. 1986), based on the theory surrounding the forest dieback in Germany (Ulrich 1983), as well a model to describe the acidification of surface waters in the Nordic countries (Kämäri and Posch 1987).

Several model evaluations and comparisons have been carried out over time. A thorough comparison of the three above-mentioned American models can be found in Eary et al. (1989) and Rose et al. (1991a, b). Sixteen (mostly European) forest-soil-atmosphere models have been compared in Tiktak and van Grinsven (1995); and a recent comparison of three models (MAGIC, SAFE and VSD; see below) can be found in Tominaga et al. (2009). In this chapter the geochemical models VSD, MAGIC, ForSAFE and SMARTml are presented, including examples of (i) site-scale applications at intensively monitored forested plots in the UK, Germany, Switzerland and Norway, illustrating the adequacy of the model behaviour and target load calculations with some of the models, i.e. MAGIC and VSD.

8.2 Model Descriptions

The most important differences and similarities of the models VSD, MAGIC, ForSAFE and SMARTml are presented in Table 8.1.

The simplest model is the VSD model, which is the simplest extension of the steady-state SMB model (see Chap. 6). The VSD model has been used for regional-scale analyses of scenarios and the calculations of target loads (see below and also Chap. 25).

The other three models are more complex, each having their own characteristics and applicabilities. In contrast to VSD, those three models contain an explicit description of litterfall and litter decomposition, which makes it possible to use them for carbon sequestration calculations. The MAGIC model further simulates the leaching from the soil to surface water, and can consequently be used to assess effects on surface waters.

The geochemical module in the ForSAFE model has been linked to a forest growth routine to have direct coupling between growth and chemistry. For the other models growth, i.e. litter input and nutrient uptake, is an input and has to be estimated beforehand. Furthermore, ForSAFE is linked to a vegetation biodiversity model to assess effects of changes in soil chemistry on vegetation.

The fourth model, SMARTml, is a multilayer model that can be used to asses effects on groundwater quality. Different from the other models, cation exchange and anion sorption are calculated with surface complexation and speciation models instead of cation exchange equations and sorption isotherms. Thus the calibration of exchange constants and isotherms is avoided, because the surface complexation model uses the abundance of the different reactive surfaces for which generic parameter values are available. More details on each model are presented below.

8.2.1 VSD

As its name implies, VSD (Posch and Reinds 2009) is a Very Simple Dynamic model that simulates soil solution chemistry and soil nitrogen pools for natural or semi-natural ecosystems. VSD is a single-layer model with an annual time step

Table 8.1 Overview of properties of the four dynamic soil chemistry models evaluated in this chapter

Features	VSD	MAGIC	ForSAFE	SMARTml
Number of soil layers	1	1–3	1–20	1–15 soil horizons 1–30 calculation layers
Temporal resolution	Annual	Annual/ monthly	Monthly	Variable (day-year)
Base cation weathering	External input	Calibrated, or external-input	Sub-model	External input
Forest nutrient uptake	External input	External input	Sub-model	Sub-model
Runoff to precipitation ratio	External input	External input	Fixed ratio	External input (from multilayer hydrological model)
Sulfate adsorption	Not included	Langmuir isotherm	Yes	At Fe/Al hydroxides using a 2pK-DDL model
N immobilization	Fractional, fixed, or external input	Fractional, fixed, or external input	Sub-model	Sub-model
Nitrification	100 %	Fractional	Sub-model	Sub-model
Denitrification	Fractional, fixed	Fractional	None or sub-model	Sub-model
Soil N build-up controlled by C:N ratio	Yes	Yes	No	Sub-model
CO_2 degassing in surface water	Yes	Yes	Yes	No surface water
$Al(OH)_3$ precipitation in stream	Yes	Yes	Yes	No surface water
Lumped base cations (Ca^{2+}, Mg^{2+}, and K^+)	Yes	No	Yes	No
DOC dissociation model	Oliver, or simple monoprotic	Triprotic	Oliver	NICA-Donnan
DOC-Al complexation	No	Yes	No	Yes (NICA-Donnan)
Cation exchange	Gaines-Thomas or Gapon	Gaines-Thomas	Gapon	OM: NICA-Donnan clay: Donnan gel hydroxides: 2 pK-DDL
Number of cation exchange equations	2	4	1	Not relevant
Ions in soil solution charge balance	12	28	16	Full speciation of ions in the soil solution
First appearance in the peer-reviewed literature	2009	1985	1993	2011

and can be seen as the simplest extension of the simple mass balance (SMB; see Chap. 6) critical load model, and is comparable to (but simpler than) the dynamic soil acidification model SMART (De Vries et al. 1989). The element input from the atmosphere, net element uptake by vegetation, net nitrogen immobilisation and element weathering from silicates are described similar to the SMB model. VSD, however, also includes cation exchange and simulates changes in aluminium (Al), base cations (Bc, where Bc stands for Ca + Mg + K), sodium (Na), chloride (Cl), nitrate (NO_3) and sulphate (SO_4) concentrations in the soil solution, while determining the H^+ concentration by using a charge balance. Nitrification is assumed to be instantaneous and complete, and thus zero ammonium (NH_4) concentration in the soil solution. VSD contains a set of mass balance equations, describing the soil input-output relationships of ions, and a set of equations describing the rate-limited and equilibrium soil processes. The soil solution chemistry in VSD depends solely on the net element input from the atmosphere (deposition minus net uptake minus net immobilisation) and the geochemical interactions in the soil (CO_2 equilibrium, weathering of carbonates and silicates, and cation exchange). Soil interactions are described by simple rate-limited reactions (e.g. nutrient uptake and weathering), first order processes (denitrification) and by equilibrium reactions (e.g. cation exchange). VSD models the exchange of Al, H and Bc with Gaines-Thomas or Gapon equations. Solute transport is described by assuming complete mixing of the element input within one homogeneous soil compartment with a constant density and a fixed depth. VSD is a single layer soil model that neglects vertical heterogeneity. It predicts the concentration of the soil water leaving this layer (mostly the root zone). The annual water flux percolating from this layer is taken equal to the annual precipitation excess. The time-step of simulations is 1 year.

8.2.2 *MAGIC*

MAGIC is an intermediate-complexity biogeochemical model of acidification and nitrogen cycling, first developed to simulate the chemistry of acid-sensitive surface waters in the 1980s (Cosby et al. 1985a, b). The core acid-base component of the model, in common with other widely used models, accounts for inputs and outputs of major base cations and acid anions to and from one or more lumped soil pools, with inputs including deposition and weathering, and outputs including leaching and uptake by vegetation. Within the soil, a range of equilibrium reactions are simulated, between base cations and Al in solution and on soil cation exchange sites; acidity-dependent dissolution of Al; adsorption and desorption of SO_4 onto the soil; the buffering of soil solution pH by weak organic acids and Al complexes; and the speciation of inorganic carbon forms. The chemistry of soil solution leached from the soil box is used to simulate surface water chemistry, after adjustment for changes in organic acid and CO_2 concentrations, Al concentrations, and any in-stream or in-lake removal of acid anions.

Since its inception, MAGIC has been developed further to provide a more realistic representation of catchment processes, and most notably to simulate the impacts of N deposition on soil and water chemistry. These developments are summarised by Cosby et al. (2001). For nitrogen, a range of input and removal processes can be specified in the model (including forest uptake and litterfall, N fixation, organic N leaching and denitrification). Remaining available N is either accumulated in soil organic matter or leached, according to an empirical function based on observed spatial relationships between NO_3 leaching and soil C/N ratio (e.g. Gundersen et al. 1998). According to this empirical function, all available N is retained in the soil, progressively reducing soil C/N until a threshold C/N ratio is attained, at which point N 'saturation' occurs, and beyond which a linearly increasing proportion of N is leached as NO_3 as C/N ratio continues to fall (Cosby et al. 2001). This relatively simple, empirically-based approach to N simulation is currently being replaced with a more mechanistically-based model of microbial C and N cycling (Oulehle et al. 2012). However the example MAGIC application presented here is based on the simpler C/N ratio-based version of the model.

8.2.3 ForSAFE

The ForSAFE model (Fig. 8.2), a multi-layer model with a monthly time step, integrates parts of four existing models, which were merged into a single structure with closed feedbacks representing the biogeochemical cycles of water, carbon and selected nutrients in a forest ecosystem (Belyazid 2006). To model the growth of the forest cover, the processes of photosynthesis, allocation, respiration, evapotranspiration and litter production were derived from the PnET model (Aber and Federer 1992). Soil chemistry is modelled according to the dynamics developed in the SAFE model (Alveteg et al. 1995; Alveteg 1998). Litter decomposition is based on the principles developed in the DECOMP model (Wallman et al. 2006). Finally, the PULSE model for soil hydrology (Lindström and Gardelin 1992) was incorporated into ForSAFE to simulate the vertical flow of moisture in the soil. ForSAFE can be coupled to a module for estimating the composition of the ground vegetation community (VEG) which derives the ground cover for a set of plant species based on chemical (soil solution alkalinity and nitrogen content) and physical (moisture, temperature, light) inputs calculated by ForSAFE (Belyazid et al. 2011; Sverdrup et al. 2007). A more detailed description of the complete ForSAFE-VEG model system is given in Chap. 12.

8.2.4 SMARTml

SMARTml (Bonten et al. 2011) is a multi-layer model based on the single layer soil acidification model SMART2 (Kros et al. 1995), which is an extension of the dynamic soil acidification model SMART (De Vries et al. 1989). SMART is

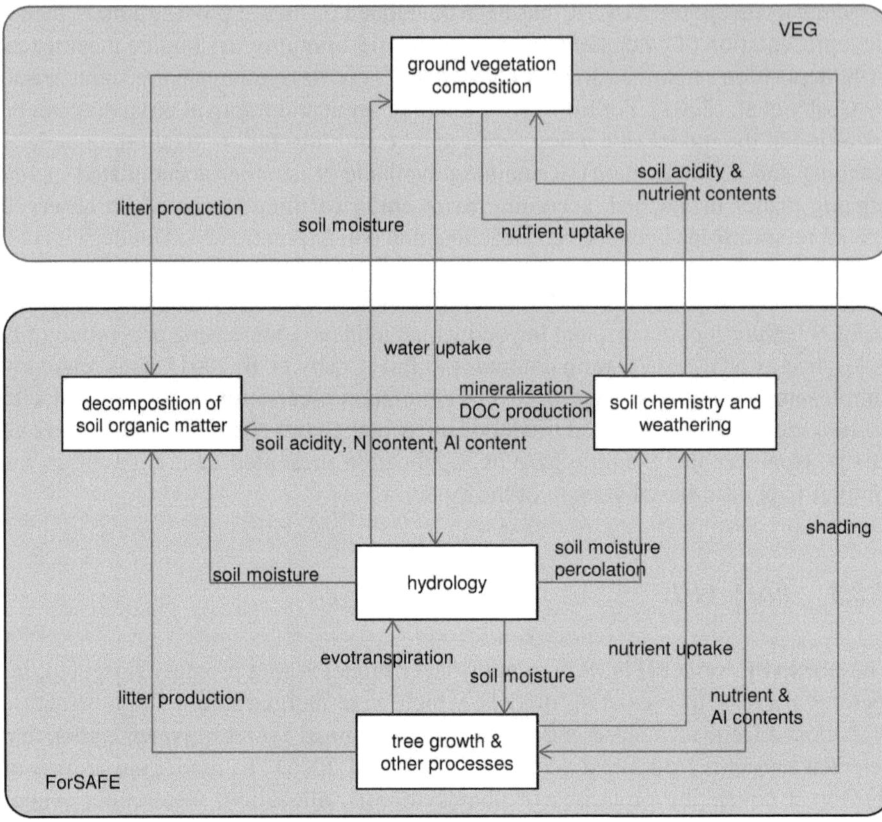

Fig. 8.2 Composition of the ForSAFE-VEG components, which together simulate a closed forest ecosystem

comparable to VSD (see above), but allows incomplete nitrification and contains equilibrium reactions for the dissolution or precipitation of carbonates and secondary Al precipitates. Consequently, apart from predictions of pH and soil solution chemistry, the model also predicts changes in solid phase characteristics depicting the acidification status, i.e. carbonate content, base saturation and readily available Al content.

Similar to SMART2, and unlike SMART and VSD, SMARTml includes nutrient cycling, through litterfall, mineralisation and uptake. The soil solution chemistry in SMARTml (and SMART2) depends on the net element input from the atmosphere (the product of deposition and filtering factor) and groundwater (seepage), canopy interactions (foliar uptake, foliar exudation), geo-chemical interactions in the soil (CO_2 equilibria, weathering of carbonates, silicates and/or secondary Al precipitates, SO_4 sorption and cation exchange) and a complete nutrient cycle (litterfall, mineralization, root uptake, nitrification and denitrification). Growth of the vegetation and litterfall are modelled by a logistic growth function, which acts as a

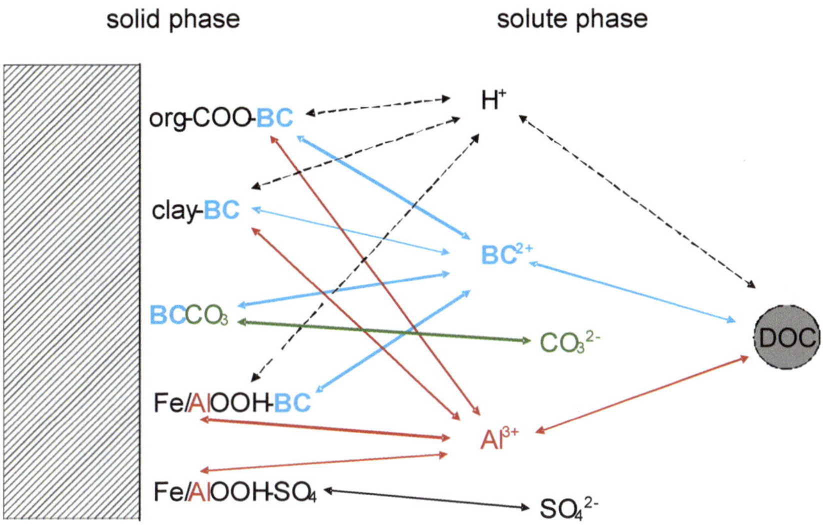

Fig. 8.3 Representation of surface complexation and speciation in the soil solution as used in SMARTml

forcing function. Nutrient uptake is only limited when there is a shortage in the total nutrient supply. In SMARTml the description of C and N mineralization is based on the VSD+model (unlike SMART and SMART2), using a four pool organic C model, where decomposition of some pools is reduced in cases of low N availability (Bonten et al. 2009). Yet, the largest difference with most other soil acidification models (see e.g. Tiktak and van Grinsven 1995) is that SMARTml uses an assemblage of surface complexation models for the calculation of cation and anion sorption instead of Gapon or Gaines-Thomas ion exchange equations for cations and Langmuir or Freundlich isotherms for sulphate and phosphate. These exchange equations and isotherms use a single exchange constant for each element to describe sorption to all sorption surfaces. This exchange constant will be some average of the exchange constants of the individual sorption surfaces, thereby being different for different soil types. Surface complexation models describe complexation by each type of reactive surface (organic matter, clay minerals, metal oxides) separately, including competition by other compounds (see Fig. 8.3). Because generic parameter sets exist for most reactive surfaces, parameterisation of surface complexation models does not depend on the soil type and only requires data on the abundance of the different reactive surfaces.

In SMARTml concentrations of all elements and pH are calculated simultaneously for each layer, using the model framework ORCHESTRA (Meeussen 2003). The assemblage of surface complexation models includes ion sorption to organic matter, clay minerals and metal (hydr)oxides (Bonten et al. 2008). Sorption to both SOM and DOM is described by the NICA-Donnan model (Kinniburgh et al. 1999), sorption to clay is calculated using a non-specific Donnan model and complexation by metal (hydr)oxides is calculated used a two-site diffusive double layer model for

Table 8.2 Generic data for the selected Swiss sites CH2038, CH2142 and CH2175. Climate and deposition data are averages of the observation period 1998–2008

Parameter	Unit	CH2038 (Frienisberg)	CH2142 (Zugerberg)	CH2175 (Lurengo)
Latitude		47.02287	47.14839	46.50471
Longitude		7.34486	8.54364	8.74813
Altitude	m asl	740	957	1550
Bedrock	–	Till	Hillside loam	Hillside loam
Soil type	–	Cambisol	Acid cambisol	Podsol-cambisol
Spruce	%	49	99	78
Larch	%	1	0	22
Beech	%	50	1	0
Stand age (in 1998)	Year	119	139	85
Stem wood	kg m^{-2}	17	33.9	21.8
Precipitation	mm yr^{-1}	1091	1507	1938
Temperature	°C	8.59	7.48	5.16
S deposition	meq $m^{-2}yr^{-1}$	42	39	41
N deposition	meq $m^{-2}yr^{-1}$	157	129	70
Bc deposition (Ca+Mg+K)	meq $m^{-2}yr^{-1}$	55	46	49

hydrous ferric oxide (HFO) (Dzombak and Morel 1990); iron and aluminium (hydr) oxides are considered the reactive metal (hydr)oxide surfaces.

8.3 Model Validations and Site Applications

8.3.1 Application of VSD and ForSAFE to Three Swiss Forest Monitoring Sites

Three Swiss forest monitoring sites were selected to model the soil solution chemistry with ForSAFE and VSD and to compare model outputs with by-weekly lysimeter data for the period 1998–2008. As the sites are part of the Swiss national critical loads program, all required model input was extracted from the national database, which holds data needed for the more complex ForSAFE model (multi-layer, monthly time resolution). Deposition, uptake and weathering fluxes needed by VSD were drawn from MAKEDEP/SAFE (Alveteg 1998, Alveteg et al. 2002) model runs. Input for these was also derived from the national database, converted to annual fluxes and average soil properties for the rooting zone. Some generic information for these sites is compiled in Table 8.2.

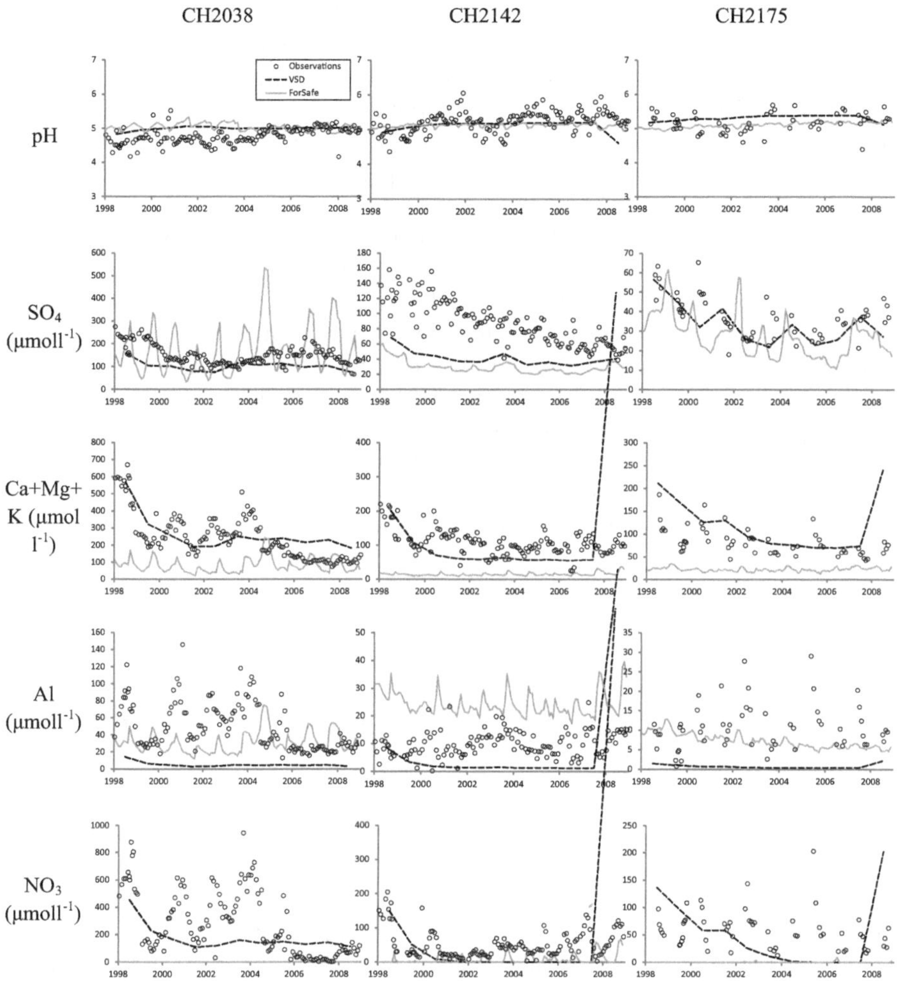

Fig. 8.4 Measured and modelled soil solution chemistry at three Swiss forest sites. Model calculations with VSD and ForSAFE

Figure 8.4 shows the observed pH values and concentrations of SO_4, NO_3, Al and base cations and the modelled values with ForSAFE and VSD. VSD models the whole rooting zone, while ForSAFE models several layers. Figure 8.4 shows only observations and ForSAFE results as close as possible to the bottom of the rooting zone.

Calculated soil solution chemistry differed substantially for the two models. VSD predicted relatively steady patterns as a result of its annual time step and single compartment—the reasonable thick rooting zone—approach. ForSAFE, on the other hand, often produced substantial intra-annual variation in the soil solution concentrations of relevant ions and divergent patterns form soil layer to soil layer.

VSD generally reproduced observed ion concentrations reasonably well. There is a weak tendency to underestimate the tracer ion concentrations, notably SO_4 in the soil solution at site CH2142. Among the nutrients, modelled base cation (Bc) concentrations often fluctuate along the lower limit of the observed range and modelled NO_3 concentrations often dip clearly below the measured range. Despite acceptable convergence of modelled and measured pH, modelled Al concentrations are substantially smaller than measured ones. However, VSD only calculates free Al^{3+}, whereas measurements also include other Al-species as Al-hydroxides and Al complexed by dissolved organic matter. The comparison of ForSAFE results with observations is complicated by the layering and intra-annual variation. The seasonality turns out to be most pronounced at site CH2038, producing roughly twice as large modelled concentration ranges of SO_4 than observed. The least seasonal influence on the solute concentration is seen at site CH2142, and modelled sulphur concentrations tend to be underestimated by the model. The concentration of base cations is underestimated particularly in the lower soil layers at all sites, and modelled NO_3 concentrations fall often markedly below the observations. In some layers modelled pH values scatter within the observed range; however, mainly upper soil layers can have higher modelled pH than observed (not shown). Compared to the range of modelled concentrations including all layers, the range of observations considering all lysimeter depths tends to often be distinctly narrower.

8.3.2 Application of MAGIC to a Long-term Forest Monitoring Site in the UK

The MAGIC model has, throughout its history, been tested against observational data from many monitoring and experimental sites (e.g. Cosby et al. 1995; Evans 2005; Kopáček et al. 2003; Majer et al. 2003; Oulehle et al. 2012; Rogora et al. 2003). Long-term data provide a particularly rigorous test of the model, and can also be used to assess and constrain uncertainties in model simulations using Bayesian calibration methods (Helliwell et al. 2014; Larssen et al. 2006; Larssen et al. 2007). The example shown here is based on the application of MAGIC to the Ladybower Forest Level II monitoring plot (Vanguelova et al. 2010). The Ladybower site is located in the Peak District, an area of Northern England which has received exceptionally high rates of historic S and N deposition, and is the most acidified Level II site in the UK. The plot is located in an area of Scots pine plantation, which was planted in 1952 on former moorland vegetation, at an altitude of 265 m on a 22° slope. The site has a mean annual temperature of 10 °C and mean annual rainfall of 1.2 m (Vanguelova et al. 2010); it has been monitored since 1996.

The MAGIC simulations shown here were originally calibrated against mean chemistry for a single year, 1999 (Evans et al. 2004), and subsequent application of the model has retained the parameter values from this original calibration. Therefore, data collected since 1999 provide an independent test of the subsequent performance of the model. As shown in Fig. 8.5a, b, MAGIC successfully reproduces

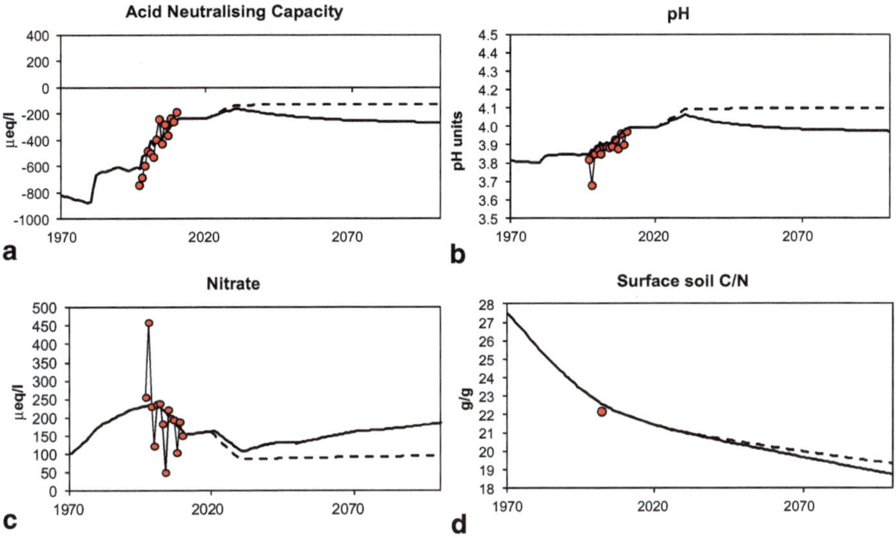

Fig. 8.5 Measured and MAGIC modelled soil solution chemistry for the Ladybower Forest Level II monitoring plot, Northern England. *Solid line* shows hindcast simulation from 1970 and forecast simulation for currently emissions legislation and replanting of 2nd rotation forest up to 2100. *Dashed lines* show simulation with felling and removal of the forest in 2020. *Red circles* show annual mean observed data for soil solution (ANC, pH and NO3) and a single observed value for the C/N ratio of the soil organic horizon

the trend in acidity (ANC and pH) over the majority of the monitoring period. Very acidic conditions in the first years of monitoring, which were not reproduced by the model, may reflect initial effects of sampler installation, as they were associated with exceptionally high NO_3 concentrations (Fig. 8.5c), or alternatively the impacts of cold winter conditions in 1995–1996 which led to a widespread pulse of NO_3 leaching (Monteith et al. 2000). This application of MAGIC did not include the effects of short-term climatic fluctuations, although more detailed (monthly time step) simulations have successfully reproduced similar short-term patterns (e.g. Sjøeng et al. 2009). In the Ladybower example shown, the model does not reproduce the observed high year-to-year variability in NO_3 concentrations, but appears to capture a general downward trend. This is attributable to decreases in N deposition at the site, while some additional variability in both acidity and NO_3 simulations is associated with forest growth and felling, as these affect deposition and uptake rates. No long-term data are available against which to evaluate the long-term predicted changes in soil C/N ratio, which the model suggests has decreased significantly over time due to soil N enrichment. This simulation does not incorporate the effects of increased C accumulation due to N deposition, however (De Vries et al. 2009; Evans et al. 2006), which would reduce the rate of change in soil C/N. The most recent version of the model (Oulehle et al. 2012) uses a different approach to N simulation, based on microbial organic matter turnover, rather than the empirical relationships between C/N ratio and N leaching used in the simulation shown here.

8.3.3 Application of ForSAFE to Two Spruce Forests in Sweden and Switzerland

ForSAFE was used to simulate an intensively monitored site in Gårdsjön, Sweden, and in Bachtel, Switzerland. Both sites are managed spruce forests, but differ in climate, atmospheric deposition levels and soils. Both sites have long records of lysimeter data. A detailed description of model inputs for Gårdsjön is given in Belyazid and Moldan (2009), and for Bachtel, in Belyazid et al. (2011). The model was calibrated for soil base saturation measurements (as well as for soil organic carbon and C/N ratio for the same year at Gårdsjön).

The model accurately simulates the standing tree biomass (excluding foliage, twigs and roots) at both sites (not shown). Because both sites have the same tree species, i.e. spruce, all differences between tree stocking are effects of differences in environmental conditions between the two sites. The model also captures the effects of the different site conditions, as. Bachtel has markedly higher stocking than Gårdsjön as expected from site conditions. During the simulated period, the Gårdsjön site has been clear-cut in 1915.

Figure 8.6 shows the observed and modelled soil solution pH and concentrations of several elements at Gårdsjön. The model reproduces the concentrations of Cl in soil solution and in the runoff water relatively well, which, Cl being a tracer, indicates that soil hydrology is appropriately modelled, assuming that Cl deposition (a model input) is well estimated. However, the episodic peaks in Cl concentrations may indicate a higher evapotranspiration under drier conditions than estimated by the model. SO_4 concentrations are slightly underestimated, and so are the concentrations of Na and nutrient base cations (Ca + Mg + K). The underestimation of SO_4 and Na concentrations can be due to an underestimation of deposition levels given that the soil hydrological transport is appropriately modelled as indicated by Cl. On the other hand, the underestimation of nutrient base cations probably lies in the biological cycle of uptake, litter fall and decomposition/mineralization. It may indicate the need to better parameterise the biota compartments, which are today simplified into three components (leaves, wood, fine roots). The inclusion of twigs and bark, which have different nutrient concentrations than the three modelled tree compartments, may be necessary to improve model performance when it comes to the cycling of macronutrients. Nitrogen concentrations in the upper soil layer are reasonably well reproduced, except for the re-occurring peaks caused by elevated evapotranspiration during dry periods. Down the soil horizon and in runoff, however, the modelled concentrations of N are lower than the measured ones, apart from the elevated evapotranspiration events. This discrepancy may reflect the fact that the dominant spruce trees take up most of their N in the uppermost layer, while they were allowed in the model to take up 40 % of their N requirements in the mineral soil. The higher uptake rates at the lower depths may cause the model to produce lower N concentrations at those depths.

At Bachtel (Fig. 8.7), Cl and SO_4 concentrations are reasonably well modelled at 20 cm depth, although the modelled Cl concentrations are slightly below the

Dep th	Soil solution [Cl] (μeq l^{-1})	Soil solution [SO_4] (μeql^{-1})	Soil solution [Na] (μeq l^{-1})	Soil solution [Bc] (μeql^{-1})	Soil solution [N] (μeq l^{-1})	Soil solution pH
5 cm	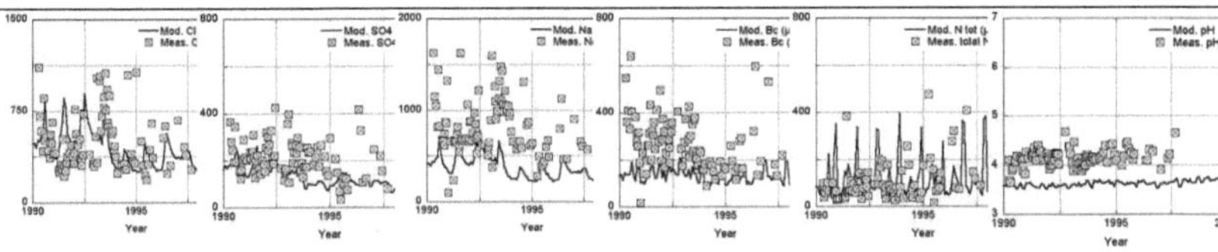					
10 cm	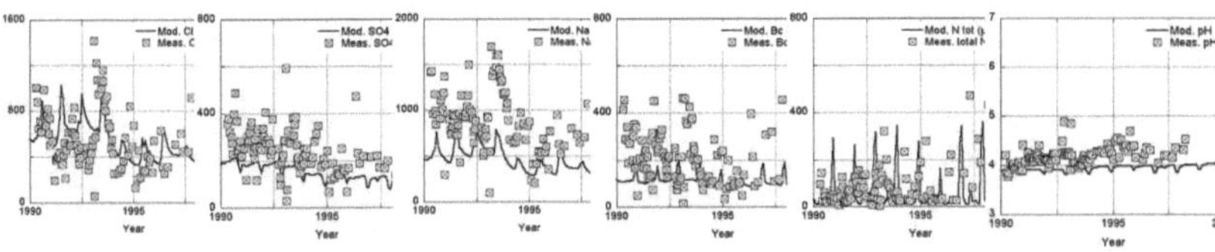					
20 cm	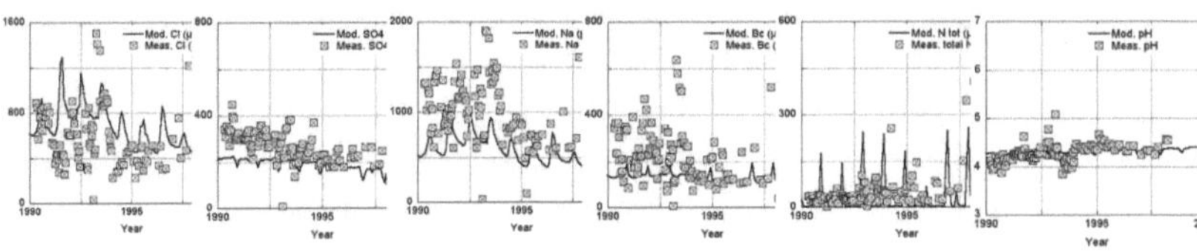					
Run off	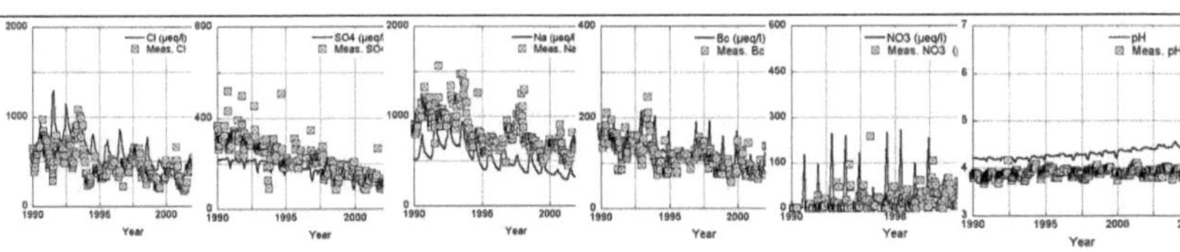					

Fig. 8.6 Measured and ForSAFE modelled concentrations of chloride, sulphate, sodium, base cations and nitrogen as well as pH in the soil solution and runoff water at Gårdsjön (SE)

measured values. However, while Cl is slightly underestimated at 100 cm depth, the modelled concentrations of SO_4 are clearly below the measurements. The measured SO_4 concentrations at 40 cm are more than double those at 20 cm. This, together with the fact that Bachtel is dominated by spruce trees with shallow roots (meaning that most water uptake takes place in shallow layers), indicates that there may be a source of S in the soil somewhere between 20 and 100 cm. This could be a large store of adsorbed SO_4 that is being released, but this is unlikely considering that S deposition had already declined from the peak level at Bachtel of above 250 meq $m^{-2}yr^{-1}$ to below 50 meq $m^{-2}yr^{-1}$ by the year 2000. The adsorbed SO_4 would have been released during the decline of deposition, as seen in the 20 cm depth comparison of measured and modelled concentrations. It indicates the possibility of a mineral source of S in the soil. Na is relatively well modelled both at 20 and 100 cm depth, although less so in the deeper soil.

Base cation concentrations are modelled within the range of the measured values, with a slight underestimation at 20 cm depth. Bachtel has a high rate of weathering, producing around 10 g Bc $m^{-2}yr^{-1}$ in the rooting zone, in the same order of

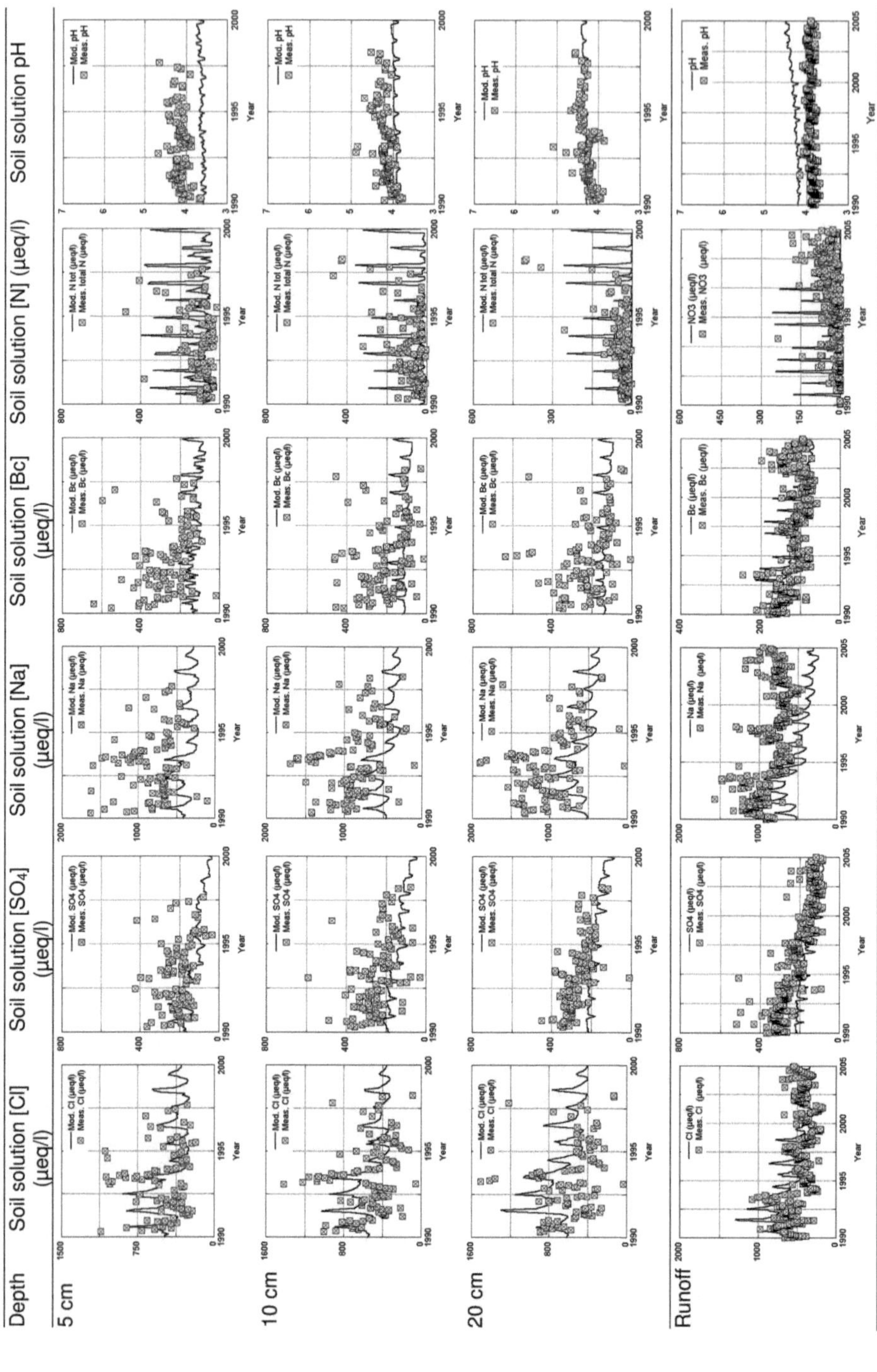

Fig. 8.7 Measured and ForSAFE modelled concentrations of chloride, sulphate, sodium, base cations, nitrogen and aluminium in the soil solution at Bachtel (CH)

magnitude as gross uptake. This means that the biochemical and the geochemical cycles contribute in equal terms to the Bc mass balance. If the model inaccurately simulates the biochemical cycle, as suspected at Gårdsjön, the effect is relatively smaller at Bachtel since weathering is large. This is probably correct since the modelled Bc concentrations are only marginally lower than the measurements at 20 cm depth (within the rooting zone) and well within the measured span at 100 cm depth (beyond the rooting zone). There is a high variability in the concentrations of Bc both in the measurements and in the model, particularly at 100 cm depth, which is not due to fluctuations in soil moisture (Cl does not vary as much). The fluctuations in Bc concentrations are due to the weathering fluxes, which triple in size between summer and winter. Nitrogen is simulated within the measured range. Al concentrations are also in the range of the measured values. Mobile Al is nearly absent from the soil solution at 100 cm depth, due to the alkaline nature of the site.

8.3.4 Application of SMARTml to a Long-term Norway Spruce Monitoring Site in Germany

The SMARTml model was applied to a Norway spruce site in Solling, Germany. This site has been extensively monitored since 1966. Monitoring included meteorology, hydrology, deposition, soil solid phase and solution chemistry as well as vegetation growth and chemistry (Meesenburg et al. 1995). For the Solling application, the model was run for the years 1968–2004. Prior to the SMARTml calculations, water fluxes, moisture contents and soil temperatures for the Solling site were calculated with the 1-D hydrological model SWAP (van Dam et al. 2008). Hydrology calculations had a temporal resolution of 5 days and were averaged per year as input for SMARTml, as it was run with a one-year stepsize. The soil profile was represented by 1 organic layer with a thickness of 7 cm, and seven mineral layers with a total thickness of 1 m. Figure 8.8 shows a comparison of modelled and measured soil solution concentrations, pH and exchangeable Ca contents for three different layers.

Hydrology calculations are validated by Cl concentrations, because Cl acts as a conservative tracer in the soil solution. Modelled and measured concentrations of Cl agree reasonably well, which suggests that hydrology was modelled correctly.

High SO_4 deposition between 1970 and 1980 causes a SO_4 peak in soil solution. The model results closely follow measured concentrations. Only for the years 1976 and 1977 the model overestimates the top of the SO_4 peak, most likely because initially sorbed SO_4 contents were overestimated. The peak of SO_4, being a major anion, is accompanied by a peak in the concentrations of major cations, for reasons of charge balance. Al, Ca and protons, i.e. the most important cations, indeed show a clear peak in both observed and simulated concentrations (and a drop in pH) and a clear drop in exchangeable Ca contents. In general, simulated trends in concentrations and contents agree with observations.

Although nitrogen species are barely complexed by reactive surfaces, they play a role in the soil charge balance and nitrate acts as an acidifying compound. Nitrogen

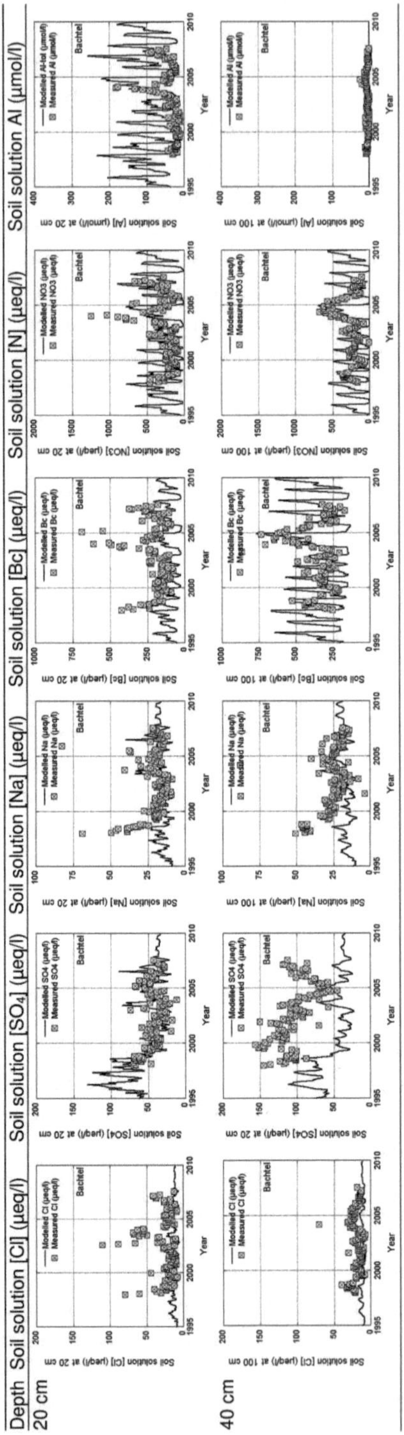

Fig. 8.8 Measured and SMARTml modelled dissolved concentrations of chloride, aluminium, calcium and nitrate, and measured and modelled pH and exchangeable calcium for the topsoil (*left*), at 40 cm (*middle*) and 90 cm depth (*right*) for a pine forest in Solling, Germany

concentrations in solution follow from the N mass balance, which consists of inputs, i.e. deposition and mineralization, and sinks, i.e. uptake and immobilisation. Although observed and simulated NO_3 show large differences between years, both show a similar increase in concentrations from 1970 to 1990 and a decrease afterwards. NH_4 concentrations were low compared to NO_3; NH_4 concentrations exceeded 0.05 meq l^{-1} only in the organic layer (data not shown).

8.4 Scenario Analyses and Target Loads

In the following section two examples with the MAGIC model are presented: one for scenario analysis and one for target load calculations at a single site. This is followed by target load calculations at multiple sites in Switzerland with the VSD/SAFE model. An early example of a European dynamic model application in the context of critical loads, their exceedances and the delay times in damage and recovery (the term target load as above was not yet defined!) can be found in Hettelingh and Posch (1994).

8.4.1 Scenario Analysis With the MAGIC Model

MAGIC has been widely used to evaluate future impacts of changes in S and N deposition on soil and surface water acidity and nitrogen status, on both site and regional scales (e.g. Wright et al. 2005). In conjunction with this, the model has also been used to evaluate the impacts of land-management, including forest management (e.g. Aherne et al. 2012; Jenkins et al. 1990; Oulehle et al. 2007) and heathland management (Evans et al. 2006), as well as various impacts of climate change over long-term, annual and episodic timescales (e.g. de Wit and Wright 2008; Evans 2005; Evans et al. 2008). The application for the Ladybower site, whose validation is described in Sect. 8.3.2 also shows projected future chemical responses to currently legislated emissions reductions (implemented by 2020), through to 2100 (Fig. 8.2). These simulations were undertaken with a ‘business as usual’ forestry scenario (with felling and replanting with the same tree species, solid line) and with the same deposition scenario but with the forest felled but not replanted (dashed line). The replanting scenario suggests that the rapid recovery from acidification observed during the last 15 years will not be sustained, with recovery ceasing (with ANC remaining negative) by 2030. This is followed by some re-acidification, which is attributable to a predicted increase in NO_3 leaching after this time. By the end of the model simulation, NO_3 concentrations are predicted to have increased back towards the peak levels observed in the 1990s (> 180 μeq l^{-1} by 2100).

The forest removal scenario highlights the impacts of forest deposition enhancement and uptake on ecosystem sensitivity to acidification and eutrophication. The

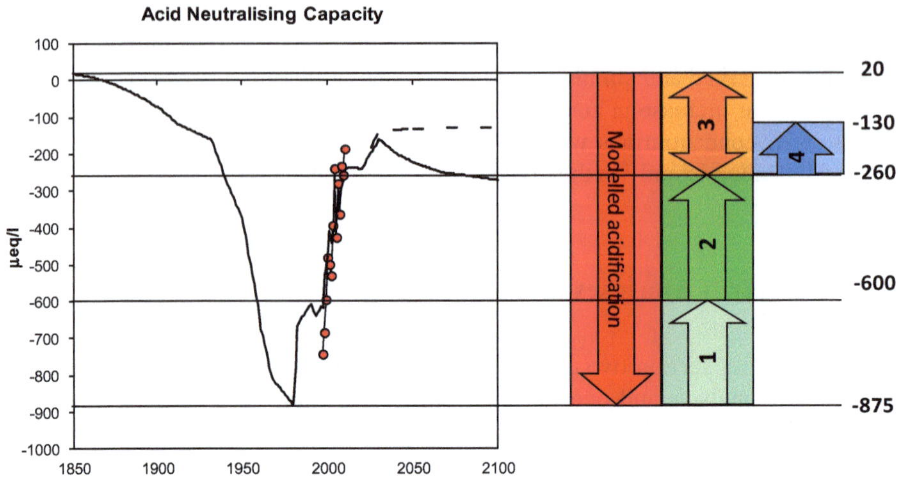

Fig. 8.9 Conceptual illustration of acidification and recovery stages relative to MAGIC modelled and observed soil solution chemistry at Ladybower, from 1850 to 2100. For explanation of *numbered arrows* see text

model suggests that removing the forest (i.e. reverting the site to pre-existing moorland) would increase soil solution ANC by 140 μeq l^{-1} (albeit still remaining negative) and pH by 0.13 units. Simulated NO_3 concentrations by 2100 would be halved, with forest removal almost halting N saturation under forecast N deposition levels. The results thus highlight the potential impact of different forestry strategies within this highly deposition-impacted area.

Figure 8.9 provides a conceptual representation of acidification and recovery for this monitoring site. The modelled pre-industrial ANC suggests that the site was always very acidification-sensitive, with an ANC of only 20 μeq l^{-1}. Acidification of the site from this time until the acidification peak in 1980, in total ~900 μeq l^{-1}, is represented by the red arrow. Subsequent recovery is represented as (a) modelled recovery prior to the start of monitoring in the mid 1990s of 275 μeq l^{-1} (arrow 1) and observed recovery during the monitoring period to the present day of 340 μeq l^{-1} (arrow 2). However, no further net recovery is simulated through the end of the simulation period, leaving a 'recovery shortfall' of 280 μeq l^{-1} between current and pre-industrial ANC levels (arrow 3). The simulation also suggests that removal of the forest plantation would reduce this recovery shortfall by half (arrow 4).

8.4.2 Target Load Calculations Using MAGIC

In the UK, most habitat classes are assigned a critical ANC threshold of zero, which the Ladybower site would remain below even after forest removal according to current emission legislation (see above). To assess the deposition reductions that

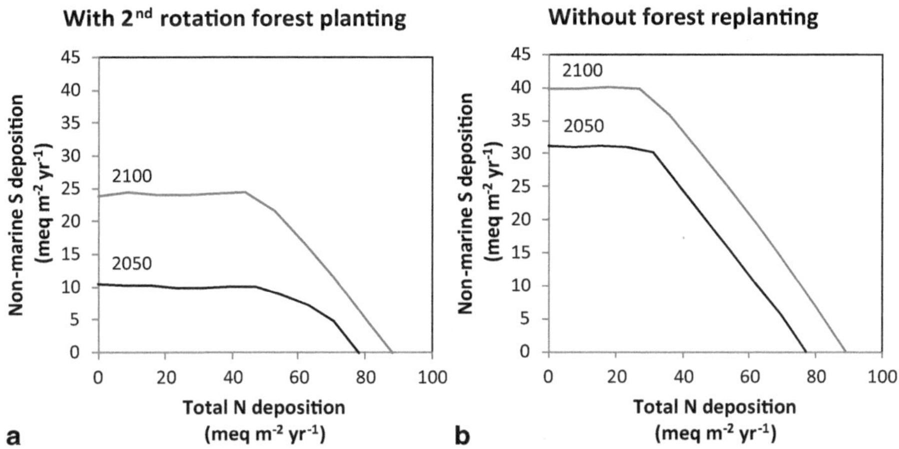

Fig. 8.10 Target load functions for the Ladybower Forest Level II monitoring plots, to achieve a target ANC value of zero by 2050 (*black line*) and 2100 (*grey line*) under **a** forest felling and replanting, and **b** forest removal

would be required to attain a specified critical threshold by a specified date, target loads can be derived. To derive target loads, the model is run iteratively in order to identify the different combinations of S and (total) N deposition that would achieve the target chemistry (in this case ANC=0) by the target year (in this case 2050 and 2100). The resulting target load function is analogous to the widely-used critical load function, which estimates the combinations of S and N deposition that would result in the desired chemistry under steady-state conditions (see Chap. 6). In damaged ecosystems, the target load is generally more stringent than the critical load, as lower deposition levels are required to achieve the critical chemical threshold at an earlier data.

Figure 8.10 illustrates the effects of both target year and forest management for the required target load. Under the forest replanting scenario, very low depositions of non-marine S (<25 meq $m^{-2}yr^{-1}$) and to a lesser extent N (<90 meq $m^{-2}yr^{-1}$) in order to achieve an ANC of zero by 2100 are required. Setting an earlier target year of 2050 would require even lower deposition levels, particularly for non-marine S. On the other hand, under the forest removal scenario, ANC=0 could be achieved with considerably higher target loads, again non-marine S deposition being most sensitive component of the target load function. This greater sensitivity of acidity to S relative to N deposition reflects the greater mobility of SO_4 within the terrestrial ecosystem compared to NH_4 or NO_3, even in this relatively highly N saturated system, is a common feature of target load functions, and contrasts with critical load functions that predict ANC under a fully N-saturated steady state condition that may take many decades or even centuries to attain.

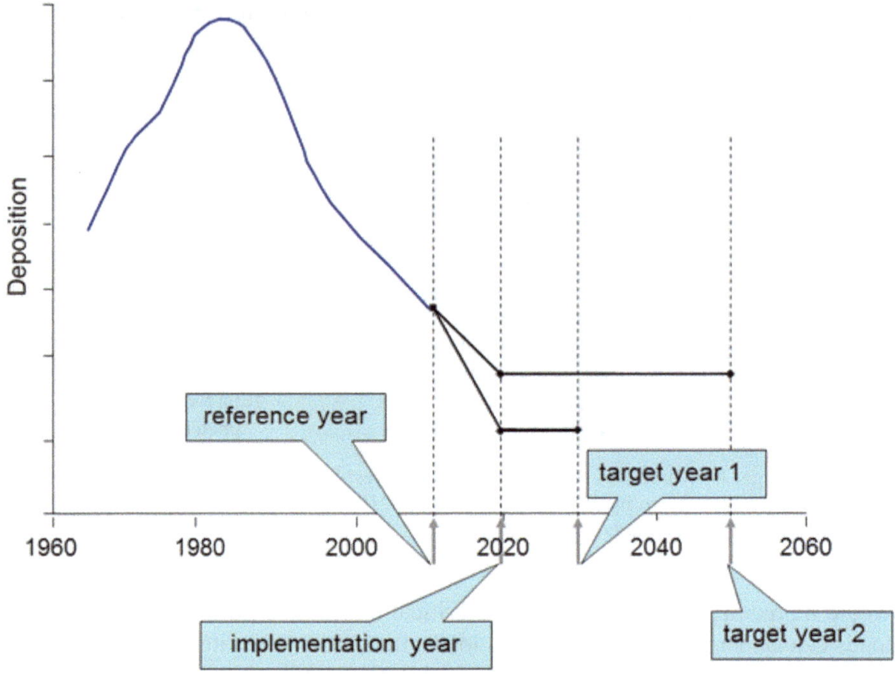

Fig. 8.11 Schematic representation of deposition paths leading to target loads by dynamic modelling, characterised by three key years. (i) The year up to which the (historic) deposition is fixed (*reference year*); (ii) the year in which the emission reductions leading to a target load are implemented (*implementation year*); and (iii) the year in which the chemical criterion is to be achieved (*target year*). Two examples of target years are shown

8.4.3 Target Load Calculations With VSD at Multiple Sites

Target loads have also been calculated on a European scale (e.g. Hettelingh et al. 2007). And here we report the dynamic modelling carried out at 260 forested sites across Switzerland (see Achermann et al. 2008) to determine target loads for acidifying deposition. The dynamic modelling was performed with the VSD model, with deposition, uptake and weathering fluxes drawn from MAKEDEP/SAFE (as in Sect. 8.3.1). The 'reference year' for the simulations was 2010 and the 'target years' addressed were 2030, 2050 and 2100 (with 'implementation year' 2020; see Fig. 8.11 for the terminology). All target loads were computed for the Al/Bc = 1 criterion.

For 246 of the 260 sites, no target load calculations are required, since the chemical criterion (Al/Bc = 1) is already achieved in 2010, or at least before 2030 (with the 3 sites described in Sect. 8.3.1 among them). For another 5 sites the chemical criterion will be met by 2050 (2100 in one case) without further emission reductions (i.e. only for 2030 a target load smaller than the critical load exists). For 4 of the remaining 9 sites target loads are unattainable, i.e. even reducing deposition to zero

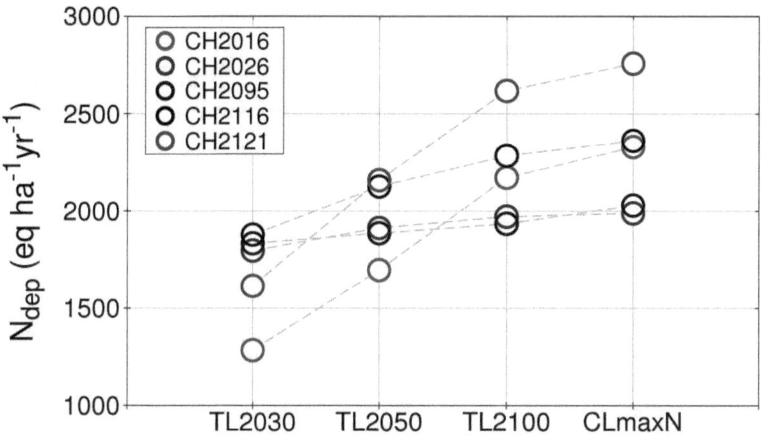

Fig. 8.12 Target loads (TLs) of acidity for target years 2030, 2050 and 2100 as well as the corresponding critical load of maximum N deposition for 5 Swiss sites

by 2020 will not lead to recovery in (one or more) target years. This leaves 5 sites for which target can be determined for all three target years; and they are displayed and compared in Fig. 8.12.

Figure 8.12 illustrates that—and this is a consequence of the definition—the later the target year the greater the target load; in fact, a critical load can be interpreted as the target load for an infinite time horizon. The difference between target loads for the same site for different target years depends on site characteristics, e.g. the size of buffers delaying equilibration, etc. For sites CH2026 and CH2095 the three target loads are quite similar and do not differ much from the critical load, whereas, e.g., for site CH2121 the 2030 target load is more than 1000 eq $ha^{-1}yr^{-1}$ smaller than the critical load. It can also be seen from Fig. 8.12 that the difference between target loads for a year far in the future (here 2100) don't differ much from the corresponding critical load.

While acidifying deposition in 2010 was already below critical loads for most of the Swiss sites, the problem of eutrophication is still prevalent: in 219 of the 260 sites the critical load of nutrient N is still exceeded in 2010, and thus further reduction in N depositions are needed.

8.5 Concluding Remarks

Over the last three decades dynamic models have been developed and applied on many sites and, increasingly—enabled by the growing power of computers—on a regional scale. The original driving force was the acidification of soils and surface waters and thus the emphasis on sulphur and base cation dynamics, which is still reflected in the core structure of most models. Furthermore, they were used for the reconstruction of past site dynamics and—after calibration—for analysing future deposition scenarios or the calculation of target loads. More recently, the emphasis has shifted to the effects of (mostly N) deposition on plant species diversity and carbon sequestration. This inspired the current efforts towards a more realistic modelling of carbon and nitrogen dynamics within existing and new dynamic models.

References

Aber, J. D., & Federer, C. A. (1992). A generalized, lumped-parameter model of photosynthesis, evapotranspiration and net primary production in temperate and boreal forest ecosystems. *Oecologia, 92,* 463–474.

Achermann, B., Rihm, B., & Kurz, D. (2008). Switzerland. In J.-P. Hettelingh, M. Posch, & J. Slootweg (Eds.), *Critical load, Dynamic modelling and impact assessment in Europe. CCE Status Report 2008* (pp. 205–210). Bilthoven: Netherlands Environmental Assessment Agency (PBL).

Aherne, J., Posch, M., Forsius, M., Lehtonen, A., & Härkönen, K. (2012). Impacts of forest biomass removal on soil nutrient status under climate change: A catchment-based modelling study for Finland. *Biogeochemistry, 107,* 471–488.

Alcamo, J., Shaw, R., & Hordijk, L. (1990). *The RAINS model of acidification.* Dordrecht: Kluwer Academic.

Alveteg, M. (1998). *Dynamics of forest soil chemistry.* PhD thesis, Lund, Sweden, Reports in Ecology and Environmental Engineering 3:1998, Department of Chemical Engineering II, Lund University.

Alveteg, M., Sverdrup, H., & Warfvinge, P. (1995). Regional assessment of the temporal trends in soil acidification in southern Sweden, using the SAFE model. *Water Air and Soil Pollution, 85,* 2509–2514.

Alveteg, M., Kurz, D., & Becker, R. (2002). Incorporating nutrient content elasticity in the MAKEDEP model. In *Reports in ecology and environmental engineering 1-2002* (pp. 52–67). Sweden: Lund University.

Belyazid, S. (2006). *Dynamic modelling of biogeochemical processes in forest ecosystems.* Doctoral Thesis. Reports in Ecology and Environmental Engineering 2006:1, Sweden: Department of chemical Engineering, Lund University.

Belyazid, S., & Moldan, F. (2009). *Use of integrated dynamic modelling to set critical loads for nitrogen deposition based on vegetation change—Pilot study at Gårdsjön.* (Report B1875). Swedish Environmental Research Institute IVL.

Belyazid, S., Sverdrup, H., Kurz, D., & Braun, S. (2011). Exploring ground vegetation change for different deposition scenarios and methods for estimating critical loads or biodiversity using the ForSAFE-VEG model in Switzerland and Sweden. *Water Air and Soil Pollution, 216,* 289–317.

Bonten, L. T. C., Groenenberg, J. E., Weng, L., & van Riemsdijk, W. H. (2008). Use of speciation and complexation models to estimate heavy metal sorption in soils. *Geoderma, 146,* 303–310.

Bonten, L., Mol, J., & Reinds, G. J. (2009). Dynamic modelling of effects of deposition on carbon sequestration and nitrogen availability: VSD plus and C and N dynamics (VSD+). In J. P. Hettelingh, M. Posch, & Slootweg J. (Eds.), *Progress in the modelling of critical thresholds, impacts to plant species diversity and ecosystem services in Europe.* CCE Status Report 2009 (pp. 69–73). Bilthoven: Netherlands Environmental Assessment Agency (PBL).

Bonten, L. T. C., Groenenberg, J. E., Meesenburg, H., & De Vries, W. (2011). Using advanced surface complexation models for modelling soil chemistry under forests: Solling forest, Germany. *Environmental Pollution, 159,* 2831–2839.

Cosby, B. J., Hornberger, G. M., Galloway, J. N., & Wright, R. F. (1985a). Modeling the effects of acid deposition: Assessment of a lumped parameter model of soil water and streamwater chemistry. *Water Resources Research, 21,* 51–63.

Cosby, B. J., Wright, R. F., Hornberger, G. M., & Galloway, J. N. (1985b). Modeling the effects of acid deposition. Estimation of long-term water quality responses in a small forested catchment. *Water Resources Research, 21,* 1591–1601.

Cosby, B. J., Wright, R. F., & Gjessing, E. (1995). An acidification model (MAGIC) with organic acids evaluated using whole-catchment manipulations in Norway. *Journal of Hydrology, 170,* 101–122.

Cosby, B. J., Ferrier, R. C., Jenkins, A., & Wright, R. F. (2001). Modelling the effects of acid deposition: Refinements, adjustments and inclusion of nitrogen dynamics in the MAGIC model. *Hydrology and Earth System Sciences, 5,* 499–517.

De Vries, W., Posch, M., & Kämäri, J. (1989). Simulation of the long-term soil response to acid deposition in various buffer ranges. *Water Air and Soil Pollution, 48,* 349–390.

De Vries, W., Solberg, S., Dobbertin, M., Sterba, H., Laubhann, D., van Oijen, M., Evans, C., Gundersen, P., Kros, J., Wamelink, G. W. W., Reinds, G. J., & Sutton, M. A. (2009). The impact of nitrogen deposition on carbon sequestration by European forests and heathlands. *Forest Ecology and Management, 258,* 1814–1823.

De Wit, H., & Wright, R. F. (2008). Projected stream water fluxes of NO_3 and total organic carbon from the Storgama headwater catchment, Norway, under climate change and reduced acid deposition. *Ambio, 37,* 56–63.

Dzombak, D. A., & Morel, F. M. M. (1990). *Surface complexation modeling: Hydrous ferric oxide*. New York: Wiley.

Eary, L. E., Jenne, E. A., Vail, L. W., & Girvin, D. C. (1989). Numerical models for predicting watershed acidification. *Archives of Environmental Contamination Toxicology, 18,* 29–53.

Evans, C. D. (2005). Modelling the effects of climate change on an acidic upland stream. *Biogeochemistry, 74,* 21–46.

Evans, C. D., Smart, S., Scott, W. A., Whittaker, J. A., Langan, S., Emmett, B. A., & Ashmore, M. (2004). Evaluation and development of dynamic models for soils and soil-plant systems. In B. A. Emmett & G. McShane (Eds.), *Terrestrial umbrella final report*. Bangor: Centre for Ecology and Hydrology.

Evans, C. D., Caporn, S. J. M., Carroll, J. A., Pilkington, M. G., Wilson, D. B., Ray, N., & Cresswell, N. (2006). Modelling nitrogen saturation and carbon accumulation in heathland soils under elevated nitrogen deposition. *Environmental Pollution, 143,* 468–478.

Evans, C. D., Reynolds, B., Hinton, C., Hughes, S., Norris, D., Grant, S., & Williams, B. (2008). Effects of decreasing acid deposition and climate change on acid extremes in an upland stream. *Hydrology and Earth System Sciences, 12,* 337–351.

Gherini, S. A., Mok, L., Hudson, R. J. M., Davis, G. F., Chen, C. W., & Goldstein, R. A. (1985). The ILWAS model: Formulation and application. *Water Air and Soil Pollution, 26,* 425–459.

Gundersen, P., Emmet, B. A., Kjønaas, O. J., Koopmans, C., & Tietema, A. (1998). Impact of nitrogen deposition on nitrogen cycling in forests: A synthesis of NITREX data. *Forest Ecology and Management, 101,* 37–55.

Helliwell, R. C., Aherne, J., MacDougall, G., Nisbet, T. R., Lawson, D., Cosby, B. J., & Evans, C. D. (2014). Past acidification and recovery of surface waters, soils and ecology in the United Kingdom: Prospects for the future under current deposition and land use protocols. *Ecological Indicators, 37B,* 396–411.

Hettelingh, J.-P., & Posch, M. (1994). Critical loads and a dynamic assessment of ecosystem recovery. In Grasman, J. & G. Van Straten (Eds.), *Predictability and nonlinear modelling in natural sciences and economics* (pp. 439–446). Dordrecht: Kluwer Academic.

Hettelingh, J.-P., Posch, M., Slootweg, J., Reinds, G. J., Spranger, T., & Tarrasón, L. (2007). Critical loads and dynamic modelling to assess European areas at risk of acidification and eutrophication. *Water Air and Soil Pollution Focus, 7,* 379–384.

Jenkins, A., Cosby, B. J., Ferrier, R. F., Walker, T. A. B., & Miller, J. D. (1990). Modelling stream acidification in afforested catchments: An assessment of the relative effects of acid deposition and afforestation. *Journal of Hydrology, 120,* 163–181.

Jenkins, A., Cosby, B. J., Ferrier, R. C., Larssen, T., & Posch, M. (2003). Assessing emission reduction targets with dynamic models: Deriving target load functions for use in integrated assessment. *Hydrology and Earth System Sciences, 7,* 609–617.

Kämäri, J., & Posch, M. (1987). Regional application of a simple lake acidification model to Northern Europe. In M. B. Beck (Ed.), *Systems analysis in water quality management* (pp. 73–84). Oxford: Pergamon Press.

Kauppi, P., Kämäri, J., Posch, M., Kauppi, L., & Matzner, E. (1986). Acidification of forest soils: Model developement and application for analyzing impacts of acidic deposition in Europe. *Ecological Modelling, 33,* 231–253.

Kinniburgh, D. G., van Riemsdijk, W. H., Koopal, L. K., Borkovec, M., Benedetti, M. F., & Avena, M. J. (1999). Ion binding to natural organic matter: Competition, heterogeneity, stoichiometry and thermodynamic consistency. *Colloids and Surfaces A-Physiochemical and Engineering Aspects, 151,* 147–166.

Kopáček, J., Cosby, B. J., Majer, V., Stuchlik, E., & Vesely, J. (2003). Modelling reversibility of Central European mountain lakes from acidification: Part I—The Bohemian Forest. *Hydrology and Earth System Sciences, 7,* 494–509.

Kros, J., Reinds, G. J., De Vries, W., Latour, J. B., & Bollen, M. J. S. (1995). Modelling the response of terrestrial ecosystems to acidification and desiccation scenarios. *Water Air and Soil Pollution, 85,* 1101–1106.

Larssen, T., Huseby, R. B., Cosby, B. J., Høst, G., Høgåsen, T., & Aldrin, M. (2006). Forecasting acidification effects using a Bayesian calibration and uncertainty propagation approach. *Environmental Science and Technology, 40,* 7841–7847.

Larssen, T., Høgåsen, T., & Cosby, B. J. (2007). Impact of time series data on calibration and prediction uncertainty for a deterministic hydrogeochemical model. *Ecological Modelling, 207,* 22–33.

Lindström, G., & Gardelin, M. (1992). Modelling groundwater response to acidification. In P. Sanden & P. Warfvinge (Eds.), *Report from the Swedish integrated groundwater acidification project* (pp. 33–36). Norrköping: Swedish Meteorological and Hydrological Institute (SMHI).

Majer, V., Cosby, B. J., Kopácek, J., Stuchlík, E., & Veselý, J. (2003). Modelling reversibility of Central European mountain lakes from acidification: Part I—The Bohemian forest. *Hydrology and Earth System Sciences, 7,* 494–509.

Meesenburg, H., Meiwes, K. J., & Rademacher, P. (1995). Long term trends in atmospheric deposition and seepage output in northwest German forest ecosystems. *Water Air and Soil Pollution, 85,* 611–616.

Meeussen, J. C. L. (2003). ORCHESTRA: An object-oriented framework for implementing chemical equilibrium models. *Environmental Science and Technology, 37,* 1175–1182.

Monteith, D. T., Evans, C. D., & Reynolds, B. (2000). Are temporal variations in the nitrate content of UK upland freshwaters linked to the North Atlantic Oscillation? *Hydrological Processess, 14,* 1745–1749.

Nikolaidis, N. P., Schnoor, J. L., & Georgakakos, K. P. (1989). Modeling of long-term lake alkalinity responses to acid deposition. *Journal of Water Pollution Control Federation, 61,* 188–199.

Oulehle, F., Hofmeister, J., & Hruška, J. (2007). Modeling of the long-term effect of tree species (Norway spruce and European beech) on soil acidification in the Ore Mountains. *Ecological Modelling, 204,* 359–371.

Oulehle, F., Cosby, B. J., Wright, R. F., Hruška, J., Kopáček, J., Krám, P., Evans, C. D., & Moldan, F. (2012). Modelling soil nitrogen: The MAGIC model with nitrogen retention linked to carbon turnover using decomposer dynamics. *Environmental Pollution, 165,* 158–166.

Posch, M., & Reinds, G. J. (2009). A very simple dynamic soil acidification model for scenario analyses and target load calculations. *Environmental Modelling and Software, 24,* 329–340.

Posch, M., Hettelingh, J.-P., & Slootweg, J. (2003). *Manual for dynamic modelling of soil response to atmospheric deposition*. (RIVM report 259101 012). Bilthoven, The Netherlands: National Institute for Public Health and the Environment.

Rogora, M., Marchetto, A., & Mosello, R. (2003). Modelling the effects of atmospheric sulphur and nitrogen deposition on selected lakes and streams of the Central Alps (Italy). *Hydrology and Earth System Sciences, 7,* 540–551.

Rose, K. A., Cook, R. B., Brenkert, A. L., Gardner, R. H., & Hettelingh, J. P. (1991a). Systematic comparison of ILWAS, MAGIC, and ETD watershed acidification models 1. Mapping among model inputs and deterministic results. *Water Resources Research, 27,* 2577–2589.

Rose, K. A., Brenkert, A. L., Cook, R. B., Gardner, R. H., & Hettelingh, J.-P. (1991b). Systematic comparison of ILWAS, MAGIC, and ETD watershed acidification models—2. Monte Carlo analysis under regional variability. *Water Resources Research, 27,* 2591–2603.

Sjøeng, A.-M. S., Wright, R. F., & Kaste, Ø. (2009). Modelling seasonal nitrate concentrations in runoff of a heathland catchment in SW Norway using the MAGIC model: I. Calibration and specification of nitrogen processes. *Hydrology Research, 40,* 198–216.

Sverdrup, H., Belyazid, S., Nihlgård, B., & Ericson, L. (2007). Modelling change in ground vegetation response to acid and nitrogen pollution, climate change and forest management at in Sweden 1500–2100 A.D. *Water Air and Soil Pollution Focus, 7,* 163–179.

Tiktak, A., & van Grinsven, J. J. M. (1995). Review of sixteen forest-soil-atmosphere models. *Ecological Modelling, 83,* 35–54.

Tominaga, K., Aherne, J., Watmough, S. A., Alveteg, M., Cosby, B. J., Driscoll, C. T., & Posch, M. (2009). Voyage without constellation: Evaluating the performance of three uncalibrated process-oriented models. *HydrologyResearch, 40,* 261–272.

Ulrich, B. (1983). Soil acidity and its relations to acid deposition. In B. Ulrich & J. Pankrath (Eds.), *Effects of accumulation of air pollutants in forest ecosystems* (pp. 127–146). Dordrecht: Reidel Publ. Co.

Van Dam, J. C., Groenendijk, P., Hendriks, R. F. A., & Kroes, J. G. (2008). Advances of modeling water flow in variably saturated soils with SWAP. *Vadose Zone Journal, 7,* 640–653.

Vanguelova, E., Benham, S., Pitman, R., Moffat, A. J., Broadmeadow, M., Nisbet, T., Durrant, D., Barsoum, N., Wilkinson, M., Bochereau, F., Hutchings, T., Broadmeadowa, S., Crow, P., Taylor, P., & Durrant Houston, T. (2010). Chemical fluxes in time through forest ecosystems in the UK—Soil response to pollution recovery. *Environmental Pollution, 158,* 1857–1869.

Wallman, P., Belyazid, S., Svensson, M. G. E., & Sverdrup, H. (2006). DECOMP—A semi-mechanistic model of litter decomposition. *Environmental Modelling and Software, 21,* 33–44.

Warfvinge, P., Holmberg, M., Posch, M., & Wright, R. F. (1992). The use of dynamic models to set target loads. *Ambio, 21,* 369–376.

Wright, R. F., Larssen, T., Camarero, L., Cosby, B. J., Ferrier, R. C., Helliwell, R. C., Forsius, M., Jenkins, A., Kopáček, J., Majer, V., Moldan, F., Posch, M., Rogora, M., & Schöpp, W. (2005). Recovery of acidified European surface waters. *Environmental Science and Technology, 39,* 64A–72A.

Chapter 9
Dynamic Geochemical Models to Assess Deposition Impacts of Metals for Soils and Surface Waters

Jan E. Groenenberg, Edward Tipping, Luc T. C. Bonten and Wim de Vries

9.1 Introduction

Elevated concentrations of heavy metals in natural soils and waters possibly deteriorate soil and water ecosystem functioning. In view of precautionary risk assessment, the concept of critical loads was developed for heavy metals, being the load below which selected receptors are protected at steady state, i.e. with inputs and outputs of metals being in equilibrium (see Chap. 7). Given the long times it takes for metal concentrations to reach steady state, the use of dynamic models should be considered to manage and evaluate the risks of metal loads in time (Lofts et al. 2007). Moreover, key processes determining the fate of metals are related to soil properties such as pH and the concentration of dissolved organic matter (DOM), both being subject to changes due to external factors such as land use change, climate change and atmospheric deposition of nitrogen. Dynamic models help to understand the complex interactions of processes due to such external factors and give insight into the timescales at which changes take effect.

Metal transfers in ecosystems are complex, but by identifying and quantifying key processes it is possible to produce useful descriptions of metal behaviour in soils and catchments with models that can be driven with limited data. Here we describe three models, which aim to capture the key processes determining the fate of metals in natural ecosystems, differing in complexity, while keeping the data demand limited. Identified key factors for the fate of metals in soils are the pH and concentrations of soil- and dissolved organic matter (SOM, DOM), which are related to sorption of metals to reactive surfaces (organic matter, oxides, clay). The models discussed in this chapter all combine predictions of soil acidity and major soil solution chemistry with heavy metal behaviour.

J. E. Groenenberg (✉) · L. T. C. Bonten · W. de Vries
Alterra Wageningen University and Research Centre, Wageningen, The Netherlands
e-mail: bertjan.groenenberg@wur.nl

E. Tipping
Centre for Ecology and Hydrology, Bailrigg, LA, UK

W. de Vries et al. (eds.), *Critical Loads and Dynamic Risk Assessments*,
Environmental Pollution 25, DOI 10.1007/978-94-017-9508-1_9

The three models, SMART2-metals, SMARTml and CHUM-AM, each have their specific purpose which is reflected in differences in spatial schematisation of the soil, soil processes incorporated and detail/complexity of the incorporated processes. The model SMART2-metals (De Vries et al. 1998) is the simplest model and was designed as the dynamic counterpart of the metal critical load models (see also Chap. 7). Because the model makes simulations with an annual time step, it only includes two soil layers and does not use an iterative procedure. Model run time is very short, which enables the model to be used for studies at large regional scales. The model CHUM-AM (CHemistry of the Uplands Model—Annual Metals) was designed to describe metal behaviour in whole catchments including predictions of stream water concentrations (Tipping et al. 2006a, b, 2007). The model considers three soil layers and uses an annual time step. The multi-layer model SMARTml (Bonten et al. 2011) was designed for site applications, includes the same processes as SMART2-metals but uses a more mechanistic process description. Furthermore, the model has a flexible time step and can be used in combination with the detailed hydrological model SWAP and is therefore able to include intra-annual variation. The number of soil layers is flexible and enables the model to predict changes at various depths including metal fluxes to groundwater and surface water.

In contrast to critical load models, dynamic model concepts can be tested on observations. In this chapter, the three above-mentioned dynamic models are presented and evaluated by comparing long-term model simulations with field (monitoring) data. Preferably, long-term (minimal several decades) monitoring data are used to evaluate model predictions. However, such long-term monitoring data are very scarce. The ability to describe the effect of land use changes on metal chemistry was evaluated with SMART2-metals using chronosequences of afforested agricultural land of different stand age. CHUM-AM was evaluated using hind-cast modelling, in which the present state is predicted by model simulations starting in the past and using constructed historical input data. Finally, SMARTml was evaluated using long-term monitoring data of a forested plot in Germany.

9.2 Modelling Approaches

Table 9.1 summarises the processes included in the three models. The SMART models with extension for heavy metals were developed primarily to simulate changes in the soil and soil solution concentrations. CHUM-AM is developed to calculate the leaching of metal from catchments to streams. SMART2-metals and CHUM-AM use an annual time step and constant hydrology. SMARTml can be combined with detailed hydrology and makes use of smaller time steps to account for intra-annual variation. Where the SMART models include canopy interactions and neglect erosion, the opposite is true for CHUM-AM. Main differences between the models are further related to the description of cation exchange and metal binding (see Table 9.1). More details on the models are given below.

Table 9.1 Overview of the key processes included in the three models

Process	SMART2-metals	CHUM-AM	SMARTml
Master species	H^+, BC^+, BC^{2+}, Al^{3+}, NH_4^+, Cl^-, NO_3^-, SO_4^{2-}, CO_3^{2-}, Ni^{2+}, Cu^{2+}, Zn^{2+}, Cd^{2+}, Pb^{2+}	H^+, Na^+, Mg^{2+}, Al^{3+}, K^+, Ca^{2+}, Fe^{3+}, NH_4^+, Cl^-, NO_3^-, SO_4^{2-}, F^-, CO_3^{2-}, $SiOH_4$, Ni^{2+}, Cu^{2+}, Zn^{2+}, Cd^{2+}, Pb^{2+}	H^+, Na^+, Mg^{2+}, Al^{3+}, K^+, Ca^{2+}, Fe^{3+}, NH_4^+, Cl^-, NO_3^-, SO_4^{2-}, CO_3^{2-}, Ni^{2+}, Cu^{2+}, Zn^{2+}, Cd^{2+}, Pb^{2+}
Application scales	Regional	Catchment	Site/plot
Number of soil layers	2 layers	3 layers	Multiple layers
Temporal resolution	Yearly	Yearly	Flexible, up to daily
Key inputs	Deposition BC, N, S, metals	Deposition BC, N, S, metals	Deposition BC, N, S, metals
Key modelled processes			
Canopy interactions	BC, N, S, metals	–	BC, N, S, metals
Uptake	BC, N, metals	–	BC, N, metals
Mineralization	BC, N, metals	C	BC, metals
Cation-exchange	Exchange-relations	Separate model for SOM[a]	Separate models for SOM[b], AlFe-oxides[c], clay[d]
Weathering	BC, metals	BC, metals	BC, metals
Erosion	–	BC, S, metals	–
Metal binding	Soil (partition function)	Separate model for SOM[a], POM[a], DOM[a]	Separate models for SOM[b], DOM[b], AlFe-oxides[c], clay[d]

[a] WHAM/Model VI
[b] NICA-Donnan
[c] Generalized Two-Layer Model
[d] Donnan-model

9.2.1 SMART2-Metals

SMART2 (Kros et al. 1995) is an extension of the dynamic soil acidification model SMART (De Vries et al. 1989), that includes the major hydrological and biogeochemical processes in the vegetation, litter and mineral soil. The model consists of two compartments, the organic layer and the mineral soil. Apart from pH, the model also predicts changes in aluminium (Al^{3+}), base cation (BC), nitrate (NO_3^-) and sulphate (SO_4^{2-}) concentrations in the soil solution and solid phase characteristics depicting the acidification status, i.e. carbonate content, base saturation and readily available Al content. The major extensions in SMART2, compared to SMART, are the inclusion of a nutrient cycle and an improved modelling of hydrology. As with SMARTml (see Chap. 8), the SMART2 model consists of a set of mass balance equations, describing the soil input-output relationships, and a set of equations describing the rate-limited and equilibrium soil processes.

In the original version of SMART, proton-base cation exchange is described with the Gaines-Thomas equation. This formulation of the exchange process assumes the exchange constant KH_{exch} to be constant over the entire pH-range. In experiments, a decreasing affinity of H^+ for adsorption is found with decreasing pH. In the case of afforestation of agricultural land the pH at the start is in general near neutral ($5.5 < pH < 7.$) but may severely decrease to pH-values as low as 3.5. Within this pH-range it is impossible to simulate the pH course accurately using a constant value for KH_{exch}. Therefore a pH-dependent exchange constant was introduced.

The model SMART2-metals has been constructed by extending SMART2 with a mass balance for metals, including all relevant processes in forest soil systems. The mass balance and processes for metals were formulated based on the following assumptions:

- Metal partitioning over the solid phase and soil solution (including DOM) is always at equilibrium.
- The soil is homogeneously mixed within each soil layer, which implies that both soil properties and concentrations of metals do not show vertical variation within a soil layer.
- The soil is in an oxidized state, reduction processes and precipitation of metals in e.g. metal sulphides are not included. The use of the model is thus limited to well drained (top) soils.
- Transport of water and solutes including metals only takes place in a vertical direction; surface runoff and bypass flow are neglected.

The most important processes which are part of the mass fluxes of heavy metals in soil-vegetation systems are included in the model. These consist of metal inputs by atmospheric deposition, direct input to the soil (fertilization, sludge amendment) or by seepage, canopy processes (filtering, foliar uptake and exudation), root uptake, adsorption/desorption of metals by soil, weathering and leaching. Canopy processes, uptake, weathering and leaching are described according to De Vries and Groenenberg (2009) (see also Chap. 7). Sorption of metals is described with the Freundlich-van Bemmelen equation (Sposito 1989). Transfer functions are used to relate the Freundlich adsorption constant to soil parameters like the organic matter content, clay content, content of Fe-oxides and soil solution parameters like the pH and the concentrations of DOM. Groenenberg et al. (2012) showed that such an empirical approach is justified for soils in the pH range 4–7.

9.2.2 CHUM-AM

The CHemistry of the Uplands Model—Annual, Metals (CHUM-AM) (Tipping et al. 2006a, b, 2007) aims to quantify metal behaviour in whole catchments with limited data by combining simple process descriptions of the soil N and S cycles, water percolation and mineral dissolution and formation, with a comprehensive model of chemical equilibria, while operating with an annual time step. CHUM-AM emphasises the competitive interactions of cations (i.e. metals, H^+, base cations, Al, Fe) with organic matter, both solid-phase and dissolved, as the principal controls on the retention and transport of metals. Application of the model requires deposition scenarios that cover several hundred years, from "pristine" conditions to the present, together with weathering inputs. The model ignores soil and stream spatial heterogeneity, cycling in vegetation, short-term variability in flow pathways, sorption by mineral components in the soil, "ageing" of metals added to soil, and bioturbation. The role of physical erosion is taken into account simply through an assumed average rate of release of soil solids.

The model considers a soil-rock profile consisting of three completely mixed layers (L1, L2, L3). Atmospherically deposited elements and species, including dry-deposited gases NH_3, NO_2 and SO_2 as well as particulate elements are assumed to be in solution when entering the plant-soil system. Extra deposition of major solutes from cloud water is taken into account. Water percolates at a constant rate from L1 to L2 to L3 and then to the stream. This involves repeated filling and emptying of the available pore space, with acquisition of solutes in input water and from decomposition and weathering reactions, attainment of new equilibrium distributions of chemical species, and outputs in drainage. Elements associated with SOM (N, S, metals) transfer annually from L1 to L2 and from L2 to L3. Annual outputs of organic matter from L1 and L2, as dissolved organic matter (DOM) and particulate organic matter (POM) in percolating water, are specified in the input file, as is the soil pCO_2. Dissolved CO_2 in percolation water from L2 enters L3, where it may be supplemented by CO_2 from carbonate dissolution. The soil takes up N and S, depending upon solution concentrations and releases the elements in proportion to their soil pools. The chemical "master species", i.e. the inorganic species on which all the chemical reactions are based, consist of major cations (H^+, Na^+, Mg^{2+}, Al^{3+}, K^+, Ca^{2+}, Fe^{3+}, NH_4^+), metal cations (Ni^{2+}, Cu^{2+}, Zn^{2+}, Cd^{2+}, Pb^{2+}), anions (OH^-, Cl^-, NO_3^-, SO_4^{2-}, F^-, CO_3^{2-}) and the neutral species $Si(OH)_4$. In L1 and L2, solutes may bind to organic matter, which may be in solid phase, in solution (DOM) or in suspension (POM). The interactions are described using the soil version of WHAM/Model VI (Tipping 1998, 2002). Account is also taken of reactions among inorganic species in solution and of precipitation and dissolution of $Al(OH)_3$, $Fe(OH)_3$ and carbonates. In L3, the same reactions are possible, together with surface complexation at the surfaces of Al and Fe oxides. Weathering inputs of major cations, metals and carbonate are described with the formulation of Schnoor and Stumm (1986), in which the rate is given by the product of a constant (k_w) and H^+ activity raised to the power n_w. Values of n_w for different solutes were taken from Stidson et al. (2002), while values of k_w (which includes the effects of mineral type and available surface area) are adjusted to match observations.

In stream water the reactants are the inorganic components, together with DOM and POM. Outgassing of CO_2 occurs in the stream, to a specified partial pressure. The lakes are assumed to be fed by a single stream, representative of the whole catchment, and to be of constant volume, completely mixed, and at CO_2 equilibrium with the atmosphere. Lake chemical compositions are determined by atmospheric deposition directly to the water surface, stream water input, lake water output, particle sedimentation and sediment-water exchange of solutes. Nitrate, sulphate and ferric iron are reduced within the accumulating sediment. Lake mixing, sedimentation and sediment chemistry are simulated.

The modelling strategy is first to deal with the anionic components Cl, N, S and DOM, which are assumed to behave independently. Next, the major cationic elements (H, Na, Mg, Al, K and Ca) are considered, with calibration of weathering rates. Finally, heavy metal behaviour is predicted, with further calibration if necessary.

9.2.3 SMARTml

SMARTml is an extension of single-layer model SMART ('ml' stands for multilayer). The largest difference with the single-layer SMART model (De Vries et al. 1989), and most other soil acidification models, is that SMARTml uses an assemblage of surface complexation models for the calculation of cation and anion sorption instead of Gapon or Gaines-Thomas ion exchange equations for cations and Langmuir or Freundlich isotherms for sulphate and phosphate (see Chap. 8). The use of surface complexation models makes it relatively easy to extend the model with additional elements, as long as complexation parameter values are available for these elements. For the application described in this chapter, SMARTml was extended with the metals Cd, Cu, Pb and Zn.

In addition to complexation also other relevant processes and inputs of metals in soils were included in SMARTml, i.e. inputs by fertilizers, atmospheric deposition and seepage. Uptake of the metals Cd, Cu, Pb and Zn was modelled similarly as in the calculation of critical loads (see Chap. 7), and described in more detail by De Vries and Groenenberg (2009). Biogeochemical processes include foliar uptake, metals in litterfall, root uptake, being the sum of maintenance uptake and growth uptake, and mineralisation of organic matter. Throughfall of metals was calculated from total deposition using a constant foliar uptake fraction. Metals in litterfall are the sum of a constant metal litterfall rate and metals in litterfall proportional to total deposition. The micronutrients Cu and Zn have a relative high constant rate while litterfall of the non-nutrients Cd and Pb is more dependent on deposition. Root uptake of metals is described as a maintenance uptake, to resupply the release by litterfall minus foliar uptake, plus a growth uptake, which is the net biomass growth times a metal content in the standing biomass. Similar to the critical load calculations, input through weathering of metals was neglected. Parameters for biochemical processes of the metals were taken from De Vries and Groenenberg (2009). Together with macro-ions the binding of metals to reactive surfaces is modelled using the assemblage model implemented in the model framework ORCHESTRA (Meeussen 2003) as described by Bonten et al. (2008). The assemblage model includes the NICA-Donnan model (Kinniburgh et al. 1999) for ion-binding to SOM and DOM, the generalized two-layer model of Dzombak and Morel (1990) for ion binding to Al and Fe-(hydr)oxides and a Donnan model for electrostatic binding to clay surfaces. SOM was assumed to consist of humic acids (HA) with a site density of 31% relative to generic HA and DOM was assumed to consist of 30% fulvic acids (FA), 30% HA and 40% non-reactive material.

9.3 Model Applications

9.3.1 SMART2-Metals Application on Chronosequences in the Netherlands and Sweden

The model SMART2 extended with metal chemistry was applied to chronosequences (sites of different age with the same land use and with similar site conditions, such as soil types) of forested soils on former agricultural land in the Netherlands and in Sweden. The advantage of the use of chronosequences is that the effects of afforestation over a longer period can be evaluated and models can be tested on a "long time" record. One should however be aware that the data of a chronosequence cannot just be interpreted as one long time series of measurements, since environmental conditions may be different for sites of different age in the same period of development.

9.3.1.1 Sites and Input Data

In the Netherlands a chronosequence of forest soils was sampled of forests planted on former arable soils around Sellingen (Römkens 1998, Römkens and Salomons 1998). The dataset consisted of four pairs of sites with forests at the age of 5, 11, 22 and 29 years (reference year 1996). Furthermore the dataset consisted of two arable soils and two reference forest soils that were more than 80 years old and not in use for agriculture before. All sites have the same type of soil, i.e. a gleyic podzol. All forested sites were planted with summer oak (*Quercus robur*), except for the sites of 5 years, of which one was planted with beech and one with ash. For a more detailed description of the sites see (Römkens 1998, Römkens and Salomons 1998). In Sweden a chronosequence of forest soils was sampled of forest soils planted on former arable soils in the southern part of Sweden (Karltun et al. 2005). The selected dataset consisted of 8 sites of forest soils with ages ranging from 6–87 years (reference year 1997). All stands were spruce on arenosols (FAO 1998).

Deposition data for S, N and metals for the Dutch chronosequence were derived for the period 1890–1995. Data for S and N deposition for 1980–1995 were derived from calculations with the DEADM model (Erisman et al. 1995). Historical deposition from 1890–1980 was reconstructed. For the period 1920–1980, deposition data were estimated from emissions in the same period. First, linear regressions were derived between emission and calculated deposition for the period 1950–1980. These regressions were used to estimate the deposition from emission data for the period

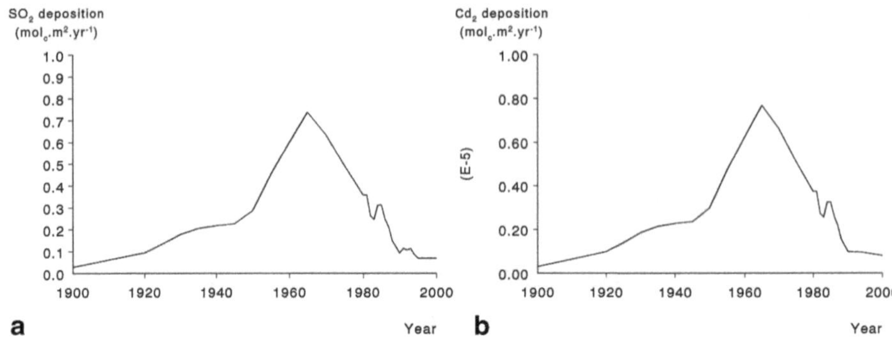

Fig. 9.1 Atmospheric deposition of SO_2 (*left*) and Cd (*right*) for the Dutch chronosequences

1920–1950. For the period 1890–1900 a background deposition for NO_x–N, NH_x–N and SO_x–S of respectively 200, 300 and 300 mol ha^{-1} yr^{-1} was used, and for the period 1900–1920 a linear increase of the deposition from the levels of 1900 to the levels of 1920 was assumed. Deposition of Cd, Cu and Zn was derived from figures for 1990, 1994 and 2000 calculated with the OPS model (Van Jaarsveld and de Leeuw 1993). Metal deposition for other years was extrapolated assuming the same trend in metal deposition as in sulphur deposition. As an example, results derived for the SO_2 and Cd deposition for the Dutch chronosequence are given in Fig. 9.1.

For the Swedish chronosequence, deposition of S, N, Cl and base cations were derived for the period 1910–1995. Data for the period 1987–1995 were taken from deposition measurements at three nearby sites (Margaretenberg, Ahla and Tjärby). Deposition figures for the period before 1987 were estimated assuming the same trend as in the Netherlands. Deposition of Cd, Cu and Zn for the years 1968, 1975, 1980, 1985, 1990 and 1995 were derived from moss surveys in the nearby Halmstad area and a regression equation relating moss contents to wet deposition (Berg et al. 1995). Cd deposition before 1968 was estimated by assuming the same trend as for SO_2.

Soil data needed in the model were derived from measured data and calculated for the 25 cm layer soil used in SMART2 by weighing the data for different depths according to their thickness and bulk density. Data on CEC and occupation of the exchange complex in the Dutch chronosequence were measured at the actual pH of the soil (unbuffered $BaCl_2$, (Deller 1983). The measured CEC refers thus to the effective CEC. In the model all CEC values were standardized to pH = 7 by linear regression of the effective CEC against pH and organic matter content of the soils. Exchange constants for H and Al with Ca were calculated for pH=7 assuming that the gap between the actual CEC and the CEC at pH = 7 is occupied by H. For the Swedish chronosequence, the CEC at pH = 7 was calculated from the organic matter and clay content of the soil (Helling et al. 1964). The cation exchange constants derived for the Dutch sites were also used for the model application at the Swedish sites.

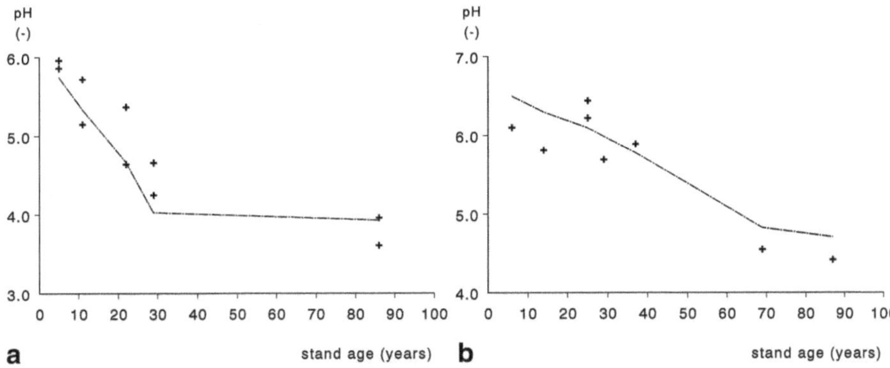

Fig. 9.2 SMART2-metals simulated (*line*) and observed (*dots*) pH for the Dutch (*left*) and Swedish (*right*) chronosequences as a function of stand age

Model parameters used for the various soils and soil vegetation combinations in SMART2 were taken from (Kros et al. 1995). Adsorption and desorption of Cd, Cu and Zn were described with transfer functions as described in Chap. 2. For the litter layer different relationships were derived, as they consist almost entirely of organic matter.

9.3.1.2 Results and Discussion

To compare model results with measured data, simulations with the model were started in the year of forest plantation and run until the latest year of measurements. The 1996 endpoint of the simulation is used to compare simulation results with measured data. In Fig. 9.2 left measured data for the pH of the stands with different stand age of the Dutch chronosequence in 1996 are presented together with the results of the model simulations for 1996 (presented as a line). The results show a clear downward trend in pH due to afforestation. The acidification of the Swedish sites (Fig. 9.2 right) is however less severe than for the Dutch sites, which is due to a lower deposition and higher buffering (higher CEC). The figure shows a good agreement of simulated and measured pH for the different stands.

Measurements and model simulations show a decreasing trend in soil cadmium concentrations with stand age for both the Dutch and Swedish chronosequences (Fig. 9.3). The decrease of Cd in soil is due to high leaching fluxes exceeding Cd inputs from deposition. The high leaching fluxes are caused by the acidification of the soil that results in a rise of Cd solution concentrations. Although Cd levels in soil are at the same level in the Netherlands and Sweden, solution concentrations in Sweden are clearly lower than in the Netherlands due to the higher pH and higher organic matter content, both resulting in a stronger sorption of Cd to the soil (Fig. 9.3). Measurements and simulations for soil Cd in the Dutch chronosequence show an almost complete depletion of the Cd pool in the upper soil for the oldest

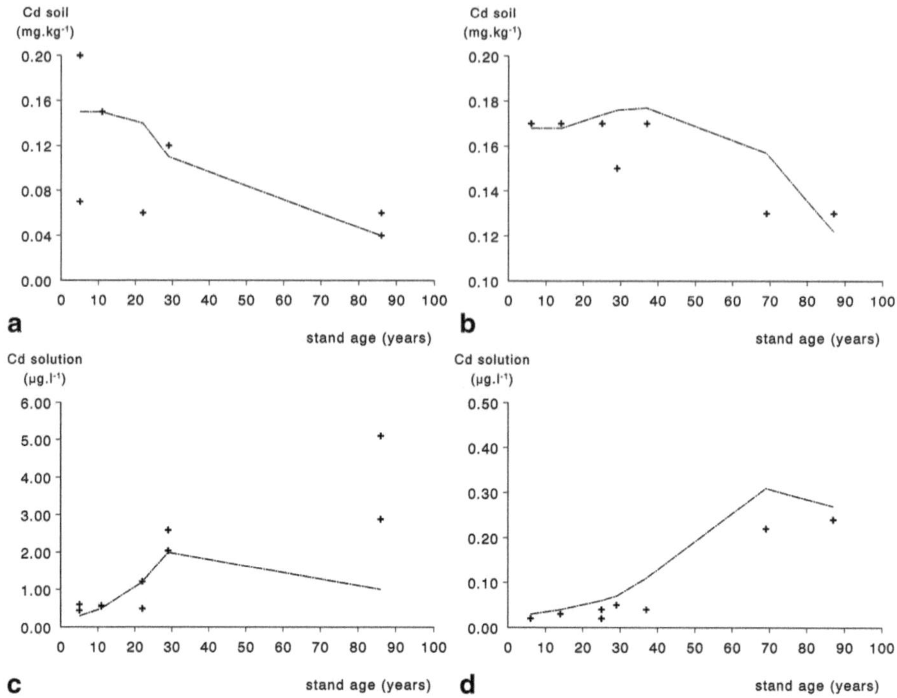

Fig. 9.3 SMART2-metals simulated (*line*) and observed (*dots*) Cd concentrations in soil and solution for the Dutch (*left*) and Swedish (*right*) chronosequences as a function of stand age

stands (Fig. 9.3 top left) due to the high leaching flux. As a result of this, the model simulates decreasing concentrations of Cd in soil solution at the end of the simulation period (Fig. 9.3 bottom left). Measured concentrations of Cd for the oldest stand are however still high. Unlike the discrepancy between model results of Cd in solution for the oldest stands, there is good agreement between model results and measurements for the younger stands.

From the results of the simulated mass balances for Cd, it appears that the biochemical cycle of Cd (uptake by vegetation, metals in litter fall and mineralization) hardly influenced the concentration of Cd in soil and soil solution. These fluxes are very small compared to inputs of metals by deposition and leaching of metal to deeper soil layers. The results also showed that accumulation in or leaching of metals from the litter layer did not have a significant effect on the concentration of Cd in the soil. The trends for the reactive soil pool and solution concentrations of Zn and Cu were comparable to those for Cd and are therefore not shown here.

9.3.2 *CHUM-AM Application to Upland Catchments in the United Kingdom*

9.3.2.1 Sites and Input Data

Tipping et al. (Tipping et al. 2006a, 2006b, 2007, 2010) applied CHUM-AM to 13 UK catchments, mostly in upland areas, and all with natural or semi-natural vegetation and soils, the predominant land use being rough grazing for sheep, with some plantation conifer forestry. The catchments vary in size from 0.3 to 36 km^2, have mean annual temperatures between 5 and 10 °C, and mean annual precipitation in the range 800–3500 mm. The surface waters comprise two lakes, two reservoirs, and nine streams, with current average pH values between 4.5 and 8.3. Most of them have suffered from acidification due to sulphur and nitrogen deposition, although recovery is now taking place.

Metal inputs to the catchments were assumed to come from either weathering or atmospheric deposition. Deposition was assumed to have been at low background levels until 1600 or later (depending upon the metal), then to have increased, accelerating to a maximum in 1960–1970, with a subsequent sharp decrease to 2000 (Tipping et al. 2006b). Temporal variations in deposition rates were estimated from contemporary measurements and lake sediment metal profiles. At sites in the Pennines (central N England), local lead mining activities in the nineteenth century caused significant deposition, especially of Zn, Cd and Pb (Tipping et al. 2010). Metals data for the catchments have been obtained from soil surveys, typically involving five representative locations, and surface water monitoring over a few years, although in one case annual average Zn concentrations are available for the period 1995–2008. In view of the paucity of data, modelling was aimed at accounting simultaneously for observed metal soil pools and annual average metal concentrations in surface water with as little adjustment as possible of metal-related parameters. Default values of the model parameter LKMA for humic acid, which is the main control on soil sorption of metals, were used for Ni, Cu and Zn. However, with the default parameter set of WHAM/Model VI, soil pools of Cd and Pb were always too low, while surface water Pb concentrations were too high. These discrepancies were rectified by modest increases in the values of the parameter LKMA describing Cd and Pb binding to soil humic acid (the same for all catchments) in order to improve agreement (Tipping et al. 2010).

9.3.2.2 Results and discussion

Figure 9.4 summarises the results of model applications. Generally, CHUM-AM provided unbiased predictions of the soil pools (Fig. 9.4 left) and surface water concentrations (Fig. 9.4 right) of Ni, Cu and Zn. Because the soil pools and stream water concentrations were reasonably well simulated for the considered metals, it can be concluded that sorption by soil organic matter is the main control

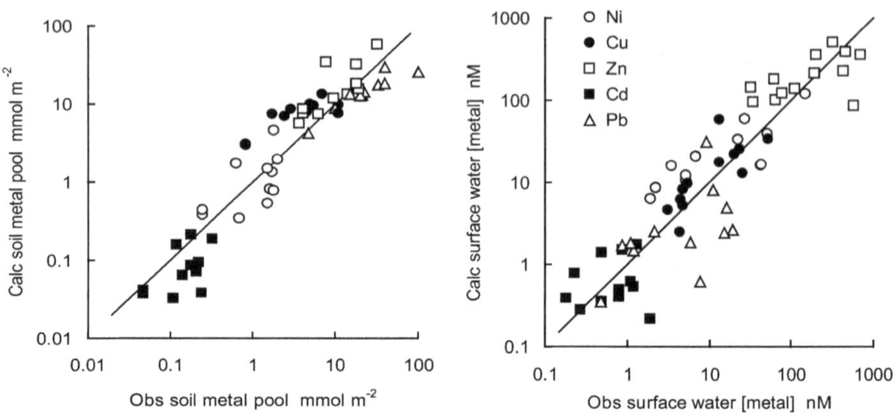

Fig. 9.4 CHUM-AM simulated and observed soil metal pools for 11 catchments and average surface water metal concentrations for 13 catchments in the UK (soil results for one catchment are absent because analyses were not performed, and for another because of artefacts due to limestone dissolution). The lines are 1:1 relationships

of atmospherically-deposited metals in these catchments, although other sorbing phases, notably the oxides of Al, Mn and Fe will also play some role (see e.g. Dijkstra et al. 2004; Bonten et al. 2008). The sorbent pools are such that substantial proportions of atmospherically deposited metals have been retained, causing enrichments of soil metal pools, which are often 10–100-fold greater than background levels. For the strongly-sorbing metals Cu and Pb, most metal deposited as a result of human activities over the last several hundred years remains held by the soils, up to 99% of Pb in some catchments. Retention of the relatively weakly-held metals (Ni, Zn and Cd) has been less, and generally more than half of each of these deposited metals was calculated to have passed through the soil. Differences in binding strength also give rise to variations in the responses of metal pools to differences and changes in soil acidity, with weakly binding metals being fairly mobile under acid conditions. However, greater co-transport of Cu and Pb with DOM was predicted as a consequence of stronger binding, and this effect countered to some extent the retention of these two metals.

9.3.3 SMARTml Validation on a Long-Term Forest Monitoring Plot in Germany

9.3.3.1 Sites and Input Data

The SMARTml metals model was applied to a Norway spruce site in Solling, Germany (see also Chap. 8 concerning the modelling of the hydrology and soil (solution)

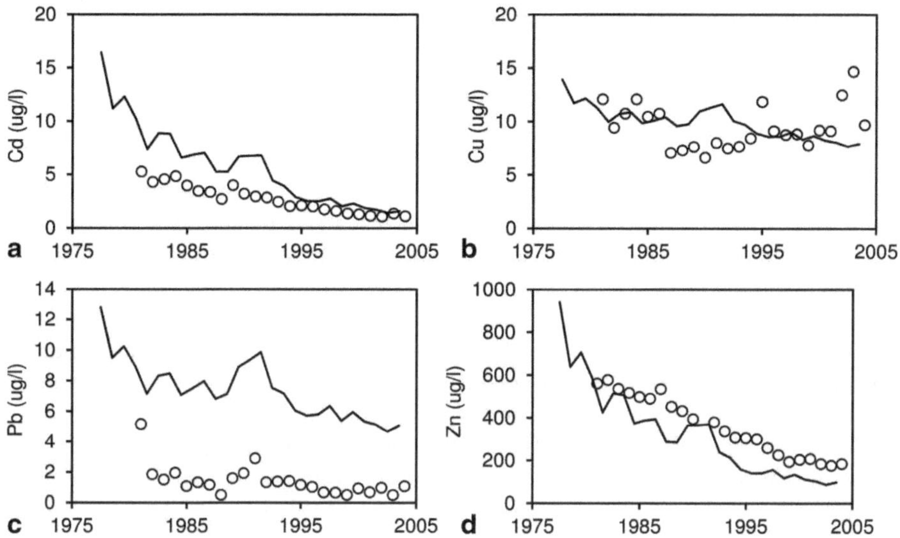

Fig. 9.5 SMARTml simulated and observed concentrations of metals (in mg l^{-1}) in the forest soil at Solling (Germany) at a depth of 90 cm; Cd (*top left*), Cu (*top right*), Pb (*bottom left*) and Zn (*bottom right*)

chemistry). This site has been intensively monitored since 1966 (Meesenburg et al. 1995). Deposition and throughfall of heavy metals have been monitored since the start of the monitoring program. Soil solution sampling of heavy metals started in 1981. For the Solling application, the model was run for the years 1968–2004.

Initial reactive metal contents in the soil were estimated from total metal contents measured in 1978 and the ratio between reactive metal contents and total contents as measured in 2001. Initial metals contents in fresh organic matter were calculated from average metal contents in litterfall and root decay for the period 1973–1978.

9.3.3.2 Results and Discussion

Modelling of the metals Cd, Cu, Pb and Zn was hampered by the limited input on initial metal contents in soil, especially for depths below 50 cm. Nevertheless, the results show that simulated concentrations match measured concentration well except for Pb where concentrations are highly overestimated (Fig. 9.5). Comparing measured and predicted partition coefficients (K_d = concentration solid phase/concentration in solution phase) show that binding of Pb to soil is clearly underestimated. Predicted partition coefficients for the metals Cd, Cu and Zn compare well with those from measurements. The predicted downward trend in concentrations of Cd and Zn match observations well and can be explained by the exhaustion of these metals in the solid phase (results not shown).

9.4 Relevance of Dynamic Modelling of Metals

To assess the risks of anthropogenic inputs of metals to soils and waters and to evaluate the effects of metal input reductions (e.g. due to emission reductions), dynamic modelling is an essential complementary tool to the steady-state critical load approach. This is because the considered systems maybe far from equilibrium and the long time it takes for metal concentrations to reach steady state. The relevance of the dynamic modelling approach was demonstrated by Ashmore et al. (2007) who applied the model CHUM-AM to five contrasting catchments, selected from the 13 catchments presented in Fig. 9.4, under a variety of future scenarios covering the period 2000–3000. This was done to provide information and understanding about the magnitude and timing of future heavy metal responses to changes in the atmospheric deposition of metals, soil and water acidity, and the carbon status of soils and waters (Ashmore et al. 2007). The scenarios permitted examination of variations in deposition of heavy metals, sulphur and nitrogen, together with changes in soil carbon cycling, including the production and leaching of dissolved organic matter. The analysis produced complex outputs, reflecting differences among both catchments and metals. The results suggest that further reductions in the deposition inputs of zinc and cadmium would have widespread beneficial effects during the twenty-first century, and the responses for these metals are expected to be similar in different catchments. In contrast, beneficial effects cannot be expected for Cu and Pb, except over much longer timescales. Some catchments will respond more quickly than others, and for Cu and Pb the responses of different catchments may not even be in the same direction under the same scenario. The differences between the two sets of metals reflect the stronger sorption of Cu and Pb to soil solids. The long timescales of recovery for Cu and Pb result from the legacy of high past deposition and slow loss rates. The results showed that it is important to ensure that metal deposition does not rise, and that soil organic matter is protected (Ashmore et al. 2007).

The results of the applications of the three dynamic models show that (long) time trends in metal concentrations can be well predicted when taking account of the most relevant processes, including metal binding to organic matter. During the long time frames that are involved before metals reach a new equilibrium, important environmental factors which strongly determine the fate of metals (pH, soil organic matter content, SOM) may change due to external factors e.g. due to land use changes and climate change. Dynamic models to predict the fate of metals should thus also be able to predict trends in soil pH and preferably SOM content, because the binding of metals is strongly dependent on these soil properties.

Acknowledgements Work on the Dutch and Swedish chronosequences was financed through the European Commission, 4th FP, contract no. FAIR-CT96-1983. We thank Mats Olsson from the Swedish University of Agricultural Sciences for providing the data of the Swedish chronosequences. Figure 9.4 is reprinted from Environmental Pollution, Vol. 158, Tipping E, Rothwell JJ, Shotbolt L, Lawlor AJ, Dynamic modelling of atmospherically-deposited Ni, Cu, Zn, Cd and Pb in Pennine catchments (northern England), Pages 1521–1529, Copyright 2010, with permission from Elsevier. Figure 9.5 is reprinted from Environmental Pollution, Vol. 159, Bonten LTC,

References

Ashmore, M., van den Berg, L., Terry, A., Tipping, E., Lawlor, A. J., Lofts, S., Thacker, S. A., Vincent, C. D., Hall, J., O'Hanlon, S., Shotbolt, L., Harmens, H., Lloyd, A., Norris, D., Nemitz, E., Jarvis, K., & Jordan, C. (2007). *Development of an effects-based approach for toxic metals*. Report to the UK Department for Environment, Food and Rural Affairs, the Scottish Executive, the National Assembly for Wales and the Department of the Environment in Northern Ireland. Contract CPEA 24. University of York.

Berg, T., Røyset, O., & Steinnes, E. (1995). Moss (*Hylocomium Splendens*) used as biomonitor of atmospheric trace element deposition: Estimation of uptake efficiencies. *Atmospheric Environment, 29,* 353–360.

Bonten, L. T. C., Groenenberg, J. E., Weng, L., & van Riemsdijk, W. H. (2008). Use of speciation and complexation models to estimate heavy metal sorption in soils. *Geoderma, 146,* 303–310.

Bonten, L. T. C., Groenenberg, J. E., Meesenburg, H., & De Vries, W. (2011). Using advanced surface complexation models for modelling soil chemistry under forests. The Solling case. *Environmental Pollution, 159,* 2831–2839.

De Vries, W., & Groenenberg, J. E. (2009). Evaluation of approaches to calculate critical metal loads for forest ecosystems. *Environmental Pollution, 157,* 3422–3433.

De Vries, W., Posch, M., & Kämäri, J. (1989). Modeling time patterns of forest soil acidification for various deposition scenarios. In J. Kämäri, D. F. Brakke, A. Jenkins, S. A. Norton, & R. F. Wright (Eds.), *Regional acidification models. Geographic extent and time development* (pp. 129–149). Berlin: Springer.

De Vries, W., Bakker, D. J., Groenenberg, J. E., Reinds, G. J., Bril, J., & van Jaarsveld, J. A. (1998). Calculation and mapping of critical loads for heavy metals and persistent organic pollutants for Dutch forest soils. *Journal of Hazardous Materials, 61,* 99–106.

Deller, B. (1983). Determination of the CEC of carbonate soils with unbuffered 0.1 M $BaCl_2$. *Zeitschrift für Pflanzenernährung Bodenkunde, 146,* 348–352.

Dijkstra, J. J., Meeussen, J. C. L., & Comans, R. N. J. (2004). Leaching of heavy metals from contaminated soils: An experimental and modeling study. *Environmental Science and Technology, 38,* 4390–4395.

Dzombak, D. A., & Morel, F. M. M. (1990). *Surface complexation modeling: Hydrous ferric oxide*. New York: Wiley.

Erisman, J. W., Potma, C., Draaijers, G., van Leeuwen, E., & van Pul, A. (1995). A generalised description of the deposition of acidifying pollutants on a small scale in Europe. *Water Air & Soil Pollution, 85,* 2101–2106.

FAO. (1998). *World reference base for soil resources*. World Soil Resources Report 84. Rome: FAO.

Groenenberg, J. E., Dijkstra, J. J., Bonten, L. T. C., De Vries, W., & Comans, R. N. J. (2012). Evaluation of the performance and limitations of empirical regression models and process based multisurface models to predict trace element solubility in soils. *Environmental Pollution, 168,* 98–107.

Helling, C. S., Chesters, G., & Corey, R. B. (1964). Contribution of organic matter and clay to soil cation exchange capacity as affected by the pH of the saturating solution. *Soil Science Society of America Journal, 28,* 517–520.

Karltun, E., Harrison, A. F., Alriksson, A., Bryant, C., Garnett, M. H., & Olsson, M. T. (2005). Old organic carbon in soil solution DOC after afforestation-Evidence from ^{14}C analysis. *Geoderma, 127,* 188–195.

Kinniburgh, D. G., Van Riemsdijk, W. H., Koopal, L. K., Borkovec, M., Benedetti, M. F., & Avena, M. J. (1999). Ion binding to natural organic matter: Competition, heterogeneity, stoichiometry and thermodynamic consistency. *Colloids and Surfaces A: Physicochemical and Engineering Aspects, 151,* 147–166.

Kros, J., Reinds, G. J., De Vries, W., Latour, J. B., & Bollen, M. J. S. (1995). *Modelling of soil acidity and nitrogen availability in natural ecosystems in response to changes in acid deposition and hydrology.* SC-DLO Report 95. Wageningen, The Netherlands.

Lofts, S., Chapman, P., Dwyer, R., McLaughlin, M., Schoeters, I., Sheppard, S., Adams, W., Alloway, B., Antunes, P., Campbell, P., Davies, B., Degryse, F., De Vries, W., Farley, K. J., Garrett, R. G., Green, A., Groenenberg, B. J., Hale, B., Harrass, M., Hendershot, W. H., Keller, A., Lanno, R., Liang, T., Liu, W.-X., Ma, Y., Menzie, C., Moolenaar, S. W., Piatkiewicz, W., Reimann, C., Rieuwerts, J. S., Santore, R. C., Sauvé, S., Schuetze, G., Schlekat, C., Skeaff, J., Smolders, E., Tao, S., Wilkins, J., & Zhao, F.-J. (2007). Critical loads of metals and other trace elements to terrestrial environments. *Environmental Science and Technology, 41,* 6326–6331.

Meesenburg, H., Meiwes, K. J., & Rademacher, P. (1995). Long term trends in atmospheric deposition and seepage output in northwest German forest ecosystems. *Water Air and Soil Pollution, 85,* 611–616.

Meeussen, J. C. L. (2003). ORCHESTRA: an object-oriented framework for implementing chemical equilibrium models. *Environmental Science Technology, 37,* 1175–1182.

Römkens, P. F. A. M. (1998). Effects of land use changes on organic matter dynamics and trace metal solubility in soils. Ph.D., Groningen, University of Groningen.

Römkens, P. F. A. M., & Salomons, W. (1998). Cd, Cu and Zn solubility in arable and forest soils: Consequences of land use changes for metal mobility and risk assessment. *Soil Science, 163,* 859–871.

Schnoor, J., & Stumm, W. (1986). The role of chemical weathering in the neutralization of acidic deposition. *Schweizerische Zeitschrift für Hydrologie, 48,* 171–195.

Sposito, G. (1989). *The chemistry of soils.* New York: Oxford University Press.

Stidson, R. T., Hamilton-Taylor, J., & Tipping, E. (2002). Laboratory dissolution studies of rocks from the Borrowdale Volcanic Group (English Lake District). *Water Air and Soil Pollution, 138,* 335–358.

Tipping, E. (1998). Humic ion-binding Model VI: An improved description of the interactions of protons and metal ions with humic substances. *Aquatic Geochemistry, 4,* 3–47.

Tipping, E. (2002). *Cation binding by humic substances.* Cambridge: Cambridge University Press.

Tipping, E., Lawlor, A. J., & Lofts, S. (2006a). Simulating the long-term chemistry of an upland UK catchment: Major solutes and acidification. *Environmental Pollution, 141,* 151–166.

Tipping, E., Lawlor, A. J., Lofts, S., & Shotbolt, L. (2006b). Simulating the long-term chemistry of an upland UK catchment: Heavy metals. *Environmental Pollution, 141,* 139–150.

Tipping, E., Yang, H., Lawlor, A. J., Rose, N. L., & Shotbolt, L. (2007). Trace metals in the catchment, loch and sediments of Lochnagar: Measurements and modelling. In N. L. Rose (Ed.), *Lochnagar: The natural history of a mountain lake* (pp. 345–373). Dordrecht: Springer.

Tipping, E., Rothwell, J. J., Shotbolt, L., & Lawlor, A. J. (2010). Dynamic modelling of atmospherically-deposited Ni, Cu, Zn, Cd and Pb in Pennine catchments (Northern England). *Environmental Pollution, 158,* 1521–1529.

Van Jaarsveld, J. A., & de Leeuw, F. A. A. M. (1993). OPS: An operational atmospheric transport model for priority substances. *Environmental Software, 8,* 91–100.

Chapter 10
Use of Combined Biogeochemical Model Approaches and Empirical Data to Assess Critical Loads of Nitrogen

Mark E. Fenn, Charles T. Driscoll, Qingtao Zhou, Leela E. Rao, Thomas Meixner, Edith B. Allen, Fengming Yuan and Timothy J. Sullivan

10.1 Introduction

Humans have perturbed the global nitrogen (N) cycle, in relative terms, to a greater degree than the much more widely recognized effects of man's activities on global carbon (C) cycling (Smil 2001). Because of widespread effects of N on the environment and on sensitive ecosystems, critical loads (CLs) of N are being increasingly used for quantifying and mapping the ecological and environmental effects of N (Hettelingh et al. 2008; Pardo et al. 2011). Critical loads are generally determined from observations or from models of varying degrees of complexity, from simple steady state to process-based dynamic models. Nitrogen CLs can be determined for aquatic and terrestrial ecosystems, both for acidification and eutrophication (N as a nutrient) effects.

M. E. Fenn (✉)
USDA Forest Service, Pacific Southwest Research Station, Riverside, CA, USA
e-mail: mfenn@fs.fed.us

C. T. Driscoll · Q. Zhou
Department of Civil and Environmental Engineering, Syracuse University, Syracuse, NY, USA

L. E. Rao
Department of Environmental Science and Center for Conservation Biology, University of California, Riverside, CA, USA

T. Meixner
Department of Hydrology and Water Resources, University of Arizona, Tucson, AZ, USA

E. B. Allen
Department of Botany and Plant Sciences and Center for Conservation Biology, University of California, Riverside, CA, USA

F. Yuan
Environmental Sciences Division, Oak Ridge National Laboratory, Oak Ridge, TN, USA

T. J. Sullivan
E&S Environmental Chemistry, Corvallis, OR, USA

W. de Vries et al. (eds.), *Critical Loads and Dynamic Risk Assessments*, Environmental Pollution 25, DOI 10.1007/978-94-017-9508-1_10

A major strength of empirical CLs is that they are based on chemical or biological response data under real-world conditions. A limitation of the empirical approach is that results may only be applied with confidence to sites with highly similar biotic and environmental conditions. Empirical CLs are important, however, for validating CL values determined with models. Advantages of simulation approaches are that ecosystem responses to alternative scenarios can be tested. These might include changes in atmospheric deposition, disturbance or climatic conditions, and responses to silvicultural treatments, grazing, fire, and other disturbances (Gimeno et al. 2009). Simulation modelling allows temporal aspects of ecosystem response in relation to CLs and CL exceedances to be evaluated, including evaluation of historical and future conditions.

Biogeochemical models such as PnET-BGC, DayCent, MAGIC, and ForSAFE-VEG have been used to simulate biogeochemical ecosystem responses to factors such as changes in air pollution, climate change, and disturbance. These models can also be applied towards estimating CLs. Such applications are described in the following sections. Empirical CLs for NO_3^- leaching were determined based on NO_3^- concentrations in streams or lakes located across N deposition gradients. The lowest N deposition at which NO_3^- leaching begins to increase above that observed at low pollution sites was taken as the empirical NO_3^- leaching CL. Nitrogen fertilization treatments were used to develop empirical N CLs for plant community changes or for increased biomass and fuel accumulation and fire risk from the growth of invasive annual grass species in desert scrub habitats.

In this chapter, we emphasize the combined approach of calculating CLs from biogeochemical simulation models in conjunction with empirical analyses. This combined approach has been applied previously to mixed conifer forests in California (Fenn et al. 2008). Herein, N CLs are based on empirical and simulated nitrate (NO_3^-) concentrations in streamwater, acid neutralizing capacity (ANC) in lakes and streamwater, changes in lake diatom assemblages or alpine vegetation communities, and fuel accumulation resulting from growth of invasive grasses in arid ecosystems. In this chapter, case studies of the combined simulation and empirical N CL approaches are described for four ecosystems including mixed conifer forests and the combination of desert scrub and pinyon-juniper woodlands in California, alpine catchments of the Rocky Mountains in Colorado and lakes in the Adirondack Mountains in New York.

10.2 California Mixed Conifer Forests

Overstory species composition in mixed conifer forests of California varies depending on elevation and site conditions, but is typically characterized by dense mixed-age stands of pine (*Pinus* spp.) and often fir (*Abies* spp.) and incense cedar (*Calocedrus decurrens* (Torr.), Florin.) in addition to broadleaved species, typically oaks (*Quercus* spp.). The climate is Mediterranean, characterized by cool wet winters and hot and dry conditions during summer. Mixed conifer forests over much of the Transverse Ranges in southern California and in the south-western Sierra Nevada

Mountains are exposed to elevated concentrations of ozone (O_3) and atmospheric N deposition from urban and agricultural emissions sources.

10.2.1 Methodologies for Empirical and Model-based CLs for Nitrate Leaching and N Trace Gas Loss

The empirical threshold for NO_3^- leaching is based on the relationship between throughfall N deposition and an empirically derived clean-site threshold value for streamwater NO_3^-. The threshold was derived by comparing temporal profiles of streamwater NO_3^- concentrations in relatively pristine low N deposition catchments against profiles for catchments receiving varying levels of chronic N deposition inputs. Based on 28 streams from relatively low pollution catchments in southern California (Fenn and Poth 1999; Riggan et al. 1985), we determined that peak NO_3^- concentrations in runoff rarely exceed 0.2 mg NO_3-N l^{-1} (14.3 μmol l^{-1}). This concentration corresponds with those reported for relatively unpolluted streams in Sweden (De Vries et al. 2007). Thus, 14.3 μmol was selected as the critical threshold value for identifying catchments that are beginning to exhibit symptoms of N excess or N saturation. The empirical N CL for NO_3^- leaching is thus the N deposition input at which this critical threshold NO_3^- concentration is exceeded (Fenn et al. 2008).

The N deposition input leading to NO_3^- leaching was determined by linear regression of streamwater NO_3^- and throughfall N deposition data collected during winter high flow events from 11 sites in the San Bernardino and southern Sierra Nevada Mountains, encompassing a broad range of annual N deposition inputs (5.7–71.1 kg $ha^{-1}yr^{-1}$).

The DayCent biogeochemistry model version 4.5 was used to simulate historical N saturation symptoms such as N trace gas emissions from soil and NO_3^- leaching and to estimate the N deposition threshold for such losses. DayCent (Del Grosso et al. 2000; Parton et al. 1998, 2001) is the daily time step version of the CENTURY model (Parton et al. 1993). It simulates the biogeochemical processes of C, N, phosphorus (P), and sulfur (S) associated with soil organic matter in multiple ecosystem types. One feature of DayCent important to this study is the improved N cycling algorithms, including the simulation of N_2O, NO, and N_2 emissions resulting from nitrification and denitrification. DayCent was parameterized for a high (Camp Paivika, CP) and low (Barton Flats, BF) deposition site in the San Bernardino Mountains located 70 km east of Los Angeles, California. See Fenn et al. (2008) for further details of model parameterization.

In conducting model runs to estimate the CL for NO_3^- leaching, N deposition was increased from an assumed background level of 0.1 kg N $ha^{-1}yr^{-1}$–3 kg N ha^{-1} yr^{-1} in 1930 and increased linearly to 10 kg N $ha^{-1}yr^{-1}$ by 1940. In 1940, N deposition was set at 19 different levels for each site ranging from 10 to 80 kg N $ha^{-1}yr^{-1}$ and maintained at these levels for simulations lasting for 120 years more in order to simulate dose/responses to increasing N deposition. Annual N deposition was set to vary randomly by $<10\,\%$. Based on simulated occurrences of peak streamwater

NO_3^- concentrations above the clean site threshold of 0.2 mg l^{-1}, we determined the N deposition rate at which CL exceedance would occur at the model sites. We included a stipulation that the critical NO_3^- leachate concentration must be exceeded at least once every 10 years on average to consider the threshold exceeded. Spurious NO_3^- exceedances that occur rarely due to extreme climatic events for example, do not indicate a catchment above the CL by our definition (Fenn et al. 2008).

10.2.2 Empirical and Model-based CLs for Mixed Conifer Forests in California

Measurements of peak NO_3^- concentrations at the 11 stream sampling sites used to estimate the empirical CL ranged from 0.07 to 107 μeq l^{-1}. Based on the linear regression ($r^2=0.98$) of streamwater NO_3^- vs. throughfall N deposition, the N deposition at which the critical threshold NO_3^- concentration (14.3 μeq l^{-1}) occurs is 17 kg N $ha^{-1}yr^{-1}$. This can be considered as the empirical threshold for NO_3^- leaching at these sites (95% confidence interval: 15–19 kg N $ha^{-1}yr^{-1}$).

The estimated N deposition thresholds (i.e., the CL) for NO_3^- leaching based on simulated NO_3^- concentrations in seepage water were 17 and 30 kg N $ha^{-1}yr^{-1}$ at CP and BF (Fig. 10.1a, b). This is based on N deposition occurring over a range of values for 30–60 years and with at least one occurrence per decade on average of a peak streamwater NO_3^- value over the 0.2 mg l^{-1} threshold. The simulated N deposition CL for NO_3^- leaching at CP (17 kg N $ha^{-1}yr^{-1}$) was equal to the empirical throughfall CL. Increasing the time period of elevated N deposition resulted in lower CL values. Reducing precipitation by 40% at CP increased the estimated CL from 17 to 20 kg N $ha^{-1}yr^{-1}$ (Fig. 10.1c). An approximate doubling of the precipitation at the much drier BF site (using CP precipitation levels) decreased the simulated CL from 30 to 7 kg N $ha^{-1}yr^{-1}$ over a 30–60 year deposition time frame (Fig. 10.1d). The simulated CL at BF increased to 16 kg $ha^{-1}yr^{-1}$ with doubled precipitation and a time frame of less than 30 years (Fig. 10.1d). The much lower simulated CL at BF compared to CP when precipitation values from CP were used is presumably due to the lower plant and microbial biomass and thus lower N demand and retention capacity at BF. Thus more N is available in soil with potential to be leached or emitted as trace gases.

We do not have sufficient field data on trace gas emissions to establish an empirical CL to protect against unacceptably high trace gas emissions. However, this CL can be estimated from the DayCent simulations. At the N deposition level equal to the empirical CL for NO_3^- leaching (17 kg N $ha^{-1}yr^{-1}$), simulated mean annual gaseous losses of N (including NO, N_2O and N_2) were ca. 1.8 kg N $ha^{-1}yr^{-1}$ at CP. If we set the acceptable gaseous N loss at 2 kg N $ha^{-1}yr^{-1}$, then the simulated CLs for trace gas loss are 20 and 10 kg N $ha^{-1}yr^{-1}$ at CP and BF, respectively (Fig. 10.1e, f). At CP and BF the percent of N deposition re-emitted as N trace gases in the year 2000 according to DayCent was 20 and 22% respectively. These results agree with European studies demonstrating that temperate forests show increased NO and N_2O fluxes when throughfall N deposition was greater than 15 kg $ha^{-1}yr^{-1}$ and that

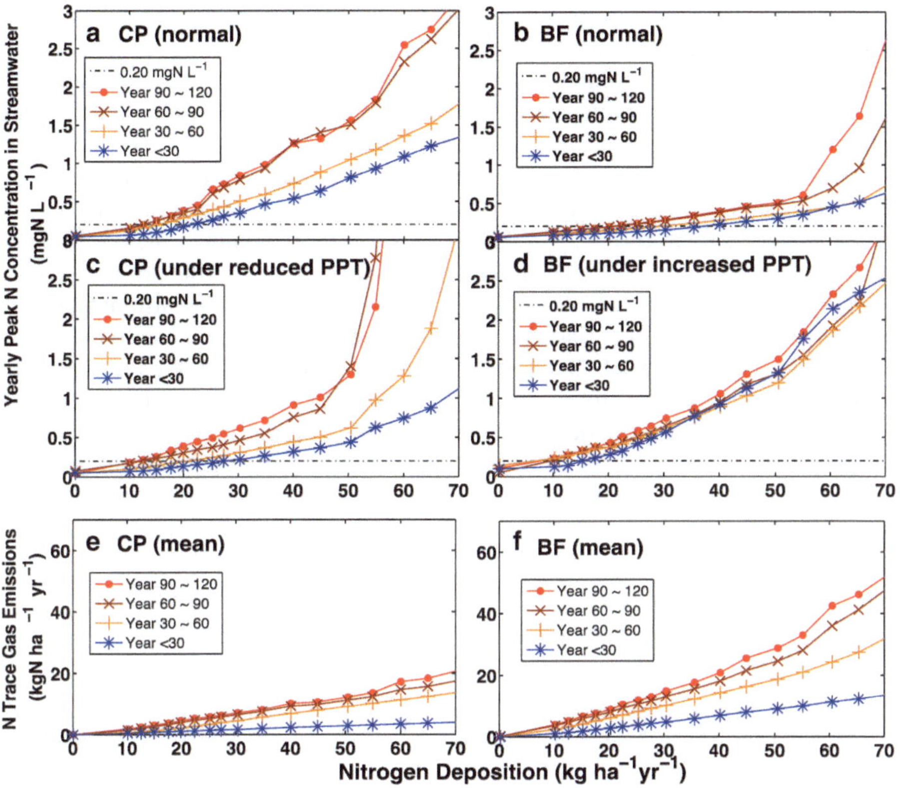

Fig. 10.1 DayCent simulated trends in annual peak NO_3–N concentrations in streamwater and annual mean gaseous emissions of N from soil that occurred within various time periods since elevated N deposition began in 1940. In these simulations N deposition since 1940 ranged from 0 to 80 kg N $ha^{-1}yr^{-1}$. The model was parameterized for the Camp Paivika (CP; plots **a**, **c** and **e**) and Barton Flats (BF; plots **b**, **d** and **f**) sites in the San Bernardino National Forest in southern California, USA. The effect of normal or altered precipitation levels on simulated nitrate leaching is shown in plots **a–d**. The prescribed N deposition values were set with < 10 % random variation. (Modified from Fenn et al. 2008)

17–20 % of the atmospherically deposited N was re-emitted as NO and N_2O (Gasche and Papen 2002; Skiba et al. 2004).

10.3 Desert Scrub and Pinyon-Juniper Woodland

Desert ecosystems downwind of urban and agricultural centers can receive N deposition as high as 30 kg $ha^{-1}yr^{-1}$ (Fenn et al. 2003). For these ecosystems, CLs cannot be set using the traditional endpoints of NO_3^- leaching or soil acidification due to the high alkalinity of the soils and the propensity of the nitrogen to be leached deep into the soil profile where it is sequestered (Walvoord et al. 2003). Instead, deposited N has a fertilization effect that promotes the production of the fast-growing ephemeral herbaceous community.

Increased production of these annual forbs and grasses has been correlated with increased frequency and intensity of fires (Brown and Minnich 1986), due in large part to the invasion of exotic annual grasses. These exotic grasses create a continuous fine fuel load that allows fires to spread rapidly through the landscape, resulting in severe damage to the ecosystem (Brooks et al. 2004). Although the amount of fine fuel necessary to carry fire in the desert is currently unknown, studies evaluating this threshold are underway, and in grassland ecosystems 1000 kg ha^{-1} of fine fuel dry mass can carry fire (Anderson 1982). Exotic grasses may also outcompete native annuals due to the plastic response of these exotics to increased resource availability (DeFalco et al. 2003; Steers and Allen 2010). Thus, CLs for southern California desert communities can be set using both the amount of N needed to increase exotic grasses and decrease native annuals, and the amount of N needed to increase herbaceous biomass production above the fire-carrying threshold. Here we provide results for both types of CL using empirical and modelling methodologies.

10.3.1 Methodologies for Empirical and Model-based CLs in View of Fire Risk

For the empirical study, four sites were selected in Joshua Tree National Park (JTNP) that span a N deposition gradient. Two sites were located in creosote bush scrub (CBS) and two sites were in pinyon-juniper woodland (PJW). Nitrogen fertilizer was applied each winter at rates of 0, 5, and 30 kg ha^{-1} for five consecutive years beginning in 2002. Nitrogen deposition along the gradient ranged from 3.4 to 12.4 kg N $ha^{-1}yr^{-1}$ as measured by ion exchange resin bulk throughfall deposition samplers (Fenn et al. 2009) and confirmed with CMAQ model simulations (Tonnesen et al. 2007). The empirical CL was determined as the lowest N treatment plus background N deposition that increased total winter biomass above the fire threshold and/or caused increased invasive grass biomass and decreased native richness. Plots were sampled for richness and percent cover of native and exotic annuals at the time of peak production. Biomass was estimated using biomass-to-percent cover regressions (see Allen et al. 2009; Rao and Allen 2010 for detailed methods).

CLs were also determined for CBS and PJW by DayCent model simulations of winter annual vegetation production under a range of soil textures, precipitation regimes, and N deposition levels (Rao et al. 2010). The model was calibrated and validated using soil profile, plant chemistry, and biomass data from the four fertilization sites in JTNP. Once parameterized for each vegetation type, winter annual biomass production under increasing N-deposition loads was modelled. Production was simulated for 100 years at each N deposition load, and the fire risk at each load was determined as the fraction of years in which the fire threshold was exceeded. The CL was defined as the N-deposition load at which the fire risk began increasing exponentially above the background fire risk (Fig. 10.2). The amount of N deposition at which fire risk no longer increased with added N was termed the fire risk

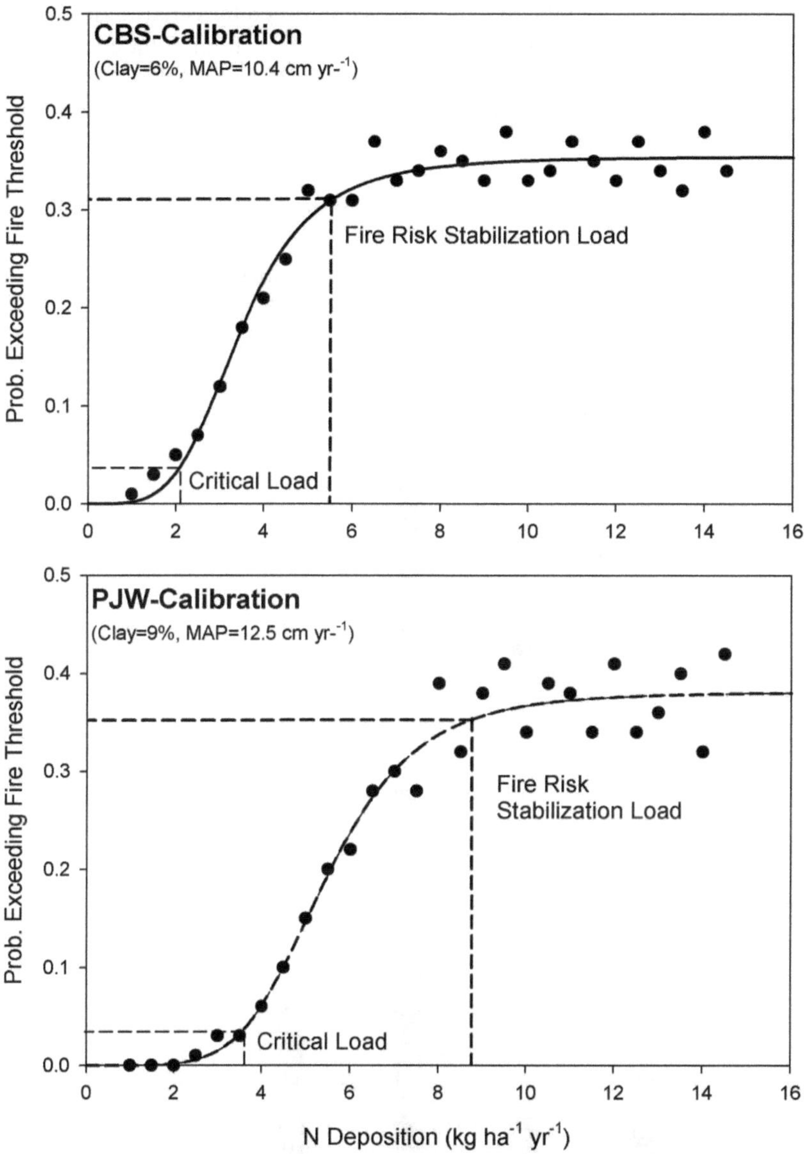

Fig. 10.2 Examples of fire risk for the model calibration sites from each vegetation type, which was calculated as the probability that annual biomass will exceed the fire threshold under a given N deposition load. For these sites, the CLs are 2.1 and 3.6 kg $ha^{-1}yr^{-1}$ of N deposition for creosote bush scrub (CBS) and pinyon-juniper woodland (PJW) respectively. The fire risk begins to level out at the stabilization load (SL), which is 5.5 and 8.8 kg $ha^{-1}yr^{-1}$ of N deposition for CBS and PJW, respectively. Between the CL and SL the fire risk changes very rapidly with increases or decreases in N deposition. *MAP* mean annual precipitation

stabilization load (SL). Above the SL, fire risk is controlled by variation in annual precipitation and does not increase with increased N deposition, although annual biomass production will still increase with added N (Rao and Allen 2010). The CLs and SLs were calculated for six precipitation regimes and six soil textures that bracketed observations from the region.

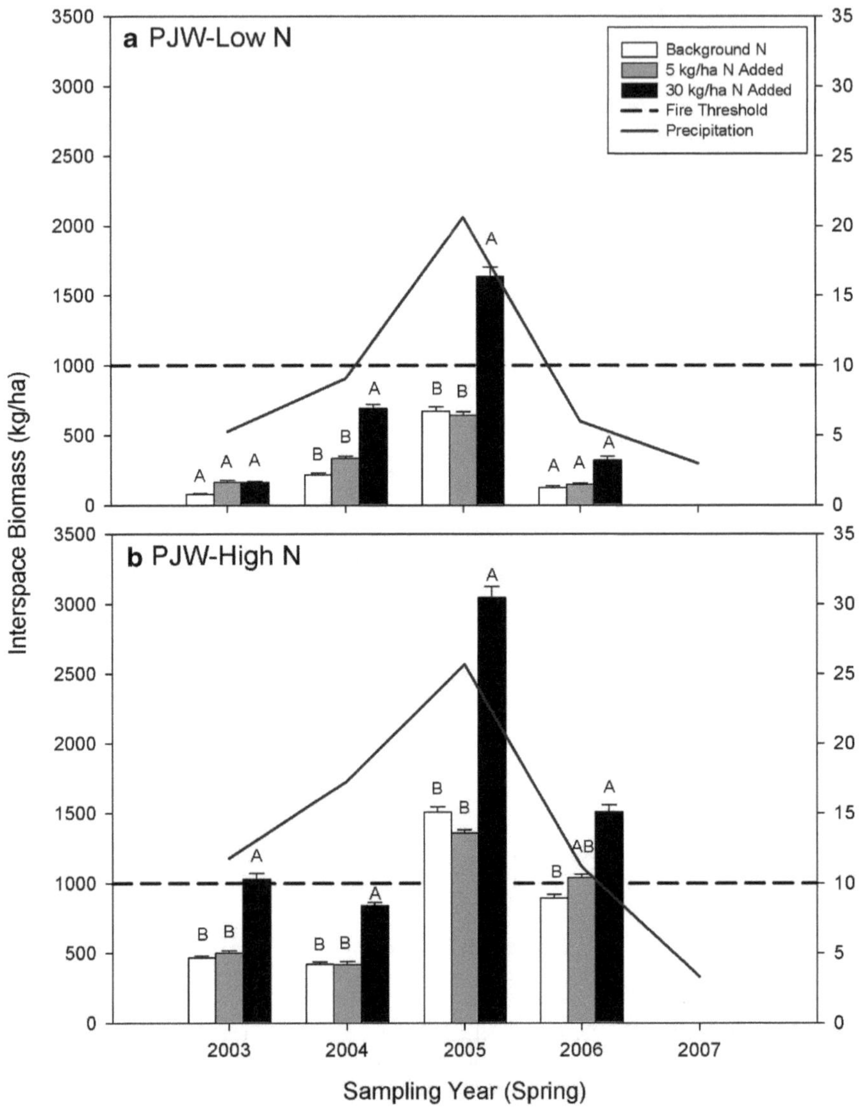

Fig. 10.3 Changes in total interspace (herbaceous) biomass with N fertilization and yearly differences in precipitation at four sites (*A* pinyon-juniper woodland (PJW) low N-deposition; *B* PJW high N-deposition; *C* creosote bush scrub (CBS) low N-deposition; *D* CBS high N-deposition). Letters above bars indicate Significant differences in soil concentrations within years. In all cases significance was determined using the post hoc test Tukey's HSD at $\alpha=0.05$

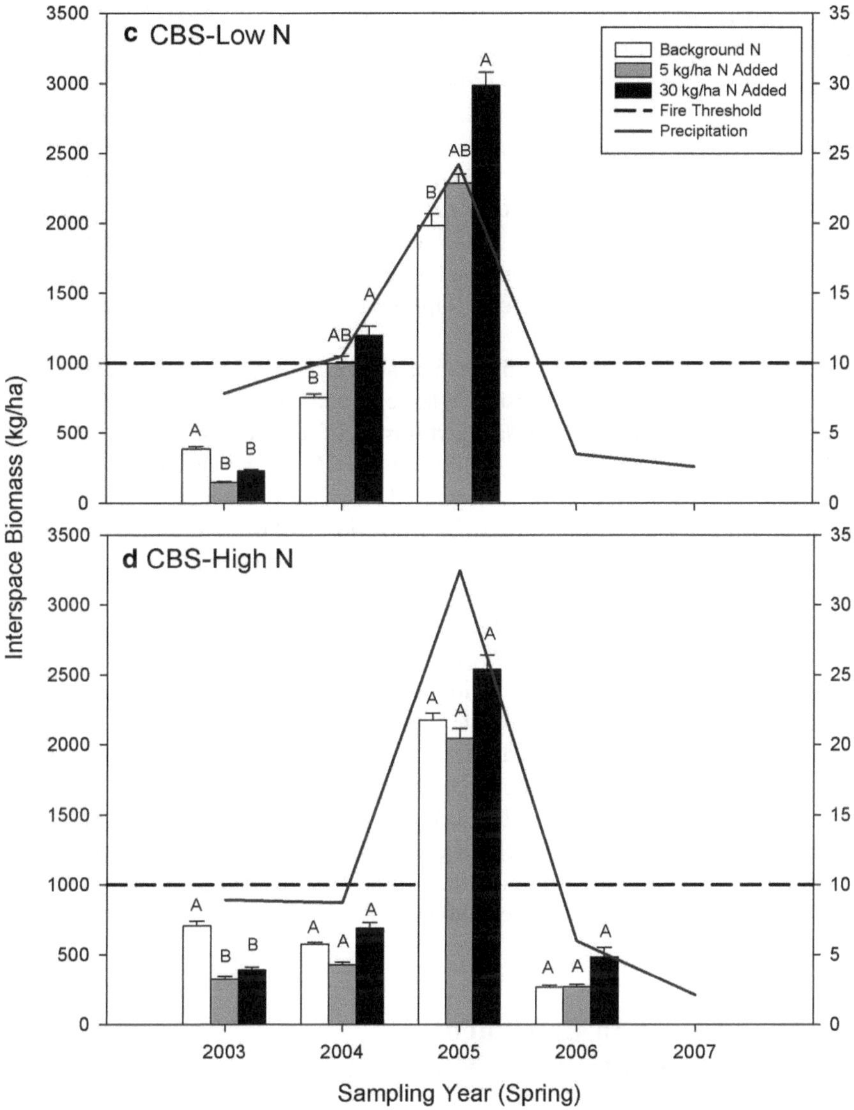

Fig. 10.3 (Continued)

10.3.2 Empirical and Model-based CLs for Desert Scrub and Pinyon-Juniper Woodland

The empirical CL calculations were somewhat confounded by the fact that both production and relative abundance of exotic and native annuals were influenced by the interaction between N fertilization and precipitation. For the CL based on fine fuel production, high precipitation resulted in biomass production above the fire threshold regardless of N fertilization level (Fig. 10.3b–d). However, in four instances under average precipitation the addition of N at 5 or 30 kg N ha^{-1} pushed the biomass above the fire threshold (Fig. 10.3a–c). Invasive grasses increased and native forbs generally decreased under both 5 and 30 kg N ha^{-1} fertilizer additions at two of the four sites in the wet year of 2005. In a drier year only 30 kg N ha^{-1} caused an increase in invasive grasses in two of the four sites. Using the background deposition value for the site with the least amount of N deposition that demonstrated an effect of fertilization, we can calculate the CL for the impact of invasive grasses on native forbs as 8 kg N ha^{-1} (5 kg N ha^{-1} added + 3 kg N ha^{-1} background deposition). The empirical CL based on fine fuel accumulation is likely at the same level, but could not be determined due to precipitation-limited production during most years of our study.

The DayCent model results indicate that above the CLs of these ecosystems, small increases in N deposition result in large increases in fire risk up to the SL. For CBS, the average SL over six soil textures and four precipitation regimes was 8.15 kg $ha^{-1}yr^{-1}$. This value is very similar to the CL for increased exotic species and decreased native species cover determined empirically through the fertilization experiment, suggesting that this level of N deposition will be detrimental to both diversity and fire dynamics. As with the empirical CLs, the modelled CLs were highly variable across different precipitation regimes (Rao et al. 2010).

The CL and SL for PJW were lower than those for CBS because PJW is more likely to burn than CBS due to the greater abundance of woody biomass (Brooks and Minnich 2006). The low N deposition levels that result in increased fire risk at PJW suggests that much of this ecosystem type in California is at high risk of fire due to the combination of increased fine fuel production caused by atmospheric N deposition and inherently high woody fuel abundance. In summary, the empirical CL for exotic grass invasion in desert ecosystems is 8 kg $ha^{-1}yr^{-1}$. Based on the DayCent simulations, the range including the lower and upper bounds of the CLs for CBS and PJW are 3.2–9.3 and 3.0–6.3 kg $ha^{-1}yr^{-1}$ respectively, depending on soil texture and precipitation regime (Rao et al. 2010).

10.4 High Elevation Catchments of the Colorado Rocky Mountains

The Front Range of the Rocky Mountains includes the high peaks of the American Cordillera that rise from the Great Plains of Colorado to elevations in excess of 3000 m in very short lateral distances (less than 30 km). Mountain areas above tree line are especially susceptible to changes in atmospheric inputs of N and other atmospheric contaminants. This susceptibility is predicated on the low biomass, and thin soils of these ecosystems. Episodic acidification and surface water ANC values below zero have been reported in high elevation catchments in the Front Range, likely due to the dilute nature of waters at the highest elevations, shallow soils and short residence times of drainage water in these ecosystems. Nutrient N effects on these systems have also been documented, and both mechanistic and empirical modelling approaches have been used to establish CLs in these high elevation catchments.

10.4.1 Methodologies for Empirical and Model-based CLs for Nitrate Leaching, Acidification and Biodiversity Impacts

Williams and Tonnessen (2000) estimated the empirical acidification CL for lakes of the Colorado Front Range based on NADP wet deposition monitoring and the combined results of the U.S. Environmental Protection Agency Western Lakes Survey and a more focused survey of 91 lakes in the Front Range. Their approach was to examine the pH, ANC, and NO_3^- concentration distributions within the two populations of lakes in connection with wet deposition data across the intermountain western states.

Baron et al. (2011a) used lake surveys and deposition data to estimate empirical CLs of N by plotting lake NO_3^- concentrations ($n=285$ for the Rocky Mountains) vs. N deposition and inferring the inflection point or threshold at which NO_3^- concentrations increased in response to N deposition. Using a different approach, Baron (2006) estimated an empirical CL from historical reconstructions of atmospheric wet deposition to 1900. She combined these hindcasts with data on the diatom communities of lakes in Rocky Mountain National Park (RMNP) to set the CL at the estimated deposition rate during the period 1950–1964 when alteration of diatom assemblages was inferred from analysis of sediment cores collected from alpine lakes in RMNP (Wolfe et al. 2003).

Using the MAGIC model Sullivan et al. (2005) estimated the N and S CLs for surface water acidification in the Colorado Front Range. The MAGIC model assumes that a catchment can be modelled as a lumped reactive soil that contains N biogeochemistry and soil geochemical exchange and weathering reactions. The CLs were estimated by simulating deposition impacts on surface water ANC and pH. Hartmann et al. (2007, 2009) simulated biology, hydrology and geochemistry for two catchments in the Front Range with the DayCent-Chem model. The study highlighted the need to capture both the biological and geochemical system response in order to properly estimate the sensitivity of alpine ecosystems to atmospheric deposition. While the study itself does not explicitly address CLs, the results include modelled estimates and current conditions of acidification and NO_3^- concentrations in both the Andrews Creek (RMNP) and Green Lakes Valley (Niwot Ridge) watersheds (Hartman et al. 2009). Based on atmospheric deposition levels at the two sites (Hartman et al. 2009), an approximation of the N CL can be determined.

Empirical and modelled CLs were also determined for plant species change and biodiversity impacts to alpine vegetation in the Rocky Mountains (Bowman et al. 2006; Sverdrup et al. 2012). The empirical CL was determined from N fertilization experiments (0, 20, 40 and 60 kg N $ha^{-1}yr^{-1}$) in an alpine dry meadow at Niwot Ridge over an 8-year period (Bowman et al. 2006). The simulated CL for biodiversity effects was produced using the ForSAFE-VEG model parameterized for a generalized site in the alpine and subalpine zones of the northern and central Rocky Mountains for the years 1750–2500 (Sverdrup et al. 2012). The ForSAFE-VEG model includes modules to simulate soil chemistry, hydrology, energy, C and N cycling, tree growth and production, geochemistry and ground vegetation (Sverdrup et al. 2012). The VEG submodel includes dynamic competition principles for interspecies interactions.

10.4.2 Empirical and Model-based CLs for the Colorado Rocky Mountains

The empirical CL for NO_3^- leaching based on a survey of lakes in the Rocky Mountains was 3.0 kg N $ha^{-1}yr^{-1}$ (Baron et al. 2011a). Baron et al. (1994) simulated stream N export with increasing N deposition in the Loch Vale Watershed in RMNP with the Century model and reported a threshold for increased N export at a N deposition of 2–3 kg N $ha^{-1}yr^{-1}$. This CL is in close agreement with the empirical CL (Baron et al. 2011a; Table 10.1). Paleolimnological hindcast studies in high elevation lakes of the Colorado Front Range (Baron 2006; Wolfe et al. 2003), the eastern Sierra Nevada of California and the Central Rocky Mountains indicate an empirical N CL of 1.4–1.5 kg N $ha^{-1}yr^{-1}$ for effects on phytoplankton communities (Saros et al. 2011).

Field surveys and N enrichment studies have shown that alpine lake NO_3^- concentrations within the range of 3–12 µmol l^{-1} are associated with increased phytoplankton biomass, changes in phytoplankton communities, and a shift to P limitation (Bergström and Jannson 2006; Elser et al. 2009; Interlandi and Kilham 2001). DayCent-Chem simulations of streamwater NO_3^- concentrations for Andrews Creek and Niwot Ridge indicated streamwater NO_3^- concentrations typically ranging from 0 to 100 µmol l^{-1}, similar to empirical data (2–60 µmol l^{-1}) for the same sites (Hartman et al. 2009). With field and simulated NO_3^- values indicative of elevated NO_3^- leaching and with NO_3^- concentrations much higher than those known to affect phytoplankton communities, we conclude that these sites (N deposition levels of 3.6 and 5.9 kg $ha^{-1}yr^{-1}$) are in exceedance of the N CL. Further DayCent-Chem simulations with atmospheric N inputs varying over the appropriate range are needed to refine the simulated N deposition CL for nutrient enrichment.

The empirical CL estimates for lake acidification of 4 kg N $ha^{-1}yr^{-1}$ as wet deposition (Williams and Tonnessen 2000) and 4 kg N $ha^{-1}yr^{-1}$ as total inorganic N deposition (Baron et al. 2011a) are within the margin of error, particularly for estimating N deposition in this region where dry deposition is purported to constitute only 15 % of the total (Baron et al. 2011a). The MAGIC model yielded an acidification CL of 7.8 kg N $ha^{-1}yr^{-1}$ at Andrews Creek based on an ANC of 20 µeq l^{-1}, and a CL of 12 kg N $ha^{-1}yr^{-1}$ for the Loch, which is not among the most acid-sensitive lakes in the region (Sullivan et al. 2005). These simulated acidification CLs are higher than the empirical acidification CL, but not all lakes are equally sensitive (Sullivan et al. 2005). If a threshold ANC of 20 µeq l^{-1} is chosen for estimating the CL, the

Table 10.1 A comparison of empirical and simulated critical loads for alpine ecosystems in the Rocky Mountains, Colorado, USA

Acidification or N enrichment CL	Empirical or modelled CL	Response indicator	CL value (kg N $ha^{-1}yr^{-1}$)	References
Catchment-level responses				
Acidification	Empirical	ANC	4	(Baron et al. 2011a; Williams and Tonnessen 2000)
	Modelled (MAGIC model)	ANC	7.8–12.0	(Sullivan et al. 2005)
N enrichment	Empirical	Change in diatom assemblages	1.4–1.5	(Baron 2006; Saros et al. 2011; Wolfe et al. 2003)
	Empirical	Surface water NO_3 concentrations	3	(Baron et al. 2011a)
	Modelled (century)	Stream N export	2–3	(Baron et al. 1994)
	Modelled (DayCent-Chem model)[a]	Surface water NO_3 concentrations	≤3.6–5.9	(Hartman et al. 2007, 2009)
Terrestrial vegetation responses				
N enrichment	Empirical	Change in individual plant species	4	(Bowman et al. 2006)
	Empirical	Plant community change	10	(Bowman et al. 2006)
	Modelled (ForSAFE-VEG model)	Long-term plant biodiversity change	1–2	(Sverdrup et al. 2012)

[a] DayCent-Chem simulations of Hartman et al. (2007, 2009) were not explicitly designed to determine a CL. Thus, the CL estimates are based on the known atmospheric deposition (kg ha^{-1} yr^{-1}) at the sites for which the model runs were parameterized.

empirical CL for acidification is not exceeded at Andrews Creek where N deposition is 3.6 kg $ha^{-1}yr^{-1}$, nor at Niwot Ridge where N deposition is 5.9 kg $ha^{-1}yr^{-1}$. These empirical data suggest that at some sites the acidification CL is higher than the proposed empirical CL value of 4 kg N $ha^{-1}yr^{-1}$.

Modelling stream ANC for determining acidification CLs with the DayCent-Chem model would benefit from model refinement. Modelling stream ANC and pH was challenging because these parameters were highly sensitive to complex silica reactions and stream pCO_2 (Hartman et al. 2007). Simulated daily ANC concentrations were also sensitive to daily discharge estimates and when discharge was overestimated ANC was underestimated and vice versa. Simulations of ANC with DayCent-Chem did not always track well with empirical data, although simulated annual volume-weighted mean ANC concentration was within 12 μeq l^{-1} of the measured mean (Hartman et al. 2009).

The empirical CL for changes to individual alpine plant species was 4 kg N $ha^{-1}yr^{-1}$, compared to 10 kg N $ha^{-1}yr^{-1}$ for overall plant community change based on an 8 year fertilization study at Niwot Ridge (Bowman et al. 2006). In comparison, the ForSAFE-VEG model estimated a biodiversity CL of 1–2 kg N $ha^{-1}yr^{-1}$ for a generalized Rocky Mountain alpine site (Sverdrup et al. 2012). The simulated N CL could be higher if different parameterization values are used for precipitation, temperature, soil chemistry, plant nutrient uptake or by altering the amount of biodiversity change considered acceptable. The lower CL derived from the ForSAFE-VEG model is presumably largely due to the much longer time scale for the impacts of N to take effect compared to the fertilization study. An additional factor is that the CL for N deposition effects on alpine plant communities is already exceeded at the study site based on the estimated ambient N deposition (6 kg $ha^{-1}yr^{-1}$; Bowman et al. 2006).

In these studies in the Colorado Rocky Mountains, both empirical and simulated CL values were higher for acidification effects than for nutrient-N effects, as is commonly reported (Hettelingh et al. 2008). The empirical N CL for surface water acidification (4 kg N $ha^{-1}yr^{-1}$) is higher than the empirical CL for nutrient effects (2–3 kg N $ha^{-1}yr^{-1}$ for elevated NO_3^-; 1.5 kg N $ha^{-1}yr^{-1}$ for effects on diatom assemblages; Table 10.1). Likewise, the simulated CL for acidification using the MAGIC model (7.8 at Andrews Creek and 12.0 at the Loch) is higher than the simulated CL for nutrient N based on DayCent-Chem (<3.6 kg N $ha^{-1}yr^{-1}$; Table 10.1). However, in the latter case, N deposition inputs were not varied to explicitly determine the CL. Note that only in the case of the empirical N enrichment CL (Baron 2006) was a biological indicator used (alteration of algal assemblages) to estimate the CL for surface waters or catchment-level responses (Table 10.1). Both empirical and modelling approaches for determining CLs are useful, and are complimentary and reinforcing each other.

10.5 Adirondack Lake Watersheds in New York State

The Adirondack Mountain region is arguably among the regions of the U.S. most highly impacted by acidic deposition (Driscoll et al. 1991). The Adirondack Park encompasses 2.4 million ha in northern New York, with about 1 million ha of publicly owned state lands and 1.4 million ha of private lands. The region is a unique landscape of forested uplands and wetlands, and includes approximately 2800 lakes (>0.2 ha). The Adirondacks receive elevated inputs of atmospheric S, NO_3^- and ammonium (NH_4^+) deposition (Driscoll et al. 1991). In 1984–1987 the Adirondack Lakes Survey determined that 26% of 1469 surveyed lakes had pH <5.0, 26% had ANC <0 μeq l^{-1}, and 50% had ANC <50 μeq l^{-1} (Kretser et al. 1989). As a result, there are 128 lakes in the Adirondacks that are included on the New York State Section List because of impairment to aquatic life attributed to acidic deposition (based on the criterion of pH <6.0; www.dec.ny.gov/chemical/31290.html). The acidic conditions in acid-sensitive Adirondack lakes are due to elevated leaching of sulfate (SO_4^{2-}) and NO_3^- in surface waters relative to the supply of base cations from the watershed (Driscoll et al. 1991). There is considerable interest in the application of models and the development of empirical relationships to determine CLs or total maximum daily loads (TMDLs) with respect to lake acidity in this region.

The objective of this case study was to illustrate the use of a dynamic model (PnET-BGC) in the calculation of CLs for lake watersheds in the Adirondack region of New York. We examine the tradeoffs of controls of (SO_4^{2-}) deposition versus NO_3^- deposition. Note that PnET-BGC is capable of simulating the effects of changes in NH_4^+ deposition, but for this chapter we chose to evaluate changes in (SO_4^{2-}) and NO_3^- deposition, the focus of current emissions control programs for acidic deposition in the U.S. (USEPA 2009). We have observed only minor differences in the response of Adirondack soils and surface waters between equivalent changes in NO_3^- and NH_4^+ deposition in simulations with PnET-BGC. We also compare and relate the use of dynamic models with empirical approaches to calculate CLs.

10.5.1 Methodologies for Model-based CLs for Nitrate Leaching and Acidification

PnET-BGC was applied to 20 lakes of the Adirondack Long-Term Monitoring (ALTM) Program (Driscoll et al. 2003) to evaluate CLs of acidity and N for the region. These lakes and their watersheds exhibit a range of characteristics and sensitivity to acidic deposition. For the 20 ALTM lake-watersheds, we show model simulations that compare results of the dynamic model PnET-BGC with empirical approaches that have been used to assess CLs of surface waters. To illustrate the use of model calculations for the determination of CLs, we show results from Jockeybush Lake (43°18'N, 74°35'W), a thin till drainage lake with a forested catchment. Jockeybush Lake has been monitored for lake chemistry since 1992 (Driscoll et al. 2003). It is chronically acidic and has a mean $SO_4^{2-} = 92\,\mu eq\,l^{-1}$, $NO_3^- = 17\ \mu eq\ l^{-1}$, $ANC = 4.5\ \mu eq\ l^{-1}$ and $pH = 5.4$.

Model Description PnET-BGC is a comprehensive forest-soil-water model that links a C, N and water balance model, PnET-CN (Aber et al. 1997), with a biogeochemical model, BGC (Gbondo-Tugbawa et al. 2001). The model performs well in small, high-elevation watersheds where detailed site data are available to constrain inputs and parameter values, but regional scale applications have also been conducted (Chen and Driscoll 2004, 2005; Zhai et al. 2008).

PnET-BGC extends the simulations of PnET-CN to include the cycling of all major elements (i.e., C, N, P, S, Ca, Mg, K, Na, Al, Cl, Si). Both major biotic and abiotic processes are represented in PnET-BGC, including atmospheric deposition, canopy interaction, CO_2 fertilization, litterfall, forest growth, root uptake, snowpack accumulation and loss, routing of water along hydrologic flowpaths, soil organic matter dynamics, nitrogen mineralization and nitrification, mineral weathering, wetland biogeochemical processes, chemical reactions involving solid and solution phases, and surface water processes. The model can be run on a time step specified by the user; a monthly time step was used for this study.

PnET-BGC requires input parameters related to the site: meteorological conditions, atmospheric deposition, element weathering, soils and vegetation and land-disturbance history. Basic soil parameters needed for PnET-BGC include properties such as soil mass, cation exchange capacity, cation exchange constants and anion adsorption constants. Vegetation is characterized in PnET-BGC using the major forest cover types represented at the study site and the element stoichiometry associated with these cover types (Gbondo-Tugbawa et al. 2001).

Model Calibration and Application Following previous research (Zhai et al. 2008), model runs for Jockeybush Lake were started in the year 1000 AD, and run under constant pre-industrial deposition and no land disturbance until 1850 to achieve steady-state and evaluate "background" (i.e., pre -1850) conditions. Changes in atmospheric deposition and land disturbance events were initiated after 1850. The model was run from 1850 through 2010 based on measured values of atmospheric deposition and reconstructions of historical deposition from emission records (Fig. 10.4a, b; Driscoll et al. 2001, Zhai et al. 2008). For the period 1978–2010, we estimated atmospheric deposition from measured wet deposition values at Huntington Forest (44°00'W, 74°13'N; NADP NY20) in the central Adirondacks. For the period before 1850, we assumed that atmospheric deposition was 10% of current values. For the period 1978–1900, we used empirical relations between atmospheric anthropogenic emissions for the atmospheric source area of the Adirondacks and the measured atmospheric deposition data for Huntington Forest (1978–2010) to reconstruct deposition based on historical emission records for the period. For the period between 1850 and 1900, we assumed a linear increase in atmospheric deposition from background values. To estimate total deposition from wet deposition we assume a fixed dry to wet deposition ratio based on dry deposition measurements at Huntington Forest. To extrapolate these estimates of atmospheric historical deposition for Huntington Forest to Jockeybush Lake watershed, we used the spatial deposition model of Ito et al. (2002). Model simulations continued to the year 2200 with a series of forecasts which included a range of deposition scenarios from current to "background" deposition for SO_4^{2-}, and NO_3^- individually and in combination. This range corresponds to 0–100% decreases in anthropogenic atmospheric SO_4^{2-} and NO_3^- deposition. Future scenarios involved a 10 year ramp from current values to the level of deposition of interest and continued simulation at this deposition level to 2200 (Fig. 10.4a, b). This range of values enabled evaluation of tradeoffs associated with reductions in sulfur dioxide, and nitrogen oxides to achieve ecosystem recovery from acidic deposition.

10.5.2 Empirical and Model-based CLs for Adirondack Lake Watersheds

The model hindcast for Jockeybush Lake suggests that prior to the advent of acidic deposition, the ANC was approximately 60 µeq l^{-1} (Fig. 10.4c–e). With increases

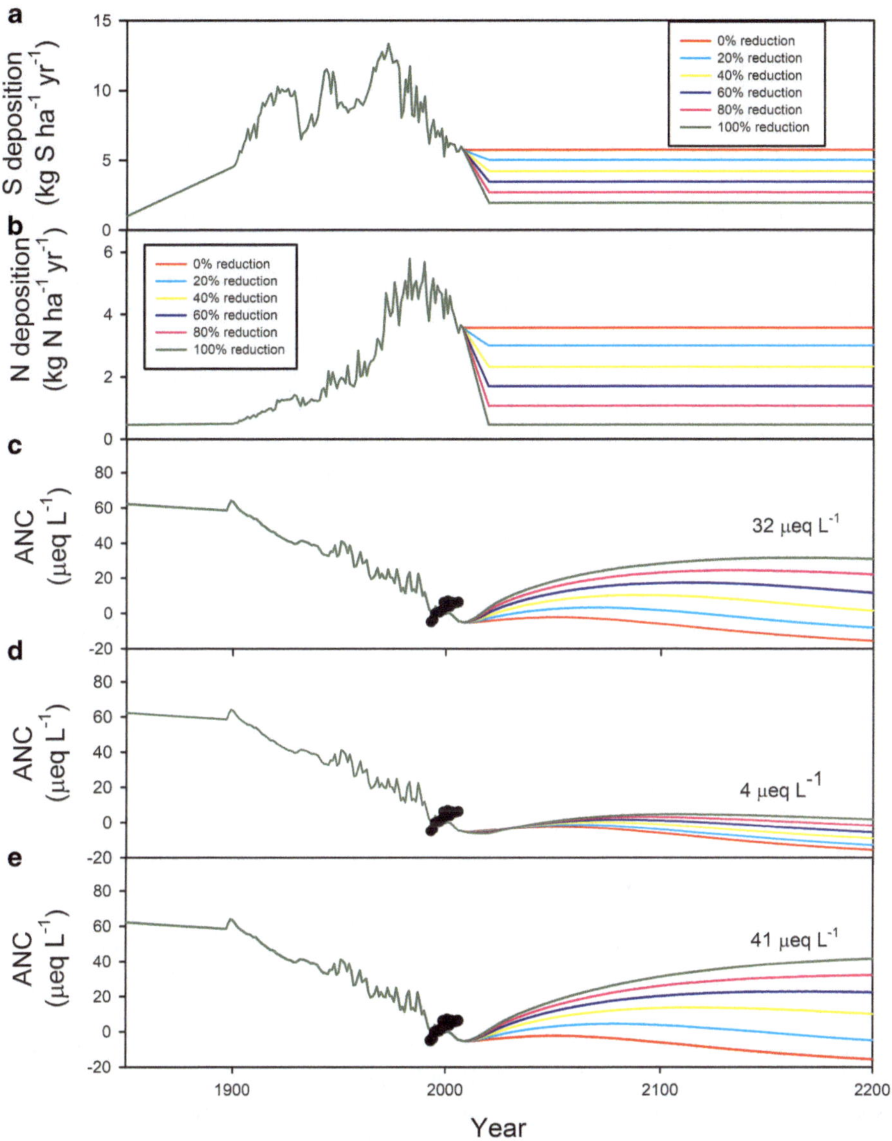

Fig. 10.4 Time-series of model inputs of atmospheric sulfur (S) (**a**) and nitrate (NO_3^-) (**b**) deposition for Jockeybush Lake and model simulations of lake acid neutralizing capacity (ANC) hindcast (1950–2010) and model projections (2010–2200) under a suite of decreases in S (**c**), NO_3^- (**d**), and S+NO_3^- (**e**) deposition, corresponding to 0–100% decreases in anthropogenic S and NO_3^- deposition. Measured values of ANC are shown as *black dots*

in atmospheric SO_4^{2-} and NO_3^- deposition through the twentieth Century, model simulation indicate that ANC decreased to approximately zero (−4 μeq l^{-1}). The simulated loss of ANC for the lake is similar to regional values reported previously (Zhai et al. 2008).

Model forecasts for Jockeybush Lake were conducted for a series of future scenarios of SO_4^{2-} and NO_3^- deposition ranging from no reduction from current deposition (no reduction) to complete elimination of anthropogenic deposition and decreases to "background" (pre-anthropogenic) deposition (100 % reduction; Fig. 10.4a, b). A range of recoveries of ANC are projected coinciding with the range of deposition scenarios (Fig. 10.4c–e). Projections of lake ANC responses are initially relatively rapid following decreases in SO_4^{2-} loading but not to decreases in NO_3^- loading. The rate of ANC increase subsequently decreases as the watershed approaches steady-state with respect to the lower input of atmospheric SO_4^{2-} or NO_3^- deposition. Note under the lower projections of decreases in atmospheric deposition (e.g., 0, 20 %) following an initial period of partial recovery, the lake ANC ultimately decreases. The reason for this projected re-acidification is two-fold. First, limited decreases in atmospheric deposition are not large enough to reverse the net loss of exchangeable calcium and magnesium from the soil, and the continued acidification of soil ultimately contributes to the acidification of the drainage water. Second, the PnET-BGC simulates an aggrading forest. As the forest ages, its demand for N decreases and NO_3^- leaching increases. This stand age-dependent retention of N and leaching of NO_3^- also contributes to the re-acidification of lake water.

Model projections suggest that under no additional controls of SO_4^{2-} or NO_3^- deposition (0 % reduction) the ANC of Jockeybush Lake will continue to acidify to −15 μeq l^{-1} by 2200 (Fig. 10.4e). In contrast, if anthropogenic atmospheric SO_4^{2-} deposition is eliminated (100 % reduction) with no change in atmospheric NO_3^- deposition, the ANC is projected to increase to 32 μeq l^{-1} by 2200 (Fig. 10.4c). Note that even under the scenario of complete elimination of atmospheric SO_4^{2-} deposition, Jockeybush Lake does not recover its pre-anthropogenic ANC (~60 μeq l^{-1}) by 2200. This net loss of ANC is due to soil acidification associated with the net depletion of exchangeable calcium and magnesium in soil from historical acidic deposition.

The projected lake responses to decreases in atmospheric NO_3^- deposition were similar to SO_4^{2-} responses, but considerably more muted. If anthropogenic atmospheric NO_3^- deposition is eliminated (100 % reduction) with no change in atmospheric SO_4^{2-} deposition, the ANC is projected to increase only to 4 μeq l^{-1} by 2200. This analysis indicates that controls on atmospheric SO_4^{2-} deposition are more effective in facilitating increases in ANC than decreases in NO_3^- deposition in Adirondack watersheds.

The projections of lake responses to decreases in future SO_4^{2-} and NO_3^- in combination show greater increases in ANC than decreases in SO_4^{2-} alone (up to 41 μeq l^{-1} for 100 % reduction; Fig. 10.4e). These simulations suggest that there are benefits associated with simultaneous controls of SO_4^{2-} and NO_3^- in achieving increases in ANC.

Comparison of Model Results with Approaches for Empirical Critical Loads In this analysis, we consider two approaches for evaluating empirical vs. modelled CLs for NO_3^- leaching and acidification by using simulation data for the 20 ALTM lakes. The empirical CL for NO_3^- leaching is based on field observations of surface water NO_3^- concentrations which typically exhibit a "dogleg" pattern in response to increases in atmospheric N deposition (Aber et al. 2003). The "dogleg" pattern is manifested by low surface water concentrations of NO_3^- at lower levels of atmospheric N deposition and at a critical load some waters show a pattern of increases in NO_3^- with increases in deposition. This leaching response is highly variable in watersheds receiving elevated N deposition due to historical land disturbance, and vegetation and land cover characteristics (Aber et al. 2003; Baron et al. 2011b). These empirical observations are typically obtained from surface water data for many sites along a spatial gradient of atmospheric N deposition. This "dogleg" pattern of watershed response to variations in N deposition has been reported for Europe and the eastern and western U.S. (Baron et al. 2011b) and has been used in the determination of empirical CLs for surface waters (Pardo et al. 2011). For example, Aber et al. (2003) observed this pattern of NO_3^- response to variation in atmospheric N deposition for surface waters in the northeastern U.S, including the Adirondacks, and found that the critical load of N above which elevated leaching of NO_3^- could occur was 8 kg N $ha^{-1}yr^{-1}$. Aber et al. (2003) observed maximum surface water NO_3^- concentrations of 30–40 μmol l^{-1} at the highest observations of N deposition of 10–12 kg N $ha^{-1}yr^{-1}$. The assumption that is inherent in the use of these empirical relations for CLs is that the observed spatial pattern of NO_3^- leaching is indicative of the temporal response that would occur in response to changes in N deposition at a given site (i.e. space-for-time substitution; e.g., Aber et al. 2003).

We used results of PnET-BGC simulations for the 20 ALTM watersheds to evaluate whether the pattern of NO_3^- leaching with changes in N deposition was similar to empirical observations, such as those reported by Aber et al. (2003). We averaged time-series of lake NO_3^- simulations over 10-year intervals for each of the ALTM lakes over the period of hindcasts (1850–2010; 16 observations per site). Thus, the results of PnET-BGC runs capture both spatial (20 sites) and temporal (time series hindcast) simulated NO_3^- response to changes in N deposition.

In contrast to the "dogleg" pattern typical of empirical observations, PnET-BGC simulations generally showed a linear pattern of increase in NO_3^- concentrations in response to increases in N deposition to Adirondack lake-watersheds (Fig. 10.5; according to the equation, NO_3^- (μmol l^{-1})$=2.9\times$ total N deposition (kg N $ha^{-1}yr^{-1}$)$+2.7$; $r^2=0.55$). As for the empirical data (Aber et al. 2003; Baron et al. 2011a) there was considerable variability around this "linear" relationship. Some of the simulation periods for some of the ALTM sites corresponded with the time period shortly after a land disturbance event (e.g., cutting, blowdown; noted in Fig. 10.5 as land disturbance). PnET-BGC simulates elevated NO_3^- leaching associated with land disturbance events and these observations are depicted as outliers in Fig. 10.5. Given this approximately linear pattern of NO_3^- leaching in response to changes in N deposition from PnET-BGC simulations, a CL for N could be estimated by specifying an acceptable level of leaching. For example, Gundersen et al.

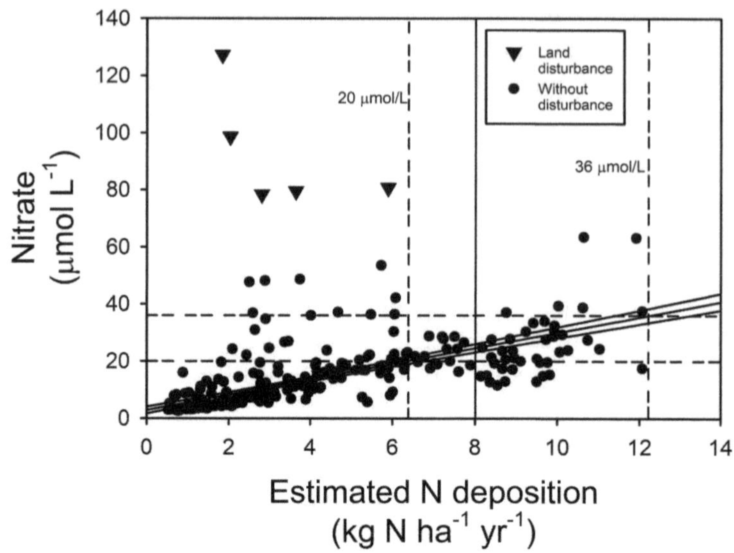

Fig. 10.5 Simulated response of 10 year average concentrations of nitrate to changes in atmospheric NO_3^- deposition over the 1850–2010 hindcast period for 20 Adirondack lake watersheds. Outlier observations are shown which represent simulated conditions shortly after land disturbance events (e.g., clear-cutting, blowdowns). A linear regression of simulated nitrate as a function of N deposition is shown with 95 % confidence interval is shown (NO_3^- (µmol l^{-1})=2.9 × total N deposition (kg N $ha^{-1}yr^{-1}$)+2.7; r^2=0.55). Using this regression, CLs are estimated for lake nitrate of 20 and 36 µmol l^{-1} (shown with *dashed vertical lines*) at 6.4 kg N $ha^{-1}yr^{-1}$ and 12.2 kg N $ha^{-1}yr^{-1}$, respectively. A *vertical line* is also shown at 8 kg N $ha^{-1}yr^{-1}$ to indicate the level of atmospheric deposition where elevated nitrate leaching approximately occurs from empirical spatial observations

(2006) proposed a threshold for annual mean surface water NO_3^- of 36 µmol l^{-1} that is thought to be indicative of a "leaky" watershed (based on data from North America and Europe). More recently, Fenn et al. (2011) suggested a threshold of 20 µmol l^{-1} for US watershed in the determination of empirical CLs. Invoking these NO_3^- concentration limits to the PnET-BGC-derived relationship for NO_3^- leaching response to varying atmospheric N deposition of Adirondack lake-watersheds, a CL of 6.4 kg N $ha^{-1}yr^{-1}$ and 12.2 kg N $ha^{-1}yr^{-1}$ were obtained for NO_3^- thresholds of 20 µmol l^{-1} and 36 µmol l^{-1}, respectively (Fig. 10.5) in general agreement with the value of 8 kg N $ha^{-1}yr^{-1}$ suggested for northeastern U.S. surface waters (Pardo et al. 2011).

Empirical approaches have also been used to evaluate CLs of acidity for surface waters. The empirical F-factor has been used in steady-state CL applications (e.g., USEPA 2009). The F-factor is the ratio of the change over time in the sum of concentration of base cations (C_B) divided by the change in $SO_4^{2-} + NO_3^-$ concentration (C_A) relative to the pre-industrial steady-state water chemistry condition.

Invoking the F-factor assumes that the supply of base cations from the watershed and atmospheric deposition to surface waters is a linear function of $SO_4^{2-} + NO_3^-$ concentrations which are derived from atmospheric deposition. To examine this assumption we plot simulated time series trajectories of ANC through hindcast simulations of (a), the sum of base cations (b) and soil percent base saturation (c) for three of the ALTM watersheds (Carry Pond, Arbutus Pond and Constable Pond (Fig. 10.6). The three sites were selected as examples to show a range of responses across the 20 ALTM lake watersheds simulated. The trajectories show a decrease in ANC (Fig. 10.6a) and an increase in C_B (Fig. 10.6b) with increasing concentrations of $SO_4^{2-} + NO_3^-$. After an initial period of relatively steady soil base saturation, soils acidify (i.e., loss of base saturation) with increases in concentrations of $SO_4^{2-} + NO_3^-$ (Fig. 10.6c). Following peak concentrations of $SO_4^{2-} + NO_3^-$, the lake watersheds show limited change in ANC with subsequent decreases in $SO_4^{2-} + NO_3^-$. There is a generally "linear-like" change in C_B with changes in $SO_4^{2-} + NO_3^-$; however, all lake watersheds show a hysteresis pattern where the waters have lower concentrations of C_B for a given concentration of $SO_4^{2-} + NO_3^-$ in the recovery leg of the trajectory, as compared with the acidification leg (Fig. 10.6b). Examining the slopes of change in C_B as a function of change in $SO_4^{2-} + NO_3^-$ ($\Delta C_B/C_A$; F-factor), this pattern is evident (Table 10.2). Initially, at the onset of acidic deposition, the F-factor is high due to the accelerated leaching of base cations from soil exchange sites. Over time, soil exchangeable base cations become depleted, the soil and water acidify in response to additional acidic deposition and F-factor values decrease. This loss of soil exchangeable base cation limits the recovery of ANC following decreases in acidic deposition.

Note that the three example lake watersheds have distinct patterns of ANC recovery. Carry Pond is increasing in ANC in response to decreases in lake $SO_4^{2-} + NO_3^-$ concentrations. Of the 20 ALTM lake watersheds simulated, 7 exhibit this type of recovery response. Arbutus Lake shows no change in ANC in response to decreases in lake $SO_4^{2-} + NO_3^-$ concentrations. Of the 20 ALTM lake watersheds simulated, 3 exhibit this type of recovery response. Finally Constable Pond is decreasing in ANC (continuing to acidify) in response to decreases in lake $SO_4^{2-} + NO_3^-$ concentrations. Of the 20 ALTM lake watersheds simulated, 10 are continuing to acidify under existing inputs of acidic deposition.

This analysis using PnET-BGC suggests that the assumption of a linear response of C_B to changes in $SO_4^{2-} + NO_3^-$, which is fundamental to many of the steady-state approaches to CLs for acidity, does not depict the dynamics of soil acid-base chemistry under changing acidic deposition. The non-linear behaviour of watershed base cation supply in response to changes in acidic deposition is a critical consideration in calculations of CLs for acidity. There are increasingly well-recognized uncertainties and limitations associated with the widely used simple steady-state CL modelling approaches for acidity (Curtis et al. 2001; Watmough et al. 2005). Such concerns have led to the more wide-spread use of process-based models that include the dynamics of base cation depletion in calculations of CLs for acidity (Rapp and Bishop 2009).

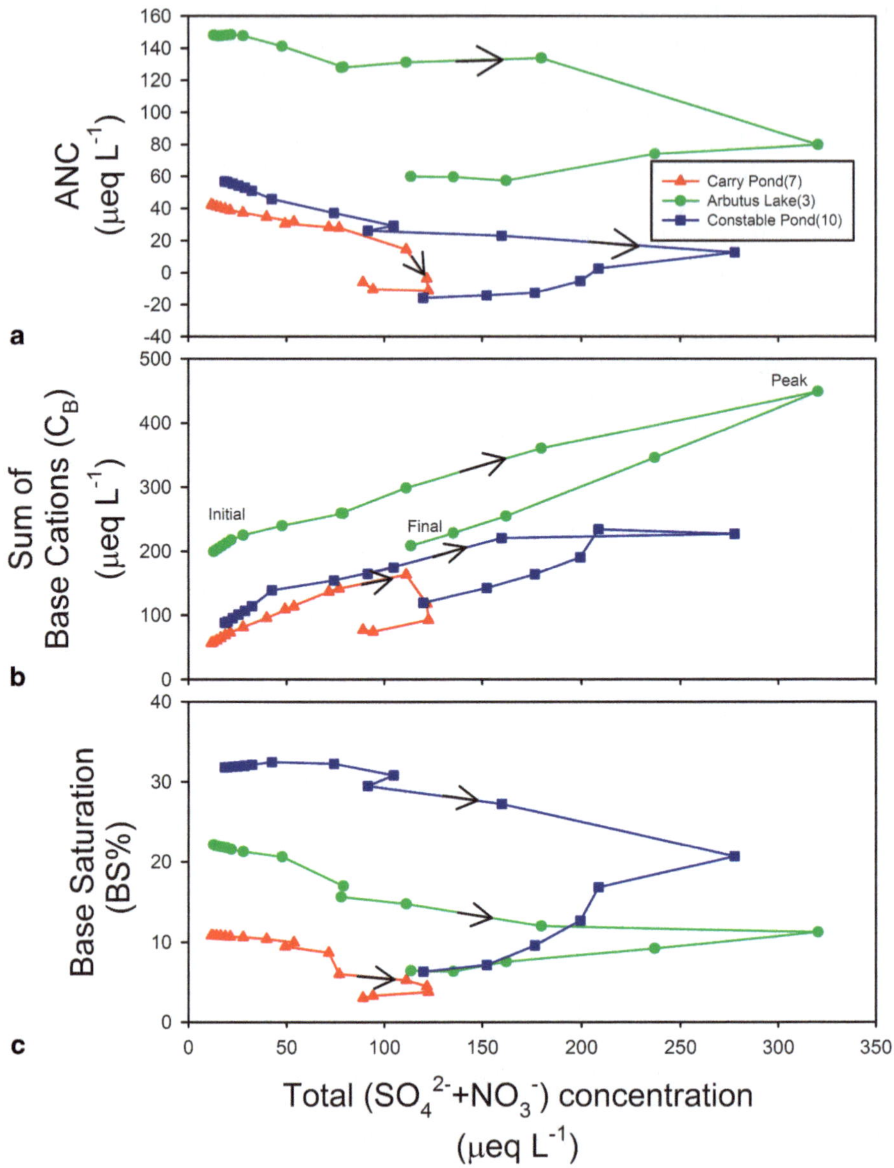

Fig. 10.6 Time series trajectories of PnET-BGC simulations of **a** acid neutralizing capacity (ANC); **b**, the sum of base cations (C_B), and **c** soil percent base saturation as a function of sulfate (SO_4^{2-}) plus nitrate (NO_3^-) concentrations for three Adirondack Long Term Monitoring (ALTM) watersheds (Carry Pond, Arbutus Pond and Constable Pond). Numbers in parenthesis refer to the number of 20 ALTM lakes simulated that show increases in ANC in response to decreases in acidic deposition (*7*); no change in ANC in response to decreases in acidic deposition (*3*); and further decreases in ANC in response to decreases in acidic deposition (*10*)

Table 10.2 Slopes of segments of time series of simulated hindcasts of the ratio of the change in sum of base cations to the change in the concentrations of sulfate plus nitrate ($\Delta C_B/\Delta C_A$: F-factor) for three Adirondack Long-Term Monitoring Lakes (see Fig. 10.6b)

ALTM lake	Carry pond	Arbutus lake	Constable pond
Initial slope (eq/eq)	1.74	2.06	1.88
Slope at peak C_A (eq/eq)	0.66	0.98	0.82
Slope at end of time series of C_A hindcast (eq/eq) (current conditions)	0.53	0.96	0.78

Discussion and Conclusions

The case studies included in this chapter illustrate the value of the combined approach for assessing CLs of N. Empirical CLs are needed to document ecosystem responses to N deposition under ambient conditions and as a reality check for simulation models. Biogeochemical models, on the other hand, provide the capability of evaluating ecosystem responses, including temporal dynamics, and CL exceedances under a variety of scenarios. Models can suggest how long it may take to reach various states of restoration. Models can also predict how climate change, disturbance, and land use scenarios will affect the CL. For example, DayCent simulations were used to evaluate the effectiveness of various combinations of decreased N deposition and prescribed fire intervals in reducing the impacts of excess N accumulation in forests of southern California (Gimeno et al. 2009).

Another advantage of biogeochemical models is that CLs can be estimated for a variety of ecological processes, pools or parameters. For example, CLs can be determined for acidification of soils or surface waters, NO_3^- leaching, stimulation of nitrification, nitrogenous trace gas emissions, biomass accumulation, or any other ecological parameter or process of interest that a model can effectively simulate. The ForSAFE-VEG model can estimate N CLs to protect plant biodiversity (Sverdrup et al. 2012).

Simulation models allow for hindcasting, forecasting, and the capability to evaluate a much wider combination of factors and environmental conditions than can be accomplished with empirical data or manipulative experiments. However, this can be an iterative process in that until further development, models may lack sufficient capability to effectively simulate the effects of some environmental factors that may influence the CL. Ecosystems are affected by multiple biotic and abiotic stressors which may include more than one type of pollutant, changing land use, or management approaches, in addition to a changing climate. Thus, empirical CLs determined from spatial gradient studies, N addition experiments, or long-term studies at one or a few sites may provide insights and CL estimates under multiple stress conditions with greater confidence than present models can provide. For example, in California mixed conifer forests, the CL for NO_3^- leaching or for effects on fine root biomass (Fenn et al. 2008) are likely affected by O_3 exposure; N deposition, including the changing ratio of reduced to oxidized N; fire suppression; and drought stress. Current models generally lack the functionality to estimate the CL as affected by these multiple co-occurring stressors. Thus, empirical CLs based on field observations may provide the best estimates of the N CL under these conditions. This points to the need of continuing model development for understanding these complex interactions and their implications for CLs and ecosystem protection.

Acknowledgments This research was funded in part by a National Science Foundation (USA) grant (NSF DEB 04–21530) and by New York State Energy Research and Development Authority.

References

Aber, J. D., Ollinger, S. V., & Driscoll, C. T. (1997). Modelling nitrogen saturation in forest ecosystems in response to land use and atmospheric deposition. *Ecological Modelling, 101,* 61–78.

Aber, J. D., Goodale, C. L., Ollinger, S. V., Smith, M.-L., Magill, A. H., Martin, M. E., Hallett, R. A., & Stoddard, J. L. (2003). Is nitrogen deposition altering the nitrogen status of Northeastern forests? *Bioscience, 53,* 375–389.

Allen, E. B., Rao, L. E., Steers, R. J., Bytnerowicz, A., & Fenn, M. E. (2009). Impacts of atmospheric nitrogen deposition on vegetation and soils in Joshua Tree National Park. In R. H. Webb, L. F. Fenstermaker, J. S. Heaton, D. L. Hughson, E. V. McDonald, & D. M. Miller (Eds.), *The Mojave desert: Ecosystem processes and sustainability* (pp. 78–100). Las Vegas: University of Nevada Press.

Anderson, H. E. (1982). *Aids to determining fuel models for estimating fire behavior.* (General Technical Report INT-122). Ogden, Utah: USDA Forest Service Intermountain Forest and Range Experiment Station.

Baron, J. S. (2006). Hindcasting nitrogen deposition to determine an ecological critical load. *Ecological Applications, 16,* 433–439.

Baron, J. S., Ojima, D. S., Holland, E. A., & Parton, W. J. (1994). Analysis of nitrogen saturation potential in Rocky Mountain tundra and forest: Implications for aquatic systems. *Biogeochemistry, 27,* 61–82.

Baron, J. S., Driscoll, C. T., Stoddard, J. L., & River, E. E. (2011a). Empirical critical loads of atmospheric nitrogen deposition for nutrient enrichment and acidification of sensitive US Lakes. *Bioscience, 61,* 602–613.

Baron, J. S., Driscoll, C. T., & Stoddard, J. L. (2011b). Inland surface waters. In L. H. Pardo, M. J. Robin-Abbott, & C. T. Driscoll (Eds.), *Assessment of N deposition effects and empirical critical loads of nitrogen for ecoregions of the United States. General technical report NRS-80* (pp. 209–228). Newtown Square: U.S. Department of Agriculture, Forest Service, Northern Research Station.

Bergström, A.-K., & Jannson, M. (2006). Atmospheric nitrogen deposition has caused nitrogen enrichment and eutrophication of lakes in the northern hemisphere. *Global Change Biology, 12,* 635–643.

Bowman, W. D., Gatner, J. R., Holland, K., & Wiedermann, M. (2006). Nitrogen critical loads for alpine vegetation and terrestrial ecosystem response: Are we there yet? *Ecological Applications, 16,* 1183–1193.

Brooks, M. L., & Minnich, R. A. (2006). Southeastern deserts bioregion. In N. G. Sugihara, J. W .V. Wagtendonk, K. E. Shaffer, J. Fites-Kaufman, & A. E. Thode (Eds.), *Fire in California's ecosystems* (pp. 391–414). Berkeley: University of California Press.

Brooks, M. L., D'Antonio, C. M., Richardson, D. M., Grace, J. B., Keeley, J. E., DiTomaso, J. M., Hobbs, R. J., Pellant, M., & Pyke, D. (2004). Effects of invasive alien plants on fire regimes. *Bioscience, 54,* 677–688.

Brown, D. E., & Minnich, R. A. (1986). Fire and changes in creosote bush scrub of the western Sonoran desert, California. *American Midland Naturalist, 116,* 411–422.

Chen, L., & Driscoll, C. T. (2004). An evaluation of processes regulating spatial and temporal patterns in lake sulfate in the Adirondack region of New York. *Global Biogeochemical Cycles, 18,* GB3024, doi:10.1029/2003GB002169.

Chen, L., & Driscoll, C. T. (2005). Regional assessment of the response of the acid-base status of lake watersheds in the Adirondack region of New York to changes in atmospheric deposition using PnET-BGC. *Environmental Science & Technology, 39,* 787–794.

Curtis, C. J., Reynolds, B., Allott, T. E. H., & Harriman, R. (2001). The link between the exceedance of acidity critical loads for freshwaters, current chemical status and biological damage: A re-interpretation. *Water, Air & Soil Pollution: Focus, 1,* 399–413.

De Vries, W., Kros, J., Reinds, G. J., Wamelink, G. W. W., Mol, J., van Dobben, H., Bobbink, R., Emmett, B., Smart, S., Evans, C., Schlutow, A., Kraft, P., Belyazid, S., Sverdrup, H. U., van Hinsberg, A., Posch, M., & Hettelingh, J.-P. (2007). Developments in deriving critical limits and modelling critical loads of nitrogen for terrestrial ecosystems in Europe. (Report 1382). Wageningen, the Netherlands: Alterra Wageningen UR.

DeFalco, L. A., Bryla, D. R., Smith-Longozo, V., & Nowak, R. S. (2003). Are Mojave Desert annual species equal? Resource acquisition and allocation for the invasive grass *Bromus madritensis* subsp rubens (Poaceae) and two native species. *American Journal of Botany, 90,* 1045–1053.

Del Grosso, S. J., Parton, W. J., Mosier, A. R., Ojima, D. S., Kulmala, A. E., & Phongpan, S. (2000). General model for N_2O and N_2 gas emissions from soils due to denitrification. *Global Biogeochemical Cycles, 14,* 1045–1060.

Driscoll, C. T., Newton, R. M., Gubala, C. P., Baker, J. P., & Christensen, S. W. (1991). Adirondack mountains. In D. F. Charles (Ed.). *Acidic deposition and aquatic ecosystems: Regional case studies* (pp. 133–202). New York: Springer-Verlag.

Driscoll, C. T., Lawrence, G. B., Bulger, A. J., Butler, T. J., Cronan, C. S., Eagar, C., Lambert, K. F., Likens, G. E., Stoddard, J. L., & Weathers, K. C. (2001). Acidic deposition in the northeastern United States: Sources and inputs, ecosystem effects, and management strategies. *Bioscience, 51,* 180–198.

Driscoll, C. T., Driscoll, K. M., Roy, K. M., & Mitchell, M. J. (2003). Chemical response of lakes in the Adirondack region of New York to declines in acidic deposition. *Environmental Science & Technology, 37,* 2036–2042.

Elser, J. J., Andersen, T., Baron, J. S., Bergström, A. K., Jansson, M., Kyle, M., Nydick, K. R., Steger, L., & Hessen, D. O. (2009). Shifts in lake N:P stoichiometry and nutrient limitation driven by atmospheric nitrogen deposition. *Science, 326,* 835.

Fenn, M. E., & Poth, M. A. (1999). Temporal and spatial trends in streamwater nitrate concentrations in the San Bernardino Mountains, southern California. *Journal of Environmental Quality, 28,* 822–836.

Fenn, M. E., Haeuber, R., Tonnesen, G. S., Baron, J. S., Grossman-Clarke, S., Hope, D., Jaffe, D. A., Copeland, S., Geiser, L., Rueth, H. M., & Sickman, J. O. (2003). Nitrogen emissions, deposition, and monitoring in the western United States. *Bioscience, 53,* 391–403.

Fenn, M. E., Jovan, S., Yuan, F., Geiser, L., Meixner, T., & Gimeno, B. S. (2008). Empirical and simulated critical loads for nitrogen deposition in California mixed conifer forests. *Environmental Pollution, 155,* 492–511.

Fenn, M. E., Sickman, J. O., Bytnerowicz, A., Clow, D. W., Molotch, N. P., Pleim, J. E., Tonnesen, G. S., Weathers, K. C., Padgett, P. E., & Campbell., D. H. (2009). Methods for measuring atmospheric nitrogen deposition inputs in arid and montane ecosystems of western North America. In A. H. Legge (Ed.), *Developments in environmental science, Vol. 9: Air quality and ecological impacts: Relating sources to effects* (pp. 179–228). Amsterdam: Elsevier.

Fenn, M. E., Lambert, K. F., Blett, T. F., Burns, D. A., Pardo, L. H., Lovett, G. M., Haeuber, R. A., Evers, D. C., Driscoll, C. T., & Jeffries, D. S. (2011). *Setting limits: Using air pollution thresholds to protect and restore U.S. ecosystems.* (Issues in Ecology, Report Number 14). Washington: Ecological Society of America.

Gasche, R., & Papen, H. (2002). Spatial variability of NO and NO_2 flux rates from soil of spruce and beech forest ecosystems. *Plant Soil, 240,* 67–76.

Gbondo-Tugbawa, S. S., Driscoll, C. T., Aber, J. D., & Likens, G. E. (2001). Evaluation of an integrated biogeochemical model (PnET-BGC) at a northern hardwood forest ecosystem. *Water Resources Research, 37,* 1057–1070.

Gimeno, B. S., Yuan, F., Fenn, M. E., & Meixner, T. (2009). Management options for mitigating nitrogen (N) losses from N saturated mixed conifer forests in California. In A. Bytnerowicz, M. J. Arbaugh, A. R. Riebau, & C. Andersen (Eds.), *Wildland fires and air pollution. Developments in environmental science* (Vol. 8, pp. 425–455). Amsterdam: Elsevier.

Gundersen, P., Schmidt, I. K., & Raulund-Rasmussen, K. (2006). Leaching of nitrate from temperate forests—effects of air pollution and forest management. *Environmental Reviews, 14,* 1–57.

Hartman, M. D., Baron, J. S., & Ojima, D. S. (2007). Application of a coupled ecosystem-chemical equilibrium model, DayCent-Chem, to stream and soil chemistry in a Rocky Mountain watershed. *Ecological Modelling, 200,* 493–510.

Hartman, M. D., Baron, J. S., Clow, D. W., Creed, I. F., Driscoll, C. T., Ewing, H. A., Haines, B. D., Knoepp, J., Lajtha, K., Ojima, D. S., Parton, W. J., Renfro, J., Robinson, R. B., van Miegroet, H., Weathers, K. C., & Williams, M. W. (2009). *DayCent-Chem simulations of ecological and biogeochemical processes of eight mountain ecosystems in the United States.* (Report 2009–5150). U.S. Geological Survey Scientific Investigations.

Hettelingh, J. P., Posch, M., & Slootweg, J. (2008). *Critical load, dynamic modelling and impact assessment in Europe. CCE status report 2008.* (RIVM Report 500090003). Bilthoven, The Netherlands: Coordination Centre for Effects, National Institute for Public Health and the Environment.

Interlandi, S. J., & Kilham, S. S. (2001). Limiting resources and the regulation of diversity in phytoplankton communities. *Ecology, 82,* 1270–1282.

Ito, M., Mitchell, M. J., & Driscoll, C. T. (2002). Spatial patterns of precipitation quantity and chemistry and air temperature in the Adirondack region of New York. *Atmospheric Environment, 36,* 1051–1062.

Kretser, W. J., Gallagher, J., & Nicolette, J. (1989). *Adirondack lake survey, 1984–1987: An evaluation of fish communities and water chemistry.* Ray Brook: Adirondack Lakes Survey Corporation.

Pardo, L. H., Fenn, M. E., Goodale, C. L., Geiser, L. H., Driscoll, C. T., Allen, E. B., Baron, J. S., Bobbink, R., Bowman, W. D., Clark, C. M., Emmett, B., Gilliam, F. S., Greaver, T. L., Hall, S. J., Lilleskov, E. A., Liu, L., Lynch, J. A., Nadelhoffer, K. J., Perakis, S. S., Robin-Abbott, M. J., Stoddard, J. L., Weather, K. C., & Dennis, R. L. (2011). Effects of nitrogen deposition and empirical nitrogen critical loads for ecoregions of the United States. *Ecological Application, 21,* 3049–3082.

Parton, W. J., Scurlock, J. M. O., Ojima, D. S., Gilmanov, T. G., Scholes, R. J., Schimel, D. S., Kirchner, T., Menaut, J. C., Seastedt, T., Moya, E. G., Kamnalrut, A., & Kinyamario, J. I. (1993). Observations and modeling of biomass and soil organic-matter dynamics for the grassland biome worldwide. *Global Biogeochemical Cycles, 7,* 785–809.

Parton, W. J., Hartman, M., Ojima, D., & Schimel, D. (1998). DAYCENT and its land surface submodel: Description and testing. *Global Planet Change, 19,* 35–48.

Parton, W. J., Holland, E. A., Del Grosso, S. J., Hartman, M. D., Martin, R. E., Mosier, A. R., Ojima, D. S., & Schimel, D. S. (2001). Generalized model for NO_x and N_2O emissions from soils. *Journal of Geophysical Research: Atmospheres, 106,* 17403–17419.

Rao, L. E., & Allen, E. B. (2010). Combined effects of precipitation and nitrogen deposition on native and invasive winter annual production in California deserts. *Oecologia, 62,* 1035–1046.

Rao, L. E., Allen, E. B., & Meixner, T. (2010). Risk-based determination of critical nitrogen deposition loads for fire spread in southern California deserts. *Ecological Application, 20,* 1320–1335.

Rapp, L., & Bishop, K. (2009). Surface water acidification and critical loads: Exploring the F-factor. *Hydrology & Earth System Sciences, 13,* 2191–2201.

Riggan, P. J., Lockwood, R. N., & Lopez, E. N. (1985). Deposition and processing of airborne nitrogen pollutants in Mediterranean-type ecosystems of southern California. *Environmental Science & Technology, 19,* 781–789.

Saros, J. E., Clow, D. W., Blett, T., & Wolfe, A. P. (2011). Critical nitrogen deposition loads in high-elevation lakes of the western US inferred from paleolimnological records. *Water, Air & Soil Pollution, 216,* 193–202.

Skiba, U., Pitcairn, C., Sheppard, L., Kennedy, V., & Fowler, D. (2004). The influence of atmospheric N deposition on nitrous oxide and nitric oxide fluxes and soil ammonium and nitrate concentrations. *Water, Air & Soil Pollution: Focus, 4,* 37–43.

Smil, V. (2001). *Fritz Haber, Carl Bosch and the transformation of world food production.* Cambridge: The MIT Press.

Steers, R. J., & Allen, E. B. (2010). Post-fire control of invasive plants promotes native recovery in a burned desert shrubland. *Restoration Ecology, 18,* 334–343.

Sullivan, T. J., Cosby, B. J., Tonnessen, K. A., & Clow, D. W. (2005). Surface water acidification responses and critical loads of sulfur and nitrogen deposition in Loch Vale watershed, Colorado. *Water Resources Research, 41,* W01021.

Sverdrup, H., McDonnell, T. C., Sullivan, T. J., Nihlgård, B., Belyazid, S., Rihm, B., Porter, E., Bowman, W. D., & Geiser, L. (2012). Testing the feasibility of using the ForSAFE-VEG model to map the critical load of nitrogen to protect plant biodiversity in the Rocky Mountains region, USA. *Water, Air & Soil Pollution, 223,* 371–387.

Tonnesen, G., Wang, Z., Omary, M., & Chien, C. J. (2007). *Assessment of Nitrogen Deposition: Modeling and Habitat Assessment.* (CEC-500–2005-032. http://www.energy.ca.gov/2006publications/CEC-500-2006-032/CEC-500-2006-032.PDF). California Energy Commission, PIER Energy-Related Environmental Research.

USEPA. (2009). *Risk and Exposure Assessment for Review of the Secondary National Ambient Air Quality Standards for Oxides of Nitrogen and Oxides of Sulfur.* EPA-452/P-09–004a.

Walvoord, M. A., Phillips, F. M., Stonestrom, D. A., Evans, R. D., Hartsough, P. C., Newman, B. D., & Striegl, R. G. (2003). A reservoir of nitrate beneath desert soils. *Science, 302,* 1021–1024.

Watmough, S. A., Aherne, J., & Dillon, P. (2005). Effect of declining lake base cation concentration on freshwater critical load calculations. *Environmental Science & Technology, 39,* 3255–3260.

Williams, M. W., & Tonnessen, K. A. (2000). Critical loads for inorganic nitrogen deposition in the Colorado Front Range, USA. *Ecological Application, 10,* 1648–1665.

Wolfe, A. P., van Gorp, A. C., & Baron, J. S. (2003). Recent ecological and biogeochemical changes in alpine lakes of Rocky Mountain National Park (Colorado, USA): A response to anthropogenic nitrogen deposition. *Geobiology, 1,* 153–168.

Zhai, J., Driscoll, C. T., Sullivan, T. J., & Cosby, B. J. (2008). Regional application of the PnET-BGC model to assess historical acidification of Adirondack lakes. *Water Resources Research, 44,* W01421.

Chapter 11
Field Survey Based Models for Exploring Nitrogen and Acidity Effects on Plant Species Diversity and Assessing Long-Term Critical Loads

Ed C. Rowe, G. W. Wieger Wamelink, Simon M. Smart, Adam Butler, Peter A. Henrys, Han F. van Dobben, Gert Jan Reinds, Chris D. Evans, Johannes Kros and Wim de Vries

11.1 Introduction

Policies aimed at decreasing the environmental damage caused by deposition of atmospheric pollutants have made extensive use of the concept of critical load, defined as a "quantitative estimate of an exposure to one or more pollutants below which significant harmful effects on specified sensitive elements of the environment do not occur according to present knowledge" (Nilsson and Grennfelt 1988). Acid pollution is mainly caused by excessive amounts of sulphur (S) and nitrogen (N) deposition. Nitrogen can also affect ecosystems through eutrophication, so N deposition is assessed in relation to both the acidity critical load and the critical load for nutrient nitrogen ($CL_{nut}N$). Critical loads have mainly been established for different habitats using empirical data, as described in Chaps. 4 and 5. However, effects of N deposition can be cumulative and persistent, with a build-up of plant-available N in soil over decades or even centuries. Evidence from long-term experiments (>4 years) is preferred for setting the $CL_{nut}N$, but for changes over longer timescales it may be informative to use dynamic models to explore likely outcomes. In this chapter we describe two model chains, based on similar principles, which have been applied to the task of predicting changes in environmental conditions resulting from nutrient-N and acidity deposition, and resultant changes in floristic

E. C. Rowe (✉) · C. D. Evans
Centre for Ecology and Hydrology, Environment Centre Wales, Bangor, UK
e-mail: ecro@ceh.ac.uk

G. W. W. Wamelink · H. F. van Dobben · G. J. Reinds · J. Kros · W. de Vries
Alterra Wageningen University and Research Centre, Wageningen, The Netherlands

S. M. Smart · P. A. Henrys
Centre for Ecology and Hydrology, Lancaster Environment Centre, Lancaster, UK

A. Butler
Biomathematics & Statistics Scotland, JCMB, The King's Buildings, Edinburgh, Scotland, UK

W. de Vries et al. (eds.), *Critical Loads and Dynamic Risk Assessments*,
Environmental Pollution 25, DOI 10.1007/978-94-017-9508-1_11

composition. These model chains have been used to explore scenarios of change in pollutant load and other factors, and to define critical loads.

Nitrogen and acidity deposition disturb ecosystems, in that they are external influences that move the system away from its present state. Ecosystems that can return to their former state after a disturbance ceases are said to be resilient (see e.g. Gunderson 2000). However, if the disturbance is large or prolonged, an ecosystem may move towards a different state, from which it will not be able to return without a new external influence (Ludwig et al. 1997). Resilience in response to N deposition could result from e.g. incorporation of N in recalcitrant soil organic material, or loss of N from the ecosystem through leaching or denitrification. In ecosystems where plant production is limited by N availability, deposition of N tends to increase productivity and litterfall. Species that are taller and more competitive are able to exploit these conditions better than small-growing and stress-tolerant species, and therefore tend to become increasingly dominant in habitats affected by N deposition. This causes a loss of species-richness, and the shorter-growing species that are lost are often of greater nature conservation value. The main underlying mechanism is a decrease in ground-level light availability beneath the taller canopy and fallen litter (Hautier et al. 2009). In temperate habitats, soil pH is often positively correlated with species-richness, and the inference is often made that acidification due to N deposition also reduces species-richness.

Several soil-vegetation models are used at present in Europe to predict plant species composition as a function of atmospheric deposition of N and acidity. These models consist of a chain from dynamic models of biogeochemical change in soil and vegetation, to floristic models of environmental suitability for plant species or assemblages. The soil-vegetation models applied in these chains predict changes in environmental conditions resulting from changes in N and S deposition and other drivers such as climate.

In this chapter, we describe the use of field survey-based species models driven by dynamic soil models to: (i) predict plant species composition or diversity as a function of atmospheric N deposition; and (ii) calculate critical N loads in relation to an unacceptable plant species diversity change. The environmental suitability models that are used define a realised niche for the species or assemblage. Examples considered in this chapter are based on empirical relationships between species occurrence (MOVE, PROPS, MultiMOVE) or values for a plant diversity indicator based on Red List criteria (NTM3, Wamelink et al. 2003b) and multiple axes that define climate, soil and light conditions, derived from large datasets of field relevés. More specifically, this chapter describes two parallel model chains, similar in concept, that are being applied to explore pollution scenarios and calculate critical loads for acidity and nutrient-N. The model chains applied in the Netherlands and the UK were described in De Vries et al. (2010), respectively, as SMART2-(SUMO2)-MOVE/NTM and MAGIC-GBMOVE. Since that review, some of the component models have been updated or replaced with alternatives, and the model chains are now better described as SMART2-(SUMO2)-PROPS/NTM3 and MADOC-MultiMOVE. In this chapter these updates to component models are described, and applications of these model chains are demonstrated.

The set-up of this chapter is as follows: In Sect. 11.2 we outline the modelling approach and describe its implementation in two currently-used model chains. Examples of the application of these model chains, illustrating the model validation status and the use of the models in critical load assessments, are presented in Sect. 11.3. In Sect. 11.4 we describe recent developments of the models used, and the development of a new database of field survey data from across Europe. Conclusions are drawn in Sect. 11.5.

11.2 Methods

11.2.1 Model Chains: Linking Biogeochemistry to Niche Models

As outlined in the introduction, the model chains explored in this chapter consist of dynamic models of biogeochemical processes in soil and vegetation, which are used to drive static or regression models of environmental suitability for species or species-assemblages (Fig. 11.1). The model chains currently used are not dynamically integrated, in that they do not allow for feedbacks from species composition to biogeochemical processes.

The model chains considered are SMART2(-SUMO2)-MOVE/NTM3 and MADOC-MultiMOVE, which were developed in the Netherlands (NL) and the United Kingdom (UK), respectively. These model chains are comparable in that they consist of:

- a soil model, SMART2 (NL) or MADOC (UK) that simulates the cycling of nutrients in the soil and predicts soil acidity and N availability, and
- a set of multiple regression equations relating species presence to abiotic factors that define realized niches for individual species of vascular plants (MOVE, NL) or vascular plants, bryophytes and terrestrial lichens (MultiMOVE, UK). The NTM3 model (NL) directly relates a nature conservation value indicator to abiotic factors.

Firstly, the biogeochemical models simulate the effects of N and S pollution on soil conditions expressed in terms of soil N availability (e.g. soil C/N ratio, extractable N, mineralisable N) and soil acidity (e.g. pH or base saturation). Next, transfer functions are used (in most configurations of the model chains) to convert soil conditions into estimates of mean scores for floristic trait-means that can be used as proxy variables for environmental conditions. Lastly, multidimensional regression relationships between the prevalence of plant species and the set of floristic trait-means and other environmental axes are used to determine likely suitability for the species or habitat-class.

The intermediate step, using transfer functions to generate floristic trait-means, has advantages and disadvantages. The occurrence of species can be seen as in-

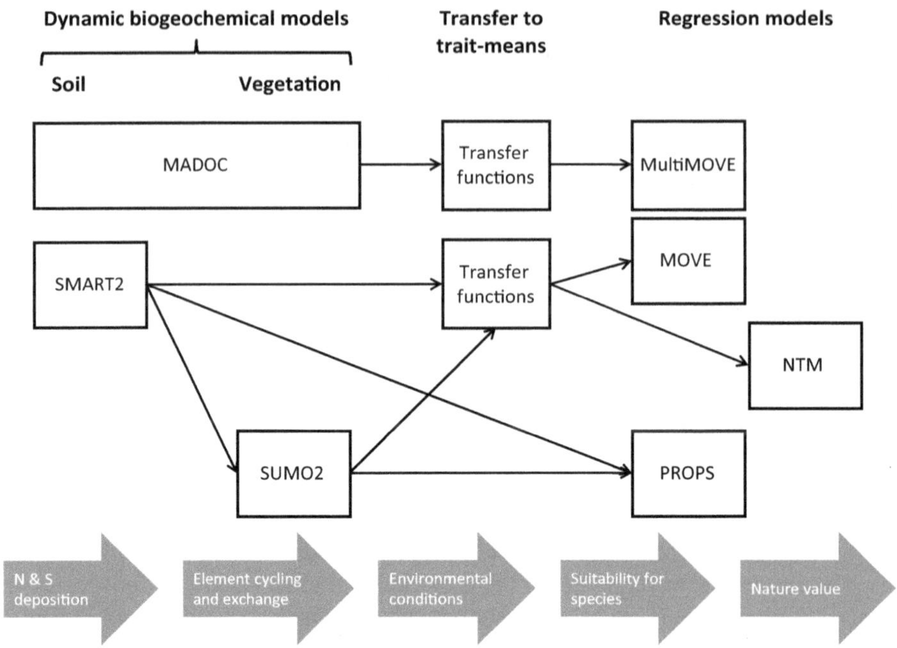

Fig. 11.1 Biogeochemical-niche model chains for exploring acidity and nitrogen effects on species and habitats: concepts and links. Outputs from dynamic models of acidity and N cycling are used, together with indicators of fixed site conditions, to solve regression models for environmental suitability for species or species-assemblages. Some chains use transfer functions to convert abiotic variables simulated by the biogeochemical models into indicators of environmental conditions that are based on species trait-means, such as mean 'Ellenberg N' score

tegrating environmental conditions over time (Diekmann 2003), and is less susceptible to measurement error than direct biophysical measurements, in particular measurements of soil nutrient status. Indicator-value means can be calculated from most existing floristic datasets, whereas comparatively few relevés have synlocated biophysical measurements. Using trait-means therefore allows niche models to be established on the basis of much larger training datasets. On the other hand, the use of an intermediate predictive step has the potential to introduce bias and increase uncertainty (Wamelink and van Dobben 2003; Wamelink et al. 2005).

11.2.2 Dynamic Biogeochemical Models Relating N & S Deposition to Soil Conditions

Effects of changes in N and S deposition on soil variables such as pH, C/N ratio or N availability are simulated by SMART2 (Kros 2002) or MADOC (Rowe et al. 2014a). The MADOC model is a dynamic integration of the VSD ion exchange model (Posch and Reinds 2009), the N14C model of plant production and soil organic matter (Tipping et al. 2012) and a simplified version of the DyDOC model of dissolved organic carbon (Michalzik et al. 2003). Both SMART2 and MADOC include the major hydrological and biogeochemical processes in the soil compartment to calculate the long-term effects of atmospheric deposition of NO_x, NH_y, SO_x and base cations (BC^{2+}) on soil solution chemistry and bulk soil properties. They consist of a set of mass-balance equations, describing the soil input-output relationships, and a set of equations describing the rate-limited and equilibrium soil processes. As well as predicting pH, the models predict changes in aluminium, base cation, ammonium, nitrate and sulphate concentrations in the soil solution and solid phase. In both SMART2 and MADOC, litterfall, mineralization, root uptake and immobilization are modelled explicitly. SMART2 has an internal simplified growth module which can be used as an alternative to SUMO2. The stand-alone version of SMART2 has been used for the calculation of critical loads and target loads.

11.2.3 Vegetation Biogeochemistry and Succession

The SMART2 and MADOC models provide dynamic quantitative descriptions of soil conditions, for example pH and plant-available N, which can then be used to drive species models. An important aspect of habitat suitability for species is, however, provided by above-ground vegetation structure, which determines light availability at different heights. The SMART2 and MADOC models include simple routines to simulate plant biomass, which can be used to provide an approximate measure of canopy height and light conditions (Rowe et al. 2011b). The SUMO2 model simulates vegetation structure in more detail; it is a process-based model that simulates biomass growth under given soil, climate and management conditions (Wamelink et al. 2009a, c). The basis of the model is a maximum growth, determined by temperature and CO_2 concentration in the air, which is then reduced using a series of linear and non-linear reduction factors. These reduction factors depend on the availability of water, N, phosphorus (P), calcium (Ca), magnesium (Mg) and potassium (K). The SUMO2 model distinguishes five functional plant types (climax trees, pioneer trees, shrubs, dwarf shrubs and herbs) that compete for light and nutrients. Their competitive balance is based on traits, i.e. canopy height and biomass of roots and leaves per functional type. Management is simulated as biomass removal by mowing, grazing, cutting, fire or sod cutting. The proportions of total biomass in the five functional plant types determine the successional stage (e.g. pioneer, grassland, heathland, forest). SUMO2 can be coupled to niche models e.g. NTM3 or MOVE through vegetation structure and soil chemical conditions (pH and nutrient availability) simulated by SMART2. To simulate soil biogeochemistry, litterfall and litter quality (N, P, K, Ca and Mg content) are crucial input terms, so SMART2 and SUMO2 are dynamically linked. The vegetation type generated by SUMO2 is also used as an input by the SMART2 model.

11.2.4 Niche Models

The MOVE (Latour and Reiling 1993; Van Adrichem et al. 2010) and MultiMOVE (Butler 2010) models describe niches for individual species in relation to the environment at a given site. The NTM model instead describes niches for plant community types; this discussion mainly refers to species' niches, but similar considerations apply for community types. Different variables are used to represent the environment in terms of climate, soil and vegetation structure, as described below. The models are fitted by relating the occurrence of a species to different explanatory axes using multiple logistic regression or similar statistical techniques. The main environmental axes that are relevant to effects of N pollution are: light availability, since more productivity and litterfall increase ground-level shading, which is a major cause of species loss (Hautier et al. 2009); soil pH (also affected by acidity deposition); and nutrient availability. Different models have used different direct or proxy measurements to represent these axes.

The models represent the realised niche, i.e. they account for competitive exclusion as expressed in the training data. The models are empirical, in that they characterise the conditions where a species has been observed. However, to deduce that they will work in the opposite direction, i.e. that under a given favourable combination of environmental conditions the species will occur, is to neglect biological delays (see Posch et al. 2003) for an explanation of chemical and biological recovery delays). A species may not colonise a site even when conditions are favourable, if it is not present locally and/or is limited in its ability to disperse and establish. When driven using biogeochemical models, delays to chemical recovery can be simulated, but the assumption of no delays in biological response may be problematic. For this reason, the outputs from niche models are generally referred to as 'suitability' for the species rather than predicted prevalence. Issues with equilibrium or instantaneous modelling species responses are discussed further in Sect. 11.4.4. A related issue is that species occurrence data may not reflect the full range of suitable habitats for a species due to delays in establishment (Svenning and Skov 2004).

The models MOVE and NTM3 (Schouwenberg et al. 2001; Wamelink et al. 2003b) are based on response curves, in which the probability of plant species (MOVE) or nature conservation value or plant community occurrence (NTM3) is determined by vegetation structure and the abiotic site conditions: groundwater depth, soil pH and N availability. The models were fitted using 160,000 vegetation relevés (Schaminee et al. 2012) that were labelled in terms of vegetation type by an automated procedure (van Tongeren 2000). The probability of occurrence in MOVE is derived using second-order logistic regression based on presence/absence data, representing species occurrence along environmental gradients (Van Adrichem et al. 2010). In NTM3, niche models were fitted separately for four vegetation classes. In MOVE, vegetation class and geographic region were included as explanatory variables. The abiotic factors that govern habitat suitability are: groundwater table depth, soil pH, Cl and N availability, including all interactions and geographical region and vegetation type, if significant. These are derived from floristic data using transfer functions, as outlined in the following section. Since MOVE and NTM3 focus on more than one abiotic factor, the curves are multi-dimensional. A total of 914 species have MOVE niche models.

The MultiMOVE model (Butler 2010) and its earlier version GBMOVE (Smart et al. 2010b) are also based on statistical relationships between occurrence and environmental factors, established using large relevé datasets, in this case from the UK. The data used to derive the niche models were assembled from a variety of sources (e.g. Roy et al. 2000) and consisted of 32,263 vegetation relevés. The environmental factors considered were: % soil moisture, soil pH, soil nutrient availability and vegetation height (which are derived from trait-means as described below), and three long-term annual average climate variables: annual precipitation, July maximum temperature and January minimum temperature. Interaction terms are also included, to quantify the extent to which a species' response on one gradient is conditioned by another gradient (c.f. Pakeman et al. 2008). The MultiMOVE model is a small ensemble based on three separate modelling techniques rather than just one, which can decrease the likelihood of bias (Araujo and New 2007). For one of the ensemble members (that derived using a Generalised Linear Model) a parsimonious model is determined for each species from the full set of explanatory terms and interactions by backward selection using Akaike's Information Criterion. The same set of explanatory variables is also used in the second ensemble member (that derived using a Generalised Additive Model; Hastie and Tibshirani 1986). The third ensemble (that derived using Multivariate Adaptive Regression Splines; Friedman 1991) undertakes model selection automatically as part of the algorithm for parameter estimation. A model average is calculated from the output of each of the three techniques applied to each species, without applying any weighting. Using multiple techniques generates a more robust projected distribution of habitat suitability scores, because it takes advantage of the benefits of each technique while moderating the influence of their individual shortcomings, such as a tendency for non-parametric models to overfit to the training data (Randin et al. 2006; Smart et al. 2010a). MultiMOVE niche models have been derived for 1342 UK plant and lichen species.

11.2.5 Transfer Functions From Soil Conditions to Trait-Means

The niche models described above were trained using prevalence data from large floristic datasets, many of which lack associated environmental data, e.g. from soil samples taken from the relevé location. For this reason, these models make use of indicator or trait scores that have been allocated to many European species (Ellenberg et al. 1992; Hill et al. 2000). These indicator values describe the typical or optimum location of plant species along an environmental axis, and have been derived for about 2720 central-European vascular plants (Ellenberg et al. 1992). Since species' environmental optima can vary across their geographical range, modified indicator values have been developed for some regions (e.g. Ertsen et al. 1998;

Hill et al. 2000). The environmental axes for which indicator values have been defined are annual mean temperature (E_T), annual temperature range or continentality (E_K), moisture (E_W), soil pH (E_R), nutrient availability (E_N), light availability (E_L) and salinity (E_S). The typical height for species is also represented in trait databases, for example using the scoring system developed by Grime et al. (1988).

The mean trait value for species that are present in a relevé is an indicator of environmental conditions at the site. The MOVE and MultiMOVE models both make use of mean E_N to represent nutrient availability and mean E_R to represent soil pH. The MultiMOVE model also makes use of the mean E_W score to represent soil moisture conditions. Since vegetation structure and height are often not reliably and objectively recorded alongside relevé data, canopy height is also represented in MultiMOVE using a mean trait score, in this case mean typical height (Grime et al. 1988). To ensure independence of explanatory and response variables, the MultiMOVE model for each species was fitted to mean trait scores calculated without including that species. Currently MultiMOVE does not include salinity, and consequently species with coastal distributions are not modelled, although work is in progress to include E_S to represent saline influence as an additional explanatory environmental axis (Jarvis et al. in prep.).

To make use of trait means within predictive model chains, it is necessary to establish relationships between trait-means and biophysical measurements. This gives an empirical basis to the use of trait-means, and also allows the outputs from biogeochemical models to be converted into trait-means for use in driving niche models. This translation requires a training dataset where soil conditions have been measured alongside floristic recording.

The relationship between soil pH and E_R is relatively uncontroversial, and simple transfer functions between E_R and soil measurements have been established, e.g.

$$pH = 3.1 + 0.53 E_R \tag{11.1}$$

in MOVE (van Dobben et al. 2006), or

$$E_R = 0.529 - 0.025 MC + 1.67\, pH - 0.106\, pH^2 - 0.00566 C \tag{11.2}$$

in MultiMOVE (Smart et al. 2010b), where MC is soil moisture (g water 100 g^{-1} fresh soil), pH is soil pH as measured in a slurry of 10 g fresh soil with 25 g water, and C is soil carbon concentration (g organic C 100 g^{-1} dry soil). For soil moisture conditions, and for vegetation height, relatively simple transfer functions are also used in MultiMOVE (Rowe et al. 2011b; Smart et al. 2010b):

$$E_F = \frac{1}{0.55}\left(In \frac{MC}{100 - MC} + 3.27 \right) \tag{11.3}$$

and

$$mean\ Grime\ height\ score = \max\{1, 1.17\ In(Height) - 1.22\} \tag{11.4}$$

where *MC* is as defined above and *Height* is the measured vegetation height in cm. The transfer function between height and Grime height score was obtained by regression of median values for the Grime height classes against their scores, whereas the other transfer functions used in these model chains were established by fitting statistical models to empirical data.

Transfer functions from soil measurements to E_N have proved more problematic, due to difficulties with defining measurements that are relevant to soil fertility. The E_N axis is often termed the nitrogen axis; and Ellenberg was not clear about whether the intended meaning was nitrogen availability or nutrient availability more generally. Subsequent studies have associated this axis with plant productivity (Hill and Carey 1997; Schaffers and Sykora 2000), which is likely to be related to N availability, at least where N is limiting or co-limiting. The prevalence of N limitation in terrestrial ecosystems is debated (e.g. Vitousek et al. 2010), but the assumption is usually made, in the model chains reviewed here, that increasing N availability increases plant productivity. The challenge has been to find soil measurements that consistently indicate N availability (Wamelink et al. 2012b). Initially, for GB-MOVE, relationships were based on total soil C and N concentration and/or ratio, e.g. (Smart et al. 2010b):

$$E_N = \exp(0.7751 - 0.00006MC^2 - 0.000099MC^2 - 0.1475C + 0.2639pH - 0.01684pH^2 + 0.1908N) \quad (11.5)$$

where N is soil total N concentration (g N 100 g^{-1} dry soil). However, a more recent study (Rowe et al. 2011a) suggests that, while soil C concentration and/or moisture content represent a primary component of variation in fertility (as represented by E_N) from relatively infertile, moist, organic soils to relatively fertile, dry, mineral soils, there is a component of this variation which is better captured by measuring plant-available N (Rowe et al. 2011a):

$$E_N = 7 + 0.77pH - 4.84\log_{10} MC + 5.66N_{mbl}^{0.333} \quad (11.6)$$

where N_{mbl} is mineralisable N (mg g^{-1} dry soil), measured as total mineral N after flushing with 1 mg N l^{-1} and 4-week incubation.

The MOVE model also uses a transfer function to estimate mean E_N based on N availability (Ertsen et al. 1998):

$$N_{av} = 0.64E_N + c_v \quad (11.7)$$

Where N_{av} is the N availability as measured by mineralisation (kmol ha^{-1}yr^{-1}) as described in Ertsen et al. (1998) and the value of c_v is 5.01 for grass, 4.29 for heath, 5.92 for coniferous forests and 6.19 for deciduous forests.

11.2.6 Niche Models from Direct Soil Measurements

In the past 10 years, much effort has been put into the collection and collation of data that can be used to derive niche models directly in relation to environmental measurements (Wamelink et al. 2012b). This has allowed the development of the PROPS model as a successor to MOVE. It is based on the same principles, but avoids the uncertainty introduced in using a transfer function to translate abiotic measurements into mean trait-scores, which can be considerable (Schouwenberg et al. 2001; Wamelink et al. 2002, 2003a). An uncertainty analysis of the SMART2-SUMO2-MOVE4 model was described by Wamelink et al. (2011). Uncertainty was introduced in the most important model parameters, as well as in two important sets of inputs; a soil map and a groundwater map. We evaluated how the uncertainty propagates through the model chain, and identified the major sources of the uncertainty in the end results, i.e. the predicted number of plant species. The main source of uncertainty were three regression functions that translate the soil model results for groundwater table, soil pH and nitrogen availability the into Ellenberg indicator values E_F, E_R and E_N, respectively. Compared to these regression equations, other sources of uncertainty were almost negligible on average, although at some of the 1500 sites examined, the soil model parameters and the input maps also contributed significantly to the uncertainty.

The PROPS model is still under construction. The PROPS model, as shortly described by Reinds et al. (2012), is based on a dataset of over 40,000 vegetation relevés with measured abiotic soil parameters from across Europe, and is therefore an advance from MOVE4, which was only calibrated using relevés from the Netherlands and based on Ellenberg indicator values. Ideally, all relevés would be related to field measurements of abiotic conditions, but these data were only available for 6000 relevés (pH) or 1000 relevés (NO_3). Therefore, for the first version of the model the pH measurements from the field were used, but instead of measured NO_3 we used either modelled NO_3 concentrations or log-transformed N deposition values. Niche models for each species were fitted to data using a multiple logistic regression technique, including all interactions. To guarantee that the occurrence probability is a value between 0 and 1, the logit-function of the probability y is used and approximated (fitted) by a quadratic polynomial in the environmental variables $x_1, \ldots, x_n$ (n=number of environmental variables) according to:

$$z = \text{logit}(y) = \log\frac{y}{1-y} = a_0 + \sum_{i=1}^{n} a_i \cdot x_i + \sum_{i=1}^{n}\sum_{j\geq i}^{n} a_{i,j} \cdot x_i \cdot x_j \tag{11.8}$$

with constant, linear, interaction and quadratic regression coefficients a_0, a_i and $a_{i,j}$, ($i=1,\ldots,n; j=i,\ldots,n$), respectively. Some of the variables were log-transformed (e.g. N deposition) and all of them standardized by subtracting the mean and dividing by the standard deviation, i.e. $x=(x_{raw}-x_{mean})/x_{stdev}$, where x_{raw} is the original (log-transformed) variable. From Eq. 11.8 the probability y is obtained as:

$$y = \frac{1}{1+\exp(-z)} \tag{11.9}$$

where z is the quadratic polynomial.

Several combinations of environmental factors were tested to fit the species response curves, including measured pH, modelled NO_3 concentration, modelled N availability, (log) N deposition, precipitation and annual mean temperature. The predictions and interactions ultimately selected were measured soil pH, annual mean temperature and log N deposition. Annual temperature per relevé was derived from the CRU meteorological data set (Mitchell et al. 2004). Deposition data were obtained from the atmospheric transport model of EMEP/MSC-W (Tarrasón et al. 2007). Soil pH was measured at the sites. This PROPS model was used in a European wide application by (Reinds et al. 2012) to investigate impacts of climate change, N and S deposition on plant species diversity indices (see also Chap. 24).

The NTM3 (Natuur Technisch Model/Vegetation Evaluation Model; Wamelink et al. 2003b, see also Chap. 3) model is based on a similar idea. It is a regression model in which the biodiversity index (see following section) is described as a function of abiotic conditions: groundwater level, soil pH and N availability and vegetation structure type (grassland, heathland, deciduous forest, pine forest). The model uses spline functions to fit these relations, and includes all two- and three-way interactions (Wamelink et al. 2003b). The percentages of explained variance of the responses are of the order of 40–60 %.

11.2.7 From Niche Model Outputs to Nature Value and Critical Loads

In this section we describe methods that have been applied to aggregate outputs from species models into an index of nature value, or to obtain such a value directly. We also consider model-based methods for defining critical loads for species or assemblages.

To aggregate species predictions into a measure of nature value or habitat quality, it is necessary to select species that are relevant for the habitat. This already implies scoring or weighting the species (as included or not) and some authors have recommended further weighting schemes, based on inclusion in lists of positive and negative indicator species (Rowe et al. 2011c) or on the scarcity and/or decline of the species (van Dobben and Wamelink 2009).

In NTM3, values for a biodiversity index are calculated directly from the abiotic conditions, using a regression approach similar to that used to define the niche models. The regressions used are based on a calibration set of 160,000 relevés made in the Netherlands. The abiotic conditions of the relevés were calculated as the averages of E_F, E_R and E_N for species that were present. The biodiversity value of a relevé was based on the conservation value of the species as defined using Red List criteria (Mace and Stuart 1994), i.e. rareness of the species, the trend of the species (for 5 × 5 km grids) and the international importance of the species. A value was calculated for all species, including species not on the Red List. For example, invasive species, i.e. those that have shown recent increases in range size, get a negative value due to their positive trend. The biodiversity value of a relevé was calculated as the mean of the values of the species, but corrected for the number of species to prevent high values for relevés with many common species; see also Wamelink et al. (2003b).

The concept of critical load has been useful in pollution science and policy, and some efforts have been invested in using the biogeochemical-floristic model chains described here to define critical loads. There are theoretical reasons for thinking that model-based critical loads may be more reliable than critical loads defined using reviews of experimental and survey data. The delays in chemical response to and recovery from N deposition may be considerable, and effects may not be revealed in comparatively short-term experiments. Modelling critical loads can also be an alternative to laborious and costly field assessments.

Critical values for abiotic factors (e.g. N availability, nitrate concentration C/N or soil pH) can be determined for each vegetation type, either directly or from species-specific information, and subsequently used by inverting the coupled soil model to back-calculate the critical loads for N and acidity. In addition, effects of N deposition can be evaluated by models, based on the occurrence of plant species, mostly focused on target species and or Red List species. In the approach of Van Dobben et al. (2006), the 20- and 80-percentiles of these frequency distributions were used as critical limits, i.e. the range between these percentiles was considered as the optimal range for each vegetation type.

11.3 Results and Discussion

11.3.1 Model Testing

Testing of the model chains described in this chapter has mainly been confined to tests of individual (sub-)models. Biogeochemical models such as MADOC and its organic matter model N14C have been tested against biogeochemical observations of the mean age of soil organic carbon (using radiocarbon dating, Tipping et al. 2012) or surface water chemistry (Rowe et al. 2014b). Predictions of vegetation biomass using SMART2-SUMO2 have been tested against time-series measurements (Wamelink et al. 2009a), e.g. from unfertilised grassland sites at Ossekampen in the Netherlands (Elberse et al. 1983) and Park Grass in the UK (Jenkinson et al. 1994; Smart et al. 2005). These are long-term experimental sites, where one or more hay cuts are made each year and harvested biomass recorded.

At Ossekampen, the measured biomass responded to inter-annual variation in rainfall and temperature (Fig. 11.2a). Effects of yearly differences in temperature and rainfall were not included in the model run. Instead long-term averages were used, which explains the absence of variation in the model run. The high measured biomass in the first year is probably caused by the former agricultural use of the site. Both the measured and the simulated biomass show a decrease over the years, due to the yearly removal of nutrients in biomass. The results for Park Grass show that the harvested biomass is reasonably well simulated by SUMO2 (Fig. 11.2b).

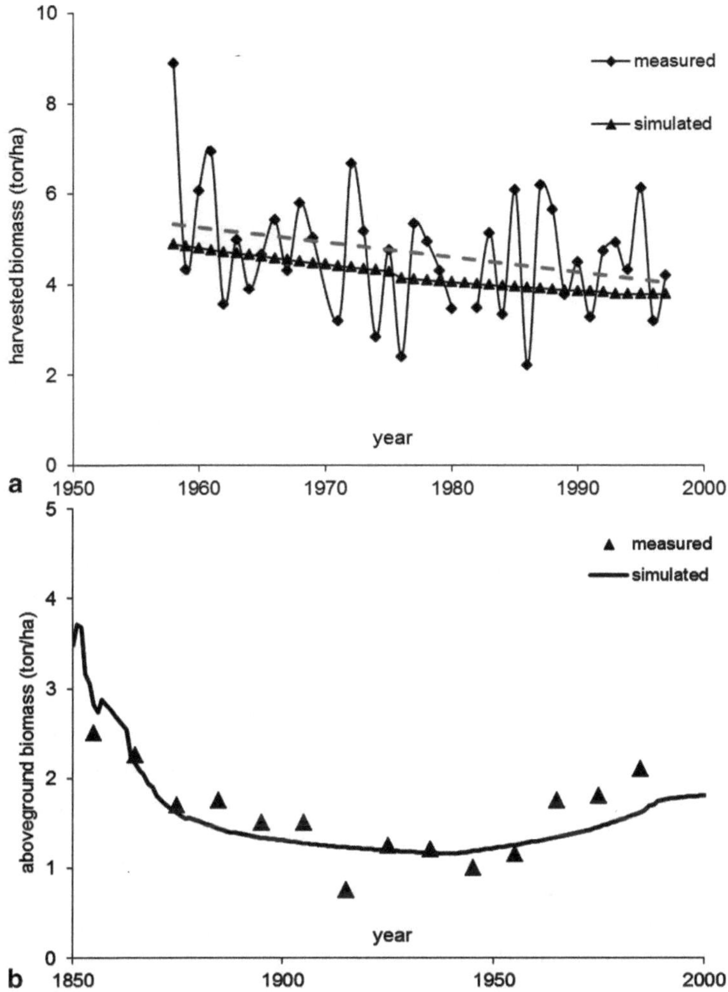

Fig. 11.2 Measured and simulated above-ground biomass by SMART2-SUMO2 at. **a** Ossenkamp and **b** Park Grass

Model results show a reduction in biomass harvest between 1850 and 1900, due to exhaustion of the soil. Then there is a stabilisation of the harvest when the effect of N deposition compensates for the nutrient exhaustion. In the last years there is an increase of the harvest due to the further increase in deposition since 1950. The effect of the N deposition since approximately 1960 is, however, underestimated, which could be due to the lack of temperature rise in the model. The SMART2-SUMO2 model was also validated on a heathland and a forest site in the Netherlands, and on a European scale for forests, including the effects of climate change for temperature, precipitation and carbon dioxide effects (Wamelink et al. 2009a).

The other major components of the model chains, the niche models describing environmental suitability for individual species, have mainly been assessed vi-

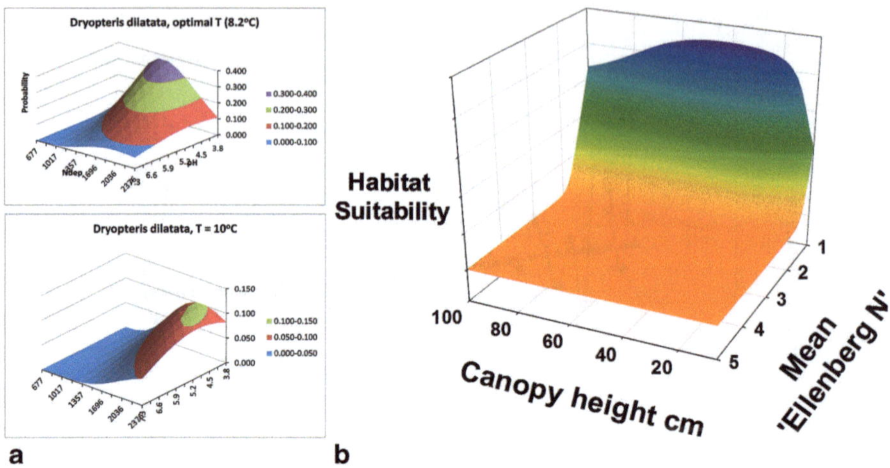

Fig. 11.3 Illustrations of multiple-dimension niche models in relation to selected environmental axes: **a** response simulated by PROPS for *Dryopteris dilatata* in relation to soil pH and N deposition rate (mol $ha^{-1}yr^{-1}$) for optimal (*top*) and sub-optimal (*bottom*) temperature; **b** response simulated by MultiMOVE for *Rhynchospora alba* in relation to canopy height and mean 'Ellenberg N' score, at mean values for other environmental variables observed for this species

sually. There is a close correspondence between these empirical models and the ecological requirements of species as described in standard floras (e.g. Fig. 11.3). GBMOVE niche models were also tested on independent survey data (Smart et al. 2010a). Performance was poorer when predictions were based on mean indicator values calculated from soil variables. This was partly attributable to soil moisture being derived from soil texture in the test dataset. Model tests using independent data tend to give a harsher assessment of model performance than tests based on subsets of the training data (Newbold et al. 2010).

The ability of GBMOVE to reproduce species compositional turnover in response to succession as a driver has also been tested. The target community was secondary woodland established at the Broadbalk Wilderness at Rothamsted. This woodland developed over the course of 20–40 years since abandonment of arable land (Harmer et al. 2001) and has been assigned in the UK National Vegetation Classification (Rodwell 1991–2000) to W8d, the *Hedera helix* sub-community of *Fraxinus excelsior-Acer campestre-Mercurialis perennis* woodland. The aim was to test whether GBMOVE, driven by soil biogeochemistry and a forced incremental increase in canopy height, could reproduce the target woodland community type starting from an unimproved grassland on the same starting soil type. Trajectories of change in species composition were modelled using soil biogeochemical outputs from MAGIC (Cosby et al. 1985) to drive changes in soil pH and C:N ratio from 2001 to 2050. Canopy height was incremented annually following a sigmoidal forest growth curve ending at an average canopy height of 15 m in 2050. The projection was considered successful, since the target community was reproduced with a reasonable degree of accuracy (Fig. 11.4). When the habitat suitability indices for the projected assemblage in 2050 (Fig. 11.4b) were matched against the National Vegetation Classification as a frequency table, a 62.3 % Jaccard similarity (Jaccard 1912) was found between the model community and W8d.

Tests of the full model chains, for example using changes in drivers such as acidity and N deposition to simulate changes in habitat suitability for species and comparing these against observed changes in species occurrence or abundance, have been limited. This is partly because there are remarkably few datasets available in which species have been recorded in sufficient detail and for a long enough period to detect the effects of drivers such as N deposition.

The MADOC and MultiMOVE components of the UK model chain have recently been finalised, and testing of the full chain from deposition scenarios to floristic responses is only now underway. However, a previous version of the UK model chain using MAGIC as the biogeochemical model driving GBMOVE, has been shown to reproduce observed species composition.

The observed data consisted of species lists per plot whereas outputs from GBMOVE provided the probability of each species being present in each plot. The GBMOVE outputs were compared against the observed data by using the estimated GBMOVE probabilities to simulate lists of species for each plot. These simulated lists were then compared against the actual list of observed species. The process, which was the same as that used to create so-called pseudo-quadrats from a frequency table as described in Tipping et al. (2013), was performed separately for each plot. The number of simulated plots was made equal to the number of plots in the observed dataset. This statistical model is fully described in Smart et al. (2005). The percentage of species correctly predicted was calculated as 100 × (number of species on both observed and simulated lists)/(number of species on observed list).

To assess the influence of uncertainty in the calibration equations relating soil properties to mean Ellenberg scores, predictions of species composition based on soil C/N and pH generated by MAGIC were compared with predictions based on observed mean Ellenberg scores. These comparisons were carried out for control plots at the long-term continuous Park Grass hay experiment at Rothamsted (unimproved neutral grassland) and the Hard Hills grazing and burning experiment at Moorhouse National Nature Reserve (ombrogenous bog), described in Smart et al. (2005). The results for Rothamsted indicated that when mean Ellenberg scores based on observed species composition were used as input to GBMOVE, on average 67 % of species observed were actually predicted. However, when predictions were based on MAGIC simulations of soil C/N and pH as input to GBMOVE, the percentage of all species correctly predicted decreased substantially (Fig. 11.5a). The main reason for the poor performance at Rothamsted appears to be that observed soil changes were inconsistent with observed vegetation changes. This is most likely due to sampling practices in the experimental plots, avoiding a thin mat of persistent litter that had developed in the O horizon over the course of the experiment. This resulted in C/N measurements indicating a higher fertility in the rooting zone than that encountered by at least some of the more shallow rooting species present. Also at Moorhouse, both the predictions from MAGIC linked to GBMOVE and the predictions based solely on observed mean Ellenberg values did not compare well

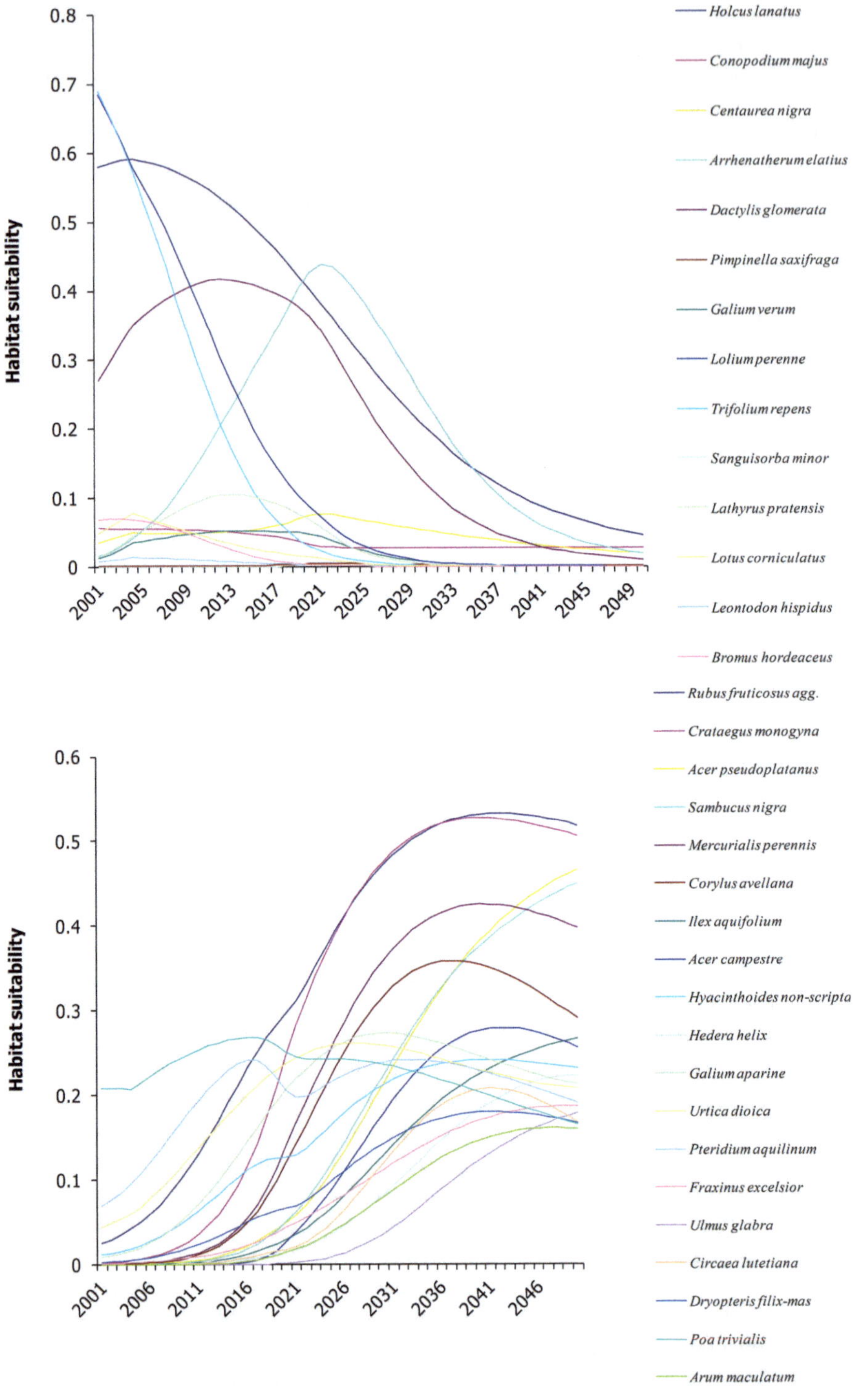

Fig. 11.4 Projected change in habitat suitability at Rothamsted, UK, based on modelling soil change between 2001 and 2050 coupled with incremental increases in canopy height using MAGIC

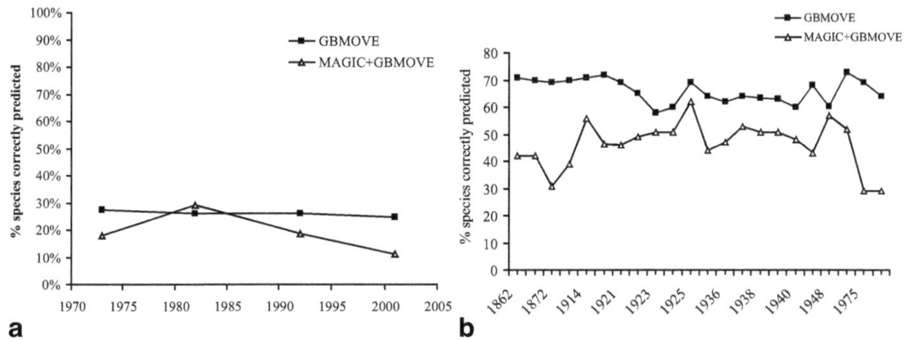

Fig. 11.5 Percentage of species correctly predicted (see text) in **a** the three Park Grass control plots; and **b** Moor House, by GBMOVE (*squares*) using observed mean Ellenberg scores only, and by the MAGIC-GBMOVE model chain (*triangles*)

with the species actually observed in the control plots, although the key dominants in the vegetation were predicted to be present by both models (Fig. 11.5b). When predicted species lists for both GBMOVE and MAGIC+GBMOVE were examined, key absences included a range of bryophytes. Possibly, bryophytes are more responsive to direct deposition effects, and less so to changes in soil chemistry, making the approach less useful for lower plants.

At Moorhouse, observed versus predicted species changes over time were summarised as slope coefficients for each species in a linear regression line relating (i) observed abundance over the years and (ii) predicted habitat suitability from MAGIC+GBMOVE across the same time period. Despite considerable scatter there was a positive correlation between observed change in species frequency and predicted change in habitat suitability (Fig. 11.6). A chi-square test of observed versus predicted directions of change was significant ($p=0.016$). While the correlation between observed and predicted slopes was also significant, predicted rates of change covered a narrower range than observed species changes. Such behaviour often arises when looking at the output from mechanistic environmental models, and is likely to reflect the fact that:

- observational data contain recording error, which means that observational data will tend to be more variable than the (unknown) true values; and
- models generally omit some components of genuine variation, which means that model outputs will tend to be less variable than the true values.

Statistical modelling of the observational data, using linear mixed models, could be used to try and disentangle these effects by quantifying the relative importance of recording error and different components of variability, but such an analysis has not yet been attempted.

and GBMOVE. **a** species present in neutral grassland control plots at the start of the projection, **b** species occurring within the local 10 km^2. (Preston et al. 2002) that were projected to increase in habitat suitability as canopy height increased

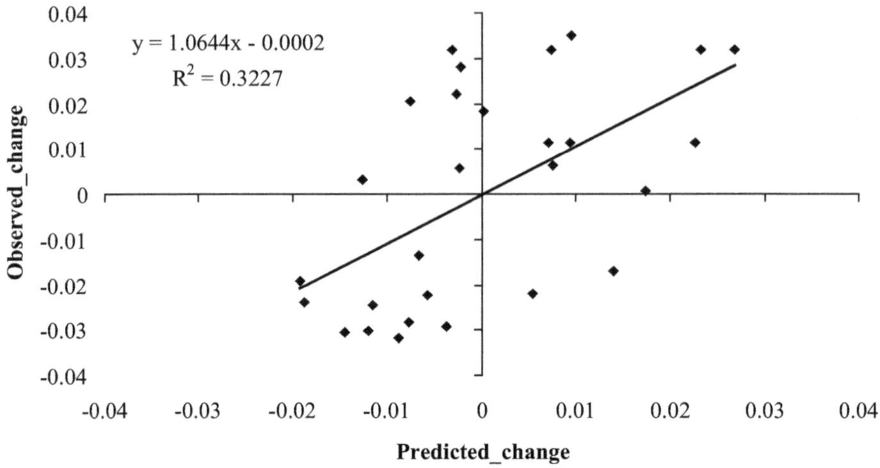

Fig. 11.6 Predicted versus observed change in individual species in the Moorhouse Hard Hills control plots. Predicted change is the slope coefficient of a linear regression on occurrence probabilities predicted by MAGIC+GBMOVE for each year between 1973 and 2001. Observed change is the slope coefficient of a linear regression on % frequency in sample plots in each survey year (Pearson correlation coefficient = 0.568, p = 0.002)

Attempts have also been made to validate MOVE4 (Van Adrichem et al. 2010; Wamelink et al. 2012a). This was done by dividing field data into classes with the same abiotic characteristics and then comparing the field data per class with model predictions. Per class we tested if the number of species that had predicted chances outside the uncertainty range of the field data was higher than could be expected based on coincidence. For most of the tested classes this was the case, indicating that the predictions were not very good, although predictions for common species and for species absences were better. The validation also showed that many data are needed to be able to validate MOVE4 properly.

11.3.2 Model Applications

The model chains described in this chapter have a wide range of potential applications, since they predict likely effects of interacting environmental drivers (including but not restricted to pollutant load) on habitat suitability for individual species. As well as being directly useful in biodiversity assessment, predicted effects on species can be related to species-specific ecosystem services. The biogeochemical submodels used in the chains also predict intermediate variables that are relevant for policy development and scenario analysis. Three example applications are described here: determining long-term critical loads for individual species and habitats; spatial mapping of biodiversity value; and exploring trade-offs between biodiversity and carbon sequestration.

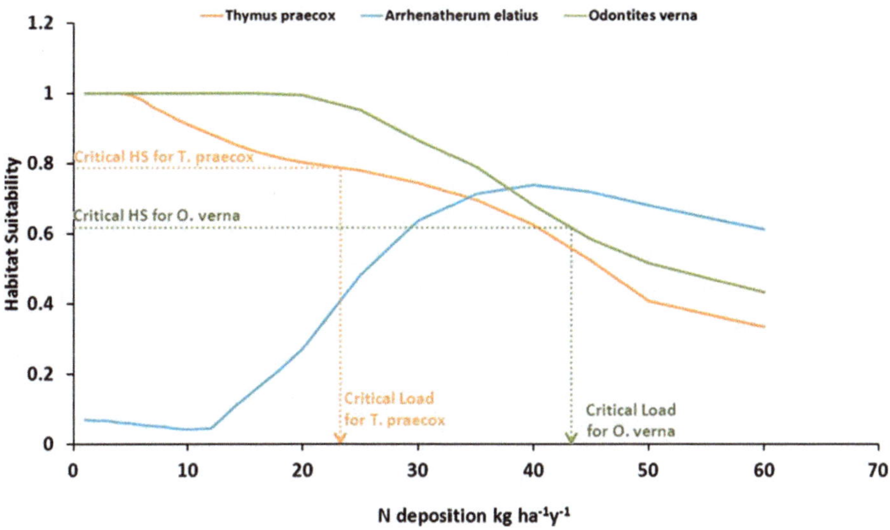

Fig. 11.7 Method for determining species-specific critical N load, by comparing critical Habitat Suitability (HS) for the species, derived from occurrence data as described in the text, against HS simulated by the MADOC-MultiMOVE model for different N loads

Model-based Critical Loads: Empirical critical loads for N ($CL_{nut}N$) have been defined based on ecosystem responses to increased N deposition over periods of the order of one decade (for experiments) to one century (for surveys) (Bobbink and Hettelingh 2011). However, chronic N deposition is likely to increase N stocks in slowly turning over soil organic pools, resulting in a very persistent increase in plant-available N, and consequent long-term effects on habitat suitability for species. Such long-term effects are difficult to assess using current observations, but the model chains described in this chapter can be used to determine the likely eventual effects of sustained deposition.

As an example, the long-term habitat suitability (HS) for three species that are used as indicator species for UK dune grassland was assessed under different N deposition scenarios (Fig. 11.7). The site and model application used for this assessment is described in Rowe et al. (2011b). For each species, a threshold HS was defined as the minimum HS at which the species occurred in the UK Countryside Survey dataset. The critical load can be defined as the N deposition rate which caused HS to decline below this threshold. Applying the precautionary principle, the critical load for *Thymus praecox* (approximately 22 kg N $ha^{-1}yr^{-1}$) would be used in this case. For this habitat, *Arrhenatherum elatius* is considered a negative condition indicator. The degree to which undesirable species should be included in critical load assessments is debatable; some authors judge biodiversity loss as occurring whenever alien or untypical species invade, whereas others view such species as harmful only when they cause declines in native species (Davis et al. 2011). If negative condition indicator species are to be included, the critical load for a habitat could also be defined as the point at which habitat suitability for the negative indicator species increases above the threshold for its occurrence.

Spatial Mapping of Biodiversity Value: The PROPS model has been applied to forests in Europe. A desired vegetation list was defined for each forest type based on EUNIS classes (http://eunis.eea.europa.eu/). The probability of occurrence of these species was calculated by PROPS and transferred into presence-absence data. For each plot the Simpson index was then calculated, based on the predicted species composition. Historic N and S deposition data were taken from Schöpp et al. (2003). Scenarios of N and S deposition were obtained from the atmospheric transport model of EMEP/MSC-W (Tarrasón et al. 2007). For 2020, two emission scenarios were used reflecting current legislation (CLE) and maximum (technically) feasible reductions (MFR) (Amann et al. 2007), with deposition assumed constant after 2020. As a climate scenario we used the IPCC SRES A1-scenario (Nakícenovíc et al. 2001). As a null-scenario we used past climate assuming no further climate change. More information can be found in Reinds et al. (2012) and also in Chap. 24 of this book.

Trade-offs between Biodiversity and Carbon Sequestration: As an extension of the biodiversity mapping study described above, the NTM3 model was used in conjunction with SMART2-SUMO2 to evaluate N and S deposition scenarios in terms of their effects on both biodiversity and C sequestration (Wamelink et al. 2003b). The soil model SMART2-SUMO2 can be used to produce spatial pictures of soil pH and N availability as a function of (changes in) N and S deposition, and hydrological models and scenarios can produce pictures of future groundwater levels. These can be combined by the NTM3 model to produce projections of (potential) (botanical) biodiversity under given scenarios as shown in Fig. 11.8.

The SUMO2 model is able to simulate carbon-sequestration in forest in living biomass, whereas SMART2 can simulate carbon stocks in soil organic matter. In a study for the Netherlands we compared the effects of climate change and N deposition on both carbon sequestration and plant diversity (Wamelink et al. 2009b), as shown in Fig. 11.9.

When N deposition increases, carbon sequestration increases in the Netherlands (for the year 2020). However, the plant diversity index in grasslands (and heathland; results not given) decreased dramatically, indicating a loss of rare (Red List) species. In another run we estimated the potential plant diversity index for all natural areas for two different policy scenarios. Results are given for 2020 (Fig. 11.8a) and differences per scenario were calculated (Fig. 11.8b). The results indicate where high and low potential biodiversity is located (high dune and clay areas, low the middle and the south; areas with high N deposition). Especially the difference map locates vulnerable areas which could be improved either by policy changes affecting the vulnerable areas or mitigating management changes.

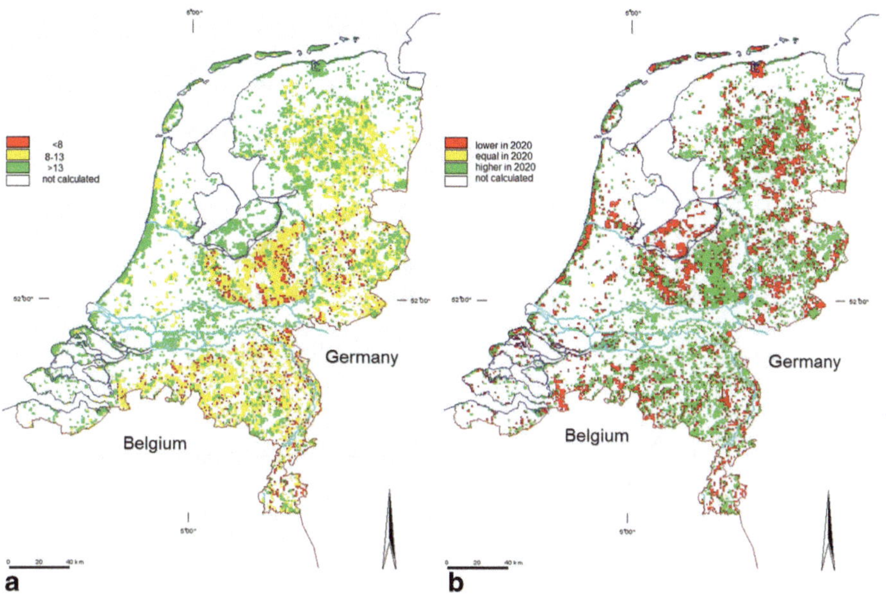

Fig. 11.8 Potential botanical diversity inferred by the NTM3 model in 2020 for **a** all vegetation types; and **b** the change in 2020 compared to the situation in 1995. (Wamelink et al. 2003b)

11.4 Outlook

The model chains described in this chapter are the result of considerable investment in development of biogeochemical and niche models, model integration, and collation of data for setting up and testing models. Much progress has been made, but the models continue to be developed in the light of new observations and theoretical developments. In this section we outline some key areas where development efforts are likely to be focused in the near future.

11.4.1 Model Choice

Several models have been applied to simulating soil, vegetation and/or floristic change in response to acid and nitrogen deposition. These vary in scope, emphasis, detail in terms of processes, number of parameters and timestep. Ideally, a choice would be made as to the best model for exploring a given question or scenario, based on accuracy and parsimony. However, models have mainly been developed and applied within a single country, and it remains difficult to make fair comparisons of accuracy and parsimony among the available models.

It is increasingly clear that models that use soil total C/N ratio as the only indicator of N availability to plants are unlikely to be accurate. The latest soil-vegetation biogeochemistry models include several conceptual pools of soil organic matter that

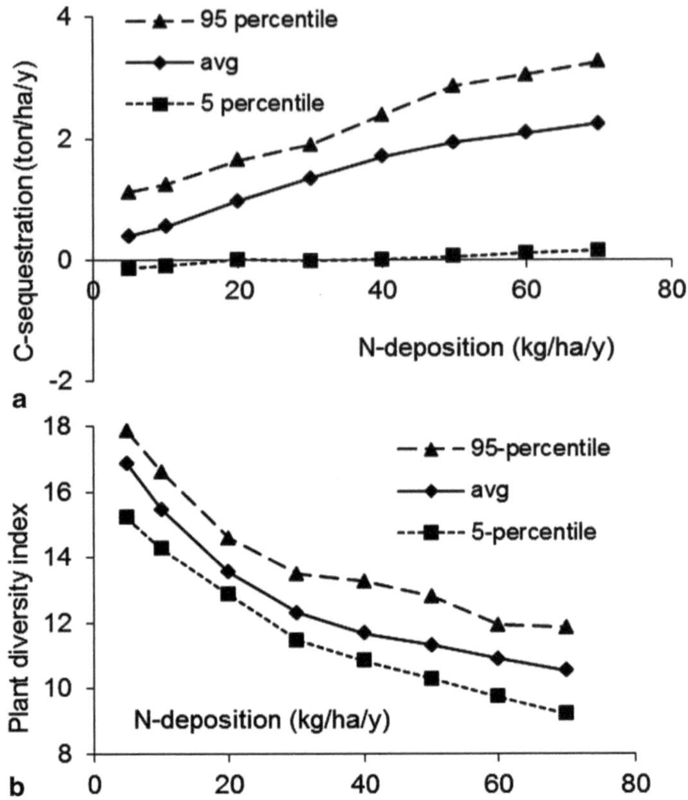

Fig. 11.9 Simulated carbon sequestration by deciduous forests (average for the Netherlands; *left*) and the plant diversity index for grassland (average for the Netherlands; *right*) as a result of N deposition in 2020. (Wamelink et al. 2009a)

vary in turnover rate. Other processes which it might be necessary to account for include: limitations to productivity by other elements, effects of anaerobiosis on C and N cycling and plant growth, and effects of management on vegetation structure. Adding more process detail to models generally improves accuracy, but there is a trade-off—adding detail inevitably creates a requirement for more parameters, making the model less general and increasing the number of measurements needed to set it up. The appropriate amount of detail is ultimately a question for scientists (Carpenter 2003; de Angelis and Mooij 2003), but calculating measures of model parsimony such as Akaike's information criterion (Akaike 1974) can aid with model comparisons.

The question of which indicator of nutrient availability to use remains difficult. Mean E_N remains useful as a time-integrated measure of habitat fertility that is independent of abiotic measurements, but is limited in geographic scope, since E_N scores have not been assigned to species that occur beyond Central and Northern Europe. Measurements based on soil total N (including soil C/N ratio) are of limited value since much of this N is not plant-available within the short- or medium-term.

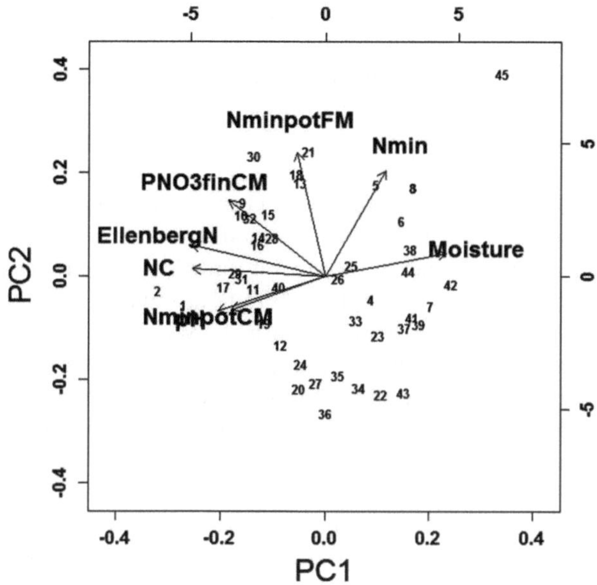

Fig. 11.10 Ordination of samples and measurements (of mean Ellenberg N and biogeochemical indicators) on first and second principal components, showing that mineralisable N (NminpotFM) explains a component of the variation in Ellenberg N that is orthogonal to the main axis of variation related to moisture content and soil total N/total C ratio (NC). (From Rowe et al. 2011a)

Instantaneous extractions, using for example 1-Molar KCl to measure NO_3 and NH_4 concentrations in soil, are highly subject to sampling error, since the mineral-N concentration is likely to fluctuate with rainfall and mineralisation events. Lysimeters for extracting soil solution require a long set-up time, and the N concentration in the sample may be affected if there are delays between extraction and analysis. Strong ion exchange resins can be used to obtain time-integrated samples, but it is not easy to relate such a sample to a known volume or mass of soil. Measurements of mineralisable N can be useful in distinguishing components of 'fertility' that are inherent to the soil type from the effects of N pollution (Fig. 11.10), but there is no standard analytical method. Plant responses to N are likely to be affected by other nutrients such as phosphorus, especially under wet circumstances (Wassen et al. 2005). The variety of measurements possible means that even where soil fertility measurements have been taken alongside floristic data, datasets with consistent fertility measurements are rarely large enough to derive niche models.

11.4.2 Database Developments

The challenge of applying models to larger geographic regions is being addressed through the setting up a database of relevés and associated chemical observations for major parts of Europe, as was done for the Netherlands (Wamelink et al. 2012b). The database will include, where available, measurements of soil pH, nitrate concentration, phosphate concentration, base cations and C/N ratio. Based on that information (plus climatic and deposition data) response functions for European species will be estimated (following Wamelink et al. 2005) and then could be used to parameterize PROPS (or any similar model) further and to a larger extent. At present (2014) about 40,000 relevés are available, which is only a small fraction of what is known to exist (Dengler et al. 2011). Abiotic conditions are known, at least for one variable (mostly soil pH), for about 11,000 relevés in the current database. These data were used to set up the PROPS model as a successor to MOVE4. The aim was to replace transfer functions using trait-means by species responses directly based on field measurements, e.g. pH and nitrate or mineral N concentration. This work will continue as more data become available. Where nitrate measurements are unavailable, values may be estimated for relevés based on the indicator values of the present species. Temperature and precipitation data from weather stations can be used, or these inputs can be calculated in the same way as for nitrate. Input for the model will be derived directly from model runs by e.g. VSD+ (pH, nitrate or C/N) or scenarios for N deposition and climate change (temperature, precipitation). Thus PROPS will be able to simulate the combined effects of N deposition and climate change on the occurrence of plant species on a European scale. In turn, the trait-means used in NTM3 will also be replaced by direct relationships.

11.4.3 Model Validation

Considerable work is still needed to inform the component models and model-chains with data from a broader set of conditions, to generate more confidence in model predictions. The interpretation of raw probabilities predicted by niche models such as MOVE4 and MultiMOVE can be difficult, since differences among species in overall prevalence can mean the raw probabilities are not directly comparable. A significant problem in the interpretation of raw probability outputs is that the range of values for a species is partly a function of species prevalence in the dataset. The consequence is that species probabilities cannot be readily compared. Approaches in which the raw probability is rescaled by known prevalence in the training data (Albert and Thuiller 2008) are promising.

11.4.4 Delays in Biological Response

Model predictions of species composition do not currently incorporate delays in the biological response of species to changes in chemical conditions. These time lags may be due to limits to dispersal distance, absence from the local species-pool, seed dormancy, and generation time. If a species is not present in the seed bank, even when circumstances become favourable for its growth it may not colonise the site for years or decades. If the distance to the nearest source of propagules is larger than the dispersal distance, intermediate generations may be necessary. This requires suitable habitat in the area to travel through, and sufficient time for the species to grow and reproduce in this habitat, which may be several decades in the case of some tree species. All species will show these effects to some extent, so the use of equilibrium niche models alone will always underestimate response times. The dispersal model DIMO has been developed to simulate these biological delays (Wamelink et al. 2007). However, effective dispersal modelling often requires species-specific parameterization, which can be costly to achieve for a wide range of species (Boulangeat et al. 2012; Catterall et al. 2012). The difficulty of predicting actual colonization or extinction probabilities is one reason why the outputs of niche models such as MultiMOVE are interpreted and applied as habitat suitability functions. Thus a high value is not treated as a direct prediction of occurrence but as prediction that conditions are highly suitable for the species, whether it is present or not. Model outputs remain useful despite this change in interpretation. For example the model may be applied to a nature reserve, where a species is not present, but was known to occur in the past and where habitat suitability projections can help estimate the likelihood of successful reintroduction.

Conclusions

Models should be assessed on their accuracy (whether predictions match observations) and their parameter-efficiency or parsimony (whether accuracy is achieved without over-parameterisation). Accuracy generally improves when process detail or spatio-temporal resolution is increased, for example, when processes are simulated in multiple soil layers rather than a single layer. The integration of different models can also increase accuracy, although the aim is usually to expand the scope of the overall model, such as extending models of soil biogeochemistry to also simulate vegetation and floristics. These factors provide a strong impetus for models to grow in size and detail, and can lead to models that require a large number of parameters, but that perform poorly outside the domain in which they were developed (Rowe et al. 2014a). Formal tests and comparisons of model parsimony remain rare. Both relatively simple and relatively detailed models can be valuable (Carpenter 2003; de Angelis and Mooij 2003), but the advantages and shortcomings of both need to be recognised.

The effects of N and S deposition on soil acid-base dynamics and the exchange of ions between solution and sorbed forms are largely understood. Remaining uncertainties mainly relate to variation in ion selectivity coefficients among soil types, which in some cases is not well characterised, and to base cation weathering rates. However, these terms can usually be calibrated using present-day observations of base saturation in the exchangeable pool and/or soil solution ion concentrations. Effects on biotic processes are more difficult to predict than effects on abiotic processes, due to factors such as resource-limitations that reduce response to N fertilisation, or interactions among trophic levels, particularly competition between decomposers and plants for soluble N. The prediction of N availability to plants remains a key uncertainty in the model chains described here. This difficulty is compounded by a lack of agreement on the best method for measuring plant-available N, which means that predictions cannot be simply tested against standard measurements.

The main uncertainty in the use of the model chains, however, relates not to biophysical processes but to the interpretation of their outputs. The models predict environmental suitability for particular species or assemblages, but policy makers generally require indicators that can be related to targets such as that included in the European Union's biodiversity strategy, to "halt the loss of biodiversity and the degradation of ecosystems services in the EU by 2020" (European Comission 2011). The NTM3 model provides a biodiversity endpoint indicator which is understandable and accessible to policy makers, but the data demands of this model are large, and the criteria used to translate environmental suitability for species and assemblages into a biodiversity value are specific to the Netherlands. Other studies have proposed methods based on indicator-species lists that have been previously defined for specific habitats (Rowe et al. 2011c), or on weighting of species using Red-List criteria (van Dobben and Wamelink 2009).

Despite the uncertainties discussed above, the accuracy of the biogeochemical models is improving as they are applied to an increasing number of experimental sites and treatments. The PROPS and MultiMOVE models have a solid empirical basis in very large floristic datasets. Progress is also being made with testing the entire soil-vegetation-floristic model chains and extending these to new regions. The model chains described in this chapter are likely to become increasingly useful for assessing N and S pollution scenarios and estimating critical loads.

References

Akaike, H. (1974). A new look at the statistical model identification. *IEEE Transactions on Automatic Control, 19,* 716–723.

Albert, C. H., & Thuiller, W. (2008). Favourability functions versus probability of presence: Advantages and misuses. *Ecography, 31,* 417–422.

Amann, M., Asman, W., Bertok, I., Cofala, J., Heyes, C., Klimont, Z., Schöpp, W., & Wagner, F. (2007). *Cost-effective emission reductions to meet the environmental targets of the thematic strategy on air pollution under different greenhouse gas constraints.* NEC Scenario Analysis Report Nr. 5. IIASA, Laxenburg, Austria.

Araujo, M. B., & New, M. (2007). Ensemble forecasting of species distributions. *Trends in Ecology & Evolution, 22,* 42–47.

Bobbink, R., & Hettelingh, J. P. (Eds.). (2011). *Review and revision of empirical critical loads and dose-response relationships: Proceedings of an expert workshop, Noordwijkerhout, 23–25 June 2010.* Co-ordination Centre for Effects, Bilthoven, The Netherlands.

Boulangeat, I., Gravel, D., & Thuiller, W. (2012). Accounting for dispersal and biotic interactions to disentangle the drivers of species distributions and their abundances. *Ecology Letters, 15,* 584–593.

Butler, A. (2010). *The GBMOVE ensemble (MultiMOVE): A technical summary.* Biomathematics & Statistics Scotland, Edinburgh. Report to Centre for Ecology and Hydrology, Lancaster, UK.

Carpenter, S. R. (2003). The need for fast-and-frugal models. In C. D. Canham, J. J. Cole, & W. K. Lauenroth (Eds.), *Models in ecosystem science* (pp. 455–460). Princeton University Press, Princeton, USA.

Catterall, S., Cook, A. R., Marion, G., Butler, A., & Hulme, P. E. (2012). Accounting for uncertainty in colonisation times: A novel approach to modelling the spatio-temporal dynamics of alien invasions using distribution data. *Ecography, 35,* 901–911.

Cosby, B. J., Wright, R. F., Hornberger, G. M., & Galloway, J. N. (1985). Modelling the effects of acid deposition: Assessment of a lumped parameter model of soil water and streamwater chemistry. *Water Resources Research, 21,* 51–63.

Davis, M. A., Chew, M. K., Hobbs, R. J., Lugo, A. E., Ewel, J. J., Vermeij, G. J., Brown, J. H., Rosenzweig, M. L., Gardener, M. R., Carroll, S. P., Thompson, K., Pickett, S. T. A., Stromberg, J. C., Del Tredici, P., Suding, K. N., Ehrenfeld, J. G., Grime, J. P., Mascaro, J., & Briggs, J. C. (2011). Don't judge species by their origins. *Nature, 474,* 153–154.

De Angelis, D. L., & Mooij, W. M. (2003). In praise of mechanistically rich models. In C. D. Canham, J. J. Cole, & W. K. Lauenroth (Eds.), *Models in ecosystem science* (pp. 63–82). Princeton University Press, Princeton, USA.

De Vries, W., Wamelink, W., van Dobben, H., Kros, H., Reinds, G.-J., Mol-Dijkstra, J., Smart, S., Evans, C., Rowe, E., Belyazid, S., Sverdrup, H., Van Hinsberg, A., Posch, M., Hettelingh, J.-P., Spranger, T., & Bobbink, R. (2010). Use of dynamic soil-vegetation models to assess impacts of nitrogen deposition on plant species composition and to estimate critical loads: An overview. *Ecological Applications, 20,* 60–79.

Dengler, J., Jansen, F., Glockler, F., Peet, R. K., De Caceres, M., Chytry, M., Ewald, J., Oldeland, J., Lopez-Gonzalez, G., Finckh, M., Mucina, L., Rodwell, J. S., Schaminee, J. H. J., & Spencer, N. (2011). The Global Index of Vegetation-Plot Databases (GIVD): A new resource for vegetation science. *Journal of Vegetation Science, 22,* 582–597.

Diekmann, M. (2003). Species indicator values as an important tool in applied plant ecology—A review. *Basic and Applied Ecology, 4,* 493–506.

Elberse, W. T., Van Den Bergh, J. P., & Dirven, J. G. P. (1983). Effects of use and mineral supply on the botanical composition and yield of old grassland on heavy-clay soil. *Netherlands Journal of Agricultural Science, 31,* 63–88.

Ellenberg, H., Weber, H. E., Dull, R., Wirth, V., Werner, W., & Paulissen, D. (1992). Zeigerwerte von pflanzen in mitteleuropa: 2nd ed. *Scripta Geobotanica, 18,* 1–258.

Ertsen, A. C. D., Alkemade, J. R. M., & Wassen, M. J. (1998). Calibrating Ellenberg indicator values for moisture, acidity, nutrient availability and salinity in the Netherlands. *Plant Ecology, 135,* 113–124.

European Comission. (2011). Biodiversity: Commission announces new strategy to halt biodiversity loss within ten years. http://europa.eu/rapid/press-release_IP-11-526_en.htm. Accessed 31 July 2014.

Friedman, J. H. (1991). Multivariate adaptive regression splines. *Annals of Statistics, 19,* 1–67.

Grime, J. P., Hodgson, J. G., & Hunt, R. (1988). *Comparative plant ecology: A functional approach to common British species*. London: Unwin Hyman.

Gunderson, L. H. (2000). Ecological resilience—In theory and application. *Annual Review of Ecology and Systematics, 31,* 425–439.

Harmer, R., Peterken, G., Kerr, G., & Poulton, P. (2001). Vegetation changes during 100 years of development of two secondary woodlands on abandoned arable land. *Biological Conservation, 101,* 291–304.

Hastie, T., & Tibshirani, R. (1986). Generalized additive models. *Statistical Science, 1,* 297–310.

Hautier, Y., Niklaus, P. A., & Hector, A. (2009). Competition for light causes plant biodiversity loss after eutrophication. *Science, 324,* 636–638.

Hill, M. O., & Carey, P. D. (1997). Prediction of yield in the Rothamsted Park grass experiment by Ellenberg indicator values. *Journal of Vegetation Science, 8,* 579–586.

Hill, M. O., Roy, D. B., Mountford, J. O., & Bunce, R. G. H. (2000). Extending Ellenberg's indicator values to a new area: An algorithmic approach. *Journal of Applied Ecology, 37,* 3–15.

Jaccard, P. (1912). The distribution of flora in the alpine zone. *New Phytologist, 11,* 37–50.

Jarvis, S. J., Rowe, E. C., Henrys, P. A., Smart, S. M., Jones, M. L. M., & Garbutt, A. (in prep.) Empirical realised niche models for coastal plant species. *Journal of Coastal Conservation.*

Jenkinson, D. S., Potts, J. M., Perry, J. N., Barnett, V., Coleman, K., & Johnston, A. E. (1994). Trends in herbage yields over the last century on the Rothamsted long-term continuous hay experiment. *Journal of Agricultural Science, 122,* 365–374.

Kros, J. (2002). Evaluation of biogeochemical models at local and regional scale. PhD thesis. Wageningen University, The Netherlands.

Latour, J. B., & Reiling, R. (1993). A multiple stress model for vegetation ('move'): A tool for scenario studies and standard-setting. *Science of the Total Environment, 134,* 1513–1526.

Ludwig, D., Walker, B., & Holling, C. S. (1997). Sustainability, stability, and resilience. *Conservation Ecology, 1,* 7.

Mace, G. M., & Stuart, S. N. (1994). Draft IUCN Red List Categories, Version 2.2. *Species, 21–22,* 13–24.

Michalzik, B., Tipping, E., Mulder, J., Lancho, J. F. G., Matzner, E., Bryant, C. L., Clarke, N., Lofts, S., & Esteban, M. A. V. (2003). Modelling the production and transport of dissolved organic carbon in forest soils. *Biogeochemistry, 66,* 241–264.

Mitchell, T. D., Carter, T. R., Jones, P. D., Hulme, M., & New, M. (2004). A comprehensive set of high-resolution grids of monthly climate for Europe and the globe: The observed record (1901–2000) and 16 scenarios (2010–2100). Tyndall Centre for Climate Change Research, Norwich, UK.

Nakícenovíc, N., Alcamo, J., Davis, G., De Vries, B., Fenhann, J., Gaffin, S., Gregory, K., Grübler, A., Jung, T. Y., Kram, T., Emilio la Rovere, E., Michaelis, L., Mori, S., Morita, T., Pepper, W., Pitcher, H., Price, L., Riahi, K., Roehrl, A., Rogner, H.-H., Sankovski, A., Schlesinger, M. E., Shukla, P. R., Smith, S., Swart, R. J., Van Rooyen, S., Victor, N., & Dadi, Z. (2001). *Special report on emission scenarios: Intergovernmental panel on climate change*. Cambridge University Press, Cambridge, UK.

Newbold, T., Reader, T., El-Gabbas, A., Berg, W., Shohdi, W. M., Zalat, S., El Din, S. B., & Gilbert, F. (2010). Testing the accuracy of species distribution models using species records from a new field survey. *Oikos, 119,* 1326–1334.

Nilsson, J., & Grennfelt, P. (1988). *Critical loads for sulphur and nitrogen.* Report 188:15. Copenhagen: UNECE/Nordic Council of Ministers.

Pakeman, R. J., Reid, C. L., Lennon, J. J., & Kent, M. (2008). Possible interactions between environmental factors in determining species optima. *Journal of Vegetation Science, 19,* 201–208.

Posch, M., & Reinds, G. J. (2009). A very simple dynamic soil acidification model for scenario analyses and target load calculations. *Environmental Modelling & Software, 24,* 329–340.

Posch, M., Hettelingh, J.-P. & Slootweg, J. (2003). *Manual for dynamic modelling of soil response to atmospheric deposition* (RIVM report 259101 012). Bilthoven: National Institute for Public Health and the Environment.

Preston, C. D., Pearman, D. A., & Dines, T. D. (2002). *New plant Atlas of the British and Irish flora.* New York: Oxford University Press, Oxford, UK.

Randin, C. F., Dirnbock, T., Dullinger, S., Zimmermann, N. E., Zappa, M., & Guisan, A. (2006). Are niche-based species distribution models transferable in space? *Journal of Biogeography, 33,* 1689–1703.

Reinds, G. J., Bonten, L., Mol-Dijkstra, J. P., Wamelink, G. W. W. & Goedhart, P. (2012). Combined effects of air pollution and climate change on species diversity in Europe: First assessments with VSD+ linked to vegetation models. In M. Posch & J. P. Hettelingh (Eds.), *CCE status report 2012* (pp. 49–61). Co-ordination Centre for Effects, Bilthoven, The Netherlands.

Rodwell, J. S. (Ed.). (1991–2000). *British Plant Communities* (Vol. 1–5). Cambridge University Press, Cambridge, UK.

Rowe, E. C., Emmett, B. A., Smart, S. M., & Frogbrook, Z. L. (2011a). A new net mineralizable nitrogen assay improves predictions of floristic composition. *Journal of Vegetation Science, 22,* 251–261.

Rowe, E. C., Jones, M. L. M., Henrys, P. A., Smart, S. M., Tipping, E., Mills, R. T. E., & Evans, C. D. (2011b). Predicting effects of N pollutant load on plant species based on a dynamic soil eutrophication indicator. CCW Science Report 977. Countryside Council for Wales, Bangor, UK.

Rowe, E. C., Smart, S. M., & Emmett, B. A. (2011c). Defining a biodiversity damage metric and threshold using Habitats Directive criteria. In W. K. Hicks, C. P. Whitfield, W. J. Bealey, & M. A. Sutton (Eds.), *Nitrogen deposition and Natura 2000: Science and practice in determining environmental impacts* (pp. 136–141). COST.

Rowe, E. C., Tipping, E., Posch, M., Oulehle, F., Cooper, D. M., Jones, T. G., Burden, A., Monteith, D. T., Hall, J., & Evans, C. D. (2014a). Predicting nitrogen and acidity effects on long-term dynamics of dissolved organic matter. *Environmental Pollution, 184,* 271–282.

Rowe, E. C., Wright, D. G., Bertrand, N., & Reis, S. (2014b) Fusing and disaggregating models, data and analysis tools for a dynamic science-society interface. In Riddick, A. T., Kessler, H., & Giles, J. R. A. (Eds) Integrated Environmental Modelling to Solve Real World Problems: Methods, Visions and Challenges. *Geological Society, London, Special Publications,* 408, http://dx.doi.org/10.1144/SP408.1.

Roy, D. B., Hill, M. O., Rothery, P., & Bunce, R. G. H. (2000). Ecological indicator values of British species: An application of Gaussian logistic regression. *Annales Botanici Fennici, 37,* 219–226.

Schaffers, A. P., & Sykora, K. V. (2000). Reliability of Ellenberg indicator values for moisture, nitrogen and soil reaction: A comparison with field measurements. *Journal of Vegetation Science, 11,* 225–244.

Schaminee, J. H. J., Hennekens, S. M., & Ozinga, W. A. (2012). The Dutch national vegetation database. *Biodiversity and Ecology, 4,* 201–209.

Schöpp, W., Posch, M., Mylona, S., & Johansson, M. (2003). Long term development of acid deposition (1880–2030) in sensitive freshwater regions in Europe. *Hydrology and Earth System Sciences, 7,* 436–446.

Schouwenberg, E. P. A. G., Houweling, H., Jansen, M. J. W., Kros, J., & Mol-Dijkstra, J. P. (2001). *Uncertainty propagation in model chains: A case study in nature conservancy.* Alterra-rapport 001, ISSN 1566–7197. Wageningen, Netherlands.

Smart, S., Evans, C., Rowe, E., Wamelink, W., Wright, S., Scott, A., Roy, D., Preston, C., Hill, M., Rothery, P., Bullock, J., Moy, I., Emmett, B., & Maskell, L. (2005). *Atmospheric nitrogen pollution impacts on biodiversity: Phase 1—Model development and testing.* Final report to DEFRA, JNCC and English Nature. Centre for Ecology and Hydrology, Lancaster, UK.

Smart, S. M., Henrys, P. A., Scott, W. A., Hall, J. R., Evans, C. D., Crowe, A., Rowe, E. C., Dragosits, U., Page, T., Whyatt, J. D., Sowerby, A., & Clark, J. M. (2010a). Impacts of pollution and climate change on ombrotrophic *Sphagnum* species in the United Kingdom; analysis of uncertainties in two empirical niche models. *Climate Research, 45,* 163–177.

Smart, S. M., Scott, W. A., Whitaker, J., Hill, M. O., Roy, D. B., Critchley, C. N., Marini, L., Evans, C., Emmett, B. A., Rowe, E. C., Crowe, A., Le Duc, M., & Marrs, R. H. (2010b). Empirical realised niche models for British higher and lower plants—Development and preliminary testing. *Journal of Vegetation Science, 21,* 643–656.

Svenning, J.-C., & Skov, F. (2004). Limited filling of the potential range in European tree species. *Ecology Letters, 7,* 565–573.

Tarrasón, L., Fagerli, H., Jonson, J. E., Simpson, D., Benedictow, A., Klein, H., Vestreng, V., Aas, W., & Hjellbrekke, A.-G. (2007). *Transboundary acidification, eutrophication and ground level ozone in Europe in 2005.* EMEP Report 1/2007. Oslo: Norwegian Meteorological Institute.

Tipping, E., Rowe, E. C., Evans, C. D., Mills, R. T. E., Emmett, B. A., Chaplow, J. S., & Hall, J. R. (2012). N14C: A plant-soil nitrogen and carbon cycling model to simulate terrestial ecosystem responses to atmospheric nitrogen deposition. *Ecological Modelling, 247,* 11–26.

Tipping, E., Henrys, P. A., Maskell, L. C., & Smart, S. M. (2013). Nitrogen deposition effects on plant species diversity: Threshold loads from field data. *Environmental Pollution, 179,* 218–223.

Van Adrichem, M. H. C., Wortelboer, F. G., & Wamelink, G. W. W. (2010). *MOVE: MOdel for terrestrial VEgetation version 4.0.* (WOT werkdocument 153). Wageningen: Wettelijke Onderzoekstaken Natuur & Milieu.

Van Dobben, H., & Wamelink, W. (2009). A Red-List-based biodiversity indicator and its application in model studies in the Netherlands. In J.-P. Hettelingh, M. Posch, & J. Slootweg (Eds.), *CCE status report 2009: Progress in the modelling of critical thresholds, impacts to plant species diversity and ecosystem services in Europe* (pp. 77–81). ICP Modelling and Mapping Co-ordination Centre for Effects, Bilthoven, The Netherlands.
Van Dobben, H. F., Van Hinsberg, A., Schouwenberg, E., Jansen, M., Mol-Dijkstra, J. P., Wieggers, H. J. J., Kros, J., & De Vries, W. (2006). Simulation of critical loads for nitrogen for terrestrial plant communities in The Netherlands. *Ecosystems, 9,* 32–45.
Van Tongeren, O. (2000). *Programma ASSOCIA gebruikershandleiding en voorwaarden.* Wageningen: Data-analyse Ecologie.
Vitousek, P. M., Porder, S., Houlton, B. Z., & Chadwick, O. A. (2010). Terrestrial phosphorus limitation: Mechanisms, implications, and nitrogen-phosphorus interactions. *Ecological Applications, 20,* 5–15.
Wamelink, G. W. W., & Van Dobben, H. F. (2003). Uncertainty of critical loads based on the Ellenberg indicator value for acidity. *Basic and Applied Ecology, 4,* 515–523.
Wamelink, G. W. W., Joosten, V., Van Dobben, H. F., & Berendse, F. (2002). Validity of Ellenberg indicator values judged from physico-chemical field measurements. *Journal of Vegetation Science, 13,* 269–278.
Wamelink, G. W. W., Van Dobben, H. F., & Berendse, F. (2003a). Apparently we do need phytosociological classes to calibrate Ellenberg indicator values! *Journal of Vegetation Science, 14,* 619–620.
Wamelink, G. W. W., Ter Braak, C. J. F., & Van Dobben, H. F. (2003b). Changes in large-scale patterns of plant biodiversity predicted from environmental economic scenarios. *Landscape Ecology, 18,* 513–527.
Wamelink, G. W. W., Goedhart, P. W., Van Dobben, H. F., & Berendse, F. (2005). Plant species as indicators of soil pH: Replacing expert judgement with measurements. *Journal of Vegetation Science, 16,* 461–470.
Wamelink, G. W. W., Van Der Greft, J., Jochem, R., Prins, A. H., Van Dobben, H. F., & Grashof-Bokdam, C. (2007). Dispersion of plant species in a scattered landscape on a regional scale; a modeling approach. In R. G. H. Bunce, R. H. G. Jongman, L. Hojas, & S. Weel (Eds.), *25 years of landscape ecology: Scientific principles in practice*. Proceedings of the 7th IALE World Congress part 2 (pp. 1089–1090).
Wamelink, G. W. W., Van Dobben, H. F., & Berendse, F. (2009a). Vegetation succession as affected by decreasing nitrogen deposition, soil characteristics and site management: A modelling approach. *Forest Ecology and Management, 258,* 1762–1773.
Wamelink, G. W. W., Van Dobben, H. F., Mol-Dijkstra, J. P., Schouwenberg, E., Kros, J., De Vries, W., & Berendse, F. (2009b). Effect of nitrogen deposition reduction on biodiversity and carbon sequestration. *Forest Ecology and Management, 258,* 1774–1779.
Wamelink, G. W. W., Wieggers, H. J. J., Reinds, G. J., Kros, J., Mol-Dijkstra, J. P., Van Oijen, M., & De Vries, W. (2009c). Modelling impacts of changes in carbon dioxide concentration, climate and nitrogen deposition on carbon sequestration by European forests and forest soils. *Forest Ecology and Management, 258,* 1794–1805.
Wamelink, G. W. W., Akkermans, L. M. W., Brus, D. J., Heuvelink, G. B. M., Mol-Dijkstra, J. P., & Schouwenberg, E. P. A. G. (2011). *Uncertainty analysis of SMART2-SUMO2-P2E-MOVE4. The Nature Planner soil and vegetation model chain* (WOT-Report 108). Wageningen: Alterra.
Wamelink, G. W. W., Van Adrichem, M. H. C., & Goedhart, P. W. (2012a). Validatie van MOVE4. Wageningen: Wettelijke Onderzoekstaken Natuur & Milieu, WOt-werkdocument 311.
Wamelink, G. W. W., Van Adrichem, M. H. C. Van Dobben, H. F. van, Frissel, J. Y., Den Held, M., Joosten, V., Malinowska, A. H., Slim, P. A., & Wegman, R. M. A. (2012b). Vegetation relevés and soil measurements in the Netherlands: A database. *Biodiversity and Ecology, 4,* 125–132.
Wassen, M. J., Venterink, H. O., Lapshina, E. D., & Tanneberger, F. (2005). Endangered plants persist under phosphorus limitation. *Nature, 437,* 547–550.

Chapter 12
Use of an Integrated Soil-Vegetation Model to Assess Impacts of Atmospheric Deposition and Climate Change on Plant Species Diversity

Salim Belyazid, Harald U. Sverdrup, Daniel Kurz and Sabine Braun

12.1 Introduction

The 1999 Gothenburg protocol to abate acidification, eutrophication and ground-level ozone within the Convention on Long-range Transboundary Air Pollution (LRTAP) has the scope of decreasing emissions of N and S substantially when fully implemented. The setting of national environmental goals for air pollution is linked to critical loads, which are based on either empirical assessments (see Chaps. 4 and 5), on steady-state models (Chap. 6) or (simple) dynamic models in which the causal chain of dose to effects is reversed (see Chaps. 8 and 10). Steady-state models do not allow predicting the temporal response of ecosystems, in terms of, e.g., impacts on plant species diversity, to deposition scenarios. This requires the use of the dynamic integrated soil-vegetation models. Such models can also be used to assess critical loads, while accounting for differences in sensitivity to perturbation depending on their current state and recent history.

At present, national environmental goals for biodiversity are defined in relatively vague and static terms, and the environmental effects on plant species diversity, originating from acid deposition, eutrophication and climate change, have been investigated independently. In order to give the national authorities a starting point from which to develop more quantitative environmental goals for linking policy to critical loads for biodiversity, a more integrated dynamic approach is needed. In Chap. 11, an approach is described in which the dynamics in soil chemistry is

S. Belyazid (✉) · H. U. Sverdrup
Department of Chemical Engineering, Lund University, Lund, Sweden
e-mail: salim@belyazid.com

D. Kurz
EKG Geoscience, Bern, Switzerland

S. Braun
Insitut für Angevandte Pflanzenbiologie, Schönenbuch, Basel, Switzerland

W. de Vries et al. (eds.), *Critical Loads and Dynamic Risk Assessments*,
Environmental Pollution 25, DOI 10.1007/978-94-017-9508-1_12

included but the impacts on plant species diversity are assessed by a static regression model, implying that the model predictions refer to the potential vegetation that is likely to occur at a new equilibrium. In this chapter, we describe an integrated dynamic soil and plant species diversity (vegetation) model for Europe, i.e. the ForSAFE-Veg model chain. This model chain consists of: (i) the ForSAFE model, aimed at the dynamic simulation of changes in soil chemistry, soil organic matter, hydrology and tree biomass growth in relation to changes in environmental factors (described in Chap. 8) and (ii) the Veg submodel, which simulates changes in the composition of the ground vegetation in response to changes in biotic and abiotic factors such as light intensity at the forest floor, temperature, grazing pressure, soil moisture, soil pH and alkalinity in addition to competition between species based on height and root depth (described in this chapter). The study presented in this chapter is an illustration of the model chain application. Emission reductions established under European air quality agreements will be re-evaluated during the coming years, and models for predicting pollution impacts on biodiversity components can strengthen the case for stricter adherence to effects-based critical loads for multiple components of the ecosystems (Martinson et al. 2005; Nilsson and Grennfelt 1988; Sverdrup et al. 2005; Sverdrup and Warfvinge 1988).

12.2 The ForSAFE-Veg Model

The ForSAFE-Veg system is an integrated numerical model, based on a systems analysis of the whole forest ecosystem. It consists of a modular assembly of subsystems. The model is used in scenario mode for analysing the system responses to input changes with respect to nitrogen inputs, climate change parameters, nutrient status and management (Belyazid 2006; Belyazid et al. 2006, 2010; Wallman et al. 2005).

12.2.1 Theory

In the ForSAFE-Veg system, the interaction between the geochemistry and the reactions of the plants are built into a feedback system (Belyazid et al. 2006; Sverdrup et al. 2007). Parts of this feedback system constitute ecological competition between plants. The plants compete in the model for essential nutrients and energy in order to gain strength and take a maximum amount of territory for themselves. In nature, plants compete for nitrogen, phosphorus, water and light in particular, according to different strategies. The physiological effect of N on plants is modified depending on the competition at any given moment (Ellenberg et al. 1992; Falkengren-Grerup 1992; Flückiger and Braun 1998; Landolt 1977). However, N is only one of several strong factors affecting competition. A simple response diagram alone cannot be used, as it does not incorporate the effect of other factors nor include competition in a way that makes it readable. The causal parameter is therefore defined as the nitrogen system input, the effect parameters are ecological and have several manifestations. Studies that record field occurrence of plants try to define 'optimal' environmental ranges for plants (De Vries et al. 2010; Rowe et al. 2005).

In reality, plants are in competition with other plants for the same room in the territory, and many of them occupy sub-optimal parts of their own potential range, because that is what they can get in the competition with other plants (Ellenberg et al. 1992; Landolt 1977). Plants do not associate because they belong together, but their coexistence is caused rather from their competitive expression getting similar outputs under certain conditions. As soon as the basic conditions change substantially, the plants in the associations may seem to fly apart as their competition expression change differently and then they may seem to create new plant associations. Thus, the composition of a plant association is an ecosystem output that varies with time. Models based on optimal ranges are likely to generate forecasts with strong biases. Biodiversity depends on many variables. In terrestrial ecosystems these include components such as plants, soil microorganisms, soil micro-fauna, insects, animals. All these live in a feedback web where single chemical limits no longer contain enough of the system's properties for a critical load determination. Aspects of structure and habitat quantity and quality are also vital in defining the state of the system (Fenn et al. 2003; Körner 2003; Lambers et al. 1998). We will be concentrating here on the ground vegetation components.

12.2.2 Model Approach

The organization of the ForSAFE-Veg system in terms of model modules and inputs to the model is given in Fig. 12.1. In a forested landscape, nitrogen, acidity, forest management and climate change have roughly equal importance for the biodiversity, and none can be ignored without significant implications. ForSAFE-Veg thus handles the effects of all these drivers on biodiversity. A full description of ForSAFE is available in Belyazid (2006), Sverdrup et al. (2007) and Belyazid et al. (2011b). Here we give a summary description.

Figure 12.2 shows the causal loop diagram, describing the causal chains and feedbacks included in the model, illustrating the effect of forest management, atmospheric pollution and climate change on the whole system. The effects from acidity, nitrogen, temperature, water availability, and management on plant group competition strength can be read from the diagram through the way the variable "external forces" affects the favourable conditions of different plants. The causal loop diagram also shows clearly why the "single cause approach" has been running into significant problems as a viable pollution assessment method as there is no single trail of cause, but rather a whole system of them. The critical load is exceeded at the time when the input of pollution has transgressed the limit where it will eventually lead to significant shifts in vegetation composition, abundance or the entry or departure of one or more plant groups. This will in many cases be well before such effects are actually observed, because of system delays. Thus, waiting for effects to be observed in the field is not in line with a precautionary approach. With such an approach, the countermeasure will always be too late with respect to a preventive strategy. The present model does not include ozone impacts on trees or ground vegetation. The long-term phosphorus mass balance and the impact of phosphorus on biodiversity is not yet included in the model. To a smaller degree the model output also depends on the trajectory of the climate change scenario. The model is able

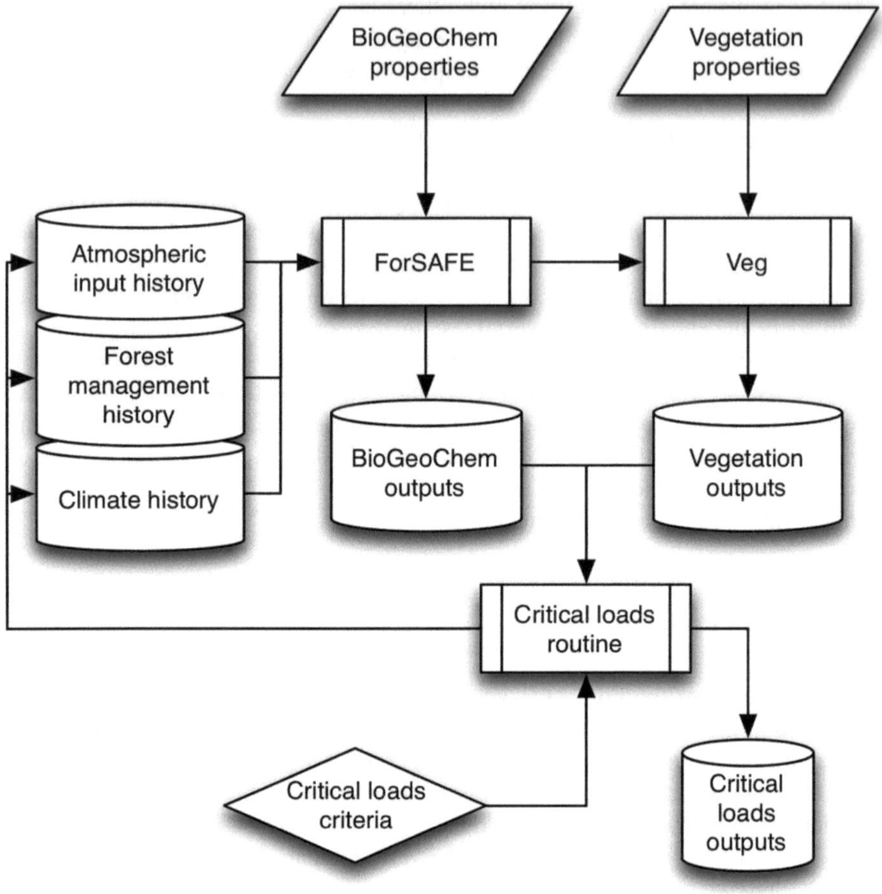

Fig. 12.1 Organization of the ForSAFE-Veg system in terms of model modules and inputs to the model

to illustrate goal conflicts in forestry, acidifying pollution and nitrogen mitigation policies, nature conservation policies and climate change mitigation policies.

The integrated ForSAFE-Veg model includes the following ecosystem components:

- Tree vegetation layer
- Ground vegetation layer
- Soil chemistry and geochemical processes
- Soil stocks and cycling of nutrients (Ca, Mg, K, P, N)
- Cycling of carbon; growth, litterfall, soil organic matter, decomposition and solute transport
- Soil hydrology and energy balance
- Effluents from the system: carbon dioxide, DOC, solution chemistry

The ForSAFE core is based on mass balances arranged as difference equations, which are solved numerically. Difference equations are preferred over differential

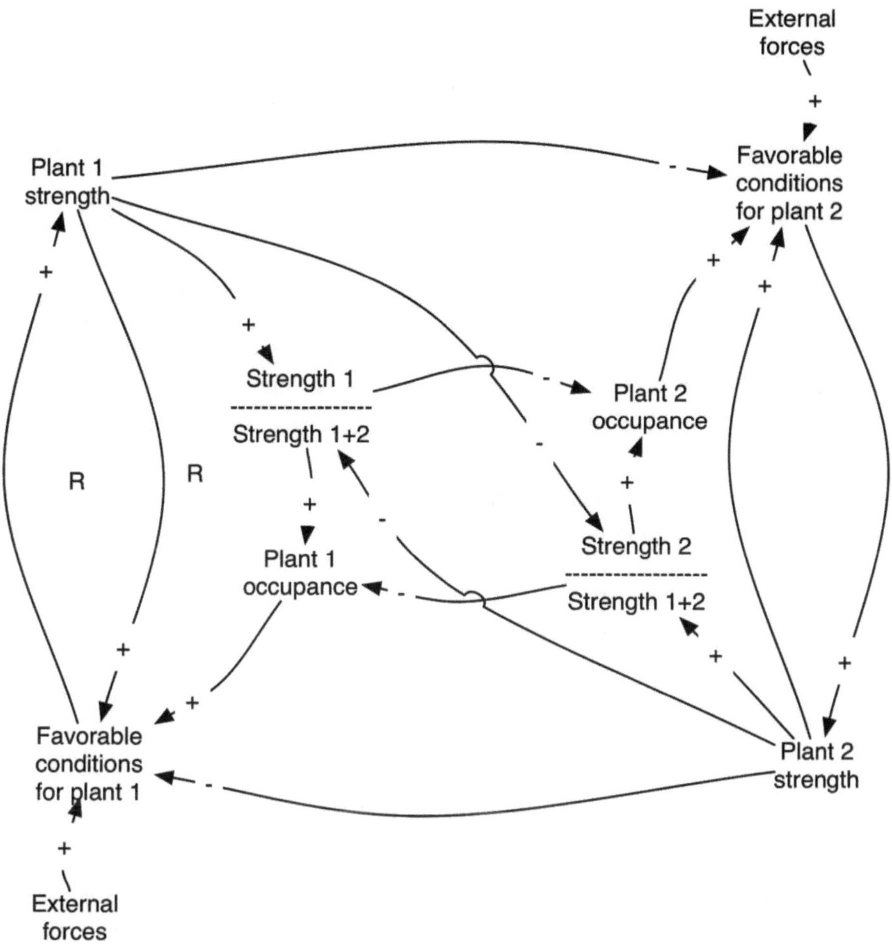

Fig. 12.2 Causal loop diagram visualizing the competition in the model

equations because of the possibility to use faster numerical solvers and improve model performance. The model operates with all systems in multiple communicating soil layers. In order to synthesize an integrated multiple stress system, ForSAFE-Veg provides simultaneous predictions of ecosystem effects of climate change, soil acidification and eutrophication with vegetation changes on forest growth, cations, acidity, nitrogen and carbon cycling. ForSAFE contains fully mechanistic nitrogen- and carbon cycle sub-models as well as predictions of forest growth under production management based on current knowledge. The ground vegetation composition in a forest or field is determined by a number of drivers in the submodel Veg (Belyazid 2006; Belyazid et al. 2011b; Sverdrup et al. 2007):

- Soil solution nitrogen concentrations (kmol m^{-3})
- Soil chemistry in terms of acidity, antagonists and co-agents: ($[H^+]$, $[BC^{2+}]$, $[Al^{3+}]$)
- Soil water activity (soil moisture volume fraction, m^3 of water per m^3 of soil volume)

- Site soil temperature (°C),
- Light reaching the ground (μmol photons $m^{-2}s^{-1}$)
- Grazing by ungulents (moose units km^{-2})
- Forest fires (Removal of plant groups, conversion of biomass to available water-soluble inorganic ions (K, Ca, Mg, Na, SO_4, P, but most N escapes to the atmosphere)
- Forest management (biomass removal containing C, Ca, Mg, K, P and N, planting, harvest)

An important feature of the ForSAFE-Veg model is that the plants have to compete for their share of the territory following specific plant strengths based on the response functions to the different drivers just listed. The drivers act to assign a completion strength to each given plant group. The competition is made up of the following elements where the individual plant group has feedback on the drivers, affecting partly itself as well as other plants.

- The above-ground competition strategy of the plant group for capturing light and preventing others from getting depends on plant height and its shading capacity.
- The root competition strategy is for capturing nutrients.
- Each plant group has a predefined water requirement niche.
- Each plant group has a predefined N niche.
- Exposure to soil chemistry (H^+, Al^{3+} in relation to BC^{2+}), expressed through root distribution in different soil depths, and the ability to take up nutrients competitively under different chemical conditions.
- Uptake of Ca, Mg and K is a necessary requirement, where lack is limiting, but excess is not growth promoting.

The competition strength is the weight each group has when the territory claim assigned to a plant group is determined. The plant group then must use this and the passage of time to take that much of the territory. The plant groups are assumed to be groups of plants with identical responses to all parameters. Indicator plant species are identified for each group. In this chapter a selection of tree species, as seedlings and as mature trees, are considered in addition to ground vegetation groups.

The change over time in the occupancy fraction of a species is modelled as:

$$\frac{dx}{dt} = \frac{k}{T} \cdot (X_{eq} - X) \tag{12.1}$$

where X is the fraction of the ground occupied at present, X_{eq} is the fraction the plant should have according to its strength at equilibrium, k is a constant and T is the delay time. The strengths illustrated in Fig. 12.2 are expressed in this way. The equilibrium occupancy is given by the ratio of the strength of plant group i relative to the sum of all strengths present:

$$X_{eq} = S_t / \sum_{all} S_j \tag{12.2}$$

The strength of a species is computed from a number of both promoting and retarding scaling factors (Fig. 12.3):

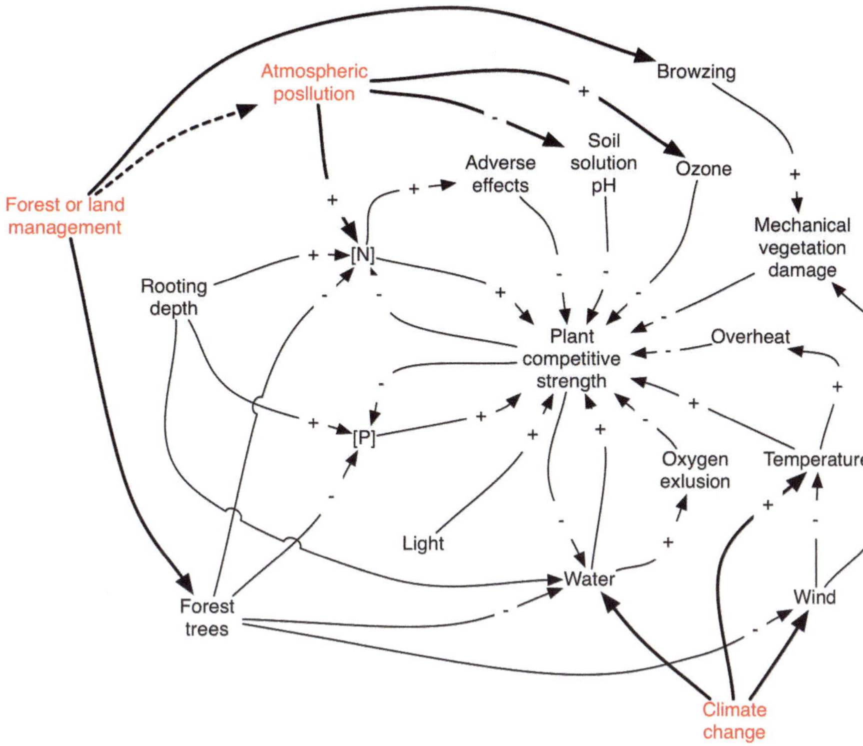

Fig. 12.3 Causal loop diagram for the elements of plant strength used in competition between the plants on the ground

$$S = f_1(pH) \cdot f_2(N) \cdot f_3(water) \cdot f_4(light) \cdot \ldots \tag{12.3}$$

The effect of pH is retarding at very high and low values with a neutral, no-effect piece in the middle, basically reflecting the curve in Fig. 12.4. The same type of curve is fitted to the other parameters as well. The plant group parameterization for this study was carried out in two steps. It is an integral part of the development process within Veg, and the setting of the parameters automatically calls up the response functions for a reasonability test. In order to parameterize, the ecologists involved fully understood the feedbacks of the model and the implication of the formulation of the response functions. The parametric responses to drivers presented in Figs. 12.4 and 12.5 are defined in mathematical terms below. For example, the N response combines a promotion function and a retardation function:

$$f_2(N) = a_0 \cdot \frac{[N]^{w}+}{k_+ + [N]^{w+}} \cdot \frac{k\_}{k\_ + [N]^{w-}} \tag{12.4}$$

where $[N]$ is the N solution concentration, w_+ denotes the slope of the promotion section, k_+ determines the threshold of $[N]$ where positive response starts, k_- the

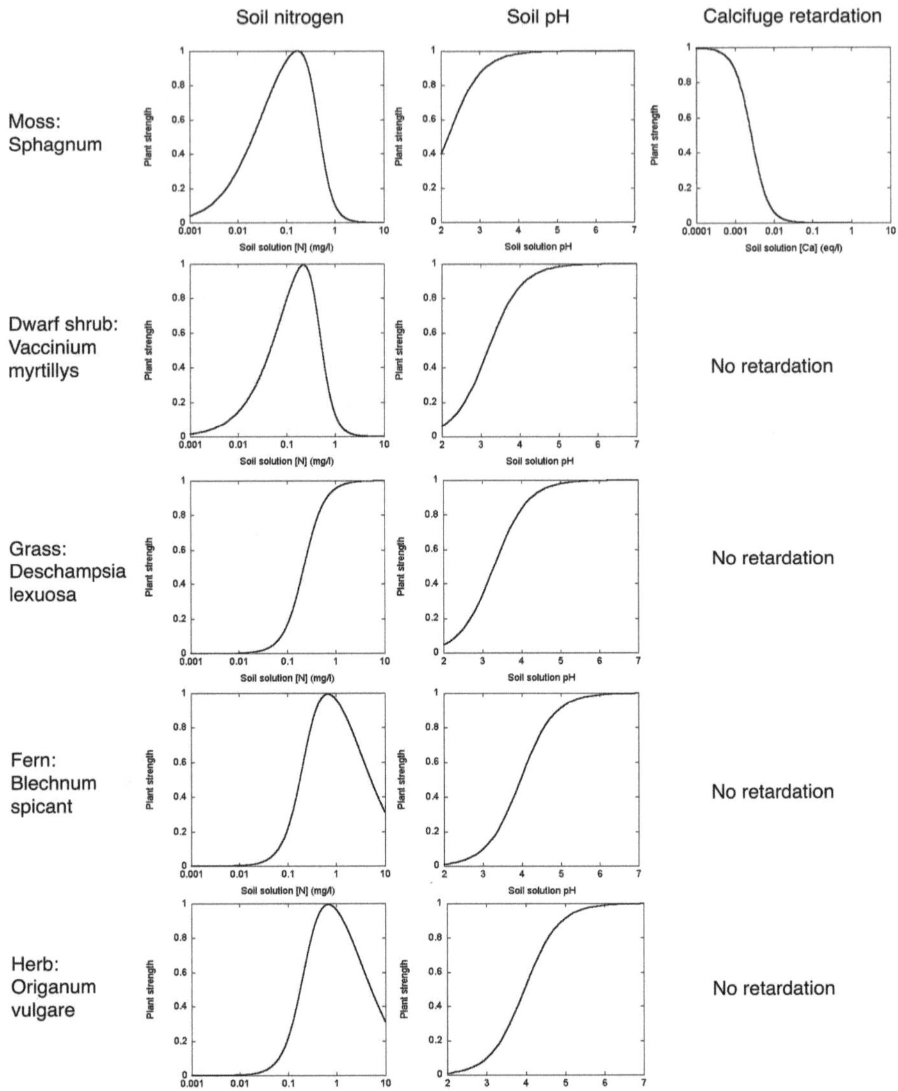

Fig. 12.4 Examples of parametric plant responses to soil solution drivers

threshold [*N*] where the retardation slope starts, w_{-} the slope of the retardation function, and a_0 is a normalization factor.

The pH response is given in Fig. 12.4, and is expressed as a function of [H^+] in the soil solution (Belyazid et al. 2011a, b). The calcifugicity retardation function is defined for a limited number of plants which show signs of decline when their uptake function is impaired by elevated Ca^{2+} concentrations (Sverdrup et al. 2007). The plant responses to soil moisture, air temperature and ground-level photosynthetically active radiation (PAR) are extrapolated between minimal, optimal and

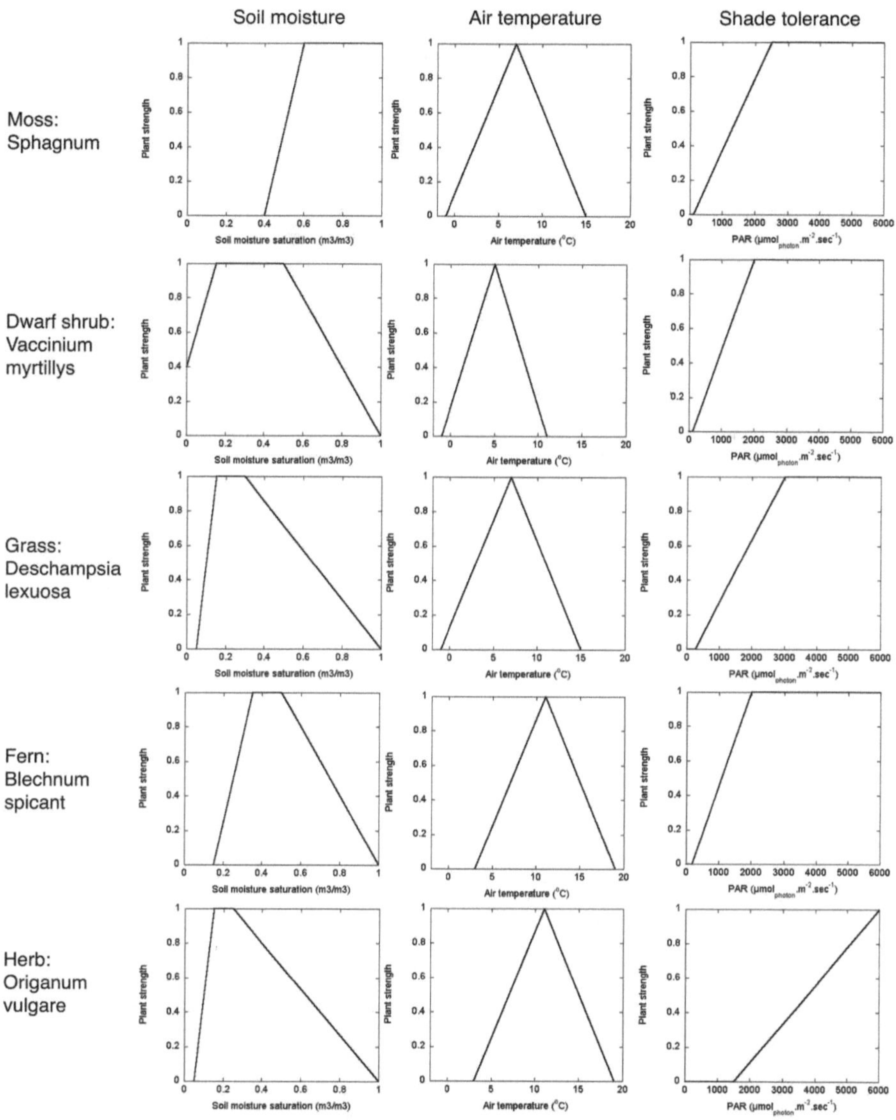

Fig. 12.5 Examples of plant responses to soil moisture, temperature and shading

saturation thresholds (we assume plants do not have any light saturation threshold). Besides the parametric responses described above, each plant has two competitive strategies. The first is shading above ground, and is expressed in the shading height of the plant, i.e. the elevation of the plant organ which is able to shade out other plants. The second is the root distribution over soil depth, which denotes the soil depth that the root plants can reach to have access to water and nutrients. The list of plant parameters that was developed for Sweden is shown in Table 12.1, where 42 indicator plants were parameterised.

Table 12.1 Parameters for ForSAFE-VEG vegetation module

Latin name	t	[N]					$[H^+]$			W			T			I		H	Root	kP	kG
	years	a0	k+	w+	K−	w−	Kbc/al	kbc	kph	Min	Top	Max	Min	Top	Max	Min	Max	(m)	class		
Cladonia lichen	20	1	0.01	1	0.003	3	0.07	0	1050	−0.2	0.05	0.25	−2.5	5.5	13.5	500	2500	0.05	0	0.1	0.7
Hylocomium mosses	20	1	0.03	1	–	–	0.07	150,000	1050	0.05	0.15	0.35	−1	7	15	100	2500	0.02	0	3	0
Mnium mosses	20	1	0.3	2	–	–	0.4	0	6000	0.15	0.25	0.6	0	8	16	50	2500	0.02	0	3	0
Sphagnum moss	20	1	0.03	1	0.1	3	0.01	150,000	150	0.4	0.6	1	−1	7	15	100	2500	0.02	0	1	0
Calluna vulgaris	30	1.4	0.2	1	3	3	0.2	0	3000	−0.25	0.15	0.4	−1	7	15	500	5000	0.25	2	1	0.7
Empetrum nigrum	15	1.6	0.03	1	0.003	3	0.2	150,000	3000	−0.2	0.1	0.4	−1.5	6.5	14	500	5000	0.1	1	1	0
Erica tetralix	15	1.6	0.3	1	0.03	3	0.4	0	6000	0.2	0.35	0.6	0	8	16	1000	5000	0.15	1	1	0
Vaccinium myrtillus	10	1.6	0.1	1	0.1	3	0.1	0	1500	−0.1	0.15	0.5	−1	5	11	100	2000	0.3	1	1	2.3
Vaccinium vitis-idea	15	1.6	0.03	1	0.003	3	0.35	0	5250	−0.2	0.1	0.45	−1.5	4.5	10.5	500	4000	0.15	1	1	0.7
Agrostis capillaries	10	1	0.5	2	–	–	0.2	0	3000	0.05	0.15	0.5	3	11	19	750	4000	0.25	2	3	2.3
Brachio-podium pennatum	5	1	20	2	–	–	6	0	3500	0.1	0.2	0.35	3	11	19	1000	3500	0.5	1	3	9

Table 12.1 (continued)

Latin name	t	[N]					[H⁺]			W			T			I		H	Root	kP	kG
	years	a0	k+	w+	K−	w−	Kbc/al	kbc	kph	Min	Top	Max	Min	Top	Max	Min	Max	(m)	class		
Bromus benekenii	5	1	20	2	–	–	12	0	180,000	0.1	0.2	0.4	5	13	21	250	3000	0.6	2	30	9
Cala-magrostis arundinasius	5	1	0.5	2	–	–	1.8	0	20,800	0.1	0.2	0.4	2	10	18	750	3500	0.5	2	3	0.67
Deschamp-sia cespitosa	5	1	0.5	2	–	–	0.2	0	3000	0.15	0.35	0.6	3	11	19	1000	5000	0.35	2	3	0
Deschamp-sia flexuosa	5	1	0.05	2	–	–	0.13	0	1950	0.05	0.15	0.3	−1	7	15	250	3000	0.2	2	3	2.3
Festuca ovina	10	1.4	0.02	2	10	1	0.1	0	1500	−0.25	0.05	0.25	3	11	19	1500	5000	0.1	1	30	0.67
Milium effusum	5	1	20	2	–	–	8	0	150,000	0.15	0.45	0.6	5	15	20	250	3000	0.5	2	3	9
Molinia caerulea	5	1	1	2	–	–	0.2	0	3000	0.2	0.3	0.45	5	13	21	1000	5500	0.4	2	30	2.3
Nardus stricta	10	1.2	0.05	2	10	1	0.2	150,000	3000	0.15	0.25	0.4	0	8	16	1500	5000	0.15	2	1	0
Poa nemoralis	5	1	5	2	–	–	8	0	120,000	0.05	0.1	0.2	2	10	20	1250	5000	0.4	2	3	9
Dryopteria dilata coll	20	1	0.5	2	–	–	2	0	30,000	0.1	0.3	0.5	3	11	19	150	2500	0.4	2	1	2.3
Pteridium	20	1	0.5	2	–	–	12	0	180,000	0.05	0.2	0.3	2	8	18	750	3250	0.5	2	1	0
Aconitum lycoctonum	20	1	5	2	–	–	10	0	150,000	0.25	0.55	0.9	2	6	10	1000	5000	1	2	1	0

Table 12.1 (continued)

Latin name	t	[N]					$[H^+]$			W			T			I		H	Root	kP	kG
	years	a0	k+	w+	K−	w−	Kbc/al	kbc	kph	Min	Top	Max	Min	Top	Max	Min	Max	(m)	class		
Allium ursinum	2	1	20	2	–	–	40	0	600,000	0.25	0.2	0.6	4	12	20	250	5000	0.25	2	30	0
Anemone nemorosa	10	1	0.5	2	–	–	0.8	0	12,000	0.2	0.3	0.4	2	10	18	250	3500	0.15	1	3	2.3
Antennaria diocia	5	1	0.01	2	–	–	0.1	0	1500	0.05	0.1	0.2	0	6	12	2000	5500	0.01	1	1	0
Arnica montana	5	1	0.01	2	–	–	0.6	0	9000	0.05	0.1	0.2	7	15	20	2000	5500	0.01	1	1	0
Epilobium augustifo-lium	5	1	1	2	–	–	2	0	30,000	0.15	0.2	0.3	0	8	20	1750	5500	0.8	2	3	32
Galium odorata	3	1	5	2	–	–	1.2	0	18,000	0.15	0.25	0.4	3	11	19	250	3000	0.15	1	1	0.67
Geranium sylvestrum	3	1	1	2	–	–	1.8	0	27,000	0.15	0.25	0.4	2	10	14	500	3000	0.5	2	3	9
Hepatica nobilis	20	1	1	2	–	–	8	0	120,000	0.15	0.25	0.4	2	10	18	375	3000	0.5	1	3	0
Mercurialis perennis	5	1	5	2	–	–	2	0	30,000	0.1	0.25	0.4	5	15	20	1000	5000	0.5	1	1	0
Origanum vulgare	20	1	0.5	2	30	1	10	0	150,000	0.05	0.15	0.25	4	12	20	1500	6000	0.04	2	3	0.67
Oxalis acetocella	2	1	0.5	2	–	–	0.2	0	3000	0.1	0.2	0.4	0	8	18	250	3000	0.05	1	1	0
Trientalis	2	1	0.5	2	10	1	0.2	0	3000	0.1	0.2	0.4	2	10	18	250	3000	0.15	1	1	0.67

Table 12.1 (continued)

Latin name	t	[N]					$[H^+]$			W			T			I		H	Root	kP	kG
	years	a0	k+	w+	K−	w−	Kbc/al	kbc	kph	Min	Top	Max	Min	Top	Max	Min	Max	(m)	class		
Trifolium repens	5	1	1	0	–	–	1.3	0	19,500	0.2	0.35	0.4	5	15	25	1250	5500	0.3	2	1	32
Urtica dioica	5	1	5	2	–	–	10	0	150,000	0.15	0.25	0.45	2	10	20	500	5000	0.8	1	3	0
Norway spruce	100	1	0.3	2	30	1	0.33	0	5000	0.1	0.4	0.9	5	15	20	400	700	0.25	1	3	0.7
Sitka spruce	110	1	0.1	2	3	1	0.07	0	1050	0.15	0.45	0.9	2	12	17	600	700	0.25	1	1	0.7
Scots pine	150	1	1	2	100	1	0.28	0	4730	0.05	0.3	0.8	3	13	18	1200	2296	0.2	2	1	0.7
Larch	70	1	1	2	30	1	0.5	0	7500	0.05	0.25	0.8	6	16	25	400	700	0.2	2	3	0.7
Birch	60	1	1	2	100	1	0.25	0	4000	0.15	0.45	0.9	2	12	17	800	1600	0.2	2	1	9
Beech	120	1	3	2	300	1	0.22	0	3500	0.15	0.45	0.7	6	16	30	320	600	0.25	3	3	2.3
Oak	160	1	3	2	300	1	0.2	0	3000	0.05	0.3	0.7	6.5	16.5	35	600	800	0.2	3	3	9
Ash	80	1	1	2	–	–	0.25	0	4000	0.15	0.45	0.7	7	17	35	600	1600	0.25	2	3	9
Norway maple	80	1	3	2	–	–	0.25	0	4000	0.05	0.3	0.7	5.5	15.5	25	160	280	0.2	3	3	9
Myrica qale	10	1	1	2	–	–	0.8	0	12,000	0.25	0.35	0.6	3	7	18	1500	4000	0.6	2	1	0.67
Rhododen-dron toment	10	1	0.03	2	–	–	0.2	150,000	3000	0.25	0.35	0.5	−1	5	9	1000	3500	0.5	2	1	0
Rubus idaeus	5	1	1	2	–	–	1	0	15,000	0.15	0.25	0.4	2	10	18	1500	5000	0.8	2	3	9
Salix lanata	30	1	0.5	1	0.1	3	1	0	9000	0.15	0.35	0.6	−2	2	6	1000	4000	1.2	3	1	2.3
Salix myrsinifolia	30	1	0.5	2	–	–	0.5	0	9000	0.15	0.35	0.6	−1	5	11	1000	4000	1.2	3	1	9

12.3 Indicators for Change in Ground Vegetation Composition

12.3.1 Defining Excessive Change in a Plant Community Composition

Three parameters are crucial to estimating whether a change in the composition of the ground vegetation due to N deposition is acceptable or not:

- The reference population under a given reference deposition, against which eventual changes in the composition of the vegetation are evaluated.
- The target population, i.e. the segment of the ground vegetation community for which change is evaluated.
- The limit of acceptable change, which is the magnitude of divergence of the plant community from the reference population beyond which change in the target population is unacceptable.

For the method testing presented here, future N deposition were used based on a scenario whereby emissions of oxidized and reduced nitrogen are reduced in Europe through the application of best available control technologies (Maximum Feasible Reductions, MFR). MFR is adopted as the reference N deposition, with the corresponding vegetation communities as the reference populations. The target population is defined as the entire plant community. A critical limit of 5% difference is adopted, meaning that total differences in areal cover of plants are acceptable up to a limit of 5% of the site area.

12.3.2 Defining a One-Dimensional Indicator for Changes in Vegetation

Unlike classical chemical indicators, such as base saturation for acidity or nitrogen concentrations for eutrophication, which are one-dimensional, using plant communities as indicator requires tracking multiples species simultaneously. The multiple plant occurrences need to be simplified into a single variable that can be tracked over time and on which a limit can be set to identify excessive change. Another difference with the classical critical load method is that instead of defining a fixed critical values (e.g. Bc/Al = 1), using the plant community as indicator requires a steadily changing reference level, as plant populations are affected by other drivers than N deposition. To solve this problem, the first step is distilling the relative occurrences of different plants to area cover, and tracking the amount of ground area that is different between different scenarios of N deposition over time. In this way, it is possible to produce a single number over time of the magnitude of the difference in the plant community between two deposition scenarios, expressed in the area that is different between a given scenario and the chosen reference scenario. We decided

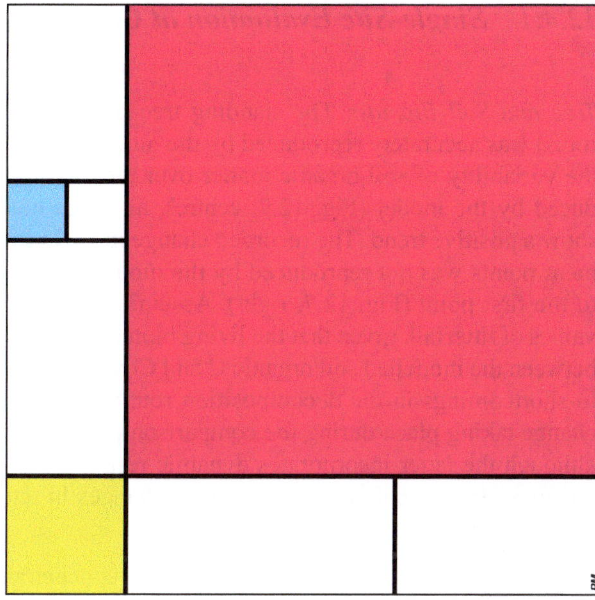

Fig. 12.6 The above 'composition' divides the canvas into non-overlapping monochrome sub-areas. If one colour expands, it would replace another colour partly or entirely. (Inspired by Piet Mondrian's 'Composition with Red, Blue and Yellow' from 1930)

to refer to the area difference between two plant communities as Mondrians, in reference to the famous artwork produced by Piet Mondrian in the 1930s and 1940s (Fig. 12.6). In Mondrian's compositions, each colour in confined to a well-defined area of the canvas, and the entire canvas is occupied by one colour or another. In the same way we model the ground vegetation as species covering a given area. In Mondrian's canvas, if a colour would expand over its given sub-area, it would have to do it at the expense of another colour. In the same way, if a plant expands its cover, it will be at the expense of other plants at the site which will have to retreat. When a site is simulated under different conditions, the mosaic of plant area cover will look different between different scenarios. The amount of area that is different between two scenarios is referred to as M (Mondrians). $1M$ means that 1 % of the area is covered by a chosen target plant group of interest. If the target plant group is the entire plant community, Mondrian units become equivalent to the classical Bray-Curtis similarity index.

12.4 Evaluation of Model Performance

Below we describe an extensive evaluation of the model at a single site (Gårdsjön) and at multiple sites (32 Swiss sites + 16 Swedish sites). The model has also been favourably evaluated at six intensively monitored sites in Switzerland (Aschau, Lurengo, Bachtel, Frienisberg, Olsberg and Wengernalp) as summarized in Belyazid and Braun (2009).

12.4.1 Single-Site Evaluation at Gårdsjön

Tree and Soil Biomass The standing tree biomass (excluding foliage, twigs and roots) was accurately reproduced by the model (Fig. 12.7, left). On the other hand, the variability of soil organic matter over the measurement period was not reproduced by the model (Fig. 12.7, centre), although both measurements and model show a positive trend. The recorded change in C/N ratio between the four measurement points was not reproduced by the model, although the model was calibrated to the first point (Fig. 12.7, right). Assuming that the model produces reasonable values of litterfall, given that the living biomass is well reproduced, the discrepancy between the modelled soil organic C and C/N ratios and the measured values points to shortcomings in the decomposition routine. Possibly the effect of the climatic change taking place during the comparison period was not captured by the model, although the latter incorporates dynamic responses to changes in temperature and moisture. In particular, the response to changes in temperature may be underestimated by the model.

Soil Chemistry The model reproduces the concentrations of chloride in the soil solution and in the runoff water relatively well. Sulphate concentrations are slightly underestimated by the model, and so are the concentrations of sodium and nutrient base cations (Ca^{2+}, Mg^{2+}, K^{+}). Nitrogen concentrations in the upper soil layer are reasonably well reproduced, apart from the reoccurring peaks caused by elevated evapotranspiration during dry periods. Down the soil horizon and in runoff, however, the modelled concentrations of N are lower than the measured ones, (see Figure 8.6 in Chap. 8, where also the differences between modelled and measured concentrations are explained).

Ground Vegetation Composition Belyazid and Moldan (2009) show the model's prediction of ground cover for different plant groups versus the surveyed plants at the site (Fig. 12.8). The model reproduces the dominance structure of the vegetation observed at Gårdsjön, predicting a *Vaccinium myrtillus* dominated community, with mosses as the subdominant plant type (Fig. 12.9). On the other hand, the model predicts the presence of herbs and grasses (*Deschampsia*) that were not recorded in the survey at G2, but appear in the survey of neighbouring plots, where it covers more than 15 % of the area. This difference can be due to the surveying method where only the dominant type of each survey square is reported. On the other hand, the model reproduces the occurrence of each species in relation to all the other modelled plants, thus predicting the presence of some plants that although not reported in the survey may occur at the site.

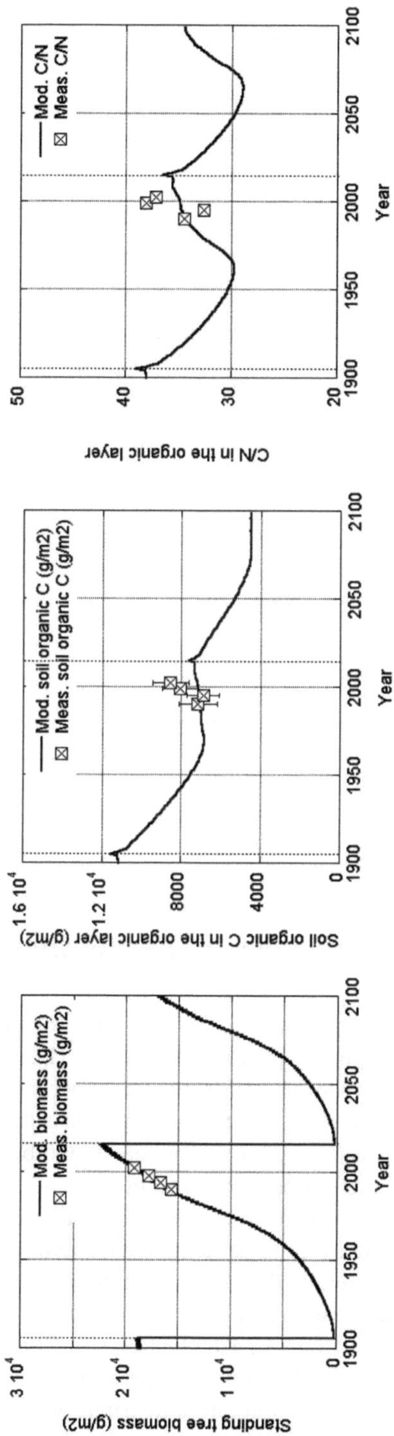

Fig. 12.7 Modelled and measured values of tree biomass, humus soil organic carbon and C/N ratio at the Gårdsjön site

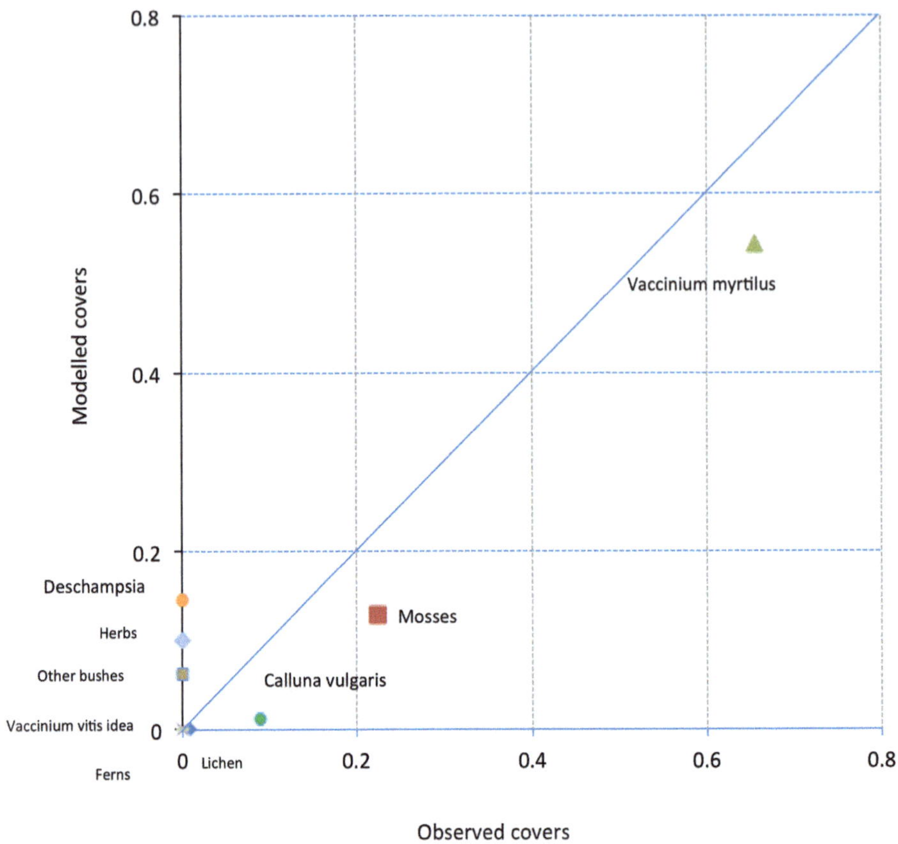

Fig. 12.8 Observed and modelled ground area covers of plant groups at the Gårdsjön site

12.4.2 Multi-Site Evaluation at 32 Swiss Sites

Soil chemical data were available for 32 Swiss sites for the period of 1998–2008, and from two or three depths depending on the site (IAP data), on a biweekly or monthly resolution. The model produces one value per month. It was therefore not possible to match each modelled point to a single measured point in time. Instead, for the one-to-one comparisons, the comparison shown below puts the medians of the measurements over the entire monitored period and for each measured depth against the corresponding medians of the modelled data for each site and depth (all sites plotted together). The same set of data is used in the cumulative distributions comparisons, which also include the ranges of both measured and modelled data.

A comparison of model results and observations for the concentration of Cl, SO_4, Na, Bc and NO_3 are given in Fig. 12.10. Overall, soil Cl concentrations are acceptably estimated by the model, although a regression shows that the model slightly underestimates the values. The cumulative distributions show that around 80% of

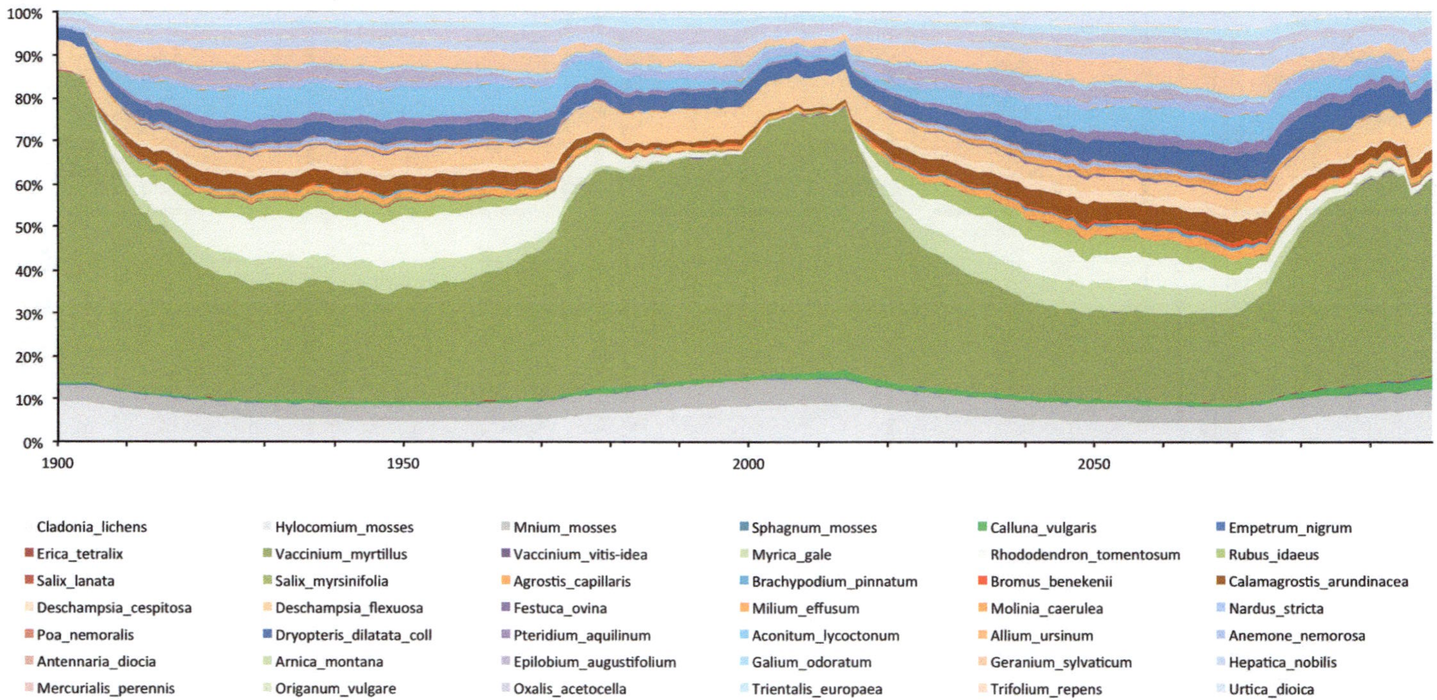

Fig. 12.9 Plant relative ground cover between 1900 and 2100 at Gårdsjön. The Y-axis represents the total area over a 1 m^2, and the fractions that each plant occupies within this m^2. Vaccinium myrtillus is dominant at the site, but retreats markedly after the two clear cutting events (in 1905 and 2015)

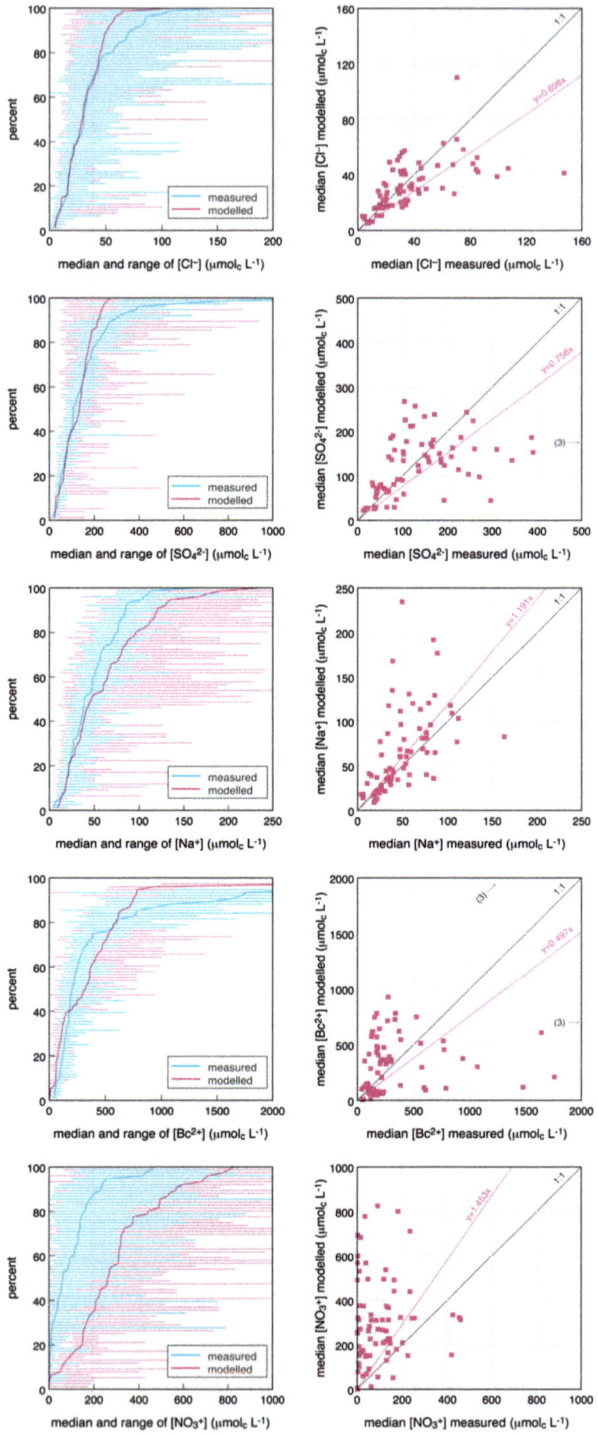

Fig. 12.10 Modelled and measured concentrations of Cl^- (*top*), SO_4^{2-} (*below top*), Na^+ (*middle*) Bc^{2+} (*above bottom*) and NO_{3-} (*bottom*). *Left*: cumulative distributions and ranges. *Right*: site- and depth-specific medians. For BC^{2+}, six points are outside the range shown in the one-to-one diagram, and not included in the regression

the simulated Cl concentrations are well within the range of the measurements. This indicates that the hydrology is relatively well modelled at the sites (Fig. 12.10 top). The simulated SO_4 concentrations vs. the measurements behave similarly to those of Cl, as both elements are mostly steered by deposition and hydrology (in the absence of any significant soil sources of S). Sulphur has a wider spread of values and higher ranges of concentrations (Fig. 12.10, below top). Sodium (Na) is fairly well estimated by the model, with a slight overestimation of the medians (Fig. 12.10 middle). The cumulative distributions show however that the modelled values have a wider spread than the measurements. Na concentrations give an indication of the accuracy of the weathering rates, since the only sources of Na are weathering and atmospheric deposition, and since the deposition is satisfactory considering the above discussion about Cl and S. Although a regression indicates that Bc concentrations are underestimated by the model (Fig. 12.10 above bottom), it may be an artefact of the few extreme points with high measured values and low modelled data. The majority of the points indicate rather that the modelled Bc concentrations are above the measured values. Bc is affected by the biochemical cycle of uptake, litterfall and decomposition, and an underestimation of the concentrations may be due to discrepancies in the biological tissue content of Bc. The cumulative distribution graph shows that the spread of the measured and modelled data is comparably large. Nitrate concentrations are grossly overestimated by the model (Fig. 12.10 bottom). As discussed above for the site-specific model evaluation, the model fails to retain inorganic N other than through uptake by trees. The lack of N retentions in the soil and forest floor vegetation produces higher modelled soil concentrations than in reality. This is a serious limitation of the model and needs to be remedied. The process behind this discrepancy is probably mostly dependent on the biological retention and not the chemical speciation of N in the soil solution.

12.5 Scenario Analysis of Changes in Climate, Deposition and Forest Management

A set of simulations was carried out with the aim of identifying the contribution of climate, deposition and forest management (harvest intensities) in affecting soil chemistry and eventually the composition of the plant community. The simulations were performed on a set of 48 Swedish sites, and the discussion below refers to the median of the time trends of each of the discussed indicators. The two adopted climate scenarios were a future trend with no deviation from today's climate, and a future trend according to the IS92a scenario. For deposition, three scenarios were used: Maximum Feasible Reduction (MFR), Current Legislation (CLE) and No Control (NOC). MFR and CLE diverge after the year 2010, while NOC remains at the 1980s level. Finally, two harvest scenarios were adopted, the first assuming conventional stem-only harvesting and the second whole-tree harvesting, with 80–90% of the above-ground wood and 50% of the foliage harvested. The possible combinations of these scenarios result in 12 simulations. The scenario combinations are denoted with

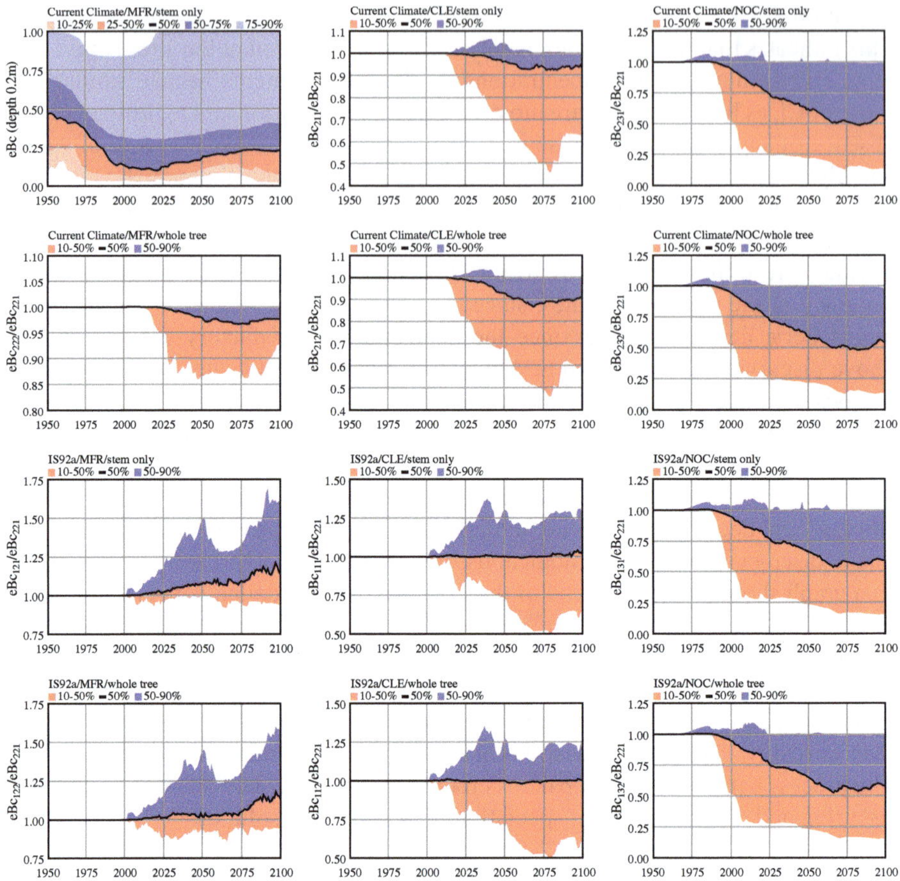

Fig. 12.11 Median and ranges of responses of soil base saturation (BS) at 48 Swedish sites due to climate, atmospheric deposition and harvest scenarios

the indices of the three drivers in this order: climate-deposition-harvest. Scenario 1-1-1 refers to changing climate, CLE deposition, and stem-only harvest. Scenario 2-3-1 refers to unchanging climate, NOC deposition, and stem-only harvest.

12.5.1 Effects of on Soil Chemical Indicators

To visualise the changes over time of different chemical and biological indicators, the diagrams in Fig. 12.11 present trends of the median of all 48 sites over time as well as given percentiles. Scenario 2-2-1, assuming unchanging climate, MFR deposition and stem only harvest, is adopted as the base scenario. For each indicator, a trend over time under the base scenario is given when relevant. The other scenarios are normalised with respect to the base scenario. The base scenario leads to a steady but slow recovery in soil base saturation (BS) over the coming 100 years, that will start to level off towards the end of the 2000s (Fig. 12.11, first figure, top left). If deposition after 2010 follows CLE instead of MFR, the simulations predict a slight loss in BS (Fig. 12.11, second figure, top centre), which will be in median less than 10 % lower than under MFR (both scenarios assume unchanging climate and stem-only harvest). On the other hand, if deposition would have remained at NOC from the 1980s onwards, BS would be around half the level in 2100 as compared to MFR. Whole-tree harvesting has a quite negligible (negative) effect on BS (Fig. 12.11, second row, first column). If climate change is simulated, still keeping MFR as the deposition level, BS would recover by around 20 % by 2100, exclusively due to climate change's effects (Fig. 12.11, third row, first column). The main reason for this simulated boost in BS under climate change is found in the latter's effect on weathering rates and is discussed below. The negative effect of CLE on BS in comparison to MFR is cancelled out by the positive effect of climate change (Fig. 12.11, third row, second column). However, the positive contribution from climate change on BS would not be enough to mitigate the negative effect from NOC, and BS would be in that case slightly less than 50 % below the level of the base scenario.

Weathering is substantially boosted by a change in climate (Fig. 12.12), primarily due to the positive effect of higher soil temperatures. Climate change will increase weathering rates in median by around 30 % by 2100, with a range from 20 to 50 %. Atmospheric deposition according to CLE and NOC cause a reduction in weathering due to the weathering inhibition by Al and Bc concentrations in soil solution. While the inhibitory effect of CLE on weathering is marginal, NOC causes a significant reduction by around 20 %, as it increases Bc mobility (reducing BS) and Al solubility in the soil solution. Both Bc and Al in the soil solution have a retarding effect on weathering rate. When combined with climate change, NOC will reduce weathering in an initial phase (before mid-2000) but will be counteracted by climate change afterward, resulting in a net increase in weathering rates by around 10 % in median, within a range of − 10 % to more than 40 %.

The median Bc/Al ratio declined following the historical increase in acid deposition prior to the 2000s, reaching the critical value of 1 in the 1980s. After emissions are reduced according to MFR, Bc/Al recovers and by the 2050s 80 % of the sites will have a Bc/Al above the critical value of 1 (Fig. 12.13). After the 2050s, while the median remains relatively stable, the 25 percentile will fall below the critical value of 1, probably due to forestry practices (16 clearcuts take place during this period). Atmospheric deposition has the strongest effect on the Bc/Al, with CLE lowering the median Bc/Al by around 40 % and NOC by 90 % in comparison to the reference scenario. While a change in climate can help reverse the negative effect of CLE deposition on the Bc/Al during the second half of the 2000s, it is unable to significantly alter the effect of NOC (Fig. 12.13). Whole-tree harvesting has a negligible negative effect on the median Bc/Al, yet for 25 % of the sites it can reduce the Bc/Al by more than 10 %.

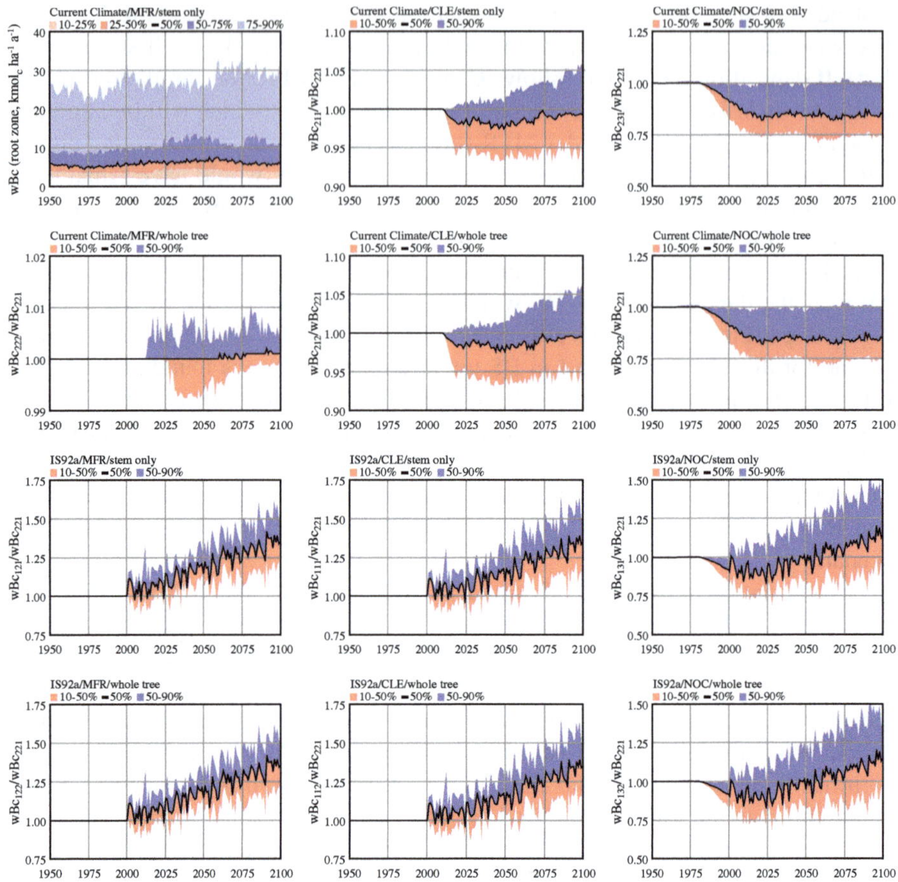

Fig. 12.12 Weathering rates at 48 Swedish sites in response to climate, atmospheric deposition, and harvest scenarios

The N leaching flux follows both the increase and then the decrease in atmospheric N deposition during the second half of the 1900s (Fig. 12.14). MFR will lead to a steady decline in N leaching at all sites, levelling off towards the end of the 2000s at lower levels than in the 1950s. CLE causes a doubling in the median of N leaching, and for 10 % of the sites N leaching reaches up to 1000 times by 2100 (being sites with virtually no N leaching under MFR, meaning that even a small increase in leaching can result in a manifold order of magnitude augmentation). Whole-tree harvesting causes a marginal decrease in N leaching. Climate change is expected to cause a doubling of the median of N leaching, and a two orders of magnitudes increase for 10 % of the sites. Climate change will also aggravate the increased N leaching caused by NOC when the two are combined, but the two factors are non-additive.

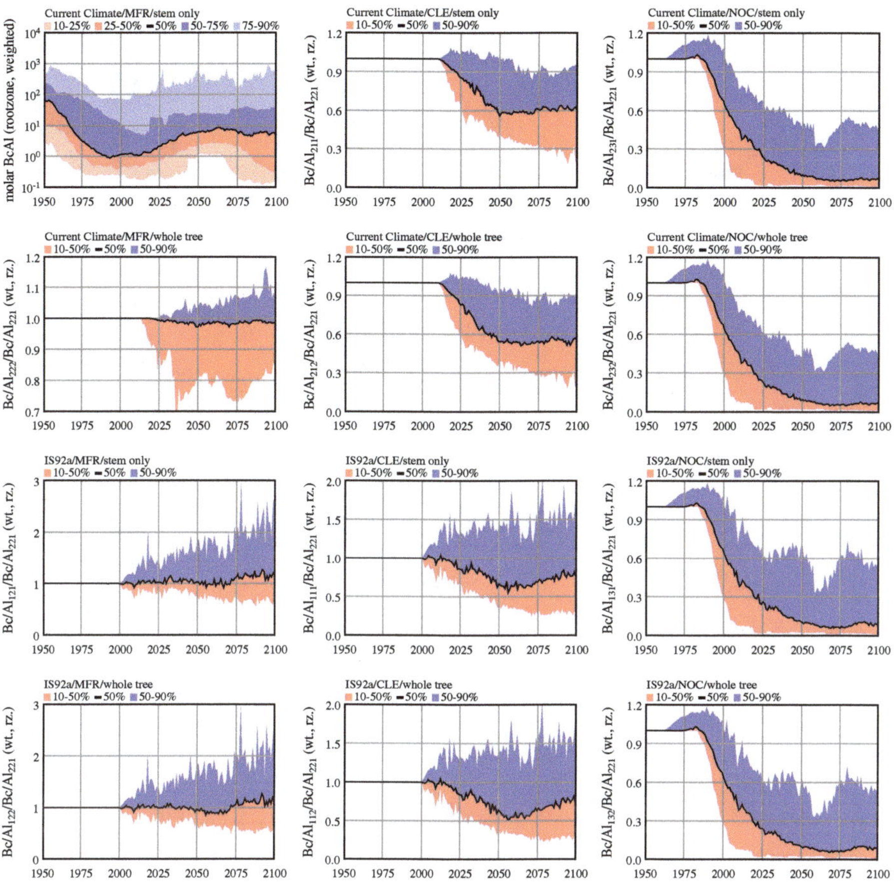

Fig. 12.13 Temporal trends of the Bc/Al ratio at the 48 Swedish sites under 12 scenarios

12.5.2 Effects on Ground Vegetation Composition

At each site, the plant community is composed of a multitude of plants present simultaneously at the site (theoretically 42 plants can be represented in Sweden, and 89 in Switzerland). At Gårdsjön for example, the plant community composition varies over time, with changing relative ground covers of present plants and invasions and losses of other species over time which are driven by clear-cuts as well as changes in climate and in atmospheric deposition (Fig. 12.9; Belyazid and Moldan 2009). It is not evident to visualise the plant communities of the 48 simulated Swedish sites. In the analysis below we focus on the area difference at each site as explained in Sect. 12.3.2. The ground vegetation communities established under MFR, no climate change and stem-only harvest are adopted as the reference communities for the multiple driver analysis presented here. Increasing future atmospheric deposition from MFR to CLE leads to around 7*M* difference in

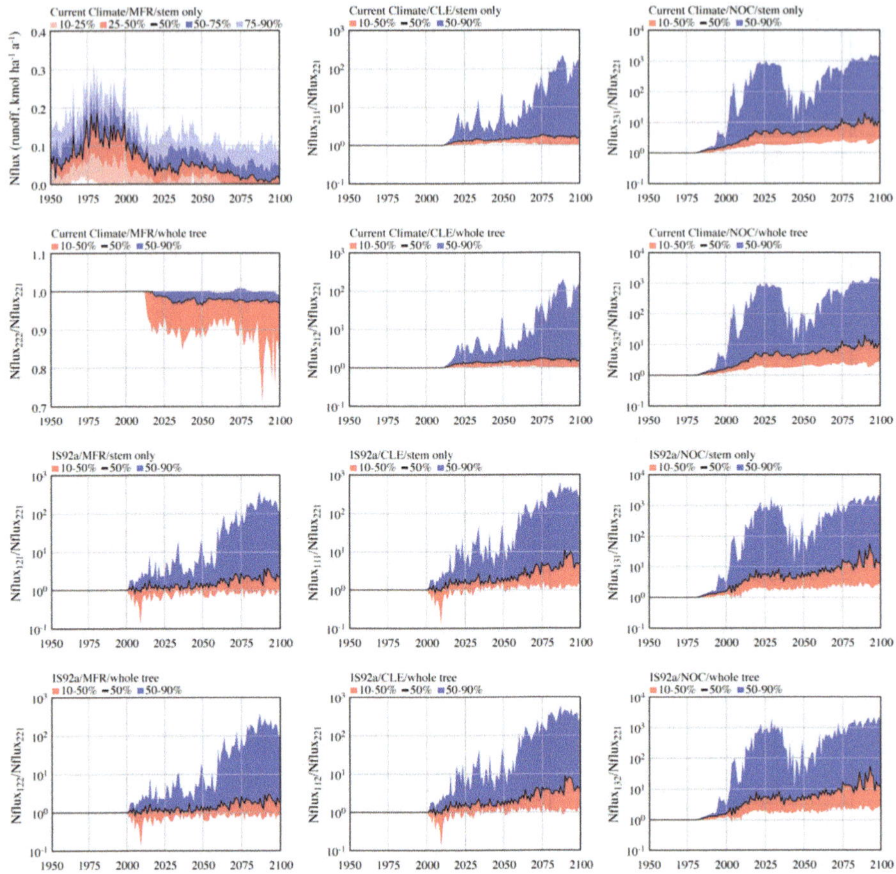

Fig. 12.14 Median changes in N leaching fluxes and relative deviations from the base scenario (current climate/MFR deposition/stem-only harvest)

the median of the simulated sites by 2100, while NOC leads to a 25*M* difference. The changes caused by CLE and NOC increase steadily over time, reflecting the increased accumulation of N in comparison to MFR. While the plant community median difference levels off under CLE (but causing a 20*M* difference at 5 % of the sites) NOC drives a continuously growing difference (Fig. 12.15).

Climate change alone causes a 35*M* difference in median of the simulated plant communities. When combined with CLE deposition, the difference is marginal, while NOC drives a further 5*M* difference. This may be due to the fact that climate changes itself leads to an important mobilisation of inorganic N, reducing the relative importance of atmospheric deposition in controlling the availability of N in the soil solution. Whole-tree harvesting has the least effect on the plant communities, causing only around 1*M* difference in the absence of climate change and negligible under climate change.

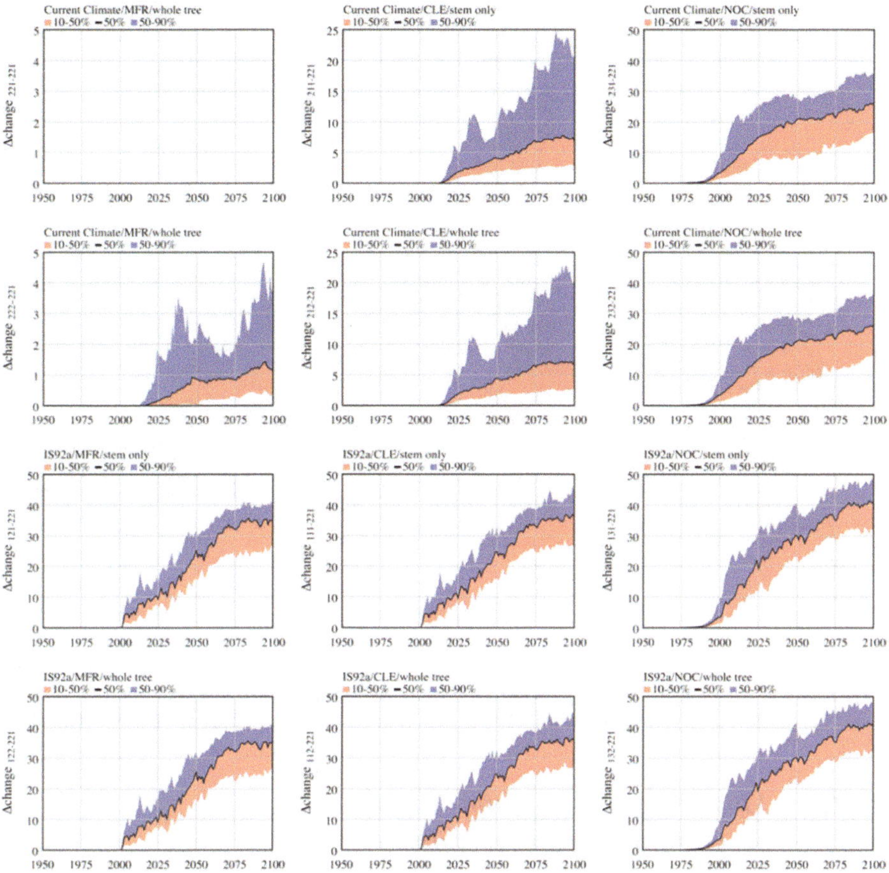

Fig. 12.15 Median of differences in area cover of ground vegetation communities due to different climate, deposition and harvest intensity scenarios

12.6 Regional Risk Assessments of N Deposition on Ground Vegetation Cover by Dynamic Modelling and Assessing Critical Load Exceedances

A regional assessment of N deposition effects of ground vegetation and related critical loads of N was performed by making use of the Swedish national terrestrial database for critical loads consisting of 640 sites. In order to calculate the critical loads of N deposition based on its effect on the modelled vegetation communities, background N deposition is used as the reference deposition.

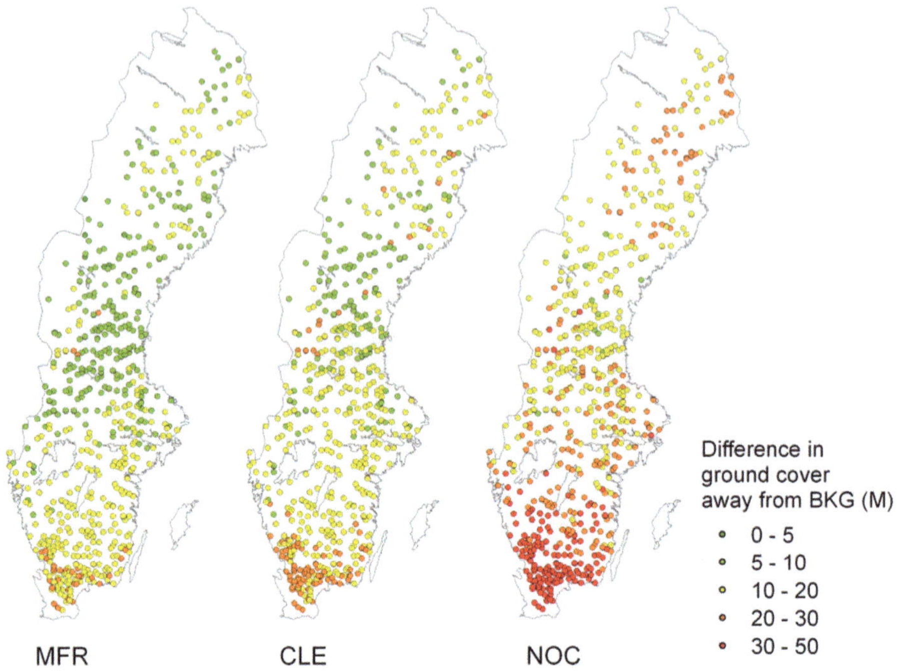

Fig. 12.16 Impacts of N deposition according to the Maximum Feasible Reduction (MFR), Current Legislation (CLE) and No Control (NOC) scenario on plant responses at 640 Swedish sites

12.6.1 Modelling N Deposition Impacts on Ground Vegetation Cover

As with the 48 sites, the impact of N deposition on ground vegetation cover was estimated for three scenarios, i.e. Maximum Feasible Reduction (MFR), Current Legislation (CLE) and No Control (NOC). The divergence in the modelled plant communities caused by MFR, CLE and NOC is shown in Fig. 12.16. All deposition scenarios cause higher divergence in the simulated plant communities in the south of the country, where deposition is higher under any of the scenarios. N deposition under the NOC scenario leads to over 50*M* divergence in the majority of the southern sites and in a substantial fraction of sites in the rest of the country, implying that more than half the communities' areal composition would have changed, if the LRTAP Convention had not come into force.

12.6.2 Modelling Critical N Load Exceedances in View of Impacts on Ground Vegetation Cover

Critical loads of N deposition were simulated under three different critical limits/assumptions: (1) the reference community was defined as under hypothetical back-

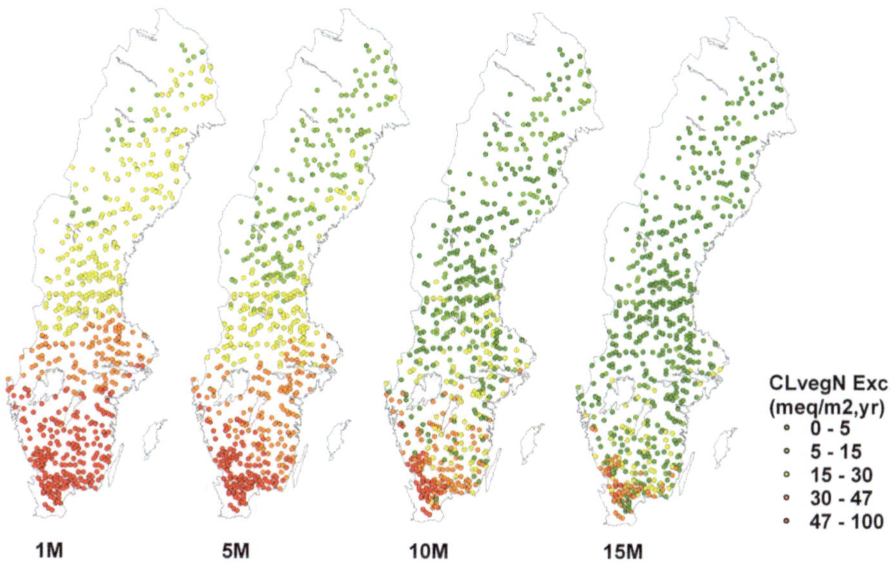

Fig. 12.17 Exceedances of critical loads of N defined for different critical limits of vegetation areal cover change at 640 Swedish sites. One Mondrian unit *M* is equivalent to 1 % of area cover by plants

ground deposition; (2) the target population was defined as the entire community; and (3) the critical loads were estimated for four different critical limits (see Belyazid et al. 2011a) for a detailed description of the assumptions adopted). An important implication of assumption 2 is that any change in any constituent of the plant community is proportional in importance to the reference occupancy of that plant. This means that a change in a marginal plant is reflected as marginal, while a change in a dominant plant can be more strongly reflected in the CL estimation.

Figure 12.17 shows the exceedances of CLE deposition to different critical loads defined for gradually increasing critical limits (i.e., increasingly more tolerant definitions of the level of harmful effect). With the acknowledged limitations of this method of weighing the plants in a community (Van Dobben et al. 2010; see also Chap. 3), the model can produce sensible results (Fig. 12.17). Depending on the critical limit of change which is tolerated for the critical load (i.e. the level of change which is still below a significant harmful effect), different critical loads are produced, and the lower the critical limit (the more stringent the definition of a harmful effect is), the lower the critical load and the higher the critical load exceedance. With a critical limit of 1*M* (i.e. a change of 1 % of the total area cover of a plant community), CLE deposition exceeds the critical load in nearly all of Sweden, with the exception of some of the mountainous ecosystems in the northeast of the country. On the other side of the tested scale, CLE deposition exceeds the critical load of N only in parts of southern and south-western Sweden, when the critical limit is set to 15*M*.

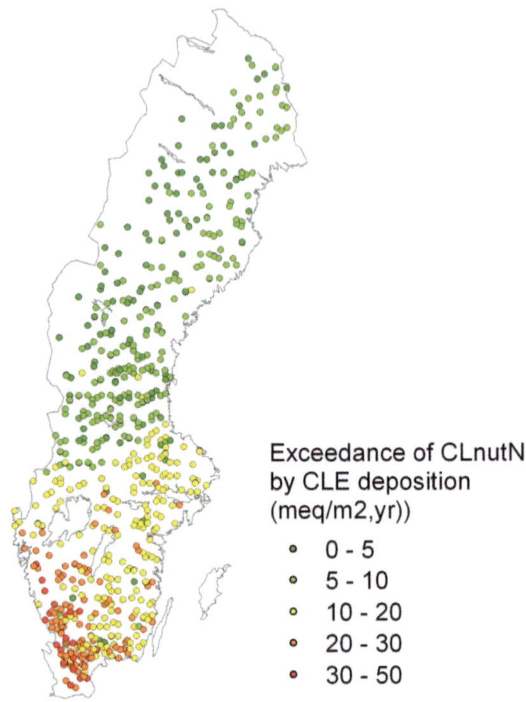

Fig. 12.18 Exceedances of nutrient N critical loads based on a limit of 0.3 mgN l^{-1} in runoff below the rooting zone at 640 Swedish sites

Calculated critical loads of N, based on a critical nutrient N concentration of 0.3 mg N l^{-1} in runoff water (Fig. 12.18), are comparable to that estimated at 10*M* based on changes in plant communities. Roughly, this comparison means that if atmospheric deposition of N is not allowed to exceed the level causing runoff N concentrations to reach or exceed 0.3 mgN l^{-1}, it will not cause more than 10*M* change in forest floor plant communities.

Conclusions

The pilot study presented here shows that it is possible to simulate the link between atmospheric deposition and the ensuing biological impact, in this case the composition of the plant community, while at the same time keeping track of the concurrent effects of changes in climate and land use. The multiple driver analysis presented above indicates that climate change and atmospheric deposition have comparably important effects on soil chemistry, and that these effects are not additive. The intensity of forest harvesting has a lesser effect on soil chemistry.

On the other hand, the ground vegetation responds most strongly to the two climate scenarios tested, and only marginally to harvest intensities. The difference in response between soil chemistry and plant community is due to the fact that plants are directly affected by temperature and moisture in addition to soil chemistry. It is therefore essential to include climate change in any future simulations of ecosystem responses to atmospheric deposition. The presented methodology provides a tool to investigate the delay times governing plant communities' responses to change in environmental conditions. Because of this, it has been shown that the adoption of the Maximum Feasible N-emission Reduction scenario (MFR) as the reference deposition, with its historically accumulated N during the elevated deposition period prior to emissions reductions, is not appropriate. Instead, a reference level should be used that allows the reference population to reach a stationary level reflecting low ecosystem N loads. The present simulations are limited to a set of specific sites. The method presented needs to be tested at a wider geographical scale to investigate its implications on the estimation of critical loads on a level which is meaningful for the LRTAP Convention. In Chap. 13 an application in France is described. Finally, in order to establish critical loads specific to desired ecosystem traits, it is noted that target plant populations and the protection limits need to be specified for different ecosystem types that are relevant for this trait.

Acknowledgments This study would not have been possible without the generous support of the Swiss Federal Office for the Environment (FOEN) and the Swiss National Focal Centre for modelling and mapping activities under the UNECE LRTAP Convention. The Swedish critical loads effort has profited significantly from the Swiss cooperation and support. The study used models and datasets developed within the Swedish long-term research programs ASTA 1997–2006 and SUFOR 1996–2005, supported by the MISTRA research foundation in Sweden. This work was jointly financed by FOEN, the Swedish Environmental Protection Agency (Naturvårdsverket—NV) and the Coordination Centre for Effects (CCE located at RIVM, The Netherlands). The authors would like to thank specifically Jean-Paul Hettelingh from the CCE for his continuous and constructive input, Beat Achermann from FOEN, and Titus Kyrklund from NV for their support and constructive input to the work and report.

References

Belyazid, S. (2006). *Dynamic modelling of biogeochemical processes in forest ecosystems*. Doctoral Thesis. Reports in Ecology and Environmental Engineering 2006:1. Sweden: Department of chemical Engineering, Lund University.

Belyazid, S., & Braun, S. (2009). *Estimating the effects of whole tree harvest and ash recycling at selected forest stands in Switzerland with the ForSAFE model*. Report for the Swiss Institute for Applied Plant Physiology.

Belyazid, S., & Moldan, F. (2009). *Use of integrated dynamic modelling to set critical loads for nitrogen deposition based on vegetation change—Pilot study at Gårdsjön*. Report B1875. Swedish Environmental Research Institute IVL.

Belyazid, S., Westling, O., & Sverdrup, H. (2006). Modelling changes in forest soil chemistry at 16 Swedish coniferous forest sites following deposition reduction. *Environmental Pollution, 144*, 596–609.

Belyazid, S., Bailey, S., & Sverdrup, H. (2010). Past and future effects of atmospheric deposition on the forest ecosystem at the Hubbard Brook Experimental Forest: Simulations with the dynamic model ForSAFE. In G. Hanrahan (Ed.), *Modelling of pollutants in complex environmental systems* (pp. 357–377). Chicago: ILM Publications.

Belyazid, S., Sverdrup, H., Kurz, D., & Braun, S. (2011a). Exploring ground vegetation change for different deposition scenarios and methods for estimating critical loads or biodiversity using the ForSAFE-VEG model in Switzerland and Sweden. *Water, Air, and Soil Pollution, 216,* 289–317.

Belyazid, S., Kurz, D., Braun, S., Sverdrup, H., Rihm, B., & Hettelingh, J. P. (2011b). A dynamic modelling approach for estimating critical loads of nitrogen based on plant community changes under a changing climate. *Environmental Pollution, 159,* 789–801.

De Vries, W., Wamelink, G. W. W., van Dobben, H., Kros, J., Reinds, G. J., Mol-Dijkstra, J. P., Smart, S. M., Evans, C. D., Rowe, E. C., Belyazid, S., Sverdrup, H. U., van Hinsberg, A., Posch, M., Hettelingh, J.-P., Spranger, T., & Bobbink, R. (2010). Use of dynamic soil-vegetation models to assess impacts of nitrogen deposition on plant species composition: An overview. *Ecological Applications, 20,* 60–79.

Ellenberg, H., Weber, H. E., Düll, R., Wirth, V., Werner, W., & Paulißen, D. (1992). Zeigerwerte der Pflanzen in Mitteleuropa. *Scripta Geobotanica 18,* 1–248.

Falkengren-Grerup, U. (1992). *Mark och floraförändringar i sydsvensk adellövskog.* Rapport 4061. Naturvårdsverket.

Fenn, M. E., Haeuber, R., Tonnesen, G. S., Baron, J. S., Grossman-Clarke, S., Hope, D., Jaffe, D. A., Copeland, S., Geiser, L., Rueth, H. M., & Sickman, J. O. (2003). Nitrogen emissions, deposition, and monitoring in the western United States. *BioScience, 53,* 391–403.

Flückiger, W., & Braun, S. (1998). Nitrogen deposition in Swiss forests and its possible relevance for leaf nutrient status, parasite attacks and soil acidification. *Environmental Pollution, 102,* 69–76.

Körner, C. (2003). *Alpine plant life: Functional plant ecology of high mountain ecosystems.* Heidelberg: Springer.

Lambers, H., Stuart Chapin, E., & Pons, T. (1998). *Plant physiological ecology.* Berlin: Springer.

Landolt, E. (1977). *Ökologische Zeigerwerte zur Schweizer Flora.* Zürich: ETH.

Martinson, L., Alveteg, M., Kronnäs, V., Sverdrup, H., Westling, O., & Warfvinge, P. (2005). A regional perspective on present and future soil chemistry at 16 Swedish forest sites. *Water, Air, & Soil Pollution, 162,* 89–105.

Nilsson, J., & Grennfelt, P. (1988). *Critical loads for sulphur and nitrogen.* Report from a Workshop held at Skokloster Sweden March 19–24 1988. Miljø rapport 1988: 15. Copenhagen Denmark Nordic Council of Ministers.

Rowe, E. C., Moldan, F., Emmett, B. A., Evans, C. D., & Hellsten, S. (2005). *Model chains for assessing impacts of nitrogen on soils, waters and biodiversity: A review. Centre for Ecology and Hydrology.* Contract report project no C02887 for DEFRA (UK).

Sverdrup, H. U., & Warfvinge, P. G. (1988). Assessment of critical loads of acid depostion on forest soils. In J. Nilsson & P. Grennfelt (Eds.), *Critical loads for sulphur and nitrogen. NORD 1988:97* (pp. 81–129). Copenhagen: Nordic Council of Ministers.

Sverdrup, H., Stjernquist, I., Thelin, G., Holmqvist, J., Wallman, P., & Svensson, M. (2005). Application of natural, social, and economical sustainability limitations to forest management, based on Swedish experiences. *Journal of Sustainable Forestry, 21,* 147–176.

Sverdrup, H., Belyazid, S., Nihlgård, B., & Ericson, L. (2007). Modelling change in ground vegetation response to acid and nitrogen pollution, climate change and forest management at in Sweden 1500–2100 A.D. *Water, Air, and Soil Pollution: Focus, 7,* 163–179.

Van Dobben, H., Hettelingh, J. P., De Vries, W., Slootweg, J., & Reinds, G. J. (2010). Plant species diversity indicators for impacts of nitrogen and acidity and methods for their simulation: an overview. In J. Slootweg, M. Posch, & J. P. Hettelingh (Eds.), CCE status report 2010.

Wallman, P., Svenssson, M., Sverdrup, H., & Belyazid, S. (2005). ForSAFE—An integrated process-oriented forest model for long-term sustainability assessments. *Forest Ecology and Management, 207,* 19–36.

Chapter 13
Evaluation of Plant Responses to Atmospheric Nitrogen Deposition in France Using Integrated Soil-Vegetation Models

Anne Probst, Carole Obeidy, Noémie Gaudio, Salim Belyazid, Jean-Claude Gégout, Didier Alard, Emmanuel Corket, Jean-Paul Party, Thierry Gauquelin, Arnaud Mansat, Bengt Nihlgård, Sophie Leguédois and Harald U. Sverdrup

13.1 Introduction

Anthropogenic activities have significantly increased the emissions to the atmosphere of reactive nitrogen (Nr, Galloway et al. 2003), mostly in its reduced (NH_4^+) and oxidized (NO_3^-) form. These nitrogen species are dispersed in the atmosphere and lead to atmospheric deposition, causing an increase in the acidification and eutrophication of terrestrial ecosystems (De Schrijver et al. 2008; De Vries et al. 2007; Galloway et al. 2003; Tarrasón and Nyíri 2008). In addition to sulphur, they

A. Probst (✉) · C. Obeidy · N. Gaudio · A. Mansat · S. Leguédois
Université de Toulouse ; INP, UPS ; EcoLab (Laboratoire Ecologie Fonctionnelle et Environnement), ENSAT, Castanet Tolosan, France
e-mail: anne.probst@ensat.fr

CNRS ; EcoLab ; Castanet-Tolosan, France

S. Belyazid
Belyazid Consulting & Communication AB, Malmö, Sweden

J.-C. Gégout
AgroParisTech, UMR 1092 INRA-AgroParistech, Laboratoire d'Etude des Ressources Forêt-Bois (LERFoB), Nancy, France

D. Alard · E. Corket
UMR BioGeco Université Bordeaux 1, Talence, France

J.-P. Party
Sol-Conseil, Strasbourg, France

T. Gauquelin
IMBE, Aix-Marseille University, Provence, France

B. Nihlgård · H. U. Sverdrup
Department of Chemical Engineering, University of Lund, Lund, Sweden

W. de Vries et al. (eds.), *Critical Loads and Dynamic Risk Assessments*, Environmental Pollution 25, DOI 10.1007/978-94-017-9508-1_13

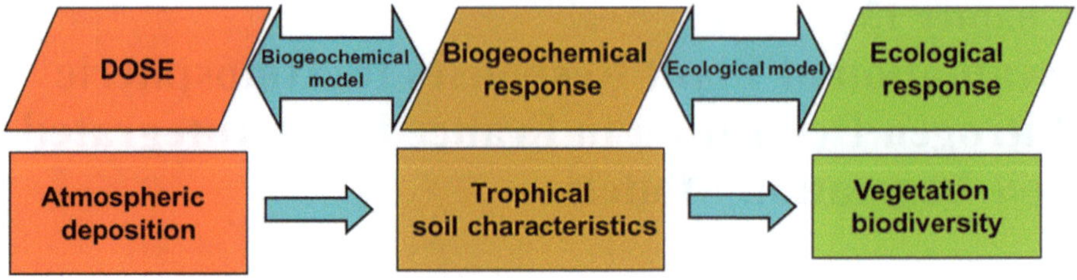

Fig. 13.1 Approach and hypothesis of the coupled biogeochemical-ecological modelling

contributed to forest dieback (Friedrich et al. 2012; Hutchinson et al. 1998; Probst et al. 1995) in eastern and northern countries over Europe in the eighties and also to the increased occurrence of nitrophilous species (Galloway et al. 2003).

Similarly to the other European countries, the semi-natural ecosystems in France are under the influence of N deposition due to long-range transboundary air pollution (Rogora et al. 2006; Tarrasón and Nyíri 2008), which is considered one of the main factors influencing the decrease in vegetation diversity in Europe (De Schrijver et al. 2008; De Vries et al. 2007).

Within the 1979 Convention on Long-range Transboundary Air Pollution (LRTAP), critical loads (see Chap. 1) are used as a tool to help regulate the air pollution over Europe and evaluate its impact on (forest) ecosystems (Nilsson and Grennfelt 1988; Posch 2002; Probst et al. 2008; Probst and Leguédois 2008). For nitrogen, the critical load models already developed (steady-state or dynamic geochemical models, empirical models, see Hettelingh et al. 2008) were obviously not adequate to reach objectives in terms of vegetation biodiversity conservation namely regarding ecological richness. Indeed, they took into account the influence of nitrogen deposition on vegetation only in an indirect manner by applying an assumed critical limit of N concentration in soil solution (allowable drainage flux; see Chap. 2) or in terms of a recommended tolerable maximal deposition based on expert advice. In France, these two kinds of approaches, i.e. the use of a biogeochemical steady-state model and empirical critical loads, lead to a major difference in critical load evaluation, which was estimated to 400 eq $ha^{-1}yr^{-1}$ for 50 % of French forest ecosystems in a range of value of 200–1500 eq $ha^{-1}yr^{-1}$ (Probst and Leguédois 2007).

Following the 25th meeting of the ICP Modelling and Mapping of the Working Group on Effects under the LRTAP Convention, a new approach was proposed based on coupling biogeochemical dynamic models with ecological integrated models (De Vries et al. 2007), in order to better take into account the impact of N deposition on biodiversity. This new approach was based on the hypothesis that N deposition will lead to a modification on biogeochemical characteristics of the soils (e.g. pH, C/N ratio). These changes in trophic conditions were supposed to cause changes in ecological response of the vegetation community (Fig. 13.1). The dose (atmospheric N deposition) was the origin of the biogeochemical response and as a cascade process provoked the ecological response (Belyazid et al. 2006;

Sverdrup et al. 2007; Wallman et al. 2005). To evaluate nitrogen critical loads, dynamic models have to be run in inverse mode, i.e. it is necessary to calculate the maximum deposition that will not cause the exceedance of biogeochemical critical limits that might imply significant changes in ecological richness (Fig. 13.1). Thus it was required: (i) to define objectives of vegetation biodiversity protection; (ii) to determine the trophic limits to reach the protection objectives using an ecological model; (iii) then, using a biogeochemical model, to assess the maximum atmospheric deposition, i.e. the critical load, that led to exceedance of trophic limit conditions.

In Europe (United Kingdom, Netherlands, Switzerland, Sweden) various kinds of coupled models have been developed and tested: SMART2-SUMO-MOVE/NTM, MAGIC-SUMO-GBMOVE (see Chap. 11), BERN (Chap. 14) and ForSAFE-VEG (Belyazid et al. 2006, 2011a, b, Sverdrup et al. 2007, 2012). However, a synthesis of these pioneer work indicated that there is a need for adaptation and up-scaling of the models beyond the regions for which dose-response relationships have been parameterized, to make them generally applicable (De Vries et al. 2010). It proved challenging to extrapolate the use of these ecological models to areas for which no calibration is available, since:

- the geographical extension of vegetation species is limited;
- for a given species, the ecological response to trophic conditions could vary over the European scale territory (Diekmann and Lawesson 1999). Moreover, concerning the biogeochemical module, it is necessary to evaluate the sensitivity and the applicability at the country scale taking into account the variables and parameters needed as input.

The assessment of these possibilities is a fundamental step before applying this modelling approach at the country scale.

In this chapter, we propose as a first approach to apply the two biogeochemical models developed and used for critical load evaluation over Europe and northern America, ForSAFE (Wallman et al. 2005) and VSD$^+$ (Bonten et al. 2009) and their coupled module VEG, on selected French ecosystems. The specific aims were:

- to test the feasibility of applying two dynamic models (ForSAFE and VSD$^+$) on forest sites with different environmental conditions (characterized by different climatic conditions, soil physico-chemical properties such as C/N ratio and pH, nitrogen deposition, vegetation types…)
- to build a vegetation table composed of main forest species covering various ecological conditions typical of the very large biodiversity of French territory, to run the VEG module;
- to validate and run the coupled biogeochemical-vegetation models: VSD$^+$-VEG and ForSAFE-VEG;
- to evaluate the variability of the model responses in term of species richness and composition predicted under current nitrogen atmospheric deposition and to discuss their sensibility.

13.2 Methods/Approach

13.2.1 Description of the Study Sites

As the French ICP-Forests network (called RENECOFOR, the national network for long-term monitoring of forest ecosystems) has already been used for a considerable number of investigations, forest stands are very well documented (Gandois et al. 2010a, b; Hernandez et al. 2003; Party 1999). In this study, four documented forest sites (CHS41, EPC08, SP57 and PM40c) belonging to this network (Fig. 13.2 and Table 13.1) were selected to run the VSD$^+$ and ForSAFE-VEG models. The mean N deposition and climatic characteristics are listed in Table 13.2. The two sites from the north-eastern part of France received the highest nitrogen inputs and were dominated by spruce (EPC08) and silver fir (SP57), whereas oak and pine are the dominant species in CHS41 and PM40c, respectively. Acid soils such as cambic podzol (EPC 08), dystric cambisol (SP 57), gleyic podzol (PM40c) and stagnic luvisol (CHS41), were also addressed (Gandois et al. 2010a; Hernandez et al. 2003).

13.2.2 Models Description and Calibration

The two models VSD$^+$ and ForSAFE were selected since they are mechanistic models proved to be efficient in simulating the biogeochemical response to nitrogen deposition in soils. They are both linked to a VEG-module, which allows testing the forest vegetation response to nitrogen deposition (Belyazid et al. 2006, 2011a; Posch and Reinds 2009).

13.2.2.1 The VSD$^+$ Model

The VSD$^+$ model is an extension of the simpler VSD model (Posch and Reinds 2009), which was primarily developed to simulate acidification processes in soils. The extensions of VSD$^+$ concerns primarily the inclusion of the carbon (C) and N cycles (Bonten et al. 2009, 2011) to make it more suitable for applications in climate change (C sequestration) and biodiversity studies. Still, VSD$^+$ is a simple model designed for applications on a (large) regional (European) scale. Thus certain processes (e.g. sulphate sorption and aluminium complexation) were neglected, and the input data requirements were kept to a minimum. For a more detailed description of the VSD$^+$ model see Chap. 8. VSD$^+$ is a single layer model, and here a soil depth of 30 cm was chosen to represent the rooting depth of the vegetation.

The input data for VSD$^+$ were mainly extracted from the French RENECOFOR network, and more precisely from the four forest sites described previously, and from the French critical load database. To ensure regional application of VSD$^+$,

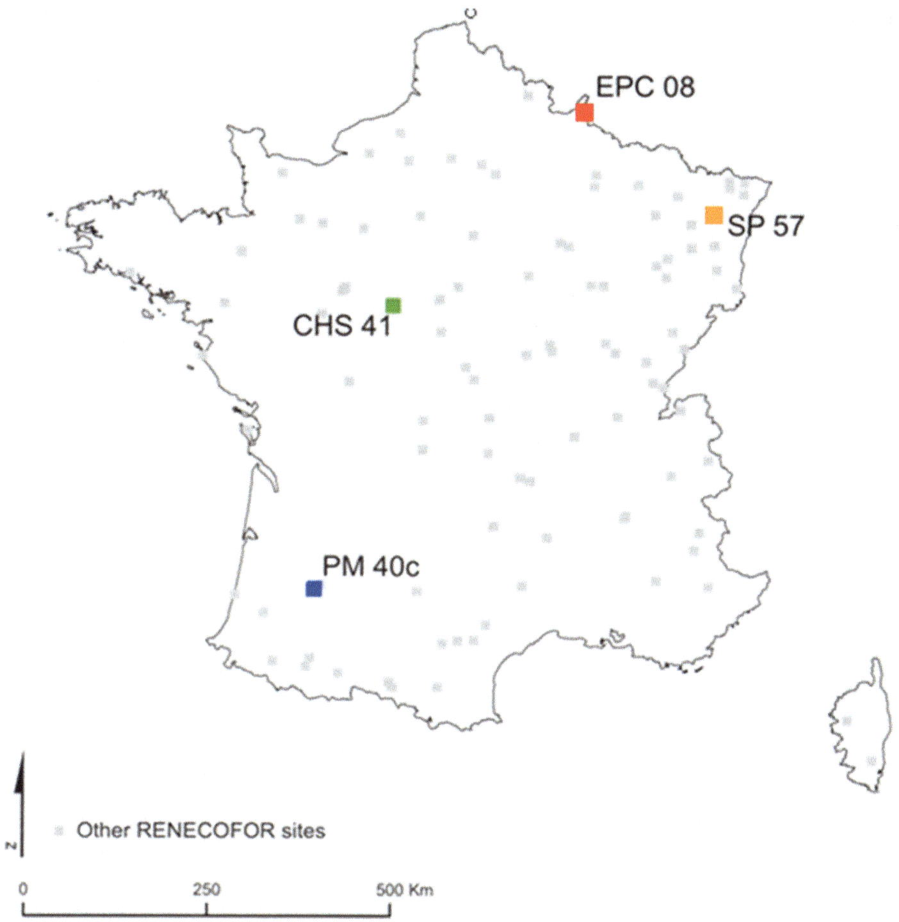

Fig. 13.2 Location of the selected RENECOFOR forest sites

EMEP deposition data (NO_x, NH_y, S) were used except for base cations deposition data (Ca, Mg, K and Na) that came from the RENECOFOR sites.

The initial state of the model corresponded to the early state of the present-day woodland. At this stage, the initialization of the model required data for the following parameters: C/N ratio, C and N pools, base saturation etc. Since no measured initial data were available, the model generated those data from (present-day) observations, using a Bayesian calibration method (Posch and Reinds 2009). The factors for mineralization ($rf_{\_min}$), nitrification ($rf_{\_nit}$) and denitrification ($rf_{\_denit}$) due to moisture and temperature were computed by the MetHyd model developed with VSD$^+$ (Posch and Reinds 2010). Mean vegetation growth and litter production were considered as constant over time. Indeed, a growth function could be calculated in different ways depending on data available on biomass production and harvest. In

Table 13.1 Input data to VSD⁺ for the selected RENECOFOR sites. (Brêthes and Ulrich 1997)

Parameter [unit]	CHS41	EPC08	SP57	PM40c
Period simulated	1900–2100	1960–2100	1941–2100	1980–2100
Thickness [cm]	0.3	0.3	0.3	0.3
Bulk density [g cm^{-3}]	1.183	1.004	1.38	1.024
Theta[a] [m m^{-1}]	0.3051	0.3776	0.2279	0.2133
CEC [meq kg^{-1}]	39.6	52.22	26.67	32.56
Temperature[a] [°C]	11.05	8.609	8.307	12.87
Percolation[a] [m yr^{-1}]	0.1759	0.586	0.3729	0.584
Ca weathering [eq $m^{-3}yr^{-1}$]	0.00774	0.00304	0.00026	0.00016
Mg weathering [eq $m^{-3}yr^{-1}$]	0.0018	0.00038	0.0018	0
K weathering [eq $m^{-3}yr^{-1}$]	0.00146	0.00042	0.0013	0.00014
Na weathering [eq $m^{-3}yr^{-1}$]	0.00988	0.00766	0.00048	0.00012
$Bsat_{obs}$[b] [eq $m^{-3}yr^{-1}$]	0.1854	0.039	0.1348	0.3402
$Cpool_{obs}$[b] [g m^{-2}]	5732	4937	2240	4544
$Npool_{obs}$[b] [g m^{-2}]	299.8	294.4	119.5	179.5
$CNrat_{obs}$[b] [g g^{-1}]	19	17	18,75	26
pH_{obs}[b]	4,4	4,1	4,1	4.45
Dominant species	Quercus petraea (L.)	Picea abies (L.)	Abies alba (Mill.)	Pinus pinaster (Ait.)

[a] Calculated with the MetHyd model
[b] Measured values in 1995

Table 13.2 Open field N depositions (meq $m^{-2}yr^{-1}$), annual rainfall (mm) and annual air temperature (°C) for the selected RENECOFOR sites (mean over 1993–2008)

Site code	Mean NO_x (meq $m^{-2}yr^{-1}$)	Mean NH_y (meq $m^{-2}yr^{-1}$)	Mean annual rainfall (mm)	Mean air temperature (°C)
CHS41	17.6	26.1	766	11.4
EPC08	42.4	64.9	1398	8.8
PM40C	18.2	22.5	938	9
SP57	36.9	44.2	1353	13

the future, as described in the VSD⁺ manual, a logistic growth function might be used to take into account human activities like wood uptake and harvest, as well as tree growth. Validation results of the model outputs are shown in Fig. 13.3.

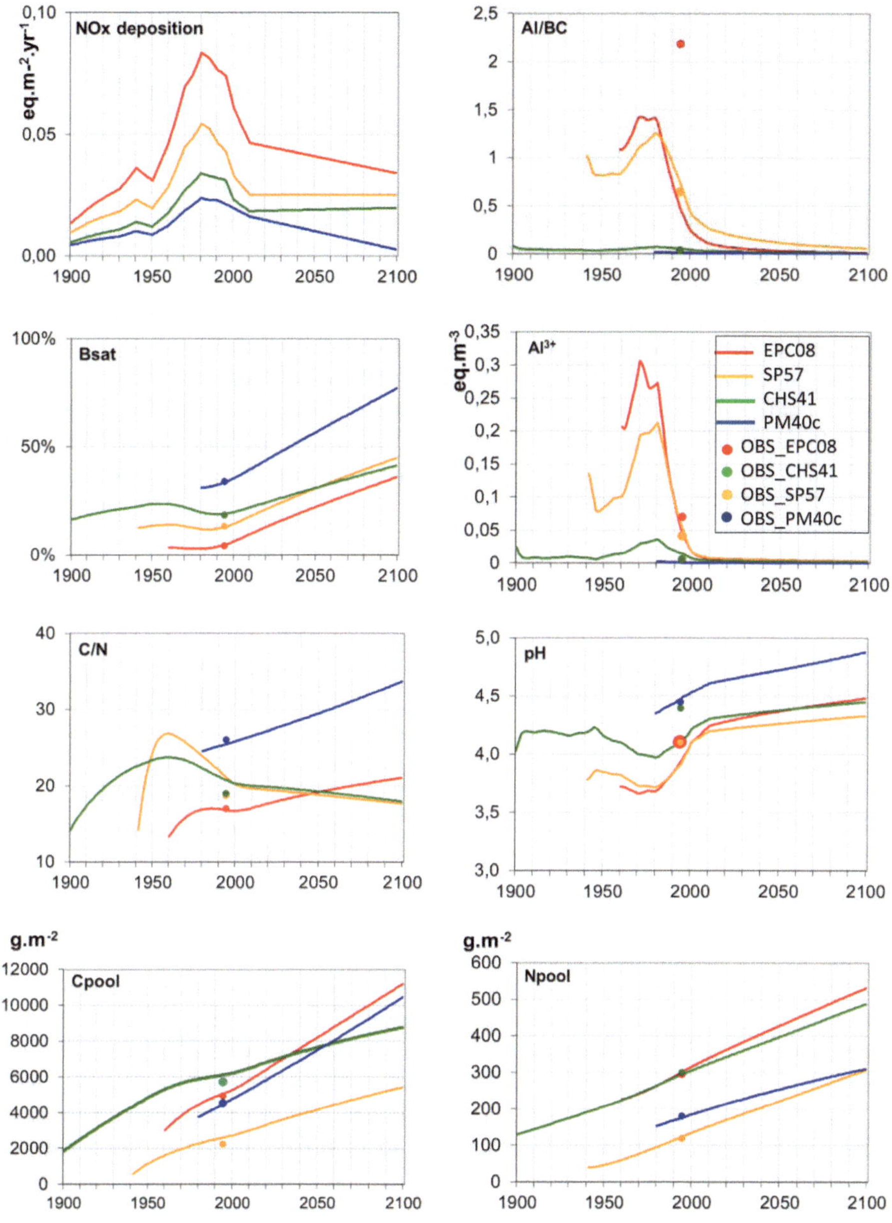

Fig. 13.3 VSD+ output simulations (*continuous lines*) for the selected forested sites from the RENECOFOR network, and corresponding simulated EMEP nitrogen deposition. When available, the measured data (1995) are marked (*circles*)

13.2.2.2 The ForSAFE-VEG Model Chain

ForSAFE: ForSAFE is a mechanistic biogeochemical model developed in Sweden (Wallman et al. 2005) and regularly improved (Belyazid et al. 2006, 2011b). The main goals of the model are to simulate (i) the cycles of carbon, nitrogen, base cations, and water in forest ecosystems; (ii) the soil acidity and aluminium mobility (Al^{3+}). ForSAFE is composed of four modules dedicated to specific soil and forest processes. Used at a monthly stepwise, ForSAFE simulates base cation weathering rate, soil water content, element concentrations in soil solution, flux of nitrogen and base cations accumulated in plants, mineralisation and decomposition rates of the litter, photosynthesis and vegetation growth (Belyazid 2006). Tree biomass growth is simulated by a module derived from the PnET model (Aber and Federer 1992). Photosynthesis is determined by the nitrogen content in leaves and by the amount of light reaching the site, then corrected by water and nutrients availability. The tree biomass is simulated as three tree compartments (leaves, wood and roots). This module provides information on:

- the element flux absorption to the soil chemistry module;
- the water absorption to the hydrology module; and
- the litter flux to the decomposition module.

To simulate the soil chemistry behaviour, the SAFE model has been adapted and linked to the other modules (Alveteg et al. 1995). In the chemistry module, the fluxes from weathering and from cationic exchange processes, the aluminium mobilisation, as well as the soil solution chemistry balance, are simulated. The hydrological module was adapted from the PULSE model (Lindström and Gardelin 1992) which simulates the vertical transfer of soil solution from one layer to another and the evapotranspiration. Finally, the model DECOMP was included to simulate the litter decomposition in relation with chemical and hydrological soil conditions (Wallman et al. 2006).

Input and output data in ForSAFE are more numerous than in VSD$^+$ since ForSAFE takes into account more complex processes, and also considers multiple soils layers. Consequently, this allows a better control of the simulations at each site, but on the other hand applications over more sites are limited since more input data are needed, and their availability is not so frequent at a large scale. Moreover, ForSAFE has no convenient graphic interface and basic data should be integrated. Nevertheless, some scripts were developed recently to facilitate the use of input data on several forest sites.

VEG: VEG is a module simulating the composition of the plant communities in terrestrial ecosystems (Sverdrup et al. 2007). VEG reconstitutes the abundance of different kinds of understorey plants in a specific site, depending on the response of the plants to temperature and surrounding light, soil water content, soil solution pH, and nitrogen and base cation concentrations in soil solution. Therefore, the data required to run VEG are related to all those parameters, which can be implemented directly in the VEG-table or by the way of ForSAFE.

ForSAFE model was run on three of the previously selected sites (except PM40c because of the lack of soil solution data). ForSAFE-VEG was then applied using the French VEG-database to perform trials on one of the forest sites (CHS41).

13.2.3 The French Vegetation Database

A French VEG-table (see examples in Table 13.3) was set up by a consortium of French and international experts to take into account the diversity of plant species in the French forest ecosystems. More than 230 species were added to the initial plant list set up in Sweden and Switzerland. This database is now included in a VEG European database. The extension of the species list for France was set up during a dedicated workshop with vegetation experts in October 2009. Relevant species were chosen to represent the various French forest ecosystems on the basis on expert knowledge. For each plant added to the plant list which was already documented for Sweden and Switzerland, the VEG-parameters were estimated compiling several sources of data (i.e. Rameau et al. 1989) and the new species were documented. For some parameters, the link between existing databases and the VEG-parameters needed a scale calibration (such as "nitrogen classes based on C/N ratio", "pH" and "temperature", which were obtained from EcoPlant database (Gégout et al. 2005). The main parameters (see Table 13.3) were determined as follows:

- The delay time expressed in years, based on average generation time and lifespan, was drawn from the French flora (Rameau et al. 1989, 2008) and expert opinions.
- The classes characterizing vegetation that is promoted by enhanced nitrogen availability were based on C/N values extracted from the EcoPlant database (Gégout et al. 2005) and adapted to the VEG classes. For the missing species of EcoPlant, the information was found in the French flora and using the Ellenberg N parameter (Bardat et al. 2004; Ellenberg et al. 1992; Ellenberg 1988; Julve 1998; Louvel et al. 2013).
- The classes characterizing vegetation whose growth is inhibited by enhanced nitrogen availability and the water and light response classes were deduced from the French flora.
- The lowest pH value was from the EcoPlant database and from the French flora when missing.
- The minimum temperature (the lowest annual average temperature when the plant can start taking ground) was extracted from the EcoPlant database and from the French flora when missing.
- The effective shading height was deduced from the French flora. For trees and shrubs, the height was considered only for seedling with a standard height of 0.1 m.

Table 13.3 Examples of VEG parameters in the new French vegetation table

Latin name	k+	k-	w+	k(Ca)	k(pH)	Wmin	Wtop	Wmax	Tmin	Ttop	Tmax	Lmin	Lmax	hmin	Root class	k(Grazing)
Lycopodium annotinum	0.10	0.30	2.00	0.00	3.87	0.15	0.35	0.62	−1.00	7.00	15.00	12.50	208.33	0.10	1.00	0.00
*Sphagnum nemorum_*	0.17	0.30	2.00	51.74	2.18	0.40	0.60	1.00	−1.00	7.00	15.00	8.33	208.33	0.06	0.00	0.00
Vaccinium myrtillus	0.36	0.10	2.00	0.00	3.18	0.10	0.15	0.39	−1.00	5.00	11.00	8.33	166.67	0.30	2.00	2.30
Carex pilulifera	0.22	0.10	2.00	0.00	4.32	0.05	0.15	0.39	−1.00	7.00	15.00	20.83	250.00	0.10	1.00	2.30
Sesleria albicans	0.22	0.10	2.00	0.00	6.37	0.10	0.25	0.51	3.00	11.00	19.00	125.00	500.00	0.20	2.00	0.70
Polytrichum formosum	017	0.10	2.00	0.00	4.95	0.10	0.15	0.39	0.00	8.00	15.00	8.33	208.33	0.03	0.00	0.00
Leucobryum glaucum	0.01	0.10	1.00	0.00	3.48	0.10	0.15	0.39	2.00	10.00	18.00	8.33	208.33	0.02	0.00	0.00
Pinus pinaster	0.20	0.10	2.00	0.00	3.42	0.05	0.15	0.39	−1.00	7.00	15.00	50.00	183.33	0.10	3.00	0.70
Pinus uncinata	0.20	0.10	2.00	0.00	4.17	0.15	0.25	0.51	4.00	12.00	20.00	83.33	216.67	0.10	3.00	0.70
Salix cinerea	5.43	0.10	2.00	0.00	4.89	0.15	0.25	0.51	4.00	12.00	20.00	66.67	333.33	0.10	3.00	9.00
Larix decidua	0.87	0.10	2.00	0.00	4.37	0.10	0.20	0.45	7.00	15.00	23.00	66.67	208.33	0.10	3.00	9.00
Hypnum cupressiforme	0.14	3.00	1.00	0.00	3.37	0.14	0.22	0.47	7.70	10.10	12.50	16.67	221.67	0.02	0.00	0.00
Vaccinium uliginosum	0.37	1.00	2.00	0.00	3.37	0.14	0.22	0.47	2.00	2.00	4.80	33.33	248.83	0.10	2.00	9.00
Erica cinerea	0.20	1.00	2.00	0.00	3.84	0.04	0.14	0.38	10.60	12.10	13.60	33.33	248.83	0.30	2.00	0.70
Arctostaphyllos uva ursi	0.35	1.00	2.00	0.00	3.97	0.04	0.14	0.38	2.10	7.00	12.00	25.00	235.25	0.50	2.00	0.70
Erica_tetralix	0.32	1.00	2.00	0.00	4.07	0.21	0.28	0.54	4.10	10.00	12.00	33.33	248.83	0.30	2.00	0.70
Danthonia decumbens	0.74	1.00	2.00	0.00	4.49	0.09	0.18	0.43	10.20	11.90	13.50	33.33	249.25	0.10	1.00	0.70
Polytrichum juniperinum	0.74	1.00	2.00	0.00	3.37	0.04	0.14	0.38	2.00	2.00	7.60	25.00	235.25	0.01	0.00	0.00

- The browsing based on the palatability characteristics of the species was extracted from the literature, the pastoral floras (Boulanger et al. 2009; Bruneton 2005; Dorée 1995; Gusmeroli et al. 2007; Morellet and Guibert 1999) and expert advices.

13.3 Results and Discussion

13.3.1 VSD⁺ Output and Validation

Figure 13.3 shows the trends in C and N pool, C/N ratio, base saturation (Bsat), Al^{3+}, Al/BC ratio and pH under the influence of the evolution of nitrogen deposition over 200 years (1900–2100) depending on the early state of the present-day woodland in the sites.

The forecasted C and N pool trends in the four selected sites show a good agreement with observed values. The C and N pools were both continuously increasing, despite the decrease in N deposition by the nineties; nevertheless for the two sites that correspond to the youngest forests (EPC08 and PM40c), the C pool increased faster related to the rapid growth of the coniferous trees probably stimulated by the "fertilizing nitrogen input". Indeed, the highest nitrogen deposition corresponded to the period of trees plantation. This increasing pattern is in accordance with the literature (Bonten et al. 2009). However, this result could also be related to some specific inputs such as modifying factors of mineralization (rf_{-min}), nitrification (rf_{-nit}) and denitrification (rf_{-denit}) that are known to largely influence the C/N ratio. These parameters were poorly documented, and values used here may require further verification. Further studies are needed in order to improve these input estimates, especially by using adapted French meteorological data (e.g. missing sunshine data) to calibrate these parameters using the MetHyd model.

At all the sites, the C/N ratio shows an increasing trend, even before the period of increasing N deposition at the beginning of the simulations. This may be interpreted as a significant N uptake for the growth of young vegetation. In PM40c, the less significant increase may be linked to the available N content in soil following the high N deposition in this period (around 1981). The C/N ration increased in EPC08 and PM40c in agreement with the significant increase in C pool, but the ratio decreased from 1980 in CHS41 and SP57. These latter forest sites are characterized by soils with a low nitrogen mineralization (Brêthes and Ulrich 1997).

The increasing N deposition led to the acidification of the four forest soils, as shown by the simulated pH and base saturation pattern. From 2000, pH and Bsat increased at the same time as N deposition decreases. Regarding acidification process, the impact of N deposition seemed more important than the influence of the vegetation and the soil type, as already shown by Moncoulon et al. (2007). With a lower N deposition level, PM40c displayed the lowest perturbations. However, the model underestimated the soil pH values for all sites (except for PM40c), but base saturation remained overall well predicted.

13.3.2 ForSAFE Output and Validation

Before applying the ForSAFE-VEG platform, ForSAFE was validated on the selected French forest sites. The assessment of model performance consisted of the comparison between predicted and observed values (Belyazid et al. 2006, 2011a). The concentrations of Na^+ and Cl^- ($\mu eq\ l^{-1}$) were used to validate the hydrological ForSAFE module since they were powerful tracers in those types of soils (Fig. 13.4). Validation was performed on about the first 20–30 cm. A non-seasonal lag time between modelled and measured data was observed at some sites as well. The observed model behaviour discrepancy between sites might be explained by the differences in deposition levels between sites (i.e. at EPC08, Na and Cl deposition was twice that measured at CH41). Indeed, some improvements are needed in ForSAFE, including an improved representation of hydrological processes linked to water horizontal fluxes and to topography. Accounting for the mineralization of twigs and barks could further help in improving the low values that are forecasted for base cations in the soil solution (data not shown). For that purpose, a new version of ForSAFE is currently under development and validated on various sites and data.

13.3.3 Comparison Between and Validation of ForSAFE-VEG and VSD⁺-VEG

Plants from the French VEG-list (Table 13.3) have been classified into main group types (grasses, herbs, mosses, lichens and trees). As a comparison purpose, ForSAFE-VEG and VSD$^+$-VEG were run on the CHS41 RENECOFOR site, the site with the most documented dataset. The observed and modelled data of the ground cover for the different plant groups were compared on Fig. 13.5.

To validate VEG model performance, the list of species predicted by the VEG model was compared with the species observed in the field relevés (if available). The first validation was established on site CHS41 for the year 2005. Among about one hundred species that ForSAFE was able to predict for such a site, only 23 species were observed in the site relevés: *Teucrium scorodonia, Melampyrum pratense, Deschampsia flexuosa, Lonicera periclymenum, Holcus mollis, Atrichum undulatum, Polytrichum formosum* were modelled and observed in this site with similar abundance. Most of them were typical plants of acidophilus oak grove characterizing this forest site. The model output was encouraging in term of ability to predict in a good way the plant response. However, about ten species observed in the field were not predicted by the model, for some of them obviously because they were note inside the VEG table. Finally, more than 80 species predicted by ForSAFE were not observed in the field relevés. Among them, some might have a good probability of occurring in the field, but others not at all because of their known temperature range. For example, some mountainous species like *Cicerbita alpina, Arctostaphylos uva ursi* or Atlantico-mediterranean species like *Quercus suber*, or

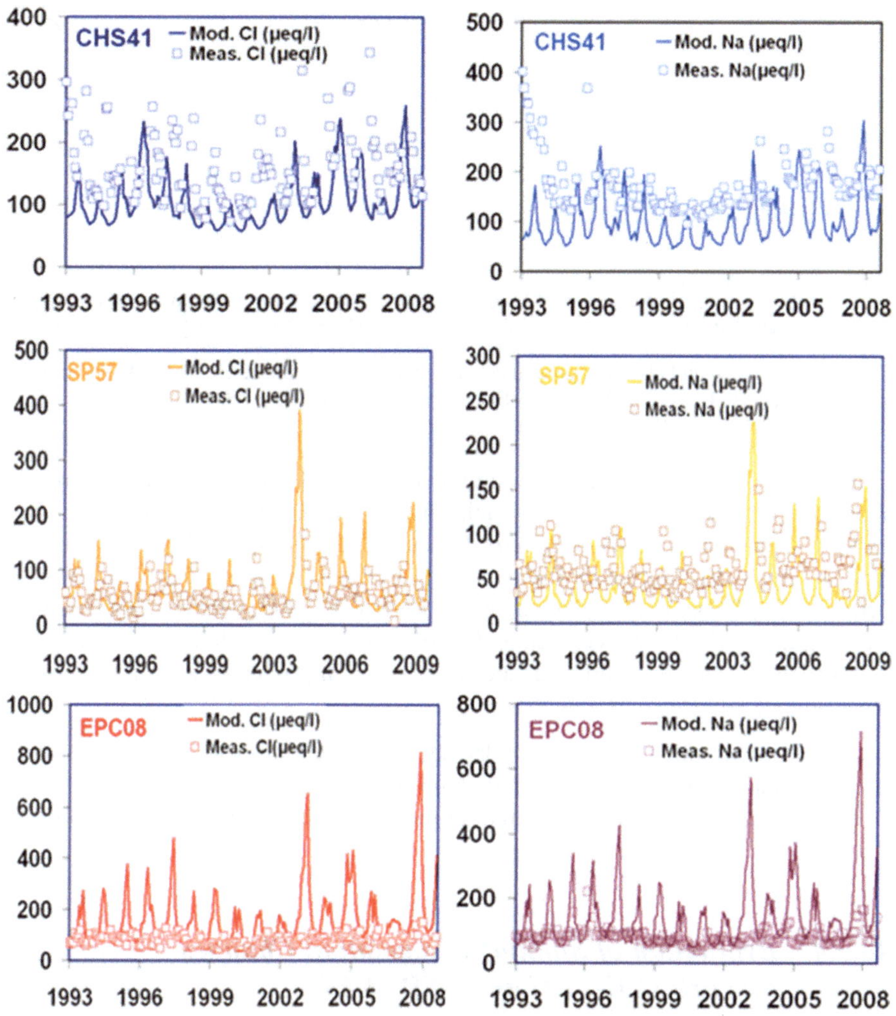

Fig. 13.4 Predicted (*continuous lines*) and measured (*squares*) Na and Cl concentrations in soil solutions between 1993 and 2009 for the selected RENECOFOR sites (CHS41, EPC08 and SP57)

more typical Mediterranean species like *Pistacia sp.*, were mostly improbable to be found in this environmental condition. The presence of these species in the outputs indicated that improvements were needed, and this could be improved by a more thorough validation on some typical sites of the RENECOFOR network since relevés were available. This could help also to improve the parameterization of the VEG-table in order to better predict the occurrence and abundance of the species, particularly as regard to their belonging to large biogeographical regions in Europe.

Unlike ForSAFE-VEG, the VSD$^+$-VEG model only predicted the percentage of occupancy of plant species and not the number of species that were matched by the model and observed in the site. For this reason, only the 17 species predicted

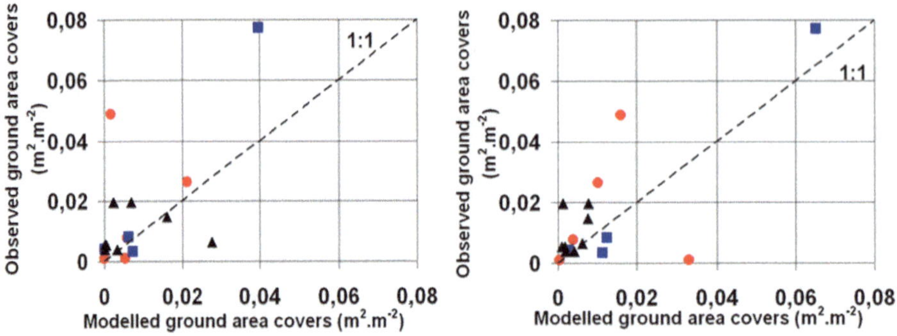

Fig. 13.5 Modelled and observed ground cover of plant groups: grasses: ● herbs: ■, and mosses: ▲ for the site CHS41 using FORSAFE-VEG (*left*) and VSD$^+$-VEG (*right*)

by ForSAFE were used to compare to the VSD$^+$-VEG model outputs. The results showed (Fig. 13.5) that the two models underestimated the occurrence of certain forbs or herbs (such as *Hedera helix L* listed in the group) and grasses (such as *Holcus mollis L.*). Moreover, the presence of certain mosses (such as *Dicranella heteromalla Hedw.*) was overestimated by ForSAFE-VEG, whereas others are underestimated by VSD$^+$-VEG (such as *Polytrichum formosum Hedw.*) (Fig. 13.5).

From these trials, it could be deduced that some improvements were still needed regarding:

- pH and C/N, which are the main driving parameters influencing the occurrence of vegetation species (Gégout et al. 2005) for both models;
- the adjustment of some parameters in the VEG-table such as pH and temperature.

13.3.4 Evaluation of Simulated Vegetation Changes by ForSAFE-VEG and VSD$^+$-VEG

Figure 13.6 shows the simulated changes in the occurrence of ground vegetation plant groups (% of all simulated plant groups) as predicted by the two models from 1900 to 2100 in the CHS41 site. ForSAFE-VEG model results include variations in the computed occurrence of the plant groups. Herbs and grasses were computed to be dominant between 1920 and 1930. This could be related to the modelled behaviour of nitrogen and carbon pools and to site reforestation as mentioned in (Ponce et al. 1998). During the period between 1968 and 2002, the simulated plant response was related to the high increase in N deposition that occurred from 1960 to 1980 (Fig. 13.3). Particularly, the relative moss occurrence was computed to increase significantly from 4% in 1930 to 21% in 1984. This indicated that this vegetation group is sensitive to N deposition. Indeed, from the literature (Armitage et al. 2011; Poikolainen et al. 2009; Xiao et al. 2010) it turns out that, mosses (*Hypnum* and *Leucobyrum*), grasses (*Molinia* and *Calamagrostis*), forbs or herbs (*Hedera* and *Impatiens*) and ferns (*Dryopteris* and *Pteridium*) are described most sensitive to changes in nitrogen deposition. The computed change in species occurrence was particularly obvious for *Hypnum* (5–10%) from 1900 to 1980, which corresponded to the maximum of N deposition (Fig. 13.7). ForSAFE-VEG simulations were performed on various RENECOFOR sites showing a general increasing trend for *Hypnum* cover in relation to NO_x deposition. However, this trend was not uniform for all sites and there is no straightforward relationship between the magnitude of N deposition and moss cover response. It is likely that site-specificity (e.g., biological activities, climate) is a confounding factor that needs to be addressed in more detail in the model to sufficiently reflect the monitored pattern of the given species.

On the other side for VSD^+-VEG model, no significant changes were simulated in the modelled occurrence of plant groups during the same period (Fig. 13.6). This indicates that the model was less sensitive than ForSAFE-VEG to simulate tenuous vegetation changes. Nevertheless, further investigations are needed in order to compare the performance of the two models on other sites.

13.3.5 Model Improvements Needed and Outlook

VSD^+-VEG is a simple model tailored to requirements of regional/national scale simulations of impacts to N deposition, but regional simulations remain to be tested and validated. Results from a first run with VSD^+ on the selected sites correspond rather well with reality, but we must be cautious that only one period of dataset was available for validation. This good agreement between simulated and measured soil data was encouraging, but further validation is needed. However, the following issues turned up that need to be addressed:

- the estimation of some of the key input parameters, like the "growth function" to better reflect C and N dynamics in the soil-plant system;
- meteorological inputs need to be adapted to French conditions to determine reduction functions for mineralization and (de)nitrification using the MetHyd module;
- the pH of French soils needs to be validated using recent soil samplings that were performed on the French ICP forest network. We must keep in mind that the underestimation of simulated soil pH (as observed on Fig. 13.3 particularly for the considered site CHS41) may have influenced the VSD^+-VEG outputs.

A tentative comparison of simulation results between the ForSAFE-VEG and VSD^+-VEG model yields the following findings (Fig. 13.6). For this, simulations were validated against times series of data. Obviously ForSAFE-VEG was more sensitive to predict vegetation changes under nitrogen deposition. But the numerous input data required to run ForSAFE-VEG might stand in the way of its application on a large country scale and, moreover, this might induce large uncertainties in model output (Wallman et al. 2005). The advantage of this modelling approach

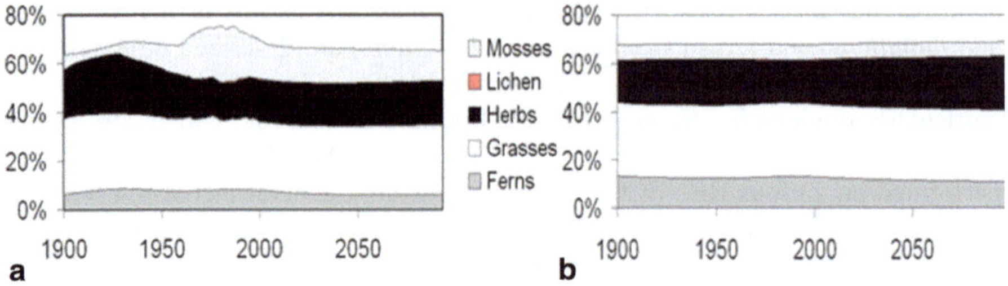

Fig. 13.6 Variation of the percentage cover of plant dominant groups at the CHS41 RENECOFOR site during the period 1900 and 2100 as simulated by (**a**) the ForSAFE-VEG and (**b**) VSD+-VEG model

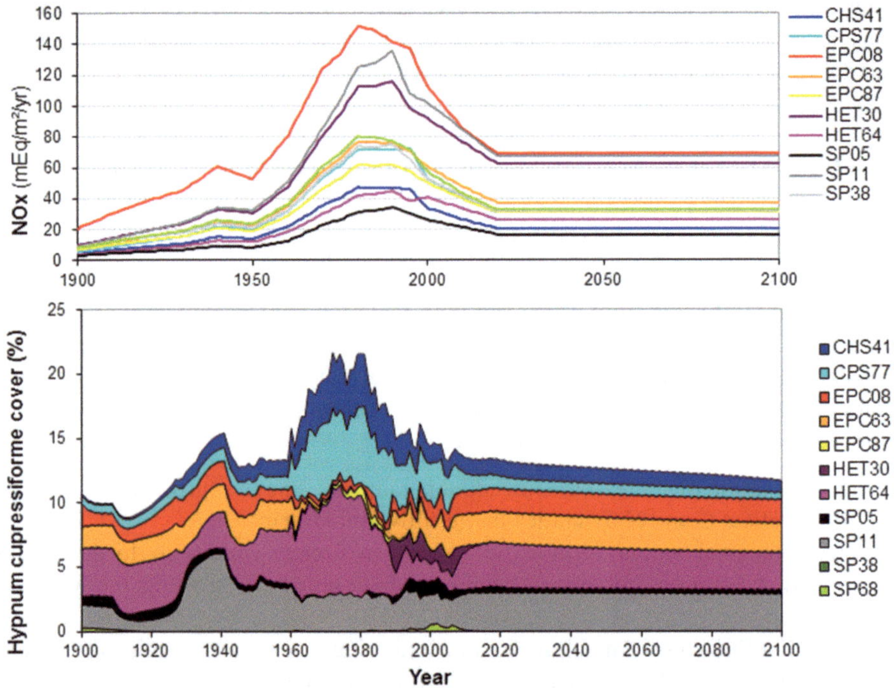

Fig. 13.7 Evolution of the EMEP modelled NO_x deposition (*top*) and of the cover percentage of *H. cupressiforme* (*bottom*) as simulated by ForSAFE-VEG model at the RENECOFOR sites CHS41, EPC87, CPS77, EPC63, HET30, SP05, SP11, SP38 and HET64, between 1900 and 2100

is the long-term assessment of N deposition impact on the vegetation diversity. More investigations are also needed in order to improve the validation of the treatment of hydrology in ForSAFE (Belyazid et al. 2006). Indeed, validation of the nitrogen outputs in soil solution and vegetation changes must be extended to other forest sites. Moreover, management practices such as clear-cutting or thinning have been recently implemented in ForSAFE-VEG platform since the run on 15 French ICP-Forests sites showed that these prove an adequate footprint of the temporal variability of pH and nitrogen outputs. The relatively minor influence of nitrogen deposition could be explained by the strong interaction between nitrogen deposition and the canopy, as modelled by the "Canopy budget" models (Staelens et al. 2008). Thus, the amount of nitrogen reaching the soil (i.e. nitrogen having an impact on soil nitrogen characteristics) could be significantly different from the amount of nitrogen contained in bulk deposition, which was the input factor considered here. Also for some sites the influence of dry deposition could be significant and must be better taken into account in future simulations.

Recently, the VEG table has been revised at the European scale by scientific experts and some of the ecological parameters were simplified. We propose also to test a new modelling approach based on the definition of ecological and functional types (under progress). With this approach, we aim to verify the hypothesis that, the vegetative response of functional groups to site specific N deposition might complete current site specific findings regarding the interaction between species, in the context of environmental specific station parameters.

Conclusions

France is known for its large variety of landscape in semi-natural conditions. The ICP-Forests network has registered long time series of N atmospheric deposition and a large set of environmental parameters. Thanks to these data, it is possible to investigate the response of nitrogen deposition at those forest sites under various site specific environmental conditions using coupled biogeochemical and vegetation models.

These trials on ForSAFE-VEG and VSD$^+$-VEG models and validations using measured data on RENECOFOR forest sites have allowed identifying their limitations, but they also showed that it was realistic to simulate the link between N deposition and under storey plant diversity. The two models have shown their ability to predict changes in soil parameters under the influence of N deposition. Changes in the vegetation cover according to the considered groups are better predicted by ForSAFE-VEG model chain than with VDS$^+$-VEG, since the response was more obvious for nitrogenous species like *Hypnum cupressiforme*. Mosses are particularly reactive and showed the highest increase in cover percentage compared to other vegetation in classes, indicating a high sensitivity to atmospheric N deposition. The under- or overestimation in predicting the presence of certain mosses plant cover using the two models indicated that improved calibration is needed. The ForSAFE-VEG simulated change over time of the dominant ground plant groups could be related to both changes in site environmental conditions, and in particular to the influence of the maximum wet N deposition. However, it is likely that the simulated vegetation response to nitrogen deposition becomes even more obvious at selected sites when dry deposition is taken into account. Finally, the French VEG-table has been used to enlarge the European version and to increase their robustness. Ecological parameters, in addition to C/N ratio, are currently under quantification using the EcoPlant database data (Gégout et al. 2005).

The performance of the ForSAFE-VEG and VSD$^+$-VEG model on selected sites showed that:

- it is realistic to simulate the link between N deposition and vegetation diversity, and
- it is possible to improve the simulations by implementing processes and influencing factors in the geochemical models and the VEG database.

Further investigations, using other sites, are needed to compare the performance of the two models. However, the modelling of critical loads of nitrogen for forest and grasslands ecosystems based on coupled geochemical-ecological approaches is a promising way forward.

Acknowledgments We are grateful to the French partners Manuel Nicolas and Marc Lanier from the National Forest Office who helped in providing basic data from the ICP-Forest network RENECOFOR. Thanks also go to Jean-Luc Dupouey from INRA (Nancy), Vincent Boulanger (ONF) and Jean-François Picard (INRA Nancy) for providing palatability expertise and E. Bortoluzzi (EcoLab, Toulouse) for preparing the initial work on ForSAFE. We also thank Jaap Slootweg from the CCE at RIVM (Bilthoven, NL) and Luc Bonten from Alterra (Wageningen NL) for fruitful exchanges about VSD$^+$. The French EPA (ADEME) and the Swedish EPA are particularly thanked for their financial support and work facilitation. This work was done within the PRIMEQUAL-PREDIT "Valeriane Project : Vulnérabilité des écosystèmes à l'azote d'origine atmosphérique. Vers un modèle de charges critiques en N intégrant la protection de la biodiversité végétale du territoire français" ADEME-CNRS convention 09 62C 0073.

References

Aber, J. D., & Federer, C. A. (1992). A generalized, lumped-parameter model of photosynthesis, evapotranspiration and net primary production in temperate and boreal forest ecosystems. *Oecologia, 92,* 463–474.

Alveteg, M., Sverdrup, H., & Warfvinge, P. (1995). Regional assessment of the temporal trends in soil acidification in southern Sweden, using the SAFE model. *Water, Air, & Soil Pollution, 85,* 2509–2514.

Armitage, H. F., Britton, A. J., Woodin, S. J., & van der Wal, R. (2011). Assessing the recovery potential of alpine moss-sedge heath: Reciprocal transplants along a nitrogen deposition gradient. *Environmental Pollution, 159,* 140–147.

Bardat, J., Bioret, F., Botineau, M., Boullet, V., Delpech, R., Géhu, J.-M., Haury, J., Lacoste, A., Rameau, J.-C., Royer, J.-M., Roux, G., & Touffet, J. (2004). *Prodrome des végétations de France.* (Coll. Patrimoines naturels, 61). Paris: Muséum national d'histoire naturelle.

Belyazid, S. (2006). Dynamic Modelling of biogeochemical processes in forest ecosystems. Doctoral Thesis. Reports in Ecology and Environmental Engineering 2006:1, Sweden, Department of chemical Engineering, Lund University.

Belyazid, S., Westling, O., & Sverdrup, H. (2006). Modelling changes in forest soil chemistry at 16 Swedish coniferous forest sites following deposition reduction. *Environmental Pollution, 144,* 596–609.

Belyazid, S., Sverdrup, H., Kurz, D., & Braun, S. (2011a). Exploring ground vegetation change for different deposition scenarios and methods for estimating critical loads or biodiversity using the ForSAFE-VEG model in Switzerland and Sweden. *Water, Air, & Soil Pollution, 216,* 289–317.

Belyazid, S., Kurz, D., Braun, S., Sverdrup, H., Rihm, B., & Hettelingh, J. P. (2011b). A dynamic modelling approach for estimating critical loads of nitrogen based on plant community changes under a changing climate. *Environmental Pollution, 159,* 789–801.

Bonten, L., Mol, J., & Reinds, G. J. (2009). Dynamic modelling of effects of deposition on carbon sequestration and nitrogen availability: VSD plus and C and N dynamics (VSD^{+}). In J. P. Hettelingh, M. Posch, & J. Slootweg (Eds.), *Progress in the modelling of critical thresholds, impacts to plant species diversity and ecosystem services in Europe* (pp. 69–73). CCE Status Report 2009.

Bonten, L., Mol-Dijkstra, J., & Reinds, G. J. (2011). Validation of VSD^{+} and critical loads for nutrient N. In M. Posch, J. Slootweg, & J.-P. Hettelingh (Eds.), *Modelling critical thresholds and temporal changes of geochemistry and vegetation diversity: CCE Status Report 2011* (pp. 49–52). Bilthoven, The Netherlands: RIVM Report 680359003.

Boulanger, V., Baltzinger, C., Said, S., Ballon, P., Picard, J. F., & Dupouey, J. L. (2009). Ranking temperate woody species along a gradient of browsing by deer. *Forest Ecology and Management, 258,* 1397–1406.

Brêthes, A., & Ulrich, E. (1997). *RENECOFOR—Caractéristiques pédologiques des 102 peuplements du réseau.* Fontainebleau, France: Office National de Forêts, Département des Recherches Techniques.

Bruneton, J. (2005). *Plantes toxiques: végétaux dangereux pour l'Homme et les animaux.* Ed. Médicales internationales.

De Schrijver, A., Verheyen, K., Mertens, J., Staelens, J., Wuyts, K., & Muys, B. (2008). Nitrogen saturation and net ecosystem production. *Nature, 451,* E1-E1.

De Vries, W., Kros, J., Reinds, G. J., Wamelink, G. W. W., Mol, J., van Dobben, H., Bobbink, R., Emmett, B., Smart, S., Evans, C., Schlutow, A., Kraft, P., Belyazid, S., Sverdrup, H. U., van Hinsberg, A., Posch, M., & Hettelingh, J.-P. (2007). *Developments in deriving critical limits and modelling critical loads of nitrogen for terrestrial ecosystems in Europe.* (Report 1382). Wageningen, the Netherlands: Alterra Wageningen UR.

De Vries, W., Wamelink, G. W. W., van Dobben, H., Kros, J., Reinds, G. J., Mol-Dijkstra, J. P., Smart, S. M., Evans, C. D., Rowe, E. C., Belyazid, S., Sverdrup, H. U., van Hinsberg, A., Posch, M., Hettelingh, J.-P., Spranger, T., & Bobbink, R. (2010). Use of dynamic soil–vegetation models to assess impacts of nitrogen deposition on plant species composition: An overview. *Ecological Applications, 20,* 60–79.

Diekmann, M., & Lawesson, J. (1999). Shifts in ecological behaviour of herbaceous forest species along a transect from northern central to North Europe. *Folia Geobotanica, 34,* 127–141.

Dorée, A. (1995). *Flore pastorale de montagne: graminées, légumineuses et autres plantes des pâturages.* Éditions Boubée.

Ellenberg, H., Jr. (1988). Eutrophierung-Veränderungen der Waldvegetation-Folgen für den Reh-Wildverbiss und dessen Rückwirkungen auf die Vegetation. *Schweiz Z Forstwes, 139,* 261–282.

Ellenberg, H., Weber, H. E., Düll, R., Wirth, V., Werner, W., & Paulißen, D. (1992). *Zeigerwerte von Pflanzen in Mitteleuropa.*

Friedrich, U., Von Oheimb, G., Kriebitzsch, W. U., Schlesselmann, K., Weber, M. S., & Haerdtle, W. (2012). Nitrogen deposition increases susceptibility to drought—Experimental evidence with the perennial grass *Molinia caerulea* (L.) Moench. *Plant Soil, 353,* 59–71.

Galloway, J. N., Aber, J. D., Erisman, J. W., Seitzinger, S. P., Howarth, R. W., Cowling, E. B., & Cosby, B. J. (2003). The nitrogen cascade. *Bioscience, 53,* 341–356.

Gandois, L., Probst, A., & Dumat, C. (2010a). Modelling trace metal extractability and solubility in French forest soils by using soil properties. *European Journal of Soil Science, 61,* 271–286.

Gandois, L., Nicolas, M., VanderHeijden, G., & Probst, A. (2010b). Importance of biomass net uptake for a trace metal budget in a forest stand (North-Eastern part of France). *Science of The Total Environment, 408,* 5870–5877.

Gégout, J.-C., Coudun, C., Bailly, G., & Jabiol, B. (2005). EcoPlant: A forest site database linking floristic data with soil and climate variables. *Journal of Vegetation Science, 16,* 257–260.

Gusmeroli, F., Della Marianna, G., Puccio, C., Corti, M., & Maggioni, L. (2007). Indici foraggeri di specie legnose ed erbacee alpine per il bestiame caprino. *Quaderno SOZOOALP*, 4, 73–82.

Hernandez, L., Probst, A., Probst, J. L., & Ulrich, E. (2003). Heavy metal distribution in some French forest soils: Evidence for atmospheric contamination. *Science of The Total Environment, 312,* 195–219.

Hettelingh, J. P., Posch, M., & Slootweg, J. (2008). *Critical load, dynamic modelling and impact assessment in Europe. CCE status report 2008.* (RIVM Report 500090003). Bilthoven, The Netherlands: Coordination Centre for Effects, National Institute for Public Health and the Environment.

Hutchinson, T. C., Watmough, S. A., Sager, E. P. S., & Karagatzides, J. D. (1998). Effects of excess nitrogen deposition and soil acidification on sugar maple (*Acer saccharum*) in Ontario, Canada: An experimental study. *Canadian Journal of Forest Research, 28,* 299–310.

Julve, P. (1998). Indexe écologique et chorologique de la flore de France. http://philippe.julve.pagesperso-orange.fr/catminat.htm.

Lindström, G., & Gardelin, M. (1992). Modelling groundwater response to acidification. In P. Sanden & P. Warfvinge (Eds.), *Report from the Swedish integrated groundwater acidification project* (pp. 33–36). Norrköping, Sweden: Swedish meteorological and hydrological institute. SMHI.

Louvel, J., Gaudillat, V., & Poncet, L. (2013). *EUNIS, European Nature Information System, Système d'information européen sur la nature. Classification des habitats. Traduction française. Habitats terrestres et d'eau douce.* Paris: MNHN-DIREV-SPN, MEDDE.

Moncoulon, D., Probst, A., & Martinson, L. (2007). Modeling acidification recovery on threatened ecosystems: Application to the evaluation of the Gothenburg protocol in France. *Water Air & Soil Pollution Focus, 7,* 307–316.

Morellet, N., & Guibert, B. (1999). Spatial heterogeneity of winter forest resources used by deer. *Forest Ecology and Management, 123,* 11–20.

Nilsson, J., & Grennfelt, P. (1988). *Critical loads for sulphur and nitrogen*. Report from a Workshop held at Skokloster Sweden March 19–24 1988. Miljø rapport 1988: 15. Copenhagen Denmark Nordic Council of Ministers.

Party, J. P. (1999). *Acidification des sols et des eaux de surface des écosystèmes forestiers français: facteurs, mécanismes et tendances*. Strasbourg: Louis Pasteur University.

Poikolainen, J., Piispanen, J., Karhu, J., & Kubin, E. (2009). Long-term changes in nitrogen deposition in Finland (1990–2006) monitored using the moss *Hylocomium splendens*. *Environmental Pollution, 157,* 3091–3097.

Ponce, R., Ulrich, E., & Garnier, F. (1998). *RENECOFOR—Essai de synthèse sur l'histoire des 102 peuplements du réseau.* Office National des Forêts, Département des Recherches Techniques.

Posch, M. (2002). Impacts of climate change on critical loads and their exceedances in Europe. *Environmental Science & Policy, 5,* 307–317.

Posch, M., & Reinds, G. J. (2009). A very simple dynamic soil acidification model for scenario analyses and target load calculations. *Environmental Modelling & Software, 24,* 329–340.

Posch, M., & Reinds, G. J. (2010). Appendix C. MetHyd—A meteo-hydrological pre-processor for VSD^+. In J. Slootweg, M. Posch, & J. P. Hettelingh (Eds.), *Progress in the modelling of critical thresholds and dynamic modelling, including impacts on vegetation in Europe. CCE Status Report 2010* (pp. 175–177).

Probst, A., & Leguédois, S. (2007). France. In J. Slootweg, M. Posch & J. P. Hettelingh (Eds.), *European critical load and dynamic modelling and impact assessment in Europe. CCE Status Report 2007* (pp. 139–145).

Probst, A., & Leguédois, S. (2008). France. In J. P. Hettelingh, J. Slootweg, & M. Posch (Eds.), *Critical load, dynamic modelling and impact assessment in Europe. CCE Status Report 2008* (pp. 133–140).

Probst, A., Lelong, F., Viville, D., Durand, P., Ambroise, B., & Fritz, B. (1995). Comparative hydrochemical behaviour and element budgets of the Aubure (Vosges Massif) and Mont-Lozère (Massif Central) spruce forested catchments. In G. Landmann & M. Bonneau (Eds.), *Forest decline and atmospheric deposition effects in the French mountains* (pp. 203–225). Berlin: Springer.

Probst, A., Moncoulon, D., Leguédois, S., Party, J. P., & Dambrine, E. (2008). Qu'a apporté le réseau pour le calcul des charges critiques en polluants atmosphériques en France? *Rendez-Vous Techniques de l'ONF (hors série), 4,* 77–81.

Rameau, J. C., Mansion, D., Dumé, G., Timbal, J., Lecointe, A., Dupont, P., & Keller, R. (1989). *Flore forestière française: Guide écologique illustré, Tome 1, Plaines et collines*. Paris: Institut pour le développement forestier.

Rameau, J. C., Mansion, D., Dumé, G., Lecointe, A., Timbal, J., Dupont, P., & Keller, R. (2008). *Flore forestière française: Guide écologique illustré, Tome 2, Montagnes*. Paris: Institut pour le développement forestier.

Rogora, M., Mosello, R., Arisci, S., Brizzio, M., Barbieri, A., Balestrini, R., Waldner, P., Schmitt, M., Stahli, M., Thimonier, A., Kalina, M., Puxbaum, H., Nickus, U., Ulrich, E., & Probst, A. (2006). An overview of atmospheric deposition chemistry over the Alps: Present status and long-term trends. *Hydrobiologia, 562,* 17–40.

Staelens, J., Houle, D., de Schrijver, A., Neirynck, J., & Verheyen, K. (2008). Calculating dry deposition and canopy exchange with the canopy budget model: Review of assumptions and application to two deciduous forests. *Water, Air, & Soil Pollution, 191,* 149–169.

Sverdrup, H., Belyazid, S., Nihlgård, B., & Ericson, L. (2007). Modelling change in ground vegetation response to acid and nitrogen pollution, climate change and forest management at in Sweden 1500–2100 A.D. *Water Air & Soil Pollution Focus, 7,* 163–179.

Sverdrup, H., McDonnell, T. C., Sullivan, T. J., Nihlgård, B., Belyazid, S., Rihm, B., Porter, E., Bowman, W. D., & Geiser, L. (2012). Testing the feasibility of using the ForSAFE-VEG model to map the critical load of nitrogen to protect plant biodiversity in the Rocky Mountains region, USA. *Water, Air, & Soil Pollution, 223,* 371–387.

Tarrasón, L., & Nyíri, Á. (2008). *Transboundary acidification, eutrophication and ground level ozone in Europe in 2006.* (EMEP status report). Norwegian Meteorological Institute.

Wallman, P., Svenssson, M., Sverdrup, H., & Belyazid, S. (2005). ForSAFE—An integrated process-oriented forest model for long-term sustainability assessments. *Forest Ecology and Management, 207,* 19–36.

Wallman, P., Belyazid, S., Svensson, M. G. E., & Sverdrup, H. (2006). DECOMP—A semi-mechanistic model of litter decomposition. *Environmental Modelling & Software, 21,* 33–44.

Xiao, H. Y., Tang, C. G., Xiao, H. W., Liu, X. Y., & Liu, C. Q. (2010). Stable sulphur and nitrogen isotopes of the moss Haplocladium microphyllum at urban, rural and forested sites. *Atmospheric Environment, 44,* 4312–4317.

Chapter 14
Use of an Empirical Model Approach for Modelling Trends of Ecological Sustainability

Angela Schlutow, Thomas Dirnböck, Tomasz Pecka and Thomas Scheuschner

14.1 Introduction

At least since the Convention on Biological Diversity was signed, models are sought that describe the response of ecosystems, and particularly the response of biodiversity, to geochemical and climate change. The BERN (Bioindication for Ecosystem Regeneration towards Natural conditions) model is one of these. A large number of surveys on vegetation and their site conditions (about 40,000 relevés) form the basis of the BERN model. Therefore one can call the model "empirical"; on the other hand, the BERN model is a deterministic model too, since it simulates changes in plant community composition depending on changes in soils and climate.

The BERN model is based on the concept of a plant community. According to Tüxen (1957), p. 151, the plant community "is a species assemblage with an interactive structure determined by site properties in which all elements effect each other. The plant community is a self-regenerating and self-regulating system, which, in competition for space, nutrients, water and energy, has reached a sociological dynamic balance and is characterized by dynamic harmony between site, production and all aspects of life".

According to this, the investigation of anthropogenic impacts on vegetation has to start with a definition of the natural balance between site and vegetation. Balanced state parameters are the basis for the definition of the reference-state site classification. The vegetation found under natural, balanced condition, which can be described with reference-state parameters, is defined as the "reference community".

A. Schlutow (✉) · T. Scheuschner
OEKO-DATA Strausberg, Strausberg, Germany
e-mail: angela.schlutow@oekodata.com

T. Dirnböck
Environment Agency Austria, Vienna, Austria

T. Pecka
Institute of Environmental Protection—National Research Institute, Warsaw, Poland

W. de Vries et al. (eds.), *Critical Loads and Dynamic Risk Assessments*,
Environmental Pollution 25, DOI 10.1007/978-94-017-9508-1_14

And a comprehensive set of reference communities build a baseline against which changes can be measured.

The reference community approach is comparable to the niche concept of species. The definition of the fundamental niche after Hutchinson (in Shugart 1984) describes the n-dimensional hyperspace of all site factors under which a species can survive neglecting any interaction with other species (Burrows 1990; Dierschke 1994; Martin 2002). The realized niche of a species is a sub-space of its fundamental niche, determined by the interaction with all other species occurring at the site. Species interactions, such as competition, facilitation or coexistence are complex and therefore very difficult to describe with a model (Wamelink and van Dobben 2003). However, it is possible to predict the possibility for a natural plant community to occur (Lortie et al. 2004), since it represents the final solution of a long-term competitive balance between co-occurring species (Callaway 1995). The BERN model describes the realized niche of plant communities with mathematical formulas, taking advantage of the great amount of knowledge available about the qualitative relation between site properties and plant communities (see Sect. 14.2). The BERN model uses the approach, suggested by Zadeh (1978), of a fuzzy correlation of site types with natural plant species composition. The degree of the fuzzy correlation is determined by a distribution function of possibilities (ranging between 0 and 1) of plant community occurrence, depending on one or more site parameters.

In this chapter we describe the BERN model, present some applications and discuss the pros and cons of the model.

14.2 Methods

14.2.1 Mathematical Concept

The realized niches of a natural reference plant community show the reference state parameter ranges of a site type. The following site parameters are used in the BERN model to characterise reference site types (in the shape of trapezoidal functions; see below):

- Soil water content at field capacity [m^3 m^{-3}]
- Base saturation [%]
- pH value (in H_2O)
- C/N ratio [g g^{-1}]
- Climatic water balance [mm/vegetation period]: precipitation minus potential evaporation
- index of continentality: De Martonne-Index = precipitation in vegetation period/ mean temperature in vegetation period + 10
- Length of vegetation period [d yr^{-1}]: number of days of the year with an average daily temperature above 10 °C

- Available energy from solar radiation [kWh $m^{-2}yr^{-1}$] during the vegetation period: depends on latitude, slope, aspect, cloudiness, and the shading caused by overlapping vegetation layers and their coverage in the plant communities
- Temperature [°C]: The trapezoid function was defined by the following indicators: minimum (frost hardiness), minimum and maximum of optimum (beginning and ending of photosynthesis) and maximum (heath hardiness)

The distribution function of the possibility ("DFP") of a plant species' occurrence depending on site parameters has to be heuristically derived since a theoretical distribution is not available (Roberts 1986).

The BERN model derives the niches of those plant species, which mainly constitute the community, i.e. the constant plant species, which are by definition, the characteristic species and all attendant species that can be found with a similar abundance in more than 70 % of all vegetation relevés representing the plant community at the same ranges of the state parameters. The assemblage of constant plant species of a community does not vary significantly within a climatic region or at a short time scale, if the site state parameters do not vary significantly in space or time.

The fuzzy thresholds of the reference state parameters for a plant species' occurrence, which are constant in more than one natural reference plant community, are determined by fusing the sets of the site parameter ranges of these communities to obtain the parameters describing the DFP. These parameter ranges of the species are inserted into the database of the BERN model. A DFP is described by a trapezoid with four characteristic corner points: (1) minimum, 10-percentile of occurring site factor range, (2) minimum of the optimum plateau, 40-percentile, (3) maximum of the optimum plateau, 60-percentile, (4) maximum, 90-percentile.

At present, there are 1940 plant species catalogued and described with their parameters ranges in the BERN model.

We denote the trapezoidal possibility distribution function (DFP) of plant species a related to site factor x_i ($i=1,\ldots,n$)—describing their fuzzy relation scaled between 0 and 1—with $\pi_a(x_i)$. In order to consider all factors of a site type—and as site factors are not independent from each other—the fuzzy conditions of the single factors are combined in the BERN model by using the minimum-operator (Dubois and Prade 1997):

$$\pi_a(\mathbf{x}) = \min\{\pi_a(x_1), \pi_a(x_2), \ldots, \pi_a(x_n)\} \quad (14.1)$$

This multivariate possibility or membership function for each plant species represents the realized niche and describes the n-dimensional subset of the site parameters $\mathbf{x}=(x_1, x_2, \ldots, x_n)$ for which the plant species exists (Zhang 1994).

The DFP for a whole plant community should be defined in a way that the possibility space of the plant community reaches the highest values at the point where most constant species have their highest possibility values (Fig. 14.1). Such a combination is implemented in the BERN model by using the non-convex algebraic gamma minimum operator (Dubois and Prade 1997) with the possibility of the combination A_γ defined as:

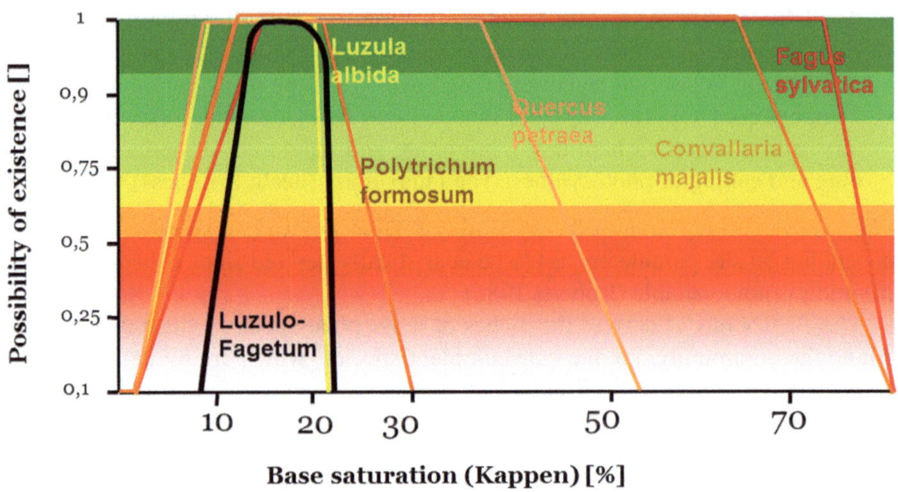

Fig. 14.1 The derivative of a community's possibility function (*black line*) for a single site parameter (here: base saturation) through the construction of the minimum range, using the possibility functions of the constant species (*brown*, *yellow*, *red* and *beige lines*)

$$A_\gamma(\mathbf{x}) = \left(\prod_{a=1}^{m} \pi_a(\mathbf{x})\right)^{\gamma} \left(1 - \prod_{a=1}^{m} \left(1 - \pi_a(\mathbf{x})\right)\right)^{1-\gamma} \tag{14.2}$$

where m is the number of (constant) species; and $\gamma = 0.2$ in the BERN model.

Since the DFP of a community is defined by the DFPs of its constant plant species, the function to the n-dimensional vector of all relevant site parameters of one site type is applied.

For use in other models—e.g. critical load calculations (see Chap. 6) using exactly defined quantities like critical limits (see Chaps. 2 and 3)—requires to convert the fuzzy quantity of site parameters for a plant community into a classical sharp quantity by assessing a so called α-level-quantity (Dubois and Prade 1997). Therefore a threshold of possibility has to be given. All elements with contribution values above this threshold are elements of the α-level-quantity. According to the definition of critical loads, the maximum stress loads of one or more pollutants are met (but not exceeded) as long as changes in the structure or/and functionality of specified sensitive elements of the environment do not occur. Therefore a biologically determined critical load can be obtained from the threshold value of the possibility function of the natural reference plant community. The parameters in the BERN model for which critical thresholds for the preservation of plant communities can be estimated are similar to the parameters used in the "Simple Mass Balance" (SMB) method for critical load computations (see Chap. 6), e.g. C/N ratio, base saturation, pH value. A reasonable threshold value is the degree of possibility at the intersection point of the optimum plateau border line with the site gradient for

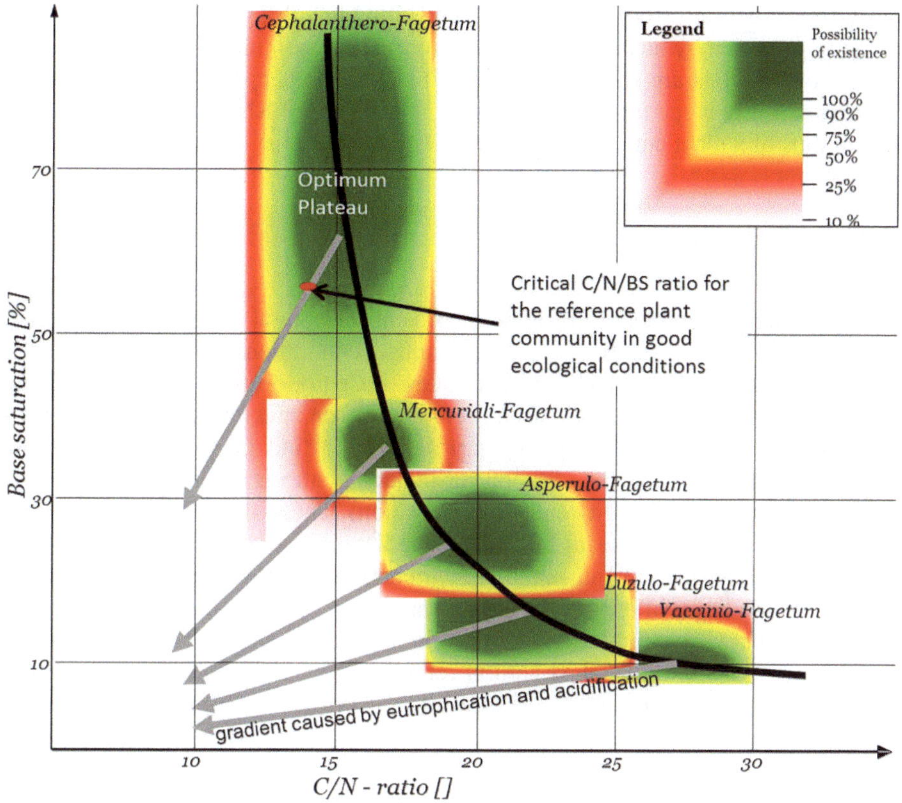

Fig. 14.2 (1) Distribution of the possibility function of all plant communities of the habitat type: natural woodland in the planar-subatlantic region with a plane relief, groundwater distance >2 m, (2) an obviously regular arrangement of the natural plant communities, which demarcates an indirect proportional connection between the base saturation and C/N-ratio at the optima of possibility ranges (gradient of nutrient balance = *black line*), (3) assessment of the actual site gradient trend after acidification and eutrophication (gradient of nutrient imbalance = *grey lines*), (4) the point of the critical limit, where the possibility degree function along the edge of the optimum plateau meets the gradient of nutrient imbalance

nutrient imbalance with decreasing C/N-ratio and decreasing base saturation caused by eutrophication and acidification (see Fig. 14.2). At this value of the site parameter (critical limit), the natural reference plant community has just the maximum possibility of its occurrence.

Figure 14.2 shows several natural plant communities and their possibilities of occurrence. For illustrative purposes only two parameters were used to display the variation of possibilities for different site conditions, but actually the realized niche is a 9-dimensional space. The straight grey lines indicate a quite severe shift in base saturation and C/N ratio resulting in nutrient imbalances depending on nitrogen deposition (partly intensified by sulphur deposition). On the other hand, the black curve represents a shift of the site factors resulting in nutrients balanced on another level (e.g. caused by litter raking in forests or by basic ash input), where the natural plant communities are able to adapt to soil chemical changes. By using the modelled possibility value for different plant communities BERN is able to define limit conditions of a plant community and thus its critical limits.

14.2.2 Databases

There are large numbers of historical records about natural and semi-natural communities on reference sites in a balanced nutrient condition. In order to create the BERN model data bases such reference sites of Central Europe were selected and analysed. The BERN model is based preferentially on sites and vegetation surveys with relative low stress. Since it is very unlikely to find sites in Europe with no anthropogenic influence, the criterion for including sites is the balance of site factors (e.g. nutrient availability and base saturation). These site properties may vary in time but as soon as a steady state is reached at a higher or lower trophic level a new (potentially) natural and suitable plant community may immigrate (Ellenberg 1992). The outcome of this modelling and analysing provides the possibility range for a natural plant community's existence in dependence on site parameter ranges. Over time additional published natural reference plant communities are added to the database. Succession phases as a result of short-term transient conditions and gradual variations in space in overlapping boundary ecotones are not included in the BERN model. This is necessary to be in line with the vital concepts of this model and in order to keep this approach applicable. Only non- or extensively utilized communities are considered. Therefore the indicator value of change is only weakly forced by anthropogenic influences, e. g. air pollution. These reference plant communities are discrete, repeatable vegetation units according to Clements (1916).

Currently, the database contains 343 reference plant communities of woodland and 238 reference plant communities of grassland, heaths, fens and swamps. Forty-nine near-natural afforested wood communities and 43 site typical agricultural weed communities are also included (Anders et al. 2002; Dierschke 1985a, b; Ellenberg 1996; Grabherr and Mucina 1993; Härdtle 1984, 1989, 1995a, b; Hartmann and Jahn 1967; Hofmann 1969; Horvat et al. 1974; Hundt 1964; Issler 1942; Jakucs 1961; Kevey and Borhidi 2005; Klapp 1965; Knollová and Chytrý 2004; Mahn 1959, 1965; Matuszkiewicz 1962; Michalko 1986; Moravec and Neuhäusl 1976; Mucina et al. 1993; Oberdorfer 1957, 1992–1998; Passarge 1964; Passarge and Hofmann 1968; Pott 1994; Preising 1953; Preising et al. 1990a, b, 1997; Rolecek 2005; Scamoni 1960; Schmidt et al. 2002; Schubert 1960, 1991; Schubert et al. 1995; Slobodda 1982, 1987; Soó 1964; Succow 1974, 1988; Succow and Joosten 2001; Tüxen 1937, 1957, 1958; Tüxen and Westhoff 1963; Willner 2002; Wolfram 1996).

These datasets indicate the reference site types and its natural reference vegetation in Germany, Austria, Poland, Slovakia, Czech Republic, Hungary, Switzerland,

Greece, Serbia, Macedonia, Albania, Croatia, Bosnia, Slovenia and Bulgaria. For these areas more than 40,000 records were evaluated with vegetation relevés after Braun-Blanquet (1961) and additional information about site factors. Both sets of information are supposed to be collected or carried out before the massive increase of deposition in Europe in the second half of the twentieth century.

14.2.3 Validation

The first part of the model validation was done on 72 sites of the ICP Forest network (Nagel et al. 2010). The overall match between measured ranges of values for the site factors and modelled or predicted ranges was quite satisfying. Especially the topsoil C/N ratio showed a good agreement. Almost 100% of the measured values on the plots lay within the modelled ranges of the BERN model.

The level of agreement was higher if the measured base saturation was high. Several plots reached levels of base saturation (e.g. $BS < 4\,\%$) which are unlikely to be found under undisturbed conditions. Therefore the BERN model cannot properly predict the vegetation composition since the natural competition processes are dominated by poly-ecological plant species. This behaviour was anticipated and confirms the proper functionality of the BERN model.

The comparison of measured and modelled pH values showed a similar behaviour. The highest measured pH values in the root zone lay within the ranges of the modelled values, but the lowest measurements were outside the BERN ranges.

The results of a dynamic modelling exercise (Nagel et al. 2010) on 11 representative Level-II-plots allow some general conclusions:

In general there is no significant linear connection between elevated nitrogen deposition and changes in the C/N ratio (see Table 14.1). This was expected since the development of the C/N ratio depends on several other factors as well. The relation between C/N ratio and deposition was clearer on plots with high deposition (>25 kg N $ha^{-1}a^{-1}$) with starting values for the C/N ratio and base saturation on levels of an almost undisturbed condition.

The loss of typical constant plant species in the shrub and herb layer (reference species) was not restricted to the elevated deposition. The clearer (multivariate) relation was identified between changes in biodiversity and a group of site factors (C/N ratio, base saturation and length of the vegetation period).

An analysis on 283 plots in a region in the North-East of Germany (Nagel et al. 2010) showed that there is no significant negative dependency between the number of plant species (including unconstant species) and shifts in C/N ratio, pH value, base saturation or climatic parameters. In contrary, the number of species rose in times of nitrogen saturation after phases of limited nitrogen availability. The same happened on nitrogen saturated sites if the tree or shrub layer was thinned. In both cases plant species that are not a constant member of a natural plant community, benefitted most from the disturbances.

Table 14.1 Results of the dynamic VSD/BERN modelling on 11 Level-II-plots. (Nagel et al. 2010)

Independent factor	Dependent factor	R^2
Change N deposition (1920–2000)	Change of C/N (reference to 2000)	0.13
Change N deposition (1920–2000)	C/N (2000) and BS (2000)	0.47
C/N (reference)	Change of C/N(reference to 2000)	0.91
BS (reference)	Change of BS (reference to 2000)	0.85
Change C/N (reference to 2000)	Change of BS(reference to 2000)	0.96
Change N deposition (1920–1985)	Relative qualitative loss (%)—current species compared with reference species list	0.44
Change of C/N (reference to 2000)		0.61
Change of BS (reference to 2000)		0.62
C/N (2000) and BS (2000)		0.68
Change C/N (reference to 2000) and change vegetation time (reference to 2000)		0.88
Change BS (reference to 2000) and change vegetation time (reference to 2000)		0.85

14.3 Results

14.3.1 Evaluation of Plant Responses to Nitrogen and Sulphur Deposition

From historical (modelled or measured) time series of the abiotic site parameters (e.g. base saturation, pH and C/N ratio and climate factors) the BERN model derives the changes of the existence possibility of the pristine plant community. The trend either could be the decrease of naturalness and vitality, or the development to similar natural plant communities on another trophic level (in forests: e.g. nutrient losses through litter raking, humus degradation after clear cut etc.; in grassland: nutrient losses through mowing, soil compression on pastures through stepping etc.) taking into account the cultivation history (in forests: main tree species, time of clearcut, afforestation; in grassland: type of management and intensity). The time series of the future development of C/N ratios, pH, base saturation and climatic parameters, which could be modelled by geochemical dynamic and climate change models, are based on assumptions like scenarios of deposition, and planned changes of ecosystem management.

In order to test dynamic modelling of vegetation changes at selected sites in Austria the soil acidification model VSD+ (Bonten et al. 2010) and BERN was calibrated for two permanent soil-vegetation plots from the ICP Integrated Monitoring (IM) site "Zöbelboden". Based on VSD+ geochemical outputs, vegetation development was modelled with the BERN model. Deposition and soil data was taken from two intensive monitoring plots, which are typical for the two site types of the area and was allocated to the respective permanent plots.

For the parameter setting for VSD+ we used modelled deposition of N and S, which is approximately double the throughfall deposition measured at the sites. Deposition of base cations was taken from Van Loon et al. (2005) for the year 2000. An increase of 70 % was assumed from 1880 to 2000 and 50 % from 1970 to 2000. A further decrease of 10 % until 2009 was estimated from the throughfall data. The sites are characterized by a high proportion of coarse fraction, but the chemical soil parameters were analysed from the fine fraction (<2 mm). This fact was taken into account by reducing the soil depth by the respective depth of the coarse fraction. All further parameters (CEC, base saturation, etc.) were calculated with the reduced soil depth. Soil water capacity was taken from measured values. Base cation uptake as well as C and N input from above and below-ground tree biomass was modelled with GrowUp (Bonten et al. 2012). The GrowUp model is a tool for computing forest growth, nutrient uptake and litterfall. The management was defined by a standard forest yield model and field data. Tree species specific biomass data were not changed but the C and N input had to be decreased by 70 % in order to be in the measured range (see discussion below). The following pairs of parameters were calibrated with VSD studio: cation GAPON exchange constants for aluminium/hydronium and base cations, the starting carbon pool and C/N ratio, weathering rates for calcium and magnesia. The simulation period was set to 1900–2050.

The BERN model uses the output from VSD+ (particularly the C/N ratio and the base saturation). In order to evaluate BERN model runs, observations from the years 1992, 2005 and 2010 were used. We restricted the species to those occurring in the understory and scaled the sum of the cover values to one in the modelled and the observed data. The BERN-model coupled to the VSD+ first links the existing plant community to the pristine plant community, and second models the dominance structure of the respective species related to this community (see Fig. 14.3). Thus, the potential species pool is pre-filtered to achieve an ecologically reference species pool to start with. The BERN model is explicitly predicting the potential tree species composition and the potential community. For both plots, subplot 01 and subplot 44, the *Helleboro nigri-Fagetum* is predicted as the most possible community with *Fagus sylvatica* followed by *Picea abies* as the most possible species. For subplot 44 a higher proportion of additional deciduous tree species are predicted. This is well in line with the potential communities and the tree species composition of the plot.

The understory species composition is also predicted and compared with the balanced (homeostatic) competition relationships in a natural community. The dominant species of subplot 44 such as *Carex alba* or *Calamagrostis varia* are also modelled as being dominant. On the other hand a number of subordinate species are predicted which do not occur. As a result, the Czekanowski index is still quite low, and for subplot 01 not satisfactory. The forest of subplot 01 is an old spruce

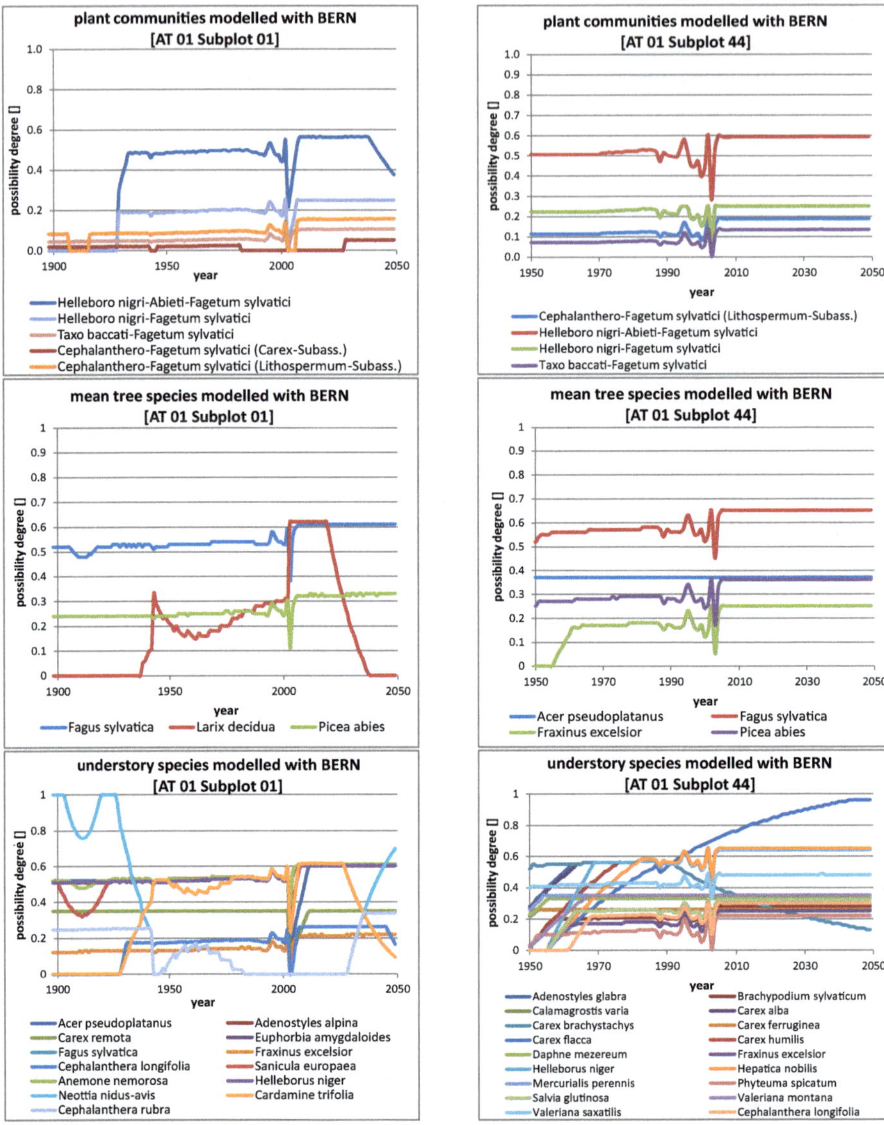

Fig. 14.3 Results of BERN community's (*top*), tree species (*middle*) and understory species (*low*) occurrence of two plots of the ICP-IM site "Zöbelboden" (*left*: AT01_01; *right*: AT01_44)

plantation and not dominated by the potential tree species. The effects of management on understory species composition are rather difficult to predict.

We also modelled the "Neuglobsow" plot. It is located in a dense forest in the low deposition region in the North-East of Germany, next to the oligotrophic lake "Stechlinsee". The soil properties are those of a podsol cambisol. The climate is Baltic-maritime with moderate warm summers and mild winters. The beech dominated plant community corresponds to *Avenello flexuosae-Fagetum sylvatici*.

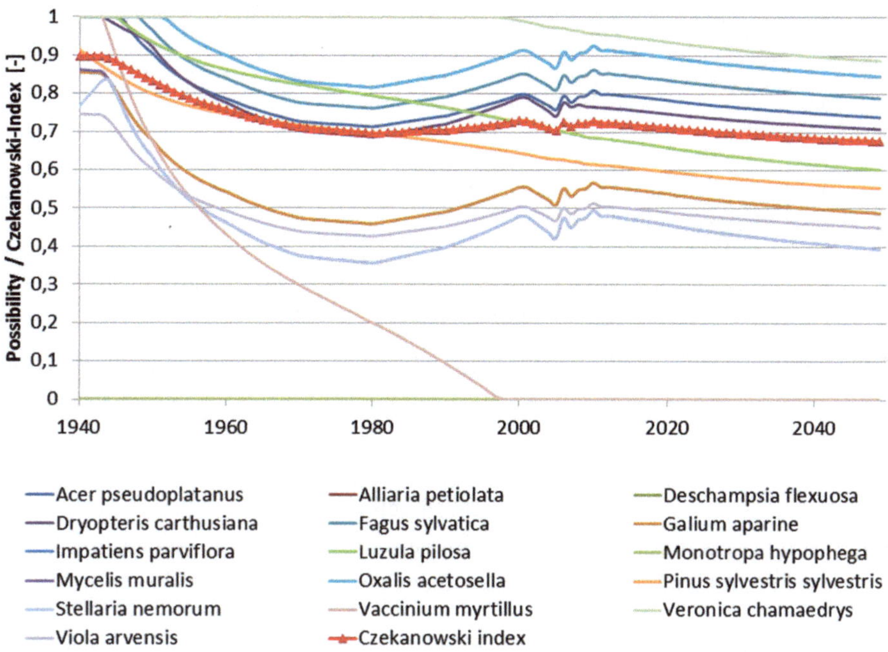

Fig. 14.4 Modelled trends of possibility of existence of the occurring plant species at the IM site "Neuglobsow" including an analysis of the recent and potential plant composition (= Sørensen index)

The soil profile was analysed and a tree leaf analysis was done (Schulte-Bisping 2012) according to the manual of the ICP IM. Also water and matter fluxes were measured, and the parameters of water retention curves used in the SVAT model were determined (Schulte-Bisping 2012). These input data was fed into the VSD+ model. The resulting time series of C/N ratio and pH were validated with measured values and imported into the BERN model.

The modelled trends of the possibility of existence for each plant species found in the plant survey in the year 2010 showed a good agreement with the potentially possible reference species. One indicator is the Czekanowski index which shows the similarity (0-1) of the current plant species composition regarding the abundances with the chosen reference (see Fig. 14.4).

In order to interpret the gradient of Czekanowski index in Fig. 14.4, an analysis of the deposition time series in comparison with the critical load might be helpful. The crucial question to understand the development is the proper choice of receptors and the related critical limits for calculating a critical load to protect the plant species diversity (see Fig. 14.5).

In this unhydromorphic soil the long-term exceedance of the critical load for nutrient nitrogen, $CL_{nut}N$, led to nitrogen saturation in the root zone. Keeping in mind the results of the modelling with BERN and the stagnant trend of the Sørensen index it seems that the most sensitive critical limit ($[N]_{crit} = 1$ mg l^{-1}, according to the Mapping Manual, www.icpmapping.org) is not valid for this site. This makes sense

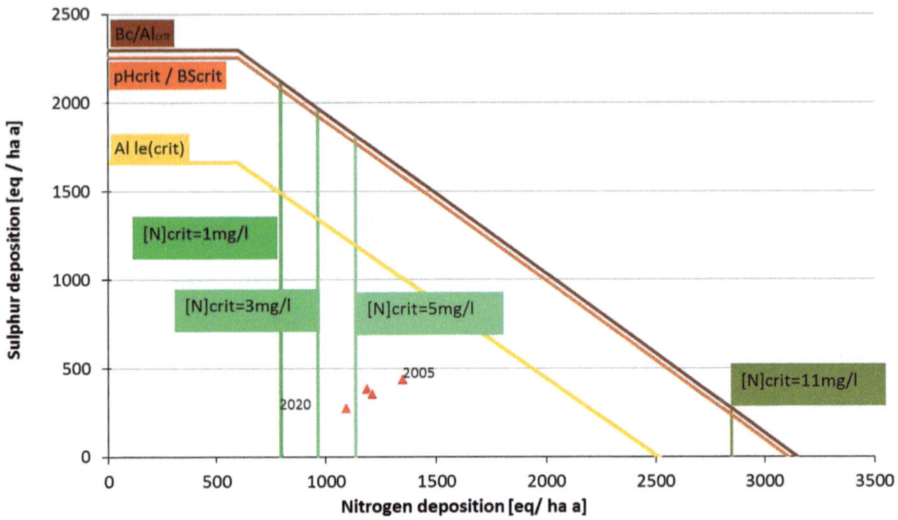

Fig. 14.5 Critical load functions for IM site "Neuglobsow" with different critical limits; deposition of S and N for the years 2005–2007 (measured) and 2020 (trend based on CLE-scenario)

since elevated nitrogen leaching into the groundwater might have only little effect on the species richness in this case. The drinking water quality limit ($[N]_{crit}=11$ mg l^{-1}) on the other side of the range might also not the best choice for the same reasons. The limit of 3 mg l^{-1} nitrogen for $[N]_{crit}$ aims on the protection of the fine roots and shall prevent increased sensitivity against frost and fungal diseases.

Most fitting in this case is the consideration of the essential base cation to nitrogen ratio. As long as this Bc/N-ratio meets the site suitability for a plant species composition is not disturbed by nutrient imbalance. This method cannot be described in detailed here due to space limitations (but see Schlutow in Balla et al. 2013).

The most crucial point is that it is important to include plant specific Critical Limits in order to preserve the plant species richness in a sustainable way. Such critical limits, particularly Bc/N, C/N, base saturation, pH, can be derived by the BERN model and are widely used by the German NFC of the ICP Modelling and Mapping.

14.3.2 Evaluation of Plant Responses Under Geochemical and Climate Change

Climate change, in particular air temperature and precipitation changes, may have significant impact on plant vegetation period and water regime in forest ecosystems. They can change soil chemistry properties and can be one of the most important drivers forcing changes in plant communities.

The scope of the study was an assessment of future sulphur and nitrogen deposition scenarios and climate change impact on plant associations. The VSD+ and BERN models were tested on forest monitoring plot No. 208 (ICP-Forests Poland) located in north-east of Poland. Plot No. 208 is located on poor sandy soils (haplic arenosol) with low groundwater level and covered by semi-dry *Peucedano-Pinetum* forest association (Matuszkiewicz 2001). It is a *Pinus silvestris* dominant 63 years old forest stand with appearance of *Betula pendula*, *Quercus robur* and *Populus tremula* trees. The ground vegetation is typical for this plant association, the largest cover being *Vaccinium myrtillus* and *Vaccinium vitis-idaea*, with appearance of *Trientalis europaea*, *Chimaphila umbellata*, *Festuca ovina*, *Calamagrostis arundinace* and *Carex ericetorum*. Most species have very low or low soil and water requirements, including low nitrogen demand (Zarzycki et al. 2002). It is the most typical oligotrophic and acidic coniferous forest association type in Poland.

Current climate data and base cation depositions were taken from the ICP-IM station "Wigry", located in the same region (Krzysztofiak 2010). Soil parameters (cation exchange capacity, initial values for carbon pool, C/N ratio and base saturation) were derived by analysing soil depth horizons. Model calibration was done for the exchange constants for aluminium/hydronium and base cations.

The MAGICC/SCENGEN model (Wigley 2008) was used to obtain the climate scenario. Calculations of average air temperature and precipitation under the A2 climate change (CC) scenario until 2050 were carried out for the whole of Poland (Mill et al. 2008). Calculation results for NE Poland, showed an increase in both the average air temperature (2010: 6.87 °C, 2025: 7.17 °C, 2025: 7.87 °C) and precipitation (2010: 684 mm, 2025: 685 mm, 2050: 695 mm). The average air temperature increase may extend the vegetation period by ca. 10 days in 2050.

For further VSD+/BERN calculations, CLE and CLE + CC scenarios were considered. Simulations started in the year 1920. Model results—base saturation, pH and C/N—from VSD+ for the 1920–2050 periods were used as an input to the BERN model. Modelling results of the BERN model with respect to the possibility degree for the existence of the plant community in the year 2000 (*Peucedano-Pinetum sylvetris*) and the potential natural forest communities at the site are displayed in Fig. 14.6.

With decreasing site suitability for *Peucedano-Pinetum sylvetris* (due to decreasing C/N-ratio) also the competition capacity decreases, resistance against environmental factors like windfall, breakage caused by snow, frost, insects and pests as well as a decline of the recovery capabilities. With decreasing site suitability for *Pinus sylvestris*, also the risk for the number of pine trees within the stock increases which couldn't reach their optimal age for economic utilization. If one calculates the inverse value of the possibility degree, one gets an index for latent mortality. On the other hand the vitality of *Q. petraea* and *Q. robur* doesn't change significantly. A possibly consequence could be to bring in some oak species for mixing the stands in order to increase the resistance against changes of environmental site factors.

Finally, the increasing temperature and precipitation under climate change may force the migration of *Peucedano-Pinetum sylvetris* to *Querco roboris-Carpinetum betuli*. In a situation where the temperature increase is moderate, a migration to *Potentillo albae-Quercetum petraece-roboris* is also possible.

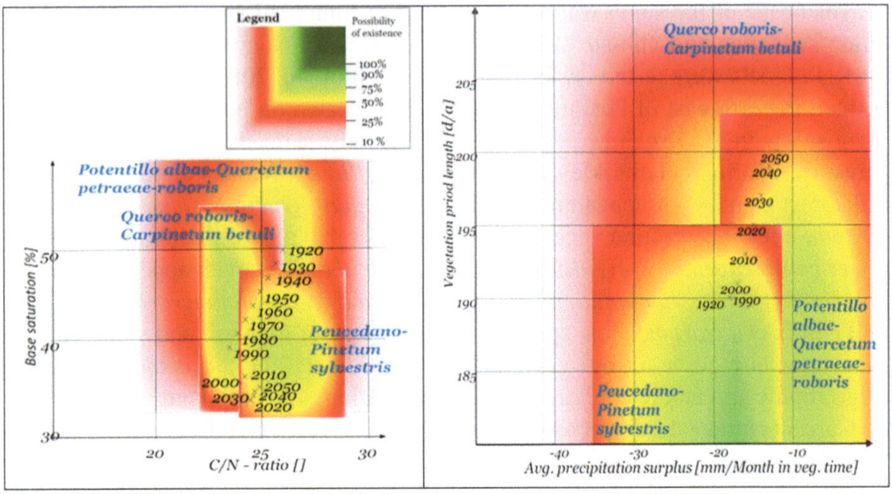

Fig. 14.6 Polish plot 208; Scenario 2: Dynamic of possibility for self-regeneration and sustainable existence of site-typical natural plant communities, depending on changes of base saturation and C/N-ratio (*left*) and climate (*right*)

14.3.3 Using the BERN Model to Assess Critical Loads for Natura 2000 Habitat Types in Germany

Critical load computations are already a standard procedure to assess impacts of air pollution in the connected cultural landscape in NATURA 2000 habitats in Germany. Especially the critical load of nitrogen (aspects of eutrophication and acidification) became an indicator for sensitivities of habitat types. In order to create a harmonized and standardized procedure in Germany the Federal Highway Research Institute (BASt) funded a project (Balla et al. 2013) with the aim to create a list with typical critical loads of N for most of German NATURA 2000 habitat types.

The critical load computations followed the Simple Mass Balance (SMB) approach according to the Mapping Manual (www.icpmapping.org) using plant-specific critical limits derived by the BERN model. The typical critical vegetation C/N ratio, $CN_{crit(phyto)}$, was used to derive the N immobilization rates (according to Klap et al. 1997). The typical critical base saturation (BS_{crit}) in the soil which the plant community demands as a minimum was used for (a) deriving the acceptable leaching of Acid Neutralizing Capacity according to the Mapping Manual (www.icpmapping.org), (b) estimating the critical N concentration in the leaching water with focus on avoiding nutrient imbalances. This includes the critical Bc/N ratio linked to the vegetation (see Schlutow in Balla et al. 2013).

The described method of choosing plant community specific critical limits and including information from the Mapping Manual for critical limits of Bc/Al_{crit}, pH_{crit}, $Al_{le(crit)}$ led to a list of 1990 records containing the information about the

Table 14.2 Aggregated CL(N) (Minimum of $CL_{nut}N$ und $CL_{max}N$) [kg N $ha^{-1}yr^{-1}$] ranges according SMB/BERN modelling for Natura 2000 habitat types. (Schlutow in Balla et al. 2013)

LRT	CL(N)	LRT	CL(N)	LRT	CL(N)	LRT	CL(N)	LRT	CL(N)
2120	10–12	4060	7–19	6431	15–77	8160	5–15	9190	8–14
2130	7–12	4070	18	6432	11–28	8210	7–23	91D0	7–28
2140	7	5110	11–14	6440	16–58	8220	4–25	91E0	6–28
2150	10–11	5130	8–30	6510	12–43	8230	6–17	91F0	11–31
2160	9–19	6110	7–33	6520	17–26	9110	10–21	91G0	15–24
2170	5–11	6120	11–21	7110	5–23	9120	8–17	91T0	4–13
2180	4–19	6212	13–28	7120	5–23	9130	9–22	91U0	6–18
2310	8–26	6213	12–30	7140	6–26	9140	15–27	9410	11–27
2320	12–13	6214	13–35	7150	7–17	9150	13–23	9420	11–36
2330	5–24	6230	10–24	8110	11–23	9160	14–21	9430	12–29
4010	6–16	6240	14–33	8120	10–23	9170	11–23		
4030	8–21	6410	12–36	8150	5–22	9180	8–30		

modelled $CL_{nut}N$ and $CL_{max}N$ and typical site factors and type of vegetation of the NATURA 2000 habitat types (see Table 14.2).

In order to validate the project results, the modelled ranges were compared with the empirical critical loads ($CL_{emp}N$) according to Bobbink and Hettelingh (2011). Almost one third of the typical combination types of site factors and vegetation composition have no analogy in the recent list of EUNIS codes of the empirical critical loads. In about 80 % of the matching cases the modelled critical load of the BASt project is within the range of the $CL_{emp}N$. In 5 % of the matching cases the modelled critical load is lower and in 15 % the empirical critical load is more sensitive. In both cases the variations can be explained by different ways of inclusion of uptake and removal of N by biomass in dependence of difference land use and cultivation.

Discussion and Conclusions

The application of the BERN model was carried out with the following objectives:

- Determination of reference communities together with their reference site parameters;
- Comparison of the currently observed site parameters with the reference parameters; determination of the similarity to the pristine conditions;
- The determination of the regeneration target and its regeneration potential;
- Dynamic modelling of vegetation changes based on sulphur and nitrogen depositions as well as climate change scenarios by coupling the BERN model to a dynamic soil models for assessing the risk of (1) structural changes of natural reference plant communities (2) biodiversity losses;

- The derivation of the critical loads for eutrophying and acidifying depositions based on biologically inspired critical limits (indicated as C/N ratio, base saturation, pH for plant's and/or community's existence possibility).

The main difference of BERN to some other vegetation models is the focus on plant communities and on the constant plant species of the natural or semi-natural communities. The BERN model includes reference types of vegetation, which are the plant communities with a self-regenerating and self-regulating ability in the sense of Tüxen (1957). Even if this concept is not undisputable in the current ecological discourse it is foundational for this approach and crucial for the practical application of this vegetation model. This concept of natural or semi-natural communities includes the definition and description of a so-called natural or pristine composition of plant species. This composition is meant to be a result of evolutionary processes and shall define the best-fitting solution of ecological competition according to a certain site factor configuration. In forests, as an example, the natural communities are more resilient and less sensitive than pure stands of trees. This resilience has the effect that plant communities with saturated species composition in a competitive equilibrium do not change their abundance continuously along a site factor gradient. They only change or enter a transition stage when certain critical limits are violated (Dierschke 1994). Deriving vegetation depending critical limits via the BERN model is based on this concept. These limits are the edges of the realized niches of plant communities as a minimum of the intersection of the niches of the constant plant species. The fundamental niches of several plant species might be large, and thus could be the reason for "neutral theory" which neglects the niche definition at all. But the minimal overlap of the realized plant niches results in most cases in a rather small range of preferences (= realized niche) for plant communities.

The deduction of development targets based on the BERN model has the current condition as a starting point. Predicted time series of abiotic parameters allow the prediction of the vegetation development potential. The target must be identical with one or more of the reference vegetation types in a balanced condition of nutrients and other abiotic parameters which were identified and included in the BERN model. The meaning of current or near-past condition is not necessary defined by time intervals. More important is the inclusion of the naturalness in order to have information about the reference condition. By including information from various regions, a wide range of site factors can be linked to different configurations of the vegetation in a good ecological condition.

Particularly climate changes are hardly reversible. The climate condition expected in the future might in Central Europe may already exist in southern Europe and support a natural plant community from there, which could serve as reference condition. The aim of applying the BERN model and link it to soil chemical models or include climate scenarios is not to get back to the past but to have a target which is capable of being in a good ecological condition.

This idea describes already the major limitation of the BERN model. The description of a plant community that might reach a good ecological condition with respect to the occurring site factors is fixed. Therefore, chances that this predicted plant community will appear cannot be modelled. The availability of genetic material for returning of the already extinct species is hardly predictable. Assuming that there are several physical and ecological barriers for natural dispersion of species from other regions the probability of an ecological reconfiguration cannot be described by the BERN model, but only the possibility of existence.

Another problem is the immigration of "invasive species" into pristine communities. Hobbs et al. (2009) describe the issue: "Some changes will result in hybrid systems retaining some original characteristics as well as novel elements, whereas larger changes will result in novel systems, which comprise different species, interactions and functions." Jackson and Hobbs (2009) reason: "Ecological restoration efforts should aim to conserve and restore historical ecosystems where viable, while simultaneously preparing to design or steer emerging novel ecosystems to ensure maintenance of ecological goods and services." A first attempt to handle this topic with the BERN model was done by including (planted and managed) forest plant communities. Additional research is required on this topic.

Current work focusses on the improvement of the modelling of vegetation changes due to sulphur and nitrogen depositions as well as climate change scenarios in the dynamic BERN model, especially regarding "damage delay times" and "recovery delay times" (= hysteresis effect) of ecosystems.

References

Anders, S., Beck, W., Bolte, A., Hofmann, G., Jenssen, M., Krakau, U.-K., & Müller, J. (2002). *Ökologie und Vegetation der Wälder Nordostdeutschlands–Einfluss von Niederschlag-sarmut und erhöhtem Stickstoffeintrag auf Kiefern-, Eichen-, und Buchen-Wald- und For-stökosysteme des nordostdeutschen Tieflandes*. Eberswalde: Verlag Dr. Kessel Oberwinter.

Balla, S., Uhl, R., Schlutow, A., Lorentz, H., Förster, M., & Becker, C. (2013). Untersuchung und Bewertung von straßenverkehrsbedingten Nährstoffeinträgen in empfindliche Biotope. (Research and development study FE 84.0102/2009). On Behalf of German Federal Highway Research Institute (BASt), in press.

Bobbink, R., & Hettelingh, J.-P. (2011). Review and revision of empirical critical loads and dose-response relationships: Proceedings of an expert workshop, Noordwijkerhout, 23–25 June 2010. (Report 680359002/2011). Bilthoven: Coordination Centre for Effects, National Institute for Public Health and the Environment.

Bonten, L., Posch, M., & Reinds, G. J. (2010). The VSD+ soil acidification model—Model description and user manual. Wageningen and Bilthoven: Aterra and CCE.

Bonten, L., Mol-Dijkstra, J. P., Wiegger, R., & Reinds, G. J. (2012). *GrowUp: A tool for computing forest growth, nutrient uptake and litterfall*. Wageningen: Aterra.

Braun-Blanquet, W. (1961). *Pflanzensoziologie*. Stuttgart: Ulmer.

Burrows, C. J. (1990). *Processes of vegetation change*. London: Unwin Hyman.

Callaway, R. M. (1995). Positive interactions among plants. *Botanical Review, 61,* 306–349.

Clements, F. E. (1916). *Plant succession*. (Publication #242). Washington: Carnegie Institute Washington.

Dierschke, H. (1985a). Pflanzensoziologische und ökologische Untersuchungen in Wäldern Süd-Niedersachsens. II. Syntaxonomische Übersicht der Laubwald-Gesellschaften und Gliederung der Buchenwälder. *Tuexenia, 5,* 491–521.

Dierschke, H. (1985b). Experimentelle Untersuchungen zur Bestandesdynamik von Kalkmagerrasen (Mesobromion) in Südniedersachsen. I. Vegetationsentwicklung auf Dauerflächen

1972–1984. In K.-F. Schreiber (Ed.), *Sukzession auf Grünlandbrachen* (pp. 9–24). Paderborn: Münsterische Geographische Arbeiten 20.
Dierschke, H. (1994). *Pflanzensoziologie*. Stuttgart: Ulmer.
Dubois, D., & Prade, H. (1997). The three semantics of fuzzy sets. *Fuzzy Sets and Systems, 90,* 141–150.
Ellenberg, H., Jr. (1992). Folgen der verbesserten Verfügbarkeit von Stickstoff als Nähr-stoff für Flora und Fauna in Mitteleuropa. Habilitationsschrift Univ. Hamburg.
Ellenberg, H. (1996). *Vegetation Mitteleuropas mit den Alpen* (5. ed.). Stuttgart: Ulmer.
Grabherr, G., & Mucina, L. (1993). *Die Pflanzengesellschaften Österreichs, Teil II: Natürliche waldfreie Vegetation*. Stuttgart: Gustav-Fischer-Verlag Jena.
Härdtle, W. (1984). Vegetationskundliche Untersuchungen in Salzwiesen der Ostholsteini-schen Ostseeküste (Heft 48). Kiel: Mitteilungen der AG Geobotanik in Schleswig-Holst. und Hamburg.
Härdtle, W. (1989). Potentiell Natürliche Vegetation–Ein Beitrag zur Kartierungsmethode am Beispiel der Topographischen Karte 1623 Owschlag (Heft 40). Kiel: Mitteilungen der AG Geobotanik in Schleswig-Holst. und Hamburg.
Härdtle, W. (1995a). Vegetation und Standort der Laubwaldgesellschaften (Querco-Fagetea) im Nördlichen Schleswig-Holstein (Heft 48). Kiel: Mitteilungen der AG Geobotanik in Schleswig-Holst. und Hamburg.
Härdtle, W. (1995b). Zur Systematik und Synökologie artenarmer Buchenwälder (Flatter-Gras-/Sauerklee-Buchenwälder) in Schleswig-Holstein. *Tuexenia, 15,* 45–51.
Hartmann, F. K., & Jahn, G. (1967). Waldgesellschaften des mitteleuropäischen Gebirgsraumes nördlich der Alpen. Jena.
Hobbs, R. J., Higgs, E., & Harris, J. A. (2009). Novel ecosystems: implications for conservation and restoration. *Trends in Ecology and Evolution, 24,* 599–605.
Hofmann, G. (1969). Zur pflanzensoziologischen Gliederung der Kiefernforsten des nordostdeutschen Tieflandes. *Feddes Repertorium, 80,* 401–412.
Horvat, I., Glavac, V., & Ellenberg, H. (1974). *Vegetation Südosteuropas*. Stuttgart: Gustav-Fischer-Verlag Jena.
Hundt, R. (1964). *Die Bergwiesen des Harzes, Thüringer Waldes und Erzgebirges*. Stuttgart: Gustav Fischer Verlag.
Issler, E. (1942). *Vegetationskunde der Vogesen. Pflanzensoziologie* (Vol. 5). Jena: G. Fischer Verlag .
Jackson, S. T., & Hobbs, R. J. (2009). Ecological restoration in the light of ecological history. *Science, 325,* 567–569.
Jakucs, P. (1961). *Die phytozönologischen Verhältnisse der Flaumeichen–Buschwälder Südostmitteleuropas*. Budapest: Verlag der Ungarischen Akademie der Wissenschaften.
Kevey, B., & Borhidi, A. (2005). The acidophilous forests of the Mecsek and their relationship with the Balkan-Pannonian acidophilious forests. *Acta Botanica Hungarica, 47,* 273–368.
Klap, J. M., De Vries, W., Erisman, J. W., & van Leeuwen, E. P. (1997). Relationships between forest condition and natural and anthropogenic stress factors on the European scale; pilot study. (Report 150). Wageningen, the Netherlands: DLO Winand Staring Centre for Integrated Land, Soil and Water Research.
Klapp, E. (1965). *Grünlandvegetation und Standort–nach Beispielen aus West-, Mittel- und Süddeutschland.* Berlin: Verlag Paul Parey.
Knollová, I., & Chytrý, M. (2004). Oak-hornbeam forests of the Czech Republic: Geographical and ecological approaches to vegetation classification. *Preslia, 76,* 291–311.
Krzysztofiak, L. (Ed.). (2010). Raport o stanie środowiska przyrodniczego Stacji Bazowej Wigry w roku hydrologicznym 2009. Krzywe: Zintegrowany Monitoring Środowiska Przyrodniczego.
Lortie, C. J., Brooker, R. W., Choler, P., Kikvidze, Z., Michalet, R., Pugnaire, F. I., & Callaway, R. M. (2004). Rethinking plant community theory. *Oikos, 107,* 433–438.
Mahn, E. G. (1959). Vegetations- und standortskundliche Untersuchungen an Felsfluren, Trocken- und Halbrocken rasen Mitteldeutschlands. Uni Halle.

Mahn, E. G. (1965). *Vegetationsaufbau und Standortsverhältnisse der kontinental-beeinflussten Xerothermrasengesellschaften Mitteldeutschlands*. (Abhandlungen der Sächsischen Akademie der Wissenschaften zu Leipzig). Berlin: Akademie-Verlag.

Martin, K. (2002). *Ökologie der Biozönosen*. Berlin: Springer.

Matuszkiewicz, W. (1962). *Zur Systematik der natürlichen Kiefernwälder des mittel- und osteuropäischen Flachlandes* (pp. 145–186). Stolzenau: Mitt. Flor.-soziol. Arbeitsgemeinschaft (Tüxen).

Matuszkiewicz, W. (2001). *Przewodnik do oznaczania zbiorowisk roślinnych Polski* (Guide for determining plant associations of Poland). Warszawa: PWN.

Michalko, J. (1986). *Geobotanická Mapa CSSR*. Bratislava: Vydavatel'stvo Slovenskej Akadémie Vied.

Mill, W., Kacprzyk, W., & Schlama, A. (2008). Climate change impact on critical loads of acidity and eutrophicaiton for forest ecosystems in Poland, according to Gothenburg Protocol revision process (in Polish). (Report to the Ministry of Environment). Warszawa: Institute of Environmental Protection.

Moravec, J., & Neuhäusl, R. (1976). Geobotanická mapa České Socialistické Republiky: mapa rekonstruované přirozené vegetace. Karte 1: 1.000.000 mit 4 Nebenkarten. Academia Prag.

Mucina, L., Grabherr, G., & Wallnhöfer, S. (1993). *Die Pflanzengesellschaften Österreichs, Teil III*. Jena: Gustav Fischer Verlag.

Nagel, H. D., Schlutow, A., Kraft, P., Scheuschner, T., & Weigelt-Kirchner, R. (2010). Modellierung und Kartierung räumlich differenzierter Wirkungen von Stickstoffeinträgen in Öko-systeme im Rahmen der UNECE-Luftreinhaltekonvention Teilbericht II: Das BERN-Modell – ein Bewertungsmodell für die oberirdische Biodiversität. (UBA-Texte 08/2010. FE-Vorhaben 205 85 239 im Auftrag des Umweltbundesamtes).

Oberdorfer, E. (1957). Süddeutsche Vegetationsgesellschaften. Jena.

Oberdorfer, E. (Ed.). (1992–1998). *Süddeutsche Vegetationsgesellschaften*. (Teil I, 4. Aufl. 1998, Teil II, 3. Aufl. 1993, Teil III, 3. Aufl. 1993, Teil IV 1992). Jena: Gustav-Fischer-Verlag.

Passarge, H. (1964). Pflanzengesellschaften des nordostdeutschen Flachlandes I. Jena.

Passarge, H., & Hofmann, G. (1968). Pflanzengesellschaften des nordostdeutschen Flachlandes. II. Jena.

Pott, R. (1994). *Die Pflanzengesellschaften Deutschlands*. Stuttgart: Ulmer.

Preising, E. (1953). Süddeutsche Borstgras- und Zwergstrauch-Heiden (Nardo-Callunetea). In *Mitteilungen der floristisch-soziologischen Arbeitsgemeinschaft–Band 4* (pp. 112–123).

Preising, E., Vahle, H. C., Brandes, H., Hofmeister, H., Tüxen, J., & Weber, H. E. (1990a). Die Pflanzengesellschaften Niedersachsens–Bestandsentwicklung, Gefährdung und Schutzpro-bleme: Salzpflanzengesellschaften der Meeresküsten und des Binnenlandes. (Heft 20/7). Hannover: Naturschutz und Landschaftspflege Niedersachsens.

Preising, E., Vahle, H. C., Brandes, H., Hofmeister, H., Tüxen, J., & Weber, H. E. (1990b). Die Pflanzengesellschaften Niedersachsens–Bestandsentwicklung, Gefährdung und Schutzprobleme: Wasser- und Sumpfpflanzengesellschaften des Süßwassers. (Heft 20/8). Hannover: Naturschutz und Landschaftspflege Niedersachsens.

Preising, E., Vahle, H. C., Brandes, H., Hofmeister, H., Tüxen, J., & Weber, H. E. (1997). Die Pflanzengesellschaften Niedersachsens–Bestandsentwicklung, Gefährdung und Schutzprobleme: Rasen-, Fels- und Geröllgesellschaften. (Heft 20/5). Hannover: Naturschutz und Landschaftspflege Niedersachsens.

Roberts, D. W. (1986). Ordination on the basis of fuzzy set theory. *Vegetatio, 66,* 123–131.

Rolecek, J. (2005). Vegetation types of dry-mesic oak forests in Slovakia. *Preslia, 77,* 241–261.

Scamoni, A. (1960). *Waldgesellschaften und Waldstandorte–dargestellt am Gebiet des Diluviums der Deutschen Demokratischen Republik* (3. ed.). Berlin: Akademie-Verlag.

Schmidt, P. A., Hempel, W., Denner, M., Döring, N., Gnüchtel, B., Walter, B., & Wendel, D. (2002). Potentielle natürliche Vegetation Sachsens mit Karte 1:200.000. Dresden: Sächsisches Landesamt für Umwelt und Geologie.

Schubert, R. (1960). *Die zwergstrauchreichen azidiphilen Pflanzengesellschaften Mitteldeutschlands*. Jena: VEB Gustav Fischer.

Schubert, R. (1991). *Lehrbuch der Ökologie*. Jena: Verlag Fischer.
Schubert, R., Kotz, W., & Hilbig, S. (1995). *Bestimmungsbuch der Pflanzengesellschaften Mittel- und Nordostdeutschlands*. Jena: Fischer.
Schulte-Bisping, H. (2012). Data set of IM plot Neuglobsow from Georg-August University Göttingen, Büsgen-Institut, Dep.: Ökopedologie der Gemäßigten Zonen, unpublished.
Shugart, H. H. (1984). *A theory of forest dynamics*. New York: Springer.
Slobodda, S. (1982). Pflanzengesellschaften als Kriterium zur ökologischen Kennzeichnung des Standortsmosaiks. *Arch. Nat.schutz Landsch.forsch, 22,* 79–101.
Slobodda, S. (1987). *Pflanzengesellschaften und ihre Umwelt* (2. ed.). Leipzig: Urania-Verlag.
Soó, R. (1964). *Die regionalen Fagion-Verbände und Gesellschaften Südosteuropas*. Budapest: Verlag der Ungarischen Akademie der Wissenschaften.
Succow, M. (1974). Vorschlag einer systematischen Neugliederung der mineralbodenwasserbeeinflussten wachsenden Moorvegetation Mitteleuropas unter Ausklammerung des Gebirgsraumes. *Feddes Repertorium, 85,* 57–113.
Succow, M. (1988). *Landschaftoekologische Moorkunde*. Jena, GDR: Gustav Fischer.
Succow, M., & Joosten, H. (2001). *Landschaftsökologische Moorkunde* (2. ed.). Stuttgart: Schweizerbart´sche Verlagsbuchhandlung.
Tüxen, R. (1937). Die Pflanzengesellschaften Nordwestdeutschlands. (Mitt. flor.-soz. Arb.gem. Niedersachsen 3). Hannover.
Tüxen, R. (1957). Entwurf einer Definition der Pflanzengesellschaft (Lebensgemeinschaft). (Mitteilungen der Floristisch-Soziologischen Arbeitsgemeinschaft (6/7)).
Tüxen, R. (1958). Pflanzengesellschaften oligotropher Heidetümpel Nordwestdeutschlands. (Veröff. D. Geobotanischen Instituts Rübel, 33). Zürich.
Tüxen, R., & Westhoff, V. (1963). Saginetea maritimae, eine Gesellschaftsgruppe im wechselhalinen Grenzbereich der europäischen Meeresküsten. (Mitt. flor.-soz. Arbgemeinsch. 1963, N. F. 10). Stolzenau/Weser.
Van Loon, M., Tarrason, L., & Posch, M. (2005). Modelling base cations in Europe. EMEP Technical Report MSC-W 2/2005. Oslo.
Wamelink, G. W. W., & van Dobben, H. F. (2003). Uncertainty of critical loads based on the Ellenberg indicator value for acidity. *Basic and Applied Ecology, 4,* 515–523.
Wigley, T. M. L. (2008). MAGICC/SCENGEN 5.3: User Manual (version 2). http://www.cgd.ucar.edu/cas/wigley/magicc/UserMan5.3.v2.pdf. Accessed 31 Jan 2014.
Willner, W. (2002). Syntaxonomische Revision der südmitteleuropäischen Buchenwälder. *Phytocoenologia, 32,* 337–453.
Wolfram, C. (1996). Die Vegetation des Bottsandes. Mitt. Arb.gem. Geobot. Schlesw.-Holst. *Hambg, 51,* 111.
Zadeh, L. A. (1978). Fuzzy Sets as a basis for a theory of possibility. *Fuzzy Sets and Fuzzy Systems, 1,* 3–28.
Zarzycki, K., Trzcińska-Tacik, H., Różański, W., Szeląg, Z., Wołek, J., & Korzeniak, U. (2002). *Ecological indicator values of vascular plants of Poland.* (Biodiversity of Poland Vol. 2). Kraków: W. Szafer Institute of Botany, Polish Academy of Sciences.
Zhang, J.-T. (1994). A combination of fuzzy set ordination with detrended correspondence analysis: One way to combine multi-environmental parameters with vegetation data. *Vegetatio, 115,* 115–121.

Part IV
Critical Loads and Dynamic Model Applications on a Regional Scale

Chapter 15
Assessment of Critical Loads of Sulphur and Nitrogen and Their Exceedances for Terrestrial Ecosystems in the Northern Hemisphere

Gert Jan Reinds, Maximilian Posch, Julian Aherne and Martin Forsius

15.1 Introduction

Critical loads of sulphur (S) and nitrogen (N) have been used in Europe as an indicator of ecosystem sensitivity to acidification and eutrophication under the Convention of Long-range Transboundary Air Pollution (LRTAP) within the United Nations Economic Commission for Europe (UNECE) and the Thematic Strategy on Air Pollution (TSAP) of the European Union. Several early studies determined low-resolution critical loads of N and acidity for terrestrial ecosystems at the European (De Vries et al. 1994; Kuylenstierna et al. 1998), Southeast-Asian (Hettelingh et al. 1995), Northern Asian (Bashkin et al. 1995; Semenov et al. 2001) and global scales (Bouwman et al. 2002; Kuylenstierna et al. 2001). High-resolution critical loads have also been mapped for Northern Asia (Reinds et al. 2008) and Canada (Aherne and Posch 2013) and the Arctic region (Forsius et al. 2010). In this chapter we describe global databases and subsequently use them to compute high-resolution critical loads of acidity and eutrophication for terrestrial ecosystems in the Northern Hemisphere.

Several studies have shown that the choice of the chemical criterion and critical limit can have a strong influence on critical load values (e.g. Aherne et al. 2001; Hall et al. 2001; Reinds et al. 2008), for Europe, the UK and Ireland, and on critical

G. J. Reinds (✉)
Alterra Wageningen University and Research Centre, Wageningen, The Netherlands
e-mail: gertjan.reinds@wur.nl

M. Posch
Coordination Centre for Effects (CCE), RIVM, Bilthoven, The Netherlands

J. Aherne
Environmental and Resource Studies, Trent University, Peterborough, ON, Canada

M. Forsius
Finnish Environment Institute (SYKE), Helsinki, Finland

W. de Vries et al. (eds.), *Critical Loads and Dynamic Risk Assessments*,
Environmental Pollution 25, DOI 10.1007/978-94-017-9508-1_15

load exceedances (e.g. Holmberg et al. 2001 for Finland). Similarly, De Vries et al. (2007) have shown that the widely used nutrient nitrogen limit of 0.2–0.4 mg N l^{-1}, related to vegetation changes in forests, is applicable for Scandinavia but not for Western Europe. Therefore, variations in the regional patterns of critical loads caused by different chemical criteria are also explored in this chapter.

15.2 Methods and Data

15.2.1 The Critical Load Model

Critical loads were computed with the so-called Simple Mass Balance (SMB) model, which links deposition to a chemical variable (the 'chemical criterion') in the soil or soil solution, which can be associated with ecosystem effects; and the violation of a specified 'critical limit' of this variable is linked to ecosystem damage. A low critical load implies high ecosystem sensitivity to deposition, and *vice versa*. The critical load of acidity for S and N is not a single value, but a trapezoidal function in the N-S-deposition plane characterised by three quantities, $CL_{max}S$, $CL_{min}N$ and $CL_{max}N$. In contrast, the eutrophying effect of N is defined by a single quantity, $CL_{nut}N$. The derivation of these quantities is described in Chap. 6.

15.2.2 Chemical Criteria

Critical limits are based on dose-response relationships between chemical characteristics and ecosystem functioning. A critical load equals the deposition to an ecosystem compartment (e.g. soil, groundwater, plant) at steady state that does not exceed the selected critical limit, thus preventing 'significant harmful effects on specified sensitive elements (e.g. plant roots) of the environment' (Nilsson and Grennfelt 1988).

To date mostly soil chemical criteria (e.g. nitrate and aluminium (Al) concentrations or Al to base cation (Bc=Ca+Mg+K; BC=Bc+Na) ratios in soil solution) have been used to derive critical loads with steady-state models. The relationship between these critical limits and the 'harmful effects' is one of the largest sources of uncertainty (see Chap. 2). In this chapter we used a critical Al/Bc ratio, a critical ANC value and a critical Al concentration to investigate the effect of the choice of criterion on the (patterns in) critical loads of acidity. The Al/Bc ratio is the most commonly used criterion, which is set to protect plant roots against toxic effects of Al, taking into account the protective effect of base cations, while the ANC criterion is specified to avoid base cation depletion (long-term soil stability). The Al criterion is also set to protect plant roots against Al toxicity.

For a given site, a fixed relationship exists between each of these criteria, referred to hereafter as 'equivalent criteria'. The widely used critical limit of $Al/Bc=1$ leads

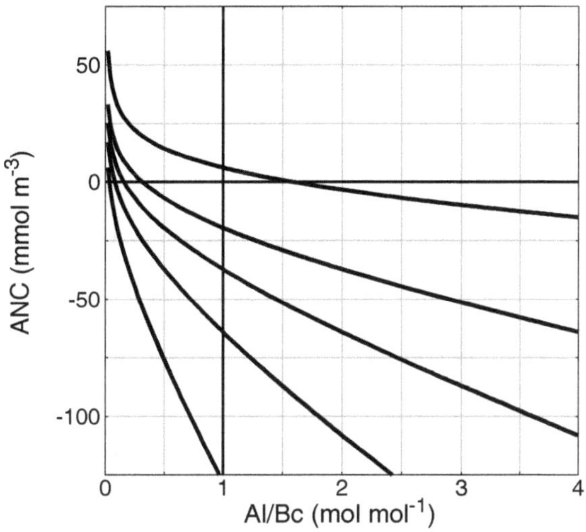

Fig. 15.1 Steady-state functional relationship between ANC and molar Al/Bc ratio for $[Bc]=1$ (*top curve*), 5, 10, 20 and 50 mmol m^{-3} (*bottom curve*) using standard soil parameters; see also Reinds et al. (2008)

to (strongly) negative ANC values in the soil solution, except for soils with very low Bc concentration (Fig. 15.1). The decrease in ANC, equivalent to $Al/Bc=1$, with increasing Bc concentrations is due to the fact that under increasing Bc concentration the Al concentration has to increase to maintain a constant Al/Bc ratio, and the increasing Al concentration leads to a decreasing ANC. In this study, we used a critical Al/Bc ratio of 1 mol mol^{-1}, a critical ANC concentration of zero, and a critical Al concentration of 0.2 eq m^{-3}. An Al concentration of 0.2 eq m^{-3}corresponds to $Al/Bc=1$ at a Bc concentration of 0.3 eq m^{-3}.

Further, acceptable N concentrations, the critical limit for $CL_{nut}N$, were set to using the acceptable concentrations of 0.2 mg N l^{-1} for coniferous forest and 0.3 mg N l^{-1} for deciduous forests and other (semi-)natural vegetation (Mapping Manual; www.icpmapping.org) to avoid changes in vegetation composition (see also Chap. 2).

15.2.3 Exceedance Calculations

Critical load exceedances were computed using the methods described in Chap. 6 and modelled depositions for the year 2000. For map displays, we used the average accumulated exceedance (AAE), i.e. the area-weighted exceedance in each $0.5° \times 0.5°$ longitude-latitude grid cell.

Deposition of S and N on a global scale was obtained from Dentener et al. (2006). In that study, 23 models were used to compute depositions of SO_2, NO_x and NH_3 to arrive at mean model ensemble estimates. Dentener et al. (2006) showed that there was a reasonable to good correlation between the mean of the model results and measurements.

15.2.4 Geographical Databases

The required input data for critical load calculations consist of spatial information describing climatic and hydrological variables, base cation deposition and weathering, nutrient uptake and N transformations derived by combining global maps of precipitation, temperature, soils, land cover and forest growth. To cover the entire geographical area of interest, several thematic maps were combined:

a. *Land cover*: The Global Land Cover 2000 map at 1 km resolution (Bartholome et al. 2002) was used; critical loads were determined and mapped for forests (EUNIS code 'G') and (semi-)natural vegetation ('D': mires, bogs and fens, 'E': natural grasslands and 'F': heathland, scrub and tundra).
b. *Soils*: The European Soil Database v2 map (JRC 2006) at a scale 1:1 M was used for Europe including the entire Russian territory, Belarus and Ukraine. For the other regions, the less detailed FAO 1:5 M soil map (FAO 1988) was used.
c. *Forest growth*: Average forest growth for European countries was derived from the European Forest Institute (EFI) database, containing growth data for a variety of species and age classes in about 250 regions in Europe (Schelhaas et al. 1999). This map was combined with a map of 74 administrative regions in Russia for which forest stock data were provided (Alexeyev et al. 2004). For other Commonwealth of Independent States (CIS) countries, Cyprus and Turkey a map was constructed that delineated forest growth regions where data were available.

Overlaying these maps and merging polygons with common soil, vegetation and region characteristics within blocks of $0.05° \times 0.05°$ resulted in about 10.8 M computational units with a total area of 20.2 M km^2. The determination of critical loads was limited to computational units larger than 1 km^2 reducing the total number to 4.7 M, representing 91 % of the study area. Data and results are presented as maps showing the median values on a $0.5° \times 0.5°$ longitude-latitude grid. In contrast, for critical loads the 5th percentiles are shown to indicate the most sensitive ecosystems. The (sub-)arctic region with shallow permafrost was excluded from modelling and represented as 'no data' areas (see below).

The soil maps are composed of associations; each map polygon representing one soil association consisting of several soil typological units (soil types) that each occupy a known percentage of the soil association, but with unknown location within the association. Soil topological units are classified into more than 200 soil types (European Soil Bureau Network 2004), with associated attribute data such as soil texture, parent material class and drainage class. Six texture classes (including peat) are defined, based on clay and sand content (FAO-UNESCO 2003). The drainage classes, which are used to estimate the denitrification fraction, were derived from the dominant annual soil water regime (European Soil Bureau Network 2004; FAO-UNESCO 2003).

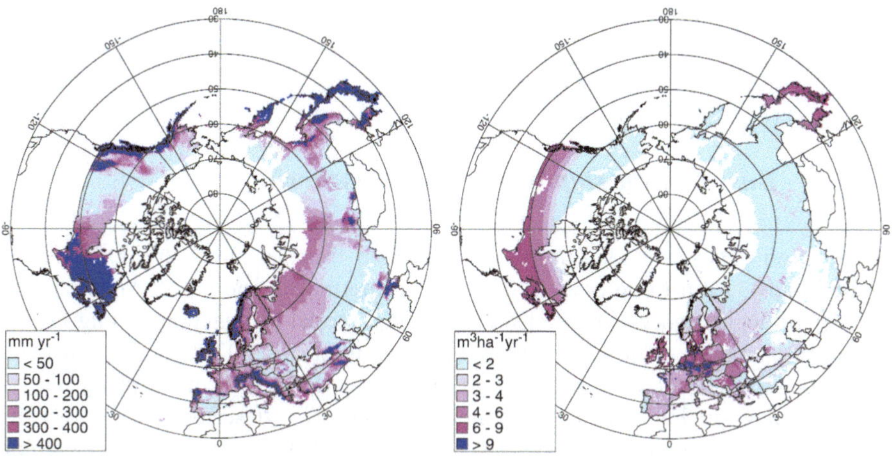

Fig. 15.2 Grid cell median annual runoff (1961–1990 average; *left*) and grid cell median forest growth (*right*)

15.2.5 Meteorology and Hydrology

The annual water flux leaving the soil at the bottom of the rooting zone (runoff or precipitation surplus) is required to compute the concentration and leaching of compounds. The bottom of the rooting zone was set at 50 cm, except for lithosols which were assumed to have a soil depth of 10 cm only. The leaching rate was estimated from meteorological data and soil properties. Long-term (1961–1990) average monthly temperature, precipitation and cloudiness were derived from a high resolution European database (New et al. 1999) that contained monthly values for the years 1901–2001 for land-based grid-cells of 10'× 10' (approx. 15 × 18 km in central Europe). For regions outside Europe, a coarser 0.5° × 0.5° global database was used (New et al. 2000).

Evapotranspiration and runoff were calculated with a sub-model used in the IMAGE global change model (Leemans and Van den Born 1994) following the approach by Prentice et al. (1993). A short description of this model can be found in Reinds et al. (2008), details in Reinds et al. (2001).

The SMB model implicitly assumes free draining soils. Large areas of the Northern Hemisphere, however, have shallow permafrost; and thus soils in areas with an average monthly temperature below zero for at least 8 months of the year were excluded. This area (see Forsius et al. 2010) corresponded well with the areas with shallow permafrost (FAO-UNESCO 2003).

Leaching fluxes varied from less than 100 mm $year^{-1}$ in arid regions such as central Spain, central Turkey, central Canada and large parts of the southern CIS states to >300 mm $year^{-1}$ in areas with high precipitation such as along the west coast of Europe and along many mountain ranges, along the Pacific rim and the eastern and western provinces of Canada (Fig. 15.2). Uncertainties in leaching flux are associated with the reliability of the climate data; values in western Europe are more certain than those in the other regions, as the density of rainfall stations used to estimate the grid rainfall is much higher.

15.2.6 Base Cation Deposition and Weathering

Base cation deposition for Europe was obtained from an atmospheric dispersion model for BC (Van Loon et al. 2005). For northern Asia and North America, calcium (Ca) deposition was taken from the global model of Tegen and Fung (1995) based on estimates of soil Ca content (Bouwman et al. 2002). Magnesium (Mg) and potassium (K) deposition outside Europe was derived from Ca using regression equations (including distance to the nearest coast) as described by Reinds et al. (2008). The distance to coast was taken from a NASA dataset (http://oceancolor.gsfc.nasa.gov/DOCS/DistFromCoast/).

Median Bc deposition shows a strong north-south gradient in Europe with values >600 eq $ha^{-1}year^{-1}$ in southern Europe and the southern parts of the CIS states, caused by high dust input from nearby desert areas, and very low values of <100 eq $ha^{-1}year^{-1}$ in the northern part of the study area (Fig. 15.3). The mapped region shows a reasonably consistent spatial pattern despite being based on two data sources.

Weathering of base cations was computed as a function of parent material and texture, corrected for temperature as described in the Mapping Manual (www.icp-mapping.org). Very low base cation weathering rates (>150 eq ha^{-1} $year^{-1}$) were estimated in most of Scandinavia where poor soils and low temperatures prevail (Fig. 15.3). The same holds for large parts of Northern Russia. Very low weathering rates were also estimated in central and western Spain in areas dominated by acid dystric regosols developed on granites. High weathering rates (>1000 eq ha^{-1} $year^{-1}$) were confined to regions with soils developed on volcanic materials and especially in areas dominated by calcareous soils that occupy parts of Spain, France, Hungary, large parts of Turkey and many areas in e.g. Kazakhstan, Uzbekistan and Turkmenistan. In Canada and Alaska estimated weathering rates were low, except for the central Prairie region, and mountainous regions in the west.

15.2.7 Nutrient Uptake, Nitrogen Immobilization and Denitrification

The net growth uptake of Bc and N by forests was computed by multiplying the estimated annual average growth of stems and branches with the element contents of Bc and N in these compartments taken from an extensive literature review by Jacobsen et al. (2002). Forest growth in EU countries was taken from the European Forest Information Scenario (EFISCEN) model (Schelhaas et al. 2007). Forest growth for the rest of Europe was derived from the EFI database (Schelhaas et al. 1999), which provides measured growth data for about 250 regions in Europe for

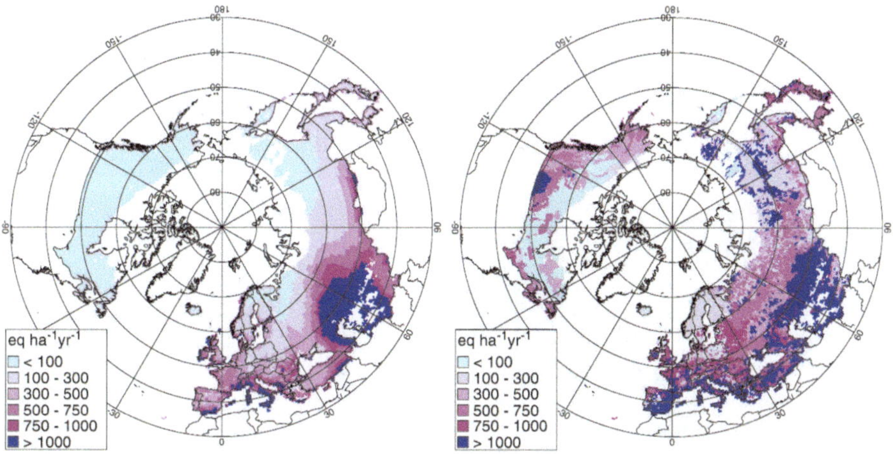

Fig. 15.3 Grid cell median base cation deposition (*left*) and grid cell median base cation weathering rate (*right*)

various species and age classes. Forest growth for Russia was estimated from data by Alexeyev et al. (2004), who compiled statistical data on growing stock and areas of stocked land from available data sources, for 74 administrative regions within Russia (see Reinds et al. 2008). For the other CIS states, growth rates were obtained from Prins and Korotkov (1994), who provided growing stock per hectare; average forest growth in these regions was based on an assumed average stand age of 60 years. For Turkey, growth rates were supplied by the Turkish ICP Forests National Focal Centre for 30 species and two forest states (degraded and non-degraded). For Cyprus a crude approximation of an average growth rate of 0.8 m^3 ha^{-1} was made based on the average standing biomass of 43 m^3 ha^{-1} given by FAO (2000) and assuming an average stand age of 60 years. Forest growth for Japan was taken from Cannell (1982), and forest growth in Korea was assumed to equal growth in Japan. Due to the lack of detailed information, growth rates for Canada and Alaska were estimated as a function of latitude starting at 6 m^3 ha^{-1} $year^{-1}$ south of 50 °N linearly deceasing to zero north of 60 °N.

Net uptake of N and base cations in forest ecosystems was determined from nutrient contents and growth rate. Median growth rates for forests show the well-known pattern in Europe; high growth rates are found in central Europe where climate, site quality and intensive management allow highly productive forests. Low growth rates in Europe were confined to arid regions such as central Spain, parts of France and Turkey and to areas with low temperature and poor site quality (shallow, poor soils) such as northern Scandinavia. In the CIS territory growth rates were generally low (1–3 m^3 ha^{-1} $year^{-1}$), although relatively high forest productivity can be found in the areas west of the Ural mountains and in Georgia. Very low growth rates (<0.5 m^3 ha^{-1} $year^{-1}$) occur in arid regions of central Asia. Growth rates for Russia are somewhat uncertain as they were derived indirectly from growing stock data. For the other CIS states and Cyprus, growth data do not represent the spatial variability as only one growth rate per country could be assigned. Although the

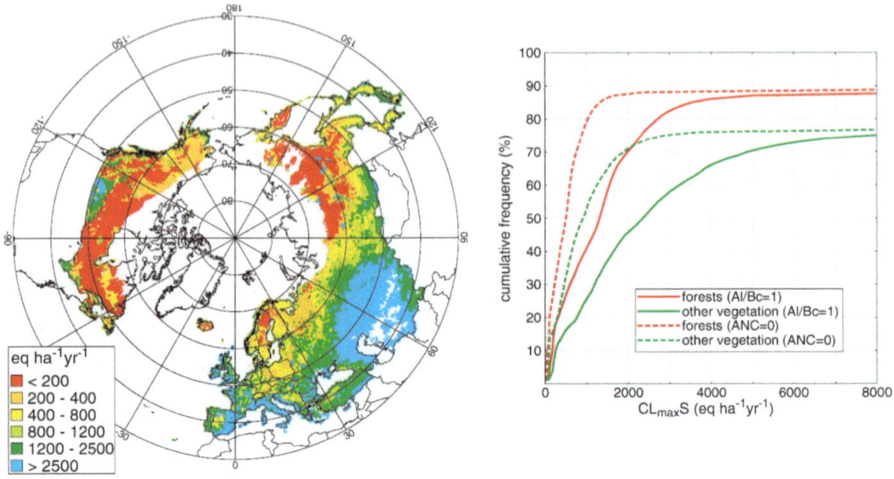

Fig. 15.4 5th percentile $CL_{max}S$ for the *Al/Bc* = 1 criterion per grid cell (*left*) and cumulative frequency distributions of $CL_{max}S$ (*right*) for forests and (semi-)natural vegetation, both for *Al/Bc* = 1 (*solid lines*) and *ANC* = 0 (*dashed lines*)

European data are based on a large data set, it is clear that some border effects occur, probably due to the fact that for some countries (e.g. Ukraine and Romania) the area-representation is relatively poor. In Canada the assumed linear decrease in growth with latitude can be clearly discerned.

The denitrification fraction, f_{de}, required to compute the denitrification flux (see Chap. 6), was estimated as a function of soil drainage status (Reinds et al. 2001) and varies between 0.1 for well drained soils to 0.8 for peaty soils. The long-term net N immobilization was set at 1 kg N ha^{-1} year^{-1}, which is at the upper end of the estimated annual accumulation rates for the build-up of stable C–N compounds in soils (see Mapping Manual; www.icpmapping.org).

15.3 Results

15.3.1 Critical Loads of Acidity

The critical load of acidity, $CL_{max}S$, is computed as the sum of BC input through weathering and deposition minus the removal of Bc by uptake and minus the critical ANC leaching based on the chemical criterion and its critical limit (see Chap. 6). High critical loads are found in areas with high BC weathering and/or BC deposition such as along the Mediterranean coast and the southern parts of the CIS states (Fig. 15.4). Furthermore, high critical loads are found in areas with high acidity

Table 15.1 Regional statistics (5th percentile, median and 95th percentile) for the critical load of acidity ($CL_{max}S$; in eq $ha^{-1}year^{-1}$) based on three critical chemical criteria: $Al/Bc = 1$ mol mol^{-1}, $ANC = 0$ and $[Al] = 0.2$ eq m^{-3}

Region	*Al/Bc*			*ANC*			*[Al]*		
	5th perc.	Median	95th perc.	5th perc.	Median	95th perc.	5th perc.	Median	95th perc.
Central Europe	526	2517	4516	248	1018	1958	432	1311	2603
Russia	169	1348	4191	96	599	9931	157	901	2635
Other CIS states	1527	3017	6957	611	1229	3665	901	1567	3829
Scandinavia	279	637	2750	97	305	1104	545	889	2740
Western Europe	443	3066	5549	228	1183	2461	751	2005	4869
Southern Europe	1170	3061	7010	465	1220	3030	811	1803	3657
Japan & Korea	629	2118	4023	272	837	1554	1012	2228	4231
Canada	63	735	1700	33	117	693	141	757	2366
Alaska	234	1187	2092	88	469	829	172	666	3018

leaching (areas with high runoff) such as along the coast of north-western Europe. Such high acidity leaching due to high base cation (especially Mg) input from sea salts is considered inappropriate by some authors, who have used alternative criteria, e.g. pH in Ireland (Aherne et al. 2001) and Ca/Al ratio in the UK (Hall et al. 2001). Low critical loads are found in areas with low weathering rates associated with coarse soils on acid parent material such as central Spain, Poland and the Precambrian Shield in Canada, and/or regions with low temperatures (Scandinavia, northern Russia and Canada).

The cumulative frequency distributions of $CL_{max}S$ for forests and (semi-)natural vegetation (grasslands/heathlands/tundra), both for the $Al/Bc = 1$ and $ANC = 0$ criteria, show that natural vegetation generally have higher critical loads because they occur on richer soils: average computed weathering rates for these soils were about twice as high as those for forests soils. More than 20% of these ecosystems are located on calcareous soils. Critical loads using the ANC=0 criterion are considerably lower those based on $Al/Bc = 1$ (see also Fig. 15.1). Table 15.1 lists the 5th percentile, median, and 95th percentile for $CL_{max}S$ per region for non-calcareous soils.

For the most sensitive ecosystems, expressed by the 5th percentile, the Al/Bc and Al criteria yield similar critical loads (in most regions). In areas with poor sandy soils and low atmospheric Bc inputs such as Scandinavia, Russia, Canada and Alaska, critical loads based on $Al/Bc = 1$ are very low, with Al concentration (far) below the critical value of 0.2 eq m^{-3}; in these regions critical load of acidity based on *[Al]* is thus often higher, see also Holmberg et al. (2001). Critical loads based on $ANC = 0$ are lower than those for $Al/Bc = 1$ and show that in regions with little buffering capacity, it will be difficult to attain this criterion.

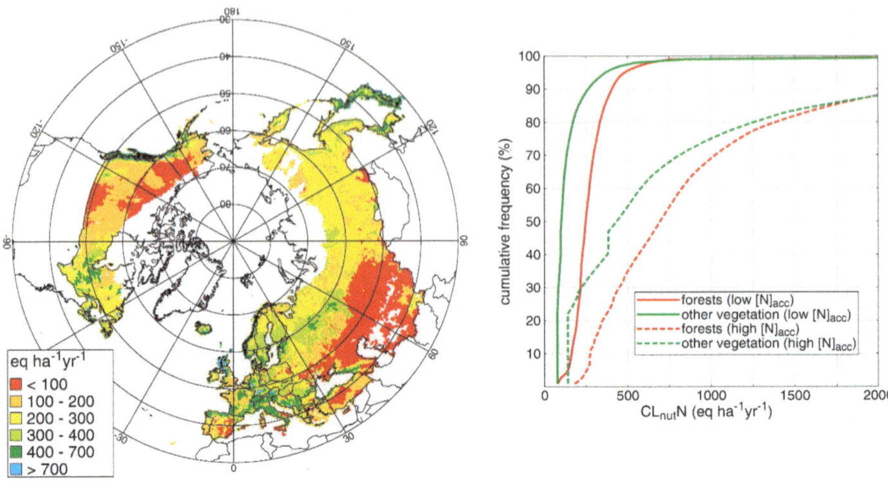

Fig. 15.5 Median $CL_{nut}N$ for the low $[N]_{acc}$ criterion per grid cell (*left*) and cumulative frequency distribution of $CL_{nut}N$ (*right*) for forests and (semi-)natural vegetation both for the low (*solid lines*) and high $[N]_{acc}$ criteria (*dashed lines*)

15.3.2 Critical Loads of Nutrient N

The critical load of nutrient N, $CL_{nut}N$, is computed from N uptake, immobilisation, denitrification and an acceptable N leaching. High critical loads were confined to regions dominated by grass and heathlands with high runoff, leading to high N leaching rates such as Ireland, the western parts of the UK, Norway, northern Spain, the Adriatic coast, the west coast of Canada and the Kamchatka peninsula (Fig. 15.5). Low critical loads were estimated in arid regions where N leaching is low, such as in the southern part of the CIS states, most of Turkey and parts of Central Europe and central and western Canada as well as Alaska.

If denitrification is low, which is often the case in forest soils, $CL_{nut}N$ mainly consists of N removal through leaching, uptake and immobilization. In general, $CL_{nut}N$ is typically computed with a critical limit ($[N]_{acc}$) of 0.2–0.4 mg N l^{-1}which is considered to be representative for forests; concentrations in soil solution above this limit may lead to changes in understory vegetation in forests (Mapping Manual; www.icpmapping.org). However, a study on critical limits of N (De Vries et al. 2007) revealed that such low limits are mostly related to vegetation changes in Scandinavia, but effects in western Europe probably occur at higher values. Based on literature data and on model results from the Netherlands, the authors suggested higher limits of about 3–6 mg N l^{-1} for Western Europe.

Comparing the cumulative frequency distributions of $CL_{nut}N$ for forests, using the acceptable concentrations in the range of 0.2–0.3 mg N l^{-1} (see Sect. 15.2.2), with $CL_{nut}N$ using a limit of 3 mg N l^{-1} for forests and 3.5 mg N l^{-1} for (semi-)natural vegetation (although specifically suggested for non-Nordic forests only) shows an obvious increase in $CL_{nut}N$ with higher concentrations. The median value for forest

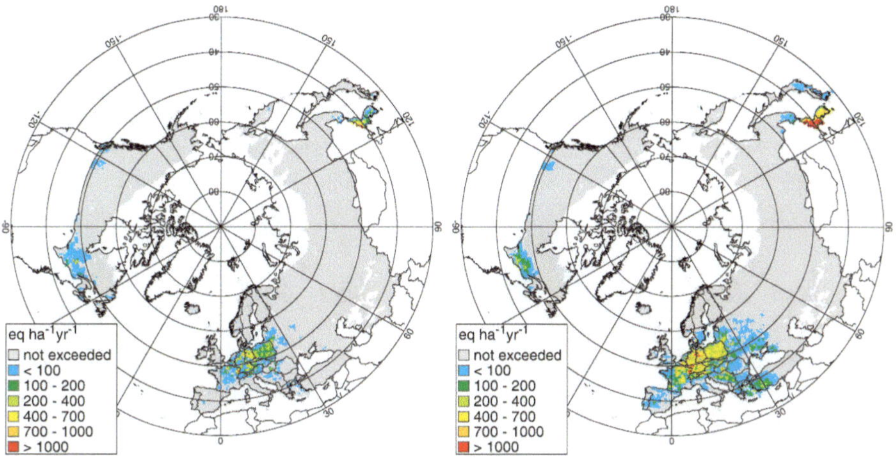

Fig. 15.6 Exceedance (AAE) of the critical load of acidity for the *Al/Bc* = 1 criterion (*left*) and the *ANC* = 0 criterion (*right*)

is 700 eq ha^{-1} year^{-1}, which is almost three times as high as the median value based on the lower acceptable leaching concentrations; for (semi-)natural vegetation the difference is even higher (400 versus 100 eq ha^{-1} year^{-1}; Fig. 15.5). Large differences in the 5th percentile critical load from both criteria are found in regions with a high precipitation surplus such as Ireland, UK and the Northern Alps. In areas with a low leaching rate, such as the eastern part of Germany, and southern and central France, the increase in the 5th percentile critical load is less than 50 % (not shown). This demonstrates that increasing the acceptable N concentration by a factor of 10 does not always mean a commensurate increase in critical load of nutrient N.

15.3.3 Exceedances

Exceedances of critical loads of acidity under 2000 deposition were observed in Europe, southern Canada (predominately southern Ontario and Quebec), Korea and Japan (Fig. 15.6). High exceedances were estimated in Central Europe and Korea due to the high deposition of acidity. Although the critical loads for the *ANC* = 0 criteria are generally much lower than those for *Al/Bc* = 1, the area were critical loads are exceeded for both criteria coincides across much of the mapped region. Only in southern and eastern Europe, the ANC criterion leads to a significant increase in the exceeded area compared with the Al/Bc criterion, as the preservation of cation pools is a stricter criterion than the protection of plant roots.

The areas with estimated exceedances of critical load of nutrient N coincided with the area of exceedances of critical load of acidity, but also included large parts of the CIS region (Fig. 15.7). Highest exceedances were estimated in areas with low leaching and/or low N uptake (Central Europe) and in areas with a high N deposition. Large parts of Scandinavia were exceeded only under the more stringent N (lower) criterion.

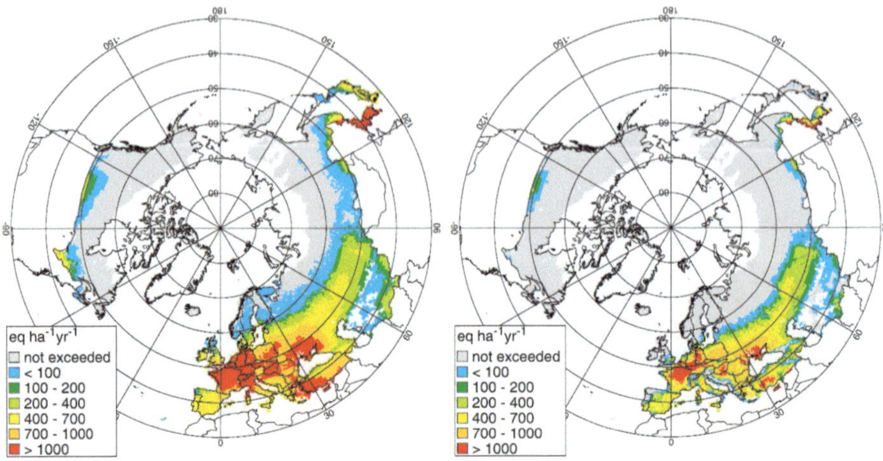

Fig. 15.7 Exceedance (AAE) of the critical load of nutrient N for the low (*left*) and high (*right*) acceptable N leaching criterion

Discussion and Conclusions

Combining available databases on soil, land cover, climate and forest growth resulted in a detailed map with almost 11 M receptors for Europe, Northern Asia, Canada and Alaska with spatially highly disaggregated critical load calculations. The spatial pattern in critical loads showed similarity with those in Kuylenstierna et al. (2001), but since their empirical approach was based on soil sensitivity alone (determined by weathering rate), they did not include the influence of N removal or precipitation excess on critical load patterns.

Critical loads depend on the chosen chemical criterion and its limit value; critical loads based on $Al/Bc = 1$ or $[Al] = 0.2$ eq m^{-3} are higher than critical loads based on $ANC = 0$. This is also evident from the equivalent criteria, i.e. $Al/Bc = 1$ leads to positive ANC in soils with a very low base cation concentration only (Fig. 15.1). The ANC leaching criterion provides a general protection against Bc depletion; however, studies providing limits for ANC leaching related to effects on terrestrial ecosystems are lacking.

Critical loads of nutrient N for Western Europe using an acceptable N concentrations of 3 mg N l^{-1} in forest soil solution, are much higher than critical loads based on the generally applied limits of 0.2–0.3 mg N l^{-1}, more representative for Northern Europe. This sensitivity to the N criterion was previously reported by De Vries et al. (1994). Across western Europe, the median critical load for nutrient N with the higher acceptable N leaching limit was about 1000 eq ha^{-1} year^{-1} (= 14 kg Nha^{-1} year^{-1}). This is in better agreement with the empirical critical loads of 10–15 kg ha^{-1} year^{-1} (Bobbink and Hettelingh 2011) related to vegetation changes in forests compared with 500 eq ha^{-1} year^{-1} based on a limit of 0.2 mg N l^{-1}; see also Holmberg et al. (2013). The strong dependence of critical load on precipitation shows the limitations of the SMB model for the assessment of nutrient N critical loads, as it assumes a relationship between nitrate concentration and plant species sensitivity, whereas N availability may be more important (De Vries et al. 2007). Moving to more elaborate models that compute critical N loads based on limits for e.g. pH and N availability related to occurrence of plant species (see e.g. Posch et al. 2011; Van Dobben et al. 2006) may improve estimates of critical loads for nutrient N.

In this chapter we have only investigated the dependence of critical loads on different soil chemical criteria. Uncertainty in critical load estimates is associated with model and parameter uncertainty (see, e.g. Skeffington et al. 2007; Zak and Beven 1999). A study for a forested catchment in Belgium showed that various local and general parameter estimates of nutrient removal by forests resulted in a wide range in estimated critical loads of nutrient N (Bosman et al. 2001). Others have shown that different methods of estimating weathering rates can yield widely varying results (Hodson and Langan 1999). Model structure is another important source of uncertainty. The SMB model by definition is a very simple model of reality; the denitrification sub-model and the fact that we assume a homogenous soil layer contribute to uncertainty. Further, previous studies have shown that there is a strong effect of the depth at which the chemical criterion should be met especially when using the Al/Bc criterion (De Vries et al. 1994). Despite significant reductions in S and N emissions during the last two decades, critical loads are still exceeded in large parts of western and central Europe. Only further reductions of (especially N) emissions will achieve the long-term political objective of the EU, i.e. non-exceedance of critical loads.

Acknowledgements We thank Ina Tegen (Leibniz Institute for Tropospheric Research, Leipzig, Germany) for providing the calcium deposition data for Asia and our colleague Lex Bouwman for pointing out this data source. We also thank the Turkish ICP-Forest NFC for supplying forest growth rates in Turkey, and Frank Dentener for kindly providing the global S and N deposition data.

References

Aherne, J., & Posch, M. (2013). Impacts of nitrogen and sulphur deposition on forest ecosystem services in Canada. *Current Opinion in Environmental Sustainability, 5,* 108–115.

Aherne, J., Farrell, E. P., Hall, J., Reynolds, B., & Hornung, M. (2001). Using multiple chemical criteria for critical loads of acidity in maritime regions. *Water Air and Soil Pollution: Focus, 1,* 75–90.

Alexeyev, V. A., Markov, M. V., & Birdsey, R. A. (2004). Statistical data on forest fund of russia and chaning of forest productivity in the second half of the XX century. St. Peterburg: Ministry of Natural Resources of Russian Federation, St. Petersburg Research Institute of Forestry and St. Petersburg Forest Ecological Center.

Bartholome, E., Belward, A. S., Achard, F., Bartalev, S., Carmona-Moreno, C., Eva, H., Fritz, S., Gegoure, J. -M., Mayaux, P., & Stibig, H. -J. (2002). GLC 2000. Global Land Cover mapping for the year 2000. Project status November 2002. (EUR 20524 EN). Ispra, Italy: European Commission, Joint Research Centre.

Bashkin, V. N., Kozlov, M. Y., Priputina, I. V., Abramychev, A. Y., & Dedkova, I. S. (1995). Calculation and mapping of critical loads of S, N and acidity on ecosystems of the Northern Asia. *Water Air and Soil Pollution, 85,* 2395–2400.

Bobbink, R., & Hettelingh, J.-P. (Eds.). (2011). *Review and revision of empirical critical loads and dose-response relationships*. Proceedings of an expert workshop, Noordwijkerhout 23–25 June 2010 Bilthoven, The Netherlands.

Bosman, B., Remacle, J., & Carnol, M. (2001). Element removal in harvested tree biomass: Scenarios for critical loads in Wallonia, South Belgium. *Water Air and Soil Pollution: Focus, 1,* 153–167.

Bouwman, A., Van Vuuren, D., Derwent, R., & Posch, M. (2002). A global analysis of acidification and eutrophication of terrestrial ecosystems. *Water Air and Soil Pollution, 141,* 349–382.

Cannell, M. G. R. (1982). *World forest biomass and primary production data*. New York: Academic Press.

De Vries, W., Reinds, G. J., & Posch, M. (1994). Assessment of critical loads and their exceedance on European forests using a one-layer steady-state model. *Water Air Soil Pollution, 72,* 357–394.

De Vries, W., Kros, J., Reinds, G. J., Wamelink, W., Van Dobben, H., Bobbink, R., Emmett, B., Smart, S., Evans, C., Schlutow, A., Kraft, P., Belyazid, S., Sverdrup, H., van Hinsberg, A., Posch, M., & Hettelingh, J. -P. (2007). Developments in modelling critical nitrogen loads for terrestrial ecosystems in Europe. (Alterra Report 1382). Wageningen, The Netherlands.

Dentener, F., Drevet, J., Lamarque, J. F., Bey, I., Eickhout, B., Fiore, A. M., Hauglustaine, D., Horowitz, L. W., Krol, M., Kulshrestha, U. C., Lawrence, M., Galy-Lacaux, C., Rast, S., Shindell, D., Stevenson, D., Van Noije, T., Atherton, C., Bell, N., Bergman, D., Butler, T., Cofala, J., Collins, B., Doherty, R., Ellingsen, K., Galloway, J., Gauss, M., Montanaro, V., Müller, J. F., Pitari, G., Rodriguez, J., Sanderson, M., Solmon, F., Strahan, S., Schultz, M., Sudo, K., Szopa, S., & Wild, O. (2006). Nitrogen and sulfur deposition on regional and global scales: A multimodel evaluation. *Global Biogeochemical Cycles, 20,* GB4003.

European Soil Bureau Network. (2004). European Soil Database (v 2.0). (EUR 19945 EN). European Soil Bureau Network and European Commission.

FAO. (1988). *Soil map of the World, wrevised legend. World soil resources report 60*. Rome: FAO.

FAO. (2000). Global forest resources assessment 2000, Main Report. (FAO Forestry Paper 140). Rome.

FAO-UNESCO. (2003). *Digital soil map of the world and derived soil properties, CD-ROM*. Rome: FAO.

Forsius, M., Posch, M., Aherne, J., Reinds, G. J., Christensen, J., & Hole, L. (2010). Assessing the impacts of long-range sulfur and nitrogen deposition on arctic and sub-arctic ecosystems. *Ambio, 39,* 136–147.

Hall, J., Reynolds, B., Aherne, J., & Hornung, M. (2001). The importance of selecting appropriate criteria for calculating acidity critical loads for terrestrial ecosystems using the simple mass balance equation. *Water Air and Soil Pollution: Focus, 1,* 29–41.

Hettelingh, J. -P., Sverdrup, H., & Zhao, D. (1995). Deriving critical loads for Asia. *Water Air and Soil Pollution, 85,* 2565–2570.

Hodson, M. E., & Langan, S. J. (1999). Considerations of uncertainty in setting critical loads of acidity of soils: the role of weathering rate determination. *Environmental Pollution, 106,* 73–81.

Holmberg, M., Mulder, J., Posch, M., Starr, M., Forsius, M., Johansson, M., Bak, J., Ilvesniemi, H., & Sverdrup, H. (2001). Critical loads of acidity for forest soils: Tentative modifications. *Water Air and Soil Pollution: Focus, 1,* 91–101.

Holmberg, M., Vuorenmaa, J., Posch, M., Forsius, M., Lundin, L., Kleemola, S., Augustaitis, A., Beudert, B., de Wit, H. A., Dirnböck, T., Evans, C. D., Frey, J., Grandin, U., Indriksone, I., Krám, P., Pompei, E., Schulte-Bisping, H., Srybny, A., & Váňa, M. (2013). Relationship between critical load exceedances and empirical impact indicators at integrated monitoring sites across Europe. *Ecological Indicators, 24,* 256–265.

Jacobsen, C., Rademacher, P., Meesenburg, H., & Meiwes, K. J. (2002). Gehalte chemischer Elemente in Baumkompartimenten. Bonn: Niedersächsische Forstliche Versuchsanstalt Göttingen; im Auftrag des Bundesministeriums für Verbraucherschutz, Ernährung and Landwirtschaft (BMVEL).

JRC. (2006). The European soil data base, Distribution version v2.0. Joint Research Centre, European Commission, Ispra, Italy.

Kuylenstierna, J. C. I., Hicks, W. K., Cinderby, S., Cambridge, H., van der Hoek, K. W., Erisman, J. W., Smeulders, S., & Wisniewski, J. R. (1998). Critical loads for nitrogen deposition and their exceedance at European scale. *Environmental Pollution, 102,* 591–598.

Kuylenstierna, J. C. I., Rodhe, H., Cinderby, S., & Hicks, K. (2001). Acidification in developing countries: Ecosystem sensitivity and the critical load approach on a global scale. *Ambio, 30,* 20–28.

Leemans, R., & Van den Born, G. J. (1994). Determining the potential distribution of vegetation, crops and agricultural productivity. *Water Air and Soil Pollution, 76,* 133–161.

New, M., Hulme, M., & Jones, P. D. (1999). Representing twentieth century space-time climate variability. Part 1: Development of a 1961–1990 mean monthly terrestrial climatology. *Journal of Climate, 12,* 829–856.

New, M., Hulme, M., & Jones, P. D. (2000). Representing twentieth century space-time climate variability. Part 2: Development of 1901–1996 monthly grids of terrestrial surface climate. *Journal of Climate, 13,* 2218–2238.

Nilsson, J., & Grennfelt, P. (1988). Critical loads for sulphur and nitrogen. Report from a Workshop held at Skokloster Sweden March 19–24 1988. Miljø rapport 1988: 15. Copenhagen Denmark Nordic Council of Ministers.

Posch, M., Aherne, J., & Hettelingh, J. -P. (2011). Nitrogen critical loads using biodiversity-related critical limits. *Environmental Pollution, 159,* 2223–2227.

Prentice, I. C., Sykes, M. T., & Cramer, W. (1993). A simulation model for the transient effects of climate change on forest landscapes. *Ecological Modelling, 65,* 51–70.

Prins, K., & Korotkov, A. (1994). The forest sector of economies in transition in Central and Eastern Europe. *Unasylva, 179,* 3–10.

Reinds, G. J., Posch, M., & De Vries, W. (2001). A semi-empirical dynamic soil acidification model for use in spatially explicit integrated assessment models for Europe. (Alterra rapport 84). Wageningen (Netherlands): Alterra.

Reinds, G. J., Posch, M., De Vries, W., Slootweg, J., & Hettelingh, J. -P. (2008). Critical loads of sulphur and nitrogen for terrestrial ecosystems in Europe and northern Asia using different soil chemical criteria. *Water Air and Soil Pollution, 193,* 269–287.

Schelhaas, M. J., Varis, S., Schuck, A., & Nabuurs, G. J. (1999). EFISCEN's European Forest Resource Database. European Forest Institute, Joensuu, Finland: EFI.

Schelhaas, M. J., Eggers, J., Lindner, M., Nabuurs, G. J., Pussinen, A., Paivinen, R., Schuck, A., Verkerk, P., Van der Werf, D., & Zudin, S. (2007). Model documentation for the European Forest Information Scenario Model (EFISCEN 3.1.3) Wageningen, Netherlands: Alterra report 1559.

Semenov, M., Bashkin, V., & Sverdrup, H. (2001). Critical loads of acidity for forest ecosystems of North Asia. *Water Air and Soil Pollution, 130,* 1193–1198.

Skeffington, R. A., Whitehead, P. G., Heywood, E., Hall, J. R., Wadsworth, R. A., & Reynolds, B. (2007). Estimating uncertainty in terrestrial critical loads and their exceedances at four sites in the UK. *Science of the Total Environment, 382,* 199–213.

Tegen, I., & Fung, I. (1995). Contribution to the atmospheric mineral load from land surface modification. *Journal of Geophysical Research, 100,* 18707–18726.

Van Dobben, H. F., van Hinsberg, A., Schouwenberg, E. P. A. G., Jansen, M. J. W., Mol-Dijkstra, J. P., Wieggers, H. J. J., Kros, J., & De Vries, W. (2006). Simulation of critical loads for nitrogen for terrestrial plant communities in the Netherlands. *Ecosystems, 9,* 32–45.

Van Loon, M., Tarrason, L., & Posch, M. (2005). Modelling base cations in Europe. EMEP Technical Report MSC-W 2/2005. Oslo.

Zak, S. K., & Beven, K. J. (1999). Equifinality, sensitivity and predictive uncertainty in the estimation of critical loads. *Science of the Total Environment, 236,* 191–214.

Chapter 16
Critical Load Assessments for Sulphur and Nitrogen for Soils and Surface Waters in China

Lei Duan, Yu Zhao and Jiming Hao

16.1 Introduction

As economic development and energy consumption increases rapidly, acid deposition and thus environmental acidification caused by huge sulphur dioxide (SO_2) and nitrogen oxides (NO_x) emissions (Zhang et al. 2009) has become one of the major environmental concerns in China (Hao et al. 2007; Larssen et al. 2006; Tang et al. 2010). The acid rain issue was first reported in the late 1970s, with the lowest annual average pH of rainwater in southwest China, especially near big cities such as Chongqing and Guiyang (Galloway et al. 1987; Zhao and Sun 1986; Zhao et al. 1988). The areas affected by acid rain in China have extended northwards from the south of Yangtze River in the middle of 1980s to the whole of eastern China at present. The chemical composition of acid rain in China is generally different from that in Europe, with a lower pH value and higher sulphate, calcium and ammonium concentrations (Larssen et al. 2006). Huge emissions of alkaline fly ash from anthropogenic sources such as cement industry and intensive construction activities, as well as soil dust from wind-blow from north China (Lei et al. 2011; Zhu et al. 2004) lead to the very high concentration of calcium relative to sulphate in precipitation, and build the high acid neutralizing capacity (ANC) of atmospheric particulate matter (PM) during washout.

In order to control acid rain and sulphur dioxide pollution, China designated areas where acid rain or serious SO_2 pollution occur or may occur as Acid Rain Control Zones and Sulphur Dioxide Pollution Control Zones (Hao et al. 2001a). Recently, much effort has been put into implementing flue gas desulphurization technology and phasing out small units in the power sector, in order to achieve the national goal of a 10% reduction in the SO_2 emission between 2005 and 2010.

L. Duan (✉) · J. Hao
School of Environment, Tsinghua University, Beijing, P. R. China
e-mail: lduan@tsinghua.edu.cn

Y. Zhao
School of the Environment, Nanjing University, Nanjing, P. R. China

W. de Vries et al. (eds.), *Critical Loads and Dynamic Risk Assessments*,
Environmental Pollution 25, DOI 10.1007/978-94-017-9508-1_16

However, the benefits of SO_2 emission reduction on acidification prevention during 2005–2010 will almost be counteracted by the increased N emissions (Zhao et al. 2009). The Chinese government has set compulsory targets to further reduce SO_2 emissions by 8%, and reduce NO_x emission by 10% by 2015. Emission and deposition abatement will be a big challenge for China and requires policy development and technology investments.

As a widely accepted scientific basis for guiding emission abatement strategies, the critical load concept (Nilsson and Grennfelt 1988) was also applied in China, for example, to designate Acid Rain Control Zones and Sulphur Dioxide Pollution Control Zones (Hao et al. 2001a). Critical loads will be more widely applied in related policy making in the future. The aim of this chapter is to give an overview of critical load assessments for sulphur and nitrogen for China. It includes a regional-scale assessment for soils, based on an extended Steady-State Mass Balance (SSMB) method (Zhao et al. 2007) and a site-scale assessment for about 100 surface waters in southern and north-eastern China, based on the MAGIC model (Cosby et al. 1985).

16.2 Mapping Critical Loads for Sulphur and Nitrogen for Soils in China Using the Extended SSMB Method

16.2.1 Extended Critical Load Function for Acidification

Critical loads of acidifying S and N for terrestrial ecosystem can be calculated with the SSMB method (Posch et al. 1995), see also Chap. 6), as shown in Eqs. (16.1)–(16.4):

$$CL(S) + (1 - f_{de})CL(N) = BC_{dep} + BC_w - BC_u + (1 - f_{de})(N_i + N_u) - ANC_{l,crit} \quad (16.1)$$

$$CL_{max}(S) = BC_{dep} + BC_w - BC_u - ANC_{l,crit} \quad (16.2)$$

$$CL_{min}(N) = N_i + N_u \quad (16.3)$$

$$CL_{max}(N) = N_i + N_u + \frac{CL_{max}(S)}{1 - f_{de}} \quad (16.4)$$

where the subscripts dep, w, u and i refer to deposition, weathering, uptake and immobilization, respectively, BC is the sum of base cations (BC = Ca + Mg + K + Na), f_{de} is the denitrification fraction, and $ANC_{l,crit}$ is the critical leaching of ANC, determined by the critical chemical value. Calculated by this method, the critical load function can be plotted as a line with generally negative slope on an x-y plane defined by N and S deposition. Thus the SSMB model defines the maximum S and N deposition acceptable for a certain ecosystem with a given ecosystem protection criterion.

When defining critical loads via Eq. (16.1), which is called critical load equation, it is implicitly assumed that all the terms on the right-hand side, including BC deposition, are quantified as long-term steady-state rates. Chloride deposition is neglected here due to very low levels in China. However, it is difficult to define a long-term steady-state BC deposition that is only from natural sources in areas such as in China, where the current BC deposition is high, and a considerable fraction is of anthropogenic origin. In a more general view, if the BC deposition is considered variable instead of constant in the critical load calculation, Eq. (16.1) can be rewritten as:

$$CL(S) + (1 - f_{de})CL(N) - BC_{dep} = BC_w - BC_u + (1 - f_{de})(N_i + N_u) - ANC_{l,crit} \tag{16.5}$$

Therefore the critical load of S and N can be approximated by linear functions of the BC deposition. This 'extended' Sulphur(S)-Nitrogen(N)-Base cation(BC) critical load function is shown in Fig. 16.1, taking changes in deposition of S, N and BC into account. As can be seen, both $CL_{max}S$ and $CL_{max}N$ would decrease with the reduction of BC deposition. However, $CL_{min}N$, which is associated only with N immobilization and vegetation uptake will remain constant, regardless of the BC deposition.

In the extended critical load function, each single point with coordinates of (N_D, BC_D, S_D) represents a deposition status. If current S and N deposition lies below the shaded surface, the deposition is acceptable, otherwise, the critical load is exceeded and ecosystem damage is expected at some time in the future. Therefore, measures to reduce S and/or N deposition are needed to prevent soil acidification. Taking point P1 in Fig. 16.1 as example, there are three kinds of measures to reduce acid deposition to meet the critical load: route a (reducing N only), route b (reducing both N and S) and route c (reducing S only). The choice depends ultimately on a multidisciplinary consideration of environment, economy and technology. Moreover, if BC deposition declines in the future, more efforts will be required to prevent soil acidification and ecosystem damage, because the critical load of S and N will be lower. For instance, when BC deposition levels are reduced from P1 to P1*, more S and/or N deposition should be reduced. If current acid deposition does not exceed the critical load (e.g. point P2), attention should still be paid to the BC deposition: The critical load at current BC deposition level is not exceeded (P2), but it may become exceeded if alkaline dust emissions are controlled and BC deposition declines (P2*). This illustrates the potential extent of future acidification induced by reduced BC deposition, if S and N deposition are not reduced correspondingly.

There are different chemical criteria for the calculation of the critical alkalinity leaching ($ANC_{l,crit}$), and the choice of the chemical criterion influences the critical load calculation (Posch et al. 1995; see also Chaps. 2 and 6). The critical molar

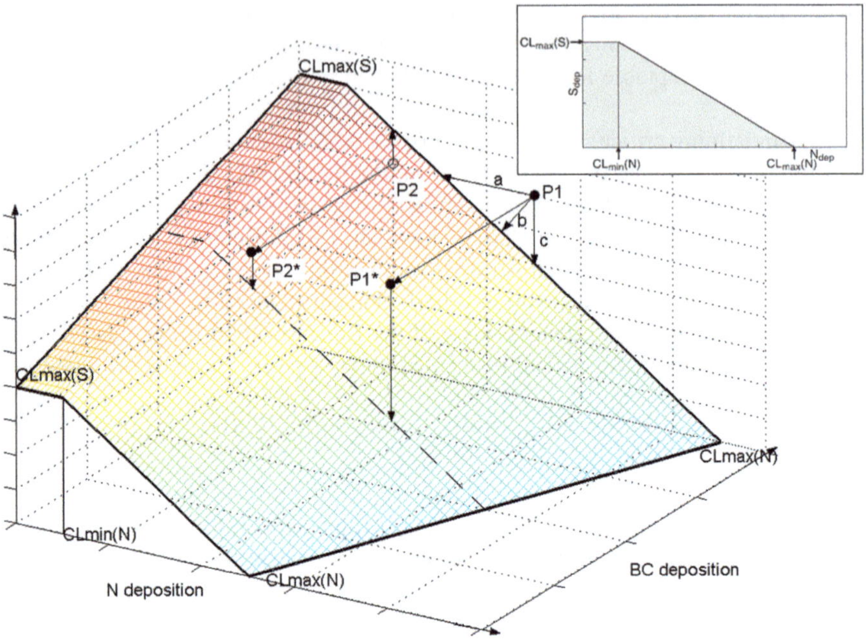

Fig. 16.1 Extended S-N-BC critical load function. To avoid exceedance, depositions of *S*, *N*, and *BC* should be limited *below* the shaded surface. Critical loads of *S* and *N* decrease if *BC* deposition is decreased whereas $CL_{min}N$ remains constant. More effort to reduce *S* and/or *N* deposition has to be made to avoid acidification as *BC* deposition is reduced, regardless of whether the critical load is currently exceeded (route P1 to P1*) or not (route P2 to P2*). The small insert diagram shows the traditional *S-N* critical load function. (Posch et al. 1995)

ratio of base cations to Al, $(Bc/Al)_{crit}$, at which fine root damage might occur in the soil, is often used as the chemical criterion for forests. Under the assumption that concentrations of Al ($[Al^{3+}]$) and hydrogen ions ($[H^+]$) in soil water are linked by an empirical equation following (Reuss et al. 1990):

$$K^* = [Al^{3+}]/[H^+]^{\alpha} \tag{16.6}$$

$ANC_{l,crit}$ can be calculated as:

$$-ANC_{l,crit} = 1.5\frac{BC_{dep} + BC_w - BC_u}{(BC/Al)_{crit}} + Q^{1-\frac{1}{\alpha}}[1.5\frac{BC_{dep} + BC_w - BC_u}{(BC/Al)_{crit} \cdot K^*}]^{\frac{1}{\alpha}} \tag{16.7}$$

where K^* is the dissolution constant, α an empirical exponent, and Q the runoff (percolation flux). For $\alpha = 3$, the gibbsite equation, governing the dissolution of a solid aluminium hydroxide phase (such as gibbsite), is obtained and the respective equation for $ANC_{l,crit}$ is often used in Europe.

In addition to the acidifying aspect of nitrogen, the effects of nitrogen deposition on the nutrient status (eutrophication) of an ecosystem should be considered when determining critical loads. Critical load for nutrient nitrogen can be calculated as (Posch et al. 1995; see also Chap. 6):

$$CL_{nut}(N) = N_i + N_u + \frac{N_{l,crit}}{1 - f_{de}} \tag{16.8}$$

where $N_{l.crit}$ is the critical leaching of N.

16.2.2 Catchment Application of Extended SSMB Method

To demonstrate the extended critical load function it was applied to a typical acidic catchment, TieShanPing (TSP), located on a sandstone ridge 25 km northeast of Chongqing city, at 450 to 500 m a.s.l. in the southeast of the Sichuan basin, China (Larssen et al. 2006). The input variables and parameters for S and N critical load calculation, including deposition, weathering rate and vegetation uptake of base cations (BC_{dep}, BC_w, and BC_u), immobilization rate and vegetation uptake of N (N_i and N_u), water flux through root zone of soil (Q), $(Bc/Al)_{crit}$ and pH-pAl equilibrium coefficients (α and log$K$*), were obtained either directly from field monitoring or indirectly by modelling and calculations based on monitored data. A summary of input data used for the critical load calculation at the TSP site is shown in Table 16.1.

Based on Eqs. (16.5) and (16.7), the extended S-N-BC critical load function was determined for the TSP catchment, described as a surface in Fig. 16.2. To draw the surface, the characteristic values, e.g., $CL_{max}S$, $CL_{max}N$ under current BC deposition, $CL_{max}S^*$, $CL_{max}N^*$ under assumed natural BC deposition (25 % of current deposition), and $CL_{min}N$, were calculated with Eqs. (16.2)–(16.4) and are labelled in the figure. As can be seen, the critical loads are highly dependent on the BC deposition. If BC deposition is cut by 75 %, i.e. to natural background, critical loads of S and N decline by 58 %.

Average N, BC and S deposition of 2002 and 2003 at the TSP site were 2.87, 8.47, and 9.91 keq ha^{-1} yr^{-1} (Point A in Fig. 16.2). With these deposition values, the critical load was not exceeded, although very close to the current S deposition. The high BC deposition therefore mitigates exceedance of the critical loads. The critical loads would be reached if BC_{dep} decreased to 7.81 keq ha^{-1} yr^{-1}, i.e. 92 % of current value (Point C), and would be far exceeded if BC_{dep} decreased by 75 % (Point B) (with S and N deposition unchanged). Based on these results, acidification effects at the TSP site are a realistic risk in the future.

This case study in a typical forested catchment indicates that the extended critical load function can be used by policy makers to set more reasonable acidity control strategies in the future. The method also highlights for policymakers the "competition" between emission control of particulate matter driven by human health targets and potential increased regional acid deposition from such measures.

Table 16.1 Input data for calculating critical loads for the TieShanPing catchment

Input data	Unit	Value	Remarks
BC_D	keq ha^{-1} yr^{-1}	8.47	Based on throughfall data monitored
BC_w	keq ha^{-1} yr^{-1}	0.60	Calculated with the PROFILE model (Warfvinge and Sverdrup 1992)
BC_u	keq ha^{-1} yr^{-1}	0.25	Duan et al. 2004
N_u	keq ha^{-1} yr^{-1}	0.21	Duan et al. 2004
N_i	keq ha^{-1} yr^{-1}	0.17	Total amount of soil N divided by period of soil formation (de Vries et al. 1994)
f_{de}	–	0.8	Geng and Sun 1999
$(Bc/Al)_{crit}$	–	2.0	Sverdrup and Warfvinge 1993
α	–	1.59	Linear regression between pH and pAl^{3+} in soil water
$\log_{10}K^*$	–	3.11	Linear regression between pH and pAl^{3+} in soil water
Q	m^3 ha^{-1} yr^{-1}	5220	Monitoring data

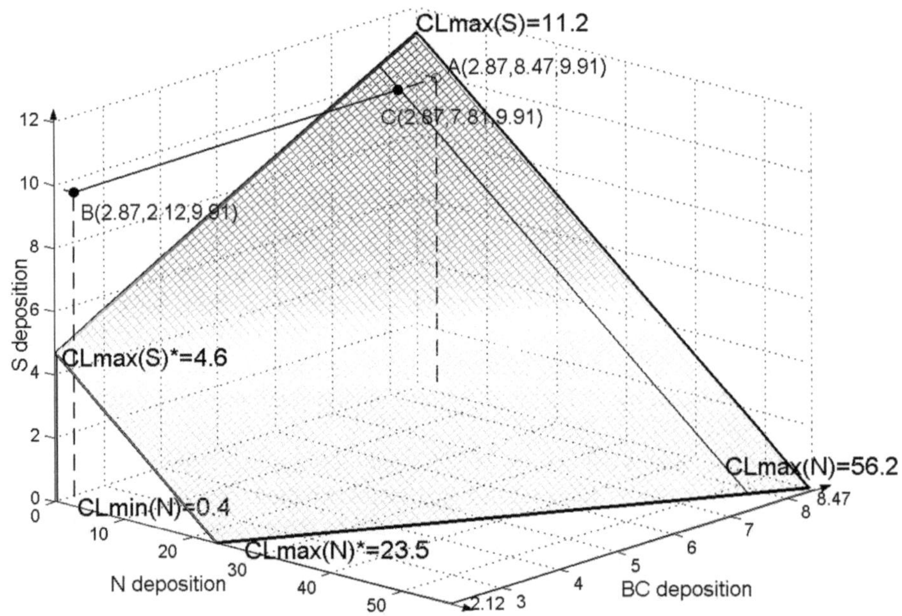

Fig. 16.2 The *S-N-BC* critical load functions for the TieShanPing site. Deposition unit: keq ha^{-1} yr^{-1}. CL_{max}*S/N* and CL_{max}*S*/N** represent the maximum *S/N* critical load under current *BC* deposition and 25 % of current *BC* deposition, respectively

16.2.3 Weathering Rates of Soils in China

As an important parameter in critical load calculations and soil acidification simulations, weathering rates of soils in China were studied using different methods (Duan et al. 2002) such as the mass balance approach (White 1995), the soil mineralogical classification (Duan et al. 2000; Nilsson and Grennfelt 1988), the total analysis correlation (Sverdrup 1990), the PROFILE model (Warfvinge and Sverdrup 1992; see also Chap. 6), the MAGIC model (Cosby et al. 1985; see also Chap. 8) and a simulated leaching experiment (Duan et al. 2002). The comparison of various methods shows that the weathering rates of soils estimated by the PROFILE model coincide well with those from other methods, such as the dynamic modelling (by MAGIC) and modified leaching experiments. Therefore, the PROFILE model was used in this study to calculate weathering rates of major soil groups in China.

For the present study, soil groups are particularly appropriate units for calculating weathering rates because the soil series within a group generally have similar properties. In total, there are 46 groups of soils in China. In the east of China, from south to north, there are belts of Latosol, Lateritic red earth, Red earth, Yellow earth, Yellow-brown earth, Brown forest earth, Dark brown forest earth and Podzolic soil. This spatial distribution of soils is called the Humid-oceanic Belt Spectrum. This contrasts with soil belts in the west of China, running from north to south form the Arid-inland Belt Spectrum, which include Chestnut soil, Brown soil, Sierozem and Desert soil. Between the two spectrums is the Interim Belt Spectrum, which contains areas of Cinnamon soil, Dark loessial soil, Chernozem Chestnut soil, Gray-cinnamon soil, Gray forest soil and Black soil, stretching northeastward from the Loess Plateau to the west of the Da-xing-an-ling Mountains (Xiong and Li 1987). An approximate correspondence of the Chinese soil taxonomy system to other international systems is given in Duan et al. (2002).

In recent years, samples of all major soil types have been collected and analyzed for physicochemical properties and mineralogy. Weathering rates of all soils were calculated by the PROFILE model. Since the soil samples within a group may have different physicochemical and mineralogical properties, and hence different weathering rates, the smallest weathering rate was assigned to the group. Based on the weathering rates of every soil group, a map of weathering rates for Chinese soils was compiled and is shown in Fig. 16.3. As can been seen, the weathering rates are very low (usually lower than 1 keq ha^{-1} yr^{-1}) for Latosol, Lateritic Red Earth, Red Earth, Yellow Earth, and Yellow-Brown Earth in south China and Dark Brown Forest Soil, Black Soil, and Podzolic Soil in northeast China, and very high for other soil groups in west China. The content of weatherable minerals in soil is the most important factor in determining the spatial distribution of weathering rates. In east China, especially in southeast China, where intensive chemical weathering occurred because of the high temperature and abundant rainfall, the weatherable material in soils has almost been consumed during the pedogenic processes. In western China, on the contrary, where historical chemical weathering was quite weak, there remain plenty of weatherable minerals, even limestone in the soil. In addition to the soil

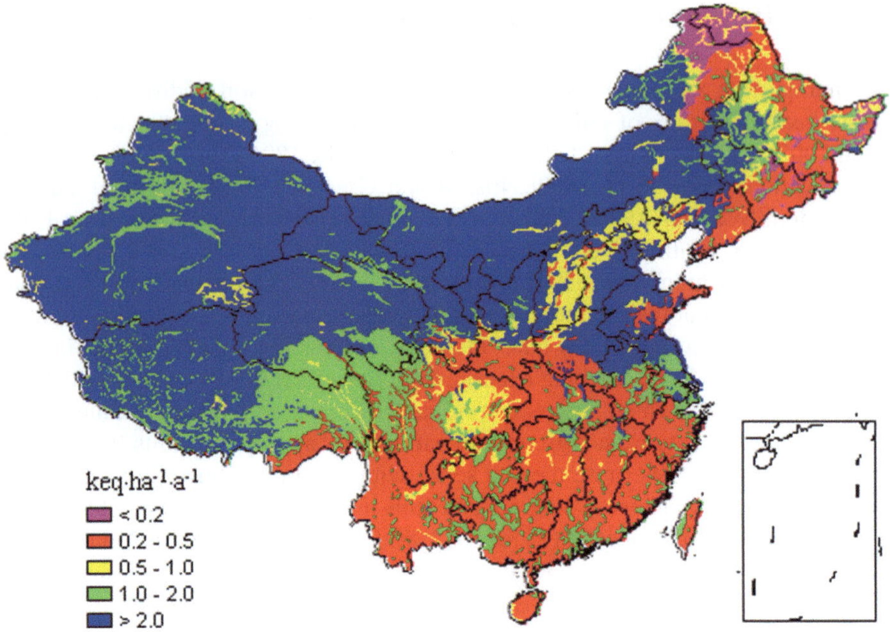

Fig. 16.3 Weathering rates in China

mineralogy, the distribution of weathering rates in China depends on several other variables such as soil texture, water content and temperature. For example, the reason why the soil weathering rate in northeast China is lower than that in southeast China can be attributed to the difference in temperature.

16.2.4 Vegetation Uptake of Nitrogen and Base Cations in China

The uptake of nitrogen and base cations by vegetation may be an important source of soil acidity. Assuming that whole-tree harvesting (stems and branches) was practiced in forests, the present growth uptake of N and base cation can be derived by multiplying the annual increase in biomass with the element contents in the various compartments according to

$$X_u = K_s X_s + K_b X_b \tag{16.9}$$

where X_u is the net uptake of element X (X=Ca, Mg, K, Na, or N), K_s and K_b are the average growth rate of stem and branches, respectively, and X_s and X_b are the content of element X in stem and branches, respectively. For steppe, meadow and desert ecosystems, the uptake rates of nitrogen and base cations were calculated by

$$X_u = rK_a X_a \tag{16.10}$$

where X_a is the content of element X in the above-ground part of grass, K_A is the growth rate of the above-ground part, and r is the grass utilization ratio with the values of 0.5, 0.6 and 0.4 for steppe, meadow and desert, respectively (Liao and Jia 1996). Although the uptake rates of nitrogen and base cations by crops such as rice, wheat and corn are very high, the loss of these elements from farming soil is generally overcompensated by agricultural practices such as frequent fertilisation, with the excess fertilizer being lost from the farming systems in drainage. Consequently, it is assumed that the net uptake by farming ecosystems is zero.

A literature review on average growth rates and element contents of major plant communities in China was carried out by Duan et al. (2004). Based on the dataset, uptake rate of nitrogen and base cations was estimated for each major vegetation type in China (vegetation classification according to Wu 1980), and maps of uptake rates were compiled (shown in Figs. 16.4 and 16.5 for nitrogen and base cations, respectively). As can be seen from the map (Fig. 16.4), the uptake rates of N vary significantly between grids because of the difference in climatic and soil conditions. In general, the uptake rates of N are low in the southeast part of China, including Northeast, North, East, and South China. In contrast, the uptake rate is high in the north-western part of China. Agricultural crops are the major vegetation type in the south-eastern part of China, with some coniferous forests, whose uptake rate of nitrogen is very low, in isolated areas in the mountains. Vegetation types with high nitrogen uptake rate cannot exist on the acid soils in the subtropical/tropical region in southeast China. Soil acidification may inhibit forest growth, resulting in low uptake rates.

In the north-western part of China, the nitrogen uptake rate of vegetation is much higher than that in the southeast, except in non-vegetated areas such as Salina, Gobi, and snow covered areas on the Tibetan Plateau and in north-western China, where the uptake rates of nitrogen and base cations are near zero. Since soil nitrogen is generally high in these arid/semi-arid regions, the N content of plants is also high, and shows a relatively small variation between different plants. The productivity of vegetation therefore becomes the major factor in determining the nitrogen uptake rate. From the southeast to the northwest there exist deciduous forest (and scrub), steppe (and meadow), and desert zones, with decreasing productivities and therefore relatively low nitrogen uptake rates. The distribution of nitrogen uptake rates matches well with that of vegetation growth rates, indicating that hydrothermal factors are of the highest importance in determining the nitrogen uptake rate of vegetation.

The distribution pattern of base cation uptake rates is quite similar to that of nitrogen uptake rates (Fig. 16.5). However, significant differences exist. With the exception of some regions in southernmost China (with tropical monsoon forests and broad-leaved evergreen forests) and in the south of Northeast China (temperate meadows and deciduous scrubs) with very high uptake rates of both nitrogen and base cations, the areas with high base cation uptake rates do not coincide with those with high nitrogen uptake rates. This difference depends mainly on the distribution of calcium uptake rates. The vegetation types with the highest calcium uptake rates

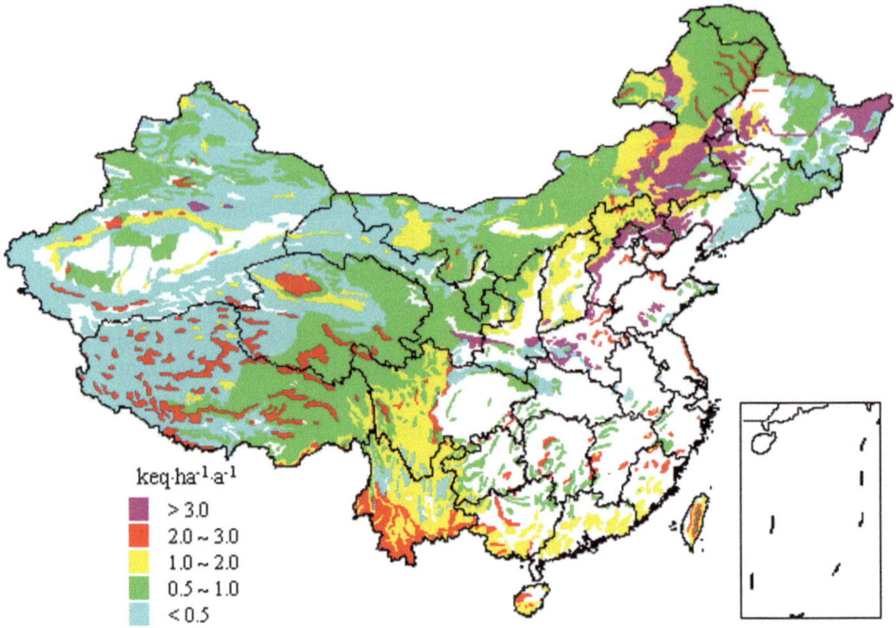

Fig. 16.4 Vegetation uptake rates of nitrogen in China (*white* areas mean zero uptake)

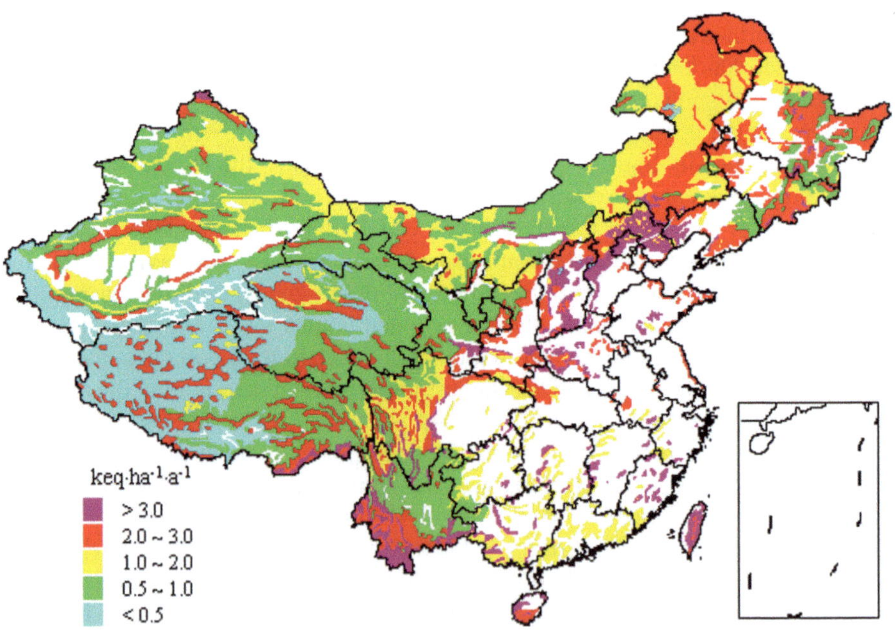

Fig. 16.5 Vegetation uptake rates of base cations in China (*white* areas mean zero uptake)

are generally distributed over calcareous soils in the temperate arid/semiarid zone in the northwest of China, which indicates that soil conditions play an important role in determining the distribution of calcium uptake rates. However, the calcium uptake rates of some temperate coniferous and broad-leaved forests located on low-calcium acid soils in the east humid zone are slightly higher (0.5–1.5 keq ha^{-1} yr^{-1}) because of the high growth rate of plants.

In summary, the base cation uptake rate is very high (>2 keq ha^{-1} yr^{-1}) in some tropical areas in southernmost China, where perfect hydrothermal conditions result in very high vegetation productivity, and in sub-humid areas in the northwest of North China, where soil conditions are favourable for base cation enrichment in vegetation. The uptake rate is also high (1–2 keq ha^{-1} yr^{-1}) in large areas in North-east China, and in isolated areas in Northwest China. However, the uptake rate of base cation is very low (<0.5 keq ha^{-1} yr^{-1}) in large areas in the subtropical zone in the east of China, and in the arid region in the west of China and on the Tibetan Plateau.

16.2.5 Modelling Atmospheric Base Cations in China: Emission, Transport and Deposition of Calcium and Magnesium

The atmospheric emissions of Ca^{2+} and Mg^{2+} (K^+ and Na^+ are not included due to their tiny shares in BC emissions) from major anthropogenic sources in China were estimated based on the emission rate and chemical composition of particulate matter (PM):

$$E_{i,j,k} = E_{\mathrm{PM},j,k} w_{i,j,k} \tag{16.11}$$

where *i*, *j*, and *k* stand for the species of base cations, regions, and sectors, respectively, *E* is the emission level, and *w* is the fraction of BCs to total PM emission by mass.

Zhao et al. (2011) estimated that the overall emissions were about 5.97 and 0.24 Mt yr^{-1} for calcium and magnesium, respectively, in the year 2005. Seven dominant source sectors were included: power, cement, iron & steel, lime production, brick making, other industrial sources, and residential coal combustion. The estimates were much higher than those for Europe of 750–800 kt yr^{-1} in 1990 (Lee and Pacyna 1999), highlighting the importance of including the (future) reductions of anthropogenic BC emissions in the assessment of critical loads their and exceedances in countries like China, where base cations play a much larger role in atmospheric chemistry and ecological effects. Cement and lime production were the leading sources of Ca^{2+}, responsible for 88 % of total emissions in 2005.

In addition to anthropogenic emissions, the base cation emission from wind-blown dust particles from arid and semi-arid regions in northern China was roughly estimated to about 2.08 Mt yr^{-1} for Ca and 0.68 Mt yr^{-1} for Mg (Zhu et al. 2004). Using a multi-layer, dynamic Eulerian model (Duan et al. 2007; Zhao et al. 2011),

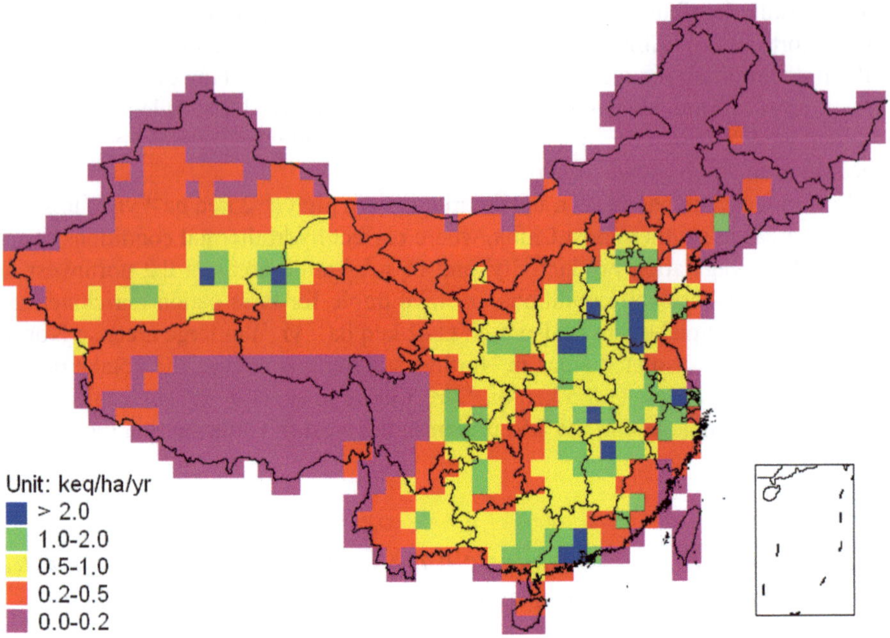

Fig. 16.6 Simulated deposition of *BCs* ($Ca^{2+} + Mg^{2+}$) for 2005

the long-range transport and deposition of base cation was estimated. Result shows that the total deposition of calcium in China was about 5.29 Mt yr^{-1} (with 1.54 Mt yr^{-1} as wet deposition and 3.84 Mt yr^{-1} as dry deposition), of which anthropogenic sources contributed 67.1 % (92.4 % of wet deposition and 57.6 % of dry deposition). The base cations emitted from natural sources are mainly deposited in the northwest of China, while anthropogenic sources greatly affected the deposition in the southeast. As shown in Fig. 16.6, the calcium deposition was mostly higher than 0.5 keq ha^{-1} yr^{-1}, or even higher than 1 keq ha^{-1} yr^{-1} in the southeast of China, where acid deposition was also very high. The distribution pattern of Mg depositions was similar with that of Ca depositions, however with values about one order of magnitude lower.

16.2.6 Mapping Critical Loads for Sulphur and Nitrogen for Soils in China

Parameters in addition to weathering rate, growth uptake and atmospheric deposition include critical BC/Al ratio, Al dissolution coefficients (see Eq. 16.8), critical N leaching, N immobilization rate, and denitrification fraction are necessary for critical load calculations. Critical BC/Al ratios for Chinese vegetation were selected from Sverdrup and Warfvinge (1993). The N immobilization of major soil groups in China was estimated as the total amount of soil N divided by period of soil formation (Hao et al. 2003). Other parameters were derived on the basis of CLRTAP (2013).

Based on the extended SSMB method, a map of critical loads for S and N for China was developed and is shown in Fig. 16.7. Here the critical load of N was calculated as the minimum of the maximum critical load of acidifying N (Eq. 16.4) and nutrient N (Eq. 16.8). As can be seen from Fig. 16.7a, the S critical loads in northern and north-western China were generally higher than 2 keq ha^{-1} yr^{-1}, due to high weathering rates and natural deposition of base cations, while the values could be lower than 0.2 keq ha^{-1} yr^{-1} in the northeast, where low temperatures, and thus low weathering rates, occur, and in the south, where both low weathering rates (due to low content of weatherable minerals) and high vegetation uptake of base cations occur. In contrary, as shown in Fig. 16.7b, the N critical load (often driven by the critical nitrogen leaching criterion) is relative lower in the northwest with poor vegetation uptake of N, and higher in the south with considerable denitrification.

16.3 Estimating Critical Loads of Sulphur for 100 Surface Waters in Southern and North-Eastern China Using the MAGIC Model

Although decades of acid deposition have apparently not resulted in surface water acidification in China, some surface waters may have the potential trend of becoming acidified, especially those in southern China (Duan et al. 2011). The dynamic acidification model MAGIC (Cosby et al. 1985) was applied to selected surface waters in southern and north-eastern China to evaluate the impact of acid deposition and to determine their critical loads of S deposition. Altogether about 100 surface waters were considered including ponds, lakes, streams and rivers with sufficient data for modelling. Except some small mountain ponds with catchment sizes less than 1 km^2, most other watersheds have larger areas. Most of the surface waters have an ANC much higher than 200 eq l^{-1}, with the lowest values occurring in a few mountain ponds in south-western China, such as the water on the top of the Jinyun mountain (269 eq l^{-1}) and the Emei mountain (237 eq l^{-1}).

Input data required for the MAGIC calibration procedure include atmospheric deposition chemistry, catchment-scale lumped soil characteristics for depth, CEC, exchangeable base cation fractions and bulk density, and present surface water chemistry. The model was calibrated for each catchment to get a set of suitable parameters for the prediction process, including cation exchange selectivity coefficients and weathering rates. The calibration simulations were started in 1920, assuming the historic deposition sequence of SO_4^{2-} was proportional to the national coal consumption (and thus anthropogenic SO_2 emissions). The year 1988 was set as the 'present' year, since observations of water chemistry were from that year. The simulated present chemistry of surface waters accorded fairly well with the observed water chemistry (Hao et al. 2001b).

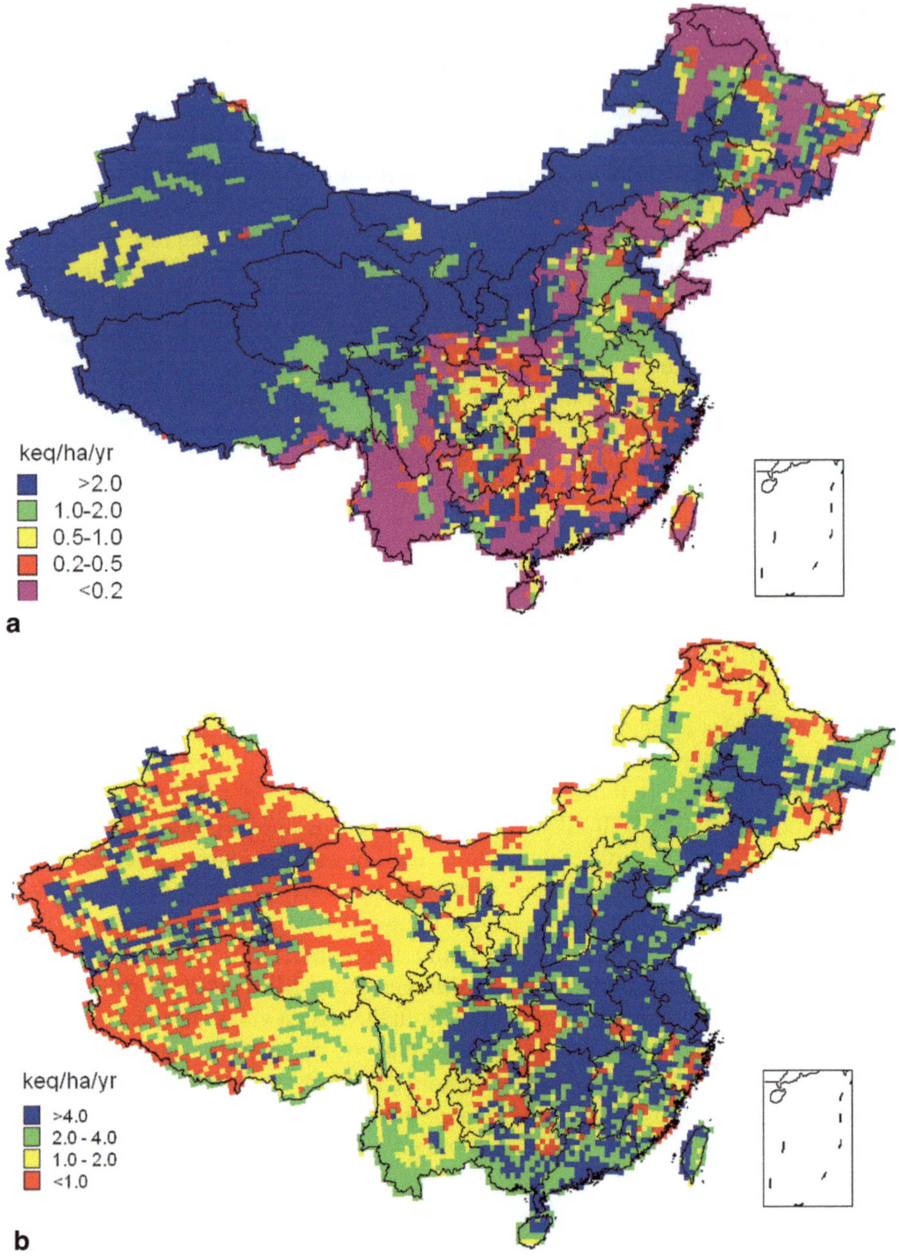

Fig. 16.7 The 5-percentile critical loads of *S* and *N* in China. The spatial resolution is 36 km by 36 km (same as for atmospheric dispersion modelling results)

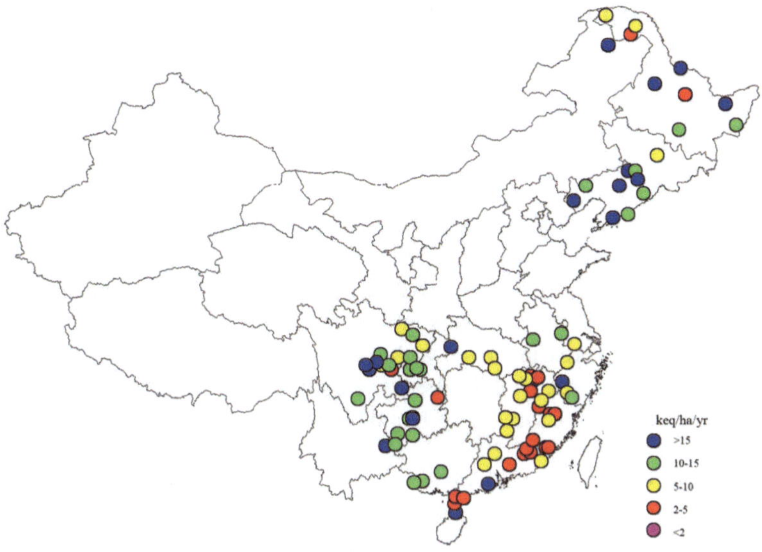

Fig. 16.8 Critical loads of *S* for some surface waters in China

Based on the calibrated parameters, the MAGIC model was then run to estimate critical load of S by adjusting the 'future' S deposition to decrease the water ANC over time to the critical value, ANC_{crit}, required to protect a defined biological target within the ecosystem. $ANC_{crit} = 50$ eq l^{-1} was adopted, and the year 2100 was taken as the target year, as this was long enough for all 100 watersheds to reach a new steady state.

The critical loads of S for all 100 waters are shown in Fig. 16.8. As can be seen, most surface waters were not sensitive to acid deposition, with critical loads of S for these waters being relatively high. On the other hand, surface waters in southern China, especially those in Fujian, Jiangxi and Guangdong provinces with average values around 5 keq ha^{-1} yr^{-1}, were more susceptible to acidification than those in north-eastern China. Among all modelled surface waters/catchments, a few small ponds, such as those on top of the Jinyun mountain and the Emei mountain, were the most sensitive to acid deposition with critical loads of 1.84 and 3.70 keq ha^{-1} yr^{-1}, respectively. Because of the considerable ANC remaining in most of the 100 surface waters, it was not likely that acidification might occur in the near future.

Critical loads for surface waters in China were much higher in comparison with the critical loads for soils. The most important reason might be the large size of the lakes (and catchments), which generally have a much higher buffering capacity than headwaters. Another reason might be the relatively old and deep soil, which can provide a buffering capacity even when the top layer is acidified. Therefore, the acidification of soils may be a precondition for surface waters acidification in China.

16.4 Application of Critical Loads in Future Emission Abatement

China is facing challenges of increasing energy consumption and related regional air pollution. Recently, much effort has been put into implementing flue gas desulphurization technology and phasing out small units in the power sector, in order to achieve the national goal of a 10% reduction in the SO_2 emissions from 2005 to 2010. The national emissions of SO_2, NO_x, PM and NH_3 in 2005 were estimated to be 30.7, 19.6, 31.3 and 16.6 Mt respectively (Zhao et al. 2009). Implementation of existing policy will lead to reductions in SO_2 and PM emissions, while NO_x and NH_3 emissions will keep increasing, even with tentative measures. National SO_2 emissions were estimated to decrease by 8.3% between 2005 and 2010, close to the national target of 10%, while the emissions of NO_x and NH_3 were estimated to increase by 24 and 26%, respectively (Zhao et al. 2009). Using the emission inventory, sulphur and nitrogen depositions were simulated with the Models-3/Community Multiscale Air Quality (CMAQ) system (V4.4) (Byun and Ching 1999), and results were compared with the critical load map derived above to obtain the exceedance of critical loads over the country. In 2005, the critical load for acidification for sulphur deposition only, was exceeded in 28% of the country's territory, mainly in eastern, central and southern China. The exceeded area will decrease to 26%, after the implementation of current plans for emission reduction. However, the areas with exceedance of the critical load for nitrogen (combining eutrophication and acidification effects) over the country were estimated to increase from 12% in 2005 to 15% in 2010 along with increasing NO_x and NH_3 emissions (Zhao et al. 2009). Moreover, when combining the acidification effects of sulphur and nitrogen, the areas with exceedance of critical loads were estimated to be constant at 28% from 2005 to 2010, implying that the benefits of SO_2 emission reduction during 2005–2010 will by and large be counteracted from increased N emissions (Zhao et al. 2009). N emission and deposition abatement will be a big challenge for China, and require appropriate policy developments and technology investments.

Control of particulate matter (PM) emissions is another important concern for China's acidification abatement strategy in the future. China's aggressive strategies to address simultaneously two of its largest air pollution challenges—soil acidification caused by S and N compounds, and health impacts because of human exposure to PM—may be in conflict. National emissions of the major acidifying pollutants are predicted for 2020, integrating both current and anticipated control policies. Similarly, emissions of PM, including base cations (BCs) Ca^{2+} and Mg^{2+}, are projected to 2020. Two emission scenarios of BCs are evaluated: a base case applying existing PM control policy to 2020 (base scenario), and a more realistic case including anticipated tightened controls (control scenario). Depositions of acidifying compounds and BCs are simulated from the estimated emission scenarios using CMAQ and a multi-layer dynamic Eulerian model (described in Sect. 16.2.5), respectively. The results were then evaluated in terms of exceedances of critical load for S. Under the current high BC deposition, the area exceeding $CL_{max}S$ covered only 15.6% of mainland China, with a total exceedance of 2.2 MtS in 2005 (Fig. 16.9a). These values decrease by 2020 in the base scenario, implying a partial recovery from soil

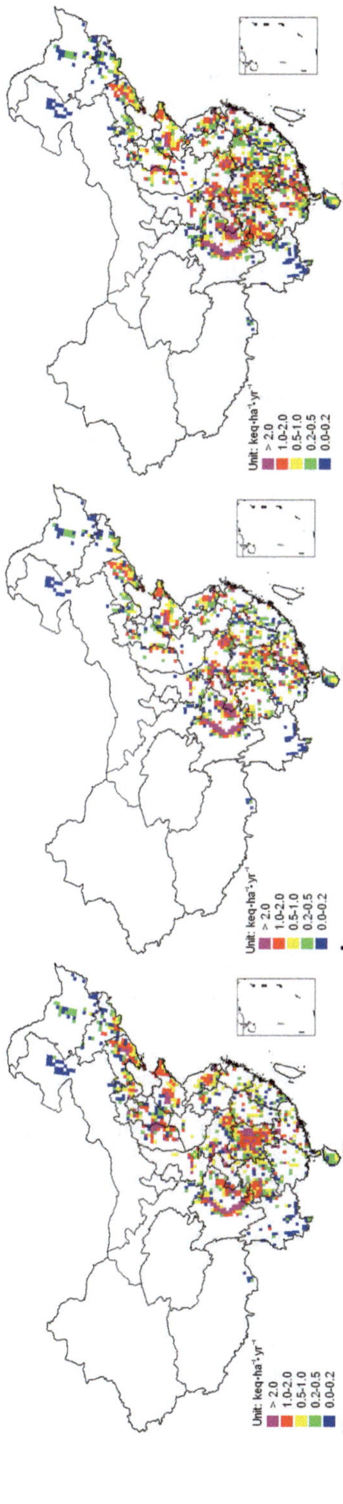

Fig. 16.9 Exceedances of critical load of acidification: **a** 2005; **b** 2020 base scenario; **c** 2020 control scenario (spatial resolution: 36 × 36 km^2) (*white* areas mean non-exceedance)

acidification if S and N are controlled as anticipated, but PM is not (Fig. 16.9b). Under the more realistic PM control scenario, however, the respective estimates are 17.9 % and 2.4 MtS (Fig. 16.9c). This represents an increased risk of acidification compared to 2005 due to simultaneous abatement of acid-neutralizing BCs (Zhao et al. 2011).

The very high concentrations of anthropogenic BCs in China indicate that PM abatement will have potentially very strong (chemical) implications for the environment. China's current program of emission controls, taken as a whole, is unlikely to achieve its long-standing goal of reduced acidification. Unanticipated side effects of the control of primary PM and thus BCs, particularly from the cement and lime industries, may completely counteract the benefits to regional acidification of reduced emissions of acid precursors, including large-scale abatement of SO_2 achieved since 2006. Despite this, ongoing PM control efforts must be continued in China because of the benefits of reduced aerosol pollution and avoided associated damages to public health. This suggests that policy-makers may have little choice but to pursue even more stringent SO_2 and NO_x controls. A "multi-pollutant control methodology", combing measures for S, N and PM reduction, is thus in high demand for future Chinese acidification mitigation. Wider consideration of the interactions of a variety of anthropogenic atmospheric pollutants may be required for control policies to be efficient and indeed effective.

References

Byun, D. W., & Ching, J. K. S. (1999). *Science algorithms of the EPA models-3 community multiscale air quality model (CMAQ) modeling system*. (EPA/600/R-99/030). Washington, DC: U.S. Environmental Protection Agency, Office of Research and Development.

CLRTAP. (2013). *Manual on methodologies and criteria for modelling and mapping critical loads and levels and air pollution effects, risks and trends*. (UNECE convention on long-range transboundary air pollution). www.icpmapping.org. Accessed Nov 2013.

Cosby, B. J., Hornberger, G. M., Galloway, J. N., & Wright, R. F. (1985). Modeling the effects of acid deposition: Assessment of a lumped parameter model of soil water and streamwater chemistry. *Water Resource Research, 21,* 51–63.

De Vries, W., Reinds, G. J., & Posch, M. (1994). Assessment of critical loads and their exceedance on European forests using a one-layer steady-state model. *Water, Air, & Soil Pollution, 72,* 357–394.

Duan, L., Xie, S. D., Zhou, Z. P., & Hao, J. M. (2000). Critical loads of acid deposition on soil in China. *Water, Air, & Soil Pollution, 118,* 35–51.

Duan, L., Hao, J. M., Xie, S. D., Zhou, Z. P., & Ye, X. M. (2002). Determining weathering rates of soils in China. *Geoderma, 110,* 205–225.

Duan, L., Huang, Y. M., Hao, J. M., Xie, S. D., & Hou, M. (2004). Vegetation uptake of nitrogen and base cations in China and its role in soil acidification. *Science of the Total Environment, 330,* 187–198.

Duan, L., Lin, Y., Zhu, X. Y., Tang, G. G., Gao, D. F., & Hao, J. M. (2007). Modeling atmospheric transport and deposition of calcium in China (in Chinese). *Journal of Tsinghua University (Science and Technology), 47,* 1462–1465.

Duan, L., Ma, X. X., Larssen, T., Mulder, J., & Hao, J. M. (2011). Response of surface water acidification in Upper Yangtze River to SO_2 emissions abatement in China. *Environmental Science & Technology, 45,* 3275–3281.
Galloway, J. N., Zhao, D. W., Xiong, J. L., & Likens, G. E. (1987). Acid rain: a comparison of China, United States and a remote area. *Science, 236,* 1559–1562.
Geng, Y. Q., & Sun, X. Y. (1999). Nitrification and denitrification of forest soil in lower mountains in Beijing (in Chinese). *Journal of Beijing Forest University, 21,* 38–43.
Hao, J. M., Duan, L., Zhou, X. L., & Fu, L. X. (2001a). Application of a LRT model to acid rain control in China. *Environmental Science & Technology, 35,* 3407–3415.
Hao, J. M., Ye, X. M., Duan, L., & Zhou, Z. P. (2001b). Calculating critical loads of sulfur deposition for 100 Chinese surface waters using MAGIC model. *Water, Air, & Soil Pollution, 130,* 1157–1162.
Hao, J. M., Qi, C. L., Duan, L., & Zhou, Z. P. (2003). Evaluating critical loads of nutrient nitrogen on soils in China using the SMB method (in Chinese). *Journal of Tsinghua University (Science and Technology), 43,* 849–853.
Hao, J. M., He, K. B., Duan, L., Li, J. H., & Wang, L. T. (2007). Air pollution and its control in China. *Frontier of Environmental Science and Engineering in China, 1,* 1–14.
Larssen, T., Lydersen, E., Tang, D. G., He, Y., Gao, J. X., Liu, H. Y., Duan, L., Seip, H. M., Vogt, R. D., Mulder, J., Shao, M., Wang, Y., Shang, H., Zhang, X., Solberg, S., Aas, W., Okland, T., Eilertsen, O., Angell, V., Li, Q., Zhao, D., Xiang, R., Xiao, J., & Luo, J. (2006). Acid rain in China. *Environmental Science Technology, 40,* 418–425.
Lee, D. S., & Pacyna, J. M. (1999). An industrial emission inventory of calcium for Europe. *Atmospheric Environment, 33,* 1687–1697.
Lei, Y., Zhang, Q., He, K. B., & Streets, D. G. (2011). Primary anthropogenic aerosol emission trends for China, 1990–2005. *Atmospheric Chemistry and Physics, 11,* 931–954.
Liao, G., & Jia, Y. (1996). *Rangeland resources of China* (in Chinese). Beijing: China Science and Technology Press.
Nilsson, J., & Grennfelt, P. (1988). *Critical loads for sulphur and nitrogen.* Report from a workshop held at Skokloster Sweden March 19–24 1988. Miljø rapport 1988: 15. Copenhagen Denmark Nordic Council of Ministers.
Posch, M., de Smet, P. A. M., Hettelingh, J.-P., & Downing, R. J. (1995). *Calculation and mapping of critical thresholds in Europe. Status Report 1995.* (RIVM report 259101004). Bilthoven: Coordination Center for Effects, National Institute of Public Health and the Environment.
Reuss, J. O., Walthall, P. M., Roswall, E. C., & Hopper, R. W. E. (1990). Aluminum solubility, calcium-aluminum exchange, and pH in acid forest soils. *Soil Science Society of America Journal, 54,* 374–380.
Sverdrup, H. U. (1990). *The kinetics of base cation release due to chemical weathering.* Sweden: Lund University Press.
Sverdrup, H., & Warfvinge, P. (1993). *The effect of soil acidification on the growth of trees, grass and herbs as expressed by the (Ca^+ Mg^+ K)/Al ratio.* (Reports in Ecology and Environmental Engineering 1993: 2). Lund University, Department of Chemical Engineering II.
Tang, J., Xu, X. B., Ba, J., & Wang, S. F. (2010). Trends of the precipitation acidity over China during 1992–2006. *Chinese Science Bulletin, 55,* 1800–1807.
Warfvinge, P., & Sverdrup, H. (1992). Calculating critical loads of acid deposition with PROFILE. A steady-state soil chemistry model. *Water, Air, & Soil Pollution, 63,* 119–143.
White, A. F. (1995). Chemical weathering rates of silicate minerals in soil. *Reviews in Mineralogy, 31,* 407–461.
Wu, Z. Y. (Ed.). (1980). *Chinese vegetation* (in Chinese). Beijing: Science Press.
Xiong, Y., & Li, Q. K. (Eds.). (1987). *Chinese soils* (2nd ed.) (in Chinese). Beijing: Science Press.
Zhang, Q., Streets, D. G., Carmichael, G. R., He, K. B., Huo, H., Kannari, A., Klimont, Z., Park, I. S., Reddy, S., Fu, J. S., Chen, D., Duan, L., Lei, Y., Wang, L. T., & Yao, Z. L. (2009). Asian emissions in 2006 for the NASA INTEX-B mission. *Atmospheric Chemistry and Physics, 9,* 5131–5153.

Zhao, D. W., & Sun, B. Z. (1986). Air pollution and acid rain in China. *Ambio, 15,* 2–5.

Zhao, D. W., Xiong, J. L., Xu, Y., & Chan, W. H. (1988). Acid rain in southwestern China. *Atmospheric Environment, 22,* 349–358.

Zhao, Y., Duan, L., Larssen, T., Hu, L. H., & Hao, J. M. (2007). Simultaneous assessment of depositions of base cations, sulfur and nitrogen using an extended critical load function for acidification. *Environmental Science & Technology, 41,* 1815–1820.

Zhao, Y., Duan, L., Xing, J., Larssen, T., Nielsen, C. P., & Hao, J. M. (2009). Soil acidification in China: Is controlling SO_2 emissions enough? *Environmental Science & Technology, 43,* 8021–8026.

Zhao, Y., Duan, L., Lei, Y., Xing, J., Nielsen, C. P., & Hao, J. M. (2011). Will PM control undermine China's efforts to reduce soil acidification? *Environmental Pollution, 159,* 2726–2732.

Zhu, X. Y., Duan, L., Tang, G. G., Hao, J. M., & Dong, G. X. (2004). Estimation of atmospheric emissions of base cations in China (in Chinese). *Journal of Tsinghua University (Science and Technology), 44,* 1176–1179.

Chapter 17
Assessment of Critical Loads of Acidity and Their Exceedances for European Lakes

Chris J. Curtis, Maximilian Posch, Julian Aherne, Jens Fölster, Martin Forsius, Thorjørn Larssen and Filip Moldan

17.1 Introduction

Lakes and rivers provided some of the earliest evidence of the effects of acidic deposition on natural ecosystems. Early warnings of the potential impacts of 'acid rain' came from Scandinavian scientists such as Odén (1968) and Jensen and Snekvik (1972), drawing attention to the acidification of lakes and rivers in Sweden and Norway, respectively. By the 1980s, the decline and loss of fish populations in Scandinavian surface waters provided one of the most visible signs of acidification and a compelling political driver for international action on transboundary air pollution. In concert, palaeolimnological studies of lake sediments provided some of the strongest evidence linking chronic acidic deposition to surface water acidification (Fig. 17.1, Battarbee et al. 1985; Flower and Battarbee 1983). Critical loads

C. J. Curtis (✉)
School of Geography, Archaeology and Environmental Studies, University of the Witwatersrand, Johannesburg, South Africa
e-mail: christopher.curtis@wits.ac.za

M. Posch
Coordination Centre for Effects (CCE), RIVM, Bilthoven, The Netherlands

J. Aherne
Environmental and Resource Studies, Trent University, Peterborough, ON, Canada

J. Fölster
Swedish University of Agricultural Sciences, Uppsala, Sweden

M. Forsius
Finnish Environment Institute (SYKE), Helsinki, Finland

T. Larssen
Norwegian Institute for Water Research (NIVA), Oslo, Norway

F. Moldan
IVL Swedish Environmental Research Institute, Gothenburg, Sweden

W. de Vries et al. (eds.), *Critical Loads and Dynamic Risk Assessments*, Environmental Pollution 25, DOI 10.1007/978-94-017-9508-1_17

Fig. 17.1 The Round Loch of Glenhead, Galloway, Scotland, is perhaps the most intensively studied site in palaeolimnology. It provided key evidence linking changes in reconstructed lake acidity, based on preserved diatom assemblages, with patterns of fossil fuel combustion and acid deposition. Despite some recovery from acidification since the 1980s, it remains acidified and continues to exceed critical loads beyond 2020. (Photo: C.J. Curtis)

for freshwaters were developed in response to the growing political controversy, when the most efficient methods of addressing acidic deposition issues (then primarily associated with sulphur (S)) at the international level were in dispute (see Chap. 1).

For freshwaters, the Steady-State Water Chemistry (SSWC) model was initially used to generate critical load and exceedance maps of S for Scandinavia and the UK (CLAG 1995; Henriksen et al. 1992, 1995). The SSWC model was employed because of its modest data requirements, needing only water chemistry and an estimate of annual runoff to calculate a critical load, and an estimate of S deposition to determine whether the critical load was exceeded. The key to the simplicity of the model is the assumption that the sulphate anion is mobile in catchments (Henriksen 1984; Seip 1980), so that input fluxes of sulphate as acidic deposition will equal output fluxes of sulphate in the runoff from a catchment. Empirical relationships are employed to derive the original, pre-industrial leaching rate of base cations from a lake or stream catchment, using only measured water chemistry. A critical acid neutralising capacity (ANC) limit is specified to ensure an acceptable level of protection to a selected biological indicator, and converted into a critical flux in runoff to derive the critical load (Henriksen et al. 1992). Details of the model are provided in Chap. 6.

While S deposition declined dramatically during the 1980s and 1990s across much of Europe, scientists in various countries had begun to notice that nitrate concentrations in acid-sensitive lakes appeared to be increasing from their very low 'pre-acidification' levels (e.g. Brown 1988; Grennfelt and Hultberg 1986; Henriksen and Brakke 1988; Sullivan et al. 1997). During the same period, emissions of oxidized forms of nitrogen (N) from vehicles and industry along with ammonia emissions from intensive agriculture had at best remained constant (INDITE 1994).

Furthermore, in the late 1980s and early 1990s the concept of N saturation was proposed (Aber et al. 1989; Aber 1992; Ågren and Bosatta 1988; Dise and Wright 1995; Skeffington and Wilson 1988; Stoddard 1994). The hypothesis states

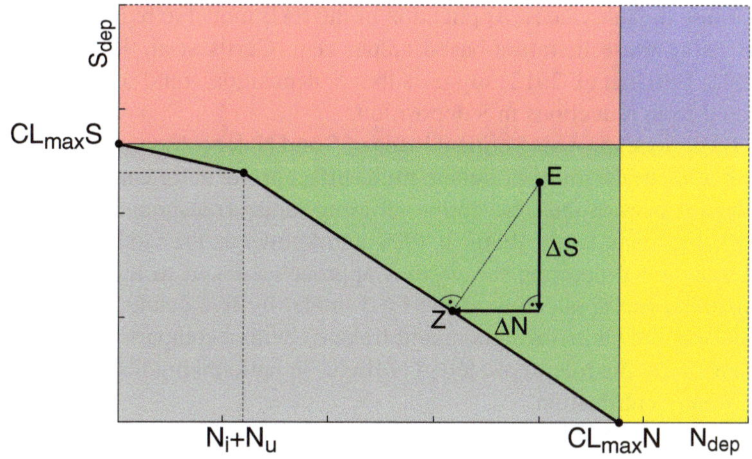

Fig. 17.2 Piece-wise linear critical load function (CLF) of acidifying *N* and *S* for a lake defined by its catchment properties. For a given deposition pair (N_{dep}, S_{dep}) the critical load exceedance is calculated by adding the *N* and *S* deposition reductions needed to reach the CLF via the shortest path (E → Z): $Ex = \Delta S + \Delta N$. The *grey* area below the CLF denotes deposition pairs resulting in non-exceedance of critical loads. If a deposition pair is located in the *green* area (such as *E*), non-exceedance can be achieved by reducing *N* or *S* deposition (or both); in the *red* (*yellow*) area S_{dep} (N_{dep}) has to be reduced to achieve non-exceedance; and in the *blue* area both N_{dep} and S_{dep} have to be reduced

that in N-limited terrestrial ecosystems, atmospheric N deposition can be assimilated until biological N demand is met entirely by internal supply through re-mineralization and nitrification. In this situation, nitrate leaching into surface waters increases. Although this process may be interpreted as a decline in retention of atmospheric inputs, rapid cycling of deposited N has been demonstrated using stable isotope methods which show that most leached nitrate from semi-natural catchments has been microbially produced (Curtis et al. 2012). A critical load model for N and S therefore has to account for the steady-state rates of N retention, removal and leaching to determine the potential contribution of enhanced nitrate leaching to acidification.

The principal freshwater model incorporating linked critical loads of N and S is the First-order Acidity Balance (FAB) model (Henriksen and Posch 2001; Posch et al. 1997, 2012), which is analogous to the Simple Mass Balance (SMB) model for terrestrial ecosystems (see Chap. 6). The FAB model employs a charge balance for N and S, which includes the base cation leaching from the SSWC model (see above) together with a critical chemical criterion, to construct a 'critical load function' (CLF; Fig. 17.2). This criterion is set to protect a chosen biological indicator, commonly brown trout (*Salmo trutta*) using a dose-response relationship between ANC and probability of occurrence, derived for Norwegian lakes (Lien et al. 1996; Lydersen et al. 2004). A description of the FAB model can be found in Chap. 6.

A requirement for the FAB model is the parameterisation of steady-state rates for the key N retention processes, which has largely been based on simplified models and literature default values. Application of the FAB model to regional and national datasets using these literature-based values (e.g. Curtis et al. 1998, 2000; Kaste et al. 2002; Posch et al. 2012) suggests that N deposition could offset any recovery anticipated from reductions in S deposition.

The FAB model provided critical loads of S and N deposition which were used in the negotiation of the multi-pollutant, multi-effect protocol for emissions reductions under the auspices of the Convention on Long-range Transboundary Air Pollution (LRTAP), signed in Gothenburg in 1999 and known as the Gothenburg Protocol. In this chapter we describe the various approaches taken to the collation of national datasets and application of the FAB model by five countries (Finland, Norway, Sweden, the United Kingdom and Ireland), with a comparison of critical load exceedances and the role of modelled N fluxes in national predictions of continued surface water acidification.

17.2 National Applications of the FAB Model in Europe

17.2.1 Finland

Critical loads of acidity and their exceedances for Finnish lakes have previously been determined by Henriksen et al. (1990), Forsius et al. (1992), Kämäri et al. (1993) and Posch et al. (1997, 2012).

Site Selection and Sampling: During 1987, a country-wide lake survey was conducted under the Finnish Research Project on Acidification (HAPRO; see Kauppi et al. 1990). A random sample of 987 lakes, from the ~56,000 lakes in Finland >1 ha (with ~15,700>10 ha), were sampled during autumn overturn and analysed for all major ions in the laboratories of the (then) National Board of Waters and the Environment. Statistical procedures for lake selection, sampling protocols, analytical methods and quality control procedures are described in detail by Forsius et al. (1990). An analysis and discussion of the lake water chemistry is given by Kortelainen et al. (1989) and Kämäri et al. (1991).

North of 66.13° (7340 km in the Finnish coordinate system) only lakes >10 ha were sampled, owing to logistical limitations and the very large number of small lakes in Finnish Lapland (Forsius et al. 1990). Thus, the data set was supplemented by a comprehensive survey of lakes sampled around the same time by the Water and Environment District Office of Lapland. This combined set of lakes constitutes the basis for the assessment of critical loads for surface waters in Finland under the LRTAP Convention (Posch et al. 1997). In the current study, only lakes north of the 7340-line and <10 ha were selected from the Lappish lake set. Moreover, the number of (small) lakes was limited so that their proportion was the same as that south of the 7340-line, resulting in a total of 1066 lakes (see Aherne et al. 2012). Not all variables were available for the additional Lappish lakes, e.g., total organic carbon

Table 17.1 FAB model parameters employed in Finland

Variable	Description	Value	Source
ANC_{limit}	ANC limit	Variable	$ANC_{limit}=ANC_{oaa}+(10.2/3)\cdot TOC$, with ANC_{oaa} (organic acid adjusted ANC)=8 μeq l^{-1} (for brown trout) (Lydersen et al. 2004)
F	F-factor	Variable	$F=1-\exp(-[BC^*]_0/B)$ where B is estimated as 131 μeq l^{-1} based on paleolimnological and water chemistry data of Finnish lakes (Posch et al. 1993)
$[SO_4^*]_0$	Pre-industrial sulphate conc.	Variable	Derived from 1880 modelled EMEP background sulphate deposition (Schöpp et al. 2003) and discharge: $[SO_4^*]_0=S_{dep,1880}/Q$
N_{upt}	N uptake in the catchment	Variable	Based on pine, spruce and broad-leaved (mostly birch) harvest data and element contents (see Aherne et al. 2012)
N_{imm}	N immobilisation	0.5 kg N ha^{-1} yr^{-1}	Mapping Manual (www.icpmapping.org)
f_{de}	Denitrification factor	$0.1+0.7f_{peat}$	Mapping Manual (www.icpmapping.org)
s_N	In-lake mass transfer coefficient for N	6.5 m yr^{-1}	Kaste and Dillon (2003)
s_S	In-lake mass transfer coefficient for S	0.5 m yr^{-1}	Baker and Brezonik (1988)

(TOC) was estimated from a linear regression with measured chemical oxygen demand (COD; Kortelainen 1993).

Derivation of FAB Datasets: The FAB model methoCy for Finland was initially described by Henriksen et al. (1993) and Posch et al. (1997) with recent advances described by Posch et al. (2012). The base cation fluxes were estimated with the SSWC model using the observed sea-salt corrected (chloride as tracer) base cation concentrations. Long-term normals for discharge (runoff) were obtained from the FINESSI project (www.finessi.info), which provided data for all of Finland on a 10 km × 10 km grid resolution (see Aherne et al. 2008). Pre-acidification lake nitrate was set to zero. A description of the FAB model parameters for Finland is provided in Table 17.1. Nitrogen removal in harvested biomass from Finnish forests was based on three species/groups: pine, spruce and broad-leaved (mostly birch). Business-as-usual (BAU) projections of forest growth and harvesting were obtained from the MELA model (Kärkkäinen et al. 2008; Redsven et al. 2004). The projections assumed no change in policies and climate (i.e., continued cuttings at present-day level). The removal (harvest) of N was estimated using element concentrations from a compilation of Nordic sources (Bringmark 1977; Finér 1989; Finér and Brække 1991; ICP Integrated Monitoring 2004; Mälkönen 1977). For further details see Aherne et al. (2012).

Table 17.2 FAB model parameters employed in Norway

Variable	Description	Value	Source
ANC_{limit}	ANC limit	Variable	Henriksen and Posch (2001), Lydersen et al. (2004)
F	F-factor	Variable	$F=\sin((\pi/2)Q[BC^*]_t/S)$, where $S=400$ meq $m^{-2}yr^{-1}$ is the base cation flux at which $F=1$ (Henriksen and Posch 2001)
$[SO_4^*]_0$	Pre-industrial sulphate conc.	Variable	Derived from 1880 modelled EMEP background sulphate deposition (Schöpp et al. 2003) and discharge: $[SO_4^*]_0=S_{dep,1880}/Q$
N_{upt}	N uptake in the catchment	0–50 meq m^{-2} yr^{-1}	Forest inventory data from Norwegian Forest and Landscape Institute (Frogner et al. 1994).
N_{imm}	N immobilisation	0.5 kg N ha^{-1} yr^{-1}	Mapping Manual (www.icpmapping.org)
f_{de}	Denitrification factor	0.1	Larssen et al. (2008)
s_N	In-lake mass transfer coefficient for N	5 m yr^{-1}	Dillon and Molot (1990)
s_S	In-lake mass transfer coefficient for S	0.5 m yr^{-1}	Baker and Brezonik (1988)

17.2.2 Norway

Site Selection and Sampling: The official Norwegian database for critical loads for surface waters is based on a 0.5° latitude by 1° longitude grid, with each grid square divided into 16 sub-grids (Henriksen 1998). The surface water chemistry within a sub-grid was derived from available observations for lakes and rivers, which includes results from the national 1500-lake survey conducted in 1995 (Skjelkvåle et al. 1996). The chemistry of the lake that was judged to be the most typical was chosen to represent the grid. If there were wide variations within a sub-grid, the most sensitive area was selected if it was >25% of the grid's area. Sensitivity was evaluated on the basis of water chemistry, topography and bedrock geology. Geology was determined from the geological map of Norway (1:1 million) prepared by the Norwegian Geological Survey.

Derivation of FAB Datasets: The methodology for Norway was described by Henriksen (1998) and later updated by Larssen et al. (2005, 2008). The base cation fluxes were estimated with the SSWC model using the observed sea-salt corrected (chloride as tracer) base cation concentrations. Mean annual runoff data were taken from runoff maps prepared by the Norwegian Water Resources and Energy Directorate. Land type characteristics (lake area, catchment area, forest area, bare rock area) were measured from maps. A description of the FAB model parameters for Norway is provided in Table 17.2.

A variable ANC limit as described by Henriksen and Posch (2001) was used, but adjusted for the strong acid anion contribution from organic acids after Lydersen et al. (2004). For the F-factor, the sine function of Brakke et al. (1990) has typically been used, but in recent applications $[BC]_0^*$ has instead been taken from hindcasts from MAGIC model runs used for calculating target loads (Larssen et al. 2005). Nitrogen removal in harvested biomass was estimated by Frogner et al. (1994) and mapped for the whole of Norway according to forest cover and productivity. All uptake rates were kept constant and assume constant removal from harvest and no change from climate, eutrophication or other factors. The denitrification factor (f_{de}) was kept constant at 0.1. In-lake mass transfer coefficients were kept constant at 5 m yr^{-1} for N and 0.5 m yr^{-1} for S, chosen as the mid-value of the ranges proposed by Dillon and Molot (1990) and Baker and Brezonik (1988), respectively.

17.2.3 Sweden

Site Selection and Sampling: The Swedish critical load lakes are part of the national surveillance monitoring of lakes sampled during 2007, 2008 and 2009 (Grandin 2007). They include 2410 lakes >1 ha selected by a stratified random selection and thus representing the lake ecosystems of the whole area of Sweden. The nine largest lakes were excluded to avoid overlap of lake catchments. Lake water was sampled in the centre of the lake during autumn circulation and analysed for water chemistry. Lakes affected by liming ($n=458$) were corrected by using the average Ca:Mg ratio from non-limed reference lakes within 20 km distance and the Mg concentration of the liming agent (Fölster et al. 2011).

Derivation of FAB Datasets: The critical loads were calculated using the FAB model as described in Henriksen et al. (1993), Posch et al. (1997) and Rapp et al. (2002), with some modifications described below (Table 17.3). The base cation leaching used in the FAB-model was taken from (MAGIC) simulated BC concentration in 2100 under the Cost Optimised Baseline (COB) deposition scenario. Thus the F-factor for estimating the weathering rate was not used. The year 2100 was used instead of 1860 for steady-state, since simulations indicated that the BC concentration of 1860 was not possible to reach within a reasonable time. The estimates of N immobilisation were based on Gundersen et al. (1998). Nitrogen immobilisation was set to 100% for deposition up to 2 kg N ha^{-1} yr^{-1}, 50% for deposition from 2 kg N ha^{-1} yr^{-1} up to 10 kg N ha^{-1} yr^{-1} and zero for deposition exceeding 10 kg N ha^{-1} yr^{-1}. In addition to this, leaching of organic N calculated from the lake concentration of Total Organic Nitrogen (TON) was regarded as non-acidifying. The chemical threshold, ANC limit, was calculated for each lake individually corresponding to a change in pH of 0.4 units from reference conditions estimated using the MAGIC model (Moldan et al. 2004). This threshold is used as a definition of acidification in the Swedish Environmental Quality Criteria and for the fulfilment of 'Good Ecological Status' within the EU Water Framework Directive (Fölster et al. 2007). Where insufficient data for MAGIC were available, parameters

Table 17.3 FAB model parameters employed in Sweden

Variable	Description	Value	Source
ANC_{limit}	ANC limit	0–300 μeq l^{-1} (exceeded lakes)	SEPA (2010) (derived from $\Delta pH=0.4$)
F	F-factor	–	Steady-state BC is calculated by MAGIC
$[SO_4^*]_0$	Pre-industrial sulphate conc.	–	Not required when using MAGIC
N_{upt}	N uptake in the catchment	2.5–43.2 meq m^{-2} yr^{-1}	Calculated from the national monitoring of forests. Uptake depends on species composition and productivity of the forest
N_{imm}	N immobilisation	0–114.7 meq m^{-2} yr^{-1}	Gundersen et al. (1998)
f_{de}	Denitrification factor	$0.1+0.7f_{peat}$	Mapping Manual (www.icpmapping.org)
s_N	In-lake mass transfer coefficient for N	5 m yr^{-1}	Dillon and Molot (1990)
s_S	In-lake mass transfer coefficient for S	0.5 m yr^{-1}	Baker and Brezonik (1988)

used in FAB were taken from a similar lake within a database of 2900 MAGIC simulated lakes using an analogue matching procedure, i.e. the MAGIC library, a web-based acidification assessment tool (see www.ivl.se/magicbibliotek).

The 2410 lakes were distributed over 2106 of the 5 km × 5 km squares used in national FAB applications. In most cases there was one modelled lake per 5 km × 5 km square and the ecosystem area was then set to 25 km^2. For lakes within squares with more than one lake, the ecosystem area was set to 25 km^2 divided by the number of lakes within that square.

17.2.4 United Kingdom

Site Selection and Sampling: Prior to 2004, the UK freshwater critical loads dataset comprised a grid-based survey of lakes and streams judged to be the most acid-sensitive in either 10 km grid squares for acid-sensitive regions (mainly in the uplands) or 20 km grid squares in less acid-sensitive lowland regions (Curtis et al. 2000; Kreiser et al. 1995). These lakes were sampled between 1991 and 1993. In 2004, additional freshwater sites from later, intensive regional studies were incorporated into the national dataset (described in Curtis et al. 2005a) with subsequent additions to the national dataset in 2008 (Curtis and Simpson 2011; Hall 2008). The resultant composite dataset includes 1752 sites which are neither random nor grid-based, but a combination of the original gridded survey sites with spatially intensive, complementary datasets in the most impacted regions. Furthermore, the UK dataset also contains stream sites as well as standing waters, due to known acidification problems in regions without lakes, such as the North York Moors (Evans et al. 2014).

Table 17.4 FAB model parameters employed in the United Kingdom

Variable	Description	Value	Source
ANC_{limit}	ANC limit	20 or 0 $\mu eq\ l^{-1}$	20 $\mu eq\ l^{-1}$ except where site-specific MAGIC modelling or palaeolimnological reconstruction indicates a lower pre-industrial value, when zero is used (Curtis and Simpson 2011)
F	F-factor	Variable	Estimated following Brakke et al. (1990) as described by Curtis et al. (2000)
$[SO_4^*]_0$	Pre-industrial sulphate conc.	Variable	Calculated as $15+0.16[BC^*]_t$ (Henriksen and Posch 2001)
N_{upt}	N uptake in the catchment	21–42 meq $m^{-2}\ yr^{-1}$	Managed coniferous woodland=21 meq $m^{-2}\ yr^{-1}$ Managed broadleaf woodland=42 meq $m^{-2}\ yr^{-1}$
N_{imm}	N immobilisation	1–3 kg N $ha^{-1}\ yr^{-1}$	Weighted by soil type; see Curtis et al. (2000), Hall et al. (1997)
N_{de}	Denitrification flux	1–4 kg N $ha^{-1}\ yr^{-1}$	Weighted by soil type; see Curtis et al. (2000), Hall et al. (1997)
s_N	In-lake mass transfer coefficient for N	5 m yr^{-1}	Dillon and Molot (1990)
s_S	In-lake mass transfer coefficient for S	0.5 m yr^{-1}	Baker and Brezonik (1988)

Derivation of FAB Datasets: While most aspects of model parameterisation were comparable with other national applications (Table 17.4), a key difference in the UK application is the use of a constant (i.e. deposition-independent) term for denitrification, based on empirically derived values for each soil type. This decision was taken on the basis of field and laboratory studies of actual and potential denitrification rates which suggested that the published method (Posch et al. 1997) would greatly over-estimate denitrification fluxes from UK peatland soils (Curtis et al. 2006). The range of N immobilisation values is also higher than used in other regions or recommended in the Mapping Manual (www.icpmapping.org).

17.2.5 Ireland

Critical loads of acidity and exceedances for Irish lakes have been previously described by Aherne et al. (2002) and Aherne and Curtis (2003). The current study incorporated an extended number of study lakes ($n=221$) sampled during the period 2000–2010.

Site Selection and Sampling: During 1997, a survey of upland headwater lakes ($n=200$) was carried out in predominantly acid-sensitive coastal regions of Ireland; site selection was pseudo-randomly weighted on acid-sensitive regions based on mapped soil characteristics and bedrock geology. A sub-set of these lakes were re-sampled during 2007 (Burton and Aherne 2012) and 2008 ($n=139$). In addition, the current study included 'acid lakes' routinely sampled by the Irish Environmental Protection Agency (EPA) under the Water Framework Directive monitoring

Table 17.5 FAB model parameters employed in Ireland

Variable	Description	Value	Source
ANC_{crit}	ANC limit	Variable	$ANC_{limit}=ANC_{oaa}+(10.2/3)\cdot TOC$, with ANC_{oaa} (organic acid adjusted ANC) = 8 meq m^{-3} for brown trout (Lydersen et al. 2004)
F	F-factor	Variable	Estimated following Henriksen and Posch (2001) as previously described by Aherne et al. (2002) and Aherne and Curtis (2003)
$[SO_4^*]_0$	Pre-industrial sulphate conc.	Variable	Derived from 1880 modelled EMEP background sulphate deposition (Schöpp et al. 2003) and discharge: $[SO_4^*]_0=S_{dep,1880}/Q$
N_{upt}	N uptake in the catchment	31 meq m^{-2} yr^{-1}	Based on Sitka spruce (*Picea sitchensis*)
N_{imm}	N immobilisation	0.5 kg N ha^{-1} yr^{-1}	Mapping Manual (www.icpmapping.org)
f_{de}	Denitrification factor	$0.1+0.7f_{peat}$	Mapping Manual (www.icpmapping.org)
s_N	In-lake mass transfer coefficient for N	6.5 m yr^{-1}	Kaste and Dillon (2003)
s_S	In-lake mass transfer coefficient for S	0.5 m yr^{-1}	Baker and Brezonik (1988)

programme ($n=41$) and other national monitoring programmes ($n=41$). All study lakes ($n=221$) were sampled during the period 2000–2010, with an average of 4 years of chemistry data for each lake (and up to 12 observations per year for some lakes).

Derivation of FAB Datasets: The lakes ranged in size from 0.08–862.58 ha, with an average area of 28.75 ha, and average elevation of 186 m a. s. l. (max = 710.7 m a. s. l.). Catchment land cover (derived from the National Teagasc Land Cover Map 1995 [TLC95]; Green and Fealy 2010) was dominated by peatlands (mean = 53 % across all catchments), exposed rock (25 %), forest (7 %) and grasslands (15 %). Not all variables were available for the EPA lakes, i.e., dissolved organic carbon was only measured on a sub-set of the lakes, as such it was estimated from a linear regression with measured colour (available for all EPA lakes). Further details on FAB model parameters are provided in Table 17.5.

The methodology followed Aherne et al. (2002) with recent updates following Posch et al. (2012). The base cation fluxes were estimated with the SSWC model using the observed sea-salt corrected (chloride as tracer) base cation concentrations. Long-term normals for discharge (runoff) on a 1 km × 1 km grid resolution were estimated using MetHyd (a meteo-hydrological model; Slootweg et al. 2010), and monthly climate normals for rainfall volume, average temperature and sunshine hours (Met Eireann [URL: www.met.ie]). Pre-acidification lake nitrate was set to zero. Nitrogen removal in harvested biomass assumed that managed Irish forests were dominated by Sitka spruce (*Picea sitchensis*), with an average yield class of 16 m^3 ha^{-1} yr^{-1} (COFORD 1994), stem concentrations of N = 0.05 % (based on spruce in Wales; Emmett and Reynolds 1996), and a wood density of 390 kg m^{-3}. The potential net uptake of 31.0 meq m^{-2} yr^{-1} was multiplied by the percent cover of forest in each catchment (based on TLC95).

17.2.6 Other European Applications of the FAB Model

There have been other large scale applications of the FAB model within Europe, including novel applications to alpine lakes in Switzerland (Posch et al. 2007) and remote lakes in several mountainous regions of Europe under the EU EMERGE Project (Curtis et al. 2005b). Both of these studies focused on small subsets of national lake datasets and are not explored further here, except to highlight the presence of (in some cases extremely) acidified lakes exceeding FAB critical loads in various alpine regions, including the Swiss and Italian Alps, the Tyrol of Austria and Italy, the Retezat Mountains of Romania, the French and Spanish Pyrenees, the Rila Mountains of Bulgaria and the Tatra Mountains of Poland and Slovakia (Curtis et al. 2005b).

17.3 Results: Critical Loads and Exceedances for European Surface Waters

17.3.1 Critical Loads of Sulphur and Nitrogen

Since the FAB model considers the effects of S and N deposition together, there is no unique critical load value for a given site. However, separate critical loads may be derived for both S and N when considered in isolation, i.e. assuming deposition of the other species to be zero (Posch et al. 1997, 2012). For example, the 5th percentile critical loads for S alone ($CL_{max}S$) are shown in Fig. 17.3. Since this model term does not include major sinks for acidity except for a minor in-lake retention term for S, it effectively provides a map of overall sensitivity to net acid inputs. Very low values of $CL_{max}S$ are found in southern Sweden, central and northern Finland, southern Norway and localised regions of northern England, north-west Scotland and Ireland. The corresponding map for N alone ($CL_{max}N$) is similar but moderated by site-specific modelled sinks for deposited N (not shown).

Cumulative distribution functions for $CL_{max}S$ in all five countries show marked differences in terms of the distribution of critical loads (Fig. 17.4). Sweden has around 10–15 % of sites with $CL_{max}S$ close to zero and a large proportion with low critical loads of <1000 eq ha^{-1} yr^{-1}. Finland and Norway also have a large proportion of very acid-sensitive sites but fewer extreme critical loads close to zero. Ireland and the UK differ from the other countries in having a much smaller proportion of sensitive sites, and around 50 % with relatively high critical loads of >1500 eq ha^{-1} yr^{-1}.

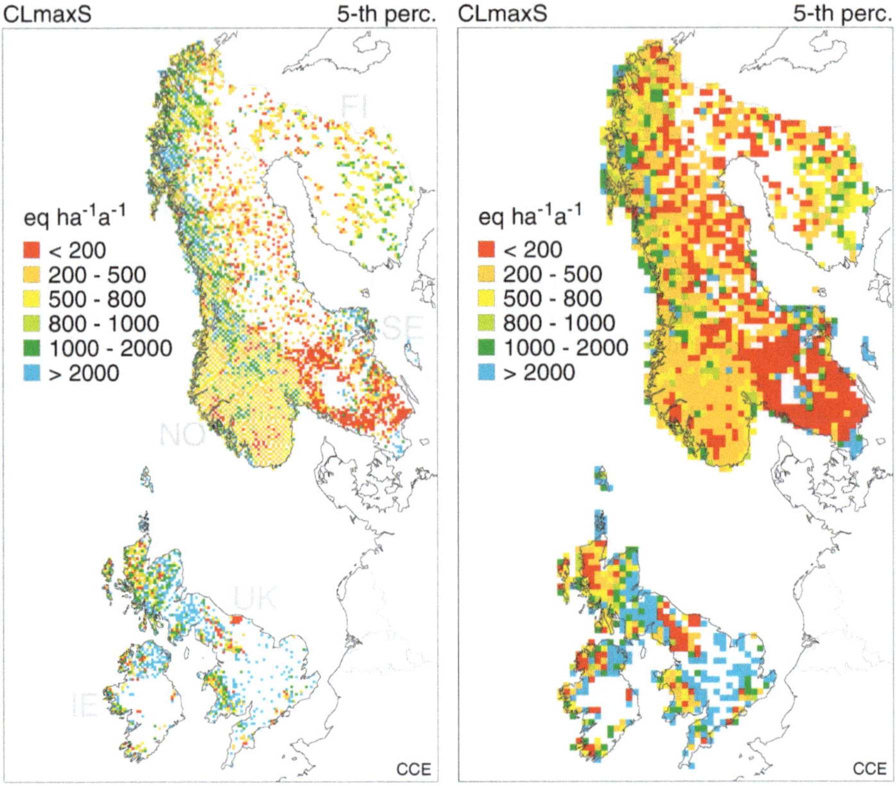

Fig. 17.3 5th percentile of $CL_{max}S$ on the 10 km × 10 km (*left*) and 25 km × 25 km (*right*) EMEP grid (the five countries are identified by their 2-letter codes)

Fig. 17.4 Cumulative distribution functions of $CL_{max}S$ for surface waters in the five countries (identified by their 2-letter codes)

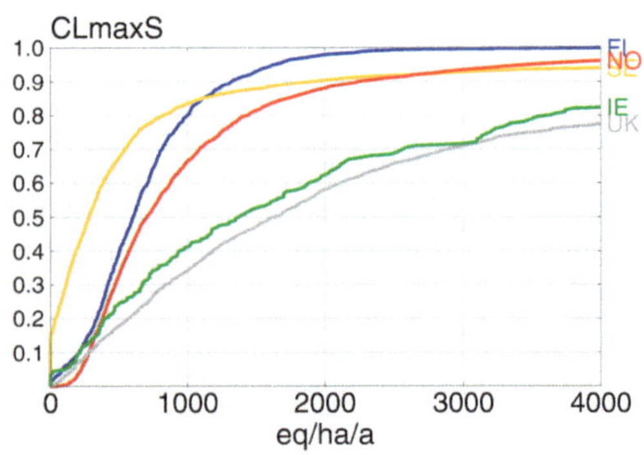

17.3.2 Critical Load Exceedances

Exceedances of critical loads are computed as an excess flux of acidity comprising contributions from both sulphate and nitrate leaching for chosen S and N depositions (see Fig. 17.2 and Posch et al. 2001 for the calculation procedure). Here, results from

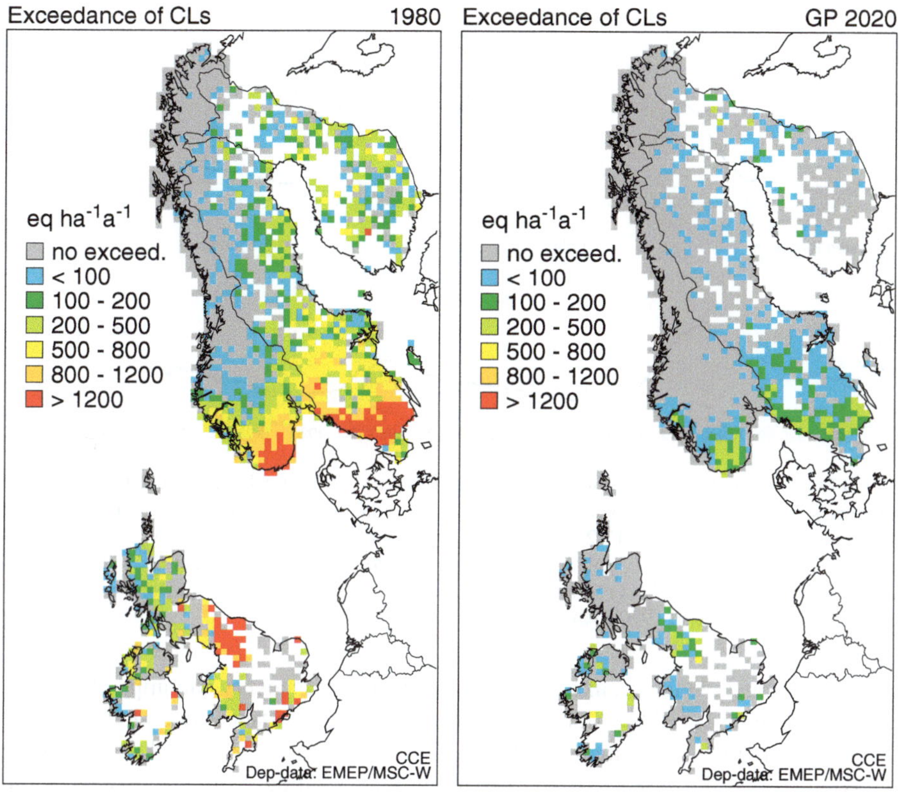

Fig. 17.5 Exceedance of acidity critical loads for surface waters in the five countries under S and N deposition for the years 1980 and 2020 (25 km × 25 km EMEP grid)

the multi-layer eulerian EMEP model (Simpson et al. 2012; Tarrasón et al. 2003) are used, consisting of relationships (source-receptor matrices) between European country emissions and depositions to specific ecosystems (forests, semi-natural vegetation, open land) in 50 km × 50 km grid cells covering Europe. Prior to 2010, historic country-specific emissions were used (see Schöpp et al. 2003 for data prior to 1990), whereas for 2020 emissions are those agreed in May 2012 in conclusion of the revision of the Gothenburg Protocol. It should be noted that there are differences between the critical load exceedances presented here for EMEP-scale deposition data, and estimates based on nationally derived depositions, which may be available at a finer spatial resolution and generated by different models.

Exceedance maps for the period of peak deposition in much of Europe (1980) and projected for 2020 on the 25 km × 25 km EMEP grid are provided in Fig. 17.5. The most severe exceedance of critical loads was observed in southern Norway, southern Sweden and parts of northern England in 1980, but exceedances were observed throughout most areas. Regions of greatest exceedance closely correspond with regions of greatest sensitivity as shown by $CL_{max}S$ (Fig. 17.3). The persistence of exceeded sites in 2020 suggests that currently planned emission reductions are insufficient to reach critical loads for all lakes, i.e. some lakes will remain acidified

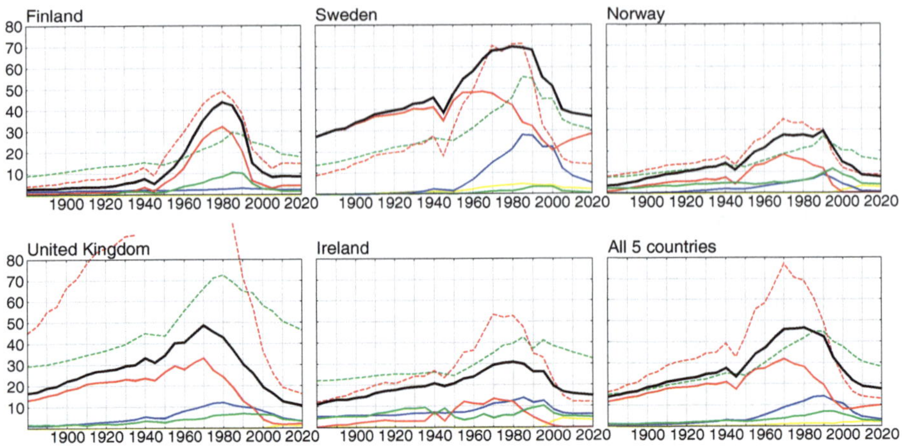

Fig. 17.6 Temporal development of percentages of lakes with critical loads exceeded in the five countries and the region as a whole (*black line*). Exceedance percentages split according to type (see coloured regions in Fig. 17.2): *red*: $S_{dep} > CL_{max}S$ but $N_{dep} < CL_{max}N$; *green*: $N_{dep} < CL_{max}N$ and $S_{dep} < CL_{max}S$; *blue*: $N_{dep} > CL_{max}N$ and $S_{dep} > CL_{max}S$; *yellow*: $N_{dep} > CL_{max}N$ and $S_{dep} < CL_{max}S$. The *dashed* lines show the total N (*green*) and S (*red*) deposition (in meq m^{-2} yr^{-1}; mean over all catchments; peaking at 165 in 1970 in the UK)

into the foreseeable future, although a partial chemical recovery and improvement in status will be achieved by 2020 as the magnitude of exceedances declines.

The temporal development of critical load exceedance from 1880 to 2020 for all countries is illustrated in Fig. 17.6, along with total deposition of excess S and N. Sulphur deposition peaked around 1970 in all countries except Sweden, where high levels persisted into the 1980s. Deposition in the UK reached a much greater and sharper peak than other countries, while Norway experienced the lowest peak in S deposition. Peaks in N deposition occurred later in all countries (1980–1990) and were much lower than peak S deposition. However, in all cases, N deposition exceeded S deposition by 2000 and this continues to 2020.

Exceedance of critical loads as a proportion of modelled sites shows a more variable peak between countries. The largest proportion of sites exceeding critical loads occurs in Sweden, peaking at about 70 % of modelled sites around 1980, which may be due to the higher level of protection selected in that country (see below). The next highest peak was found in the UK around 1970 with almost 49 % of modelled sites showing exceedance, then Finland in 1980 with 44 % exceedance. The proportion of exceeded sites reached only 31 % in Ireland (around 1980) and 29 % in Norway (around 1990). In all countries, the proportion of exceeded sites has decreased dramatically in response to declining S deposition, but all countries show exceeded sites persisting to 2020, varying from 8–11 % in Finland, Norway and the UK to 15 % in Ireland and almost 37 % in Sweden. The greatest uncertainty associated with these future exceedances is related to the long-term fate of N deposition as discussed below.

In 1980, around the time of peak exceedance for most countries, S was by far the dominant contributor to exceedance (Table 17.6). Sulphur deposition exceeded

Table 17.6 Proportion (%) of sites in exceedance categories of the Critical Load Function (see Fig. 17.2) and average deposition flux across all catchments (meq m^{-2} yr^{-1})

Year	Categories of exceedance according to CLF (%)								Deposition (meq m^{-2} yr^{-1})		
1980	Not exceeded	Total Exceeded	S or N (green)	S only (red)	N only (yellow)	S and N (blue)	Total N_{dep} > $CL_{max}N$ (blue+yellow)	Total S_{dep} > $CL_{max}S$ (blue+red)	N deposition	S deposition	N:S deposition ratio
Finland	56.3	43.7	8.6	32.4	0.0	2.7	2.7	35.1	25.8	49.2	0.52
Ireland	69.2	30.8	5.9	12.2	0.0	12.7	12.7	24.9	39.6	52.9	0.75
Norway	72.5	27.5	5.9	15.7	0.0	6.0	6.0	21.6	21.3	33.3	0.64
Sweden	30.4	69.6	1.5	42.2	4.4	21.5	25.9	63.8	45.9	71.1	0.65
UK	57.2	42.8	6.5	24.2	0.0	12.2	12.2	36.4	72.7	127.4	0.57
ALL	53.6	46.4	5.0	28.0	1.4	11.9	13.3	40.0	41.7	69.0	0.60
2020											
Finland	91.6	8.4	1.8	4.3	0.0	2.4	2.4	6.7	18.0	14.8	1.22
Ireland	85.1	14.9	5.0	0.0	3.6	6.3	10.0	6.3	32.5	12.0	2.70
Norway	92.5	7.5	4.0	0.1	2.9	0.6	3.4	0.7	15.5	8.6	1.81
Sweden	63.4	36.6	0.4	28.7	2.2	5.2	7.4	34.0	30.6	14.0	2.18
UK	89.4	10.6	3.5	2.2	1.5	3.4	4.9	5.6	46.5	16.2	2.86
ALL	82.4	17.6	2.5	10.0	2.0	3.1	5.0	13.1	28.0	12.9	2.16

$CL_{max}S$, indicating that critical loads were exceeded by S alone without reference to N deposition, in 22 (Norway) to 64 % (Sweden) of modelled sites. At that time, the relative contribution of N deposition to exceedance was much smaller than at present, as shown by N:S deposition ratios ranging from 0.52 (Finland) to 0.75 (Ireland), and taking into account that a proportion of N deposition is retained or removed in catchments before reaching surface waters. N deposition exceeded $CL_{max}N$ at only a small proportion of sites, from about 3 in Finland to 26 % in Sweden.

By 2020, major reductions in S deposition and much smaller reductions in total N deposition will mean that N deposition far outweighs S in terms of deposited acidity, with N:S deposition ratios ranging from 1.22 (Finland) to 2.86 (UK). However, the variability in modelled sinks for N deposition between countries results in a range of relative contributions to critical load exceedance. In Finland and Sweden, S deposition continues to dominate exceedance by 2020, but in Ireland and Norway N has overtaken S as the main agent of continued critical load exceedance. Relative contributions of S and N in the UK are very similar in 2020. The proportion of sites in which N deposition exceeds $CL_{max}N$ is low, ranging from about 2 % in Finland to almost 10 % in Ireland, but the future behaviour of deposited N is one of the key areas of uncertainty in the FAB model (see below).

17.4 Discussion: Limitations and Uncertainties in FAB Model Applications

17.4.1 National Approaches to Site Selection, Mapping and Setting Critical Limits

Submission of national critical load datasets to the Modelling & Mapping Programme under the LRTAP Convention allows significant flexibility in national approaches for deriving the requisite data. Some countries have used random surveys, often stratified regionally or by size class of lakes, to populate national water chemistry databases. Other countries have used additional datasets in acid-sensitive areas to supplement national random or grid-based surveys. Hence there is no consistency between countries in the degree to which national critical load datasets represent national populations of lakes. Furthermore, there are differences in the way that critical loads for lakes are represented in terms of ecosystem area.

Another difference between countries is in the choice of the critical limit, which for the SSWC and FAB models is defined as a critical ANC value. Except for Sweden, all the countries use brown trout as the indicator organism (endpoint) and a minimum ANC (critical limit) below which there is unacceptable damage. For Finland, Ireland and Norway, ANC_{oaa} takes into account the role of strong organic acids in determining this threshold, while the UK uses fixed ANC limits of 0 or 20 $\mu eq\ l^{-1}$. Sweden, on the other hand, uses ΔpH as the criterion for biological damage, in accordance with the Water Framework Directive whereby ecological status should be assessed as a change from pre-industrial conditions. It is defined as the difference between pre-industrial pH and present-day pH. Both pH values are calculated from ANC and TOC; pre-industrial ANC comes from the MAGIC model. If ΔpH exceeds 0.4 units, then the lake is considered acidified, and the critical load is exceeded. This is independent of position on the pH scale and thus can indicate ecological changes in naturally acidic systems (where biota under pre-industrial conditions were already controlled by a low pH) and not just in naturally circum-neutral lakes. The dose-response relationship between ΔpH and unacceptable biological change is in turn based on measured pH and relationships with fish, benthic fauna and periphyton for 74 lakes in southern Sweden, where pH showed a stronger predictive power than other chemical measures (Fölster et al. 2007).

The Swedish criterion of $\Delta pH = 0.4$ apparently gives lower critical loads than the ANC limit used in other countries, which also explains the drop in critical loads at the border between Norway and Sweden shown in Fig. 17.3. The implication is that Sweden has chosen to define critical loads with respect to more sensitive biota such as littoral invertebrate fauna and epiphyton in addition to fish, whereas the other four countries base their critical loads primarily on brown trout. It is also likely that a pH decline of 0.4 units in some Swedish sites would not result in breaching the critical ANC values used by other countries. Curtis and Simpson (2011) discussed the conceptual differences between the traditional critical loads approach where a fixed ANC limit allows differing degrees of change from reference conditions for different lakes, and the Water Framework Directive notion of defining ecological status relative to a fixed degree of change from reference conditions. The issue of aligning approaches under the LRTAP Convention and the EU Water Framework Directive has yet to be resolved.

17.4.2 National Approaches to the Nitrogen Mass Balance in FAB

There are four key N sink or removal terms used to calculate critical loads with the FAB model, and the national approaches for selecting them are compared below.

Net Uptake of N: All countries use similar assumptions about element concentrations in key tree species and projected harvest cycles to determine net removal of N in harvested biomass.

In-Lake Retention of N and S: All countries use similar mass-transfer coefficients to calculate in-lake retention of N and S. Previous studies using FAB found in-lake retention to be a minor sink for N and S in acid-sensitive lakes; for N potential terrestrial sinks are generally much greater (Curtis et al. 1998; Kaste et al. 2002; Kaste and Dillon 2003).

Immobilisation of N: Finland, Norway and Ireland all use a fixed value of 0.5 kg N ha^{-1} yr^{-1} for the immobilisation of N in catchment soils, which is the lower end of the suggested range in the Mapping Manual (www.icpmapping.org)

and constitutes a precautionary (worst-case nitrate leaching) approach. The UK employs a higher range of values from 1–3 kg N ha^{-1} yr^{-1} on the assumption that elevated rates of N immobilisation under anthropogenic N deposition loads may be sustainable over the long term. Sweden employs a unique approach, with a sliding scale for N immobilisation from 100 % at very low deposition to zero at higher depositions. In all cases, long-term immobilisation in FAB provides a relatively minor sink for N when compared with the upper end of the deposition range, though it can be significant in lower deposition regions.

Denitrification Fluxes: Finland, Sweden and Ireland use a first-order term for denitrification, a linear interpolation between 10 % for catchments with no peat and 80 % for catchments with 100 % peat cover. Norway uses a first-order term denitrifying a fixed 10 % of net N inputs. The UK uses a constant (i.e. deposition-independent) term from a lookup table of denitrification fluxes for each soil type, so a catchment-weighted constant value is used for each site.

For peat catchments, denitrification within the FAB model therefore provides a major sink for N deposition of up to 80 % of net inputs in Finland, Sweden and Ireland. However, it has been argued that the retention of large proportions of N deposition may be due to elevated, but possibly unsustainable, rates of N immobilisation in catchment soils (Curtis et al. 2006). One view might be that over the long term, a reduction in N immobilisation in soils could lead to an increase in denitrification fluxes as soilwater nitrate availability increases. However, for UK soils this hypothesis was tested experimentally by measuring denitrification fluxes from soils treated with ammonium nitrate solution in high and low N deposition settings. Neither untreated nor treated soils showed denitrification fluxes of the magnitude suggested by the peat fraction method (Curtis et al. 2006).

It may be concluded that the largest uncertainty within the N mass balance section of the FAB model lies in the relative importance of N immobilisation and denitrification fluxes, i.e. the future behaviour of N in terrestrial ecosystems. Nitrogen immobilisation in particular continues to present challenges at the catchment scale. The use of the FAB model to predict nitrate leaching fluxes at steady-state indicates much greater fluxes than currently observed (e.g. Curtis et al. 1998; Kaste et al. 2002). The question remains to what degree the FAB model adequately represents the real endpoint of the N saturation process, i.e. will N continue to be largely retained within terrestrial ecosystems, or will all N beyond long-term sinks specified in FAB ultimately leach as nitrate to surface waters? The latter case largely drives exceedance in Norway and Ireland. The same question has been addressed in recent modifications to the dynamic acidification model MAGIC (Oulehle et al. 2012).

17.4.3 F-Factor Issues and Lack of Steady-State

Under the original concept, a steady-state critical load should be a constant value regardless of when it is calculated. It is assumed that the calculation of critical load using contemporary water chemistry is insensitive to the stage of acidification or indeed recovery from acidification, i.e. even though water chemistry changes in response to the acidification process, the calculated critical load should not. Hence some of the national datasets still employ water chemistry survey data from the early 1990s or even 1980s, when deposition loads were much higher than they are at present.

More recently, the time-independence of the F-factor in the SSWC and FAB models has been questioned by Rapp and Bishop (2009). They used the dynamic model SAFE to compare modelled hydrochemical changes in Swedish lakes with empirical F-factors and showed that while the F-factor approach may be adequate during the acidification phase, it does not work well during the recovery phase when base cation leaching may decrease below the pre-industrial value. Critical loads based on water chemistry sampled during the recovery phase would therefore be under-estimated. However, an earlier analysis by Henriksen (1995) suggested that variations in the F-factor were only likely to be significant for critical loads in regions of high deposition, and found little effect of varying F for Norway, Finland and Sweden. Aherne and Curtis (2003) also looked at the influence of different formulations of the F-factor on SSWC model exceedance and found little impact. One reason for these differing interpretations is that the later work of Rapp and Bishop (2009) considered largely recovered systems where minor variations in critical load would mean the difference between meeting critical loads or exceedance. As a result of reservations about the F-factor approach, Sweden and Norway have more recently used the dynamic model MAGIC to calculate critical loads for lakes in their national datasets (Tables 17.2, and 17.3), while other countries continue to use different forms of the F-factor approach. Posch et al. (2012) compared SSWC/FAB and MAGIC based F-factors and found a strong correlation.

Conclusions

National scale applications of the FAB model provide maps of critical load exceedance which are highly consistent with areas of known acidification, as verified by independent chemical and biological monitoring studies, despite simplifications and uncertainties associated with steady-state models and critical limits (Chap. 2). Furthermore, many studies do show a strong relationship between increasing surface water ANC or pH and biological recovery (e.g. Hesthagen et al. 2011; Johnson and Angeler 2010; Kernan et al. 2010; Posch et al. 2012). In Finland, recovery in recruitment of perch populations matches well with reduced exceedance of critical loads (Posch et al. 2012). At long-term monitoring sites in the UK Acid Waters Monitoring Network, the re-appearance of acid-sensitive macrophytes and invertebrates has been linked to improvements in ANC with evidence that the critical ANC value of 20 $\mu eq\ l^{-1}$ does indeed represent an important ecological threshold for some species (Kernan et al. 2010). Similar results were found in a Norwegian lake study where the achievement of critical load and ANC>20 μeq^{-1} coincided with significant recovery in brown trout and invertebrate species (Hesthagen et al. 2011).

Swedish studies comparing acidified but recovering lakes with unimpacted reference lakes showed a movement of impacted phytoplankton and invertebrate assemblages towards those of reference lakes as pH of acidified lakes increased over 20 years (Johnson and Angeler 2010).

Hence the continued exceedance of FAB critical loads beyond 2020 in all five countries included here is a cause for concern as planned emission reductions under the recently revised Gothenburg Protocol do not go far enough. Critical load models like FAB evidently still have a role to play in shaping emissions policy for the protection of aquatic ecosystems. However, as emissions reductions become ever more expensive to achieve, the areas of uncertainty will face increasing scrutiny. The role of chemical and biological monitoring programmes will be critical in providing the supporting data against which model performance can be evaluated.

References

Aber, J. D. (1992). Nitrogen cycling and nitrogen saturation in temperate forest ecosystems. *Trends in Ecology & Evolution, 7*, 220–224.

Aber, J. D., Nadelhoffer, K. J., Steudler, P., & Melillo, J. M. (1989). Nitrogen saturation in northern forest ecosystems. *Bioscience, 39*, 378–386.

Ågren, G. I., & Bosatta, E. (1988). Nitrogen saturation of terrestrial ecosystems. *Environmental Pollution, 54*, 185–197.

Aherne, J., & Curtis, C. J. (2003). Critical loads of acidity for Irish lakes. *Aquatic Sciences, 65*, 21–35.

Aherne, J., Kelly-Quinn, M., & Farrell, E. P. (2002). A survey of lakes in the Republic of Ireland: Hydrochemical characteristics and acid sensitivity. *Ambio, 31*, 452–459.

Aherne, J., Posch, M., Forsius, M., Vuorenmaa, J., Tamminen, P., Holmberg, M., & Johansson, M. (2008). Modelling the hydrogeochemistry of acid-sensitive catchments in Finland under atmospheric deposition and biomass harvesting scenarios. *Biogeochemistry, 88*, 233–256.

Aherne, J., Posch, M., Forsius, M., Lehtonen, A., & Härkönen, K. (2012). Impacts of forest biomass removal on soil nutrient status under climate change: A catchment-based modelling study for Finland. *Biogeochemistry, 107*, 471–488.

Baker, L. A., & Brezonik, P. L. (1988). Dynamic model of in-lake alkalinity generation. *Water Resources Research, 24*, 65–74.

Battarbee, R. W., Flower, R. J., Stevenson, A. C., & Rippey, B. (1985). Lake acidification in Galloway: A palaeoecological test of competing hypotheses. *Nature, 314*, 350–352.

Brakke, D. F., Henriksen, A., & Norton, S. A. (1990). A variable F-factor to explain changes in base cation concentrations as a function of strong acid deposition. *Verhandlungen des Internationalen Verein Limnologie, 24*, 146–149.

Bringmark, L. (1977). A bioelement budget of an old Scots pine forest in central Sweden. *Silva Fennica, 11*, 201–209.

Brown, D. J. A. (1988). Effect of atmospheric N deposition on surface water chemistry and the implications for fisheries. *Environmental Pollution, 54*, 275–284.

Burton, A. W., & Aherne, J. (2012). Changes in the chemistry of small Irish lakes. *Ambio, 41*, 170–179.

CLAG, Critical Loads Advisory Group. (1995). *Critical loads of acid deposition for United Kingdom freshwaters*. London: ITE Edinburgh/Department of the Environment.

COFORD. (1994). Pathway to progress: A programme for forest research and development. National Council for Forest Research and Development, University College Dublin.

Curtis, C., & Simpson, G. (2011). Freshwater Umbrella: The effects of nitrogen deposition on freshwaters in the UK. (Report to DEFRA under Contract AQ0803. ECRC Research Report No. 152). London: University College London.

Curtis, C. J., Allott, T. E. H., Reynolds, B., & Harriman, R. (1998). The prediction of nitrate leaching with the first-order acidity balance (FAB) model for upland catchments in Great Britain. *Water Air and Soil Pollution, 105,* 205–215.

Curtis, C. J., Allott, T. E. H., Hughes, M., Hall, J., Harriman, R., Helliwell, R., Kernan, M., Reynolds, B., & Ullyett, J. (2000). Critical loads of sulphur and nitrogen for freshwaters in Great Britain and assessment of deposition reduction requirements with the First-order Acidity Balance (FAB) model. *Hydrology and Earth System Sciences, 4,* 1–15.

Curtis, C. J., Evans, C., Helliwell, R. C., & Monteith, D. (2005a). Nitrate leaching as a confounding factor in chemical recovery from acidification in UK upland waters. *Environmental Pollution, 137,* 73–82.

Curtis, C. J., Botev, I., Camarero, L., Catalan, J., Cogalniceanu, D., Hughes, M., Kernan, M., Kopáček, J., Korhola, A., Psenner, R., Rogora, M., Stuchlík, E., Veronesi, M., & Wright, R. F. (2005b). Acidification in European mountain lake districts: A regional assessment of critical load exceedance. *Aquatic Sciences, 67,* 237–251.

Curtis, C. J., Emmett, B. A., Reynolds, B., & Shilland, J. (2006). How important is N_2O production in removing atmospherically deposited nitrogen from UK moorland catchments? *Soil Biology and Biochemistry, 38,* 2081–2091.

Curtis, C. J., Heaton, T. H. E., Simpson, G. L., Evans, C. D., Shilland, J., & Turner, S. (2012). Dominance of biologically produced nitrate in upland waters of Great Britain indicated by stable isotopes. *Biogeochemistry, 111,* 535–554.

Dillon, P. J., & Molot, L. A. (1990). The role of ammonium and nitrate retention in the acidification of lakes and forested catchments. *Biogeochemistry, 11,* 23–43.

Dise, N. B., & Wright, R. F. (1995). Nitrogen leaching from European forests in relation to nitrogen deposition. *Forest Ecology and Management, 71,* 153–161.

Emmett, B. A., & Reynolds, B. (1996). Nitrogen critical loads for spruce plantations in Wales: Is there too much nitrogen? *Forestry, 69,* 205–214.

Evans, C. D., Chadwick, T., Norris, D., Rowe, E. C., Heaton, T. H. E., Brown, P., & Battarbee, R. W. (2014). Persistent surface water acidification in an organic soil-dominated upland region subject to high atmospheric deposition: The North York Moors, UK. *Ecological Indicators, 37,* 304–316.

Finér, L. (1989). Biomass and nutrient cycle in fertilized and unfertilized pine, mixed birch and pine and spruce stands on a drained mire. *Acta Forestalia Fennica, 208,* 3–19.

Finér, L., & Brække, F. H. (1991). Understorey vegetation on three ombrotrophic pine bogs and the effects of NPK and PK fertilization. *Scandinavian Journal of Forest Research, 6,* 113–128.

Flower, R. J., & Battarbee, R. W. (1983). Diatom evidence for recent acidification of two Scottish lochs. *Nature, 305,* 130–133.

Fölster, J., Andren, C., Bishop, K., Buffam, I., Cory, N., Goedkoop, W., Holmgren, K., Johnson, R., Laudon, H., & Wilander, A. (2007). A novel environmental quality criterion for acidification in Swedish lakes—an application of studies on the relationship between biota and water chemistry. *Water Air and Soil Pollution: Focus, 7,* 331–338.

Fölster, J., Köhler, S., von Brömsen, C., Akselsson, C., & Rönnback, P. (2011). Korrigering av vattenkemi för kalkningspåverkan—val av referenser och beräkning av osäkerheter. Institutionen för vatten och miljö (in Swedish). (Rapport 2011:1). SLU.

Forsius, M., Malin, V., Mäkinen, I., Mannio, J., Kämäri, J., Kortelainen, P., & Verta, M. (1990). Finnish lake acidification survey: Survey design and random selection of lakes. *Environmetrics, 1,* 73–88.

Forsius, M., Kämäri, J., & Posch, M. (1992). Critical loads for Finnish lakes: Comparison of three steady-state models. *Environmental Pollution, 77,* 185–193.

Frogner, T., Wright, R. F., Cosby, B. J., & Esser, J. M. (1994). *Maps of critical loads and exceedances for sulphur and nitrogen to forest soils in Norway*. (Naturens Tålegrenser Fagrapport 56). Oslo: Ministry of Environment.

Grandin, U. (2007). Strategier för urval av sjöar som ska ingå i den sexåriga omdrevsinventeringen av vattenkvalitet i svenska sjöar (in Swedish). (Rapport 2007:10). Institutionen för Miljöanalys, SLU.

Green, S., & Fealy, R. (2010). Teagasc's national landcover and habitat maps. *TResearch, 5,* 14–15.

Grennfelt, P., & Hultberg, H. (1986). Effects of nitrogen deposition on the acidification of terrestrial and aquatic ecosystems. *Water Air and Soil Pollution, 30,* 945–963.

Gundersen, P., Emmet, B. A., Kjønaas, O. J., Koopmans, C., & Tietema, A. (1998). Impact of nitrogen deposition on nitrogen cycling in forests: A synthesis of NITREX data. *Forest Ecology and Management, 101,* 37–55.

Hall, J. (2008). Status of UK critical loads and exceedances June 2008. http://cldm.defra.gov.uk/PDFs/uk_status_jun08.pdf. Accessed 1 June 2012.

Hall, J., Hornung, M., Freer-Smith, P., Loveland, P., Bradley, I., Langan, S., Dyke, H., Gascoigne, J., & Bull, K. (1997). Current status of UK critical loads data—December 1996. ITE Monks Wood, UK.

Henriksen, A. (1984). Changes in base cation concentrations due to freshwater acidification. *Verhandlungen des Internationalen Verein Limnologie, 22,* 692–698.

Henriksen, A. (1995). Critical loads of acidity to surface waters—how important is the F-factor in the SSWC method? *Water Air and Soil Pollution, 85,* 2437–2441.

Henriksen, A. (1998). *Application of the first-order acidity balance (FAB) model to Norwegian surface waters.* (Report SNO 3809-98). Oslo: Norwegian Institute for Water Research (NIVA).

Henriksen, A., & Brakke, D. F. (1988). Increasing contributions of nitrogen to the acidity of surface waters in Norway. *Water Air & Soil Pollution, 42,* 183–201.

Henriksen, A., & Posch, M. (2001). Steady-state models for calculating critical loads of acidity for surface waters. *Water Air & Soil Pollution: Focus, 7,* 375–398.

Henriksen, A., Kämäri, J., Posch, M., Lövblad, G., Forsius, M., & Wilander, A. (1990). *Critical loads to surface waters in Fennoscandia—Intra- and inter-grid variability of critical loads and their exceedance. (Nord 1990:124).* Copenhagen: Nordic Council of Ministers.

Henriksen, A., Kämäri, J., Posch, M., & Wilander, A. (1992). Critical loads of acidity: Nordic surface waters. *Ambio, 21,* 356–363.

Henriksen, A., Forsius, M., Kämäri, J., Posch, M., & Wilander, A. (1993). Exceedance of critical loads for lakes in Finland, Norway and Sweden: Reduction requirements for nitrogen and sulfur deposition. (Report 2841). Oslo, Norway: Norwegian Institute for Water Research.

Henriksen, A., Posch, M., Hultberg, H., & Lien, L. (1995). Critical loads of acidity for surface waters—Can the ANC_{limit} be considered variable? *Water Air and Soil Pollution, 85,* 2419–2424.

Hesthagen, T., Fjellheim, A., Schartau, A. K., Wright, R. F., Saksgård, R., & Rosseland, B. O. (2011). Chemical and biological recovery of Lake Saudlandsvatn, a formerly highly acidified lake in southernmost Norway, in response to decreased acid deposition. *Science of the Total Environment, 409,* 2908–2916.

ICP Integrated Monitoring. (2004). *ICP IM manual: Procedure for calculating biomass and bioelements.* Helsinki: Finnish Environment Institute (www.ymparisto.fi/default.asp?contentid=96947&lan=EN).

INDITE. (1994). Impacts of nitrogen deposition in terrestrial ecosystems. (United Kingdom Review Group on Impacts of Nitrogen Deposition on Terrestrial Ecosystems (INDITE)). London: Department of the Environment.

Jensen, K. W., & Snekvik, E. (1972). Low pH levels wipe out Salmon and Trout populations in southernmost Norway. *Ambio, 1,* 223–225.

Johnson, R. K., & Angeler, D. G. (2010). Tracing recovery under changing climate: Response of phytoplankton and invertebrate assemblages to decreased acidification. *Journal of the North American Benthological Society, 29,* 1472–1490.

Kämäri, J., Forsius, M., Kortelainen, P., Mannio, J., & Verta, M. (1991). Finnish lake survey: Present status of acidification. *Ambio, 20,* 23–27.

Kämäri, J., Forsius, M., & Posch, M. (1993). Critical loads of sulfur and nitrogen for lakes II: Regional extent and variability in Finland. *Water Air & Soil Pollution, 66,* 77–96.

Kärkkäinen, L., Matala, J., Härkönen, K., Kellomäki, S., & Nuutinen, T. (2008). Potential recovery of industrial wood and energy wood raw material in different cutting and climate scenarios for Finland. *Biomass Bioenergy, 32,* 934–943.

Kaste, Ø., & Dillon, P. J. (2003). Inorganic nitrogen retention in acid-sensitive lakes in southern Norway and southern Ontario, Canada—a comparison of mass balance data with an empirical N retention model. *Hydrological Processes, 17,* 2393–2407.

Kaste Ø. Henriksen A., & Posch M. (2002). Present and potential nitrogen outputs from Norwegian soft water lakes—an assessment made by applying the steady-state First-order Acidity Balance (FAB) model. *Hydrology and Earth System Sciences, 6,* 101–112.

Kauppi, P., Anttila, P., & Kenttämies, K. (Eds.). (1990). *Acidification in Finland.* Berlin: Springer.

Kernan, M., Battarbee, R. W., Curtis, C. J., Monteith, D. T., & Shilland, E. M. (2010). UK acid waters monitoring network 20 year interpretive report. (ECRC Research Report #141). London: University College London.

Kortelainen, P. (1993). Content of total organic carbon in Finnish lakes and its relationship to catchment characteristics. *Canadian Journal of Fisheries and Aquatic Sciences, 50,* 1477–1483.

Kortelainen, P., Mannio, J., Forsius, M., Kämäri, J., & Verta, M. (1989). Finnish lake survey: The role of organic and anthropogenic acidity. *Water Air & Soil Pollution, 46,* 235–249.

Kreiser, A. M., Patrick, S. T., Battarbee, R. W., Hall, J., & Harriman, R. (1995). Mapping water chemistry. In CLAG Freshwaters (Ed.). *Critical loads of acid deposition for United Kingdom freshwaters* (pp. 15–18). London: Department of the Environment.

Larssen, T., Høgåsen, T., & Wright, R. F. (2005). *Target loads for acidification of Norwegian surface waters.* (NIVA-Report 5099-2005). Oslo: Norwegian Institute for Water Research.

Larssen, T., Lund, E., & Høgåsen, T. (2008). Exceedance of critical loads for aciddification and nitrogen for Norway—update for the period 2002-2006. (NIVA-report 5697-2008). Oslo: Norwegian Institute for Water Research.

Lien, L., Raddum, G. G., Fjellheim, A., & Henriksen, A. (1996). A critical limit for acid neutralizing capacity in Norwegian surface waters, based on new analyses of fish and invertebrate responses. *Science of the Total Environment, 177,* 173–193.

Lydersen, E., Larssen, T., & Fjeld, E. (2004). The influence of total organic carbon (TOC) on the relationship between acid neutralizing capacity (ANC) and fish status in Norwegian lakes. *Science of the Total Environment, 326,* 63–69.

Mälkönen, E. (1977). Annual primary production and nutrient cycle in birch stands. *Communicationes Instituti Forestalis Fenniae, 91,* 1–35.

Moldan, F., Kronnäs, V., Wilander, A., Karltun, E., & Cosby, B. J. (2004). Modelling acidification and recovery of Swedish lakes. *Water Air & Soil Pollution: Focus, 4,* 139–160.

Odén, S. (1968). *The acidification of air precipitation and its consequences in the natural environment. (Energy Committee Bulletin 1).* Stockholm: Swedish Natural Sciences Research Council.

Oulehle, F., Cosby, B. J., Wright, R. F., Hruška, J., Kopáček, J., Krám, P., Evans, C. D., & Moldan, F. (2012). Modelling soil nitrogen: The MAGIC model with nitrogen retention linked to carbon turnover using decomposer dynamics. *Environmental Pollution, 165,* 158–166.

Posch, M., Forsius, M., & Kämäri, J. (1993). Critical loads of sulfur and nitrogen for lakes 1: Model description and estimation of uncertainty. *Water Air Soil & Pollution, 66,* 173–192.

Posch, M., Kämäri, J., Forsius, M., Henriksen, A., & Wilander, A. (1997). Exceedance of critical loads for lakes in Finland, Norway and Sweden: Reduction requirements for acidifying nitrogen and sulfur deposition. *Environmental Management, 21,* 291–304.

Posch, M., Hettelingh, J.-P., & De Smet, P. A. M. (2001). Characterization of critical load exceedances in Europe. *Water Air and Soil Pollution, 130,* 1139–1144.

Posch, M., Eggenberger, U., Kurz, D., & Rihm, B. (2007). *Critical loads of acidity for alpine lakes. A weathering rate calculation model and the generalized First-order Acidity Balance (FAB) model applied to alpine lake catchments. (Environmental Studies 0709).* Berne: Federal Office for the Environment.

Posch, M., Aherne, J., Forsius, M., & Rask, M. (2012). Past, present, and future exceedance of critical loads of acidity for surface waters in Finland. *Environmental Science & Technology, 46,* 4507–4514.

Rapp, L., & Bishop, K. (2009). Surface water acidification and critical loads: Exploring the F-factor. *Hydrology and Earth System Sciences, 13,* 2191–2201.

Rapp, L., Wilander, A., & Bertills, U. (2002). Kritisk belastning för försurning av sjöar. In *Kritisk belastning för svavel och kväve (In Swedish)* (pp. 81–106). Report 5174. Swedish Environmental Protection Agency.

Redsven, V., Anola-Pukkila, A., Haara, A., Hirvelä, H., Härkönen, K., Kettunen, L., Kiiskinen, A., Kärkkäinen, L., Lempinen, R., Muinonen, E., Nuutinen, T., Salminen, O., & Siitonen, M. (2004). *MELA2002 reference manual* (2nd ed.). Helsinki: Finnish Forest Research Institute.

Schöpp, W., Posch, M., Mylona, S., & Johansson, M. (2003). Long-term development of acid deposition (1880–2030) in sensitive freshwater regions in Europe. *Hydrology and Earth System Sciences, 7,* 436–446.

Seip, H. M. (1980). Acidification of freshwater—sources and mechanisms. In D. Drabløs & A. Tollan (Eds.), *Ecological impact of acid precipitation* (pp. 358–365). Oslo: Proceedings of an international conference (Sandefjord, Norway, March 11–14, 1980, SNSF Project).

SEPA. (2010). *Status, potential and quality requirements for lakes, watercourses, coastal and transitional waters* (Vol. 4) (Handbook 2007). Stockholm: Swedish Environmental Protection Agency.

Simpson, D., Benedictow, A., Berge, H., Bergström, R., Emberson, L. D., Fagerli, H., Flechard, C. R., Hayman, G. D., Gauss, M., Jonson, J. E., Jenkin, M. E., Nyiri, A., Richter, C., Semeena, V. S., Tsyro, S., Tuovinen, J.-P., Valdebenito, A., & Wind, P. (2012). The EMEP MSC-W chemical transport model—technical description. *Atmospheric Chemistry and Physics, 12,* 7825–7865.

Skeffington, R. A., & Wilson, E. J. (1988). Excess nitrogen deposition. *Environmental Pollution, 54,* 159–184.

Skjelkvåle, B. L., Henriksen, A., Faafeng, B., Fjeld, E., Traaen, T. S., Lien, L., Lydersen, E., & Buan, A. K. (1996). Regional innsjøundersøkelse 1995. En vannkjemisk undersøkelse av 1500 norske innsjøer. (Report 677/96). Oslo: Statens Forurensningstilsyn.

Slootweg, J., Posch, M., & Hettelingh, J.-P. (2010). Progress in the modelling of critical thresholds and dynamic modelling, including impacts on vegetation in Europe: CCE Status Report 2010. (RIVM Report No. 680359001). The Netherlands: Coordination Centre for Effects.

Stoddard, J. L. (1994). Long-term changes in watershed retention of nitrogen: Its causes and aquatic consequences. In L. A. Baker (Ed.). *Environmental chemistry of lakes and reservoirs* (pp. 223–284). Washington, DC: American Chemical Society.

Sullivan, T. J., Eilers, J. M., Cosby, B. J., & Vaché, K. B. (1997). Increasing role of nitrogen in the acidification of surface waters in the Adirondack mountains, New York. *Water Air & Soil Pollution, 95,* 313–336.

Tarrasón, L., Jonson, J. E., Fagerli, H., Benedictow, A., Wind, P., Simpson, D., & Klein, H. (2003). Transboundary acidification and eutrophication and ground level ozone in Europe Part III: Source-receptor relationships. (EMEP Status Report 1/03). Oslo: Norwegian Meteorological Institute.

Chapter 18
National-Scale Dynamic Model Applications for Nordic Lake Catchments

Martin Forsius, Filip Moldan, Thorjørn Larssen, Maximilian Posch, Julian Aherne, Espen Lund, Richard F. Wright and B. Jack Cosby

18.1 Introduction

The critical load (CL) approach is static; i.e. there is no consideration of ecosystem processes that require time to respond to changes in acidic deposition, climate and land management. Within terrestrial catchments and surface waters, these time lags reflect slow changes to the (often large) element pools, as well as lags associated with the perturbation of fish and invertebrate populations. In contrast, dynamic hydro-geochemical models accommodate finite pools and can be used to analyse the temporal response of ecosystems to changes in sulphur (S) and nitrogen (N) deposition, climate and forestry measures, such as the increased use of forest har-

M. Forsius (✉)
Finnish Environment Institute (SYKE), Helsinki, Finland
e-mail: martin.forsius@ymparisto.fi

F. Moldan
IVL Swedish Environmental Research Institute Ltd., Gothenburg, Sweden

T. Larssen · E. Lund · R. F. Wright
Norwegian Institute for Water Research (NIVA), Oslo, Norway

M. Posch
Coordination Centre for Effects (CCE),
RIVM, Bilthoven, The Netherlands

J. Aherne
Environmental and Resource Studies,
Trent University, Peterborough, ON, Canada

B. J. Cosby
Centre for Ecology and Hydrology,
Environment Centre Wales, Bangor, Gwynedd, United Kingdom

Department of Environmental Sciences,
University of Virginia, Charlottesville, VA, USA

W. de Vries et al. (eds.), *Critical Loads and Dynamic Risk Assessments*,
Environmental Pollution 25, DOI 10.1007/978-94-017-9508-1_18

vesting residues (e.g. Aherne et al. 2012; Larssen et al. 2010; Moldan et al. 2004; Posch et al. 2008).

European emissions of S and N oxides, and ammonia have declined by 67, 24 and 20 %, respectively, between the years 1980 and 2000 (EMEP 2004). The protocols to the Convention on Long-range Transboundary Air Pollution (LRTAP) of the United Nations Economic Commission for Europe (UNECE) and legislation of the European Union have been the key international instruments driving this development. Reductions in acidifying depositions (primarily driven by reductions in S deposition) have resulted in large-scale recovery of lakes and streams from acidification (e.g. Forsius et al. 2003; Skjelkvåle et al. 2005; Stoddard et al. 1999).

Two factors that can potentially influence the extent of chemical recovery from acidification are climate change and the increased use of forest harvesting residues for biofuel production (Aherne et al. 2008b, 2012; Wright et al. 2006). In general, modelling studies related to acidification assume a constant influence of climate. However, there is a high probability that climate will change during the coming decades, having a confounding effect on chemical recovery from acidification. Changes in temperature, precipitation and storminess can affect surface water chemistry directly, as well as indirectly through changes in vegetation and soil processes in the terrestrial catchment. Higher temperature will promote forest growth, with faster extraction of nutrients from the soil and may result in increased decomposition of soil organic matter with release of the associated N and base cations, as demonstrated in the CLIMEX experiment in Norway (Van Breemen et al. 1998). Changes in precipitation are also expected to change the concentrations of solutes leached from soils.

The confounding influence of increasing dissolved organic carbon (DOC) concentrations in surface waters has also received considerable attention (e.g., Evans et al. 2006; Vuorenmaa et al. 2006). Increasing DOC concentrations have been observed across Europe and eastern North America during the past 10–20 years (Monteith et al. 2007; Stoddard et al. 2003). It has been hypothesised that factors related to both climate change and changes in S deposition may be (partly) responsible for this increase (Evans et al. 2005; Freeman et al. 2001; Holmberg et al. 2006; Monteith et al. 2007). As such, climatic effects (and climate variability) on biogeochemical cycles should be considered when predicting the recovery of acidified soils and surface waters (Aherne et al. 2006; Forsius et al. 1997; Park et al. 2010; Wright et al. 2006).

During the past decade, biofuels have increasingly been seen as a means of contributing to the mitigation of climate change. The European Commission has presented ambitious targets for increasing the proportion of renewable energy in the total energy consumption, although many biofuel production methods compete with food production for land resources. Furthermore, some methods of biofuel production are seen as conflicting with environmental protection goals (termed 'pollutant swapping', Stevens and Quinton 2009). As a consequence, a worldwide debate on the merits and trade-offs of biofuel production is underway (see, e.g. Tilman et al. 2009).

In Finland and Sweden, with their large forest resources and highly developed forest industry, there is increased interest in forest harvesting residue removal for biofuel production. According to the long-term climate and energy strategy of the Finnish government, the target is to increase the use of renewable energy to 38 % of the energy consumption by 2020 and to 60 % by 2050 (Climate and Energy Strategy 2008). Among the agreed measures is the increased use of forest residues from 4 million m^3 in 2008 to 12 million m^3 in 2020. The Foresight Study of climate and energy policy of the government includes even more ambitious targets (Foresight Report 2009). In Sweden environmental goals are by definition equally important as production goals for setting forestry policy. Since about 1990 there has been a large increase in the harvesting of forest residues from thinning and at final harvest for energy production. There has also been increased interest to fertilize forests, especially in the northern part of the country. The demand for biomass is in part driven by the governmental goal to supply 50 % of total energy consumption from renewable resources by the year 2020. Forest harvesting influences soil and lake water quality by removing base cations (and N) from catchment soils. Long-term sustainability of forest ecosystems require that these losses are balanced by the deposition and weathering of base cations (e.g. Akselsson et al. 2007; Joki-Heiskala et al. 2003; Sverdrup and Rosén 1998) and possibly fertilization. Incentives are being put in place to limit the environmental impacts of intensified forestry such as subsidizing ash treatments from biomass burning to stimulate ash recycling in the forest.

Here we summarise the results of comprehensive national-scale applications of the Model of Acidification of Groundwater In Catchments (MAGIC) to lake catchments in Norway, Sweden and Finland under scenarios assessing the impacts of changes in atmospheric deposition, climate and forest harvesting practices.

18.2 Material and Methods

18.2.1 MAGIC Model

MAGIC (version 777) is a dynamic lumped-parameter model of intermediate complexity, developed to predict the long-term effects of acidic deposition on soil and surface water chemistry (Cosby et al. 1985a, b, 2001).

The data required to run MAGIC must be spatially and temporally aggregated ('lumped') to represent the whole catchment area and the time step (monthly or annual) of the model application. Soil physiochemical parameters are aggregated vertically by soil profile and spatially to account for different soil types within a catchment. Further details on model inputs and parameters are given in Cosby et al. (2001).

The model has been extensively applied and tested during the past three decades at many sites and in many regions around the world (Cosby et al. 2001). Further, MAGIC was used to assess the recovery of European surface waters from acidification under a deposition scenario based on full implementation of current legislated S and N emissions (CLE scenario), i.e. the Gothenburg protocol and other agreed legislation (Wright et al. 2005). Overall, the model has proven to be robust, reliable and useful in a variety of scientific and environmental management activities.

18.2.2 Norwegian Study

The Norwegian study is described in detail in Larssen et al. (2010). The lake chemistry and site data were taken from 1003 statistically-selected lakes of the 1995 Norwegian national survey (Skjelkvåle et al. 1996). The lakes were sampled in autumn 1995, and analysed for major ions. Lake and catchment characteristics were obtained from topographic maps and other national databases.

Soil data for forested areas were obtained from the NIJOS (Norwegian Institute for Soil and Forest Inventory, now part of The Norwegian Forest and Landscape Institute) national forest inventory on a 9 km × 9 km grid basis. These were aggregated (arithmetic averages weighted by soil mass) to a 12 km × 12 km critical load grid. Of the 1003 lakes, 345 were located in grid cells where forest was present). Soil data for non-forested areas (n=658; mainly upland, mountains and heathlands) came from various research projects (details given by Larssen et al. 2008). Annual net accumulation (and harvest removal) of base cations in forest biomass was assumed to be Ca^{2+} 21 meq $m^{-2}yr^{-1}$, Mg^{2+}4 meq $m^{-2}yr^{-1}$, and K^{+}4.5 meq $m^{-2}yr^{-1}$ for forests and zero for non-forested areas. Forest management was assumed not to change over time. These values come from estimated annual average removal of cations in harvested timber (data from the national forest inventory) and average base cation concentrations in tree boles for the two major species, Norway spruce (*Picea abies L.*) and Scots pine (*Picea abies L.*) (Sverdrup et al. 1990).

Deposition during the calibration year 1995 was estimated for each lake catchment from the measured lake chemistry and modelled deposition estimates for the EMEP50 (50 km × 50 km) grid, provided by the Coordination Centre for Effects (CCE; Hettelingh et al. 2008). The modelled deposition estimates were not used directly because they are not specific to each lake catchment. Factors such as orographic effects cause deposition fluxes to vary locally. Details are given by Larssen et al. (2010). Deposition sequences for S, NO_x, and NH_y during the 'historical' period 1880–2010 were provided by the CCE (Hettelingh et al. 2008). The historical deposition sequence were normalised to the deposition values calculated for the year 1995.

MAGIC was calibrated to each lake using an automatic optimiser that iteratively adjusted the values for a series of parameters such that the simulated lake chemistry for the calibration year 1995 matched the observed to within ±2 μeq l^{-1} and the simulated percent soil base saturation matched the observed for each base cation to within ±0.2% points. Acid neutralising capacity (ANC) was used to quantify lake acidification status; ANC is defined as the equivalent sum of base cations (Ca^{2+}, Mg^{2+}, Na^{+}, K^{+}) minus the equivalent sum of strong acid anions (SO_4^{2-}, NO_3^{-}, Cl^{-}).

A total of 990 lakes were successfully calibrated. For the calibrated sites, model simulations were used to generate data for four evaluation years: 1980, 1990, 2000, and 2010.

Two future scenarios of S and N deposition were used, the CLE scenario and the maximum technically feasible reduction (MFR) scenario. Both were assumed to be fully implemented by the year 2020 (linear change from 2010 to 2020). Lake ANC was estimated by running MAGIC for the calibrated lakes under the future deposition scenarios and results were evaluated for the years 2020, 2030, 2050 and 2100. Net percentage retention of N input in the lake catchment was set to the 1995 value and assumed to be constant throughout the entire modelling period. This assumption was justified on the basis of 30-year records of deposition and lake chemistry in Norway, which show no systematic trend over time in net percentage retention of N (SFT 2009).

Norway has submitted CL estimates to the CCE for each of the 2304 squares of a $0.125° \times 0.250°$ latitude-longitude grid (approximately 12 km × 12 km) covering the entire country (Henriksen and Buan 2000). The results for dynamic modelling of the 990 lakes were extrapolated to the 2304 squares in the CL grid using an analogue matching procedure in the 'MAGIC library' held at IVL Swedish Environment Institute (IVL 2007). This procedure takes the available water chemistry data for each of the 2304 grid squares and finds the closest match in the 990 lakes calibrated by MAGIC.

Brown trout population status was selected as the biological indicator of adverse effects of acidification. A 1986 survey of lake water chemistry and brown trout populations in 827 lakes in Norway showed a clear relationship between ANC and population status (Lien et al. 1996). Damaged populations were defined in the survey as those described as "sparse" or "extinct". At ANC <20 µeq l^{-1} the probability of damaged population exceeded 5 %. This was defined as the ANC_{limit} which then was used to calculate CL (Henriksen and Buan 2000).

Lydersen et al. (2004) modified the ANC-brown trout status relationship to take into account the acidifying influence of natural organic acids. They defined a modified ANC that incorporated strong organic acids, $ANC_{oaa} = ANC - (10.2/3) \cdot DOC$, where DOC is the concentration of dissolved organic carbon (mg C l^{-1}), and ANC is in µeq l^{-1}. Larssen et al. (2006) calculated a logistic regression relationship between ANC_{oaa} (organic acid adjusted ANC) and fish health, and obtained an $ANC_{ooa\ limit}$ of 8 µeq l^{-1} for brown trout. For ANC_{ooa} above this level the probability of a healthy population is >99.5 %.

18.2.3 Swedish Study

The Swedish study is described in detail in Moldan et al. (2009, 2013). The 1218 Swedish lakes were sampled in 1995, 2000 and 2005 as part of a national survey of lakes (Riksinventeringen). Description of the monitoring program and lake chemistry data are available from the Swedish University of Agricultural Sciences

(SLU; www.ma.slu.se). Lake annual runoff and catchment land use characteristics were obtained from SLU (Jens Fölster pers. comm.).

Forest soil data came from the Swedish Survey of Forest Soils and Vegetation (SK) 1983–1987, which is a regularly-repeated survey of forest vegetation and soil chemical and physical qualities, at the permanent plots of the Swedish National Forest Inventory (NFI). Soil data were lumped (arithmetic averages weighted by soil mass) for each lake catchment based on the average soil properties from the nearest 1–20 soil plots. For lake catchments on non-forested land, soil data were assigned from various research projects, including Svarén (1996).

Base cation uptake (annual net accumulation in harvested biomass) was assumed to be zero for non-forested areas. For forests we used region and forest type specific uptake rates as calculated by the ASTA program (Akselsson et al. 2007), weighted by the percent forest cover in each catchment.

Deposition sequences of S, NO_x, and NH_y for the period 1880–2010 were provided by the CCE for each EMEP50 grid square. These historical deposition values were normalised to the deposition calculated for the year 2005 as follows. First deposition of Cl^- was assumed to equal the output flux at each lake (based on concentration in lake water and discharge). The depositions of Na^+, Mg^{2+}, $Ca^{2+}$and K^+ were calculated from the deposition of Cl^- and the ionic ratios of these ions in seawater. Deposition of base cations other than from sea salt was estimated together with weathering rate during optimisation procedure when calibrating the model. Secondly SO_4^{2-} deposition was calculated; the sources for SO_4^{2-} in lake water were assumed to be (1) weathering of soil minerals or release from lake sediment, (2) deposition of seasalts, and (3) anthropogenic SO_4^{2-} deposition. Seasalt SO_4^{2-} was assumed to equal 0.103 Cl^- (ratio of these ions in seawater). Sulphate from weathering was assumed to be zero if excess SO_4^{2-} lake flux (i.e., lake SO_4^{2-} concentration multiplied by discharge minus the seasalt SO_4^{2-} flux) was less than twice the EMEP sulphate deposition (in 2005). Where excess SO_4^{2-} deposition was greater, the difference was assumed to be weathering. Finally, the deposition of NO_x and NH_y was estimated from anthropogenic SO_4^{2-} deposition (above) using the ionic ratios (NO_x/SO_4^{2-}, NH_y/SO_4^{2-}) in the EMEP50 modelled deposition data.

Model simulations were carried out for four future forest management\and climate scenarios during the period 2010–2100. All four scenarios were based on the CLE deposition scenario for S and N deposition. The CLE scenario indicated only slight decreases in future S and N deposition from levels in 2010. In addition a fifth scenario was run with a lower S and N deposition scenario (MFR) in the future.

The initial dataset comprised 1218 lakes over all of Sweden. Of these, 1151 were previously successfully calibrated using the automatic optimiser routine with N dynamics turned off, and included in the MAGIC library (www.ivl.se). A total of 858 lakes were calibrated successfully with the N dynamics turned on. These lakes fell into three categories: those with high soil C/N retain 100 % of incoming N, those with low soil C/N retain 0 % of incoming N, and those with intermediate C/N retain a varying amount of incoming N. Only the 348 lakes that fell into the last category were included in the scenario study.

A matrix of future forest management and climate scenarios was designed to cover four hypothetical future forest management and climate change combinations which spanned from present-day 'business as usual' (no change in either forest management and climate change) to 'worst case' (most extensive biomass harvesting forestry practice in combination with largest climate change). The matrix was defined as follows:

- Scenario base (no change in future climate, no change in future forest management);
- Scenario LU (no change in future climate, maximum utilisation of the forest);
- Scenario CC (maximum change in future climate, no change in future forest management);
- Scenario CCLU (maximum change in future climate, maximum utilisation of the forest).

18.2.4 Finnish Study

Model simulations were used to predict the response of 163 Finnish lake catchments to historic and future atmospheric deposition (1880–2100), climate and future tree harvesting practices (Aherne et al. 2008b; Posch et al. 2008). Deposition was assumed to follow CLE and a scenario based on MFR. Future harvesting was assumed to shift from stem-only harvesting (SOH) to whole-tree harvesting (WTH) owing to the potential increased utilisation of biomass for biofuels. Future climate (temperature and precipitation) was derived from the HadAM3 and ECHAM4/OPYC3 general circulation models under two global scenarios of the Intergovernmental Panel on Climate Change (IPCC: A2 and B2). Future scenarios for runoff were obtained from the Finnish watershed simulation and forecasting system (Vehviläinen and Huttunen 2002). Finally, the potential influence of future changes in surface water DOC concentrations on lake acid status was also explored using simple empirical models based on the relationship between observed changes in lake DOC concentrations and (a) S deposition or (b) temperature between 1990 and 2000. For each lake, change coefficients were estimated, and these coefficients were then used in two simple site-specific linear models to 'forecast' DOC under future S deposition and climate scenarios (Posch et al. 2008).

In an extended study, the environmental impact of different levels of forest residue removal under a likely future climate (IPCC A2 scenario) was evaluated at 1066 Finnish lake catchments until 2050 (Aherne et al. 2012). Forest biomass removal scenarios were derived from a management-oriented large-scale forestry model (MELA), and based on sample plot and tree data from national forest inventories (Kärkkäinen et al. 2008). The use of a large random sample of catchments covering the whole of Finland allowed for a regionally representative assessment.

Sustainability under forest harvesting scenarios and the future state of soil and surface water chemistry was evaluated using two indicators: ANC of lake water and base saturation of the catchment soils, which are key indicators for ecological impacts used in European-scale assessment work on air pollution (Hettelingh et al. 2007). Lake ANC is the most widely used chemical criterion in critical load calculations for surface waters (Henriksen and Posch 2001). Base saturation, on the other hand, is a key indicator of soil nutrient status or long-term sustainability.

18.3 Results and Discussion

18.3.1 Norwegian Study

As a result of the implementation of S emission reductions in Europe, deposition of anthropogenically-derived S in Norway has decreased by nearly 80% from peak levels around 1980 to levels that are very close to those expected for the year 2010 with full implementation of the Gothenburg protocol (Aas et al. 2008). Emissions and deposition of N, on the other hand, have declined by only about 20%. Although the role of N in freshwater acidification has increased relative to that of S, the long-term data show no systematic change in the fraction of incoming N retained in catchment soils and the lakes themselves (SFT 2009).

The highest levels of S deposition are in southernmost Norway, and it is here that the most severe acidification of freshwaters has occurred. Nationwide surveys of lake water chemistry and fish populations during the 1980s indicated that acidic deposition exceeded the critical load in about 30% of the country (Henriksen et al. 1988; Henriksen et al. 1989). The decline in S deposition since 1980 has led to gradual improvements in lake water chemistry (Skjelkvåle et al. 1998).

The MAGIC simulations suggested that substantial acidification of lakes will remain in southernmost Norway in the year 2010 despite full implementation of the CLE scenario (Fig. 18.1). Water quality in about 20% of the country will still be insufficient to support self-reproducing brown trout populations.

Brown trout is the most widely-used biological indicator of acidification status in Norwegian waters as this species is acid-sensitive, is the major inland sports fish species, and is widely distributed throughout Norway. Other fish species and organism groups are also acid sensitive and show adverse effects. Atlantic salmon (*Salmo trutta L.*) populations have been severely impacted by acidification in southern Norway; salmon is extinct in seven major salmon rivers (Overrein et al. 1980). Acidification also affects most other organism groups, and among indicator species are *Daphnia* species in the zooplankton (Havens et al. 1993) and the mayfly *Bætis rhodani* in the zoobenthos (Raddum and Skjelkvåle 1995). These species all have similar requirements for water quality with respect to acidification parameters such as ANC, pH and dissolved inorganic aluminium. Recovery of brown trout populations will thus generally mean that conditions should be satisfactory for other indicator organisms. A case study of recovery from acidification at Lake Saudlandsvatn, southernmost Norway, showed that brown trout, *Daphnia* and *Bætis* all have responded to increased pH and ANC during the last 10 years (Hesthagen et al. 2011).

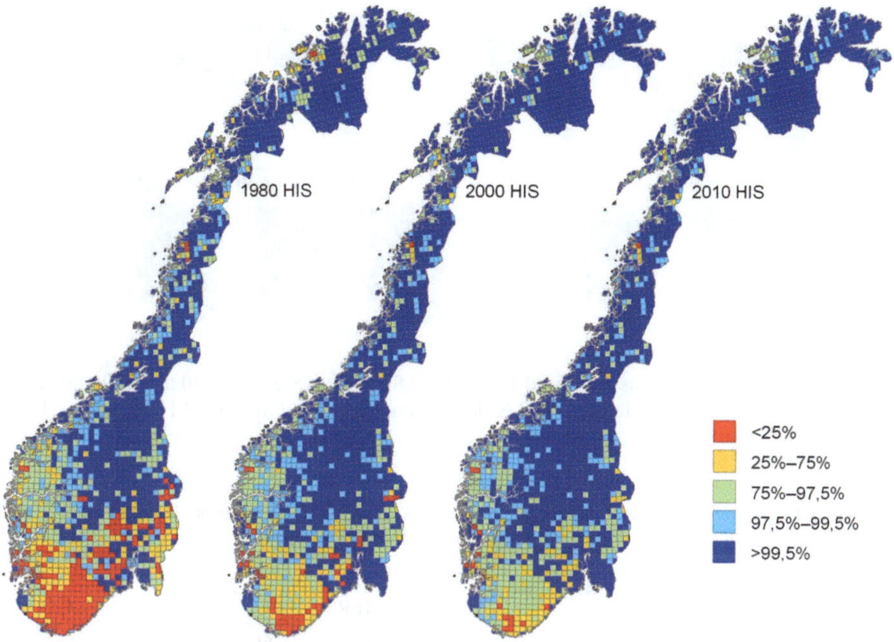

Fig. 18.1 Maps of ANCooa as simulated by MAGIC for the years 1980, 2000, and 2010. The classes represent ANCooa levels for which water quality is suitable for brown trout populations. *Red*: >75 % probability of damaged population (ANC_{ooa} < −36.7 μeq l^{-1}); *orange*: 25–75 % probability of damaged population (−36.7 to −19.8 μeq l^{-1}); *light green*: 2.5–25 % probability of damaged population (−19.8–0 μeq l^{-1}); *light blue*: 0.5–2.5 % probability of damaged population (0 to 12 μeq l^{-1}); *blue*: <0.5 % probability of damaged population (>12 μeq l^{-1}). (from Larssen et al. 2010)

Biological recovery, however, is fraught with various 'bottlenecks' such as migration rates and vulnerability to short-term acid episodes (Yan et al. 2003).

The MAGIC simulations indicated that there will be further improvements in lake water chemistry in the future (Larssen et al. 2010). Under the CLE scenario ANC_{ooa} is expected to increase modestly to the year 2020 and then very slowly throughout the rest of the century, with about 18 % of the country still affected in the year 2100 (Fig. 18.2). Under the more aggressive MFR scenario, greater improvements in ANC_{ooa} are expected, and in the year 2100 about 12 % will still be affected.

18.3.2 Swedish Study

By year 2010 almost all the expected recovery from acidification due to air pollution controls will have occurred (Table 18.1). In 1990 the average ANC was 410 μeq l^{-1}, i.e. 108 μeq l^{-1} below the modelled 1860 pre-industrial value of 518 μeq l^{-1}. By the year 2010, 20 years after 1990, the average ANC increased by 31–441 μeq l^{-1}. A

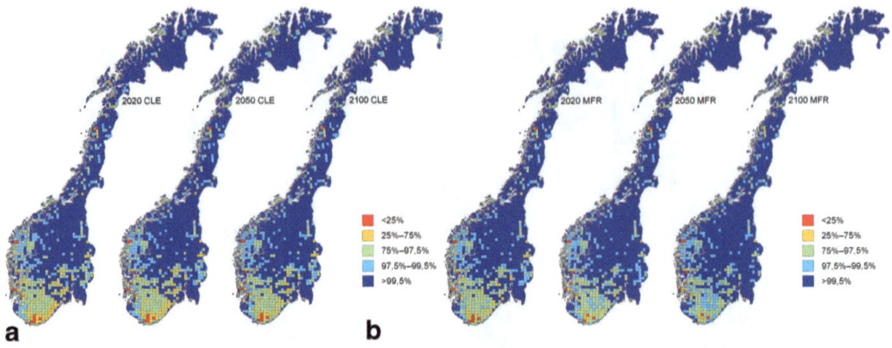

Fig. 18.2 Maps of ANC_{ooa} as simulated by MAGIC for the years 2020, 2050 and 2100 given the CLE and MRF scenarios of future S and N deposition. See legend to Fig. 18.1 for details. (From Larssen et al. 2010)

Table 18.1 Average lake water chemistry for the 348 lakes under the all modelled scenarios (from Moldan et al. 2009); $SO_4^{2-}*$ denotes the non-marine fraction of sulphate

	1860	1980	2010	2100				
$\mu eq\ l^{-1}$	All scenarios			Base	MFR	CC	LU	CCLU
$SO_4^{2-}*$	40	230	114	71	54	66	71	66
ANC	518	412	441	460	475	430	435	377

further average increase of 19 $\mu eq\ l^{-1}$ during the next 90 years was simulated under the base scenario (CLE deposition), and even more, 34 $\mu eq\ l^{-1}$, under the MFR scenario. ANC is thus expected to end up at best 43 $\mu eq\ l^{-1}$ and at worst 58 $\mu eq\ l^{-1}$ below the pre-industrial level (Table 18.1).

Under all the other future scenarios, however, the results suggest that recovery from acidification will stop and the majority of the lakes will decline in ANC during the next 90 years. Much of this decline is apparently due to the increased flux of N from catchment soils to surface waters.

The modelled lakes represent many different types of lakes distributed over the whole 450,000 km^2 area of Sweden. Therefore the variability in lake water chemistry is also considerable. In the perspective of this large variability, the changes in lake water chemistry as a whole were modest under all the future scenarios with the exception of marked increases in NO_3^- concentrations for the CC, LU and CCLU scenarios. The mean differences between years 2010 and 2100, however, were considerable. The CCLU scenario resulted in large decrease in ANC in lake water driven largely by the increase in NO_3^- at practically all lakes (Table 18.2). In this respect the results of the combined scenario CCLU was greater than the simple sum of the CC and LU scenarios due to non-linearity in the key modelled ecosystem processes.

Table 18.2 provides a comparison of modelled median changes. The response of individual lakes was, however, variable in magnitude and even in direction of change. This is to be expected given the variability of modelled lakes, variability

Table 18.2 Lake water pH and ANC, concentrations of NO_3^- and of the sum of base cations (SBC) in year 2010, and mean change from year 2010 to year 2100 for the four future scenarios. SO_4^{2-*} denotes the non-marine fraction of sulphate. Change calculated as median of 348 individual differences from 2010 to 2100. (From Moldan et al. 2009)

		Median change 2100–2010				
	Median 2010	Base	MFR	CC	LU	CCLU
pH	7.1	0.0	0.0	0.0	0.0	−0.1
ANC μeq l^{-1}	247	2	9	−11	1	−39
SO_4^{2-*} μeq l^{-1}	38	−10	−22	−15	−10	−15
NO_3^- μeq l^{-1}	1	0	−1	11	50	86
SBC μeq l^{-1}	355	7	−6	−16	25	−2

of soils and past and present acidification and land use in lake catchments, and climatic and deposition gradients across the country.

In Fig. 18.3 the changes in concentrations in years 2030 and 2100 relative to year 2010 are shown for pH, ANC, NO_3^- and SBC for the 348 lakes. For pH the results indicate that for most of the lakes only small changes are expected in the future except under the combined CCLU scenario. ANC is projected to increase under the MFR scenario, but decrease under the CC, LU and CCLU scenarios; NO_3^- is projected to increase as is SBC.

The results show in general that with MFR there will be a small amount of recovery in the future. Further under the CC and LU scenarios (and especially the CCLU scenario) there will be increased NO_3^- leaching from the soils which will offset to a lesser or greater degree the recovery from acidification. In acid-sensitive lakes re-acidification can be expected, whereas in well-buffered lakes very little change in lake chemistry (except for NO_3^-) is predicted to occur.

The scenarios LU and CC both would have an impact on soil pools of C and N. In the model, soil organic carbon was assumed to be lost as CO_2 to the atmosphere due to increased decomposition of organic matter. Nitrogen was lost from soil in two ways; via leaching of nitrate to runoff and via uptake to the trees and subsequent harvest. On average, during the next 90 years LU alone would cause a 17 % loss of soil organic C, while CC alone would cause loss of 8 %. The combination of maximum biomass harvest and climate change (CCLU) would lead to a 25 % loss of soil organic C. Similarly for N, the LU alone would cause loss of 7 % and CC alone 1 %, while the CCLU scenario would cause the largest loss of 12 %. Under all scenarios, the soil organic matter would become more N rich. The soil C/N ratio would decrease on average by 0.4 and 0.1 under base and MFR scenarios and by 1.5, 2.4 and 3.2 in the CC, LU and combined CCLU scenarios, respectively (Fig. 18.4).

On average the soil pool of exchangeable base cations (BC) was little changed after 90 years in the base scenario (Fig. 18.5). In fact, the average pool size actually increased by about 1 and 4 % in the base and MFR scenarios. That means that on average the sum of losses of BC due to present-day tree harvest practices (stems plus 30 % of branches and foliage) and leaching of BC to runoff due to the S and N deposition under the CLE scenario were largely in balance with the sum of

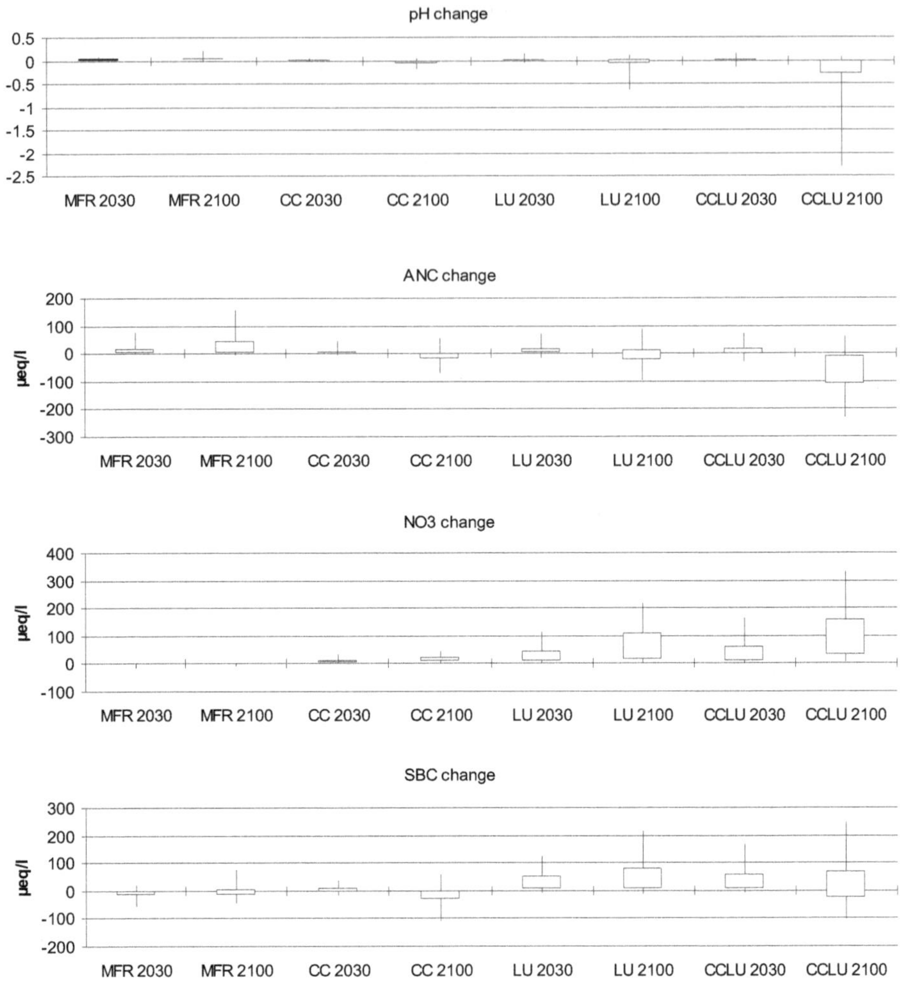

Fig. 18.3 Mean change in lake water chemistry from 2010 to 2030 and to 2100 for MFR, CC, LU and CCLU scenarios. Boxes enclose 50 % and high-low lines 90 % of 348 modelled lakes. (From Moldan et al. 2009)

BC supply from mineral weathering and deposition of base cations. Under the CC, LU and CCLU scenarios, however, soils were on average depleted of exchangeable BC by 11, 13 and 29 % respectively. The results suggested that depletion of base cations in the soil would be most severe under the CCLU scenario (Fig. 18.5). This future projected loss of soil base cations under the CC, LU and CCLU scenarios did not show a strong south-north gradient. This is because the depletion of base cations under these scenarios is driven primarily by increased biomass removal (LU scenario) and more rapid forest growth (CC scenario) and both these factors would occur over the entire country. The south-north gradient in S and N deposition apparently plays a lesser role in the future under these scenarios.

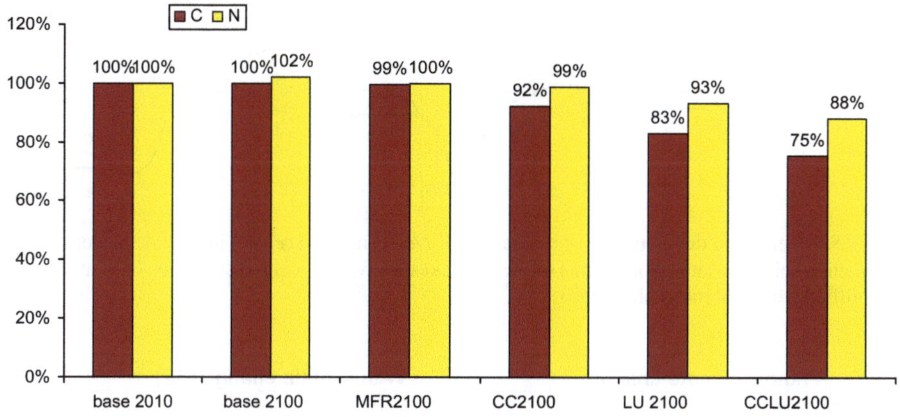

Fig. 18.4 Average change of C and N soil pools in the catchments of the 348 lakes from year 2010 to year 2100 for the four scenarios expressed relative to soil pools in 2010. (From Moldan et al. 2009)

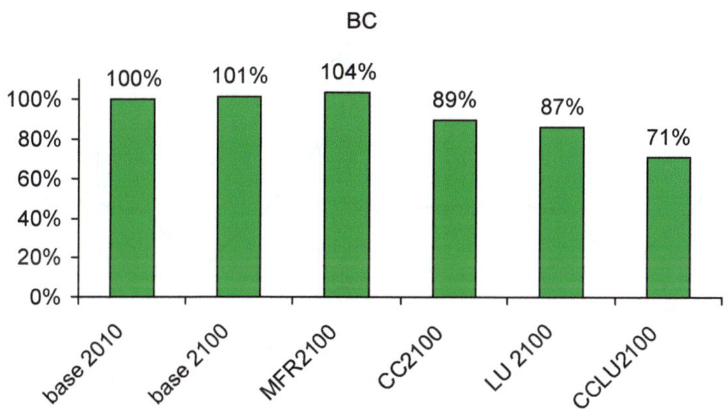

Fig. 18.5 Average pools of exchangeable base cations (Ca^{2+}, Mg^{2+}, Na^{+} and K^{+}) in catchment soils in year 2010 and under the four future scenarios in year 2100. (From Moldan et al. 2009)

18.3.3 Finnish Study

Despite the modest changes in atmospheric deposition under CLE (compared to 2010), these reductions were predicted to halt the decline in soil base saturation; however, further reductions were required to improve soil and lake water chemistry at the 163 catchments (Aherne et al. 2008b). The MFR scenario predicted a significant long-term improvement in soil base saturation leading to continued long-term recovery in surface waters (all lakes with ANC >0 by 2100, Fig. 18.6). However, under the whole-tree harvesting (WTH) scenario, significant long-term impacts (re-acidification) were predicted for soil and surface water chemistry (14 lakes with

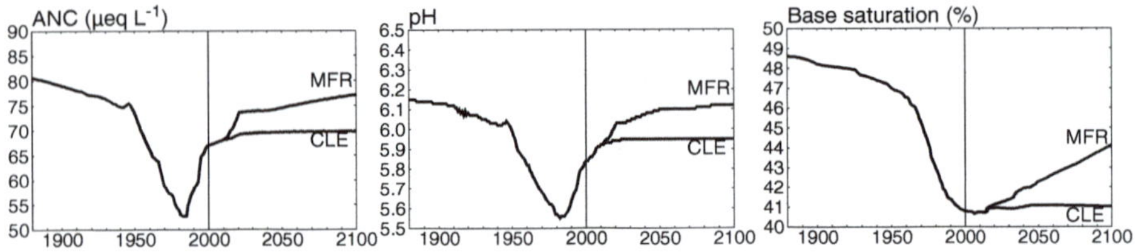

Fig. 18.6 Temporal development of median ANC (*left*) and pH (*centre*) in the lake water as well as median soil base saturation (*right*) of the 163 lakes for two deposition scenarios CLE and MFR. (Modified from Aherne et al. 2008b)

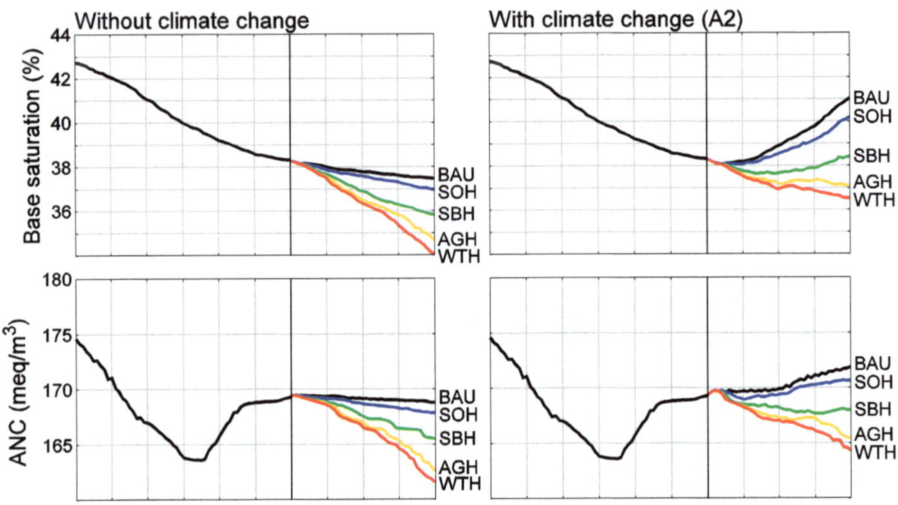

Fig. 18.7 Temporal development (1950–2050) of the median base saturation (*top*) and lake-water ANC (*bottom*) expressed as median of the 1066 catchments, under constant climate (*left*) and the A2 climate scenario (*right*) for five harvesting scenarios: BAU, SOH, SBH, AGH and WTH. (Modified from Aherne et al. 2012)

ANC <0 by 2100). Impacts of WTH or SOH (stem-only harvesting) are not shown in Fig. 18.6, but the impacts of harvesting scenarios are presented in Fig. 18.7 for the larger 1066 study lake data set, illustrating the acidifying impacts. To some extent the long-term negative impacts were reduced under MFR, indicating that increased utilisation of biofuels would necessitate implementation of large emission reductions, or soil amendment, to maintain ecosystem quality and sustainable forest growth. The current practice of stem-only harvesting (SOH) was close to the sustainable maximum harvesting under current (legislated) atmospheric deposition in Finland.

The direct influence of climate change (temperate and precipitation) on recovery was negligible, as runoff hardly changed; greater precipitation is offset by increased evapotranspiration due to higher temperatures. However, two exploratory empirical DOC models indicated that changes in S deposition or temperature could have a confounding influence on the chemical recovery of surface waters from acidification, and that the corresponding increases in DOC concentrations may offset the recovery in pH under reductions in acidifying depositions (see Posch et al. 2008).

Using the extended dataset (1066 lake catchments), 'business-as-usual' (BAU), which assumed no change in policies and climate, and four harvesting scenarios (SOH, SBH: stems plus branches harvesting, AGH: above-ground harvesting and WTH) were evaluated, each with very different effects on biomass and element removal as well as soil and water quality in Finnish forest ecosystems (Aherne et al. 2012). Consistent with the results of the earlier study using the smaller 163 catchment dataset (Aherne et al. 2008b), increased element removal in harvested biomass caused a concurrent decline in environmental quality indicators (Fig. 18.7). Only harvesting of above-ground woody biomass (SOH and SBH management scenarios) were predicted to be sustainable if increased removal was not compensated by management measures. The most intensive scenario, WTH (including the removal of stumps and roots), was predicted to double the removal of biomass, while the removal of base cations from the catchment soils would triple and the N removal would increase fourfold. Climate change was predicted to have a positive impact on future conditions by increasing the future supply of base cations from soil mineral weathering (in excess of increased uptake), partly compensating their removal due to biomass harvesting (Fig. 18.7). Additional inputs of N would however be required to ensure sustained forest growth. Nonetheless, even when considering this likely future 'benefit' of climate change, the intensive AGH and WTH scenarios resulted in a decrease in soil and water quality (Fig. 18.7).

Sustainability in terms of soil conditions can be defined as a situation where no long-term reduction in the Bc pool (base saturation) is allowed. Under present climate the majority of the study catchments showed a decrease in Bc concentration (meq $m^{-3}yr^{-1}$) and in base saturation (% yr^{-1}) at the end of the simulation period (2049–2050; Fig. 18.8). Moreover, the harvesting scenarios predicted an accelerated decline in Bc concentration and base saturation in 2050. In contrast, under the A2 climate, soil base saturation increased in many of the study catchments thus increasing the proportion of catchments where a more intensive management is sustainable (almost a third of the sites for the WTH scenario). Furthermore, under the A2 climate both soil base saturation and Bc leaching increased simultaneously (see points in the first quadrant in Fig. 18.8) due to the increased supply of Bc by weathering owing to higher temperatures (Aherne et al. 2012).

Discussion and Conclusions

The regional-scale modelling studies summarised in this chapter illustrate the dynamic aspect of the acidification and recovery processes. If the policy goal is to achieve healthy biological conditions as a restoration target, then the greater and earlier the reduction in acidic deposition, the faster this goal can be achieved. The

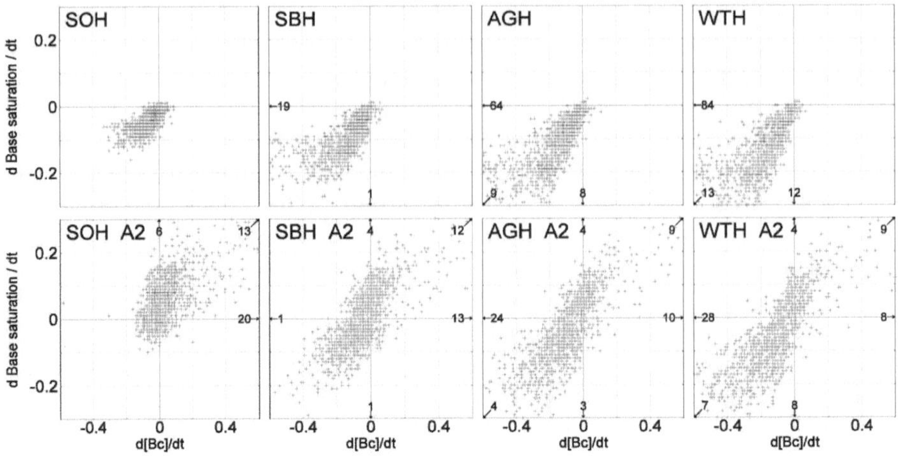

Fig. 18.8 Change from 2049 to 2050 in base cation (Bc = $Ca^{2+} + Mg^{2+} + K^{+}$) lake water concentration (meq $m^{-3}yr^{-1}$) versus change in catchment soil base saturation (% yr^{-1}) for the SOH, SBH, AGH and WTH harvesting scenarios under constant climate (*top*) and the A2 climate change scenario (*bottom*) for each study catchment (n = 1066). (Modified from Aherne et al. 2012)

replenishment of base cations on the soil exchange complex by chemical weathering is a slow process; the lower the sulphur deposition the greater the rate of replenishment. The policy choice is thus to incur the additional costs of further cuts in emissions now to achieve rapid restoration of acidified catchments and lakes, or to do nothing beyond current legislation and wait many decades for recovery of the ecosystems. This is clearly illustrated, e.g., in the Norwegian study, where the situation projected for the CLE scenario for the year 2100 was quite similar to that projected for the MFR scenario for the year 2020 (Fig. 18.2; Larssen et al. 2010). Large reductions may also be required to achieve pre-acidification levels, as illustrated in the Finnish case (see Fig. 18.6; Aherne et al. 2008b). The results for the MFR scenario clearly demonstrate that there is room for further improvement of the chemical conditions of catchments and lakes if emissions are reduced beyond 2010 levels. The MFR scenario represents maximum reduction of S and N emissions using currently-available technology. This is not a fixed parameter, however, because technological improvements are continually being made. Furthermore, future S and N emissions may decrease due to 'co-benefits' of controls on greenhouse gases. This is especially the case for CO_2, where a lower use of fossil fuels may be one of the major steps taken to reduce emissions. Reduced use of fossil fuels may thus give co-benefits in terms of reduced emissions of S and N.

Previous modelling studies on groups of acid-sensitive lakes elsewhere in Europe indicate that in most areas implementation of the CLE scenario will greatly diminish the acidification problem, but acidification with adverse biological effects will continue to occur for some lakes in southern Norway, southern Sweden, the Tatra Mountains of Slovakia, the Italian Alps and parts of the United Kingdom (Jenkins et al. 2003; Wright et al. 2005). MAGIC and other biogeochemical models of acidification have been extensively applied also in North America (see Chap. 19), and used to project the future recovery of acid-sensitive salmon rivers in Nova Scotia (Clair et al. 2004), lakes in eastern Canada and the Adirondack Mountains (Aherne et al. 2003, 2004, 2006, 2008a; Clair et al. 2007; Sullivan et al. 2007), and trout streams in the Appalachian Mountains (Sullivan et al. 2008).

The Finnish and Swedish case studies clearly show the need to consider the long-term impacts of deposition, forest biomass removal and climate change simultaneously (Aherne et al. 2008b; Aherne et al. 2012; Moldan et al. 2009; Posch et al. 2008). The harvesting scenarios resulted in very different levels of biomass and element removal. The intensive management scenarios, where forest harvest residues—such as living and dead branches, foliage, stem tops and off-cuts, stumps and roots—increasingly are removed, tended to increase base cation and N removal much more than the stem-only removal. Therefore the different forest harvesting scenarios assessed clearly had different impacts on future conditions in soils and lakes, with increasing element removal causing a consistent decline in the environmental impact indicators (see, e.g., Figs. 18.5 and 18.7). The most intensive management scenarios were predicted to lead to a significant deterioration in soil and water quality, if not compensated by management measures. Without additional nutrient inputs, only harvesting of above-ground woody biomass would ensure long-term sustainable forests.

Any modelling effort entails a degree of uncertainty, which can arise from many sources. These uncertainties also tend to increase in regional studies, where various simplifying assumptions and data interpolation methods have to be applied due to limited data availability. Sources of uncertainty include: (1) the measured data (catchment characteristics, water chemistry, soil chemistry), (2) estimated data such as historical S and N deposition, (3) the methods used to aggregate up to lake catchments or grid squares, (4) assumptions behind the model, and (5) confounding factors which might operate in the future. The greatest source of uncertainty in the dynamic modelling results are likely associated with these future confounding factors such as climate change, the long-term fate of the large store of N in catchment soils, and changes in concentrations of DOC (Aherne et al. 2004; Moldan et al. 2009; Park et al. 2010; Posch et al. 2008; Wright et al. 2006). In the regional-scale studies presented in this chapter, attempts have been made to consider several of these factors in the simulations, including effect of increased temperature on decomposition of soil organic matter and weathering rates, the effect of increased temperature on forest growth and uptake of nutrients from the soil, and the effect of increased runoff on the leaching of solutes from the soil. The results obtained clearly indicate the importance of these confounding factors on the predictions and the need for enhanced ecosystem process understanding.

Confidence in MAGIC simulations is enhanced by applications at ecosystem-scale manipulation experiments such as the Norwegian RAIN project (Wright et al. 1990) and sites with long-term datasets such as calibrated catchments (Larssen et al. 2004; Larssen 2005) and lakes (Hindar and Wright 2005). Comparison of model results using different simulation models to such unified datasets have also increased the understanding of model uncertainties and concepts (e.g. Forsius et al. 1998; Tominaga et al. 2010). Despite the many remaining uncertainties involved in dynamic model applications, the biogeochemical modelling frameworks presented in this chapter can be used for both analysis of the long-term environmental impacts of deposition and climate change scenarios, and for the development of climate-optimal and environmentally sustainable scenarios for increased use of renewable energy sources from forested ecosystems.

References

Aas, W., Solberg, S., Berg, T., Manø, S., & Yttri, K. E. (2008). *Monitoring of long-range transported air and precipitation—Atmospheric deposition, 2007. Report 1033/2008*. Oslo: State Pollution Control Authority.

Aherne, J., Dillon, P. J., & Cosby, B. J. (2003). Acidification and recovery of aquatic ecosystems in south-central Ontario, Canada: Regional application of the MAGIC model. *Hydrology and Earth System Sciences, 7,* 561–573.

Aherne, J., Larssen, T., Dillon, P. J., & Cosby, B. J. (2004). Effects of climate events on elemental fluxes from forested catchments in Ontario, Canada: Modelling drought-induced redox processes. *Water Air and Soil Pollution: Focus, 4,* 37–48.

Aherne, J., Larssen, T., Cosby, B. J., & Dillon, P. J. (2006). Climate variability and forecasting surface water recovery from acidification: Modelling drought-induced sulphate release from wetlands. *Science of the total Environment, 365,* 186–199.

Aherne, J., Futter, M. N., & Dillon, P. J. (2008a). The impacts of future climate change and sulphur emission reductions on acidification recovery at plastic lake, Ontario. *Hydrology and Earth System Sciences, 12,* 383–392.

Aherne, J., Posch, M., Forsius, M., Vuorenmaa, J., Tamminen, P., Holmberg, M., & Johansson, M. (2008b). Modelling the hydrogeochemistry of acid-sensitive catchments in Finland under atmospheric deposition and biomass harvesting scenarios. *Biogeochemistry, 88,* 233–256.

Aherne, J., Posch, M., Forsius, M., Lehtonen, A., & Härkönen, K. (2012). Impacts of forest biomass removal on soil nutrient status under climate change: A catchment-based modelling study for Finland. *Biogeochemistry, 107,* 471–488.

Akselsson, C., Westling, O., Sverdrup, H., Holmqvist, J., Thelin, G., Uggla, E., & Malm, G. (2007). Impact of harvest intensity on long term base cation budgets in Swedish forest soils. *Water Air and Soil Pollution: Focus, 7,* 201–210.

Clair, T. A., Dennis, I. F., Amiro, P. G., & Cosby, B. J. (2004). Past and future chemistry changes in acidified Nova Scotian Atlantic salmon (Salmo salar) rivers: A dynamic modeling approach. *Canadian Journal of Fisheries And Aquatic Sciences, 61,* 1965–1975.

Clair, T. A., Aherne, J., Dennis, I. F., Gilliss, M., Couture, S., McNicol, D., Weeber, R., Dillon, P. J., Keller, W., Jeffries, D. S., Page, S., Timoffee, K., & Cosby, B. J. (2007). Past and future changes to acidified eastern Canadian lakes: A geochemical modeling approach. *Applied Geochemistry, 22,* 1189–1195.

Climate and Energy Strategy. (2008). Long-term climate and energy strategy. Government report to parliament 6 Nov 2008, Ministry of employment and the economy, Helsinki, p. 130 (In Finnish with English summary). www.tem.fi/?l=en&s=2658. Accessed 24 Nov 2014.

Cosby, B. J., Hornberger, G. M., Galloway, J. N., & Wright, R. F. (1985a). Modeling the effects of acid deposition: Assessment of a lumped parameter model of soil water and streamwater chemistry. *Water Resources Research, 21,* 51–63.

Cosby, B. J., Hornberger, G. M., Galloway, J. N., & Wright, R. F. (1985b). Time scales of catchment acidification. A quantitative model for estimating freshwater acidification. *Environmental Science and Technology, 19,* 1144–1149.

Cosby, B. J., Ferrier, R. C., Jenkins, A., & Wright, R. F. (2001). Modelling the effects of acid deposition: Refinements, adjustments and inclusion of nitrogen dynamics in the MAGIC model. *Hydrology and Earth System Sciences, 5,* 499–517.

EMEP. (2004). EMEP assessment, Part I, European perspective. Oslo, Norway. www.emep.int: Norwegian Meteorological Institute.

Evans, C. D., Monteith, D. T., & Cooper, D. M. (2005). Long-term increases in surface water dissolved organic carbon: Observations, possible causes and environmental impacts. *Environmental Pollution, 137,* 55–71.

Evans, C. D., Chapman, P. J., Clark, J. M., Monteith, D. T., & Cresser, M. S. (2006). Alternative explanations for rising dissolved organic carbon export from organic soils. *Global Change Biology, 12,* 2044–2053.

Foresight Report. (2009). Towards a low-emission society in Finland [in Finnish]. Helsinki: Foresight report on climate and energy policy of the government, Valtioneuvoston kanslian julkaisusarja 28/2009.

Forsius, M., Johansson, M., Posch, M., Holmberg, M., Kämäri, J., Lepistö, A., Roos, J., Syri, S., & Starr, M. (1997). Modelling the effects of climate change, acidic deposition and forest harvesting on the biogeochemistry of a boreal forested catchment in Finland. *Boreal Environmental Research, 2,* 129–143.

Forsius, M., Alveteg, M., Jenkins, A., Johansson, M., Kleemola, S., Lükewille, A., Posch, M., Sverdrup, H., & Walse, C. (1998). MAGIC, SAFE and SMART model applications at Integrated monitoring sites: Effects of emission reduction scenarios. *Water Air and Soil Pollution, 105,* 21–30.

Forsius, M., Vuorenmaa, J., Mannio, J., & Syri, S. (2003). Recovery from acidification of Finnish lakes: regional patterns and relations to emission reduction policy. *Science of the total Environment, 310,* 121–132.

Freeman, C., Evans, C. D., Monteith, D. T., Reynolds, B., & Fenner, N. (2001). Export of organic carbon from peat soils. *Nature, 412,* 785.

Havens, K. E., Yan, N. D., & Keller, W. (1993). Lake acidification—effects on crustacean zooplankton populations. *Environmental Science and Technology, 27,* 1621–1624.

Henriksen, A., & Buan, A. K. (2000). *Critical loads and exceedances of critical loads for surface water, soils and vegetation in Norway (In Norwegian).* NIVA Report 4179-2000. Oslo: Norwegian Institute for Water Research.

Henriksen, A., & Posch, M. (2001). Steady-state models for calculating critical loads of acidity for surface waters. *Water Air and Soil Pollution: Focus, 7,* 375–398.

Henriksen, A., Lien, L., Traaen, T. S., Sevaldrud, I., & Brakke, D. F. (1988). Lake acidification in Norway—present and predicted chemical status. *Ambio, 17,* 259–266.

Henriksen, A., Lien, L., Rosseland, B. O., Traaen, T. S., & Sevaldrud, I. (1989). Lake acidification in Norway—present and predicted fish status. *Ambio, 18,* 314–321.

Hesthagen, T., Fjellheim, A., Schartau, A. K., Wright, R. F., Saksgård, R., & Rosseland, B. O. (2011). Chemical and biological recovery of Lake Saudlandsvatn, a highly acidified lake in southernmost Norway, in response to decreased acid deposition. *Science of the Total Environment, 409,* 2908–2916.

Hettelingh, J.-P., Posch, M., Slootweg, J., Reinds, G. J., Spranger, T., & Tarrasón, L. (2007). Critical loads and dynamic modelling to assess European areas at risk of acidification and eutrophication. *Water Air and Soil Pollution: Focus, 7,* 379–384.

Hettelingh, J.-P., Posch, M., & Slootweg, J. (2008). CCE status report 2008: Critical load, dynamic modelling and impact assessment in Europe. (Report No. 500090003, ISBN No. 978-90-6960-211-0). Bilthoven, The Netherlands: Coordination Centre for Effects.

Hindar, A., & Wright, R. F. (2005). Long-term records and modelling of acidification, recovery, and liming at Lake Hovvatn, Norway. *Canadian Journal Of Fisheries And Aquatic Sciences, 62,* 2620–2631.

Holmberg, M., Forsius, M., Starr, M., & Huttunen, M. (2006). An application of artificial neural net-works to carbon, nitrogen and phosphorus concentrations in three boreal streams and impacts of climate change. *Ecological Modelling, 195,* 51–60.

IVL. (2007). Description of the MAGIC library (In Swedish). http://www.ivl.se. Accessed 24 Nov 2014.

Jenkins, A., Cosby, B. J., Ferrier, R. C., Larssen, T., & Posch, M. (2003). Assessing emission reduction targets with dynamic models: deriving target load functions for use in integrated assessment. *Hydrology and Earth System Sciences, 7,* 609–617.

Joki-Heiskala, P., Johansson, M., Holmberg, M., Mattsson, T., Forsius, M., Kortelainen, P., & Hallin, L. (2003). Long-term base cation balances of forest mineral soils in Finland. *Water Air and Soil Pollution, 150,* 255–273.

Kärkkäinen, L., Matala, J., Härkönen, K., Kellomäki, S., & Nuutinen, T. (2008). Potential recovery of industrial wood and energy wood raw material in different cutting and climate scenarios for Finland. *Biomass Bioenergy, 32,* 934–943.

Larssen, T. (2005). Model prognoses for future acidification recovery of surface waters in Norway using long-term monitoring data. *Environmental Science and Technology, 39,* 7970–7979.

Larssen, T., Cosby, B. J., & Høgåsen, T. (2004). Uncertainties in predictions of surface water acidity using the MAGIC model. *Water Air and Soil Pollution: Focus, 4,* 137.

Larssen, T., Huseby, R. B., Cosby, B. J., Høst, G., Høgåsen, T., & Aldrin, M. (2006). Forecasting acidification effects using a Bayesian calibration and uncertainty propagation approach. *Environmental Science and Technology, 40,* 7841–7847.

Larssen, T., Cosby, B. J., Høgåsen, T., Lund, E., & Wright, R. F. (2008). Dynamic modelling of acidification of Norwegian surface waters. Report 5705-2008. Oslo: Norwegian Institute for Water Research.

Larssen, T., Cosby, B. J., Lund, E., & Wright, R. F. (2010). Modeling future acidification and fish populations in Norwegian surface waters. *Environmental Science and Technology, 44,* 5345–5351.

Lien, L., Raddum, G. G., Fjellheim, A., & Henriksen, A. (1996). A critical limit for acid neutralizing capacity in Norwegian surface waters, based on new analyses of fish and invertebrate responses. *Science of the total Environment, 177,* 173–193.

Lydersen, E., Larssen, T., & Fjeld, E. (2004). The influence of total organic carbon (TOC) on the relationship between acid neutralizing capacity (ANC) and fish status in Norwegian lakes. *Science of the total Environment, 326,* 63–69.

Moldan, F., Kronnäs, V., Wilander, A., Karltun, E., & Cosby, B. J. (2004). Modelling acidification and recovery of Swedish lakes. *Water Air and Soil Pollution: Focus, 4,* 139–160.

Moldan, F., Cosby, B. J., & Wright, R. F. (2009). *Modelling the role of nitrogen in acidification of Swedish lakes: future scenarios of acid deposition, climate change and forestry practices.* (IVL report B1888). Gothenburg: IVL Swedish Environmental Research Institute Ltd.

Moldan, F., Cosby, B. J., & Wright, R. F. (2013). Modeling past and future acidification of Swedish lakes. *Ambio, 42,* 577–586.

Monteith, D. T., Stoddard, J. L., Evans, C. D., de Wit, H. A., Forsius, M., Høgasen, T., Wilander, A., Skjelkvaale, B.-L., Jeffries, D. S., Vuorenmaa, J., Keller, B., Kopacek, J., & Vesely, J. (2007). Dissolved organic carbon trends resulting from changes in atmospheric deposition chemistry. *Nature, 450,* 537–540.

Overrein, L., Seip, H. M., & Tollan, A. (1980). Acid precipitation—Effects on forest and fish. Final report of the SNSF-project 1972–1980. (FR 19-80). Ås, Norway: SNSF project.

Park, J.-H., Duan, L., Kim, B., Mitchell, M. J., & Shibata, H. (2010). Potential effects of climate change and variability on watershed biogeochemical processes and water quality in Northeast Asia. *Environment International, 36,* 212–225.

Posch, M., Aherne, J., Forsius, M., Fronzek, S., & Veijalainen, N. (2008). Modelling the impacts of European emission and climate change scenarios on acid-sensitive catchments in Finland. *Hydrology and Earth System Sciences, 12,* 449–463.

Raddum, G. G., & Skjelkvåle, B. L. (1995). Critical limits of acidification to invertebrates in different regions of Europe. *Water Air and Soil Pollution, 85,* 475–480.

SFT. (2009). The Norwegian monitoring programme for long-range transported air pollutants. Annual report—effects 2008. TA-2546/2009. Oslo: The Norwegian Pollution Control Authority (SFT).

Skjelkvåle, B. L., Henriksen, A., Faafeng, B., Fjeld, E., Traaen, T. S., Lien, L., Lydersen, E., & Buan, A. K. (1996). *Regional innsjøundersøkelse 1995. En vannkjemisk undersøkelse av 1500 norske innsjøer* (Vol. 677/96). Oslo: Statens forurensningstilsyn.

Skjelkvåle, B. L., Wright, R. F., & Henriksen, A. (1998). Norwegian lakes show widespread recovery from acidification: Results of national surveys of lakewater chemistry 1986–1997. *Hydrology and Earth System Sciences, 2,* 555–562.

Skjelkvåle, B. L., Stoddard, J. L., Jeffers, J. N. R., Tørseth, K., Høgåsen, T., Bowman, J., Mannio, J., Monteith, D., Mosello, R., Rogora, M., Rzychon, D., Veselý, J., Wieting, J., Wilander, A., & Worsztynowicz, A. (2005). Regional scale evidence for improvements in surface water chemistry 1990–2001. *Environmental Pollution, 137,* 165–176.

Stevens, C. J., & Quinton, J. N. (2009). Diffuse pollution swapping in arable agricultural systems. *Critical Reviews Environmental Science and Technology, 39,* 478–520.

Stoddard, J. L., Jeffries, D. S., Lükewille, A., Clair, T. A., Dillon, P. J., Driscoll, C. T., Forsius, M., Johannessen, M., Kahl, J. S., Kellogg, J. H., Kemp, A., Mannio, J., Monteith, D. T., Murdoch, P. S., Patrick, S., Rebsdorf, A., Skjelkvale, B. L., Stainton, M. P., Traaen, T., van Dam, H., Webster, K. E., Wieting, J., & Wilander, A. (1999). Regional trends in aquatic recovery from acidification in North America and Europe. *Nature, 401,* 575–578.

Stoddard, J. L., Karl, J. S., Deviney, F. A., DeWalle, D. R., Driscoll, C. T., Herlihy, A. T., Kellogg, J. H., Murdoch, P. S., Webb, J. R., & Webster, K. E. (2003). Response of surface water chemistry to the clean air act amendments of 1990. Report EPA 620/R-03/001. North Carolina: United States Environmental Protection Agency.

Sullivan, T. J., Cosby, B. J., Herlihy, A. T., Driscoll, C. T., Fernandez, I. J., McDonnell, T. C., Boylen, C. W., Nierzwicki-Bauer, S. A., & Snyder, K. U. (2007). Assessment of the extent to which intensively-studied lakes are representative of the Adirondack region and response to future changes in acidic deposition. *Water Air and Soil Pollution, 185,* 279–291.

Sullivan, T. J., Cosby, B. J., Webb, J. R., Dennis, R. L., Bulger, A. J., & Deviney, F. A. (2008). Streamwater acid-base chemistry and critical loads of atmospheric sulfur deposition in Shenandoah national park, Virginia. *Environmental Monitoring and Assessment, 137,* 85–99.

Svarén, A. (1996). Jordmånsbildning och markkemisk övervakning i fjällområdet—en pilotstudie. (Examensarbete 1995/1996). Uppsala: Institutionen för skoglig marklära SLU.

Sverdrup, H., & Rosén, K. (1998). Long-term base cation mass balances for Swedish forests and the concept of sustainability. *Forest Ecology and Management, 110,* 221–236.

Sverdrup, H., De Vries, W., & Henriksen, A. (1990). *Mapping critical loads. Miljørapport 1990* (Vol. 14). Copenhagen: Nordic Council of Ministers.

Tilman, D., Socolow, R., Foley, J. A., Hill, J., Larson, E., Lynd, L., Pacala, S., Reilly, J., Searchinger, T., Somerville, C., & Williams, R. (2009). Beneficial biofuels—The food, energy, and environment trilemma. *Nature, 325,* 270–271.

Tominaga, K., Aherne, J., Watmough, S. A., Alveteg, M., Cosby, B. J., Driscoll, C. T., Posch, M., & Pourmokhtarian, A. (2010). Predicting acidification recovery at the Hubbard Brook experimental forest, New Hampshire: Evaluation of four models. *Environmental Science and Technology, 44,* 9003–9009.

Van Breemen, N., Jenkins, A., Wright, R. F., Beerling, D. J., Arp, W. J., Berendse, F., Beier, C., Collins, R., van Dam, D., Rasmussen, L., Verburg, P. S. J., & Wills, M. A. (1998). Impacts of elevated carbon dioxide and temperature on a boreal forest ecosystem (CLIMEX project). *Ecosystems, 1,* 345–351.

Vehviläinen, B., & Huttunen, M. (2002). *The Finnish watershed simulation and forecasting system (WSFS)*. Publication of the 21st conference of Danube countries on the hydrological forecasting and hydrological bases of water management.

Vuorenmaa, J., Forsius, M., & Mannio, J. (2006). Increasing trends of total organic carbon concentrations in small forest lakes in Finland from 1987 to 2003. *Science of the total Environment, 365,* 47–65.

Wright, R. F., Cosby, B. J., Flaten, M. B., & Reuss, J. O. (1990). Evaluation of an acidification model with data from manipulated catchments in Norway. *Nature, 343,* 53–55.

Wright, R. F., Larssen, T., Camarero, L., Cosby, B. J., Ferrier, R. C., Helliwell, R. C., Forsius, M., Jenkins, A., Kopáček, J., Majer, V., Moldan, F., Posch, M., Rogora, M., & Schöpp, W. (2005). Recovery of acidified European surface waters. *Environmental Science and Technology, 39,* 64A–72A.

Wright, R. F., Aherne, J., Bishop, K., Camarero, L., Cosby, B. J., Erlandsson, M., Evans, C. D., Forsius, M., Hardekopf, D. W., Helliwell, R., Hruska, J., Jenkins, A., Kopacek, J., Moldan, F., Posch, M., & Rogora, M. (2006). Modelling the effect of climate change on recovery of acidified freshwaters: Relative sensitivity of individual processes in the MAGIC model. *Science of the total Environment, 365,* 154–166.

Yan, N. D., Leung, B., Keller, W., Arnott, S. E., Gunn, J. M., & Raddum, G. G. (2003). Developing conceptual frameworks for the recovery of aquatic biota from acidification. *Ambio, 32,* 165–169.

Chapter 19
Critical Load Assessments and Dynamic Model Applications for Lakes in North America

Julian Aherne and Dean Jeffries

19.1 Introduction

The relationship between the acidification of surface waters and loss of biological resources provided the earliest evidence for the effects of acidic deposition in North America (Beamish and Harvey 1972; Dillon et al. 1978; Schofield 1976; Watt et al. 1979). Observations from site-specific intensive studies in the United States (US) and Canada further supported the hypothesis that declines of pH and acid neutralising capacity (ANC) in surface waters had occurred owing to acidic deposition (Dillon et al. 1987; Likens et al. 1972). In response to this early work, regional surveys of surface waters played an important role in assessing the extent and magnitude of the chemical and biological effects of acidic deposition (e.g., Landers et al. 1988). During the 1980s and 1990s, scientific and political activity on 'acid rain' focused primarily on eastern North America, owing to the spatial coincidence of acid-sensitive bedrock (Shiltz 1981) and elevated levels of acidic deposition primarily associated with emissions from coal-fired power plants in the US and smelters in Canada. Within this context, critical loads were first discussed by the US and Canada in the Memorandum of Intent on Transboundary Air Pollution (US-Canada 1983). The values suggested at that time were based largely on empirical associations between observed aquatic effects and measured deposition—subsequently refined further through the application of water chemistry models.

The impacts of acidic deposition on surface waters in Canada were a significant driver underpinning bilateral agreements on emission reductions (US-Canada 1983). The development of critical loads and (subsequent) bilateral negotiations on

J. Aherne (✉)
Environmental and Resource Studies, Trent University, Peterborough, ON, Canada
e-mail: jaherne@trentu.ca

D. Jeffries
Environment Canada, Burlington, ON, Canada

W. de Vries et al. (eds.), *Critical Loads and Dynamic Risk Assessments*,
Environmental Pollution 25, DOI 10.1007/978-94-017-9508-1_19

emission reductions (i.e., Canada-US Air Quality Agreement) initially focused on sulphur (S), as sulphate (SO_4^{2-}) deposition was recognised as the primary driver of surface water acidification. However, in recent years there has been growing concern on the role of nitrogen (N) in acidification (and ecosystem eutrophication), owing partially to the early success of S emissions control programs and subsequent decreases in SO_4^{2-} deposition (Lehmann et al. 2005; Zbieranowski and Aherne 2011). As a result, there have been significant reductions in S deposition during the last two decades, and in oxidised N since 2005 (Zbieranowski and Aherne 2011). In contrast, there has been renewed focus on the impacts of S and N deposition in western Canada and the US (Aherne and Shaw 2010; Sullivan et al. 2005) owing to increasing emissions from the oil and gas sector.

Since the early 1980s, steady-state (Chap. 6) and dynamic (Chap. 8) models have supported the assessment of impacts and development of 'acid rain' policy in North America. In Canada, critical loads have played a dominant role in the scientific assessment of impacts on surface waters. Initially steady-state models combined into an 'expert system' (later an Integrated Assessment Model [IAM] framework), underpinned estimates of critical loads (Jeffries et al. 1999); whereas, recent 'acid rain' assessments (Jeffries and Ouimet 2005) have incorporated UNECE methodologies (see Chaps. 6 and 17). Despite reductions, recent assessments indicate that critical loads of acidity (based on the SSWC model, with N leaching included) are still exceeded in eastern Canada. Subsequent assessments have incorporated linked critical loads of N and S using the FAB model, which is analogous to the Simple Mass Balance (SMB) model for terrestrial ecosystems (Chap. 6). The FAB model employs an N and S 'critical load function' (CLF), which quantifies deposition reductions required for N and S to meet the critical chemical limit (see Chap. 17; Fig. 17.2).

In this chapter we present a brief background on the history and use of steady-state models to define critical loads for surface waters in North America, and the application of dynamic models to assess the response of aquatic ecosystems to acidic deposition. In addition we present an application of the FAB model to surface waters in Canada with reference to assessments in Europe (Chap. 17).

19.2 History and Use of Steady-State and Dynamic Models in North America

19.2.1 Steady-State Models

In the US and Canada, empirical (e.g. Small and Sutton 1986) and steady-state (e.g. Henriksen 1979; Schnoor et al. 1986; Thompson 1982) models have been developed (and applied) since the 1980s to predict the impacts of acidic deposition on surface waters. Further, in Canada steady-state models have been used to define critical loads, and have underpinned all major national 'acid rain' assessments

(Environment Canada 1997, 2004; RMCC 1986, 1990). Notably, the Canada-Wide Acid Rain Strategy for Post-2000 set a long-term goal of achieving the threshold of critical loads for acidic deposition across Canada. In contrast, there is a notably paucity in the application of critical loads with respect to policy support in the US, despite the development of models such as the Direct Distribution Model (Small and Sutton 1986) and Trickle Down (Schnoor et al. 1986), both used in Canada to define critical loads (Lam et al. 1989), and early initiatives to develop 'a national critical loads framework for atmospheric deposition effects assessment' in the US (Shaffer et al. 1991; Strickland et al. 1993). One notable exception was the application of the Steady-State Water Chemistry (SSWC) model (Chap. 6) to map critical loads for 2053 surface waters in New England and eastern Canada under the auspices of the New England Governors and Eastern Canadian Premiers' 'Acid Rain Action Plan' (Dupont et al. 2005). Further, more recently there has been growing momentum in the US under the FOCUS (FOcal Center Utility Study) Project to develop a national-scale critical loads database (Blett et al. 2014).

A legacy of acid rain assessment activities in Canada has been the specification of critical loads for acidifying pollutants. Initial critical loads were set at 20 kg SO_4^{2-} ha^{-1} yr^{-1} wet deposition for south-eastern Canada; as further lake data were obtained and water chemistry models evolved and improved, critical load estimates declined. RMCC (1986) concluded that critical loads for sensitive aquatic ecosystems should be <12 kg SO_4^{2-} $ha^{-1}yr^{-1}$ sea-salt corrected wet deposition. In the 1990 Assessment (Jeffries and Lam 1993; RMCC 1990), aquatic critical loads were determined for 21 regions in south-eastern Canada using an 'expert system' of models (Lam et al. 1989). Estimates ranged from <8 to >20 kg SO_4^{2-} ha^{-1} yr^{-1} wet deposition to maintain pH 6 in 95% of the lakes. The expert system incorporated several steady-state water chemistry models, such as Trickle Down (Schnoor et al. 1986), the Cation Denudation Rate model (Thompson 1982), and the 'Site' model (Marmorek et al. 1990), later referred to as the ESSA-DFO model (Marmorek et al. 1996), which incorporated elements of the Direct Distribution Model (Small and Sutton 1986). The 1997 Assessment employed the same system (now refined and in the structure of an IAM; Lam et al. 1998) to refine the values for several eastern lake clusters and to extend the analysis to northern Alberta lakes (Jeffries et al. 1999).

More recently, aquatic critical load evaluations have employed steady-state models that were originally developed and used in Europe (Henriksen and Posch 2001). The SSWC model has been widely applied (Table 19.1) to regional data sets in Ontario (Henriksen et al. 2002), Eastern Canada (Dupont et al. 2005), all western provinces (Manitoba (Jeffries et al. 2010), Saskatchewan (Scott et al. 2010), Alberta (Gibson et al. 2010) and British Columbia (Strang et al. 2010)) and used in the most recent 'acid deposition' assessment (Environment Canada 2004). Estimated critical loads of acidity (CLA) to protect 95% of the lakes in each region generally range from <5 meq m^{-2} yr^{-1} in northern Saskatchewan and Manitoba, 10 meq m^{-2} yr^{-1} in coastal British Columbia, 15–30 meq m^{-2} yr^{-1} in Eastern Canada, and up to 60 meq m^{-2} yr^{-1} in Northern Alberta, reflecting the range in regional acid sensitivity, associated water chemistry and specified levels of protection (Table 19.1). The more recent applications (i.e., Jeffries et al. 2010; Scott et al. 2010; Strang et al.

Table 19.1 Summary of regional applications of steady-state models (Expert, FAB and SSWC) to determine critical load of acidity (CLA) for lakes (number) in Canada and the United States. The critical chemical limit, deposition year and proportion of exceeded lakes are shown

Region	Model[a]	Lakes	CLA meq $m^{-2}yr^{-1}$	Critical limit	Deposition year	Exceedance % of lakes	Source
Eastern Canada[b]	Expert	>2000	13–28[b]	pH≥6			Jeffries et al. (1999)
Ontario	SSWC	1469	34	ANC=40 μeq l^{-1}	1999	11–26%	Henriksen et al. (2002)
Ontario	FAB	285	30[c]	ANC=40 μeq l^{-1}	1995–1999 2010	59% 47%	Aherne et al. (2004)
Eastern North America[d]	SSWC	2053	<20	ANC=40 μeq l^{-1}[e]	2002	12%	Dupont et al. (2005)
Manitoba and Saskatchewan	SSWC	347	0.2–5.3[f]	Variable ANC[g]	2002	20%	Jeffries et al. (2010)
Saskatchewan	SSWC	259	2.6–5.7[h]	Variable ANC	2002	15%	Scott et al. (2010)
Alberta	SSWC	50	60	ANC=75 μeq l^{-1}	–	–	Gibson et al. (2010)
British Columbia	SSWC	72	10	Variable ANC	2005–2006	18%	Strang et al. (2010)
Northern British Columbia	SSWC	120	87	ANC=50 μeq l^{-1}	2000	0%	Krzyzanowski and Innes (2010)
United States	SSWC	>9500	–	ANC=50 μeq l^{-1}	–	–	Blett et al. (2014)

[a] Expert system of steady state models (Lam et al. 1989), Steady-State Water Chemistry (SSWC) model (Henriksen and Posch 2001), and First-order Acidity Balance (FAB) model (Posch et al. 2012)

[b] Lakes were sampled in four regional 'clusters' in Eastern Canada (Kejimkujik [Nova Scotia], Montmorency [Québec], Algoma [Ontario], and Sudbury [Ontario]); CLA range represents the four regions

[c] $CL_{max}S=30$ meq m^{-2} yr^{-1} and $CL_{max}N=90$ meq m^{-2} yr^{-1}

[d] Assessment of lakes in Quebec, New Brunswick, Nova Scotia, Newfoundland and eastern United States under the New England Governors and Eastern Canadian Premiers' 'Acid Rain Action Plan'

[e] Equivalent to pH=6 for the study lakes under the Direct Distribution Model (Small and Sutton 1986)

[f] Range represents CLA for 11 survey blocks; 5 in Manitoba and 6 in Saskatchewan

[g] Variable $ANC_{limit}=10+(10.2/3)\cdot TOC$ (Lydersen et al. 2004)

[h] Range represents the use of two models to estimate catchment runoff (Q)

2010), have incorporated a lake-specific level of protection (i.e., a variable ANC limit), adjusted for the strong acid anion contribution from organic acids after Lydersen et al. (2004). Despite the range in CLA, approximately 20 % (±5 %) of the lakes in each region exceeded critical load under total S deposition during the early 2000s. The FAB model (Henriksen and Posch 2001; Posch et al. 1997, 2012), which provides simultaneous estimates of critical loads of acidifying S and N deposition (see also Chap. 17), has not been as widely used in North America (Table 19.1). Nonetheless, under recent UNECE assessments the FAB model was applied to estimate critical loads of acidity for surface waters Canada-wide (Posch et al. 2011 and Sect. 19.3).

19.2.2 Dynamic Models

In concert with the development and application of steady-state models, it was recognised that time-dependent processes could buffer ecosystem response to acidic deposition; incorporation of these processes required time-dependent or 'dynamic' modelling frameworks (see Chap. 8). Since the 1980s, a wide-range of dynamic soil chemical models has been used in North America to predict the response of surface waters to acidic deposition. While initial dynamic models were developed both in the US (e.g. Reuss 1980) and Canada (e.g. Arp 1983) during the early 1980s, model development has been dominated by activities in the US.

Freshwater acidification models incorporate soil chemical processes, generally at the catchment scale, to describe the transfer of acidity from terrestrial to aquatic ecosystem. While many of these models are based on the same conceptual formulation with respect to soil and water acidification (Reuss and Johnson 1986), and include similar processes, their complexity varies greatly (Table 19.2). Since the early 1980s, there have been five well-established acidification models in the US (see Table 19.2), i.e. widely used and cited in the scientific literature. Their development has spanned three decades; MAGIC (Model of Acidification of Groundwater in Catchments; Cosby et al. 1985), ILWAS (Integrated Lake-Watershed Acidification Study; Gherini et al. 1985) and ETD (Enhanced Trickle-Down; Nikolaidis et al. 1988) were developed during the 1980s, followed by PnET-BGC (Photosynthesis and Evapotranspiration-Biogeochemistry; (Gbondo-Tugbawa et al. 2001), initially called PnET-BGC/CHESS (Krám et al. 1999) during the 1990s, and more recently DayCent-Chem (Hartman et al. 2007). These models represent a wide range in complexity, incorporating 2–11 soil layers, daily to annual simulation time-steps, and various sub-model or process representations (Table 19.2). Differences in model structure, such as cation exchange sub-models (e.g., Gapon versus Gaines Thomas; Table 19.2), have been shown to influence model predictions (Tominaga et al. 2009); nonetheless, all five models have been used to (successfully) simulate observed water chemistry at long-term monitoring sites. In fact, these models have been repeatedly used in many single site applications at data-rich monitoring sites (such as Hubbard Brook Experimental Forest [New Hampshire], Plastic Lake [Ontario] and

Table 19.2 Model characteristics for DayCent-Chem (DCC), Enhanced Trickle-Down (ETD), Integrated Lake-Watershed Acidification Study (ILWAS), Model of acidification of groundwater in catchments (MAGIC), and photosynthesis and evapotranspiration-biogeochemistry (PnET-BGC)

	DDC	ETD	ILWAS	MAGIC	PnET-BGC[a]
Number of soil layers	Up to 11 (including organic)	3	5	Up to 3	2 (organic and mineral)
Model time step	Daily	Daily	Daily	Annual or monthly	Monthly
Forest uptake	Submodel	External	Submodel	External	Submodel
Hydrology	Submodel	Submodel	Submodel	External	External
Soil cation exchange	Yes	First-order kinetic	Gapon and Kerr equilibrium	Gaines-Thomas equilibrium	Gaines-Thomas equilibrium
Mineral weathering	External	Kinetic process	Submodel	Calibrated or external	External
Sulphate adsorption	–	Linear isotherm	Linear isotherm	Langmuir isotherm	Yes
Dissolved organic carbon dissociation	Triprotic	–	Yes	Triprotic	Triprotic
Model description	Hartman et al. (2007)	Nikolaidis et al. (1988)	Gherini et al. (1985)	Cosby et al. (1985)	Gbondo-Tugbawa et al. (2001)

[a] Previously called PnET-BGC/CHESS (Krám et al. 1999)

Loch Vale [Colorado]) to evaluate model performance, determine model uncertainties, or predict future status under climate and emission scenarios (e.g. Aherne et al. 2008). However, only two models have been widely used for regional assessments: PnET-BGC and MAGIC, the latter being the most widely used model in North America (Table 19.3) and elsewhere (see Ferrier et al. 2003 and Chap. 18).

Regional dynamic model assessments of surface waters have primarily been applied to lakes in eastern Canada and the eastern US (notably in the Adirondack Mountains, New York), and streams in the south-eastern US (Southern Appalachians, Southern Blue Ridge, and Shenandoah National Park, Virginia). In contrast, fewer regional assessments have been carried out in western Canada and the western US (Table 19.3). Regional assessments of the response of surface waters to acidic deposition typically simulate the historic (pre-acidification) and future state of a specified chemical indicator, such as surface water ANC, pH or soil base saturation, during a 200 year period, e.g., 1850–2050. Future simulations typically employ currently agreed emissions reductions or evaluate the benefits of proposed reductions by assessing the proportion of lakes above or below a critical limit for the simulated chemical indicator (consistent with critical loads; see Table 19.1). For lakes in Ontario ($n=25$), Aherne et al. (2008) estimated the 5th percentile ANC to

Table 19.3 Summary of regional applications of dynamic models (MAGIC and PnET-BGC[a]) to assess the response of catchment soils and surface waters (lakes [L], rivers [R], streams [S]) to acidic deposition in Canada and the United States. The chemical indicator (e.g., surface water acid neutralising capacity [ANC], pH, sum of base cation [SBC] and soil base saturation [%BS]) and period of simulation are also shown

Region	Sites	Model	Indicator	Simulation period	Source
Adirondacks, New York	33 (L)	MAGIC	pH	Pre-industrial–2034	Sullivan et al. (1996)
Ontario	25 (L)	MAGIC	ANC, pH, %BS[b]	1850–2100	Aherne et al. (2003)
Atlantic Canada [c]	116 (L)	MAGIC	ANC, pH, SBC	1875–2016	Clair et al. (2003)
Nova Scotia	35 (R)	MAGIC	pH, SBC	1850–2100	Clair et al. (2004)
Southern Appalachian Mountains	130 (S)	MAGIC	Proportion in ANC classes	Pre-industrial–2040	Sullivan et al. (2004)
Adirondacks, New York	37 (L)	PnET-BGC	SO_4^{2-}, NO_3^-, pH, ANC, %BS	1850–2050	Chen and Driscoll (2005b)
New England and Maine	60 (L)	PnET-BGC	SO_4^{2-}, NO_3^-, pH, ANC, %BS	Pre-industrial–2050	Chen and Driscoll (2005a)
Nova Scotia	20 (L)	MAGIC	ANC, pH[d]	1850–2100	Whitfield et al. (2006)
Eastern Canada [e]	410 (L)	MAGIC	ANC, pH, calcium[e]	1850–2050	Clair et al. (2007)
Adirondacks, New York	70 (L)	MAGIC	Proportion in ANC classes	1850–2050	Sullivan et al. (2007)
Adirondacks, New York	44 (L)	PnET-BGC	pH, ANC, soil BS	1850–2000	Zhai et al. (2008)
Shenandoah National Park, Virginia	14 (S)	MAGIC	Proportion in ANC classes	Pre-industrial–2100	Sullivan et al. (2008)
Athabasca Oil Sands Region, Alberta	50 (L)	MAGIC	ANC, Bc:Al, %BS[f]	1900–2035	Whitfield et al. (2010)
Southern Blue Ridge Province	66 (S)	MAGIC	Proportion in ANC classes[g]	1890–2100	Sullivan et al. (2011)

[a]MAGIC model of acidification of groundwater in catchments; PnET-BGC photosynthesis and evapotranspiration-biogeochemistry
[b]ANC in 2050: 49 μeq L^{-1} (5th percentile); pH in 2050: 5.97 (5th percentile); Soil base saturation (5th percentile): 20.5 % (2050), 18.9 % (2100)
[c]New Brunswick, Newfoundland and Nova Socotia
[d]Critical limits: ANC=20 μeq L^{-1} and pH=5.4. Simulated ANC in 2100 (5th percentile): 27 μeq L^{-1}, pH in 2100 (5th percentile): 5.3
[e]Ontario, Quebec, New Brunswick, Newfoundland and Nova Scotia. Critical limit: ANC=40 μeq L^{-1}. In 2030, 10–25 % lakes have lower simulated ANC
[f]Critical limit: ANC=40 μeq L^{-1}. No impacts from acidic deposition
[g]Proportion of lakes with ANC<20 μeq L^{-1} : 29 % (2040), 23 % (2100); ANC <50 μeq L^{-1} : 73 % (2040), 73 % (2100)

be 49 μeq L^{-1} by 2050, well above the critical limit of 40 μeq L^{-1}. In contrast, Sullivan et al. (2011) predicted that 29 % of their study streams in the Southern Blue Ridge Province (n=66) had an ANC<20 μeq L^{-1} in 2040, decreasing to 23 % in 2100. Within the scientific literature, it is evident that MAGIC has played a leading role in such assessments, suggesting its level of complexity, model structure and data requirements make it a useful tool for long-term deposition scenario assessments.

Since the 1980s, the assessment of model uncertainty on predictions (i.e., prediction uncertainty) has been carried out in tandem with model development. These models (i.e., MAGIC, ILWAS, ETD and PnET-BGC) have been frequently reviewed (Eary et al. 1989), evaluated (Rose et al. 1991a, b; Tominaga et al. 2010) and critiqued (Reuss et al. 1986). Two notable systematic comparisons were carried out during the 1990s by (Rose et al. 1991a, b; MAGIC, ILWAS and ETD; Rose et al. 1991b) and more recently by Tominaga et al. (2010; MAGIC, PnET-BGC) following the methods developed by Rose et al. (1991a, b). Tominaga et al. (2010) used multiple calibrated ensemble simulations from four dynamic catchment acidification models to accommodate parameter and structural uncertainties in model predictions at Hubbard Brook Experimental Forest. Despite uncertainties, dynamic acidification models have proven to be a very useful tool in assessing the response of aquatic and terrestrial ecosystems to acidic deposition.

19.3 Critical Loads of Acidity and Exceedance for Surface Waters in Canada

In this section, critical loads and their exceedances for Canadian lakes are described with reference to similar assessments in Europe (Chap. 17: Finland, Norway, Sweden, the United Kingdom and Ireland). Critical loads were computed with the FAB model (see Chap. 6), and exceedances of critical loads were estimated using modelled total S and N deposition during 2002 and 2006 (the latter shown in Fig. 19.1). Modelled wet and dry deposition was obtained from AURAMS (A Unified Regional Air-quality Modelling System) on a 42 km × 42 km grid resolution (Zhang et al. 2002).

19.3.1 Methods and Data

Site Selection and Sampling Acid-sensitive lake catchments in Canada are located in regions with base-poor geologies overlain by thin soils, such as Nova Scotia (south and Cape Breton Island in the north), the Precambrian shield through Quebec, Ontario and northern Saskatchewan, and the coastal mountain ranges, e.g., British Columbia (see Shiltz 1981; Wolniewicz et al. 2011). In the current assessment, the study lakes were collated from existing survey data previously used in national 'acid rain' assessments in eastern Canada (Jeffries and Ouimet 2005; Jeffries et al. 1999), supplemented with recent survey data in Western Canada (Jeffries et al. 2010; Scott et al. 2010; Strang et al. 2010). While surveys were predominantly carried out in acid sensitive regions of Canada (see Fig. 19.2), survey designs included sampling strategies focused on acid sensitive lakes and stratified-random lake selection (e.g., compare Scott et al. (2010) with Jeffries et al. (2010)).

The lakes ranged in size from 0.2 to 43,100 ha, with an average area of 251.9 ha; catchment area ranged from <0.1 to 22,042 km^2, with an average area of 68.7 km^2. Catchment land cover (derived from the USGS Global Land Cover Characterization database [GLCC Version 2 Simple Biosphere Model]) was dominated by forested ecosystems. The assessment only included lakes that had all required FAB model input parameters (see Table 19.4), which was primarily dictated by lakes with catchment delineation ($n=3058$).

Derivation of FAB Datasets: The FAB model methodology to assess critical loads followed the Mapping Manual (www.icpmapping.org) with recent updates following Posch et al. (2012). Base cation flux was estimated with the SSWC model using observed sea-salt corrected (chloride as tracer) base cation concentrations. Long-term normals for catchment discharge (runoff) were estimated from meteorological data and soil properties using a model similar to MetHyd (a meteo-hydrological model; Slootweg et al. 2010). Long-term (1961–1990) average monthly temperature, precipitation and cloudiness were derived from a $0.5° \times 0.5°$ global database (Mitchell et al. 2004). Nitrogen immobilization in catchment soils was set at 0.5 kg N ha^{-1} yr^{-1} following the Mapping Manual (www.icpmapping.org). The denitrification fraction in the catchment soils was estimated as $f_{de}=0.1+0.7 \cdot f_{peat}$, where f_{peat} is the fraction of wetlands in the terrestrial catchment. Nitrogen removal in harvested biomass was set to zero following previous assessments in Canada (Jeffries and Ouimet 2005; Ouimet et al. 2006). A variable ANC limit was used, adjusted for the strong acid anion contribution from organic acids after Lydersen et al. (2004).

19.3.2 Results

Critical Loads of Sulphur and Nitrogen: The FAB model considers the effects of S and N deposition together, as such there is no unique critical load value for a given lake catchment. However, separate critical loads may be derived for both S and N when considered in isolation, i.e., assuming deposition of the other species to be zero (Posch et al. 1997; Posch et al. 2012). In the current study, in-lake retention of S was set to zero (Table 19.4); as such $CL_{max}S$ is equivalent to CLA. On a regional scale, critical loads are (typically) summarised as the 5th percentile of all lakes in the overlying deposition grid (Fig. 19.2; see also Fig. 17.3 in Chap. 17 for European assessments). The 5th percentile for $CL_{max}S$ ranged from <1 to >500 meq m^{-2} yr^{-1}, with an average of 65.5 meq m^{-2} yr^{-1} (median 41.5 meq m^{-2} yr^{-1}). The lowest

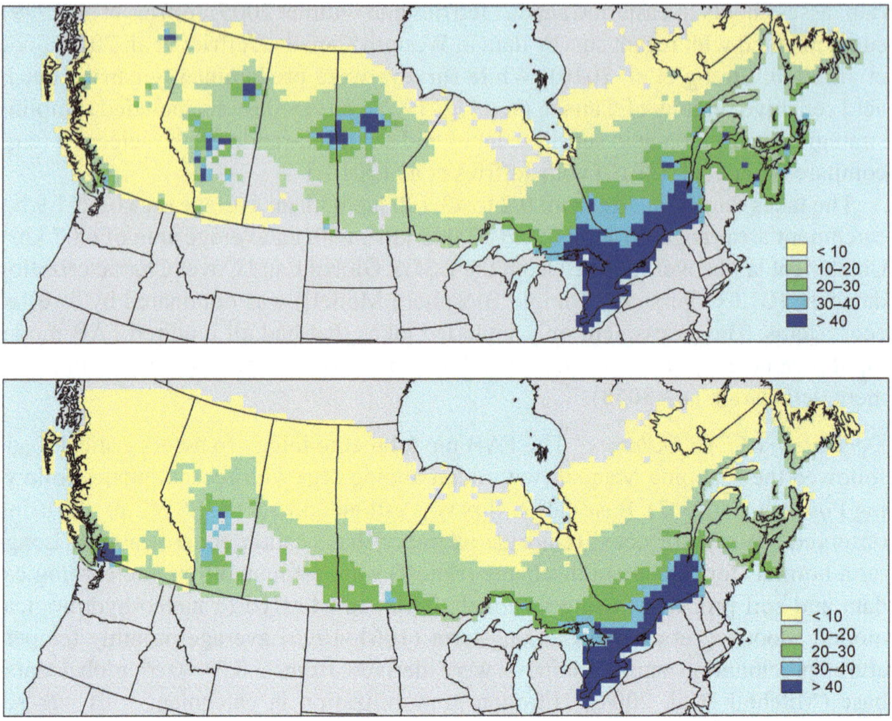

Fig. 19.1 Upper panel: AURAMS (A unified regional air-quality modelling system) modelled total (wet and dry) sulphur deposition (meq $m^{-2}yr^{-1}$) to forested catchments based on 2006 emissions. Lower panel: AURAMS 2006 total (wet and dry) nitrogen deposition (meq $m^{-2}yr^{-1}$) to forested catchments. AURAMS grid size: 42 km × 42 km (stereographic projection)

values of $CL_{max}S$ were estimated for the coastal mountain ranges in British Columbia, northern Saskatchewan and eastern Canada (e.g., Nova Scotia).

The mean $CL_{max}S$ was 82.8 meq m^{-2} yr^{-1} (median 51.8 meq m^{-2} yr^{-1}). Approximately 57 % of the lakes (weighted by catchment area) had $CL_{max}S < 100$ meq m^{-2} yr^{-1}, which was in the middle of the range for European catchments (see Fig. 17.4 in Chap. 17).

Critical Load Exceedances: The risk of 'harmful effects' is quantified by the exceedance of a critical load. Exceedance is estimated as the excess flux of acidity comprising contributions from both SO_4^{2-} and nitrate leaching, using appropriate S and N deposition datasets. Both N and S deposition can contribute to exceedance, and there are, in general, an infinite number of ways to achieve non-exceedance (i.e., a point on the CLF). To be able to evaluate exceedance, it has become customary to define it as the sum of N and S deposition reduction required to reach the CLF on the shortest path: $Ex(N_{dep}, S_{dep}) = \Delta N + \Delta S$. If (N_{dep}, S_{dep}) lies below the CLF, then there is non-exceedance, and Ex is set to zero (see Fig. 17.2 in Chap. 17, and Chap. 6 for the calculation procedure).

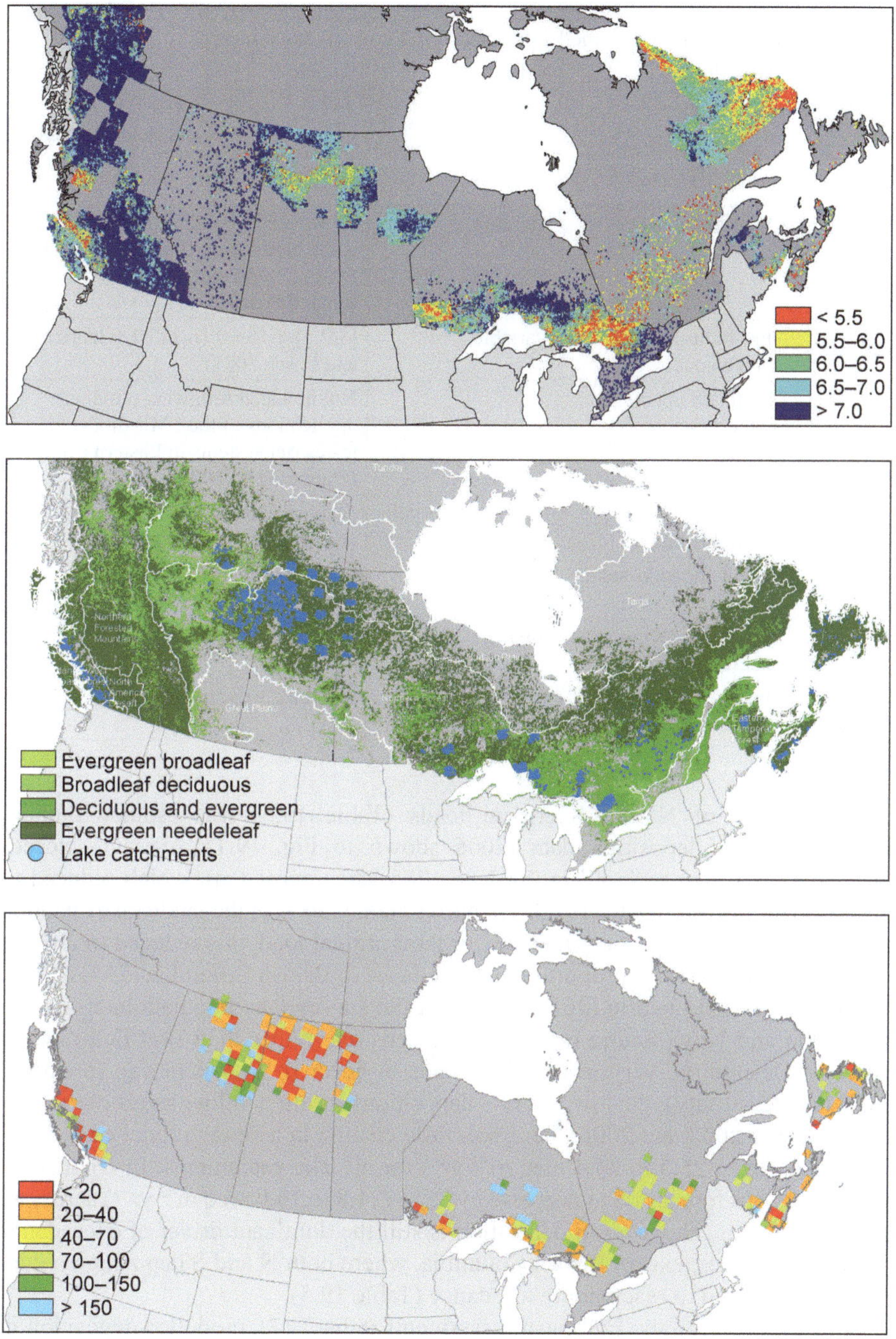

Fig. 19.2 Upper panel: Regional acid sensitivity of surface waters (lakes and streams) as indicated by pH based on >83,000 observations. Middle panel: Location of lake catchments used in the First-order Acidity Balance (FAB) model. Lower panel: 5th percentile of critical loads of acidity ($CL_{max}S$) for lake catchments (meq m^{-2}yr^{-1}) summarised on the EMEP50 grid (stereographic projection)

Table 19.4 First-order acidity balance (FAB) model parameters employed in Canada (compare Tables 17.2.1–5 in Chap. 17 for model parameters in European assessments)

	Description	Value	Source
ANC_{limit}	ANC limit	Variable	$ANC_{limit} = ANC_{oaa} + (10.2/3)\cdot TOC$, with ANC_{oaa} (organic acid adjusted ANC) = 10 meq m^{-3} (Lydersen et al. 2004)
F	F-factor	Variable	$F = \sin((\pi/2)Q[BC^*]_t/S)$, where S is the base cation flux above which $F = 1$ ($S = 400$ meq m^{-2} yr^{-1}) (Henriksen and Posch 2001)
$[SO_4^*]_0$	Pre-industrial sulphate conc.	Variable	$[SO_4^*]_0 = 15 + 0.16[BC^*]_t$ (Henriksen and Posch 2001)
N_{upt}	Nitrogen uptake in the catchment	–	Not included following previous applications of the SMB model to terrestrial systems in Eastern Canada (Ouimet et al. 2006)
N_{imm}	Nitrogen immobilisation	0.5 kg N ha^{-1} yr^{-1}	Mapping Manual (www.icpmapping.org)
f_{de}	Denitrification factor	$0.1 + 0.7\cdot f_{peat}$	Mapping Manual (www.icpmapping.org)
s_N	In-lake mass transfer coefficient for N	7.0 m yr^{-1}	Kaste and Dillon (2003)
s_S	In-lake mass transfer coefficient for S	–	Assumed to be negligible

Exceedance of N and S critical loads (Table 19.5) was estimated under 2002 and 2006 deposition data (2006 shown in Fig. 19.1). Average exceedance was 24.5 meq m^{-2} yr^{-1} in 2002 ($n = 785$ lakes exceeded), which reduced to 18.2 meq m^{-2} yr^{-1} in 2006 ($n = 621$). Approximately 26 % of the study lakes in 2002 received N and S deposition in excess of their critical load, this reduced to 20 % in 2006. There was greater proportional exceedance in British Columbia (75 %), Nova Scotia (29 %) and Ontario (26 %); note that the survey lake populations in British Columbia and Nova Scotia represent acid sensitive rather than stratified-random surveys. Despite the very low critical loads in northern Saskatchewan (owing to high acid sensitivity), the low N and S depositions resulted in low levels of exceedance. During 2002 and 2006, there was little change in average modelled S deposition, i.e., from 25.5 to 24.4 meq m^{-2} yr^{-1} for all lake catchments. In contrast, N deposition was forecasted to decrease by 35 % (24.8 to 16.0 meq m^{-2} yr^{-1}) across all lake catchments. However, S deposition is still the dominant driver of exceedance in most provinces, except British Columbia, where both N and S deposition reductions are required to reach non-exceedance (Table 19.5).

In comparison to the European assessments (Chap. 17), modelled deposition in Canada is somewhat 'inverted' when compared with 2020 deposition in Europe, i.e., 2006 N deposition in Canada is 16 meq m^{-2} yr^{-1} compared with 28 meq m^{-2} yr^{-1} in Europe in 2020. In contrast, the 2006 S deposition is 24.4 meq m^{-2} yr^{-1} compared

Table 19.5 Proportion (%) of sites in exceedance categories of the critical load function (see Fig. 17.2 in Chap. 17) and average deposition flux across all catchments (meq $m^{-2}yr^{-1}$) by province and Canada wide. The number of lakes (n) in each region is also given

Year	Region	n	Categories of exceedance according to CLF (%)								Deposition flux		
			Not exceeded	Exceeded	S or N (green)	S only (red)	N only (yellow)	S and N (blue)	N_{dep} $CL_{max}N$ (blue + yellow)	S_{dep} $CL_{max}S$ (blue + red)	N deposition	S deposition	N:S deposition ratio
2002	Alberta	398	92.5	7.5	0.5	6.8	0.0	0.3	0.3	7.0	18.6	18.5	1.01
2002	British Columbia	168	25.0	75.0	0.0	6.5	0.0	68.5	68.5	75.0	32.6	12.2	2.68
2002	Manitoba	211	90.5	9.5	0.0	9.5	0.0	0.0	0.0	9.5	4.6	14.9	0.31
2002	New Brunswick	111	85.6	14.4	6.3	8.1	0.0	0.0	0.0	8.1	22.6	25.8	0.88
2002	Newfoundland	55	81.8	18.2	1.8	14.5	0.0	1.8	1.8	16.4	8.1	11.1	0.73
2002	Nova Scotia	191	69.1	30.9	12.6	11.5	0.0	6.8	6.8	18.3	23.7	26.8	0.89
2002	Ontario	1009	60.6	39.4	60.6	20.8	0.0	0.4	0.4	21.2	45.5	46.4	0.98
2002	Quebec	82	84.1	15.9	9.8	6.1	0.0	0.0	0.0	6.1	24.8	27.9	0.89
2002	Saskatchewan	833	86.4	13.6	2.2	10.7	0.0	0.7	0.7	11.4	7.9	9.2	0.85
2002	Canada	3058	74.3	25.7	8.0	13.1	0.0	4.6	4.6	17.7	24.8	25.5	0.97
2006	Alberta	398	94.5	5.5	0.5	5.0	0.0	0.0	0.0	5.0	8.8	19.9	0.44
2006	British Columbia	168	25.0	75.0	0.0	3.0	0.0	72.0	72.0	75.0	26.2	12.4	2.10
2006	Manitoba	211	90.0	10.0	0.0	10.0	0.0	0.0	0.0	10.0	4.4	16.0	0.27
2006	New Brunswick	111	92.8	7.2	3.6	3.6	0.0	0.0	0.0	3.6	17.7	23.0	0.77
2006	Newfoundland	55	76.4	23.6	5.5	16.4	0.0	1.8	1.8	18.2	8.3	12.2	0.68
2006	Nova Scotia	191	70.7	29.3	9.4	15.7	0.0	4.2	4.2	19.9	17.0	29.8	0.57
2006	Ontario	1009	73.6	26.4	10.1	16.2	0.0	0.1	0.1	16.3	28.2	41.6	0.68
2006	Quebec	82	90.2	9.8	3.7	6.1	0.0	0.0	0.0	6.1	20.0	27.8	0.72
2006	Saskatchewan	833	87.9	12.1	0.5	11.2	0.0	0.5	0.5	11.6	5.1	9.8	0.52
2006	Canada	3058	79.7	20.3	4.4	11.4	0.0	4.4	4.4	15.9	16.0	24.4	0.65

with 12.9 meq m^{-2} yr^{-1} in Europe in 2020. This results in significant differences in the ratio of N to S deposition between Canada and Europe; the N:S ratio was 0.65 for Canada during 2006 compared with 2.16 for Europe during 2020. Atmospheric deposition in Canada is dominated by S deposition, with the N:S ratio similar to the 1980 ratio in Europe (see Chap. 17, Table 17.3). Nonetheless, the overall proportion of lakes receiving deposition in excess of their critical load in Canada in 2006 (20 %) was similar to predicted exceedance in Europe in 2020 (18 %). British Columbia is the only province in Canada with required N reductions, it had the highest N:S deposition ratio (2.68 [2002], 2.10 [2006]), which more closely resembled European countries under predicted 2020 deposition (see Chap. 17: e.g., Ireland in Table 17.3).

Uncertainties in the determination of critical loads of surface waters have been long discussed (Posch et al. 1993; Skeffington 2006); for a comprehensive examination on limitations and uncertainties in the FAB model see Chap. 17.

Conclusions

It is well established that models are useful tools to assess the impacts of acidic deposition on lake catchments. These models represent a variety of approaches, ranging from simple to complex, empirical to process-oriented, and steady-state to dynamic.

In Canada, steady-state models have been used to underpin bilateral negotiations with the US on emission reductions. Notably the Canada-Wide Acid Rain Strategy for Post-2000 sets a long-term goal of achieving the threshold of critical loads for acidic deposition across Canada. In contrast, despite initial interest in critical loads in the US, to-date they have played a limited role in supporting policy. Nonetheless, in combination with dynamic models, they have been used to assess ecosystem response to future emission reduction scenarios.

In Canada, approximately 20 % of lakes surveyed in acid-sensitive regions received modelled N and S deposition during 2006 in excess of their critical load. As such, critical load models like FAB continue to have a role to play in shaping future emission policies for the protection of aquatic ecosystems. While exceedance is still dominated by S deposition, there is a notable impact from N emissions in western Canada. Furthermore, there is concern that emissions from the Athabasca Oil Sands Region may impact acid-sensitive lakes in northern Saskatchewan.

Acknowledgements This research was undertaken, in part, thanks to funding from the Canada Research Chairs Program, and an NSERC Discovery grant. In addition, it was financially supported by the Government of Canada through the Federal Department of the Environment (Project: Forest ecosystem critical loads).

References

Aherne, J., & Shaw, P. D. (2010). Impacts of sulphur and nitrogen deposition in western Canada. *Journal of Limnology, 69,* 4–10.

Aherne, J., Dillon, P. J., & Cosby, B. J. (2003). Acidification and recovery of aquatic ecosystems in south-central Ontario, Canada: Regional application of the MAGIC model. *Hydrology and Earth System Science, 7,* 561–573.

Aherne, J., Posch, M., Dillon, P. J., & Henriksen, A. (2004). Critical loads of acidity for surface waters in south-central Ontario, Canada: Regional application of the first-order acidity balance (FAB) model. *Water Air Soil Pollution, 4,* 25–36.

Aherne, J., Futter, M. N., & Dillon, P. J. (2008). The impacts of future climate change and sulphur emission reductions on acidification recovery at Plastic Lake, Ontario. *Hydrology and Earth System Science, 12,* 383–392.

Arp, P. A. (1983). Modelling the effects of acid precipitation on soil leachates. A simple approach. *Ecological Modelling, 19,* 105–117.

Beamish, R. J., & Harvey, H. H. (1972). Acidification of La Cloche Mountain lakes, Ontario, and resulting fish mortalities. *Journal of the Fisheries Research Board of Canada, 29,* 1131–1143.

Blett, T. F., Lynch, J. A., Pardo, L. H., Huber, C., Haeuber, R., & Pouyat, R. (2014). FOCUS: A pilot study for national-scale critical loads development in the United States. *Environmental Science and Policy, 38,* 225–236.

Chen, L., & Driscoll, C. (2005a). A two-layer model to simulate variations in surface water chemistry draining a northern forest watershed. *Water Resources Research, 41*(9), 8 (W09425).

Chen, L., & Driscoll, C. (2005b). Regional application of an integrated biogeochemical model to northern New England and Maine. *Ecological Application, 14,* 1783–1797.

Clair, T. A., Dennis, I. F., & Cosby, B. J. (2003). Probable changes in lake chemistry in Canada's Atlantic Provinces under proposed North American emission reductions. *Hydrology and Earth System Science, 7,* 574–582.

Clair, T. A., Dennis, I. F., Amiro, P. G., & Cosby, B. J. (2004). Past and future chemistry changes in acidified Nova Scotian Atlantic salmon (Salmo salar) rivers: A dynamic modeling approach. *Canadian Journal of Fisheries and Aquatic Sciences, 61,* 1965–1975.

Clair, T. A., Aherne, J., Dennis, I. F., Gilliss, M., Couture, S., McNicol, D., Weeber, R., Dillon, P. J., Keller, W., Jeffries, D. S., Page, S., Timoffee, K., & Cosby, B. J. (2007). Past and future changes to acidified eastern Canadian lakes: A geochemical modeling approach. *Applied Geochemistry, 22,* 1189–1195.

Cosby, B. J., Hornberger, G. M., Galloway, J. N. & Wright, R. F. (1985). Modeling the effects of acid deposition: Assessment of a lumped parameter model of soil water and streamwater chemistry. *Water Resources Research, 21,* 51–63.

Dillon, P. J., Jeffries, D. S., Snyder, W., Reid, R., Yan, N. D., Evans, D., Moss, J., & Scheider, W. A. (1978). Acid precipitation in south-central Ontario: Recent observations. *Journal of the Fisheries Research Board of Canada, 35,* 809–815.

Dillon, P. J., Reid, R. A., & de Grosbois, E. (1987). The rate of acidification of aquatic ecosystems in Ontario, Canada. *Nature, 329,* 45–48.

Dupont, J., Clair, T. A., Gagnon, C., Jeffries, D. S., Kahl, J. S., Nelson, S. J., & Peckenham, J. S. (2005). Estimation of critical loads of acidity for lakes in northeastern United States and Eastern Canada. *Environmental Monitoring and Assessment, 109,* 275–291.

Eary, L. E., Jenne, E. A., Vail, L. W., & Girvin, D. C. (1989). Numerical models for predicting watershed acidification. *Archives of Environmental Contamination and toxicology, 18,* 29–53.

Environment Canada. (1997). *1997 Canadian acid rain assessment, volume 3: The effects on Canada's lakes, rivers and wetlands.* Ottawa: Environment Canada.

Environment Canada. (2004). *2004 Canadian acid deposition science assessment.* Ottawa: Environment Canada.

Ferrier, R. C., Wright, R. F., Jenkins, A., & Barth, H. (2003). Predicting recovery of acidified freshwaters in Europe and Canada: An introduction. *Hydrology and Earth System Science, 7,* 431–435.

Gbondo-Tugbawa, S. S., Driscoll, C. T., Aber, J. D., & Likens, G. E. (2001). Evaluation of an integrated biogeochemical model (PnET-BGC) at a northern hardwood forest ecosystem. *Water Resources Research, 37,* 1057–1070.

Gherini, S. A., Mok, L., Hudson, R. J. M., Davis, G. F., Chen, C. W., & Goldstein, R. A. (1985). The ILWAS model: Formulation and application. *Water Air and Soil Pollution, 26,* 425–459.

Gibson, J. J., Birks, S. J., Kumar, S., McEachern, P. M., & Hazewinkel, R. (2010). Inter-annual variations in water yield to lakes in northeastern Alberta: Implications for estimating critical loads of acidity. *Journal of Limnology, 69,* 126–134.

Hartman, M. D., Baron, J. S., & Ojima, D. S. (2007). Application of a coupled ecosystem-chemical equilibrium model, DayCent-Chem, to stream and soil chemistry in a Rocky Mountain watershed. *Ecological Modelling, 200,* 493–510.

Henriksen, A. (1979). A simple approach for identifying and measuring acidification of freshwater. *Nature, 278,* 542–545.

Henriksen, A., & Posch, M. (2001). Steady-state models for calculating critical loads of acidity for surface waters. *Water Air and Soil Pollution:Focus, 1,* 375–398.

Henriksen, A., Dillon, P. J., & Aherne, J. (2002). Critical loads of acidity for surface waters in south-central Ontario, Canada: Regional application of the steady-state water chemistry (SSWC) model. *Canadian Journal of Fisheries Aquatic Sciences, 59,* 1287–1295.

Jeffries, D. S., & Lam, D. C. L. (1993). Assessment of the effect of acidic deposition on Canadian lakes: Determination of critical loads for sulphate deposition. *Water Science and Technology, 28,* 183–187.

Jeffries, D., & Ouimet, R. (2005). Critical loads: Are they being exceeded? In *2004 Canadian acid deposition science assessment.* Ottawa: Environment Canada.

Jeffries, D. S., Lam, D. C. L., Moran, M. D., & Wong, I. (1999). The effect of SO_2 emission controls on critical load exceedances for lakes in southeastern Canada. *Water Science and Technology, 39,* 165–171.

Jeffries, D. S., Semkin, R. G., Wong, I., & Gibson, J. J. (2010). Recently surveyed lakes in northern Manitoba and Saskatchewan, Canada: Characteristics and critical loads of acidity. *Journal of Limnology, 69,* 45–55.

Kaste, Ø., & Dillon, P. J. (2003). Inorganic nitrogen retention in acid-sensitive lakes in southern Norway and southern Ontario, Canada—a comparison of mass balance data with and empirical N retention model. *Hydrological Processes, 17,* 2393–2407.

Krám, P., Santore, R. C., Driscoll, C. T., Aber, J. D., & Hruška, J. (1999). Application of the forest-soil-water model (PnET-BGC/CHESS) to the Lysina catchment, Czech Republic. *Abstract in International Forestry Review,* 120, 9.

Krzyzanowski, J., & Innes, J. L. (2010). Back to the basics—Estimating the sensitivity of freshwater to acidification using traditional approaches. *Journal of Environmental Management, 91,* 1227–1236.

Lam, D. C. L., Swayne, D. A., Storey, J., & Fraser, A. S. (1989). Watershed acidification models using the knowledge based systems approach. *Ecological Modelling, 47,* 131–152.

Lam, D. C. L., Puckett, K. J., Wong, I., Moran, M. D., Fenech, G., Jeffries, D. S., Olson, M. P., Whelpdale, D. M., McNicol, D. K., Mariam, Y. K. G., & Minns, C. K. (1998). An integrated acid rain assessment model for Canada: From source emission to ecological impact. *Water Quality Research Journal of Canada, 33,* 1–17.

Landers, D. H., Overton, W. S., Linthurst, R. A., & Brakke, D. F. (1988). Eastern lake survey: Regional estimates of lake chemistry. *Environmental Science & Technology, 22,* 128–135.

Lehmann, C. M. B., Bowersox, V. C., & Larson, S. M. (2005). Spatial and temporal trends of precipitation chemistry in the United States, 1985–2002. *Environmental Pollution, 135,* 347–361.

Likens, G. E., Bormann, F. H., & Johnson, N. M. (1972). Acid rain. *Environment, 14,* 33–40.

Lydersen, E., Larssen, T., & Fjeld, E. (2004). The influence of total organic carbon (TOC) on the relationship between acid neutralizing capacity (ANC) and fish status in Norwegian lakes. *Science of the Total Environment, 326,* 63–69.

Marmorek, D. R., Jones, M. L., Minns, C. K., & Elder, F. C. (1990). Assessing the potential extent of damage to inland lakes in eastern Canada due to acidic deposition. I. Development and evaluation of a simple 'site' model. *Canadian Journal of Fisheries Aquatic Science, 47,* 55–66.

Marmorek, D. R., MacQueen, R. M., Wedeles, C. H. R., Korman, J., Blancher, P. J., & McNicol, D. K. (1996). Improving pH and alkalinity estimates for regional-scale acidification models: Incorporation of dissolved organic carbon. *Canadian Journal of Fisheries Aquatic Science,* 53, 1602–1608.

Mitchell, T., Carter, T. R., Jones, P. D., Hulme, M., & New, M. (2004). *A comprehensive set of high-resolution grids of monthly climate for Europe and the globe: The observed record (1901–2000) and 16 scenarios (2001–2100).* Working Paper 55. Tyndall Centre.

Nikolaidis, N. P., Rajaram, H., Schnoor, J. L., & Georgakakos, K. P. (1988). A generalized soft water acidification model. *Water Resources Research, 24,* 1983–1996.

Ouimet, R., Arp, P. A., Watmough, S. A., Aherne, J., & Demarchant, I. (2006). Determination and mapping critical loads of acidity and exceedances for upland forest soils in eastern Canada. *Water Air and Soil Pollution, 172,* 57–66.

Posch, M., Forsius, M., & Kämäri, J. (1993). Critical loads of sulfur and nitrogen for lakes 1: Model description and estimation of uncertainty. *Water Air and Soil Pollution, 66,* 173–192.

Posch, M., Kämäri, J., Forsius, M., Henriksen, A., & Wilander, A. (1997). Exceedance of critical loads for lakes in Finland, Norway and Sweden: Reduction requirements for acidifying nitrogen and sulfur deposition. *Environmental Management, 21,* 291–304.

Posch, M., Slootweg, J., & Hettelingh, J.-P. (2011). Modelling critical thresholds and temporal changes of geochemistry and vegetation diversity: CCE Status Report 2011. RIVM Report 680359003. The Netherlands: Coordination Centre for Effects.

Posch, M., Aherne, J., Forsius, M., & Rask, M. (2012). Past, present, and future exceedance of critical loads of acidity for surface waters in Finland. *Environmental Science & Technology, 46,* 4507–4514.

Reuss, J. O. (1980). Simulation of soil nutrient losses resulting from rainfall acidity. *Ecological Modelling, 11,* 15–38.

Reuss, J. O., & Johnson, D. W. (1986). *Acid deposition and the acidification of soils and waters.* Germany: Springer-Verlag.

Reuss, J. O., Christophersen, N., & Seip, H. M. (1986). A critique of models for freshwater and soil acidification. *Water Air and Soil Pollution, 30,* 909–930.

RMCC. (1986). *Assessment of the state of knowledge on the long-range transport of air pollutants and acid deposition, Part 3: Aquatic effects.* Ottawa: Federal/Provincial Research and Monitoring Committee.

RMCC. (1990). *The 1990 Canadian long-range transport of air pollutants and acid deposition assessment report, part 4: Aquatic effects.* Ottawa: Federal/Provincial Research and Monitoring Coordinating Committee.

Rose, K. A., Brenkert, A. L., Cook, R. B., Gardner, R. H., & Hettelingh, J.-P. (1991a). Systematic comparison of ILWAS, MAGIC, and ETD watershed acidification models—2. Monte Carlo analysis under regional variability. *Water Resources Research, 27,* 2591–2603.

Rose, K. A., Cook, R. B., Brenkert, A. L., Gardner, R. H., & Hettelingh, J. P. (1991b). Systematic comparison of ILWAS, MAGIC, and ETD watershed acidification models 1. Mapping among model inputs and deterministic results. *Water Resources Research, 27,* 2577–2589.

Schnoor, J. L., Lee, S., Nikolaidis, N. P., & Nair, D. R. (1986). Lake resources at risk to acidic deposition in the eastern United States. *Water Air and Soil Pollution, 31,* 1091–1101.

Schofield, C. L. (1976). Acid precipitation: Effects on fish. *Ambio,* 5, 228–230.

Scott, K. A., Wissel, B., Gibson, J. J., & Birks, J. S. (2010). Chemical characteristics and acid sensitivity of boreal headwater lakes in northwest Saskatchewan. *Journal of Limnology, 69,* 33–44.

Shaffer, P. W., Rosenbaum, B., Holdren, G. R., Strickland, T. C., McDowell, M. K., Rosenbaum, B., Holdren, G. R., Strickland, T. C., McDowell, M. K., Papp, M., Cassell, D., Hazard, J., & Martin, B. E. (1991). *Estimating critical loads of sulfate to surface waters in the northeastern United States: A comparative assessment of three procedures for estimating critical loads of sulfate for lakes*, 600-3-91-062. US Environmental Protection Agency.

Shiltz, W. W. (1981). *Sensitivity of bedrock to acid precipitation: Modification by glacial processes*. Paper 81-14. Ottawa: Geological Survey of Canada.
Skeffington, R. A. (2006). Quantifying uncertainty in critical loads: (a) literature review. *Water Air and Soil Pollution, 169,* 3–24.
Slootweg, J., Posch, M., & Hettelingh, J.-P. (2010). Progress in the modelling of critical thresholds and dynamic modelling, including impacts on vegetation in Europe: CCE Status Report 2010. RIVM Report No. 680359001. The Netherlands: Coordination Centre for Effects.
Small, M. J., & Sutton, M. C. (1986). A direct distribution model for regional aquatic acidification. *Water Resources Research, 22,* 1749–1758.
Strang, D., Aherne, J., & Shaw, D. P. (2010). The hydrochemistry of high-elevation lakes in the Georgia Basin, British Columbia. *Journal of Limnology, 69,* 56–66.
Strickland, T. C., Holdren, G. R., Ringold, P. L., Bernard, D., Smythe, K., & Fallon, W. (1993). A national critical loads framework for atmospheric deposition effects assessment I: Method summary. *Environmental Management, 17,* 329–334.
Sullivan, T. J., Cosby, B. J., Driscoll, C. T., Charles, D. F., & Hemond., H. F. (1996). Influence of organic acids on model projections of lake acidification. *Water Air and Soil Pollution, 91,* 271–282.
Sullivan, T. J., Cosby, B. J., Herlihy, A. T., Webb, J. R., Bulger, A. J., Snyder, K. U., Brewer, P. F., Gilbert, E. H., & Moore, D. L. (2004). Regional model projections of future effects of sulfur and nitrogen deposition on streams in the southern Appalachian Mountains. *Water Resources Research, 40*(2), W02101.
Sullivan, T. J., Cosby, B. J., Tonnessen, K. A., & Clow, D. W. (2005). Surface water acidification responses and critical loads of sulfur and nitrogen deposition in Loch Vale watershed, Colorado. *Water Resources Research, 41,* W01021.
Sullivan, T. J., Cosby, B. J., Herlihy, A. T., Driscoll, C. T., Fernandez, I. J., McDonnell, T. C., Boylen, C. W., Nierzwicki-Bauer, S. A., & Snyder, K. U. (2007). Assessment of the extent to which intensively-studied lakes are representative of the Adirondack region and response to future changes in acidic deposition. *Water Air and Soil Pollution, 185,* 279–291.
Sullivan, T. J., Cosby, B. J., Webb, J. R., Dennis, R. L., Bulger, A. J., & Deviney, F. A. (2008). Streamwater acid-base chemistry and critical loads of atmospheric sulfur deposition in Shenandoah National Park, Virginia. *Environmental Monitoring and Assessment, 137,* 85–99.
Sullivan, T. J., Cosby, B. J., & Jackson, W. A. (2011). Target loads of atmospheric sulfur deposition for the protection and recovery of acid-sensitive streams in the Southern Blue Ridge province. *Journal of Environmental Management, 92,* 2953–2960.
Thompson, M. E. (1982). The cation denudation rate as a quantitative index of sensitivity of eastern Canadian rivers to acidic atmospheric precipitation. *Water Air and Soil Pollution, 18,* 215–226.
Tominaga, K., Aherne, J., Watmough, S. A., Alveteg, M., Cosby, B. J., Driscoll, C. T., & Posch, M. (2009). Voyage without constellation: evaluating the performance of three uncalibrated process-oriented models. *Hydrology Research, 40,* 261–272.
Tominaga, K., Aherne, J., Watmough, S. A., Alveteg, M., Cosby, B. J., Driscoll, C. T., Posch, M., & Pourmokhtarian, A. (2010). Predicting acidification recovery at the Hubbard Brook Experimental Forest, New Hampshire: Evaluation of four models. *Environmental Science & Technology, 44,* 9003–9009.
US-Canada. (1983). Memorandum of intent on transboundary air pollution. Report of the impact assessment working group I, Section 3 Aquatic effects.
Watt, W. D., Scott, D., & Ray, S. (1979). Acidification and other chemical changes in Halifax County lakes after 21 years. *Limnology and Oceanography, 24,* 1154–1161.
Whitfield, C. J., Aherne, J., Watmough, S. A., Dillon, P. J., & Clair, T. A. (2006). Recovery from acidification in Nova Scotia: temporal trends and critical loads for 20 headwater lakes. *Canadian Journal of Fisheries Aquatic Science, 63,* 1504–1514.
Whitfield, C. J., Aherne, J., Watmough, S. A., & Cosby, B. J. (2010). Modelling catchment response to acid deposition: A regional dual application of the MAGIC model to soils and lakes in the Athabasca Oil Sands region, Alberta. *Journal of Limnology, 69,* 147–160.

Wolniewicz, M. B., Aherne, J., & Dillon, P. J. (2011). Acid sensitivity of lakes in Nova Scotia, Canada: Assessment of lakes at risk. *Ecosystems, 14,* 1249–1263.
Zbieranowski, A. L., & Aherne, J. (2011). Long-term trends in atmospheric reactive nitrogen across Canada: 1988–2007. *Atmospheric Environment, 45,* 5853–5862.
Zhai, J., Driscoll, C. T., Sullivan, T. J., & Cosby, B. J. (2008). Regional application of the PnET-BGC model to assess historical acidification of Adirondack lakes. *Water Resources Research, 44,* W01421.
Zhang, L., Moran, M. D., Makar, P. A., Brook, J. R., & Gong, S. (2002). Modelling gaseous dry deposition in AURAMS: A unified regional air-quality modelling system. *Atmospheric Environment, 36,* 537–560.

Chapter 20
Critical Loads and Critical Limits of Cadmium, Copper, Lead and Zinc and Their Exceedances for Terrestrial Ecosystems in the United Kingdom

Jane Hall, Edward Tipping, Stephen Lofts, Michael Ashmore and Laura Shotbolt

20.1 Introduction

The soil critical load for a heavy metal at a given location is calculated first by specifying a critical limit for the metal (see also Chap. 2). The critical load is then determined by calculating the input load of the metal that, at steady state, would cause the soil metal content to equal the critical limit (see also Chap. 7). The calculations focus on inputs, outputs and concentrations of metals in the topsoil, where the greatest potential toxic effects are anticipated.

Soil metal concentrations, including known or expected toxic levels, have traditionally been expressed as the amount of metal per mass of dry soil (Merrington and Schoeters 2011; Steenbergen et al. 2005; Thakali et al. 2006). This is simple and practical, but does not lend itself to the incorporation of the growing knowledge about mechanisms of toxicity, and is therefore limited with respect to (a) the application of laboratory knowledge to the field, and (b) scientifically-defensible forecasting. Therefore, we developed critical limits based on free metal ion concentrations in soil solution (De Vries et al. 2007; Lofts et al. 2004).

The free ion approach we developed for cationic potentially-toxic metals is a simple version of models such as the Biotic Ligand Model (Niyogi and Wood 2004; Paquin et al. 2000) and WHAM-F_{TOX} (Stockdale et al. 2010), which relate toxic

J. Hall (✉)
Centre for Ecology and Hydrology, Environment Centre Wales, Bangor, UK
e-mail: jrha@ceh.ac.uk

E. Tipping · S. Lofts
Centre for Ecology and Hydrology, Lancaster Environment Centre, Bailrigg, Lancaster, UK

M. Ashmore
Environment Department, University of York, Heslington, York, UK

L. Shotbolt
Geography Department, Queen Mary University of London, London, UK

W. de Vries et al. (eds.), *Critical Loads and Dynamic Risk Assessments*,
Environmental Pollution 25, DOI 10.1007/978-94-017-9508-1_20

response to the degree of binding of the potentially-toxic metal by the organism, in effect treating the organism as a chemical reactant. By this approach, the competition effects of solution cations (including H^+, i.e. pH dependence) can be taken into account. The resulting description, although ostensibly more complicated, relates chemical exposure to effects more reliably. A misconception that should be laid to rest is that the free metal ion (or indeed any specified fraction of the metal) is the "toxic form" of the metal. Instead, the idea is that the free metal ion concentration, after taking account of the competitive effects of other cations, is a better *predictor* of toxicity than the total amount of metal in the soil.

In our simple approach, only soil solution pH is taken into account, which is possible because in most soils the concentrations of the main competing cations (Mg^{2+}, Al^{3+}, Ca^{2+}) are strongly pH-dependent. The critical free metal ion concentration is thus obtained from the soil pH. The total dissolved metal concentration in soil solution can be calculated by taking into account the binding of the metal with solution ligands, in other words calculating how much additional metal is present, in equilibrium with the free metal. This calculation is done with the speciation model WHAM (Tipping 1994, 1998), knowing or assuming concentrations of metal-binding ligands (see also Chap. 7). The most important for soils is dissolved organic matter, measured or estimated as the dissolved organic carbon concentration, [DOC]. By this means, we find the total metal concentration in soil solution at the critical limit.

The critical load refers to the steady state condition at which the soil has a free ion concentration corresponding to the critical limit (see Chaps. 2 and 7). Since this is a steady state, it is not affected by the reactive metal content of the soil solids (i.e. metal undergoing solid-solution exchange and available for uptake by biota), since that will have been forced into equilibrium with the solution by the continuous, constant, inputs and outputs of the metal. The critical load is therefore calculated in terms of the soil solution not the whole soil.

However, it is also informative to compare current observed soil conditions with those that pertain under the steady state critical load condition, i.e. to know how close the soil metal concentration is to the toxicity-based critical limit. In principle this should be done via data on the soil solution. However, when mapping exceedance of critical limits over large areas, rather than assessing an individual site, the available data for comparison is invariably the total soil metal concentration, expressed in, for example, μg metal per g soil. Therefore it is necessary to calculate the soil metal content that would be found at equilibrium with solution metal at the critical limit. This is done using pedotransfer functions, which are multiple regression equations that relate the total metal associated with the soil solids to solution pH, free metal ion concentration, and the organic matter content of the soil solids (Groenenberg et al. 2010; Tipping et al. 2003).

Critical limit exceedance occurs where the total soil metal concentration is greater than the critical limit concentration, and provides an indication of the current soil condition. Critical load exceedance occurs where current atmospheric deposition exceeds the steady-state critical load and provides an indication of areas potentially at risk when steady state is reached. It is possible for the critical load to be exceeded but not the critical limit; however, if deposition continues at current levels, exceedance of the critical limit may happen at some point in the future. It is therefore important to look at exceedance of both the critical loads and critical limits.

Critical loads have been calculated for Cd, Cu, Pb and Zn for six terrestrial habitats in the UK using pH-dependent free-ion critical limits (Lofts et al. 2004) and steady-state methods (Hall et al. 2006). This chapter focuses on the development in the methodology used in the UK since 2006, including updates to the critical limit functions and estimates of DOC; and on the results of critical load and critical limit exceedance for all four metals.

20.2 Methods

20.2.1 Critical Load Calculations

The critical loads methodology used in the UK is described in detail in Hall et al. (2006) and based on the following steady-state mass balance equation (see Chap. 7):

$$CL(M) = M_u + M_{we} + M_{le(crit)} \tag{20.1}$$

where $CL(M)$ is the critical load of a heavy metal M (g $ha^{-1}yr^{-1}$), M_u the net uptake of metal in harvestable parts of plants under critical load conditions, M_{we} the input of metal derived from mineral weathering and $M_{le(crit)}$ the flux of heavy metal M leaching from the soil layer based on the free ion critical limit for the metal.

M_{we} was set to zero so the critical loads reflected the atmospheric inputs only. M_u is an important component of the critical load calculations for managed (productive) woodlands. Uptake values were derived by Forest Research based on UK data from the ICP Forest Survey Level II plots. These plots comprise four oak stands, six beech sands, four Scots pine stands, two Norway spruce stands and four Sitka spruce stands. For national mapping the mean uptake rates for the broadleaf plots and for the conifer plots were applied in the managed woodland critical load calculations (see Table 20.1)

The other key term in the equation is the leaching flux of metal, defined as:

$$M_{le(crit)} = Q_{le} \cdot [M]_{(crit)} \tag{20.2}$$

where Q_{le} is the flux of drainage water leaching from the soil and $[M]_{crit}$ is the total concentration of metal in soil drainage water at the free-ion critical limit. $[M]_{crit}$ comprises: (a) metal bound to suspended particulate matter (SPM) calculated from the critical free ion concentration using pedotransfer functions based on soil pH and loss on ignition (LOI); (b) total dissolved metal concentration calculated from the critical free ion concentration using the Windermere Humic-Aqueous Model (WHAM) (Tipping 1994, 1998), which uses soil pH and DOC concentrations.

The derived critical loads are equivalent to the deposition rate that, at steady-state, leads to free ion concentrations equal to the free ion critical limit at the relevant

Table 20.1 Mean metal uptake rates (g $ha^{-1}yr^{-1}$) applied in national critical load calculations for managed (productive) coniferous woodland and managed (productive) broadleaved woodland. (Values based on wood cores taken from each plot and the inner wood, outer wood and bark analysed for metals. To calculate uptake rates the metal concentrations in wood were combined with estimated biomass increments over the course of rotation at each site)

Habitat	Metal uptake rates (g $ha^{-1}yr^{-1}$)			
	Cd	Cu	Pb	Zn
Managed coniferous woodland	0.67	9.27	2.98	44.1
Managed broadleaved woodland	0.17	4.42	1.81	10.0

soil pH. Critical loads were calculated and mapped for six habitat types (managed coniferous woodland, managed broadleaved woodland, unmanaged woodland, acid and calcareous grassland, dwarf shrub heath, and bog) based on the dominant soil type in each 1 × 1 km habitat grid square for the UK.

20.2.2 Critical Limit Functions

The critical limit functions (CLFs) were based on a model of pH-dependent free-ion toxicity by Lofts et al. (2004). The objective of the CLF is to describe the variation with pH of the free ion concentration that is protective of 95 % of the organisms in the ecosystem. The CLF is derived from the theory that the toxicity of a metal to an organism, expressed as the logarithm of the free ion concentration, should be a linear function of the soil solution pH (pH_{ss}) and concentrations of other cations assumed to compete with the metal for binding to the organism:

$$\log[M]_{free,toxic} = a \cdot pH_{ss} + \beta \cdot \sum[C_i] + \gamma \tag{20.3}$$

where $[M]_{free,\ toxic}$ is the concentration of free metal ion corresponding to a threshold of effect on the organism (the toxic endpoint concentration), $[C_i]$ is the concentration of a competing cation i (e.g. Mg^{2+}, Al^{3+}, Ca^{2+}), and α, β and γ are empirical coefficients. In principle, given sufficient toxicity data, the expression can be applied to organism-specific datasets, to derive organism-specific sets of α- and γ-values. Using these values, the critical limit could be calculated at a specific pH by considering the distribution of calculated log $[M]_{fre,\ toxic}$ values. In practice, the data available are insufficient for this purpose, and so Lofts et al. (2004) developed an approach to derive the CLF from a lumped toxicity dataset, comprising endpoints for plants, soil invertebrates and microbial processes together. This approach relies on two assumptions:

- The pH-dependence of metal toxicity (i.e. α) for the soil ecosystem can be obtained from a lumped toxicity dataset comprising endpoint data for multiple organisms, with the data points weighted to give equal importance to data from different organisms;
- The concentrations of competing cations $[C_i]$ in soil solutions are correlated with pH

Table 20.2 Metal-specific parameters for the empirical function that gives free metal ion concentration as a function of soil pH and organic matter content, and soil labile metal (see Eq. 20.5)

Metal	Parameter			
	a	b	c	d
Cu	−1.26	−0.63	0.93	4.99
Zn	−0.45	−0.61	0.57	0.55
Cd	−0.52	−0.60	0.60	−0.14
Pb	−1.02	−0.69	1.05	4.33

so that the CLF can be expressed as (see also Chap. 2):

$$\log[M]_{free,toxic} = a \cdot pH_{ss} + \gamma \tag{20.4}$$

An important step in the calculations is to express the endpoint metal concentrations as the free metal ion. Endpoint concentrations from soil toxicity tests are conventionally expressed as the concentration of metal added to the soil. These concentrations were converted to free metal ion, using the following expression:

$$\log[M]_{free} = a \cdot pH_{ss} + b \cdot \log OM + c \cdot \log[M]_{soil,labile} + d \tag{20.5}$$

where $[M]_{free}$ is the free metal ion in soil solution (mol dm^{-3}), *OM* is the soil organic matter content (%) and $\log[M]_{soil,\ labile}$ is the added or labile soil concentration (mol g^{-1}). Metal–specific coefficients for this equation, from Lofts et al. (2004), are given in Table 20.2.

Figure 20.1 demonstrates the principle of the CLF approach. This shows a plot of the lumped endpoint dataset for a metal (expressed as the log of the free ion concentrations) against the pH_{ss} of the soil in which the organism was tested, showing both the pH-dependence of the endpoint concentrations, and the distribution of intrinsic organism sensitivity (the vertical spread of endpoint concentrations at a given pH). The slope of a median regression of the lumped endpoints against pH (the dashed line in Fig. 20.1) gives the pH-dependency of the CLF (the coefficient α). The constant term γ is set such that the line representing the CLF (the solid line in Fig. 20.1) partitions the toxicity dataset so that, after taking the weighting of the data points into consideration, 95 % of the endpoints are effectively above the line. This line then represents the pH-dependent free ion concentration that is protective of 95 % of the ecosystem.

Critical limit functions of Cd, Zn, Cd and Pb were originally calculated by Lofts et al. (2004). Revised functions were presented by De Vries et al. (2007), following an update of databases of toxic endpoints. The functions used here were derived following a further update of the databases by Ashmore et al. (2007). This had little impact on the limits for Cd, Cu, and Pb and resulted in only a small change to the limit for Zn (see also Chap. 2, Table 2.1). Values of the function coefficients are

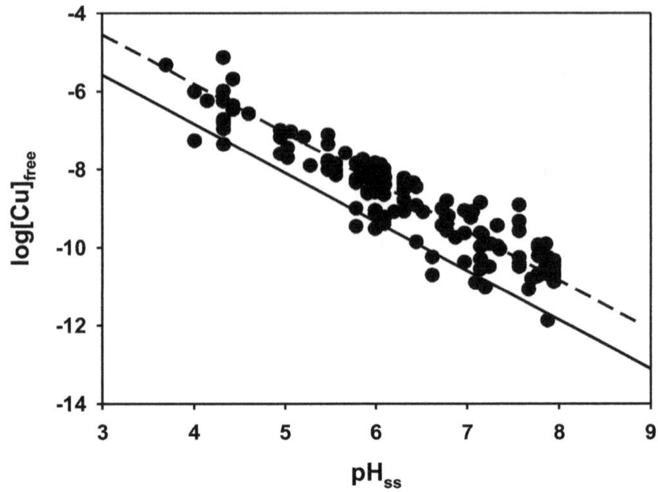

Fig. 20.1 Toxicity dataset for copper (Cu), demonstrating the derivation of a critical limit function. Endpoint concentrations of Cu, expressed as the free ion, are plotted against the soil solution pH (pH_{ss}). The dashed line indicates the median toxic free ion as a function of pH, the solid line indicates the critical limit function

Table 20.3 Coefficients of the critical limit functions (CLFs) for Cu, Zn, Cd and Pb (see Eq. 20.4)

Metal	Parameter	
	α	γ
Cd	−0.31	−6.36
Cu	−1.26	−1.80
Pb	−0.93	−3.50
Zn	−0.25	−5.07

given in Table 20.3. As the critical limits are related to soil properties they vary from one grid square to the next depending on the properties of the dominant soil type in each 1 × 1 km grid square.

20.2.3 DOC

Due to a lack of national DOC data, a relationship between DOC flux, runoff and habitat type was used by Hall et al. (2006) to estimate DOC values for each habitat type. This resulted in a DOC flux that increased with increasing runoff, and varied by habitat type. Since 2006 a targeted sampling programme, with the aim of reducing the uncertainties in the DOC estimates previously used, was carried out in the UK to obtain DOC measurements across a range of habitat types. A total of 53 sites were surveyed across a wide geographic range covering a rainfall (and runoff) gradient, with a minimum of eight sites for each of the six habitat types (acid/

Table 20.4 Average DOC concentrations by habitat for the UK based on a survey of 53 sites

Habitat	DOC mg l^{-1}	Rounded DOC values mg l^{-1}
Grassland:		20
Acid/neutral grassland	16.6	
Calcareous grassland	18.4	
Dwarf shrub heath	31.4	30
Bog	60.3	60
Woodland		25
Coniferous woodland	24.5	
Broadleaved woodland	24.1	

neutral grass, calcareous grass, bogs, heaths, deciduous woods, coniferous woods). Soil porewater DOC samples were collected monthly for a one-year period to give average DOC concentrations for each habitat type. The DOC for the bog habitat was significantly larger than the values for the other habitats. There was no significant difference between the values for the different grasslands, or for the different woodlands, so the data for these habitats were pooled, and annual average values were rounded for use in the critical limit and critical load calculations (Table 20.4).

20.2.4 Exceedance of Critical Loads

Critical loads are compared with atmospheric deposition of metals to identify the habitat areas where deposition exceeds the critical load. Exceedance results in this chapter are based on UK annual average deposition data for 2005 provided by the UK Heavy Metals Monitoring Network (Fowler et al. 2006; RoTAP 2012). Measurements of bulk deposition and heavy metal concentrations in airborne particulates were interpolated to generate deposition fields on 5 × 5 km grid for UK. The calculation of wet deposition includes treatment of the seeder-feeder effect to account for enhanced rainfall at higher altitudes. For dry deposition, the dry deposition velocity was calculated from land use maps and local wind data for each 5 × 5 km square. The exceedance calculations employed ecosystem-specific deposition for (i) grassland/moorland and (ii) woodland. The highest rates of deposition are found in central and western England, the north-west of Scotland, and upland areas of Wales; many of these are areas where rainfall is higher, and hence wet deposition is higher. The maps of deposition to woodland additionally show areas of higher deposition across south and eastern England where metal concentrations in air are greatest. Critical load exceedances were calculated on a 1 × 1 km grid (the same resolution as the critical load maps) by assuming that deposition is constant across each 5 × 5 km grid square.

20.2.5 Exceedance of Critical Limits

Exceedances of the critical limits were derived by comparing observed soil metal concentrations with the levels of soil metal calculated to be in steady-state with the critical solution concentration. Ideally, this would have been done by comparing measured and calculated (from solution critical free ion concentrations, pH, and soil properties) reactive soil metal contents, but the available measurements were only for total metal. However, Tipping et al. (2003) found significant relationships between reactive and total soil metal, and Hall et al. (2006) accordingly derived pedotransfer functions to predict the total soil metal. The critical limits, expressed as total metal concentrations, were then compared with national soil survey data (5×5 km data for England and Wales (McGrath and Loveland 1992) and Northern Ireland (Jordan et al. 2000), 10×10km data for Scotland (Macaulay Institute, Aberdeen)) to identify habitat areas, mapped on a 1×1 km grid, where the critical limits were exceeded.

20.2.6 Exceedance of Critical Limits and Critical Loads

The 1×1km maps of critical load exceedance and critical limit exceedance were combined for each habitat and each metal to generate four-class maps showing areas where:

- Neither critical loads or critical limits were exceeded
- Critical loads only were exceeded (critical limits not exceeded)
- Critical limits only were exceeded (critical loads not exceeded)
- Both critical loads and critical limits were exceeded

The percentage area of each habitat within each of the four classes was calculated from the habitat distribution maps.

20.3 Results

Heavy metal critical load values range from averages of 10–40 g $ha^{-1}yr^{-1}$ for Cd to averages an order of magnitude greater for Cu, Pb and Zn (Table 20.5). Spatial differences across the UK arise due to the dependence of the critical limits on soil pH (Sect. 20.2), and because complexation by DOM varies among metals. Maps of critical loads for Cd, Cu, Pb and Zn for UK habitats can be found in Ashmore et al. (2007). In general the lowest critical loads are found in the south and east of the UK, and the higher values in the north and west. This mainly reflects the national pattern of the critical leaching flux of metal from the soil, which is higher in north-western areas where rainfall and runoff are higher. The spatial pattern of heavy metal critical

Table 20.5 Minimum, maximum and mean critical loads for Cd, Cu, Pb and Zn by habitat type for the UK

Metal	Habitat	Critical loads (g $ha^{-1}yr^{-1}$)		
		Min	Max	Mean
Cd	Managed coniferous woodland	2	145	20
	Managed broadleaf woodland	1	116	10
	Unmanaged woodland	1	131	12
	Grassland (acid & calcareous)	1	181	22
	Heathland	1	184	36
	Bog	1	184	39
Cu	Managed coniferous woodland	14	5244	172
	Managed broadleaf woodland	9	1655	70
	Unmanaged woodland	5	1588	88
	Grassland (acid & calcareous)	3	6241	176
	Heathland	7	6205	358
	Bog	8	6350	441
Pb	Managed coniferous woodland	9	2941	153
	Managed broadleaf woodland	6	1504	61
	Unmanaged woodland	4	1445	80
	Grassland (acid and calcareous)	5	3503	176
	Heathland	7	3483	319
	Bog	12	3564	407
Zn	Managed coniferous woodland	72	2583	428
	Managed broadleaf woodland	35	2164	209
	Unmanaged woodland	27	2448	250
	Grassland (acid and calcareous)	21	3173	427
	Heathland	31	3228	691
	Bog	37	3231	733

loads contrasts with the acidity critical load maps for the UK where the more acid sensitive soils, and hence lower critical loads, are found in the north and west.

The calculations of critical load exceedance show virtually no exceedance for Cd, with just 3ha (0.001 %) of managed coniferous woodland being exceeded. There is also virtually no exceedance ($\leq$0.1 % habitat area) of the critical loads of Cu, Pb and Zn for bog and heathland habitats and only 3–11 % of the habitat area exceeds the critical load for grassland and managed coniferous woodland. However, over 50 % of the areas of managed broadleaved woodland and unmanaged (broadleaved and coniferous) woodland exceed the critical load for Cu, Pb and Zn. The area and magnitude of exceedance tends to be larger for woodland habitats due to the greater scavenging of dry deposition by trees compared to low growing vegetation. There is less exceedance for managed coniferous woodland as the critical loads for this habitat are higher due to a lower net input of metals as the removal rate of metal by harvesting is greater than for managed broadleaved woodland (Table 20.1). Where exceedances do occur they are mainly found in central, eastern and southern England where critical loads are also lower. Critical load exceedance maps are not shown here, but can be found in Ashmore et al. (2007).

Comparison of the critical limits (converted to total metal concentrations) with the available national soil concentration data shows virtually no exceedance of the critical limit for Pb for any habitat. The percentage areas of habitats with critical limit exceedance are also low for Cd with no more than 10 % exceeded for any habitat. The low exceedance of the critical limits for Pb, despite the high critical load exceedance in some habitats, is most likely due to the slow dynamics of the system, and the high affinity of Pb for organic matter, meaning that long periods of time are needed for Pb to accumulate in soils to concentrations at which the free-ion based critical limit is exceeded. Hall et al. (2006) showed that exceedance of the critical limit for Pb in the UK was much lower when it was expressed on a free-ion basis than when it was expressed on the basis of reactive soil concentration, indicating the importance of partitioning of the metal in assessing critical limit exceedance. Cu shows a higher exceedance than Cd, with a slightly higher area exceeded for grassland (18 %), 42 % exceeded for unmanaged woodland and 50 % of managed broadleaf exceeded. The largest areas of critical limit exceedance are seen for Zn (26–56 % of habitat exceeded). Maps of critical limit exceedance for Zn do not show any real spatial pattern of exceedance, although higher values appear to be associated with upland areas which may have received high deposition inputs of Zn. Maps of critical limit exceedances are published in Ashmore et al. (2007).

As the critical limits are dependent on soil pH they vary spatially across the country and exceedance depends on the spatial distribution of total metal concentrations in the soils. Some of the soil concentration data were collected around 30 years ago (e.g., McGrath and Loveland 1992) and represent both past and present inputs of metals. Although the habitat areas mapped for this study are semi-natural land use or commercial forest and the primary input of anthropogenically derived metals may be expected to be atmospheric, some areas also receive metal inputs from geological sources and sources of historical mining activity, or from agricultural application across lowland grassland areas. Such inputs may be significant, and affect critical limit exceedance, but are not considered within the critical loads evaluation.

The percentage area of each habitat in each of the four critical load and critical limit exceedance categories (Sect. 20.2.6) is given in Table 20.6 and the results are summarised below:

Cd: Critical limits are exceeded for between 1 and 10 % of the habitat areas; there are no exceedances of the critical load for any habitat. Note that the results for Cd exclude Scotland as national data on soil Cd concentrations for this area were not available.

Cu: There is very little exceedance of the critical load or limit for heathlands and bogs. The managed conifer and grassland habitats show some small areas (2–5 %) with both critical load and critical limit exceedance. However, managed broadleaved and unmanaged woodland are fairly evenly spread between the four exceedance categories, with approximately 30 % of each habitat exceeding both the critical load and critical limit.

Pb: There is virtually no exceedance of the critical load or limit for grasslands, heathlands or bogs and only 11 % of the area of managed coniferous forests with critical load exceedance. However, approximately 60 % of the areas of managed broadleaved and unmanaged woodland exceed the critical load, but < 1 % of the area of these habitats have exceedance of the critical limits.

Zn: There are more areas exceeding the critical limits and loads for this metal than the other metals above. Between 23 and 40 % of each habitat area has exceedance of the critical limit and 24–28 % of managed broadleaved woodland and unmanaged woodland also exceed the critical load, resulting in 30 % of these woodland habitats exceeding both the load and limit. Of this 30 %, three-quarters of the area also exceeds the critical load and critical limit for Cu for these woodland types.

To exemplify the methodology, mapped results for the unmanaged woodland habitat for each metal are given in Fig. 20.2. For Pb, Cu and Zn, critical loads are exceeded in central and south-eastern England where the critical loads are lower due to lower runoff. For Cu and Zn the critical limit is also exceeded in many of these areas, as would be expected based on an assumption of steady-state, but also across much of Wales, Northern Ireland and western areas of England. While this may reflect other inputs to the soils, for example, from agriculture and historical industrial/mining activity, it most likely reflects the large decrease in metal deposition across the UK since 1970, which will have led to a loss of critical load exceedance in these areas where the critical load is higher, but where the soil metal concentrations may not yet have fallen below the critical limit. The pattern observed for Cd is likely to have a similar explanation with decreased deposition leading to almost all of the UK falling below the critical load, but some areas of England and Wales still showing exceedance of the critical limit. In contrast, for Pb concentrations in soils are below the critical limit virtually everywhere, except in isolated areas of historical mining or industrial activity. As noted above, this most likely reflects the very long period required for Pb to reach soil concentrations at which the free-ion based critical limit is exceeded.

Table 20.6 Percentage area of habitats occurring in each of the four critical load and critical limit exceedance categories for each metal

Metal	Habitat	Percentage area of habitat in the following exceedance categories			
		No exceedance of critical load or critical limit	Only critical load exceeded	Only critical limit exceeded	Both critical load & critical limit exceeded
Cd[a]	Managed conifer	94	0	6	0
	Managed broadleaf	91	0	9	0
	Unmanaged wood	90	0	10	0
	Grassland	92	0	8	0
	Heathland	95	0	5	0
	Bog	99	0	1	0
	All habitats	92	0	8	0
Cu	Managed conifer	88	4	6	2
	Managed broadleaf	29	21	21	29
	Unmanaged wood	38	20	12	30
	Grassland	80	3	13	5
	Heathland	98	<0.1	2	<0.1
	Bog	99	<0.1	1	<0.1
	All habitats	81	5	8	7
Pb	Managed conifer	89	11	0.2	<0.1
	Managed broadleaf	37	62	0.2	0.4
	Unmanaged wood	41	58	0.2	0.3
	Grassland	96	3	1	<0.1
	Heathland	99	0.1	1	0
	Bog	99	0	1	0
	All habitats	87	13	0.5	0.1
Zn	Managed conifer	60	4	34	2
	Managed broadleaf	16	28	26	30
	Unmanaged wood	23	24	23	30
	Grassland	51	5	40	4
	Heathland	74	0.1	26	<0.1
	Bog	74	<0.1	26	<0.1
	All habitats	56	6	31	7

[a] Soil Cd concentration data are unavailable for Scotland and hence critical limit exceedances cannot be calculated for Scotland. Cd results in this table refer to data for England, Wales and Northern Ireland only

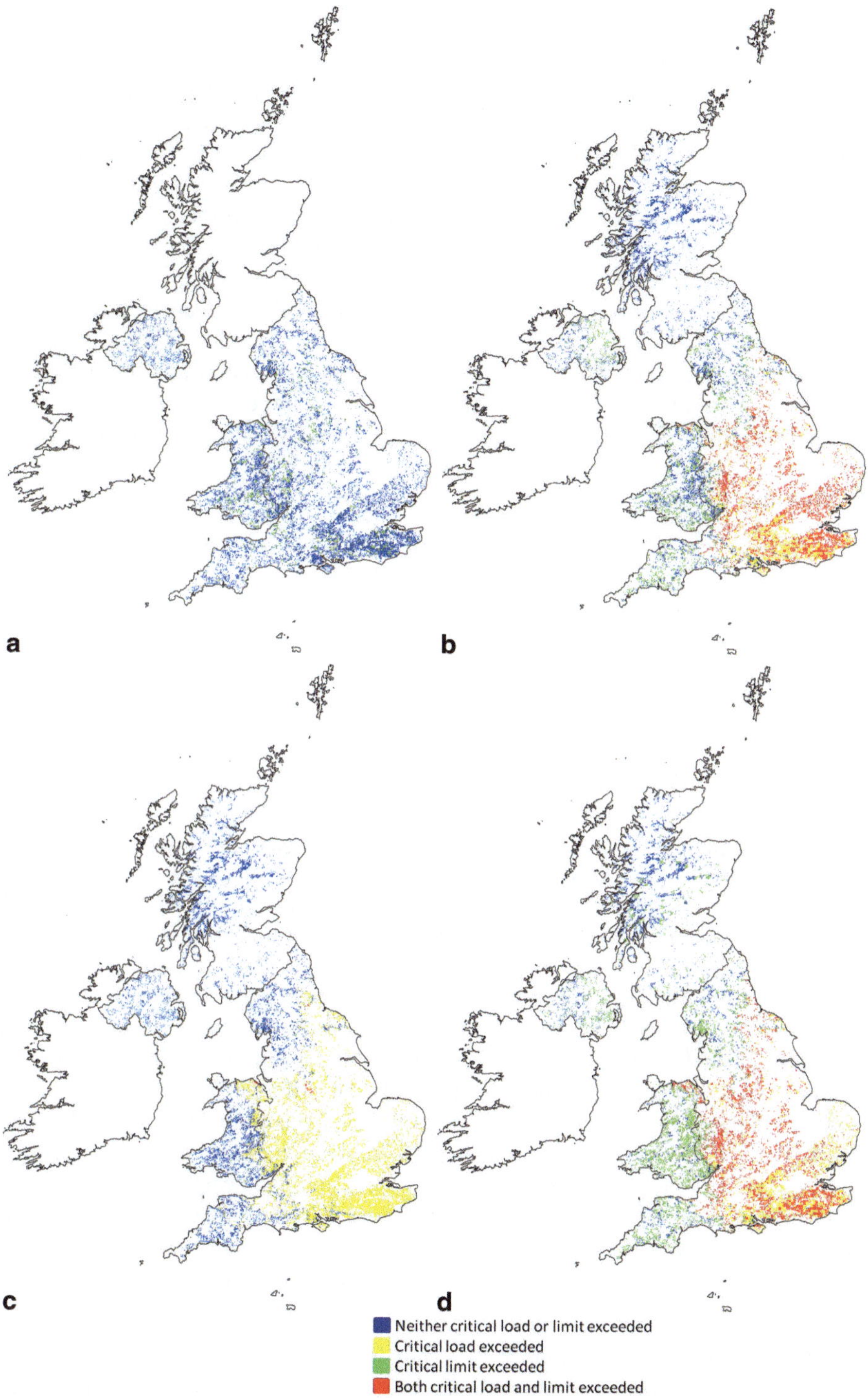

Fig. 20.2 Categories of critical load and critical limit exceedance for **a** Cd, **b** Cu, **c** Pb and **d** Zn for areas of unmanaged woodland in the UK. Note: Cd soil concentration data are unavailable for Scotland and hence this area of the map is left blank

Discussion and Conclusions

Exceedance of the critical load can be considered a potential threat to the ecosystem under steady-state conditions, while exceedance of the critical limit indicates a potential threat based on our critical limit risk model which is based on protection of 95% of species. If current deposition rates remain constant, we would predict that sites with neither critical load exceedance nor critical limit exceedance are not at risk of adverse impacts from heavy metals, and will not become so. Sites with critical load exceedance but not critical limit exceedance are currently not at risk but will become so at steady state. Sites without critical load exceedance but with critical limit exceedance may currently be impacted but could recover if deposition does not increase. Sites with exceedance of both critical load and critical limit are currently at risk of adverse impacts and this risk will not be reduced unless deposition is reduced further. Therefore, considering Table 20.6, the following conclusions can be drawn with regard to the threats from current rates of atmospheric deposition of heavy metal pollution to UK terrestrial ecosystems;

- Cadmium deposition does not pose a long-term threat, but cumulative deposition is still a minor risk factor for several habitat types;
- Copper and zinc deposition has accumulated sufficiently to be a risk factor, primarily in woodlands, and current deposition still poses a long-term threat in central and southern England;
- Lead poses a long-term threat to woodlands, even though it is not currently a significant risk factor.

The meaning of "long-term" is clearly important with respect to expected ecosystem responses (see also Chap.9). While critical load exceedance is important in indicating the potential for adverse environmental effects of atmospheric deposition of metals, it has significant limitations. The most important limitation of an assessment based on critical load exceedance only, is that the steady-state approach does not consider the dynamics of responses of environmental concentrations of metals (and hence their impacts) to increases or decreases in emissions. As demonstrated by this study, the exceedance patterns for critical loads can be quite different from those of critical limits, which show where thresholds for adverse effects are currently exceeded. A major reason for the different patterns of critical limit and critical load exceedance for the four metals is the different rates of change in the soil metal pools in relation to changes in metal deposition. In the UK, dynamic modelling studies (Tipping et al. 2006, 2010) showed that the dynamics of changes in heavy metal concentrations in soils and waters in upland catchments in response to changes in deposition are of the order of decades to centuries, or for Pb in particular, longer (see also Chap.9). This has a number of profound implications for the assessment of benefits using an approach such as critical loads that is based on steady-state assumptions:

- The benefits of measures taken now may only be experienced a long time into the future, with obvious implications for cost-benefit assessments. Unless time delay is taken into account (see also Chaps.8 and 9), the benefits of emission controls will be greatly over-estimated by using exceedance of a steady-state critical load as the criterion.

- Because of these long timescales and the fact that rates of metal deposition (in the UK) have changed significantly in the last 30 years, it is likely that there is almost nowhere in the UK in which the steady-state assumptions underlying a critical load based analysis are currently valid. This means that patterns of exceedance of critical limits for effects of metals in soils and waters now may be quite different from those of exceedance of critical loads at steady-state. For example, using the methods described above large areas of critical load exceedance are predicted for Pb (for some habitats), but only a small area of the country has soil Pb concentrations above the critical limit. In these circumstances, reductions in exceedance of critical loads would over-estimate the benefits of emission control because adverse effects are not expected where critical limits are not exceeded.

The very long timescale of response to changes in emissions of metals to the atmosphere may mean that other environmental factors and hence other policy measures may have a greater impact on metal exposure and effects over this century than reductions in metal emissions. For example, results from UK dynamic modelling (Tipping et al. 2006) suggest that effects of nitrogen deposition leading to increased nitrate leaching could significantly influence the mobility of metals and the rates of change of concentrations in soils and waters. Hence, other policies not specifically focussed on metal emissions could in principle have a greater cost-benefit in terms of reducing environmental exposures to metals, or could negate any benefits of measures to reduce metal emissions. Where there is a high metal burden in soils due to historical deposition, it is even possible that increases in stream concentrations of Pb and Cu can occur as rates of deposition of these metals decrease.

The current steady-state methods result in high critical loads in areas of high runoff allowing metal to leach from soils (into waters) as a mechanism of protecting the soil and soil organisms from the adverse impacts of metal pollution. The potential impacts of these leached metals in freshwater systems will depend on whether the metal remains bound to dissolved organic matter and/or suspended particulate matter and on the pH of the surface water. In addition, the methods take no account of possible different soil conditions in the future, nor are mixture effects (the combined toxic effects of more than one metal) included.

References

Ashmore, M., van den Berg, L., Terry, A., Tipping, E., Lawlor, A. J., Lofts, S., Thacker, S. A., Vincent, C. D., Hall, J., O'Hanlon, S., Shotbolt, L., Harmens, H., Lloyd, A., Norris, D., Nemitz, E., Jarvis, K., & Jordan, C. (2007). *Development of an effects-based approach for toxic metals*. (Report to the UK Department for Environment, Food and Rural Affairs, the Scottish Executive, the National Assembly for Wales and the Department of the Environment in Northern Ireland. Contract CPEA 24). University of York.

De Vries, W., Lofts, S., Tipping, E., Meili, M., Groenenberg, B. J., & Schütze, G. (2007). Impact of soil properties on critical concentrations of cadmium, lead, copper, zinc and mercury in soil and soil solution in view of ecotoxicological effects. *Reviews of Environmental Contamination and Toxicology, 191,* 47–89.

Fowler, D., McDonald, A. G., Crossley, A., Nemitz, E., Leaver, D., Cape, J. N., Smith, R. I., Anderson, D., Rowland, P., Ainsworth, G., Lawlor, A. J., Guyatt, H., & Harmens, H. (2006). *UK heavy metal monitoring network.* (Report to Defra: Project Number EPG 1/3/204).

Groenenberg, J. E., Römkens, P. F. A. M., Comans, R. N. J., Luster, J., Pampura, T., Shotbolt, L., Tipping, E., & De Vries, W. (2010). Transfer functions for solid-solution partitioning of cadmium, copper, nickel, lead and zinc in soils: derivation of relationships for free metal ion activities and validation with independent data. *European Journal of Soil Science, 61,* 58–73.

Hall, J., Ashmore, M., Fawehinmi, J., Jordan, C., Lofts, S., Shotbolt, L., Spurgeon, S., Svendsen, C., & Tipping, E. (2006). Developing a critical load approach for national risk assessments of atmospheric metal deposition. *Environmental Toxicology and Chemistry, 25,* 883–890.

Jordan, C., Higgins, A. J., Hamill, K. P., & Cruickshank, J. G. (2000). The Soil Geochemical Atlas of Northern Ireland, Millenium Edition. Belfast, Ireland: Agricultural and Environmental Science Division, Department of Agriculture and Rural Development.

Lofts, S., Spurgeon, D. J., Svendsen, C., & Tipping, E. (2004). Deriving soil critical limits for Cu, Zn, Cd and Pb: a method based on free ion concentrations. *Environmental Science & Technology, 38,* 3623–3631.

McGrath, S. P., & Loveland, P. J. (1992). *The soil geochemistry atlas of England and Wales*. London: Chapman & Hall.

Merrington, G., & Schoeters, I. (2011). *Soil quality standards for trace elements.* Boca Raton: CRC Press.

Niyogi, S., & Wood, C. M. (2004). Biotic ligand model, a flexible tool for developing site-specific water quality guidelines for metals. *Environmental Science & Technology, 38,* 6177–6192.

Paquin, P. R., Santore, R. C., Wu, K. B., Kavvadas, C. D., & Di Toro, D. M. (2000). The biotic ligand model: a model of the acute toxicity of metals to aquatic life. *Environmental Science &* Technology, *3,* 175–182.

RoTAP. (2012). *Review of transboundary air pollution: acidification, eutrophication, ground level ozone and heavy metals in the UK.* (Contract Report to the Department for Environment, Food and Rural Affairs). Centre for Ecology and Hydrology. http://www.rotap.ceh.ac.uk.

Steenbergen, N. T. T. M., Iaccino, F., De Winkel, M., Reijnders, L., & Peijnenburg, W. J. G. M. (2005). Development of a biotic ligand model and a regression model predicting acute copper toxicity to the earthworm Aporrectodea caliginosa. *Environmental Science & Technology, 39,* 5694–5702.

Stockdale, A., Tipping, E., Lofts, S., Ormerod, S. J., Clements, W. H., & Blust, R. (2010). Toxicity of proton-metal mixtures in the field: linking stream macroinvertebrate species diversity to chemical speciation and bioavailability. *Aquatic toxicology, 100,* 112–119.

Thakali, S., Allen, H. E., Di Toro, D. M., Ponizovsky, A. A., Rooney, C. P., Zhao, F.-J., McGrath, S. P., Criel, P., van Eeckhout, H., Janssen, C. R., Oorts, K., & Smolders, E. (2006). Terrestrial Biotic Ligand Model. 2. Application to Ni and Cu toxicities to plants, invertebrates, and microbes in soil. *Environmental Science & Technology, 40,* 7094–7100.

Tipping, E. (1994). WHAM—A chemical equilibrium model and computer code for waters, sediments, and soils incorporating a discrete site/electrostatic model of ion-binding by humic substances. *Computers and Geoscience, 20,* 973–1023.

Tipping, E. (1998). Humic ion-binding Model VI: an improved description of the interactions of protons and metal ions with humic substances. *Aquatic Geochemistry, 4,* 3–47.

Tipping, E., Rieuwerts, J., Pan, G., Ashmore, M. R., Lofts, S., Hill, M. T. R., Farago, M. E., & Thornton, I. (2003). The solid-solution partitioning of heavy metals (Cu, Zn, Cd, Pb) in upland soils of England and Wales. *Environmental Pollution, 125,* 213–225.

Tipping, E., Lawlor, A. J., Lofts, S., & Shotbolt, L. (2006). Simulating the long-term chemistry of an upland UK catchment: heavy metals. *Environmental Pollution, 141,* 139–150.
Tipping, E., Rothwell, J. J., Shotbolt, L., & Lawlor, A. J. (2010). Dynamic modelling of atmospherically-deposited Ni, Cu, Zn, Cd and Pb in Pennine catchments (northern England). *Environmental Pollution, 158,* 1521–1529.

Chapter 21
Critical Loads of Cadmium, Lead and Mercury and Their Exceedances in Europe

Jean-Paul Hettelingh, Gudrun Schütze, Wim de Vries, Hugo Denier van der Gon, Ilia Ilyin, Gert Jan Reinds, Jaap Slootweg and Oleg Travnikov

21.1 Introduction

Cadmium (Cd), lead (Pb) and mercury (Hg) are known to be transported over relatively long distances from their sources. Deposited metals may accumulate over time in soils and catchments, and then follow varying pathways to endpoints in humans and the environment. Cadmium and lead, that are emitted primarily as particulate matter, and mercury in gaseous form, can be transported over long distances. This is especially the case when emissions consist of either relatively small particles or of gaseous compounds. The latter occurs especially for anthropogenically emitted mercury, tetra-methyl lead and tetra-ethyl lead.

Under the effect-oriented programmes within the LRTAP Convention, the International Cooperative Programmes on Integrated Monitoring and Forests have found that atmospherically deposited Pb and Hg can accumulate in forested catchments in northern-central Europe, and that depositions can exceed thresholds for effects in soil organisms. The combination of Cd that is atmospherically deposited, used in fertilizer and naturally occurring in soils may accumulate to exposures of concern (TFHM 2006; WHO 2007).

J.-P. Hettelingh (✉) · J. Slootweg
Coordination Centre for Effects (CCE), RIVM, Bilthoven, The Netherlands
e-mail: jean-paul.hettelingh@rivm.nl

G. Schütze
Umweltbundesamt, Dessau, Germany

W. de Vries · G. Jan. Reinds
Alterra Wageningen University and Research Centre, Wageningen, The Netherlands

H. Denier van der Gon
Climate, Air and Sustainability, TNO, Utrecht, The Netherlands

I. Ilyin · O. Travnikov
EMEP-Meteorological Synthesizing Centre East, Moscow, Russia

W. de Vries et al. (eds.), *Critical Loads and Dynamic Risk Assessments*,
Environmental Pollution 25, DOI 10.1007/978-94-017-9508-1_21

Table 21.1 Overview of indicators used in the computation of critical loads. (adapted from Chap. 5.5 in the Modelling and Mapping Manual; www.icpmapping.org)

Receptor ecosystem	Endpoints	Heavy metals of concern	Land cover types to be considered	Indicator/critical limit	Effect number[a]
Terrestrial	Human health effects	Cd, Pb, Hg	All ecosystems	Total concentration in soil water below the rooting zone(to protect ground water)	1
		Cd	Arable	Content in food, fodder and crops	2
		Cd	Grassland	Content in grass and animal products (cows, sheep)	
	Ecosystem functions	Cd, Pb	Arable land, grassland, non-agricultural	Free ion concentration in view of effects on soil micro-organisms, plants and invertebrates	3
		Hg	Forest soils	Total concentration in humus layer in view of effects on soil micro organisms and invertebrates	
Aquatic	Ecosystem functions	Cd, Pb	Freshwaters	Total concentration in view of effects on algae, crustacea, worms, fish, top predators	4
	Human health	Hg	Freshwaters	Concentration in fish	5

[a] *1* = human health effect (drinking water) via terrestrial ecosystem, *2* = human health effect (food quality) via terrestrial ecosystems, *3* = Ecotoxicological effect on terrestrial ecosystems, *4* = Ecotoxicological effect on aquatic ecosystems, *5* = human health effect (food quality) via aquatic ecosystems

Cadmium, lead and mercury are neither essential trace elements for human beings nor for plants or animals. Uptake of these metals about certain thresholds is a threat for all organisms. It also jeopardizes human health, whereby food is the most important pathway of uptake (TFHM 2006; WHO 2007). The critical load calculations related to the protection of human health effects focus on the transport of the individual metals via soil, groundwater and/or surface waters into food or drinking water (see Table 21.1), while other pathways of uptake are neglected.

Ecotoxicological effects of Pb, Cd and Hg include visible injury of plant leaves, hampered growth or reproduction as well as changes in physiologic and microbial conversion processes. The critical concentrations in environmental compartments used in critical load calculations (critical limits) have been determined in laboratory or, to a smaller extent, field experiments. More details on the scientific reasoning of the critical limits can be found in Chap. 2 and below in Sect. 21.2.3.2.

Critical loads of heavy metals have been addressed over the past two decades for possible use in support of European air pollution abatement policies in general, and the review and possible revision of the 1998 heavy metals protocol in particular. The scientific basis of the development of critical loads of heavy metals for policy support in Europe started with the European Soil & sea QUality due to Atmospheric Deposition (ESQUAD) project (Van den Hout et al. 1999; Van den Hout 1994). ESQUAD focused on the impacts of non-acidifying depositions on the quality of European forest soils and the North Sea. Since then, methodologies were gradually improved by technical groups under the LRTAP Convention and described in a number of reports (Curlík et al. 2000; Gregor et al. 1999, Gregor et al. 1997; Schütze et al. 2003; Schütze and Hettelingh 2006). The first manuals for modelling and mapping of critical loads of heavy metals were published at the end of the last century (De Vries and Bakker 1996, 1998; De Vries et al. 1998). An update of this manual has been made in 2005 (De Vries et al. 2005), and this approach has then been adopted under the LRTAP Convention for the calculation of critical loads for cadmium, lead and mercury (Mapping Manual, www.icpmapping.org).

Results of a first European application under the effect-oriented programme of the LRTAP Convention is described in Hettelingh et al. (2002) and Slootweg et al. (2005). European Parties under the Convention collaborated under the International Cooperative Programme on Modelling and Mapping of Critical Levels and Loads and Air Pollution Effects, Risks and Trends (ICP M&M) to improve the European critical loads database held at the Coordination Centre for Effects (CCE). Model analyses of the risk of heavy metal depositions computed for European emission reduction alternatives have been carried out since 2006 to provide background information to the policy process preceding the review of the Aarhus protocol.

This chapter summarizes recent results of integrated assessment of heavy metal emissions, depositions and critical load exceedances for the target years 2010 and 2020. Emphasis is put on methods and data to compile the European critical loads database that is then used in the integrated assessment to compute critical load exceedances for selected years in Europe. Emissions of cadmium, lead and mercury in 2010 have been derived from data reflecting current legislation regarding the application of abatement techniques. For 2020 a scenario is assumed whereby emissions in Europe would be the result of a full implementation of the 1998 heavy metals protocol.

21.2 Methods and Data

This section summarizes the methodologies and data to obtain country emissions of cadmium, lead and mercury to the atmosphere, their dispersion and their deposition. Furthermore, the methodologies to derive critical loads and to compute critical load exceedances in European ecosystems in 2010 and 2020 are described.

21.2.1 Emissions of Cadmium, Lead and Mercury

Visschedijk et al. (2010) provides a detailed description of the compilation of heavy metal emissions used in this study. The emission data for 2010 and 2020 were compiled in support of the revision of the 1998 Aarhus protocol on heavy metals (UNECE 1998). Emission projections for these years were compiled as two scenarios, i.e. Current LEgislation and current ratification (CLE2010) and Full Implementation of the heavy metal protocol (FI2020).

CLE2010 emission data consist of both projected as well as officially reported emissions, but at the time of the compilation the year 2010 emissions were not yet available. The officially reported emission data for 2000 as well as country reports for the period 2003–2007, both available from the Centre on Emission Inventories and Projections (www.ceip.at), were used as basis. Not all countries report heavy metal emission data to the Centre on Emission Inventories and Projections (CEIP) of the Cooperative Programme for Monitoring and Evaluation of the Long-Range Transmission of Air Pollutants in Europe (EMEP), while others report data with significant unsubstantiated variations between years or only part of the source sectors. Using the year 2000 emission inventory (Denier van der Gon et al. 2005) as a starting point, projected emissions for 2010 were developed subject to assumptions regarding activity data such as energy use by fuel type in power plants and specific sectoral activities including combustion and process emissions in the relevant industries (e.g., iron and steel industry, non-ferrous metal industries and cement industries). The projection of relevant heavy metal emitting activities were based on a baseline scenario developed for the revision of the Gothenburg protocol (Amann et al. 2008).

FI2020 emission data reflect projected national emissions of European countries assuming that all countries implemented the 1998 heavy metal protocol in 2020. Changes as a result of full implementation do furthermore not occur in countries where more stringent legislation is already in place (such as the IPPC Directive in the EU27). The relative emission reductions per type of emission abatement measure for countries going from current legislation to full implementation of the 1998 heavy metal protocol were taken from Denier van der Gon et al. (2005). In absolute terms the emission reductions are consistent with the baseline scenario from Amann et al. (2008). It turns out that FI2020 leads to a small increase of mercury emissions. The reason is that the application of emission abatement techniques under FI2020 does not offset the emission increase caused by the growth of coal-fired

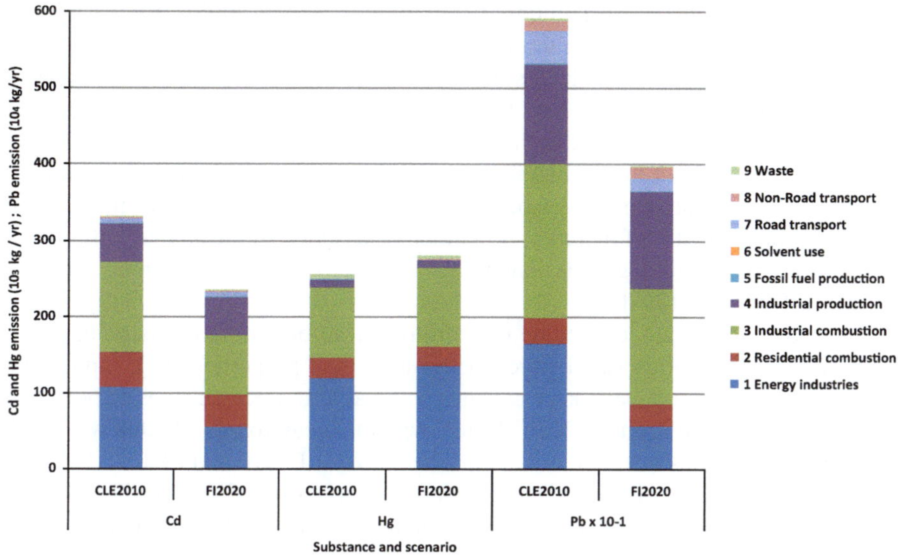

Fig. 21.1 Emissions of Cd, Hg and Pb per sector under the CLE2010 and FI2020 scenarios

power plants for which no specific additional (gaseous) mercury control measures are planned. Figure 21.1 illustrates the Cd, Pb and Hg emissions for each activity for CLE2010 and FI2020.

Figure 21.1 shows that the emission changes of Cd, Pb and Hg under FI2020 compared to CLE2010 are −29, −33 and +10%, respectively. The major emitting sectors are shown to be industrial combustion and energy industries. European mercury emissions from the latter turn out to increase by about 25 kt yr^{-1} from 255 kt yr^{-1} under CLE2010 to about 280 kt yr^{-1} under FI2020. Full implementation of the HM protocol in 2020 brings about a substantial reduction of Cd and Pb emissions in Europe. Emission reductions of Cd and Pb following the FI2020 scenario compared to CLE2010 amount to about 100 kt yr^{-1} and 1931 kt yr^{-1}, respectively. Emissions of all compounds could be further reduced, relative to CLE2010, if best available emission abatement technologies were applied (see Visschedijk et al. 2010).

21.2.2 Dispersion and Deposition of Cadmium, Lead and Mercury

Calculations of atmospheric transport and deposition resulting from CLE2010 and FI2020 emissions of Cd, Pb and Hg over the EMEP-domain[1] were carried out using

[1] The map projection of the European Parties to the LRTAP Convention is described in http://www.emep.int/grid/EMEP_domain.pdf.

the EMEP heavy metal (HM) model (Travnikov and Ilyin 2005). It is a three-dimensional eulerian model with terrain-following vertical coordinate. Vertically the model domain is split into 15 layers from the surface to about 15 km. The thickness of the lowest layer is around 70 m. The horizontal spatial resolution of the model is 50×50 km^2. The model includes the main processes governing atmospheric transport and deposition of heavy metals, such as dispersion, wet and dry ecosystem-dependent deposition, wind re-suspension of particles containing heavy metals, and atmospheric chemistry of mercury. A detailed description of the analysis of the dispersion of Cd, Pb and Hg emissions under CLE2010 and FI2020 can be found in Ilyin and Travnikov (2010).

Input data to the model include meteorological information, emission data and geophysical information (land cover distribution, soil properties etc.). Input data, except emissions, have been kept constant to be able to trace changes in pollution levels due to CLE2010 and FI2020. That is why the meteorology of 2008 was used, while land cover was assumed not to vary between CLE2010 and FI2020. Keeping meteorology constant also implies that wind re-suspension of historically accumulated heavy metals in soils is kept constant in the calculation of heavy metal depositions under CLE2010 and FI2020.

The emission data cover the territory of Europe. However, emissions from the Central Asian region, the eastern part of Russia, northern Africa and other remaining parts of Asia are needed as well for dispersion and deposition assessments. Therefore, additional emission data (Ilyin et al. 2009) were used. Other sources included emission totals for Kyrgyzstan and Kazakhstan (Denier van der Gon et al. (2005). Pb emission totals in the eastern part of Russia, in Turkmenistan, Tajikistan and Uzbekistan, in remaining parts of Asia and northern Africa were compiled from Pacyna et al. (1995) and scaled to present time. Emissions of Hg for these regions were derived from the global mercury inventory for 2005 (AMAP/UNEP 2008). No changes in Hg emission were assumed between 2005 and 2020. Global emission inventories for Cd are currently not available. That is why Cd emission data for the Asian part of the EMEP domain and for northern Africa were obtained on the basis of the global mercury inventory (AMAP/UNEP 2008). Cd emissions were assumed to be proportional to Hg emissions with a coefficient depending on the region (Ilyin and Travnikov 2010). The spatial distribution of all emissions for the Asian part of the EMEP domain and northern Africa was obtained by interpolating global gridded emissions with a $1° \times 1°$ spatial resolution into the model grid.

It should be noted that the treatment of wind re-suspension is an important element in the calculation of depositions and related critical load exceedances. The concentration of heavy metals in top soils is one of the key parameters for computing wind re-suspension. This concentration consists of a natural and of a historical component, accumulated over a long (decades to centuries) period of anthropogenic pollution. Emissions of heavy metals in Europe generally decline since about 1990. Therefore, in the long run, also soil concentrations can be expected to decrease. However over a short period, i.e. between 2010 and 2020 for our analysis, it is assumed that changes in soil concentrations are relatively minor and the resulting influence of wind re-suspension was included as a constant factor in the calculation

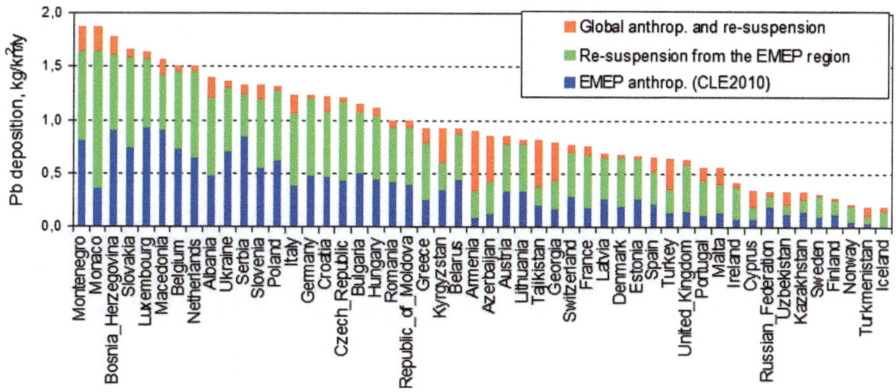

Fig. 21.2 Country-averaged deposition fluxes (kg km^{-2} yr^{-1}) of lead from European and Central Asian anthropogenic sources due to emissions under the CLE2010 scenario (*blue*), from natural sources and wind re-suspension within the EMEP region (*green*) and from both anthropogenic and other sources from non-EMEP regions in 2010

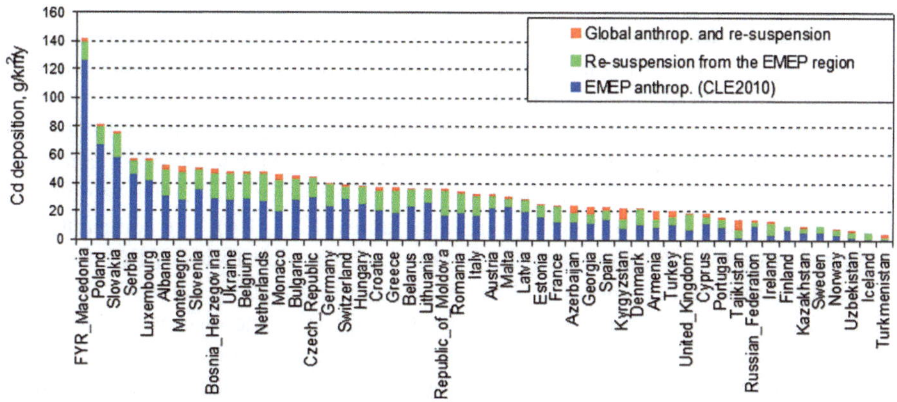

Fig. 21.3 Country-averaged deposition fluxes (g km^{-2} yr^{-1}) of cadmium from European and Central Asia anthropogenic sources due to emissions under the CLE2010 scenario (*blue*), from wind re-suspension within the EMEP region (*green*) and from both anthropogenic and other sources from non-EMEP regions in 2010

of depositions in the two target years. Wind re-suspension typically (but not always) causes the magnitude of the reduction of anthropogenic emissions, e.g. between CLE2010 and FI2020, to result in a relatively smaller decline of deposition.

For example, the contribution of wind re-suspension to Pb deposition (Fig. 21.2) amounts to about 39 % on average for Europe and central Asia as a whole, and can be as high as 75 % in Iceland. However, Cd deposition (Fig. 21.3) is largely caused by emission sources within the EMEP region.

Similar to Pb and Cd, the decline of Hg deposition in European countries is smaller than the reduction of anthropogenic emissions. The reason is the global

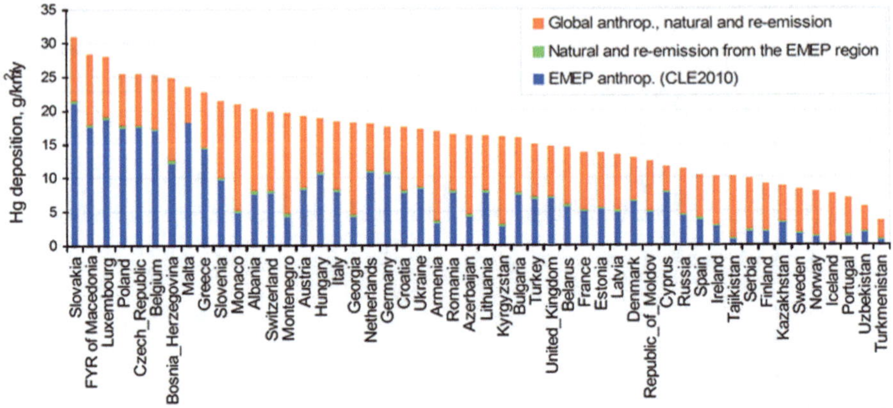

Fig. 21.4 Country-averaged deposition fluxes (g km^{-2} yr^{-1}) of mercury from European and Central Asia anthropogenic sources due to emissions under the CLE2010 scenario (*blue*), natural emission and re-emission within the EMEP region (*green*) and from both anthropogenic and other sources from non-EMEP regions in 2010

character of Hg transport through the atmosphere. Because of a long (about a year) atmospheric lifetime of Hg its spatial distribution in the global atmosphere is almost uniform compared to that of Pb and Cd.

Therefore, a significant part of Hg deposition in Europe is caused by long-range transport from anthropogenic sources in regions located outside European and Central Asian countries. This contribution is around 60 % for the EMEP region, and ranges from about 20 to 95 % in individual counties (Fig. 21.4). The contribution to Hg deposition by natural sources and re-emission from territories of European and Central Asian countries within the EMEP region (green shading in Fig. 21.4) is relatively small compared to Pb (Fig. 21.2) and Cd (Fig. 21.3) depositions from similar origins. This can be explained by the fact that natural emission and re-emission of Hg are thought to occur in gaseous elemental form, which is slowly removed from the atmosphere and quickly leaves the region due to atmospheric transport.

In contrast to natural emission and re-emission, anthropogenic emissions of Hg occur in three forms: long-lived gaseous elemental Hg, relatively short-lived gaseous oxidized and particulate Hg. Because of a relatively short atmospheric lifetime most part of gaseous oxidized and particulate mercury emitted in the EMEP region is deposited within the region.

The consequence of contributions from sources that are originating from non-EMEP regions, re-suspension, re-emissions or nature is that the share of anthropogenic within-EMEP sources varies over countries as illustrated (blue shading) in Figs. 21.2 and 21.3. The result is that the effect of emission reductions can, in some cases, be drowned out by relatively stable depositions from wind re-suspension or from sources outside of the EMEP domain.

This uncertainty propagates in the assessment of a relatively load exceedances, the changes of which cannot be uniquely attributed to the reduction of heavy metal emissions. Therefore improvements in the change of critical load exceedances (next section) as obtained between CLE2010 and FI2020, cannot always be only associated with emission reductions of specific countries. However, it is also evident that a global effort is needed to reduce anthropogenic sources of Hg emissions.

21.2.3 Critical Limits and Loads for Human Health and Ecotoxicological Effects

This section summarizes how the available information on the fate of the three metals in the environment and on effects of heavy metals on human health and biota has been translated into a methodology to calculate critical loads (see De Vries et al. 2005) and the Mapping Manual (www.icpmapping.org) for details and a complete list of references).

Critical limits for indicators of effects of cadmium, lead and mercury on the functioning of ecosystems and human health have been established for use in the computation of critical loads (see Chap. 2). Table 21.1 lists these indicators and shows that five effects are distinguished (last column). A similar method has been used to assess critical loads of other metals (Reinds et al. 2006). These are not included in this chapter because their assessment is not different from the metals addressed here, while their critical loads turn out to be hardly exceeded (Hettelingh et al. 2006).

21.2.3.1 Computing Critical Loads

Following the detailed description in Chap. 7, the critical load of a metal is calculated as the sum of the metals removed by harvest of timber or agricultural products plus an acceptable metal leaching rate, according to:

$$CL(M) = M_u + M_{le(crit)} \tag{21.1}$$

where

$CL(M)$ is the critical load of metal M (g $ha^{-1}yr^{-1}$)

M_u is the net metal uptake in harvestable biomass under critical load conditions (g $ha^{-1}yr^{-1}$)

$M_{le(crit)}$ is the critical leaching flux of metal M from the considered soil layer (g $ha^{-1}yr^{-1}$), whereby only the vertical drainage flux is considered.

Using this approach, the internal metal cycling within an ecosystem is ignored, since its influence on critical loads is relatively small, at least for Pb, Cd, Hg (De Vries et al. 2005). Because weathering of Pb, Cd, Hg causes only a minor input flux to topsoils, while uncertainties of such calculations are high, this flux was also neglected. In case of Hg, re-suspension (volatilization) of deposited Hg is ignored, because this flux is already treated as part of the atmospheric net deposition in the modelling of the dispersion of heavy metals (see above).

Critical loads of metals for terrestrial ecosystems are focused on the top soil. The soil depth to be considered in the quantification of metal fluxes depends on the receptor type and effect that is addressed (see Table 21.1). The metal net uptake is calculated from the annual yield (removal or increment) of biomass (kg $ha^{-1}yr^{-1}$) times a metal concentration (g kg^{-1}) in the harvested parts of plants. Information on annual yields can be obtained from agricultural statistics or forest growth tables. It can also be modelled by relating yields to site characteristics such as soil quality, climate and land use. The site-specific share of different crops on agricultural land has to be considered (see also Chap. 7). In general, independently from metal concentrations in soils, the median of measured metal concentrations in the harvested parts of plants from relatively unpolluted areas were used in the calculation of uptake, since there is hardly any relationship between metal contents in soil and in harvestable biomass with the exception of Cd in wheat grains (see below). These median values in general neither exceed limits for food and fodder nor phytotoxic limits. The related uptakes are therefore considered to be tolerable. However, in the calculation of the critical load of Cd regarding protection of food quality, the critical limit of 0.1 mg kg^{-1} in wheat grains has been used to be consistent with the critical limit for soil drainage water. In these calculations, which were only performed for arable land, it was assumed that all of the crop is wheat.

The tolerable leaching flux of a metal is calculated from the flux of drainage water leaching from the considered soil layer (m^3 ha^{-1} yr^{-1}) times a critical total concentration of the metal in the soil drainage water (g m^{-3}). The flux of drainage water is calculated from data on climate, land cover and soil type of the site.

As with terrestrial ecosystems, the critical load of Cd and Pb for freshwaters equals the sum of tolerable outputs by harvest within the catchment and the critical outflow from a catchment. The uptake into harvestable parts of plants is calculated in analogy to terrestrial ecosystems, while harvest in this context means in general the harvest of wood in forested catchments. In the calculation of the critical outflow of metals the lateral water flux off the catchment is multiplied with the critical total concentration in the surface water. The latter is derived from the critical limit considering the binding of metals to suspended particulate matter (SPM) and the concentration of SPM in the surface water. In calculations for lakes also net retention of metals can be considered. For a more detailed description of methods see Mapping Manual (www.icpmapping.org) and De Vries et al. (2005).

21.2.3.2 Critical Limits for Human Health Effects

Critical limits were derived for human health effects caused by the oral uptake of heavy metals via drinking water (Table 21.1; effect number 1), crops and livestock (Table 21.1; effect number 2) and fish from inland waters (Table 21.1; effect number 5).

To protect ground water that is used as drinking water (Table 21.1; effect 1) it was assumed that the Cd, Pb or Hg concentration in soil drainage water below the rooting zone should not exceed the values recommended in WHO (2004), i.e. 3, 10 and 1 μg l^{-1}, respectively. These critical limits are expressed as total concentrations in soil drainage water. For Pb this is the only pathway considered with respect to human health protection.

Food quality criteria for metals in crops or target organs (liver and kidney) of grazing animals are available for Cd, Pb and Hg (see Chap. 2). These criteria, however, need to be transferred to critical limits in soil solution to be of use for critical load calculations. A calculation from critical concentrations in crops to critical concentrations in soil solution could not be performed for Pb and Hg since these metals are hardly taken up by plant roots, making correlations between concentrations in these media impossible. Consequently, only the Cd concentration in wheat was used for addressing human health effects via food intake (Table 21.1; effect 2). The EU regulation (EG) No.466/2001 proposed a critical limit for Cd in wheat of 0.10 mg kg^{-1} fresh weight. This limit was later amended by (EC) No 629/2008 to 0.2 mg kg^{-1} fresh weight. The latter limit for Cd in wheat grain is, however, not based on effects but on the ALARA principle, i.e. "As Low As Reasonably Achievable". An effects-based food quality criterion for wheat of 0.1 mg kg^{-1} (fresh weight) was thus recommended for use under the Long-range Transboundary Air Pollution (LRTAP) Convention (Schütze et al. 2003). This limit was used in the critical load calculations for Cd, by transferring it to critical Cd concentrations in soil solution. First, soil -plant relationships were used to transform the critical limit for Cd in wheat grains to critical total Cd concentrations in soil (see also Chap. 2, Table 2.2). Next, a relationship between the total soil Cd content and the total Cd concentration in solution was used to derive critical Cd concentrations in soil solution (see also Chap. 2, Table 2.3). Following this approach the pathway of Cd to wheat leads to the lowest critical Cd concentration in soil solution compared to other food and fodder crops, thus protecting against human health effects via all possible food uptake pathways (De Vries et al. 2003).

The quantification of the risk of human health effects of Hg through fish consumption (Table 21.1; effect 5) is not related to a critical load but to a critical concentration of Hg in precipitation (mg l^{-1}) entering a catchment. This critical Hg concentration in precipitation can be calculated with a simple model from the critical Hg concentration in fish assuming a steady-state situation in the catchment (Meili 1997; Meili et al. 2003). A limit of 0.3 mg kg^{-1} fresh weight on total Hg in fish is consistent with recommendations by the US-EPA (2001) and the UN-FAO/WHO-JECFA (2003) and therefore used in the calculation. This limit is frequently exceeded in Nordic surface waters already now, although steady state is often not yet reached and Hg continues to accumulate.

21.2.3.3 Critical Limits for Ecotoxicological Effects

Critical limits for ecotoxicological effects on ecosystem functions of terrestrial (Table 21.1; effect number 3) and aquatic ecosystems (Table 21.1; effect number 4) were derived from critical limits for biota listed in Table 21.1. For areas covered by non-agricultural land (forests, semi-natural vegetation) or agricultural land (arable land and grassland), soil toxicity data for plants, invertebrates and soil micro-organisms (EC 2007) were used as basis for the derivation of critical limit functions for free metal ion concentrations (Pb, Cd) in soil solution. The critical limits have been derived by assuming that a free metal ion approach is most appropriate to account for the influence of the bioavailability of metals in view of effects on related organism groups. The bioavailability of metals does, however, not only depend on the free metal ion concentration but also on the concentration of other cations competing with the metal of concern for bio-ligands. Such relationships can be synthesised to be a function of the pH. This was taken into account in deriving critical limits as a function of pH in soil drainage water. The method of derivation of the critical limit function is described in detail in Lofts et al. (2004). The critical limits for free metal ions were translated to total critical concentrations in soil solution using the WHAM model (Tipping 1994, 1998, updated by version W6S-MTC2 in 2004). The main principles and the choice of WHAM input parameters (pH value, organic matter content of the soil, the concentrations of dissolved organic carbon and suspended particulate matter in soil solution, partial pressure of CO_2) are explained in De Vries et al. (2005) and summarized in the Mapping Manual (www.icpmapping.org) (see also Chap. 7).

Organic forest (top) soils are considered as the best understood critical receptor with respect to atmospheric Hg pollution, based on knowledge on effects on microbial processes and invertebrates. The suggested critical limit for Hg is that the concentration in the humus layer of forest soils after normalization with respect to the organic matter content should not exceed 0.5 mg (kg org)$^{-1}$ (Meili et al. 2003). The strong association of Hg with organic matter leaves virtually no free ions. Therefore the exposure of biota to Hg is controlled by the competition between biotic and other organic ligands. Furthermore the contamination of all types of organic matter is determined by the supply of organic matter relative to the supply of Hg at a given site (Meili 1991, 1997, Meili et al. 2003). As a result, the critical limit for Hg in soils is set for the organically bound Hg rather than for the free ion concentration, also in solution. The transformation to a critical total Hg concentration in soil drainage water is performed using a fractionation ratio for the Hg contamination of organic matter in solution relative to that in solids (Åkerblom 2006; Åkerblom et al. 2008; Meili et al. 2003).

In order to quantify ecotoxicological effects of Cd in surface waters the 5-percentile cut-off value of chronic toxicity from the EU risk assessment report (EC 2007) was used as critical limit. Deviating from the draft EU risk assessment no additional assessment factor was applied. For Pb the critical limit is based on Crommentuijn et al. (1997) and Skjelkvåle and Ulstein (2002). These critical limits of Cd, Pb are provided as total dissolved concentrations. The free metal ion approach could not be used to derive critical limits of Pb and Cd due to limitations in the effects database for aquatic systems.

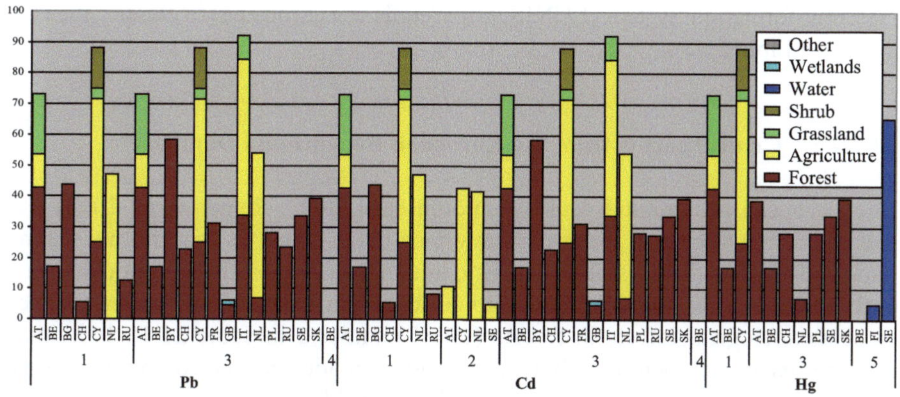

Fig. 21.5 National distribution of ecosystem types (% of country area) by effect (see Table 21.1) for critical loads of lead, cadmium and mercury (countries are identified by their 2-letter codes)

21.2.4 European Critical Loads Data

European data on heavy metals critical loads have been compiled from 18 National Focal Centres that submitted information. The CCE European background database was used for countries that did not submit data. A complete overview of the information can be found in Slootweg et al. (2005), including documentation by countries submitting critical loads data.

21.2.4.1 Critical Loads from National Data

Figure 21.5 shows the percentage of the total country area for which critical loads have been submitted for cadmium, lead and mercury. Ecosystem types are distinguished including the effect (see Table 21.1) for which critical loads have been computed. Critical loads for cadmium and lead are predominant, with an emphasis on ecotoxicological effects for terrestrial ecosystems (effect 3). Finland, Sweden and Belgium have submitted data for the human health effects for aquatic ecosystems (food quality, effect 5). Only Belgium has also submitted the ecotoxicological effects for aquatic ecosystems (effect 4). Eighteen countries submitted data on critical loads on one or more of the three heavy metals. It turned out that critical loads data of Cd were submitted by a maximum of 14 countries addressing effect 1 (submitted by 10 countries), effect 2 (submitted by 5 countries), effect 3 (submitted by 14 countries) and effect 1 (submitted by a single country). Critical loads of Pb were submitted by 10, 14 and 1 countries addressing effects 1, 3 and 4, respectively. Finally, Hg effects 1, 3 and 5 were considered by 5, 7 and 3 countries, respectively.

Forest is the dominant ecosystem considered in most of Europe, but also agricultural areas were considered by some countries. Countries applied the European

Nature Information System (EUNIS; http://eunis.eea.europa.eu) to classify the ecosystems (Table 21.2).

21.2.4.2 Critical Loads from the European Background Database

The European background database (Posch et al. 2003; Reinds et al. 2008) was used to compute critical loads of heavy metals for countries that did not submit data. Only forests (forest soils) are considered in the European background database and the calculation was thus limited to critical loads of cadmium and lead related to ecotoxicological effects in forests (effect number 3 in Table 21.1). Individual tree species are not identified, but a distinction is made between coniferous, broad-leaved (deciduous) and mixed forests. When variables were not found in existing databases, the recommendations and transfer functions from the Mapping Manual (www.icpmapping.org) were followed as closely as possible.

Input data for critical load calculations vary as a function of location and receptor (the combination of forest type and soil type). Thus an overlay of the following base maps was constructed:

a. A map with soil types at scale 1:1,000,000 for all European countries (Eurosoil 1999); except for Russia, Belarus, Ukraine and Moldova, for which the FAO 1:5,000,000 soil map (FAO 1981) was used.
b. A map of forest types, distinguishing coniferous, broad-leaved and mixed forests, taken from the harmonised European land cover map (Slootweg et al. 2005). This map is derived from the CORINE and SEI (Stockholm Environmental Institute) land cover maps.
c. A map with climate zones for Europe, derived from UNECE and EC (1996).
d. A global map of detailed elevation data (on a 30" × 30" grid) from NOAA/NGDC (Hastings and Dunbar 1998).
e. A map with EMEP 50 × 50 grid cells, in which deposition data are provided.

Overlaying these maps, merging polygons within every EMEP50 grid cell differing only in altitude, and discarding units smaller than 0.1 km^2 resulted in about 90,000 different forest-soil combinations. The soil maps are composed of so-called soil associations, each polygon on the map representing one association. Every association, in turn, consists of several soil typological units (soil types) that each covers a known percentage of the soil association. The soil typological units on the maps are classified into more than 200 soil types (Eurosoil 1999). For every soil typological unit information is available, of which soil texture classes are used here to derive other input data. Texture classes are defined in Table 6.2 of the Mapping Manual (www.icpmapping.org).

Further details regarding the derivation from this information of input variables for the computation of critical loads with Eq. 21.1 are provided in Posch and Reinds (2005).

Table 21.2 Number of ecosystems and areas for which critical loads data were submitted

Country	Country area (km^2)	EUNIS level 1	Cd		Pb		Hg	
			#ecosyst	Area (km^2)	#ecosyst	Area (km^2)	#ecosyst	Area (km^2)
Austria	83,858	Forest	503	35,822	503	35,822	503	35,822
		Agriculture	1154	9073	1154	9073	1154	9073
		Grassland	1296	16,491	1296	16,491	1296	16,491
		Total	2953	61,386	2953	61,386	2953	61,386
Belarus	207,595	Forest	9503	121,128	9503	121,128		
Belgium	30,528	Forest	1833	5237	1833	5237	1833	5237
		Water	10	9	10	9	10	9
		Total	1,843	5,246	1843	5,246	1843	5246
Bulgaria	110,994	Forest	84	48,330	84	48,330	–	–
Cyprus	9251	Forest	7438	1860	7438	1860	7438	1860
		Agriculture	13,869	3467	13,869	3467	13,869	3,467
		Shrub	10,586	2647	10,586	2647	10,586	2,647
		Total	31,893	7973	31,893	7973	31,893	7973
Finland	338,144	Water	–	–	–	–	820	16,856
France	543,965	Forest	3840	170,657	3840	170,657	–	–
Germany	357,022	Forest	101,306	101,306	101,306	101,306	101,306	101,306
		Agriculture	144,211	144,211	144,211	144,211	144,211	144,211
		Grassland	40,529	40,529	40,529	40,529	40,529	40,529
		Shrub	3205	3205	3205	3205	3205	3205
		Wetlands	659	659	659	659	659	93
		Other	93	93	93	93	93	93
		Total	290,003	290,003	290,003	290,003	290,003	290,003

Table 21.2 (continued)

Country	Country area (km²)	EUNIS level 1	Cd		Pb		Hg	
			#ecosyst	Area (km²)	#ecosyst	Area (km²)	#ecosyst	Area (km²)
Italy	301,336	Forest	436	99,327	436	99,327	–	–
		Agriculture	230	152,285	230	152,285	–	–
		Grassland	215	26,551	215	26,551	–	–
		Total	881	278,163	881	278,163	–	–
Netherlands	41,526	Forest	17,857	2900	17,857	2907	–	–
		Agriculture	12,627	19,522	12,627	19,522	–	–
		Total	30,484	22,422	30,484	22,429	–	–
Poland	312,685	Forest	88,383	88,383	88,383	88,383	88,383	88,383
Russia*	5,090,400	Forest	29,444	1,818,725	30,198	1,844,700	–	–
Slovakia	49,034	Forest	320,891	19,253	320,891	19,253	320,891	19,253
Sweden	449,964	Forest	2070	151,441	2070	151,441	5396	152,098
		Agriculture	2450	22,050	–	–	–	–
		Water	–	–	–	–	2977	293,749
		Total	4520	173,491	2070	151,441	8379	446,177
Switzerland	41,285	Forest	277	11,612	277	11,612	277	11,612
Ukraine	603,700	Agriculture	46	1,925	46	1925	–	–
United Kingdom	243,307	Forest	98,827	14,134	98,827	14,134	–	–
		Grassland	73,816	14,637	73,816	14,637	–	–
		Shrub	49,517	18,488	49,517	18,488	–	–
		Wetlands	12,494	3892	12,494	3892	–	–
		Total	234,654	51,151	234,654	51,151	–	–
All Countries	8,814,594	Grand Total	1,049,699	3,169,849	1,048,003	3,173,780	745,442	946,889

21.2.5 The Computation of Exceedances

Exceedance is defined as the non-negative difference between deposition and a critical load, i.e. $Ex = \max\{Dep\text{-}CL, 0\}$. Exceedances for 2010 were computed from national emission reductions that are consistent with current legislation (CLE2010). Exceedances for the year 2020 are the result of a scenario, named FI2020, in which national emissions are computed reflecting full implementation of abatement options as agreed in the protocol on heavy metals.

In natural areas, on broad regional scales in Europe, magnitudes of critical loads as well as depositions may vary subject to underlying data and modelling constraints that include soil characteristics, land cover and meteorology. In addition, also the surface covered by natural areas that are classified according to the EUNIS system can vary over countries. Deposition fields, on the other hand, are provided under the LRTAP Convention by EMEP in 50×50 km^2 grid cells covering Europe. More than a single exceedance can be computed in an EMEP grid cell, because the latter can include more than one EUNIS class. Moreover, since a EUNIS class can have more than one critical load, because of varying abiotic characteristics, a method was needed to assess exceedances on a regional scale. For this the Average Accumulated Exceedance (AAE) has been introduced (Posch et al. 2001). It is defined as $AAE=(A_1Ex_1+\ldots+A_nEx_n)/(A_1+\ldots+A_n)$, where A_i is the area of the i-th ecosystem in a grid cell and Ex_i its exceedance ($i=1,\ldots,n$). The AAE can be computed for any area within a country and for all natural areas of a country (or group of countries).

21.3 Results: Areas at Risk of Heavy Metal Deposition

Exceedances of critical loads of cadmium, lead and mercury have been computed (Hettelingh et al. 2006; Slootweg et al. 2007) for 2010 under current legislation (CLE2010) and for 2020 under a scenario assuming full implementation of the Aarhus protocol (FI2020). Results are shown in Fig. 21.6 for Cadmium (top), lead (middle) and mercury considering all effects listed in Table 21.1. A more detailed overview of exceedances when one effect is considered can be found in Slootweg et al. (2010). A detailed overview of the percentage of the *areas at risk*, i.e. where critical loads are exceeded by the three metals in European countries, is provided in Table 21.3.

Figure 21.6 illustrates that the critical load of cadmium is not exceeded (grey shading) in most of Europe by the atmospheric deposition of Cd. However, there are indications that the input of Cd from agricultural practices may lead to exceedances in agricultural areas. Several studies show that inputs of Cd from mineral and organic fertiliser (excluding sludge and compost) may be as important as atmospheric deposition in some European countries. The input of Cd by mineral and organic fertilisers is estimated to range between 0.4–3.3 g ha^{-1} yr^{-1} (Knappe et al. 2008), approximately 28–76 % of the total Cd input (WHO 2007).

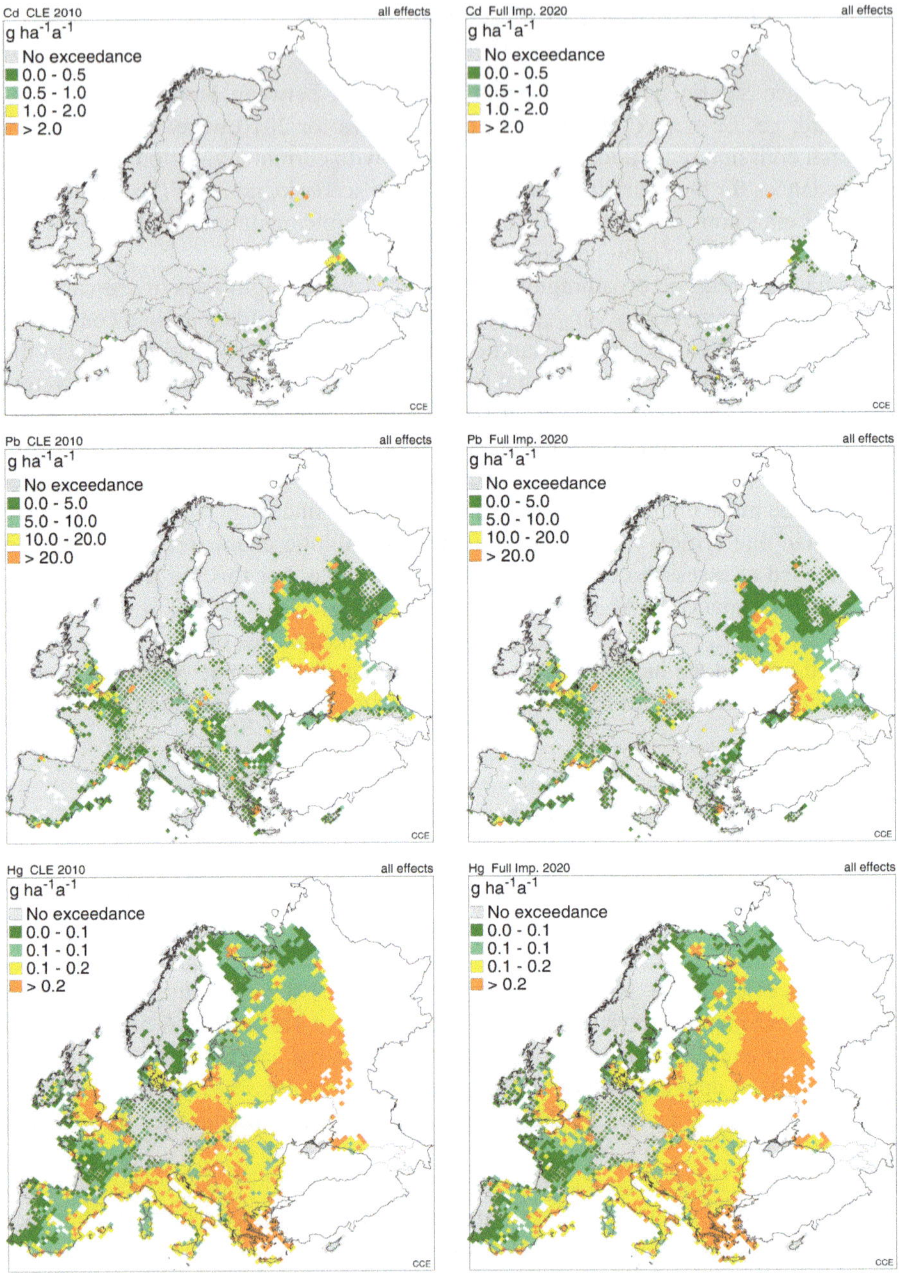

Fig. 21.6 Average accumulated exceedance (AAE in g ha^{-1}yr^{-1}) of critical loads of cadmium (top), lead (*middle*) and mercury (bottom) in 2010 (*left*) under the current legislation scenario and in 2020 (*right*) under a scenario assuming full implementation of the heavy metals protocol

Table 21.3 Percentage of ecosystem area at risk of the deposition of cadmium, lead and mercury in countries in Europe under the current legislation scenario (CLE2010) and in 2020 under a scenario assuming full implementation of the heavy metals protocol (FI2020)

Country	Cadmium		Lead		Mercury	
	CLE 2010	FI 2020	CLE 2010	FI 2020	CLE 2010	FI 2020
Albania	0.0	0.0	15.3	7.0	99.3	99.3
Austria	0.0	0.0	11.8	10.8	13.4	14.7
Belarus	0.3	0.0	9.5	3.3	100.0	100.0
Belgium	0.0	0.0	42.8	46.3	94.8	95.9
Bosnia and Herzegovina	0.0	0.0	53.2	12.1	99.8	99.8
Bulgaria	15.0	9.3	69.8	47.5	100.0	100.0
Croatia	0.0	0.0	19.1	7.0	99.8	99.8
Cyprus	1.7	0.0	33.4	30.2	3.6	4.0
Czech Republic	0.0	0.0	33.4	31.9	2.2	2.6
Denmark	0.0	0.0	0.2	0.2	99.1	99.1
Estonia	0.0	0.0	0.5	0.0	83.1	83.2
Finland	0.0	0.0	0.0	0.0	0.0	0.0
France	0.1	0.1	30.0	30.3	83.6	83.8
Germany	0.1	0.1	12.7	13.0	17.5	17.6
Greece	0.7	0.2	20.2	11.3	100.0	100.0
Hungary	0.0	0.1	31.7	21.4	100.0	100.0
Ireland	0.0	0.0	0.0	0.0	37.8	37.9
Italy	0.0	0.0	17.3	16.6	99.4	99.4
Latvia	0.0	0.0	0.0	0.0	92.4	93.1
Lithuania	0.0	0.0	0.5	0.1	98.5	99.3
Luxembourg	0.0	0.0	3.1	4.0	100.0	100.0
Macedonia	16.6	8.4	44.8	17.7	100.0	100.0
Moldova	0.0	0.0	26.8	6.8	100.0	100.0
Netherlands	0.0	0.0	29.2	30.2	93.2	93.4
Norway	0.0	0.0	0.0	0.0	13.1	14.0
Poland	0.2	0.0	7.6	5.9	99.9	99.9
Portugal	0.0	0.0	1.3	1.3	22.9	24.1
Romania	0.0	0.0	1.5	0.9	99.9	100.0
Russia	1.0	0.5	31.4	21.2	99.8	99.9
Serbia and Montenegro	1.0	0.0	28.5	4.0	99.9	99.9
Slovak Republic	0.5	0.5	26.5	25.3	73.0	78.3
Slovenia	0.0	0.0	10.4	7.3	98.0	98.0
Spain	0.6	0.7	15.5	17.4	88.7	89.7
Sweden	0.0	0.0	5.0	4.5	27.5	29.8

Table 21.3 (continued)

Country	Cadmium		Lead		Mercury	
	CLE 2010	FI 2020	CLE 2010	FI 2020	CLE 2010	FI 2020
Switzerland	0.0	0.0	3.2	3.2	30.1	27.6
Ukraine	0.0	0.0	100.0	100.0	0.0	0.0
United Kingdom	0.0	0.0	12.2	12.3	34.0	34.0
EU27	0.5	0.3	13.7	12.8	59.5	60.1
All	0.7	0.4	22.1	16.1	74.4	74.8

The deposition of lead is shown to put natural areas at risk both under CLE2010 as well as under FI2020. Table 21.3 reveals that Pb leads to computed risk in about 14% of the natural areas in the EU27 and 22% in the whole of Europe. Taking inputs by agricultural practices into account, both the magnitude of actual exceedances as well as the area where the critical load of lead is exceeded may increase. The agricultural share of Pb input is estimated to range from 8 to 33% of the total Pb input (WHO 2007) with a magnitude ranging from 1.8 to 10.4 g ha^{-1} yr^{-1} (Knappe et al. 2008). Compared to 2010, the risk of exceeding the critical load of Pb by atmospheric deposition alone in 2020 decreases from 14 to 13% in the EU27 and from 22 to 16% in the whole of Europe, respectively, under FI2020.

Finally, Fig. 21.6 shows that the risk of mercury deposition turns out to remain widespread both in 2010 and in 2020, affecting a computed percentage of natural areas in the EU27 covering about 59 and 60%[2] respectively. Areas with relatively high exceedances (red shading) can be found in most of the European countries[3]. In Europe 74–75%[2] of the area is subject to risk of Hg deposition, obviously also when it is assumed that the heavy metals protocol is fully implemented in 2020.

From Table 21.3 follows that the largest computed percentage of natural areas at risk of Cd and Pb in 2010 are found in Macedonia (17 and 45%, respectively) and in Bulgaria (15 and 70%, respectively). The computed risk of excessive Hg deposition in 2010 and 2020 turns out to affect the majority of natural areas in most countries. Last but not least, it is important to note that the critical concentration of Hg in rain water (effect 5, not included in Table 21.3) is exceeded nearly everywhere in the countries where this effect has been assessed, i.e. Belgium, Sweden and Finland (Slootweg et al. 2010).

[2] Note from Sect. 21.2.1 that emissions of Hg under FI2020 are higher than under CLE2010.

[3] Finland and Sweden focussed on the assessment of effect 5, not shown in Fig. 21.6.

Conclusions and Recommendations

Since 2005, 18 countries have computed critical loads of cadmium, lead and mercury and submitted them to the Coordination Centre for Effects. For other countries background databases have been used to compute critical loads for forest soils.

For each heavy metal, a number of well-defined human health and ecotoxicological effects were distinguished. These have been compiled into critical limit indicators and values for natural areas, classified according to the EUNIS system. These critical limit indicators enabled the computation of critical loads for terrestrial and aquatic ecosystems, covering a representative share of natural areas in Europe. The methodology enabled the computation of ecosystem-specific critical loads to protect human or environmental health. These critical loads were then compared to scenarios of ecosystem-specific depositions of the three metals in 2010 and 2020. This chapter includes a summary on how these depositions were computed from national emission reduction policies according to current legislation in 2010 and full implementation of the heavy metals protocol in 2020. The robustness of the relationship between national emissions and regional depositions is to some extent affected by the modelled treatment of re-suspension, and also by some variability in the emission data for 2010. Also, the impacts of heavy metal inputs from agricultural practices is not quantified in the exceedances described in this chapter.

Bearing these uncertainties in mind, it is shown that modelled atmospheric deposition of cadmium does not cause widespread risk in 2020, contrary to the risk of lead deposition and the significant risk caused by mercury deposition. Also, when the heavy metals protocol is fully implemented will the risk of lead deposition remain widespread in 2020, putting about 16 % of European natural areas at risk. The risk is higher for mercury affecting more than 74 and 60 % of natural area in Europe and the EU27 respectively.

References

Åkerblom, S. (2006). Anthropogenic heavy metals in organic forest soils—distribution, microbial risk assessment and Hg mobility. Doctoral thesis No. 2006:67, Uppsala Swedish University of Agricultural Sciences SLU.

Åkerblom, S., Meili, M., Bringmark, L., Johansson, K., Berggren Kleja, D., & Bergkvist, B. (2008). Partitioning of Hg between solid and dissolved organic matter in the humus layer of boreal forests. *Water, Air & Soil Pollution, 189,* 239–252.

Amann, M., Bertok, I., Cofala, J., Heyes, C., Klimont, Z., Rafaj, P., Schöpp, W., & Wagner, F. (2008). *National emission ceilings for 2020 based on the 2008 climate & energy package.* (NEC scenario analysis report Nr. 6). Laxenburg: International institute for applied systems analysis (IIASA).

AMAP/UNEP. (2008). Technical background report to the global atmospheric mercury assessment. Arctic monitoring and assessment programme/ UNEP chemicals branch.

Crommentuijn, T., Polder, M. D., & van de Plassche, E. J. (1997). Maximum permissible concentrations and negligible concentrations for metals, taking background concentrations into account. (Report no. 601501 001). Bilthoven, The Netherlands: National Institute of Public Health and the Environment.

Curlík, J., Šefcik, P., & Viechová, Z. (2000). Proceedings of UN/ECE Ad hoc international expert group on effect-based critical limits for heavy metals. Bratislava, Slovak Republik, 11–13 October 2000. Soil science and conservation research institute report.

De Vries, W., & Bakker, D. J. (1996). Manual for calculating critical loads of heavy metals for soils and surface waters. Preliminary guidelines for environmental quality criteria, calculation methods and input data. (Report 114). Wageningen, the Netherlands: DLO Winand Staring Centre.

De Vries, W., & Bakker, D. J. (1998). Manual for calculating critical loads of heavy metals for terrestrial ecosystems. Guidelines for critical limits, calculation methods and input data. (Report 166). Wageningen, the Netherlands: DLO Winand Staring Centre.

De Vries, W., Bakker, D. J., & Sverdrup, H. U. (1998). Manual for calculating critical loads of heavy metals for aquatic ecosystems. Guidelines for critical limits, calculation methods and input data. (Report 165). Wageningen, the Netherlands: DLO Winand Staring Centre.

De Vries, W., Schütze, G., Lofts, S., Meili, M., Römkens, P. F. A. M., Farret, R., De Temmerman, L., & Jakubowski, M. (2003). Critical limits for cadmium, lead and mercury related to ecotoxicological effects on soil organisms, aquatic organisms, plants, animals and humans. In Schütze et al. (Ed.), Proceedings of the Expert meeting on critical limits for heavy metals and methods for their application. Berlin 2–4 December 2002, held under the UN/ECE Convention on Long-Range Transboundary Air Pollution. UBA-Texte 47/03 Umweltbundesamt Berlin.

De Vries, W., Schütze, G., Lofts, S., Tipping, E., Meili, M., Römkens, P. F. A. M., & Groenenberg, J. E. (2005). Calculation of critical loads for cadmium, lead and mercury. Background document to a Mapping Manual on Critical Loads of cadmium, lead and mercury. (Report 1104). Wageningen, the Netherlands: Alterra.

Denier van der Gon, H. A. C., Van Het Bolscher, M., Visschedijk, A. J. H., & Zandveld, P. Y. J. (2005). Study to the effectiveness of the UNECE Heavy Metals Protocol and costs of possible additional measures; Phase I: Estimation of emission reduction resulting from the implementation of the HM Protocol. (TNO Report B & O-A R 2005/193). TNO.

EC. (2007). European union risk assessment report, cadmium oxide and cadmium metal, Part I—environment, Existing substances, 3rd Priority List. (Vol. 72). Institute for health and consumer protection, European chemicals bureau http://echa.europa.eu/documents/10162/4ea8883d-bd43-45fb-86a3-14fa6fa9e6f3. Accessed 27 Nov 2014.

Eurosoil. (1999). Metadata: Soil geographical data base of Europe v. 3.2.8.0. Ispra: Eurosoil.

FAO. (1981). FAO-UNESCO soil map of the world, 1:5 000 000. Vol. V Europe. Paris: Unesco.

Gregor, H. D., Spranger, T., & Hönerbach, F. (Eds.). (1997). *Critical limits and effects-based approaches for heavy metals and persistent organic pollutants*. United nations economic commission for Europe (UN/ECE), Convention on long-range transboundary air pollution (CLRTAP), Task force on mapping (TFM), Proceedings of a European workshop on effects-based approaches for heavy metals, bad harzburg, Germany, 3–7 Nov. 1997. Federal environmental agency (Umweltbundesamt), Berlin, UBA-Texte 5/98.

Gregor, H.-D., Mohaupt-Jahr, B., & Hönerbach, F. (Eds.). (1999). *Effects-based approaches for heavy metals*. Schwerin, Germany, 12–15 Oct. 1999: United nations economic commission for Europe (UN/ECE), Convention on long-range transboundary air pollution (CLRTAP), Task force on mapping (TFM).

Hastings, D. A., & Dunbar, P. K. (1998). Development & assessment of the global land one-km base elevation digital elevation model (GLOBE). *ISPRS Archives, 32,* 218–221.

Hettelingh, J. P., Slootweg, J., Posch, M., & Ilyin, I. (2002). Preliminary modelling and mapping of critical loads of cadmium and lead in Europe. (RIVM Report 259101011). Bilthoven (The Netherlands): CCE MSC-East Moscow and CCE Bilthoven, RIVM.

Hettelingh, J.-P., Sliggers, J., van het Bolscher, M., Denier van der Gon, H., Groenenberg, B. J., Ilyin, I., Reinds, G. J., Slootweg, J., Travnikov, O., Visschedijk, A., & De Vries, W. (2006). Heavy metal emissions, depositions critical loads and exceedances in Europe. (Report for the Ministry of VROM, Report of the VROM directorate of climate change and industry). Den Haag: VROM-DGM, Directie Klimaatverandering en Industrie. http://www.wge-cce.org/dsre

source?type=pdf&disposition=inline&objectid=rivmp:221375&versionid=&subobjectname=. Accessed 17 Nov 2014.

Ilyin, I., & Travnikov, O. (2010). Calculations of depositions of lead, cadmium and mercury for different options for the revision of the heavy metals protocol. In J. Slootweg, M. Posch, & J.-P. Hettelingh (Eds.), *Progress in the modeling of critical thresholds and dynamic modeling, including impacts on vegetation in Europe* (pp. 83–89). Coordination cenre for effects at the national institute for public health and the environment.

Ilyin, I., Rozovskaya, O., Sokovyh, V., Travnikov, O., & Aas, W. (2009). Heavy metals: Transboundary pollution of the environment. (EMEP Status Report 2/2009).

Knappe, F., Möhler, S., Ostermayer, A., Lazar, S., & Kaufmann, C. (2008). *Vergleichende Auswertung von Stoffeinträgen in Böden über verschiedene Eintragspfade*. (Forschungsbericht 203 74 275, FB 001168, UBA Texte 36/2008). Dessau: Umweltbundesamt.

Lofts, S., Spurgeon, D. J., Svendsen, C., & Tipping, E. (2004). Deriving soil critical limits for Cu, Zn, Cd and Pb: A method based on free ion concentrations. *Environmental Science & Technology, 38,* 3623–3631.

Meili, M. (1991). The coupling of mercury and organic matter in the biogeochemical cycle—towards a mechanistic model for the boreal forest zone. *Water Air & Soil Pollution, 56,* 333–347.

Meili, M. (1997). Mercury in lakes and rivers. In A. Sigel & H. Sigel (Eds.), *Mercury and its effects on environment and biology* (pp. 21–51). New York: Marcel Dekker Inc.

Meili, M., Bishop, K., Bringmark, L., Johansson, K., Munthe, J., Sverdrup, H., & De Vries, W. (2003). Critical levels of atmospheric pollution: Criteria and concepts for operational modelling of mercury in forest and lake ecosystems. *Science of The Total Environment, 304,* 83–106.

Pacyna, J. M., Scholtz, M. T., & Li, Y.-F. (1995). Global budget of trace metal sources. *Environmental Reviews, 3,* 145–159.

Posch, M., & Reinds, G. J. (2005). The European background database. In M. Posch, J.-P. Hettelingh, & J. Slootweg (Eds.), *European critical loads and dynamic modelling. CCE Status Report 2005* (pp. 33–37). Bilthoven: RIVM Report 259101016/2005, ISBN 9069601281, Coordination centre for effects, National institute for public health and the environment.

Posch, M., Hettelingh, J.-P., & De Smet, P. A. M. (2001). Characterization of critical load exceedances in Europe. *Water Air & Soil Pollution, 130,* 1139–1144.

Posch, M., Reinds, G. J., & Slootweg, J. (2003). The European background data. In M. Posch, J.-P. Hettelingh, J. Slootweg, & R.J. Downing (Eds.), *Modelling and mapping of critical thresholds in Europe. Status report 2003* Bilthoven: Coordination center for effects, RIVM Report 259101013, ISBN 90-6960-106-0.

Reinds, G. J., Groenenberg, J. E., & De Vries, W. (2006). Critical loads of copper, nickel, zinc, arsenic, chromium and selenium for terrestrial ecosystems at a European scale. A preliminary assessment. (Report 1355). Wageningen: Alterra green world research.

Reinds, G. J., Posch, M., De Vries, W., Slootweg, J., & Hettelingh, J. P. (2008). Critical loads of sulphur and nitrogen for terrestrial ecosystems in Europe and Northern Asia using different soil chemical criteria. *Water Air & Soil Pollution, 193,* 269–287.

Schütze, G., & Hettelingh, J.-P. (2006). Results of modelling and mapping of critical loads of lead, cadmium and mercury and critical concentrations of mercury in precipitation and their exceedances in Europe, Sufficiency and Effectiveness Review of the HM Protocol. (Chapter A3 15 June 2006, UNECE Convention on LRTAP).

Schütze, G., Lorenz, U., & Spranger, U. (2003). Expert Meeting on Critical Limits for Heavy Metals and Methods for their Application. Berlin: Proceedings, UBA Texte 47/2003. Umweltbundesamt Berlin.

Skjelkvåle, B. L., & Ulstein, M. (Eds.). (2002). Proceedings from the workshop on heavy metals (Pb, Cd and Hg) in surface waters—Monitoring and biological impact. March 18–20, 2002, Lillehammer, Norway. UN/ECE-CLRTAP-WGE-ICP-Waters. Norwegian institute for water research (NIVA), Oslo (Norway), ICP-Waters Report 67/2002.

Slootweg, J., Hettelingh, J.-P., Posch, M., Dutchak, S., & Ilyin, I. (Eds.). (2005). Critical loads of cadmium, lead and mercury in Europe. Bilthoven, The Netherlands, CCE-MSCE collaborative report, Netherlands environmental assessment agency. http://www.pbl.nl/sites/default/files/cms/publicaties/259101015.pdf. Accessed 27 Nov 2014.

Slootweg, J., Hettelingh, J.-P., Posch, M., Schütze, G., Spranger, T., De Vries, W., Reinds, G. J., van 't Zelfde, M., Dutchak, S., & Ilyin, I. (2007). European critical loads of cadmium, lead and mercury and their exceedances. *Water Air & Soil Pollution :Focus, 7,* 371–377.

Slootweg, J., Hettelingh, J.-P., & Posch, M. (2010). Critical loads of heavy metals and their exceedances. In J. Slootweg, M. Posch, & J.-P. Hettelingh (Eds.), Progress in the modelling of critical thresholds and dynamic modelling, including impacts on vegetation in Europe CCE Status Report 2010, RIVM report 680359001, ISBN 978-90-6960-249-3.

TFHM. (2006). Best available scientific information on the effects of deposition of heavy metals from long-range atmospheric transport sufficiency and effectiveness review of the 1998 protocol on heavy metals. UNECE convention on long-range transboundary air pollution.

Tipping, E. (1994). WHAM—A chemical equilibrium model and computer code for waters, sediments, and soils incorporating a discrete site/electrostatic model of ion-binding by humic substances. *Journal of Computers & Geoscience, 20,* 973–1023.

Tipping, E. (1998). Humic ion-binding Model VI: An improved description of the interactions of protons and metal ions with humic substances. *Aquatic Geochemistry, 4,* 3–47.

Travnikov, O., & Ilyin, I. (2005). Regional model MSCE-HM of heavy metal transboundary air pollution in Europe. (EMEP/MSC-E Technical Report 6/2005).

UNECE, & EC. (1996). Forest Condition in Europe. Results of the 1995 Survey. Brussels, Geneva.

UNECE. (1998). Protocol on heavy metals done at Aarhus, Denmark, on 24 june 1998. http://www.unece.org/env/lrtap/hm_h1.html. Accessed 27 Nov 2014.

UN-FAO/WHO-JECFA. (2003). Summary and Conclusions. Joint FAO/WHO expert commitee on food additives, sixty-first meeting, food and agriculture organization & world health organization of the United Nations, JECFA/61/SC, Rome, 10–19 June 2003.

US-EPA. (2001). *Water Quality Criterion for the Protection of Human Health: Methylmercury.* (Report EPA-823-R-01-001). Washington, DC 20460: U.S. Environmental protection agency, office of science and technology, Office of water.

Van den Hout, K. D. (Ed.). (1994). *The impact of atmospheric deposition of non-acidifying pollutants on the quality of European forests and the North sea, Main report of the ESQAD project "European Soil & sea Quality due to Atmospheric Deposition".* Den Haag: Ministry of Housing, Spatial Planning and the Environment, DG for Environmental Protection, Air and Energy Dir./640.

Van den Hout, K. D., Bakker, D. J., Berdowski, J. J. M., van Jaarsveld, J. A., Reinds, G. J., Bril, J., Breeuwsma, A., Groenenberg, J. E., De Vries, W., van Pagee, J. A., Villars, M., & Sliggers, C. J. (1999). The impact of atmospheric deposition of non-acidifying substances on the quality of European forest soils and the North Sea. *Water Air & Soil Pollution, 109,* 357–396.

Visschedijk, A., Denier van der Gon, H., Kuenen, J., & van der Burgh, H. (2010). Heavy metal emissions and reduction costs. In J. Slootweg, M. Posch, & J.-P. Hettelingh (Eds.), *Progress in the modeling of critical thresholds and dynamic modeling, including impacts on vegetation in Europe, CCE Status Report 2010* (pp. 69–82). ISBN 978-90-6960-249-3, RIVM report 680359001, Coordination Cenre for Effects at the National Institute for Public Health and the Environment, www.rivm.nl/cce.

WHO. (2004). *Guidelines for Drinking-Water Quality. Vol. 1: Recommendations.* World Health Organisation, Geneva. http://www.who.int/water_sanitation_health/dwq/GDWQ2004web.pdf. Accessed 27 Nov 2014.

WHO. (2007). Health risks of heavy metals from long-range transboundary air pollution. Copenhagen: Joint WHO/Convention task force on the health aspects of air pollution, World health organisation, Regional office for Europe.

Chapter 22
Derivation of Critical Loads of Nitrogen for Habitat Types and Their Exceedances in the Netherlands

Han F. van Dobben, Arjen van Hinsberg, Dick Bal, Janet P. Mol-Dijkstra, Henricus J.J. Wieggers, Johannes Kros and Wim de Vries

22.1 Introduction

The Netherlands has a strong policy focus on the conservation of habitat types as defined in the European Habitats Directive (EU-Directive 92/43 1992). The almost country-wide exceedance of critical loads creates the need for a standardised method to unambiguously determine the magnitude of exceedances at a local scale to support pollution abatement planning. For this aim, the empirical critical load concept outlined by e.g. Achermann and Bobbink (2003), Bobbink and Hettelingh (2011) and in Chap. 4 has a limited applicability as (a) it relies on a vegetation typology on a high level of abstraction, using broadly defined types; and (b) critical loads are given as ranges that may be rather wide e.g. 10–20 kg N $ha^{-1}yr^{-1}$. Within such a range, reduction of an actual deposition of e.g. 25 kg N $ha^{-1}yr^{-1}$ to 20 or 10 kg N $ha^{-1}yr^{-1}$makes a huge difference in terms of costs. Here we describe a method that uses a combination of a dynamic soil model and observational or expert-based abiotic responses per vegetation type to derive critical loads that (a) are unique values instead of ranges, and (b) relate to a detailed vegetation typology. The critical load values produced by this method enable a further specification of the empirical critical loads of Bobbink and Hettelingh (2011), to arrive at critical loads that are (a) sufficiently detailed to be applicable in support of Dutch policy, and (b) consistent with current international knowledge on empirical critical loads.

H. F. van Dobben (✉) · J. P. Mol-Dijkstra · H. J. Wieggers · J. Kros · W. de Vries
Alterra Wageningen University and Research Centre, Wageningen, The Netherlands
e-mail: han.vandobben@wur.nl

A. van Hinsberg
Netherlands Environmental Assessment Agency (PBL), Bilthoven, The Netherlands

D. Bal
Ministry of Economic Affairs, Den Haag, The Netherlands

W. de Vries et al. (eds.), *Critical Loads and Dynamic Risk Assessments*,
Environmental Pollution 25, DOI 10.1007/978-94-017-9508-1_22

In this chapter we first present the derivation of critical loads by the dynamic soil model SMART2 (Kros 2002) in combination with empirically derived responses of vegetation types to soil acidity, nitrogen availability and hydrology. The principle of this method is also outlined in Chap. 3 and is described in more detail in Van Dobben et al. (2004). We then present the method we used to combine the simulated critical loads per vegetation type with Bobbink and Hettelingh's (2011) empirical values per 'EUNIS' type into a table of unique critical load values per habitat type, as outlined in detail in Van Dobben et al. (2014). The results of this approach are also presented.

The focus of the present study is on the Netherlands. Reasons include that the N load in the Netherlands is high and therefore there is a need for N deposition reductions, but also because large data sets (relevés) are available containing vegetation descriptions (cf. Weeda et al. 2002), with known abiotic circumstances for a subset of these relevés (Wamelink et al. 2002, 2012). However, our method aims at a wider applicability, at least in Europe. Its core features, the soil model SMART2 (Kros 2002) and Ellenberg's indicator values (Ellenberg 1991) have been applied throughout central and north-western Europe.

22.2 Material and Methods

To characterise vegetation, we use the association in the phyto-sociological sense of Braun-Blanquet (1964), as elaborated for the Netherlands by Schaminée et al. (1995b). This is a natural vegetation unit that may be expected to have a unique critical load value. Moreover, vegetation associations (or, in some cases, sub-associations) are often used to describe targets in nature conservation, and translation tables exist between (sub-) associations and habitat types. The principle of our method is to estimate the critical abiotic conditions, i.e. the limits of occurrence in terms of soil pH and N availability for each association. These are then used in an inverse application of the dynamic soil model SMART2 (Kros 2002) to derive the N deposition at these critical limits.

22.2.1 SMART2 Inversion Procedure

SMART2 is generally used to estimate the soil conditions that are predictors for biodiversity (pH and N availability) at given deposition levels of sulphur (S) and N compounds (Kros 2002; Wamelink et al. 2003). In the present application, SMART2 is 'inverted' to SMART2^{-1}, to compute deposition levels that lead to a given pH and N availability. For this purpose we developed an iterative procedure that in successive SMART2 runs with varying deposition levels, tries to find the N deposition level that most closely approximates predefined ('requested') values for soil pH and N availability. The values for soil pH and N availability that result from this iterative search procedure are termed the 'optimised' values. In order to find a unique solution for each vegetation type, additional constraints have to be provided for SO_x deposition and the NH_y to NO_x ratio; for both we used approximate values for the period 2000–2010 (CBS et al. 2011). Besides deposition, SMART2 uses soil type, vegetation structure and hydrology (as Mean Phreatic Level in Spring [MPLS], seepage quantity and seepage quality) as inputs. In the iterative SMART2 inversion procedure these input variables are kept constant.

22.2.2 Estimation of the Critical Conditions per Association

As a starting point we used the vegetation database compiled by Schaminée et al. (1995a, 1996, 1998) and Stortelder et al. (1999) (further referred to as 'Schaminée et al. 1995'). This database contains about 400,000 so-called relevés, which are descriptions of sample plots of 1–100 m^2 in size, in terms of species presences. We made a selection of 160,209 relevés from this database that we considered representative of the Dutch natural vegetation, based on criteria described by Runhaar et al. (2002). The program ASSOCIA (Van Tongeren et al. 2008) was used for a phyto-sociological identification of these relevés. Next, a subset was made containing those relevés that could be identified on the level of association, and belong to one of the 139 associations that substantially contribute to biodiversity in the Netherlands according to criteria formulated by Bal et al. (1995). This subset, containing 46,462 relevés, was used to derive 'critical' abiotic conditions, i.e. the lowest pH and the highest N availability at which a given association can occur. However, direct measurements of abiotic conditions are lacking for the vast majority of these relevés. Therefore we estimated abiotic conditions on the basis of the species composition per relevé using Ellenberg's (1991) indicator values.

Ellenberg (1991) built a database of species responses to abiotic conditions, mostly based on expert judgement. Essentially, the optimal values of a number of abiotic factors have been estimated for a large number of species, and scored in an arbitrary nine- or twelve-point scale. Here we use these values for soil pH, N availability and groundwater level as initial estimates for the 'real' values of these variables (i.e. measurable in SI units, such as kmol $ha^{-1}yr^{-1}$ for N availability), and we used a training set consisting of relevés with known abiotic conditions to define transfer functions that relate Ellenberg indicators for acidity, nutrient availability and moisture to soil pH, N availability and groundwater level. In the rest of this chapter we will use the abbreviations E_R, E_N and E_W to denote Ellenberg's indicators for acidity, nutrient availability and moisture, respectively.

The abiotic conditions that are input to $SMART2^{-1}$, pH and N availability, are influenced by N deposition. Changes in these conditions are the direct cause of the disappearance (or loss of diversity) of certain vegetation types at increasing deposition. These losses are caused by either a too low pH value ('acidification'), or a too high N availability ('eutrophication'). Therefore, the critical pH value was estimated as the lower end of the pH range of each association. In the present study this lower end was defined as the 20-percentile (P_{20}) of E_R (Ellenberg's acidity indicator) values of the relevés belonging to a given association. The E_R value was determined for each relevé as the (unweighted) mean value of its constituent species, including mosses (Düll 1991; Siebel 1993) and lichens (Wirth 1991). Similarly, the critical N availability was estimated as the higher end of the N availability range of each association, for which the 80-percentile (P_{80}) of Ellenberg's N (N availability indicator) in a given association was taken. For all other inputs of SMART2^{-1}, we estimated mean or 'typical' values per association and soil combination, as specified below.

22.2.3 Inputs of SMART2

22.2.3.1 Soil Type

SMART2 uses a simplified typology with seven soil types: poor sand, rich sand, calcareous sand, non-calcareous clay, calcareous clay, peat, loess. We assigned each of the 139 associations included in our study to between one and three of these soil types on the basis of their ecological descriptions in Schaminée et al. (1995). Furthermore we checked these assignments by projecting a subset of 6973 relevés, whose locality was known, onto a soil map (Steur and Heijink 1991). As the critical load depends on both vegetation type and soil type, the number of critical loads that is finally estimated (227) exceeds the number of associations included in our study (139).

22.2.3.2 Vegetation Structure

SMART2 uses a simple vegetation structure typology: pine forest, spruce forest, deciduous forest, heathland, grassland. However, as spruce forest does not naturally occur in the Netherlands, this type was not used in the present study. Pine forest, deciduous forest and heathland were assigned to the associations in a straightforward way, based on each association's floristic composition. The grassland type was assigned to all associations that did not belong to any of the other types. Consequently, this type is rather heterogeneous.

22.2.3.3 Hydrology

SMART2 uses three types of hydrological input: the MPLS (in cm below soil surface), the seepage intensity (upward movement of water through the soil, in mm day^{-1}), and the ionic composition of seepage water (in six classes: rainwater, groundwater, mixed rain and groundwater, surfacewater, brackish water, seawater; Kros 2002, p. 161). In SMART2, seepage is an external source of ions. Therefore we used seepage in a rather broad sense, not only accounting for the upward movement of groundwater but also for the influence of surface water. The position of each association relative to groundwater or surface water was taken from Schaminée et al. (1995) and translated into an estimated seepage quantity according to Table 22.1. As the

Table 22.1 Assignment of seepage quantity values to vegetation types

Groundwater or surface water influence	Assigned (mm day^{-1})
No water influence	0
Periodically flooded, fen, bog hummocks	0.15
Shore vegetation, partly immersed, bog gullies	1
Immersed but temporarily dry	2
Permanently immersed	3
'Real' seepage-dependant vegetation[a]	5

[a] Used for only one vegetation type (Pellio epiphyllae-Chrysosplenietum oppositifolii)

assumed seepage quantity has a strong influence on the pH, the seepage quantities in Table 22.1 were fine-tuned, so that the optimised pH approximated the requested pH as closely as possible. Classes for seepage quality (also reflecting surface water quality) were assigned to each association also on the basis of Schaminée et al. (1995). MPLS was estimated from Ellenberg numbers as described below.

22.2.3.4 Estimation of pH and MPLS from Ellenberg Numbers

For both pH and MPLS, relevés with known values of these variables were collected and used as a training set, being equal to the one used by Wamelink et al. (2002) with a few additions. The relation between E_R and pH, and between E_W and MPLS, respectively, was determined by simple linear regression. In this regression the measured abiotic conditions were used as the response variable, while the (unweighted) mean indicator values over all species with a known indicator value in each relevé was the predictor variable. As Wamelink et al. (2002) showed that translation functions may be different per vegetation type, SMART2's vegetation structure types were included as an extra predictor variable. However, the effect of vegetation structure appeared to be non-significant in this case, which may be due to the coarse typology. Details of the regression analyses are given in Table 22.2. In calcareous soils E_R is only weakly related to pH (Schaffers and Sykora 2000), and therefore the pH was set to 7 on the calcareous soil types, irrespective of the E_R value.

22.2.3.5 Estimation of N Availability From Ellenberg N

The procedure used to derive translation functions for E_R and E_W cannot be used for E_N because measurements of N availability are very limited. Therefore we used generic data to estimate N availability for those relevés whose geographical coordinates were known, including historic deposition fields of S, NO_x and NH_y in 1950, 1960, 1970, 1980 and 1990, the soil type, and the seepage quantity and quality per $250 \times 250\ m^2$ grid cell (CBS et al. 2011). These historic data were used because nearly all relevés in the training set were collected during the period 1945–1995.

Table 22.2 Regression of MPLS (mean phreatic level in spring, in cm below soil surface, a negative number indicates flooding) and soil pH on E_W and E_R (Ellenberg's indicator numbers for humidity and acidity, respectively). Regression equation: MPLS or pH = $a_0 + a_1(E_W$ or $E_R)$

Variable	Nr of records	Percentage variance explained	a_0	a_1
			Mean ± se	Mean ± se
E_W	1537	22%	235.6±7.08	−21.57±1.02
E_R	3630	43%	3.106±0.0512	0.527±0.839

Table 22.3 Regression of N availability on logarithm of Ellenberg N and vegetation types (according to SMART2 classification)

Variable	Symbol	Estimate	±	s.e. significance
Intercept	a_0	6.191	±	0.106***
$\log(E_N)$	a_1	0.637	±	0.0608***
Vegetation type				
Grass	c_{grass}	−1.182	±	0.0394***
Heath	c_{heath}	−1.898	±	0.0842***
Coniferous	$c_{coniferous}$	−0.274	±	0.123*

Regression equation: N availability $= a_0 + a_1 \log(E_N) + c_{vegtype}$. The constants per vegetation type are relative to deciduous forest, and therefore $c_{deciduous} = 0$. Percentage explained variance = 24 %, number of records = 6911
significance: ***$p<0.001$, * $0.01<p<0.05$

We used SMART2 to estimate the N availability in the grid squares where these relevés were located as the sum of deposition and mineralisation at the points in time mentioned above. The modelled N availabilities averaged over these points in time were assumed to represent the actual N availability for the relevés.

The N availability estimated as above was regressed on the mean Ellenberg N per relevé, using the vegetation type as an additional explanatory factor. In contrast to the regression analyses for pH and MPLS, the effect of vegetation type was highly significant (see Table 22.3 for details).

22.2.4 *Combination of Simulated and Empirical Values to a Unique Critical Load per Habitat Type*

The simulation and the empirical approaches use vegetation typologies that are both different from the habitat typology. The EUNIS typology (Davies and Moss 2002; Davies et al. 2004) used in the empirical approach is rather coarse, and habitat types can be related to EUNIS types in a straightforward way, as given in Bobbink and Hettelingh (2011: Appendix 1) and Van Dobben et al. (2014: Appendix 1) and also listed in Table 22.5. The phyto-sociological typology used in the simulation approach is far more detailed than the habitat typology and the number of (sub-) association that are (part of) a given habitat type varies between zero and 19; the translation table is given in Van Dobben et al. (2014). This table is based on the definition of the habitat types (EU-Directive 92/43 1992) and their implementation for the Netherlands as given in the vegetation database SynBioSys (Hennekens et al. 2010).

As the relation of habitat type to (sub-)association is 1:n in many cases, a procedure has to be defined to combine the critical load values of different (sub-)associations. We based this procedure on the fact that in an area with a given habitat type, any of the (sub-)associations belonging to that habitat type may be present, but information on their actual presence, location and quantity is lacking. In that case the overall mean turns out to be the best approximation for the simulated critical load at a given location, i.e. the one that on average has the smallest difference with the true value. In detail the procedure to assign a simulated critical load to a habitat type is as follows:

- if the habitat type corresponds to (one or more) (sub-)associations of which critical loads depend on the soil type, the critical loads are averaged over these soil types per (sub-)association;
- if the habitat type corresponds to (one or more) sub-associations, a critical load is assigned to the corresponding association(s) as the average over its constituent sub-associations (possibly after applying step 1).
- if the habitat type corresponds to more than one association (possibly after applying step 2), a critical load is assigned to that habitat type as the average over these associations.

The approach above may or may not lead to a critical load value. Critical load values may be lacking either because there is no translation between typologies, or because there are insufficient data to derive a critical load. Empirical critical loads can also be lacking. Therefore there are four possibilities per habitat type:

1. critical loads based on both simulated and empirical values are available;
2. only critical loads based on simulated values are available;
3. only critical loads based on empirical values are available;
4. no critical loads are available.

In case 1 the final critical load should be within the empirical range so that it does not disagree with the internationally accepted values. In that case the following procedure was used: if the simulated value is within the empirical range it becomes the final critical load; if it is above the empirical range, the upper boundary of the empirical range becomes the final critical load; and if it is below the empirical range, the lower boundary of the empirical range becomes the final critical load. In case 2 the simulated value is used, and in case 3 the mean of the empirical critical load range is used (sometimes after defining a subrange according to Dutch local conditions, following the recommendations in Bobbink and Hettelingh 2011). In case 4 either no critical load is determined (some of the habitat types have no vegetation

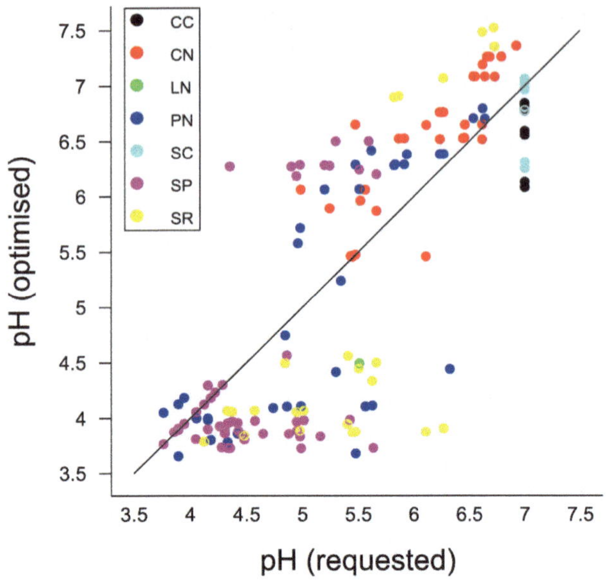

Fig. 22.1 Scatter diagram of 'requested' vs. 'optimised' pH, per soil type. The requested value is the critical (i.e., maximum tolerated) limit estimated on the basis of Ellenberg numbers, the optimised value is the closest approximation of this value by SMART2^{-1}. Note that the deposition at the critical limit is the critical load. (1:1-line also shown; soil types: *SP* nutrient poor sand, *SR* nutrient rich sand, *SC* calcareous sand, *CN* non-calcareous clay, *CC* calcareous clay, *PN* peat, *LN* loess)

and hence need no critical load), or an estimate is based on other sources, e.g. expert knowledge or other models.

22.3 Results

22.3.1 Evaluation of the Simulation Procedure

The performance of the procedure used to invert the SMART2 model was checked by comparing the 'requested' and 'optimised' values of N availability and pH. An example for pH is given in Fig. 22.1 and shows a close correspondence ($R = 0.89$) although some records have a rather large discrepancy. These discrepancies occur for all soil types except on clay and on the calcareous types; and also for all vegetation types, although mostly in grassland. For N availability the correspondence between the 'requested' and 'optimised' values is even larger ($R = 0.91$). It was checked whether the discrepancy between the 'requested' and 'optimised'values of pH influenced the final critical load value, but for this no indications were found (Fig. 22.2). The critical load mainly depends on the 'requested' N availability (i.e., the P_{80} of the estimated N availabilities of the relevés of each association) and on the vegetation type.

The simulation procedure was subjected to extensive sensitivity and uncertainty analyses whose results are summarised here, with details being presented in Van Dobben et al. (2004).

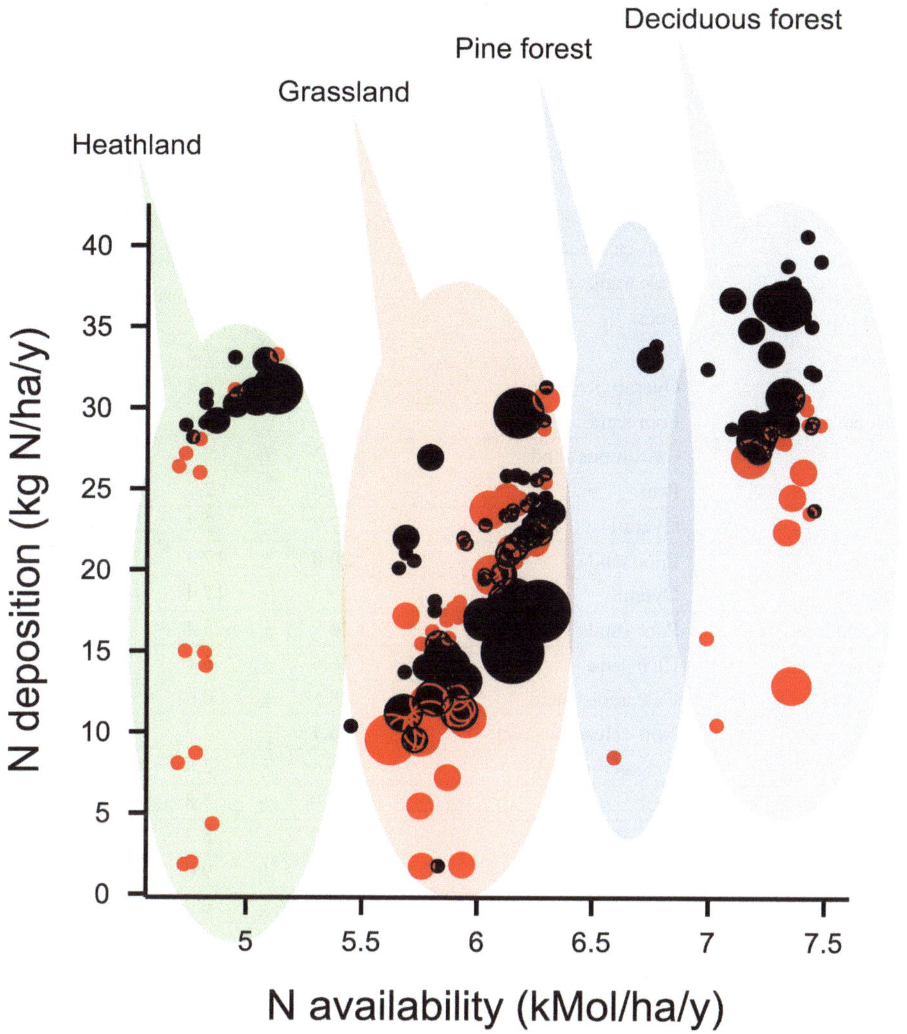

Fig. 22.2 Scatter diagram of N availability against N deposition. Points are at the P_{80} of N availability and the corresponding deposition (= critical load) per association. *Dot colour* and *size* indicate the discrepancy between the 'requested' and the 'optimised' pH (*red* = optimised > requested, *black* = optimised < requested; *largest dot* = |optimised−requested| >2, *smallest dot* = |optimised−requested| <0.5. Vegetation types are indicated by background shades

22.3.2 Simulated Critical Loads and Their Uncertainties

The simulated critical load values per (sub-)association can be found in Van Dobben et al. (2004). These critical loads are derived by adding the deposition values for NO_x and NH_y at the critical conditions, expressed in kg N $ha^{-1}yr^{-1}$. Table 22.4 summarises the critical loads per combination of vegetation type and soil type.

Table 22.4 Summary of critical loads per vegetation type and per soil type, excluding critical loads that were judged unreliable (n = number of associations)

Vegetation type	Soil type	Critical load (kg N $ha^{-1}yr^{-1}$)			
		Mean	±	s.e.	(n)
Grassland	Poor sand	12.9	±	3.3	(20)
	Rich sand	17.3	±	3.9	(16)
	Calcareous sand	20.8	±	2.4	(33)
	Non-calcareous clay	22.9	±	3.1	(21)
	Calcareous clay	23.3	±	2.3	(38)
	Loess	13.2	±	*	(1)
	Peat	19.6	±	6.2	(16)
	Overall	20.1	±	4.8	(145)
Heathland	Poor sand	25.9	±	9.1	(10)
	Calcareous sand	33.3	±	*	(1)
	Peat	29.5	±	2.7	(7)
	Overall	27.7	±	7.2	(18)
Pine	Poor sand	20.8	±	17.4	(2)
forest	Overall	20.8	±	17.4	(2)
Deciduous	Poor sand	26.9	±	5.4	(6)
forest	Rich sand	26.5	±	6.1	(7)
	Calcareous sand	28.5	±	0.6	(6)
	Non-calcareous clay	26.1	±	2.7	(9)
	Calcareous clay	36.5	±	2.8	(10)
	Peat	32.9	±	4.8	(8)
	Overall	30.0	±	5.7	(46)
All types	Poor sand	19.0	±	9.0	(38)
	Rich sand	20.1	±	6.3	(23)
	Calcareous sand	22.3	±	3.9	(40)
	Non-calcareous clay	23.9	±	3.3	(30)
	Calcareous clay	26.0	±	5.9	(48)
	Loess	13.2	±	*	(1)
	Peat	25.3	±	7.9	(31)
	Grand mean	22.9	±	6.9	(211)

In general, the estimated critical load for a large part depends upon the vegetation type, and to a lesser extent on the 'requested' N availability and the soil type. At equal abiotic conditions, the critical load for pine forest is equal to the one for grassland, but the critical load for heathland is significantly higher, and for deciduous forest significantly lower than for grassland. However, because of the difference in average abiotic conditions per vegetation type, the final critical loads increase in the order grassland < heathland < pine forest < deciduous forest (compare Table 22.4). The effect of soil type on the critical load is rather limited. Both calcareous soil types have a significantly higher critical load than poor sand, and also non-calcareous clay and peat have higher critical loads than poor sandy soils. The final ranking order of critical loads per soil type (taking account of the distribution of the vegetation types over the soil types) is poor sand < rich sand < calcareous sand < non-calcareous clay < peat < calcareous clay (compare Table 22.4).

The results of the uncertainty analysis per association can be found in Van Dobben et al. (2004). A general conclusion is that the variability around the average critical load is rather small, in the order of a standard deviation of 0.2–0.6 kg N $ha^{-1}yr^{-1}$. However, the variability around the average of critical loads at individual sites can be large, in the order of 10–20 kg N $ha^{-1}yr^{-1}$. This means that, if the true critical loads of all individual sites belonging to a given vegetation type were known, the discrepancy between the simulated value and the average of the true value over all these sites would be small, but the true values at each site can substantially deviate from the simulated value.

After the uncertainty analysis, some of the critical loads were discarded because they were judged too uncertain or otherwise unreliable (see Van Dobben et al. 2004; Appendix 2) Most notably this is the case where the standard deviation in the average critical load is larger than 5 kg N $ha^{-1}yr^{-1}$, and where the critical load is below 4 kg N $ha^{-1}yr^{-1}$. These values are not used in subsequent calculations, and were also left out of consideration in Table 22.4.

22.3.3 Comparison of Simulated and Empirical Critical Loads and Integration per Habitat Type

A comparison between simulated and empirical critical loads is challenging because of the differences in typology. However, an approximate comparison can be made by translating the EUNIS types to combinations of soil type and vegetation type. This translation can be found in Van Dobben et al. (2004), together with empirical and simulated critical load ranges (the latter based on the minimum and maximum values per (sub-)association/ soil type combination, without taking account of the uncertainty).

To make a comparison, the midpoint was determined between the lower and upper boundary of the ranges of both simulated and empirical values. Although the simulated and empirical critical loads appear to be in the same order of magnitude, the midpoints per EUNIS class are not correlated ($r = -0.02$, $P >> 0.1$). However, it should be noted that the variability in empirical critical loads between EUNIS classes is rather small; 10 out of the 22 relevant EUNIS classes have an empirical critical load midpoint of 15 N $ha^{-1}yr^{-1}$. For most EUNIS classes the ranges of the critical loads determined by both methods overlap, although the empirical values are generally slightly below the simulated ones, the mean difference of range midpoints being 3.4 kg N $ha^{-1}yr^{-1}$.

Table 22.5 gives the final critical loads derived according to the procedure outlined in Sect. 22.2.4. The table also gives a short explanation why some of the simulated critical loads were judged unreliable and what other estimates have been used instead.

Table 22.5 Final critical load values per habitat type and their derivation. The translation from Habitat type to EUNIS type is according to Appendix 1 in Bobbink and Hettelingh (2011), the translation from Habitat type to vegetation type (not given in Table) is according to Van Dobben et al. (2014). Empirical ranges are taken from Bobbink and Hettelingh (2011) and simulated values from Van Dobben et al. (2006), unless otherwise stated

Habitat-type	Description	Subtype	EUNIS type	Description	Empirical range	Simulated value	Critical load	Explanation
1110	Sandbanks which are slightly covered by sea water all the time						>34	Expert judgement
1130	Estuaries						>34	Expert judgement
1140	Mudflats and sandflats not covered by seawater at low tide						>34	Expert judgement
1160	Large shallow inlets and bays						>34	Expert judgement
1170	Reefs						>34	Expert judgement
1310	Salicornia and other annuals colonizing mud and sand	*Sagina maritima present*	A2.54	Low-mid salt marshes	20–30	20.8	21	
1310	Salicornia and other annuals colonizing mud and sand	*Sagina maritima present*	A2.55	Pioneer salt marshes	20–30	20.8	21	
1310	Salicornia and other annuals colonizing mud and sand	*Salicornia spp. dominated*	A2.54	Low-mid salt marshes	20–30	22.9	23	
1310	Salicornia and other annuals colonizing mud and sand	*Salicornia spp. dominated*	A2.55	Pioneer salt marshes	20–30	22.9	23	
1320	Spartina swards (Spartinion maritimae)		A2.54	Low-mid salt marshes	20–30	23.3	23	

Table 22.5 (continued)

Habitat-type	Description	Subtype	EUNIS type	Description	Empirical range	Simulated value	Critical load	Explanation
1320	Spartina swards (Spartinion maritimae)		A2.55	Pioneer salt marshes	20–30	23.3	23	
1330	Atlantic salt meadows (Glauco-Puccinellietalia maritimae)		A2.54	Low-mid salt marshes	20–30	22.3	22	
1330	Atlantic salt meadows (Glauco-Puccinellietalia maritimae)		A2.55	Pioneer salt marshes	20–30	22.3	22	
2110	Embryonic shifting dunes		B1.3	Shifting coastal dunes	10–20	23.6	20	
2120	Shifting dunes along the shoreline with Ammophila arenaria ("white dunes")		B1.3	Shifting coastal dunes	10–20	21.2	20	
2130	Fixed coastal dunes with herbaceous vegetation ("grey dunes")	*Acid*	B1.4	Coastal stable dune grasslands	8–15	13.1	10	Empirical subrange used: 8–10
2130	Fixed coastal dunes with herbaceous vegetation ("grey dunes")	*Grass-heath*	B1.4	Coastal stable dune grasslands	8–15	10.8	10	Empirical subrange used: 8–10
2130	Fixed coastal dunes with herbaceous vegetation ("grey dunes")	*Lime-rich*	B1.4	Coastal stable dune grasslands	8–15	17.4	15	Empirical subrange used: 10–15
2140	Decalcified fixed dunes with *Empetrum nigrum*		B1.5	Coastal dune heaths	10–20		15	Midpoint of empirical range
2150	Atlantic decalcified fixed dunes (Calluno-Ulicetea)		B1.5	Coastal dune heaths	10–20		15	Midpoint of empirical range

Table 22.5 (continued)

Habitat-type	Description	Subtype	EUNIS type	Description	Empirical range	Simulated value	Critical load	Explanation
2160	Dunes with Hippophaë rhamnoides					28.3	28	
2170	Dunes with Salix repens ssp. argentea (Salicion arenariae)					32.3	32	
2180	Wooded dunes of the Atlantic, Continental and Boreal region	*Dry, dominance of other treespecies*	G1.6	Fagus woodland	10–20	28.6	20	
2180	Wooded dunes of the Atlantic, Continental and Boreal region	*Dry, Quercus—Betula dominated*	G1.8	Acidophilous Quercus-dominated woodland	10–15	18.2	15	
2180	Wooded dunes of the Atlantic, Continental and Boreal region	*Inner fringe of dunes*				25.3	25	
2180	Wooded dunes of the Atlantic, Continental and Boreal region	*Wet*				31.3	31	
2190	Humid dune slacks	*Acid*	B1.8	Moist to wet dune slacks	10–20	18.6	15	Empirical subrange used: 10–15
2190	Humid dune slacks	*Lime-rich*	B1.8	Moist to wet dune slacks	10–20	19.5	20	Empirical subrange used: 15–20
2190	Humid dune slacks	*Open water, meso-eutrophic*					30	Expert judgement, based on simulated values of comparable vegetation types and environmental conditions

Table 22.5 (continued)

Habitat-type	Description	Subtype	EUNIS type	Description	Empirical range	Simulated value	Critical load	Explanation
2190	Humid dune slacks	*Open water, oligo-mesotrophic*	C1.16	Dune slack pools	10–20	14.0	14	Simulated value taken from Wortelboer (1990)
2190	Humid dune slacks	*Tall reed and sedge vegetation*					> 34	Expert judgement
2310	Dry sand heaths with Calluna and Genista		F4.2	Dry heaths	10–20		15	Empirical subrange used: 10–15, simulated value taken from Heil and Bobbink (1993)
2320	Dry sand heaths with Calluna and *Empetrum nigrum*		F4.2	Dry heaths	10–20		15	Empirical subrange used: 10–15, simulated value taken from Heil and Bobbink (1993)
2330	Inland dunes with open Corynephorus and Agrostis grasslands		E1.94	Inland dune pioneer grasslands	8–15	10.4	10	Empirical subrange used: 8–11
3110	Oligotrophic waters containing very few minerals of sandy plains (Littorelletalia)		C1.1	Permanent oligotrophic lakes, ponds and pools	3–10	5.9	6	Empirical subrange used: 5–10, simulated values taken from Wortelboer (1990)
3130	Oligotrophic to mesotrophic standing waters with vegetation of the Littorelletea		C1.1	Permanent oligotrophic lakes, ponds and pools	3–10		8	Expert judgement, also based on empirical subrange: 5–10
3140	Hard oligo-mesotrophic waters with benthic vegetation of Chara spp.	*In former (now enclosed) marine basins*					> 34	Expert judgement, compare 3150

Table 22.5 (continued)

Habitat-type	Description	Subtype	EUNIS type	Description	Empirical range	Simulated value	Critical load	Explanation
3140	Hard oligo-mesotrophic waters with benthic vegetation of *Chara* spp.	*In peatland ares*					30	Expert judgement, compare 3150
3140	Hard oligo-mesotrophic waters with benthic vegetation of *Chara* spp.	*On sandy soil*					8	Expert judgement, compare 3130
3150	Natural eutrophic lakes with Magnopotamion or Hydrocharition—type	*In former (now enclosed) marine basins*					> 34	Expert judgement
3150	Natural eutrophic lakes with Magnopotamion or Hydrocharition—type	*Outside former (now enclosed) marine basins*					30	Expert judgement
3160	Natural dystrophic lakes and ponds		C1.4	Permanent dystrophic lakes, ponds and pools	3–10		10	Expert judgement, also based on empirical subrange: 5–10
3260	Water courses of plain to montane levels with the Ranunculion fluitantis and						> 34	Expert judgement
3270	Rivers with muddy banks with Chenopodion rubri p.p. and Bidention p.p.						> 34	Expert judgement
4010	Northern Atlantic wet heaths with Erica tetralix	*On peat*	D2	Valley mires, poor fens and transition mires	10–15		11	Midpoint of empirical subrange used: 10–12
4010	Northern Atlantic wet heaths with Erica tetralix	*On sandy soil*	F4.11	Erica tetralix dominated wet heath lowland	10–20		17	Simulated value taken from Berendse (1988)

Table 22.5 (continued)

Habitat-type	Description	Subtype	EUNIS type	Description	Empirical range	Simulated value	Critical load	Explanation
4030	European dry heaths		F4.2	Dry heaths	10–20		15	Empirical subrange used: 10–15, simulated value taken from Heil and Bobbink (1993)
5130	Juniperus communis formations on heaths or calcareous grasslands		F4.2	Dry heaths	10–20	30.5	15	Empirical subrange used: 10–15
6110	Rupicolous calcareous or basophilic grasslands of the Alysso-Sedion albi		E1.26	Sub-Atlantic semi-dry calcareous grassland	15–25	20.1	20	
6120	Xeric sand calcareous grasslands		E1.26	Sub-Atlantic semi-dry calcareous grassland	15–25	17.5	18	
6120	Xeric sand calcareous grasslands		E2.2	Low and medium altitude hay meadows	20–30	17.5	18	
6130	Calaminarian grasslands of the Violetalia calaminariae		E1.7	Closed non-Mediterranean dry acid and neutral grassland	10–15	14.7	15	
6210	Semi-natural dry grasslands and scrubland facies on calcareous substrates		E1.26	Sub-Atlantic semi-dry calcareous grassland	15–25	21.1	21	
6230	Species-rich Nardus grasslands, on silicious substrates in mountain areas (and submountain areas in Continental Europe)	*Dry, acid*	E3.52	Heath Juncus meadows and humid Nardus stricta swards	10–20	13.7	12	Empirical subrange used: 10–12

Table 22.5 (continued)

Habitat-type	Description	Subtype	EUNIS type	Description	Empirical range	Simulated value	Critical load	Explanation
6230	Species-rich Nardus grasslands, on silicious substrates in mountain areas (and submountain areas in Continental Europe)	*Dry, lime-rich*	E1.7	Closed non-Mediterranean dry acid and neutral grassland	10–15	12.2	12	Empirical subrange used: 12–15
6230	Species-rich Nardus grasslands, on silicious substrates in mountain areas (and submountain areas in Continental Europe)	*Wet, acid*	E1.7	Closed non-Mediterranean dry acid and neutral grassland	10–15	9.6	10	
6410	Molinia meadows on calcareous, peaty or clayey-silt-laden soils (Molinion caeruleae)		E3.51	Molinia caerulea meadows	15–25	10.9	15	
6430	Hydrophilous tall herb fringe communities of plains and of the montane to alpine levels	*Dry forest edges*				26.1	26	Expert judgement, based on simulated values of comparable vegetation types and environmental conditions
6430	Hydrophilous tall herb fringe communities of plains and of the montane to alpine levels	*Wet*					>34	Expert judgement, simulated value not used because surface water (and not deposition) is main source of nitrogen
6510	Lowland hay meadows (Alopecurus pratensis, Sanguisorba officinalis)	*Alopecurus pratensis dominated*	E2.2	Low and medium altitude hay meadows	20–30	21.5	22	

Table 22.5 (continued)

Habitat-type	Description	Subtype	EUNIS type	Description	Empirical range	Simulated value	Critical load	Explanation
6510	Lowland hay meadows (Alopecurus pratensis, Sanguisorba officinalis)	*Arrhenaterum elatius dominated*	E2.2	Low and medium altitude hay meadows	20–30	19.4	20	
7110	* Active raised bogs	*Over larger extent*	D1	Raised and blanket bogs	5–10		7	Midpoint of empirical range
7110	* Active raised bogs	*Small peatbogs in heathland*	D2	Valley mires, poor fens and transition mires	10–15		11	Midpoint of empirical subrange used: 10–12
7120	Degraded raised bogs still capable of natural regeneration	*Target: 4010*					17	Value of comparable type 4010A
7120	Degraded raised bogs still capable of natural regeneration	*Target: 7110*					7	Value of comparable type 7110A
7120	Degraded raised bogs still capable of natural regeneration	*Target: 91D0*					25	Value of comparable type 91D0
7140	Transition mires and quaking bogs	*Quacking bog*	D4.1	Rich fens	15–30	16.8	17	
7140	Transition mires and quaking bogs	*Reed dominated*	D2	Valley mires, poor fens and transition mires	10–15	7.2	10	
7150	Depressions on peat substrates of the Rhynchosporion		F4.11	Erica tetralix dominated wet heath lowland	10–20		20	Empirical subrange used: 15–20; simulated values according to Berendse (1988)

Table 22.5 (continued)

Habitat-type	Description	Subtype	EUNIS type	Description	Empirical range	Simulated value	Critical load	Explanation
7210	* Calcareous fens with *Cladium mariscus* and species of the Caricion davallianae		D4.1	Rich fens	15–30		22	Midpoint of empirical range
7220	* Petrifying springs with tufa formation (Cratoneurion)						<34?	Expert judgement based on Bobbink and Lamers (1999)
7230	Alkaline fens		D4.1	Rich fens	15–30	15.8	16	
9110	Luzulo-Fagetum beech forests		G1.6	Fagus woodlands	10–20	28.0	20	
9120	Atlantic acidophilous beech forests with Ilex and sometimes also Taxus in the		G1.6	Fagus woodlands	10–20	28.7	20	
9160	Sub-Atlantic and medio-European oak or oak-hornbeam forests of the Carpinion	*On poor soil*	G1.A	Meso- and eutrophic Quercus, Carpinus, Fraxinus, Acer, Tilia, Ulmus and related woodland	15–20	30.3	20	
9160	Sub-Atlantic and medio-European oak or oak-hornbeam forests of the Carpinion	*On rich soil*	G1.A	Meso- and eutrophic Quercus, Carpinus, Fraxinus, Acer, Tilia, Ulmus and related woodland	15–20	33.7	20	
9190	Old acidophilous oak woods with Quercus robur on sandy plains		G1.8	Acidophilous Quercus-dominated woodland	10–15	18.2	15	

Table 22.5 (continued)

Habitat-type	Description	Subtype	EUNIS type	Description	Empirical range	Simulated value	Critical load	Explanation
91D0	* Bog woodland					27.5	25	Expert judgement
91E0	* Alluvial forests with Alnus glutinosa and Fraxinus excelsior (Alno-Padion, Alnion incanae, Salicion albae)	*In brook valleys*				26.1	26	
91E0	* Alluvial forests with Alnus glutinosa and Fraxinus excelsior (Alno-Padion, Alnion incanae, Salicion albae)	*In river valleys, Fraxinus and Ulmus dominated*				28.0	28	
91E0	* Alluvial forests with Alnus glutinosa and Fraxinus excelsior (Alno-Padion, Alnion incanae, Salicion albae)	*In river valleys, Salix and Populus dominated*				33.8	34	
91F0	Riparian mixed forests of Quercus robur, Ulmus laevis and Ulmus minor					29.1	29	

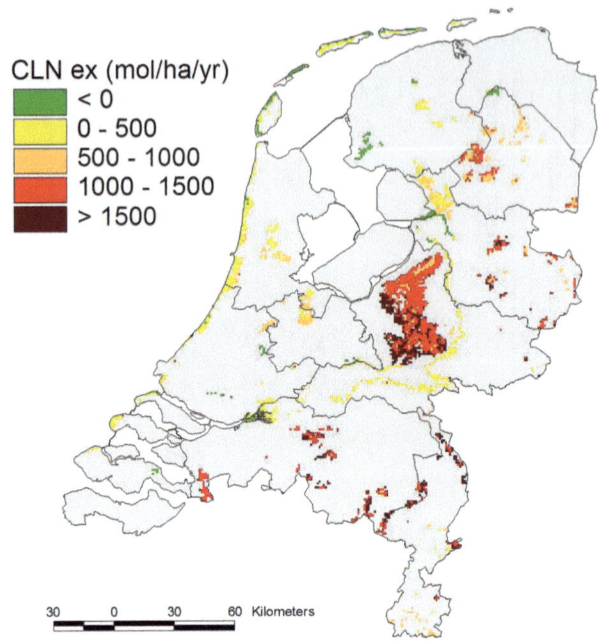

Fig. 22.3 Critical load exceedance in the Netherlands in 2010. Grid cells are 1 km^2. Critical load shown for each grid cell equals the critical load of the most sensitive Habitat type in the Natura 2000 area to which that cell belongs. Grey: open water or outside Natura 2000. CLNex = exceedance of critical load for N; 1000 mol = 14 kg N.

22.3.4 Assessment of Critical Load Exceedances for the Netherlands

Figure 22.3 gives a picture of the critical load exceedances for Natura 2000 sites in the Netherlands in 2010, based on critical load values in Van Dobben and Van Hinsberg (2008) and Dutch deposition data from CBS et al. (2011). Critical load values are from the most sensitive habitat type in the Natura 2000 area to which a grid cell belongs. Results show that low exceedances (<500 mol ha^{-1}yr^{-1}) occur near the coast, due to low N deposition values, and in areas with loamy to clayey soil, such as the northern part of the Netherlands and along rivers in the central part, due to high critical N loads. High exceedances (>1000 mol ha^{-1}yr^{-1}) occur in the forested sandy soil area in the central and southern parts of Netherlands, due to high N deposition values.

22.4 Discussion

The methodological constraints of the empirical approach are sufficiently discussed by Achermann and Bobbink (2003) and Bobbink and Hettelingh (2011). As to the simulation approach the following remarks can be made:

The uncertainty analysis presented by Van Dobben et al. (2004, 2006) showed that vegetation type and N availability strongly contribute to the variability of the simulated critical load. Of these two, N availability has by far the largest uncertainty, because it is derived from Ellenberg numbers with a transfer function that has a large amount of intrinsic uncertainly. In the calibration of the transfer function the assumption was made that the relevés in the training set have an N availability equal to the average sum of mineralisation and deposition over the period 1950–1990. In other words: for each association the assumption is made that the thus estimated N availabilities over this period represent the range of N availabilities that is characteristic for that association, or at least, that the upper end of this range (determined as the 80-percentile) is its limit of occurrence. For the present critical loads this had two consequences:

1. if the historic deposition is over or underestimated, the critical loads will be over or underestimated by the same amount; and
2. although the simulated critical loads are not derived from a comparison of an 'impacted' state with a 'reference' state, the historic relevés are assumed to represent each association in an optimally developed form. If the historic deposition has already caused a certain loss of biodiversity, this will remain unnoticed in the present method, and will ultimately lead to an over-estimation of the critical load. Even from the early nineteenth century onward, species have been lost to the Dutch flora, especially in grasslands (Weeda et al. 2002 pp. 77–79 and 175–176), and there are indications that this loss may be partly due to atmospheric deposition (Van Dam et al. 1986). If this is true, our critical loads will be biased towards values above the 'true' ones. It should be noted that in this case the same bias probably occurs in the empirical critical loads, because here the notions of biodiversity and the optimal development of vegetation types are also strongly related to the vegetation of the period just before the strong increase of deposition started (i.e. before ca. 1965).

Another consequence of the way in which E_N is translated into N availability is that it creates a certain independence of the formulation of the mineralisation process in SMART2. E.g. if SMART2 over-estimates N mineralisation, N availability at a given value of E_N is over-estimated. However, this is compensated in $SMART2^{-1}$ because the 'requested' N availability is then also over-estimated, and thus the amount of N that is to be provided by deposition is still correctly estimated.

The uncertainty analysis (Van Dobben et al. 2004, 2006) showed that the estimates of the country-wide average critical loads have a low uncertainty. However, the variability around the average of critical loads at individual sites appears to be high. This variability may have two causes: (1) it may be 'real', in which case a given association may occur at widely different N availabilities, or (2) it may be a methodological artefact, in which case the method we used to estimate N availability yields values that have a large spread at any given 'real' N availability. On the basis of the present data it is not possible to decide which of these causes mostly contributes to the local variability. In case (2) the estimated N availability comes closer to the real value if the number of observations is increased. In case (1) the variability is an intrinsic property of the response of the vegetation to N availability, possibly caused by variation in other environmental factors. If that is the case, an association can still occur at values far above the critical load, if other conditions are favourable, while if other conditions are unfavourable, it may disappear at deposition values far below the critical load. It should be noted that in that case a general reduction of deposition will have beneficial ecological effects, even if the critical load is still exceeded on a large scale, because some sites will have a critical load far above the mean value for its vegetation type. In that case any reduction in deposition will have a positive effect, and indications for this were found by De Vries et al. (2002).

The above procedure estimates critical loads that have the highest likelihood; this is a direct consequence of the averaging procedure. However, the Habitats Directive intends to '*rule out significant negative effects*'. Many scientists appear to follow a strict interpretation of the precautionary principle, where the lower end of an uncertainty range is automatically used (cf. Appendix 4 in Van Dobben and Van Hinsberg 2008). Here we use range midpoint or even upper ends if there is sufficient scientific support to do so. By using the most likely critical load, significant negative effects cannot be entirely ruled out in practice, because there will always be situations that have a higher than average sensitivity to N deposition (although lower than average situations are equally likely). This is a disadvantage that is inherent to the application of generic standards at site level. It would require a very large effort to determine for each site if its local critical load is significantly different from the generic value. Therefore we choose to determine a unique value *within* a range using a well-defined method. The reviewers of Appendix 4 in Van Dobben and Van Hinsberg (2008) considered this methodology *as a great step forward in applying science based effects thresholds in local and national environmental policy*, although they recommend discussing whether in case of doubt, the precautionary principle could be justified, i.e. to choose the lower range endpoint to exclude negative effects. However, because of the main conclusion of the review cited above we decided to adhere to the above method.

To these arguments an ecological argument can be added. In our view a Habitat type should not be considered as the sum or a mosaic of a number of vegetation types that all have to be present. Rather we consider it as a type in itself and a conservation goal as such. Of course, the more of its constituent (sub-) associations are (potentially) present, the better its state of development, but the actual presence or absence of vegetation types also depends on other local conditions besides deposition. To this the practical argument can be added that in Natura 2000 areas the location of Habitat types is known, but the location of vegetation types is usually not.

Conclusions

The following conclusions can be drawn from our study:

- a critical load in the order of 10–30 kg N $ha^{-1}yr^{-1}$ for most vegetation types in The Netherlands can be considered as a reliable number. The mean critical load resulting from our study is 23 ± 7 kg N $ha^{-1}yr^{-1}$, and almost all EUNIS classes have empirical critical loads in this range;

- by combining the empirical and the simulation approach, critical load values can be derived for Habitat types in the Netherlands that (a) are unique values and thus useful as deposition reduction goals, and (b) have a broad international support;
- it is not possible to determine critical load values on a site scale, as the uncertainty in simulated values becomes very high in that case, and empirical values are not available at that level of detail.

References

Achermann, B., & Bobbink, R. (Eds.). (2003). Empirical critical loads for nitrogen: Expert workshop, Berne, 11–13 November 2002. Swiss Agency for the Environment, Forests and Landscape.

Bal, D., Beije, H. M., Hoogeveen, Y. R., Jansen, S. R. J., & van de Reest, P. J. (1995). Handboek natuurdoeltypen in Nederland. (Rapport IKC-N 11). Wageningen: IKC-Natuurbeheer.

Berendse, F. (1988). *De nutriëntenbalans van droge zandgrondvegetaties in verband met de eutrofiëring via de lucht. Deel 1 Een simulatiemodel als hulpmiddel bij het beheer van vochtige heidevelden.* Wageningen: Centrum voor Agrobiologisch Onderzoek.

Bobbink, R., & Hettelingh, J.-P. (2011). *Review and revision of empirical critical loads and dose-response relationships: Proceedings of an expert workshop, Noordwijkerhout, 23–25 June 2010.* (Report 680359002/2011). Bilthoven: Coordination Centre for Effects, National Institute for Public Health and the Environment.

Bobbink, R., & Lamers, L. P. M. (1999). *Effecten van stikstofhoudende luchtverontreiniging op vegetaties – een overzicht.* (TCB R13). University of Nijmegen; Technische Commissie Bodembescherming (in Dutch).

Braun-Blanquet, W. (1964). *Pflanzensoziologie. Grundzüge der Vegetationskunde.* 3. Aufl. Wien: Springer.

CBS, PBL, & Wageningen UR. (2011). www.compendiumvoordeleefomgeving.nl. CBS, Den Haag PBL, Den Haag/Bilthoven and Wageningen UR, Wageningen.

Davies, C. E., & Moss, D. (2002). *EUNIS Habitat Classification.* (2001 Work Programme, Final Report to the European Environment Agency European Topic Centre on Nature Protection and Biodiversity). Centre for Ecology and Hydrology.

Davies, C. E., Moss, D., & Hill, M. O. (2004). *EUNIS Habitat Classification revised 2004.* European Environment Agency, European topic centre on nature protection and biodiversity.

De Vries, W., van Dobben, H., van Herk, C. M., Roelofs, J., van Pul, A., van Hinsberg, A., Duijzer, J., & Erisman, J. W. (2002). Effecten emissiebeleid voor verzuring op de natuur. *Arena, 8,* 105–108.

Düll, R. (1991). Zeigerwerte von Laub- und Lebermoose. *Scripta Geobotanica, 18,* 175–214.

Ellenberg, H. Jr. (1991). Ökologsiche Veränderungen in Biozönosen durch Stickstoff-eintrag. In K. Henle & G. Kaule (Eds.), *Arten- und Biotopschutzforschung für Deutschland. Berichte aus der ökologischen Forschung* (Vol. 4, pp. 75–90).

EU-Directive 92/43. (1992). *Council Directive 92/43/EEC of 21 May 1992 on the conservation of natural habitats and of wild fauna and flora.* http://eur-lex.europa.eu/LexUriServ/LexUriServ.do?uri=CONSLEG:1992L0043:20070101:EN:PDF.

Heil, G. W., & Bobbink, R. (1993). "Calluna", a simulation model for evaluation of impacts of atmospheric nitrogen deposition on dry heathlands. *Ecological Modelling, 68,* 161–182.

Hennekens, S. M., Smits, N. A. C., & Schaminée, J. H. J. (2010). *SynBioSys Nederland versie 2.* Alterra, Wageningen UR. http://www.wageningenur.nl/en/show/SynBioSys-Nederland.htm.

Kros, J. (2002). Evaluation of biogeochemical models at local and regional scale. PhD thesis, Wageningen, The Netherlands, Alterra, 2002, Alterra scientific contributions 7.

Runhaar, J., Alkemade, J. R. M., Hennekens, S. M., Wiertz, J., & van 't Zelfde, M. (2002). *Afstemming biotische responsmodules DEMNAT-SMART/MOVE* (RIVM/ Rapport 408657008/2002). Bilthoven: RIVM.

Schaffers, A. P., & Sykora, K. V. (2000). Reliability of Ellenberg indicator values for moisture, nitrogen and soil reaction: A comparison with field measurements. *Journal of Vegetative Science, 11,* 225–244.

Schaminée, J. H. J., Weeda, E. J., & Westhoff, V. (1995a). *De vegetatie van Nederland 2: plantengemeenschappen van wateren, moerassen en natte heiden.* Uppsala: Opulus Press.

Schaminée, J. H. J., Stortelder, A. H. F., & Westhoff, V. (1995b). *De vegetatie van Nederland 1. inleiding tot de plantensociologie: grondslagen, methoden en toepassingen.* Uppsala: Opulus Press.

Schaminée, J. H. J., Stortelder, A. F. H., & Weeda, E. J. (1996). *De vegetatie van Nederland 3: plantengemeenschappen van graslanden, zomen en droge heiden.* Uppsala: Opulus Press.

Schaminée, J. H. J., Weeda, E. J., & Westhoff, V. (1998). *De vegetatie van Nederland 4: plantengemeenschappen van de kust en van binnenlandse pioniermilieus.* Uppsala: Opulus Press.

Siebel, H. N. (1993). *Indicatiegetallen van blad- en levermossen.* (Report IBN 47).

Steur, G. G. L., & Heijink, W. (1991). Bodemkaart van Nederland schaal 1: 50 000. Algemene begrippen en indelingen, 4e uitgave (in Dutch). (Rapport 168). Wageningen: DLO Winand Staring Centre.

Stortelder, A. H. F., Schaminée, J. H. J., & Hommel, P. W. F. M. (1999). *De vegetatie van Nederland 5: plantengemeenschappen van ruigten, struwelen en bossen.* Uppsala: Opulus Press.

Van Dam, D., van Dobben, H. F., ter Braak, C. F. J., & de Wit, F. (1986). Air pollution as a possible cause for the decline of some phanerogamic species in The Netherlands. *Vegetatio, 65,* 47–52.

Van Dobben, H. F., & van Hinsberg, A. (2008). Overzicht van kritische depositiewaarden voor stikstof, toegepast op habitattypen en Natura 2000 gebieden. (Alterra Report 1654). Wageningen: Alterra.

Van Dobben, H., Schouwenberg, E. P. A. G., Mol, J. P., Wieggers, H. J. J., Jansen, M., Kros, J., & De Vries, W. (2004). Simulation of critical loads for nitrogen for terrestrial plant communities in The Netherlands. (Report 953). Wageningen: Alterra.

Van Dobben, H. F., van Hinsberg, A., Schouwenberg, E. P. A. G., Jansen, M. J. W., Mol-Dijkstra, J. P., Wieggers, H. J. J., Kros, J., & De Vries, W. (2006). Simulation of critical loads for nitrogen for terrestrial plant communities in the Netherlands. *Ecosystems, 9,* 32–45.

Van Dobben, H., Bobbink, R., Bal, D., & van Hinsberg, A. (2014). Critical loads for nitrogen deposition of Natura 2000 habitat types: overview of occurring in The Netherlands. (Alterra Report 2488). Wageningen: Alterra.

Van Tongeren, O., Gremmen, N., & Hennekens, S. (2008). Assignment of relevés to pre-defined classes by supervised clustering of plant communities using a new composite index. *Journal of Vegetation Science, 19,* 525–536.

Wamelink, G. W. W., Joosten, V., van Dobben, H. F., & Berendse, F. (2002). Validity of Ellenberg indicator values judged from physico-chemical field measurements. *Journal of Vegetation Science, 13,* 269–278.

Wamelink, G. W. W., ter Braak, C. F. J., & van Dobben, H. F. (2003). Changes in large-scale patterns of plant biodiversity predicted from environmental economic scenarios. *Landscape Ecology, 18,* 513–527.

Wamelink, G. W. W., van Adrichem, M. H. C., van Dobben, H. F., Frissel, J. Y., den Held, M., Joosten, V., Malinowska, A. H., Slim, P. A., & Wegman, R. M. A. (2012). Vegetation relevés and soil measurements in the Netherlands: the Ecological Conditions Database (EC). In J. Dengler, J. Oldeland, F. Jansen, M. Chytrý, J. Ewald, M. Finckh, F. Glöckler, G. Lopez-Gonzalez, R.K. Peet, & J.H.J. Schaminée (Eds.), Vegetation databases for the 21st century. *Biodiversity & Ecology, 4,* 125–132.

Weeda, E. J., Schaminée, J. H. J. & van Duuren, L. (2002). *Atlas van Plantengemeenschappen in Nederland 2: graslanden, zomen en droge heiden.* Utrecht: KNNV Uitgeverij.

Wirth, V. (1991). Zeigerwerte von Flechten. *Scripta Geobotanica, 18,* 215–237.

Wortelboer, F. G. (1990). A model on the competition between two macrophyte species in acidifying shallow soft-water lakes in The Netherlands. *Hydrobiological Bulletin, 24,* 91–107.

Chapter 23
Assessing the Impacts of Nitrogen Deposition on Plant Species Richness in Europe

Jean-Paul Hettelingh, Carly J. Stevens, Maximilian Posch, Roland Bobbink and Wim de Vries

23.1 Introduction

Critical load exceedance is sometimes viewed as a 'limited' mid-point assessment of damage to ecosystem structure and function, because exceedance does not provide information on the actual biological effect. At the same time it is argued that the worldwide increase of N deposition (Galloway et al. 2008) has many reasons to be of global concern, especially in view of impacts on plant species diversity (Bobbink et al. 2010; Phoenix et al. 2006). Therefore, information on the adverse effects of excessive N deposition is considered important as part of the justification of policies to reduce emissions of reduced and oxidized N.

Scientific support of these policies is particularly challenged to quantify effects on biological endpoints such as provisioning and production services of ecosystems. These services are the basis of human welfare and well-being (Millennium Ecosystem Assessment 2005), capable of bridging the gap between ecology and economy (Chan et al. 2012). The ensuing importance for policy development includes estimations of the global economic value of these services to be in the range of 16–54 trillion US $ yearly (Costanza et al. 1997) and, as advocated more recently, should also include non-material values (Chan et al. 2012). De Groot et al. (2012) recently gave an overview of the monetary value of ecosystem services of

J.-P. Hettelingh (✉) · M. Posch
Coordination Centre for Effects (CCE), RIVM, Bilthoven, The Netherlands
e-mail: jean-paul.hettelingh@rivm.nl

C. J. Stevens
Lancaster Environment Centre, Lancaster University, Lancaster, UK

R. Bobbink
B-Ware Research Centre, Radboud University Nijmegen, Nijmegen, The Netherlands

W. d. Vries
Alterra Wageningen University and Research Centre, Wageningen, The Netherlands

W. de Vries et al. (eds.), *Critical Loads and Dynamic Risk Assessments*,
Environmental Pollution 25, DOI 10.1007/978-94-017-9508-1_23

ten main biomes, with values ranging between 2871 US$ yr^{-1} for grasslands and 4267 US $ yr^{-1} for freshwater lakes. Therefore it is considered important to have scientific support of (European) air quality to also include impacts on endpoints such as ecosystem functions.

Maestre et al. (2012) documented causal relationships between biodiversity and multiple ecosystem functions and services in arid, semi-arid and dry sub-humid ecosystems ('drylands'). The result of this work was based on an extensive evaluation of "how the richness of perennial vascular plants ('species richness') and a range of key abiotic factors (climate, slope, elevation, and soil texture) relate to multi-functionality in 224 dryland ecosystems sampled from all continents, except Antarctica". The study concluded that species richness was of greater importance for multi-functionality than climatic and abiotic factors in drylands, consistent with findings reported for other ecosystems (see references in Maestre et al. 2012). This implies that documented relationships between N deposition and plant species richness (Bobbink et al. 1998; Bobbink and Hettelingh 2011; Emmett et al. 2007; Stevens et al. 2004; Stevens et al. 2010a, b) can be used to obtain insight in adverse impacts on the multi-functionality of terrestrial ecosystems.

This chapter describes a tentative assessment of a base year (2000) and expected future (2020) adverse effects of N deposition on plant species diversity of (semi-) natural grasslands at a European scale. The impacts in 2020 were assessed in response to two policy relevant N emission scenarios, i.e. emission reductions agreed under the revised Gothenburg Protocol (GP2020) and Maximum Feasible Reduction (MFR) technology to reduce N emissions.

23.2 Methods and Data

23.2.1 Overall Approach

The change of species richness was assessed by combining N deposition on a European scale with available relationships between N deposition (dose) and plant species richness (response) for selected habitat classes. Two kinds of D-R relationships were used, one derived from experimental N addition studies (Bobbink 2008; Bobbink and Hettelingh 2011) and another one derived from a nitrogen deposition gradient study (Stevens et al. 2010b). The focus is on natural and semi-natural grasslands as common receptor in both studies.

First, a land cover map of Europe was compiled whereby 'nature' was classified according to the European Nature and Information System (EUNIS; Davies and Moss 1999). This EUNIS classification was then used to apply available D-R relationships to appropriately selected natural and semi-natural grassland ecosystems. Ecosystem-specific depositions of total N were computed over Europe and used in the D-R relationships to yield a statistic for plant species richness in 2000 and 2020 for these European grassland ecosystems.

23.2.2 Compiling a Harmonized Land Cover Map

A harmonized land cover under the LRTAP Convention was compiled in 2005 in a collaboration between the Coordination Centre for Effects, the Stockholm Environment Institute (SEI) and the University of Leiden (Slootweg et al. 2005). The term 'harmonized' was chosen to express the need for one single land cover map to be used under the Convention both for the assessment of ecosystem-specific depositions and concentrations of air pollutants as well as for the modelling and mapping of critical loads and their exceedances.

The compilation of the harmonized map involved a comparison of the classifications used in maps from the CORINE land cover database (version 12/2000 extended coverage) of the European Environment Agency (www.eea.europe.eu), the land cover map of the Stockholm Environment Institute (SEI) (www.york.ac.uk/sei/projects/completed-projects/european-land-cover), and the Pan-European Land Cover Monitoring (PELCOM) map (Mücher et al. 2001). Generally, CORINE is considered the best available land cover database, but only part of the modelling domain of the European Monitoring and Evaluation Programme (EMEP; www.emep.int) is covered by it. The land cover map used here is a combination of CORINE, where available, and SEI data where CORINE is missing. This map has been created by the CCE as a grid map in the EMEP coordinate system with a grid size of 250×250 m^2.

This land cover map enables the analysis of critical loads, exceedances and D-R relationships for any EUNIS class and regions in the European network of protected natural areas (Natura 2000; www.eea.europa.eu/data-and-maps/data/natura-4). The D-R relationships established from N addition experiments and from N deposition gradient studies (see 23.2.3) were assigned to relevant EUNIS classes in the harmonized land cover map. This chapter addresses (semi-)natural grasslands (EUNIS class E), focussing on E1 (dry grasslands), E2 (mesic grasslands) and E3 (seasonally wet and wet grasslands).

23.2.3 Calculation of Nitrogen Deposition

Ecosystem-specific N depositions were computed on the 50×50 km^2 scale (EMEP grid) for 2000 and for 2020, using the source-receptor matrices (Amann et al. 2011; EMEP 2005) that are commonly used under the Convention on Long-range Transboundary Air Pollution (LRTAP). Depositions for 2020 were computed from two scenarios prescribing reductions of national emissions of oxidized and reduced N in that year, compared to 2000. The first scenario used was the revised Gothenburg Protocol scenario (GP2020) scenario, which follows the emissions ceilings for 2020 that were agreed under the LRTAP Convention in 2012 (UNECE 2013). The second scenario assumes that emissions will be reduced by 2020 due to the implementation of Maximum Feasible emission Reduction technology (MFR2020).

23.2.4 Derivation of Dose-Response Relationships

D-R Relationships Derived from Experimental N Addition Studies:

The basis for the derivation of D-R relationships from experimental N addition studies was an extended database of publications prepared for an assessment of empirical critical N loads, as summarized in Bobbink and Hettelingh (2011). Studies providing insights into ecosystem reactions to increases in N loads have been conducted for a variety of reasons. This has resulted in many different experimental designs. In this study, the outcomes of field addition experiments were only used if they were in accordance with the following criteria:

- N treatments only; no applications of additional nutrients or lime;
- N additions below 150 kg N $ha^{-1}yr^{-1}$;
- experimental period longer than two years, thus incorporating at least three growing seasons);
- located in Europe (west of 40°E longitude);
- including data on plant species richness (number of plant species per plot)

D-R relationships from N addition experiments for grasslands (EUNIS class E_1, E_2, E_3 and E_4) have been obtained from experimental field studies under both dry and wet conditions and with very different soil pH values (acid - calcareous) across Europe (6 countries) (Beltman et al. 2007; Berlin 1998; Bobbink 1991; Bonanomi et al. 2006; Crawley et al. 2005; Jacquemyn et al. 2003; Lüdi 1959; Tallowin and Smith 1994).

The number of plant species per control plot or N-treated plot were either directly obtained from the literature, or calculated from presented full species tables or extrapolated from given figures. Data from studies used for the derivation of D-R relationships were averaged per N treatment per investigated site to avoid pseudo-replication and overrepresentation. The used experimental N loads (in kg N $ha^{-1}yr^{-1}$) were also noted. To eliminate the effects of differential plot sizes used in the studies, and to make the outcome plot-independent, a species richness ratio (per N treatment) S_n/S_c has been defined, where S_n is the number of species in the N-treated situation and S_c their number in the control situation. This ratio expresses whether species richness in the addition experiment is equal ($S_n/S_c=1$), higher ($S_n/S_c>1$) or smaller ($S_n/S_c<1$) to the control. This species richness ratio is simple to interpret: a ratio of 0.75 indicates 25% reduction in species number after N treatment, compared with the control situation.

For each grassland addition experiment (see Fig. 23.1 for locations) the N deposition at the location and time of the experiment (around the year 2000) was derived from historical depositions based on EMEP model results (Hettelingh et al. 2008). The D-R relationships were then derived by regressing the species richness ratio on the total N load (experimental N addition plus estimated deposition at the site).

A significant negative relationship (negative exponential fit; $p<0.001$) between the species richness ratio and the total N load was found (Fig. 23.2). However,

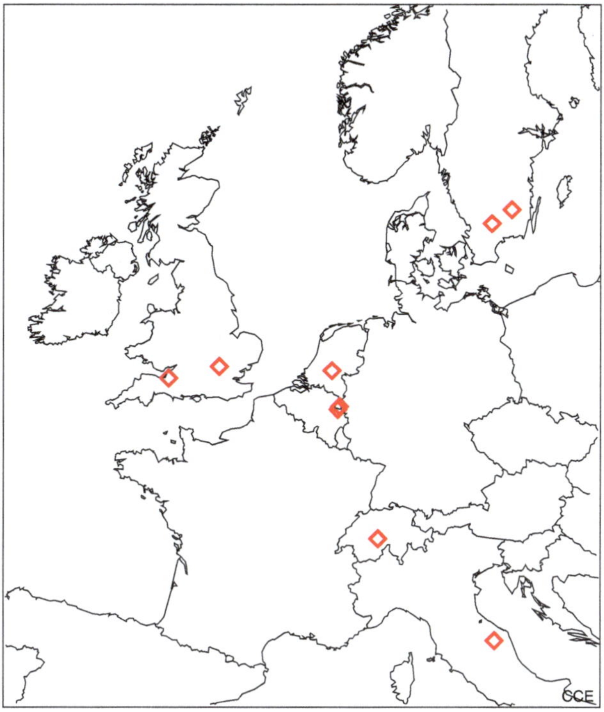

Fig. 23.1 Locations of plots of natural and semi-natural grasslands where N addition experiments were conducted

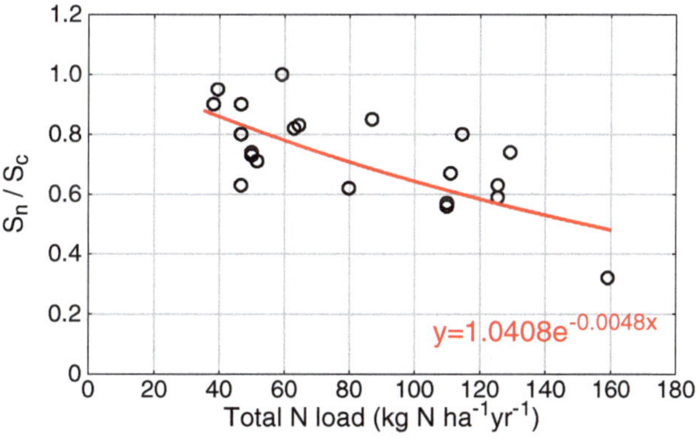

Fig. 23.2 Relation between the species richness ratio (S_n/S_c) and total N load (N addition plus background deposition) in grassland habitats (six countries; $n=22$; $p<0.001$) derived from N addition experiments

the scatter is relatively large, probably caused by the use of data from studies in very different grassland types and differences in experimental layout (especially duration).

Species richness is, in addition, probably a rather crude indicator for changes in the vegetation composition. In many studies it has been observed that nutrient-poor plant species have been outcompeted by other plants such as 'N-favouring' or 'invasive' species, without changing the overall number of species per plot (i.e. species richness) (Bobbink et al. 1998). It is very likely that the number of plant

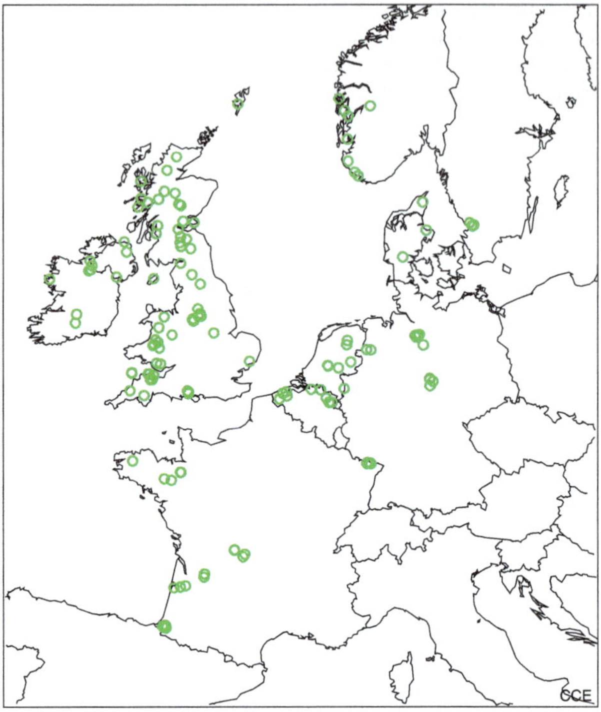

Fig. 23.3 Locations of plots of acid grasslands surveyed in the Atlantic biogeographic region of Europe. (Source: Stevens et al. 2010b)

species characteristic of unaffected vegetation, for example those that are typical to nutrient-poor environments, is a more sensitive indicator for changes in plant diversity with respect to N deposition. However, in almost all publications, plant species data have been unclassified. Therefore, it is likely that the fitted curve for plant species richness in grassland ecosystems would be steeper if data on the decrease in characteristic plant species were available from these experiments.

Dose-Response Relationship Derived from N Deposition Gradient Studies:

Stevens et al. (2010b) surveyed 153 semi-natural acid grasslands on a transect across the Atlantic biogeographic zone of Europe with total atmospheric N deposition ranging from 2.4 to 43.5 kg N $ha^{-1}yr^{-1}$, covering much of the range of deposition found in the industrialised world. The surveyed grasslands were dominated by species such as *Agrostis capillaris, Festuca ovina* and *rubra, Potentialla erecta* and *Galium saxatile*. The survey consisted of nine grasslands in Belgium, three grasslands in Denmark, twenty-five grasslands in France, twelve grasslands in Germany, eleven grasslands in Ireland, Northern Ireland and the Isle of Man, seven grasslands in the Netherlands, nine grasslands in Norway, four grasslands in Sweden and seventy-seven grasslands in Great Britain (Fig. 23.3). The large number of sites surveyed in Great Britain, based on an intensive national survey (Stevens et al. 2004) is due to the fact that natural grasslands cover a much larger area in the UK than in the other countries. All surveys were conducted between 2002 and 2007, between May and September, using a consistent methodology; none of the sites was fertilized or in the vicinity of a point source of N, and many were in areas where nature

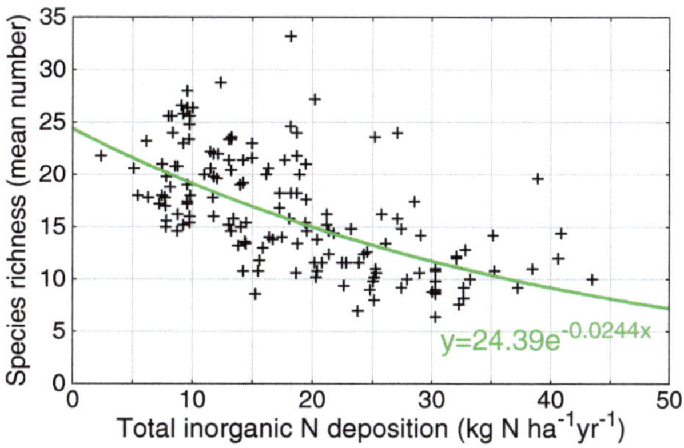

Fig. 23.4 Relationship between species richness of acid grasslands in the Atlantic region and N deposition. (adapted from Stevens et al. 2010b)

conservation policies applied. Within each site, five randomly located 4 m^2 quadrats were surveyed, avoiding areas of vegetation belonging to a different community (e.g. areas of shrub) or strongly affected by animals, tracks and paths, or in the rain shadows of trees or hedges. Within each quadrat all vascular plants and bryophytes were identified at species level, and percent cover was estimated by eye.

For all sites, well-documented deposition models were used for estimating the deposition of N. National models were used for Germany (Gauger et al. 2002), the Netherlands (Asman and van Jaarsveld 1992; Van Jaarsveld 1995, 2004) and Great Britain (NEGTAP 2001). For all other countries the EMEP-based IDEM (Pieterse et al. 2007) models were used. All models calculated deposition as a three-year average to provide a more robust estimate of longer-term N inputs.

The relationship between N deposition and species richness was best fitted with a negative exponential curve, giving an r^2 of 0.40 ($p<0.0001$, Fig. 23.4). The exponential curve implies a potentially greater loss of species richness when N deposition increases from a low level than from a high level.

23.2.5 Regionalization of Dose-Response Relationships over Europe

Regionalization of N Addition-Based Dose-Response Relationships:

The D-R relationships derived from N addition studies (Bobbink 2008; Bobbink and Hettelingh 2011) were used to estimate the loss of species richness of (semi-)natural grasslands (EUNIS class E). Uncertainty is expected to be important because of the relative scarcity of N-addition studies over Europe (see Hettelingh et al. 2011). The D-R relationships could be applied to approximately 70 million ha of grasslands.

In applying the D-R relationship derived from N addition studies, the response variable is the ratio of the plant species number in treatment over control, i.e. the model predicts the change in plant species richness compared to a reference point. Consequently, to apply the model based on N addition experiments to field conditions, we needed to set a reference point with which to compare the effects of changing deposition. The N addition experiments were predominantly carried out around the year 2000. This implies that the regionalization of D-R relationships based on N additions should be used in a relative context, i.e. to compare the change of plant species diversity between different N emission reduction scenarios as compared to 2000.

Regionalization of Gradient-Study Based Dose Response Relationships:

The harmonized European land cover map is used for a regionalized application of the gradient study based D-R function for grasslands with EUNIS codes E1, E2 and E3. The analysis was restricted to locations with precipitation between 490 and 1970 mm yr^{-1}, altitude below 800 m and soil pH below 5.5. These restrictions were applied to ensure that only grasslands with precipitation and soil pH within the range of conditions found in the original dataset were considered. The limitation of available precipitation data led to the analysis being restricted to E1, E2 and E3 grasslands areas located west of 32°E. This resulted in an area covering about 45 million ha.

23.3 Scenario Analysis of the Change of Plant Species Diversity in Europe

23.3.1 Impacts of N Addition and N Gradient Based Dose-Response Relationships

For the assessment of species richness on the European scale we normalise the D-R functions such that species richness at zero deposition corresponds to 100 %.

Computation results of species richness significantly depend on whether an N addition or an N gradient based D-R relationship is used (Fig. 23.5), because the slopes of the relationships differ between the two methods.

The regionalization of the N gradient based D-R relationship shows a broader range of percentages of species richness of grasslands in Europe than the N addition based D-R relationship. This is illustrated in Fig. 23.6 for both approaches, showing cumulative distribution functions of plant species richness in 2000 and under the GP2020 and MFR2020 scenarios. For the N addition based D-R relationship, all European grasslands have plant species diversity above 85 %, for all years and scenarios. If we focus on the most impacted 10 % of European grasslands, depositions in 2000, GP2020 and MFR2020 lead to a species richness near 96, 98 and 99 %, respectively (Fig. 23.6 left). These results illustrate that this addition based D-R function has a limited explanatory power to assess species loss in 2000. As mentioned above, most N addition experiments were carried out around the year 2000 and considerable species loss may have already occurred by then. Using 2000 as a reference point, results imply an increase in species richness of 2 % under N deposition under GP2020 and 3 % under MFR2020 in 10 % of the European grasslands. Changes in species richness are thus very low when currently available N addition based D-R relationships are extrapolated to European grasslands.

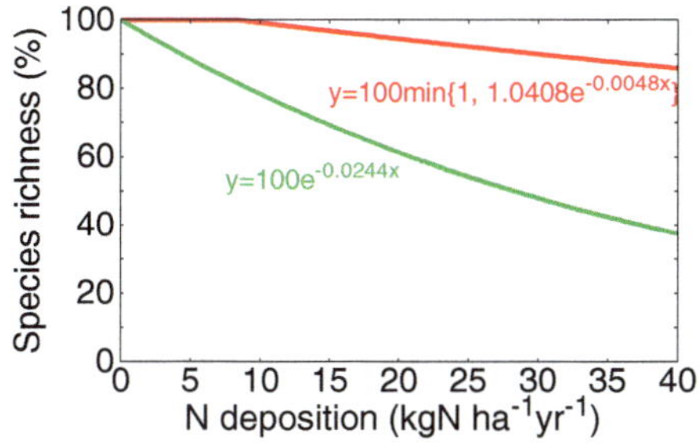

Fig. 23.5 Normalised species richness as a function of N deposition for the addition based (*red*; Fig. 23.2) and gradient based (*green*; Fig. 23.4) D-R relationships. Starting at a species richness of 100 % under zero N deposition, only 84 and 40 % of species richness remains, respectively, when deposition reaches 40 kg N $ha^{-1}yr^{-1}$

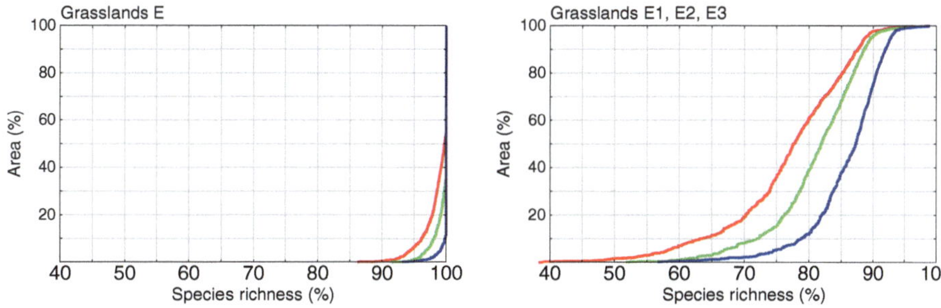

Fig. 23.6 Cumulative distribution functions of species richness (in %) in grasslands (EUNIS code E1, E2 and E3) in 2000 (*red*) and in 2020 for the revised Gothenburg Protocol (*green*) and under MFR emissions (*blue*) using the N addition based approach (*left*) and the N gradient approach (*right*). From the latter it can be seen that in 2000 the species richness in, e.g., 5 % of the area is about 60 %, i.e. reduced by 40 %

Application of the N gradient based D-R relationship shows that half of the grassland area is computed to have a species richness of more than about 78 % in the year 2000, while more than about 83 and 87 % could be achieved with N deposition under GP2020 and MFR2020, respectively (Fig. 23.6 right). Overall, the species richness percentages range between about 45 and 99 %. More than 40 % of the species richness is computed to be lost in about 5 % of the most affected European grasslands in 2000. This loss is reduced to less than 25 %, when N emissions are reduced using maximum feasible techniques.

23.3.2 Spatial Variation in Biodiversity Changes Using Gradient Based D-R Relationships

The regional variation in plant species richness over European grasslands based on application of the N gradient-based D-R relationship is presented in Fig. 23.7. Areas

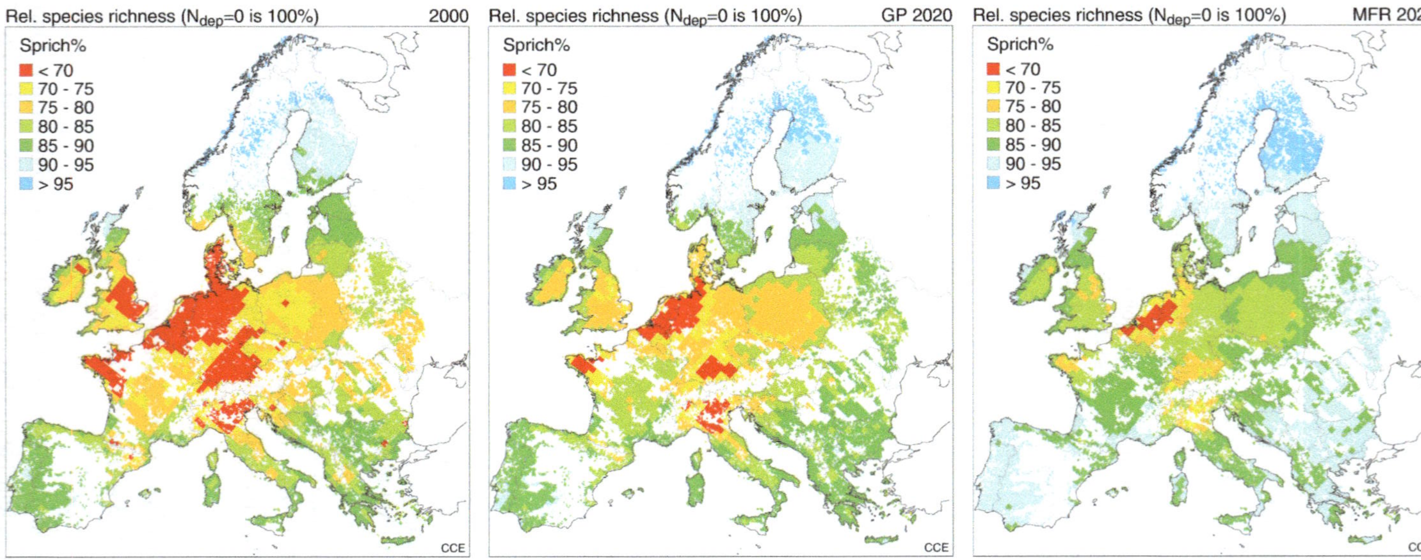

Fig. 23.7 Species richness (in %) in grasslands (EUNIS code E1, E2 and E3) applying a dose-response relationship derived from gradient studies to computed N depositions in 2000 (*left*) and in 2020 under emissions agreed under the revised Gothenburg Protocol (*centre*) and under the Maximum Feasible Reductions scenario (*right*)

with more than 90 % species richness (blue shading) are found in the north, east and south of Europe, under MFR2020 in particular. Areas with a species richness of less than 70 % (red shading) can be found in central and western Europe. Under the GP2020 and MFR2020 scenarios this area is markedly reduced. Species richness is higher than 75 % in most of the European grasslands under MFR2020, implying that the loss is reduced to less than 25 %. Note that the computed (GP2020 and MFR2020) increase in the species number (from the 2000 level) may occur later under field conditions, due to recovery delay time.

Conclusions

In this chapter dose-response relationships derived from N addition based experiments and from N deposition gradient studies are extrapolated over (semi-)natural grasslands (EUNIS class E) in Europe. The analysis shows that the range of plant species richness computed over these grasslands is smaller when the N addition based D-R relationship is used than when the N gradient based relationship is applied. Using the gradient based relationship, the loss of species richness in about 5% of the area of European grasslands is computed to reduce from about 40% in 2000 to less than about 25 % in 2020, if N emissions were reduced using maximum feasible techniques. The loss of species richness declines from about 4 % in 2000 to 2 % in MFR2020 when N addition based dose-response functions are used. As expected, both methods identify the greatest loss in areas with highest N deposition.

The large difference between both methods may be due to the fact that addition based D-R relationships depend on the N deposition at the time of experiment, when considerable species loss could already have occurred. Since gradient based D-R relationships are anchored in the deposition at the time of measurement, they are most suitable to identify the distribution in Europe of areas with plant species diversity at risk of further N deposition. Until now, the widely established method for computing exceedances of critical loads, which cause a violation of geochemical requirements for sustainability, could not convincingly communicate consequences for biology. It is suggested here that the regional use of both empirical critical loads based on N addition based methods and dose-response relationships based on N deposition gradient studies could increase the robustness of effect-oriented assessments in support of regional air pollution abatement policies.

References

Amann, M., Bertok, I., Borken-Kleefeld, J., Cofala, J., Heyes, C., Höglund-Isaksson, L., Klimont, Z., Nguyen, B., Posch, M., Rafaj, P., Sandler, R., Schöpp, W., Wagner, F., & Winiwarter, W. (2011). Cost-effective control of air quality and greenhouse gases in Europe: Modeling and policy applications. *Environmental Modelelling and Software, 26,* 1489–1501.

Asman, W. A. H., & van Jaarsveld, J. A. (1992). A variable-resolution transport model applied for NH_x in Europe. *Atmospheirc Environment, 26A,* 445–464.
Beltman, B., Willems, J. H., & Güsewell, S. (2007). Flood events overrule fertiliser effects on biomass production and species richness in riverine grasslands. *Journal of Vegetation Science, 18,* 625–634.
Berlin, G. (1998). Semi-natural meadows in southern Sweden—changes over time and the relationship between nitrogen supply and management. PhD thesis, Lund, Sweden, Lund University.
Bobbink, R. (1991). Effects of nutrient enrichment in Dutch chalk grassland. *Journal of Applied Ecology, 28,* 28–41.
Bobbink, R. (2008). The derivation of dose-response relationships between N load, N exceedance and plant species richness for EUNIS habitat classes. In J.-P. Hetterling, M. Posch, & J. Slootweg (Eds.), *Critical load, dynamic modelling and impact assessment in Europe* (pp. 63–72). Bilthoven: CCE Status Report 2008, Coordination Centre for Effects, Netherlands Environmental Assessment Agency.
Bobbink, R., & Hettelingh, J.-P. (2011). *Review and revision of empirical critical loads and dose-response relationships: Proceedings of an expert workshop, Noordwijkerhout, 23–25 June 2010.* (Report 680359002/2011). Bilthoven: Coordination Centre for Effects, National Institute for Public Health and the Environment.
Bobbink, R., Hornung, M., & Roelofs, J. G. M. (1998). The effects of air-borne nitrogen pollutants on species diversity in natural and semi-natural European vegetation. *Journal of Ecology, 86,* 717–738.
Bobbink, R., Hicks, K., Galloway, J., Spranger, T., Alkemade, R., Ashmore, M., Bustamante, M., Cinderby, S., Davidson, E., Dentener, F., Emmett, B., Erisman, J. W., Fenn, M., Gilliam, F., Nordin, A., Pardo, L., & De Vries, W. (2010). Global assessment of nitrogen deposition effects on terrestrial plant diversity: a synthesis. *Ecological Applcation,* 20, 30–59.
Bonanomi, G., Caporaso, S., & Allegrezza, M. (2006). Short-term effects of nitrogen enrichment, litter removal and cutting on a Mediterranean grassland. *Acta Oecologica, 30,* 419–425.
Chan, K. M. A., Satterfield, T., & Goldstein, J. (2012). Rethinking ecosystem services to better address and navigate cultural values. *Ecological Economics, 74,* 8–18.
Costanza, R., d'Arge, R., de Groot, R. S., Farber, S., Grasso, M., Hannon, B., Limburg, K., Naeem, S., O'Neill, R. V., Paruelo, J., Raskin, R. G., Sutton, P., & van den Belt, M. (1997). The value of the world's ecosystem services and natural capital. *Nature, 387,* 253–260.
Crawley, M. J., Johnston, A. E., Silvertown, J., Dodd, M., de Mazancourt, C., Heard, M. S., Henman, D. F., & Edwards, G. R. (2005). Determinants of species richness in the Park Grass experiment. *American Naturalist, 165,* 179–192.
Davies, C. E., & Moss, D. (1999). *EUNIS habitat classification*. Copenhagen: European Environment Agency.
De Groot, R., Brander, L., van der Ploeg, S., Costanza, R., Bernard, F., Braat, L., Christie, M., Crossman, N., Ghermandi, A., Hein, L., Hussain, S., Kumar, P., McVittie, A., Portela, R., Rodriguez, L. C., ten Brink, P., & van Beukering, P. (2012). Global estimates of the value of ecosystems and their services in monetary units. *Ecosystem Services, 1,* 50–61.
EMEP. (2005). Transboundary acidification, eutrophication and ground level ozone in Europe in 2003. EMEP Status Report 2005. (EMEP Report 1/2005). Joint CCC & MSC-W Report, The Norwegian Meteorological Institute.
Emmett, B., Ashmore, M., Britton, A., Broadmeadow, M., Bullock, J., Cape, N., Caporn, S., Carroll, J., Cooper, J., Cresser, M., Crossley, A., d'Hooghe, P., De, Lange, I., Edmondson, J., Evans, C., Field, C., Fowler, D., Grant, H., Green, E., Griffiths, B., Haworth, B., Helliwell, R., Hicks, K., Hinton, C., Holding, H., Hughes, S., James, M., Jones, A., Jones, M., Jones, M. L. M., Leake, J., Leith, I., Maskell, L., McNamara, N., Moy, I., Oakley, S., Ostle, N., Pilkington, M., Power, S., Prendergast, M., Ray, N., Reynolds, B., Rowe, E., Roy, D., Scott, A., Sheppard, L., Smart, S., Sowerby, A., Sutton, M., Terry, A., Tipping, E., Van den Berg, L., Van Dijk, N., Van, Zetten, E., Vanguelova, E., Williams, B., Williams, D., & Williams, W. (2007). Terrestrial Umbrella: effects of eutrophication and acidification on terrestrial ecosystems. Final report.

(CEH Project Number: C02613, Defra Contract No. CPEA 18). NERC/Centre for Ecology & Hydrology.
Galloway, J. N., Townsend, A. R., Erisman, J. W., Bekunda, M., Cai, Z., Freney, J. R., Martinelli, L. A., Seitzinger, S. P., & Sutton, M. A. (2008). Transformation of the nitrogen cycle: Recent trends, questions, and potential solutions. *Science, 320,* 889–892.
Gauger, T., Anshelm, F., Schuster, H., Erisman, J. W., Vermeulen, A. T., Draaijers, G. P. J., Bleeker, A., & Nagel, H.-D. (2002). *Mapping of ecosystems specific long-term trends in deposition loads and concentrations of air pollutants in Germany and their comparison with critical loads and critical levels*. Germany: Institut Fur Navigation. University of Stuttgart.
Hettelingh, J. P., Posch, M., & Slootweg, J. (2008). *Critical load, dynamic modelling and impact assessment in Europe. CCE status report 2008. (RIVM Report 500090003).* Bilthoven, The Netherlands: Coordination Centre for Effects, National Institute for Public Health and the Environment.
Hettelingh, J.-P., Posch, M., Slootweg, J., & Le-Gall, A.-C. (2011). Revision of the Gothenburg protocol: Environmental effects of GAINS scenarios developed during summer 2011. In Posch, M., J. Slootweg, & J.-P. Hettelingh (Eds.), *Modelling critical thresholds and temporal changes of geochemistry and vegetation diversity. CCE status report 2011.* Bilthoven: RIVM.
Jacquemyn, H., Brys, R., & Hermy, M. (2003). Short-term effects of different management regimes on the response of calcareous grassland vegetation to increased nitrogen. *Biological Conservation, 111,* 137–147.
Lüdi, W. (1959). Versuche zur Alpweide verbesserung auf der Schynigen Platte bei Interlaken. Interlaken: Beil. Jahresbericht. Ver. Alpengarten Schynige Platte.
Maestre, F. T., Quero, J. L., Gotelli, N. J., Escudero, A., Ochoa, V., Delgado-Baquerizo, M., García-Gómez, M., Bowker, M. A., Soliveres, S., Escolar, C., García-Palacios, P., Berdugo, M., Valencia, E., Gozalo, B., Gallardo, A., Aguilera, L., Arredondo, T., Blones, J., Boeken, B., Bran, D., Conceição, A. A., Cabrera, O., Chaieb, M., Derak, M., Eldridge, D. J., Espinosa, C. I., Florentino, A., Gaitán, J., Gabriel, G. M., Ghiloufi, W., Gómez-González, S., Gutiérrez, J. R., Hernández, R. M., Huang, X., Huber-Sannwald, E., Jankju, M., Miriti, M., Monerris, J., Mau, R. L., Morici, E., Naseri, K., Ospina, A., Polo, V., Prina, A., Pucheta, E., Ramírez-Collantes, D. A., Romão, R., Tighe, M., Torres-Díaz, C., Val, J., Veiga, J. P., Wang, D., & Zaady, E. (2012). Plant species richness and ecosystem multifunctionality in global drylands. *Science, 335,* 214–118.
Millennium Ecosystem Assessment. (2005). Ecosystems and human well-BEING: synthesis. (A report of the millennium ecosystem assessment). Washington, DC: Island. http://www.millenniumassessment.org/documents/document.356.aspx.pdf. Accessed 23 June 2014.
Mücher, C. A., Champeaux, J.-L., Steinnocher, K. T., Griguolo, S., Wester, K., Heunks, C., Winiwarter, W., Kressler, F. P., Goutorbe, J. P., Ten Brink, B., Van Katwijk, V. F., Furberg, O., Perdigao, V., & Nieuwenhuis, G. J. A. (2001). *Development of a consistent methodology to derive land cover information on a European scale from remote sensing for environmental monitoring: The PELCOM report*. (Report 178). Wageningen: Alterra.
NEGTAP. (2001). Transboundary air pollution: acidification, eutrophication and ground-level ozone in the UK. In first report of the national expert group on transboundary air pollution London: UK department of the environment, transport and the regions.
Phoenix, G. K., Hicks, W. K., Cinderby, S., Kuylenstierna, J. C. I., Stock, W. D., Dentener, F. J., Giller, K. E., Austin, A. T., Lefroy, R. D. B., Gimeno, B. S., Ashmore, M. R., & Ineson, P. (2006). Atmospheric nitrogen deposition in world biodiversity hotspots: the need for a greater global perspective in assessing N deposition impacts. *Global Change Biology, 12,* 470–476.
Pieterse, G., Bleeker, A., Vermeulen, A. T., Wu, Y., & Erisman, J. W. (2007). High resolution modelling of atmosphere-canopy exchange of acidifying and eutrophying components and carbon dioxide for European forests. *Tellus,* 59B, 412–424.
Slootweg, J., Hettelingh, J.-P., Tamis, W., & van 't Zelfde, M. (2005). Harmonizing European land cover maps. In M. Posch, J. Slootweg, & J.-P. Hettelingh (Eds.), *European critical loads and dynamic modelling: CCE report 2005* (pp. 47–62).

Stevens, C. J., Dise, N. B., Mountford, J. O., & Gowing, D. J. (2004). Impact of nitrogen deposition on the species richness of grasslands. *Science, 303,* 1876–1879.

Stevens, C. J., Thompson, K., Grime, P., Long, C. J., & Gowing, D. J. G. (2010a). Contribution of acidification and eutrophication to declines in species richness of calcifuge grasslands along a gradient of atmospheric nitrogen deposition. *Functional Ecology, 24,* 478–484.

Stevens, C. J., Duprè, C., Dorland, E., Gaudnik, C., Gowing, D. J. G., Bleeker, A., Diekmann, M., Alard, D., Bobbink, R., Fowler, D., Corcket, E., Mountford, J. O., Vandvik, V., Aarrestad, P. E., Muller, S., & Dise, N. B. (2010b). Nitrogen deposition threatens species richness of grasslands across Europe. *Environmental Pollution, 158,* 2940–2945.

Tallowin, J. R., & Smith, R. E. N. (1994). The effects of inorganic fertilisers in flower-rich hay meadows on the somerset levels. (English Nature Research Report 87). Peterborough: English Nature.

UNECE. (2013). 1999 Protocol to abate acidification, eutrophication and ground-level ozone to the convention on long range transboundary air pollution, as amended on 4 May 2012. (ECE/EB.AIR/114). UN economic council for europe, executive body for the LRTAP-convnetion.

Van Jaarsveld, J. A. (1995). Modelling the long-term atmospheric behaviour of pollutants on various spatial scales. Ph. D. Thesis, Utrecht, Universiteit Utrecht.

Van Jaarsveld, J. A. (2004). *The Operational Priority Substances model. Description and validation of OPS-Pro 4.1. (RIVM Report 500045001).* Bilthoven, the Netherlands: National Institute of Public Health and the Environment.

Part V
Integrated Assessment, Policy Applications and Synthesis

Chapter 24
Integrated Assessment of Impacts of Atmospheric Deposition and Climate Change on Forest Ecosystem Services in Europe

Wim de Vries, Maximilian Posch, Gert Jan Reinds, Luc T. C. Bonten, Janet P. Mol-Dijkstra, G. W. Wieger Wamelink and Jean-Paul Hettelingh

24.1 Introduction

Atmospheric deposition, especially of N, affects the quality of forest soils and services that forests provide, such as the habitat function for plant species and carbon sequestration through wood production. The latter can contribute to the mitigation of impacts of climate change. The concept of ecosystem services is already well established, but the linkage between N deposition and those services is relatively new. Nitrogen deposition may cause both beneficial and adverse impacts, depending on the level of N input and the type of service analysed. Until now, most of the focus in research has been on N critical loads in view of impacts on plant species diversity. Limited insight is available on physiological optimal N deposition that maximizes forest growth while safeguarding an 'acceptable' impact on plant species diversity.

The Concept of Ecosystem Services: Ecosystems provide services that are vital to human health and livelihood. Since the mid-1960s and early 1970s (e.g. Helliwell 1969), an increasing amount of information is being collected on the ecological and socio-economic value of goods and services provided by natural and semi-natural ecosystems to human society (see for example Costanza et al. 1997; Costanza et al. 2014; De Groot et al. 2002; De Groot et al. 2010; TEEB 2010). The Millennium Ecosystem Assessment has made a distinction in provisioning services, regulating services, supporting services and cultural services (Millennium Ecosystem Assessment 2005). Provisioning services are the products obtained from ecosystems, specifically food, fibre and wood/fuel, and also the provision of fresh water. Regulating services include the regulation of climate (carbon sequestration), the buffering and

W. de Vries (✉) · G. J. Reinds · L. T. C. Bonten · J. P. Mol-Dijkstra · G. W. W. Wamelink
Alterra Wageningen University and Research Centre, Wageningen, The Netherlands
e-mail: wim.devries@wur.nl

M. Posch · J.-P. Hettelingh
Coordination Centre for Effects (CCE), RIVM, Bilthoven, The Netherlands

W. de Vries et al. (eds.), *Critical Loads and Dynamic Risk Assessments*,
Environmental Pollution 25, DOI 10.1007/978-94-017-9508-1_24

filtering capacity (water purification) of the soil, ensuring an adequate water quality, and ground water and surface flow regulation affecting water quantity. Supporting services relate to the capacity of natural and semi-natural ecosystems to regulate essential ecological processes and life support systems through bio-geochemical cycles, including the provision of a habitat for a wide range of animal and plant species. Cultural services include, for example, recreation and landscape features or species with aesthetic or spiritual value. In this context, forests are among the most important ecosystems.

Nitrogen Deposition and Ecosystem Services: Nitrogen emission and deposition, in combination with S deposition, affects several ecosystem services (see Table 24.1 for details) as it has an impact on:

- Production of food and feed (crops) and wood (forests) for timber/fuel (provisioning service);
- Water quantity by affecting water uptake and thereby provisioning of fresh water by ground water recharge and runoff to surface waters (provisioning service);
- Diversity of plant species by its impact on the habitat function for wild plants and animals (supporting service, recently revised to habitat service), affecting biological diversity and related products (provisioning service);
- Carbon sequestration by affecting growth and production of non-CO_2 greenhouse gases, specifically N_2O and O_3 (climate regulating service);
- Soil buffering functions, induced by accumulation of N, depletion (leaching) of base cations and Al and changes in soil acidity (pH), thus affecting ground water and surface water quality (water quality regulating service);
- Soil biodiversity and thereby nutrient cycling and primary production (supporting service).

Within the context of critical load assessments and dynamic modelling of N and S deposition impacts, most attention has focused on the water quality regulation function of ecosystems, which is negatively impacted by N and S induced acidification. This is caused by a decrease in the soil N retention, by a decrease in the soil C/N ratio (see Chap. 2), and in the acid buffering, by depleting available BC and Al pools. Changes in the buffering of N and acidity are clearly reflected by an increase in the dissolved (soil solution) concentrations of N (mainly nitrate) and Al. More recently, the impacts of N and S deposition on the productivity of forest ecosystems have also been modelled. Furthermore, large-scale modelling has been focused on effects of N, in interaction with climate change, on the diversity of plant species.

Differential Nitrogen Deposition Impacts on Forest Ecosystems: Until a physiological optimum level is reached, forests will react to additional N inputs by an increased biomass production until, provided that other nutrients are not limiting. The physiological optimum may be beyond an ecological optimum in view of adverse impacts on soil quality and plant species diversity, as illustrated in Fig. 24.1.

Table 24.1 Major relationships between N deposition and ecosystem services as distinguished in the Millennium Ecosystem Assessment (after De Vries et al. 2014b)

Ecosystem services	Examples of nitrogen effects	Causal link with nitrogen deposition
Provisioning services		
Food/feed, including	–	–
Crops	Increase in crop production	N deposition increases crop growth in N limited systems (low N fertilizer inputs)
Wild plants and animal products	Impacts on biodiversity (based products)	N induced eutrophication and soil acidification affects plant and faunal species diversity and thereby biodiversity-based products
Timber/wood fuel	Increase in wood production	In N-limited systems, nitrogen increases forest growth and wood production; in N saturated forests, N can induce mortality
Fresh water	Impacts on ground water recharge and drainage	N induced impacts on growth and plant species diversity also affect water uptake and thereby freshwater supply (see also water quantity regulation)
Regulating services		
Air quality regulation	Decline in air quality	Nitrogen deposition is correlated with increased concentrations of ammonia (NH_3), nitrogen oxides (NO_x), ozone (O_3) and particulate matter (PM_{10} and $PM_{2.5}$), all affecting human health and ecosystems
Climate regulation Greenhouse gas balance	Increased carbon sequestration in forests	In N limited systems, N deposition increases forest growth and related tree carbon sequestration, but can enhance mortality in some species. It also can cause an increased litterfall and reduced decomposition, leading to soil carbon sequestration
–	Increased/decreased carbon sequestration in peat lands	At low N deposition, additional atmospheric N deposition may stimulate net primary productivity. At high rates of N deposition, species composition changes lead to loss of peat land forming species and changed microbial activity causing degradation of peat lands
–	Increased N_2O production	N_2O emissions increase with increased N loading
–	Decreased CH_4 consumption	Soil microbes decrease CH_4 consumption in response to increased NH_4 availability

Table 24.1 (continued)

Ecosystem services	Examples of nitrogen effects	Causal link with nitrogen deposition
–	Increased O_3 production	Increased production in tropospheric O_3 from interactions with NO_x and VOC emitted from ecosystems, which serves as GHG and can also inhibit CO_2 uptake through plant damage
Water quality regulation (water purification)	Decline in ground water and surface water (drinking water) quality	N eutrophication and N induced soil acidification causes a decrease in soil C/N ratio, in available BC and Al pools and in pH, leading to: Increasing NO_3, Cd and Al concentrations in groundwater and surface water, which may exceed drinking water quality criteria in view of human health effects
Water quality regulation (water purification)		Increased Al concentrations in acid sensitive surface waters resulting in the reduction or loss of fish (salmonid) populations and reduction of aquatic diversity at several trophic levels (acidification)
		Fish dieback by algal blooms and anoxic zones (eutrophication). Eutrophication is also affected by silica and phosphorus in estuaries
Pest/disease regulation	Increased human allergic diseases	Increasing N availability can stimulate greater pollen production, causing human allergic responses, such as hay fever, rhinitis and asthma
	Increase in forest pests	Increase in bark or foliar N concentrations can attract higher infestation rates, such as beech bark disease
Supporting services		
Nutrient cycling and primary production	Increases N inputs by litterfall reduces soil biodiversity	N induced impacts on growth/litterfall and on soil biodiversity (soil mesofauna and bacteria composition) affects decomposition, nutrient mineralization and N immobilization, and thereby impacts nutrient cycling and primary production
Cultural services		
Cultural heritage values	Impacts on culturally significant species in historically important landscapes	N deposition may change heathlands into grasslands, affecting historically important landscapes

Table 24.1 (continued)

Ecosystem services	Examples of nitrogen effects	Causal link with nitrogen deposition
Recreation and ecotourism	Impacts on recreation due to impacts on ecosystems	Nitrogen induces the increase in nitrophilic species like stinging nettles and algal blooms reducing recreational and aesthetic values of nature. Extreme examples include closed beaches due to algal blooms resulting from N-induced eutrophication in estuaries and coastal ecosystems

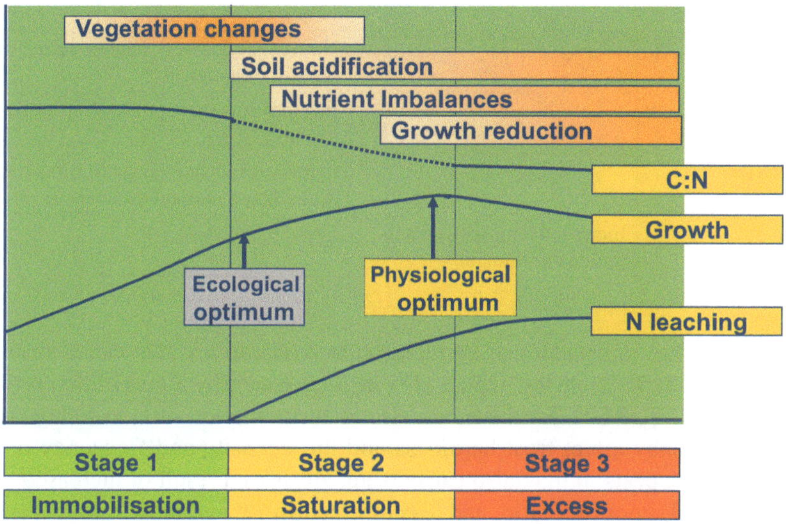

Fig. 24.1 Hypothetical relationship between the stage of N saturation and the effects on terrestrial ecosystems in terms of soil processes, vegetation changes and growth. (This figure is an update of the original figure by Aber et al. (1989), further updated by (Gundersen 1991), Aber et al. (1998) and Gundersen et al. (2006))

Below the physiological optimum for growth, the ecological optimum may already be exceeded, including gradual changes of the forest (floor) plant species composition towards more nitrophilic species (Bobbink et al. 1998; Bobbink and Hettelingh 2011; Ellenberg 1985). In forests with a continuous elevated N input, the ecosystem may approach 'N saturation' (Aber et al. 1989). In this stage, the N leaching will increase above (nearly negligible) background levels, associated with soil acidification in terms of elevated leaching of BC or Al, causing a decrease in acid neutralizing capacity. At the stage of 'N saturation' or 'N excess', the ecosystem may be destabilised by the interaction of a number of factors. Release of Al by soil

Table 24.2 Possible effects of increased atmospheric N deposition and exposure to NO_x and NH_3 on forest ecosystems (after Erisman and De Vries 2000)

Forest ecosystem compartment	Effects	
	Chemistry	Ecosystem
Soil (solution)	Increased N concentrations in soil (solution)	Increase in nitrophilous species/ decrease in biodiversity
		Increase in NO3 leaching
	Increased ratios of NH4 + and Al3 + to base cations	Inhibition of uptake (nutrient imbalances) root damage and mycorrhiza decline
Trees (foliage)	Increased N concentrations in foliage	Increased frost sensitivity
		Increased parasite injury (insects, fungi, virus)
		Nutrient deficiency absolute or relative (to N)/ discoloration
		Increased biomass production/ water demand
		Increased ratio of foliage to roots (risk of drought and nutrient deficiency)
	Increased arginine concentrations in foliage	Growth reduction

acidification and imbalances of ammonium to base cations may cause absolute or relative nutrient deficiencies, which may be aggravated by a loss of mycorrhiza or root damage (Fig. 24.1). Furthermore, strong accumulation of N in foliage (e.g. as amino acids) may affect frost hardiness and the intensity and frequency of insect and pathogenic pests. It may also cause water stress as a result of increased canopy size, increased shoot/root ratio, and loss of mycorrhizal symbioses. An overview of possible effects on forests as a result of increased atmospheric acid and N deposition and/or exposure to air pollutants is presented by Erisman and De Vries (2000), and summarised in Table 24.2.

In this chapter, impacts of changes of N and S deposition and of climate (temperature and precipitation) on forest ecosystem services are presented at the European scale in terms of:

- Soil quality, focusing on changes in the soil C/N ratio and the readily available pools of BC and Al, in view of impacts on water quality, i.e. increased dissolved NO_3 and Al concentrations in excess of critical limits;
- Climate regulation, in view of forest growth and C sequestration;
- Diversity of plant species in forest ecosystems.

We modelled the combined effects of past (1900–2010) and projected future (2010–2050) changes in climatic variables (temperature and precipitation) and deposition of N and S on these ecosystem services, focusing on European forests. With respect to forest growth and carbon sequestration, the possible limitation of base cation major nutrients, i.e. calcium, magnesium and potassium, was also investigated. Two scenarios for deposition (current legislation and maximum feasible reductions) and two climate scenarios (no change and SRES A1 scenario) were used. Results are presented in terms of time trends over the period 1900–2050.

24.2 Models, Data and Indicators to Assess Impacts on Forest Ecosystem Services

Three different models were used to assess impacts of changes in atmospheric deposition of N and S and climate change on forest ecosystem services:

- VSD (Posch and Reinds 2009, see also Chap. 8) to assess impacts on soil regulating functions and related soil water quality;
- EUgrow (De Vries and Posch 2011) to assess impacts on forest growth and carbon sequestration;
- PROPS (Reinds et al. 2012, see also Chap. 11) to assess impacts on plant species diversity.

Below, relevant details on the modelling approach are given, including the input data used for running the models.

24.2.1 VSD: Assessing Impacts on Soil Regulation Function and Related Soil Water Quality

The response of soil quality indicators, i.e. soil C/N ratios, readily available BC and Al pools and dissolved NO_3 and Al concentrations, to changes in N and S deposition and climate at the European scale for the period 1900–2050 were assessed with the VSD model (Posch and Reinds 2009). VSD consists of a set of mass balance equations, describing the soil input-output relationships, and a set of equations describing rate-limited and equilibrium soil processes. The soil solution chemistry in VSD depends solely on the net element input from the atmosphere (deposition), net uptake, net immobilisation and denitrification and the geochemical interaction in the soil (CO_2 equilibria, mineral weathering, cation exchange and internal production or organic anions). More details on VSD are given in Chap. 8. The impact of N and S deposition on soil quality, in terms of depletion of the readily available BC and Al pools, was derived as total BC and Al soil release rates minus BC and Al weathering rates (De Vries et al. 1994). Base cation depletion from the exchangeable base cation pool is directly computed by VSD. Aluminium depletion was derived by

Al leaching and runoff minus Al weathering, which was assumed as twice the BC weathering. Results are given as annual average BC and Al depletion in eq $ha^{-1}yr^{-1}$.

The impacts of the of deposition and climate scenarios on soil C/N ratio was evaluated by requiring that the C/N ratio (i) should stay constant, as the decrease of C/N potentially leads to an increase in N leaching, while an increase of C/N could imply nutrient deficiency or (ii) stays above a value of 20, being a critical value for the C/N ratio in the organic layer (see Chap. 2). The impacts of deposition and climate scenarios on the concentrations of Al, NO_3 and N in soil solution, draining to groundwater and surface water, was evaluated by comparing them with:

- a limit of 20 µmol l^{-1} for Al, being a value in soil and ground water related to potential impacts (see Chap. 2) and a NO_3 concentration in groundwater, intended for drinking water, of 50 mg NO_3 l^{-1} according to the EU Drinking Water Directive (EC 1998);
- a limit of 2 mg N l^{-1} for N in surface water in view of eutrophication effects. This value is based on an extensive study on the ecological and toxicological effects of inorganic N pollution (Camargo and Alonso 2006), (ii) an overview of in national surface water quality standards (Liu et al. 2011), and (iii) different European objectives for N (Laane 2005) indicate maximum allowable N concentrations in surface waters in the range of 1.0–2.5 mg N l^{-1}.

24.2.2 EUgrow: Assessing Impacts on Tree Growth and Carbon Sequestration

The environmental effects on forest growth, G, for every year, was modelled by the empirical model EUgrow by modifying the reference growth, G_{ref}, according to (De Vries and Posch 2011):

$$G = G_{ref} \cdot f_{climate} \cdot f_{Ndep} \cdot f_{nutlim} \tag{24.1}$$

where the f-factors describe the reduction or enhancement of growth due to climate, N deposition and nutrient limitations, respectively. The reference growth data were related to observed growth for the period 1970–1990. In our simulations we assumed that these data all refer to the year 1980; and for this year the values for $f_{climate}$, f_{Ndep} and f_{nutlim} are normalised to one.

Based on literature data, a conceptual response curve of N retention to N deposition can be hypothesized with three stages, i.e. (i) below a N deposition threshold N1, N retention stays high and growth of forests is stimulated, (ii) between N deposition thresholds N1 and N2, both the stimulation of forest growth and the N retention starts to decrease, thus reducing the C response to N input and (iii) above N deposition threshold N2, there are adverse effects on growth and there is a substantial decrease in N retention due to N saturation, soil acidification and nutrient imbalances. Field studies indicate threshold values of 10–15 and 20–50 kgN $ha^{-1}yr^{-1}$ for

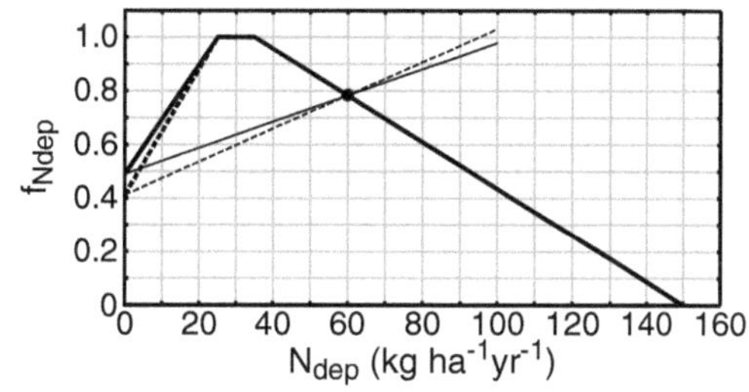

Fig. 24.2 Response functions f_{Ndep}, relating forest growth to N deposition

N1 and N2, respectively (De Vries et al. 2014a). For the impacts of N deposition on growth, we defined a hormetic response function f_{Ndep}, approximated by a trapezoidal shape (Fig. 24.2), with the following considerations:

- growth of approximately half of its maximum at zero N deposition (e.g. due to N fixation) based on Solberg et al. (2009),
- maximum growth between 25–35 kg N $ha^{-1}yr^{-1}$ based on Solberg et al. (2009)
- linear growth decrease above 35 kg N $ha^{-1}yr^{-1}$ up to a collapse of the system (no growth) at 150 kg N $ha^{-1}yr^{-1}$, based on results of Magill et al. (2004).

Apart from N, the growth of forests is limited by the availability of other nutrients, specifically phosphorous (P) and base cations (Ca, Mg, K). Due to limited data availability, P limitation is not included. Base cation limitation is included by comparing the supply of these nutrients by deposition and weathering with the demand of each nutrient, based on tree growth and required minimum content in the tree, whereby the most limiting nutrient determines the growth. Details on these calculations, the various reduction factors for climate and other nutrients, and the assessment of the input data is given in De Vries and Posch (2011).

24.2.3 PROPS: Assessing Impacts on Plant Species Diversity

The environmental effects on plant species diversity were estimated with the PROPS (probability of occurrence of plant species) model, which assesses the probability of plant species occurrence as a function of environmental factors derived from combined plant species and soil data in vegetation relevés (Reinds et al. 2012). The database for PROPS consists of about 6000 vegetation relevés at which pH and partly C/N ratio have been measured. The data base have been augmented with data on annual temperature and precipitation excess from the CRU meteorological data set (Mitchell et al. 2004) and N and S deposition from the EMEP model results (Simpson et al. 2003), using data from the grid cell corresponding to the location of the relevé. Several combinations of environmental factors were tested to fit the response curves, including pH, modelled nitrate concentration, modelled N availability, N deposition, temperature and precipitation surplus. For the simulations described here, the model with measured soil pH, interpolated temperature and log-transformed N deposition was used. The probability of occurrence of each plant species at each plot was calculated by PROPS using VSD predicted values for soil pH and model and database derived values for N deposition and temperature as described below. More information is given in Reinds et al. (2012) and in Chap. 11.

To evaluate changes in species diversity, the Bray-Curtis Index, *BCI*, was computed:

$$BCI = 1 - \tfrac{1}{2}\sum_{i=1}^{n} | x_i - y_i | \qquad (24.2)$$

where x_i and y_i ($i = 1,\ldots,n$) denote two sets of (plant) abundances, normalised to one, either at two different points in time or from two different 'measurements' (model outputs) (at the same time). This index always lies in the range between 0 and 1, and it is 1 only if the two sets are identical.

The Bray-Curtis index is an index describing the similarity of plant species composition to a reference point. Based on the probability of occurrence of each species calculated by PROPS, the Bray-Curtis index was calculated taking 2010 as the reference state, and assuming that probabilities of occurrence are linearly related to abundances. The impact of air pollution and climate change on the index was evaluated by comparing index values between scenarios, as described below.

24.2.4 Assessment of Input Data

Input data for the simulations of VSD, EUgrow and PROPS consist of spatially explicit information on (i) deposition of S and N and also base cations for VSD and (ii) climatic data. Furthermore, VSD needs input on soil and forest data, including weathering of base cations, nutrient uptake, N transformations and soil properties such as carbon pool and cation exchange capacity. The assessment of those data is described below.

Nitrogen and Sulphur Deposition Data: Model simulations were carried out for the period 1900–2050. Historic N and S deposition data were taken from Schöpp et al. (2003). Scenarios of N and S deposition were obtained from the atmospheric transport model of EMEP/MSC-W (Tarrasón et al. 2007). Two emission scenarios were used, reflecting current legislation (CLE) and maximum (technically) feasible reductions (MFR), developed for the Thematic Strategy on Air Pollution of the EU (Amann et al. 2007). The CLE scenario was used from 2010 onwards, whereas the MFR scenario was assumed to be implemented by 2020 and assumed constant thereafter. Results of the simulated N deposition for the period 1990–2050 are shown in Fig. 24.3. Results show that when MFR are implemented, N deposition on forests in 2050 is below 2000 eq $ha^{-1}yr^{-1}$ in almost the whole of Europe, except for the Netherlands and its surroundings.

Meteorological Data: Historical climate data were taken from a European data base (Mitchell et al. 2004) that contains monthly values of temperature, precipita-

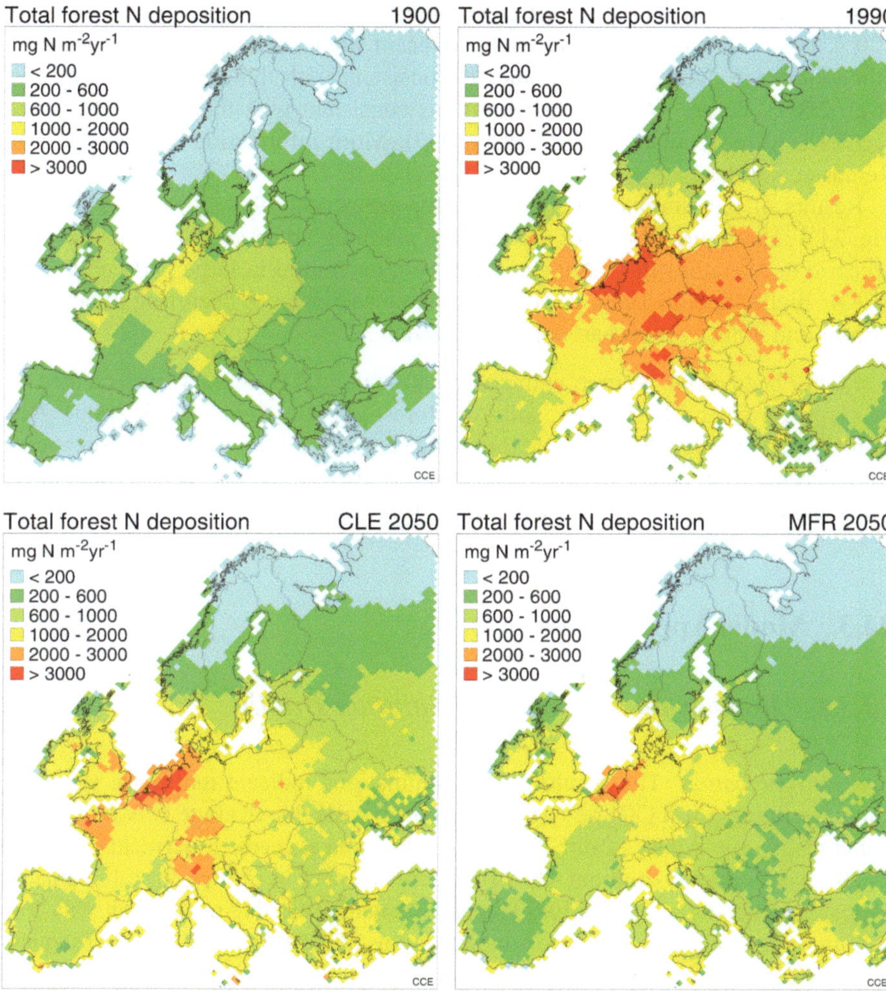

Fig. 24.3 Modelled N deposition in Europe in (from *top* left to *bottom* right) 1900, 1990, 2050 assuming Current Legislation (CLE) and 2050 assuming Maximum Feasible Reductions (MFR)

tion and cloudiness for the years 1901–2005 for land-based grid-cells west of 32°E of size 10′ × 10′ (approx. 15 × 18 km in central Europe). In the simulations 10-year averages were taken between 1900 and 1990 to smooth the climate pattern. For the future we used two climate scenarios: (1) constant climate after 1990, using the computed mean monthly temperature, precipitation and cloudiness of the period 1961–1990 as reference climate set, and (2) from 2010 onwards the A1 storyline from the Special Report on Emission Scenarios (SRES) with its related climate scenario (Nakícenovíc et al. 2001). The A1 scenario represents a future world with globalization and rapid economic growth, low population growth, and the rapid introduction of new and more efficient technologies everywhere (Strengers et al. 2005). The A1 scenario was chosen since recent studies have shown that observed climate change over the last decades is consistent with this scenario (Rahmstorf et al. 2007). Future CO_2 air concentrations consistent with the above mentioned scenarios were obtained from Carter (2007). More information can be found in Reinds et al. (2012).

Results of the simulated temperature for the period 1900–2050 are shown in Fig. 24.4. The figure illustrates that climate warming under the A1 climate change scenario accelerates after 1990. In 2050 the temperature in large parts of Europe is 2–4° higher than in 1990, with most pronounced changes occurring in southern Europe and in the eastern parts of Scandinavia.

Soil and Forest Data: A map with computational units (receptors) was created by overlaying maps on land cover, soils and forest growth (Reinds et al. 2009). This results in about 385,000 sites covering about 2.3 million km^2. Soil and forest data were derived from databases on soils and forest growth regions, as described in De Vries and Posch (2011) and Reinds et al. (2012).

24.3 Model Applications

24.3.1 Impacts on Soil and Soil Solution Quality

Soil Eutrophication: The results of our simulations indicate that there is only a slight decrease in soil C/N ratio. On a European scale, the average C/N ratio in the mineral soil only reduces from 25 to 24 and the area exceeding a critical C/N ratio of 20 stayed constant at about 4% of the forested area in Europe for all evaluated deposition and climate scenarios. The small changes are due to the relatively low N immobilization compared to available N pools and the fact that C is also sequestered in response to N deposition.

Trends in the annual average ambient NO_3 concentration (Fig. 24.5a) for the period 1900–2000 mainly reflect the trend in N deposition, increasing up to 1990, followed by a decrease since then. The impact of the MFR scenario as compared to the CLE scenario is much larger than the difference between the climate change scenario and no climate change (Fig. 24.5a). This is also in line with results by Reinds et al. (2009), but are probably influenced by neglecting the effect of climate change on N transformations in VSD.

The area exceeding a N concentration of 2 mg N l^{-1} (Fig. 24.5b)—a critical value for surface water quality (see Sect. 24.2.1) and also affecting the ground vegetation, at least in the Nordic countries (De Vries et al. 2007)—under the four scenarios drops from about 40% in 1990 to 25% in 2050 in case of the CLE scenario and to 5% in case of the MFR scenario, with very limited impacts of climate change (Fig. 24.6b).

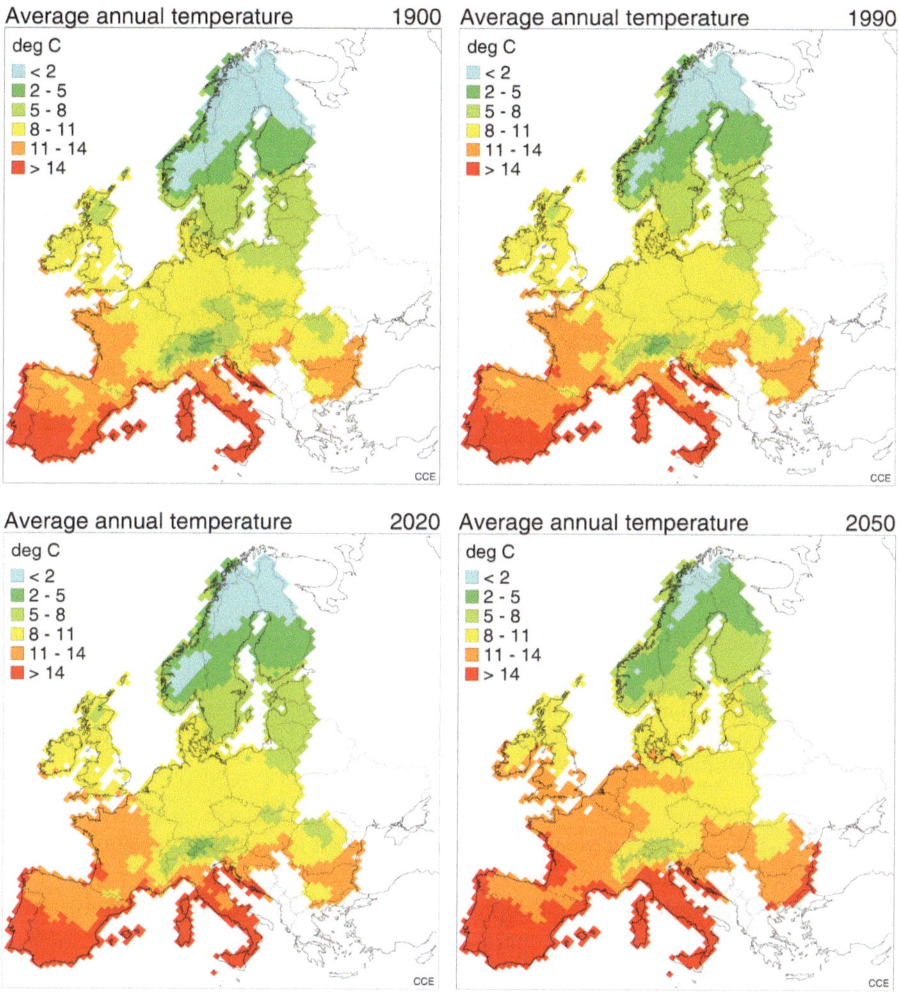

Fig. 24.4 Median grid temperature in Europe in (from *top* left to *bottom* right) 1900, 1990, 2020 and 2050, with 2020 and 2050 being predictions under the A1 climate scenario

Soil Acidification: Temporal trends in the average depletion of readily available BC and Al pools are shown in Fig. 24.6. The depletion of BC refers to the exchangeable pool (decrease in base saturation), whereas the Al depletion refers to the amounts adsorbed to organic matter and Al hydroxides. Results show that Al and BC depletion was low in the period 1900–1950, with BC depletion being approximately seven times as large as Al depletion. This is logical, since Al release in excess of weathering only occurs in quite strongly acidified systems. The large increase in both Al and BC depletion between 1950 and 1980 reflects the strong increase in N and S deposition in that period, whereas the strong decrease between 1980 and 2010 reflects the response to the large emission reductions. At its peak in

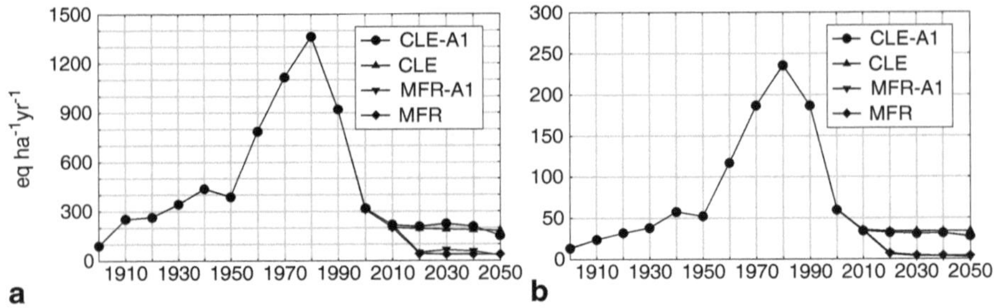

Fig. 24.5 Past and future development of **a** the annual average BC depletion and **b** Al depletion, in the period 1900–2050 in response to four scenarios for deposition and climate

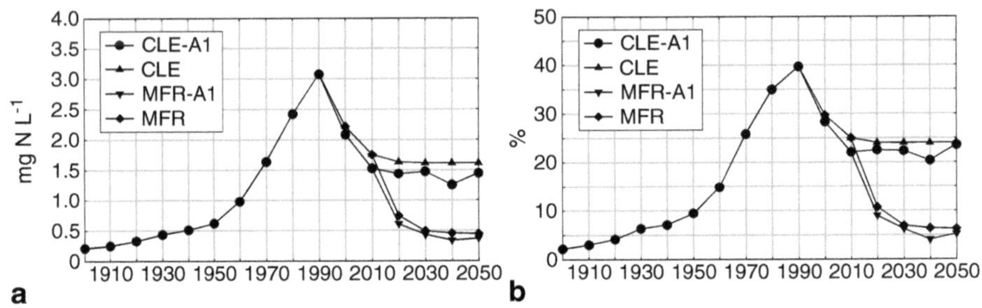

Fig. 24.6 Past and future development of **a** the annual average N concentration and **b** the area exceeding a critical N concentration of 2 mg N l^{-1}, in the period 1900–2050 in response to four scenarios for deposition and climate

1980, annual average Al and BC depletion was about ten times as high as in 1900. At present, Al depletion is only an issue in central Europe, but BC depletion is more widespread, although the rate of depletion has decreased strongly due to the emission reductions taken, and is comparable to the low values around 1900.

The simulations for the future (2010–2050) according to the four scenario combinations show that the impact of the MFR deposition scenario as compared to the CLE scenario is much larger than the difference between the A1 climate change scenario compared to no climate change (Fig. 24.6); and this is in line with results by Reinds et al. (2009). Climate change leads to higher weathering rates and N uptake in the model, but positive effects on the recovery from acidification are limited. This limited effect may be caused by neglecting the effect of climate change on N transformations (mineralization, nitrification and denitrification) in the VSD model.

Temporal trends of the annual average dissolved Al concentration (Fig. 24.7a) are comparable to those of the annual average Al depletion. Unlike N concentration, which increased up to 1990, both Al (and BC) depletion rates and dissolved Al concentrations show a decline between 1980 and 1990 (Figs. 24.6 and 24.7a). This different behaviour of N and Al reflects the still ongoing increase in N deposition between 1980 and 1990 but the large decrease in S deposition in that period. This results in ongoing eutrophication (N saturation), but decreasing acidification. The

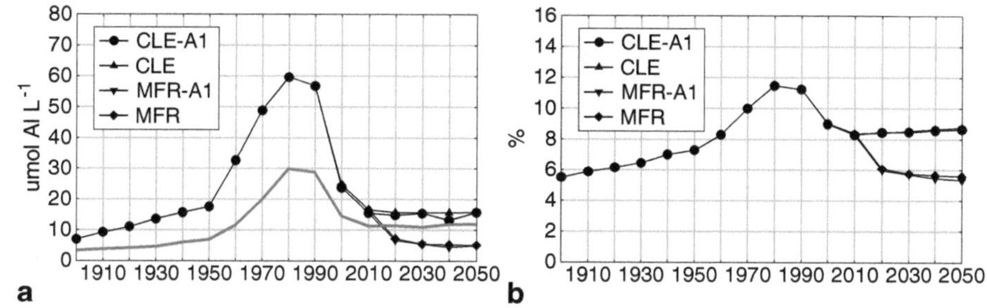

Fig. 24.7 Past and future development of **a** the annual average Al concentration and **b** the area exceeding a critical Al concentration of 20 μmol Al l^{-1}, in the period 1900–2050 in response to four scenarios for deposition and climate. The thick grey line in **a** is the 90th percentile trace of CLE-A1, indicating the extreme skewness of the Al concentration distributions

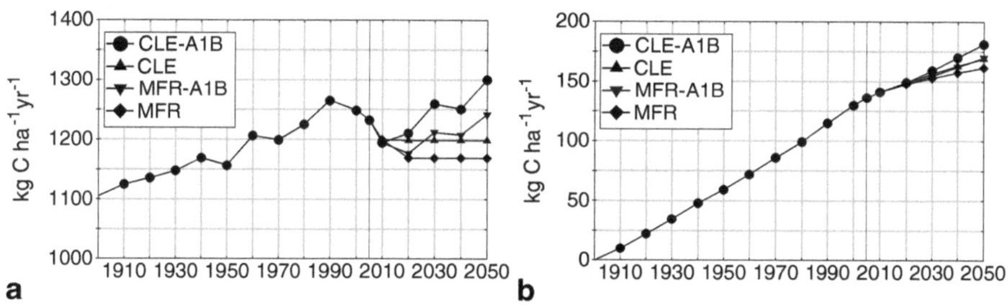

Fig. 24.8 Past and future development of **a** the mean annual European tree C sequestered and **b** the mean additional annual European tree C sequestered since 1900, in response to four scenarios for deposition and climate

impact of the MFR scenario as compared to the CLE scenario on the annual average Al concentration and the area exceeding a critical Al concentration of 20 μmol Al l^{-1} is, however, less than for the annual average N concentration (compare Fig. 24.5 and 24.7), reflecting the larger future reduction in N emissions than in S emissions. Again, the climate change impact is limited. Note that the annual average Al concentration in 1980 is thrice the critical value, whereas the area exceeding the critical limit is only near 12% (Fig. 24.7b), due to the extremely skewed distribution of [Al] (mean > 90th percentile; see Fig. 24.7a) caused by very high Al concentrations in central Europe.

24.3.2 Impacts on Forest Growth and Carbon Sequestration

Trends in tree C sequestration for the four scenario combinations of N deposition (CLE and MFR) and climate change (no climate change and A1 climate change scenario) are presented in Fig. 24.8a. Up to 2010 the curves follow the N deposition trends. Thereafter, the influence of the warmer climate can be clearly seen. Overall, the growth decline induced by strong N deposition reductions under the MFR scenario is almost entirely compensated by increased growth due to higher temperatures (compare CLE without climate change to MFR with climate change). The long-term average additional sequestered C in trees (Fig. 24.8b) shows a smoother path. By 2000 an additional—i.e. compared to what has been sequestered in 1900–average of 125 kg C $ha^{-1}yr^{-1}$ have been sequestered in trees for 100 years. Considering the long-term average additional N deposition in 2000 of approximately 5 kg N $ha^{-1}yr^{-1}$ compared to 1900, this implies an average tree response of 25 kgC kgN^{-1}. Under the CLE-A1 scenario, the average additional N deposition compared to 1900 stays near 5 kg N $ha^{-1}yr^{-1}$ between 2000 and 2050, implying that the additional sequestration of 50 kg C $ha^{-1}yr^{-1}$ in that period (from about 125–175 kg C $ha^{-1}yr^{-1}$; see Fig. 24.8b) is mainly due to climate change.

Geographic variations in trends in C sequestration in the period 1900–2050 are largely determined by the reference growth in 1980, since both N deposition and climate change are normalized to the reference growth (see Eq. 24.1). The grid-average tree C sequestered in 1900 (Fig. 24.9a), and 2050 (both for CLE-A1, Fig. 24.9c and MFR without climate change, Fig. 24.8d) are largest in areas with a high reference growth in 1980 (see Fig. 24.9b), combined with a low reduction in N deposition in either 1900 or 2050 as compared to 1980.

24.3.3 Impacts on Plant Species Diversity

When we assume constant climate and constant deposition, using the year 2010 as the reference year, the simulated Bray-Curtis index, a measure for the similarity between two species compositions, was close to one in 2050 for the whole of Europe, even though the soil pH slightly changed between 2010 and 2050 (not shown). In the scenarios with constant climate and CLE, the index decreased only slightly. With MFR depositions, but constant climate, the computed Bray-Curtis index changes significantly, showing the effect of strongly decreasing N deposition on species composition. In the scenario in which changes in both climate and N deposition were evaluated, the Bray-Curtis index decreased sharply, indicating that climate change in the PROPS model strongly changes the occurrence probability in areas with pronounced temperature change in 2050 (Fig. 24.10).

Discussion and Conclusions

Discussion: This study focused on multiple effects of changes in deposition of S and N compounds, in combination with climate change, on forest ecosystems. It included not only impacts of acidification and eutrophication on soil and soil solution quality and biodiversity, but also of atmospheric N deposition on the sequestration

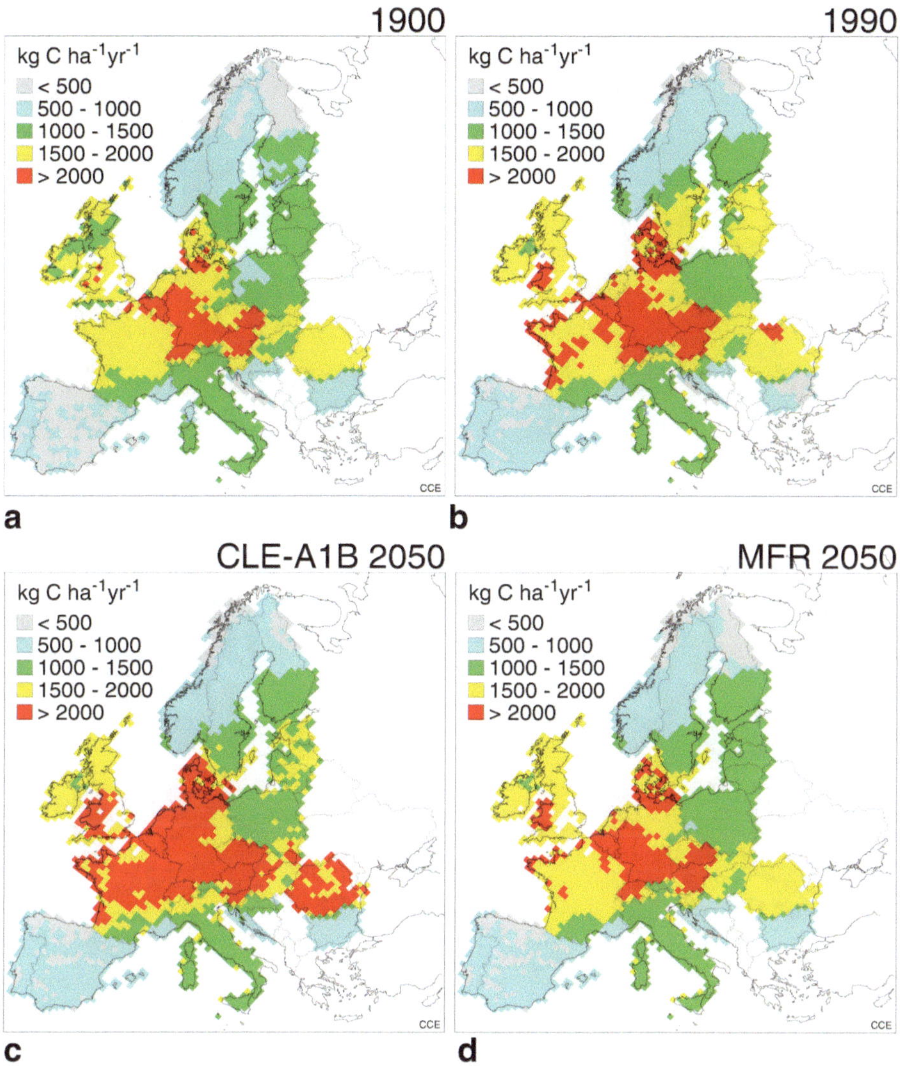

Fig. 24.9 Calculated average carbon sequestration in forests (trees) in Europe in (from *top* left to *bottom* right) 1900, 1990, 2050 under the CLE-A1 scenario and 2050 under the MFR-A1 scenario

of CO_2. A qualitative summary of the various effects is given in Table 24.3. Results of the model simulations show that adverse effects of N deposition on soil solution quality and plant species diversity occur while there are still positive effects on forest growth. Similar This is due to the assumed positive response of forest growth to N deposition up to 25 kg N $ha^{-1}yr^{-1}$ (see Fig. 24.2), whereas critical N loads for eutrophication and plant species impacts are (much) lower. This is illustrated in Table 24.4, showing areas exceeding critical loads for those effects in 1990,

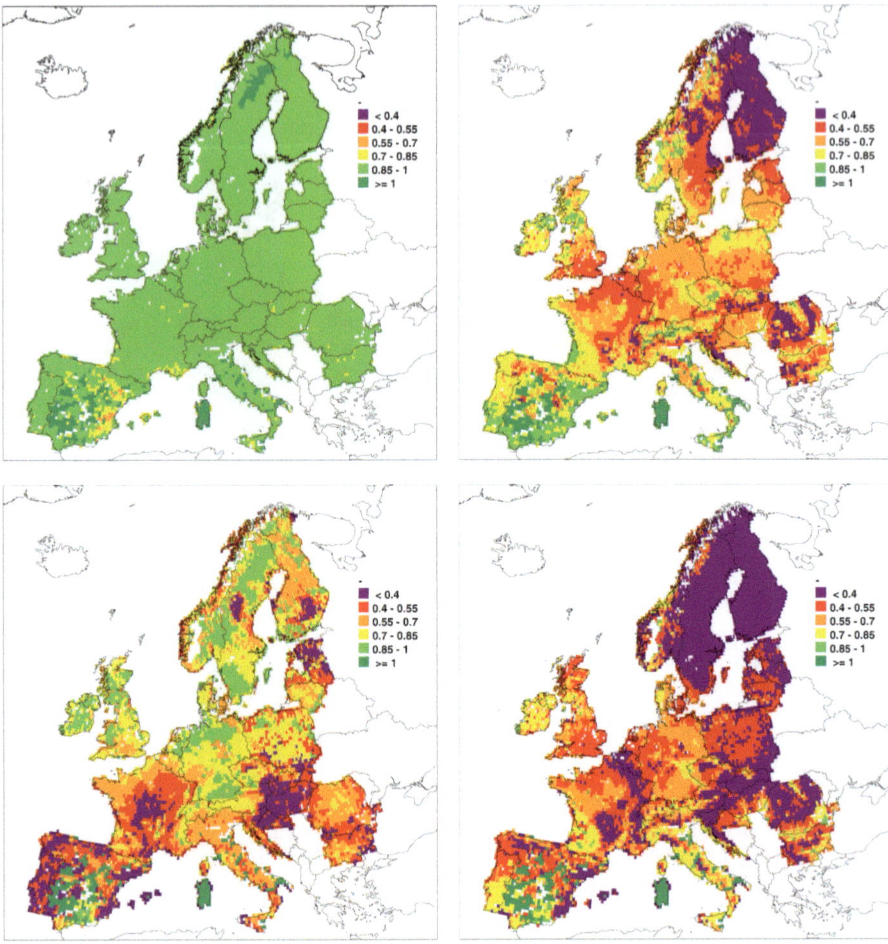

Fig. 24.10 Calculated median Bray Curtis index per for forests in Europe for 2050 in response to four scenarios of deposition and climate, i.e. (from *top* left to *bottom* right) the CLE, CLE-A1, MFR and MFR-A1 scenario

2020 and 2050 using the SMB model. Predicted areas exceeding critical N loads related to plant species diversity (using computed and empirical critical N loads; see Chap. 4 and 6) and soil and surface water quality (using a critical N concentration of 2 mg N l^{-1}) are much larger than those for forest growth (assuming a critical load of 25 kg N $ha^{-1}yr^{-1}$).

The results of the study are in line with a model study by Wamelink et al. (2009), who investigated the effect of nitrogen deposition reduction on the carbon sequestration and plant species diversity of terrestrial ecosystems in the Netherlands. They applied the dynamic soil model SMART in combination with the vegetation dynamic model SUMO and the biodiversity regression model NTM (see also Chap. 11).

Table 24.3 Simulated effects of elevated atmospheric deposition of N and S compounds and climate change (increased temperature) on soil and soil solution quality, forest growth and plant species diversity by the models VSD, EUgrow and PROPS, respectively

Impacts	Elevated atmospheric deposition of N and S compounds	Climate change (increased temperature)
Soil and soil solution quality	Increased N and Al concentrations	Slightly decreased N and Al concentrations due to higher weathering rates and N uptake in the VSD model
Forest growth	Increased growth except for areas with very high N inputs	Increased growth except for Mediterranean areas where drought stress reduces growth
Plant species diversity	Decreased plant species diversity	Decreased plant species diversity

Table 24.4 Area exceeding critical loads and Average Accumulated Exceedance (AAE) of critical N loads of European forests related to soil and surface water quality, plant species diversity and forest growth in 1990, in 2020 under the CLE scenario and in 2050 under the MFR scenario

Impacts	Area exceeding critical loads (% of total area)			Average Accumulated Exceedance of critical loads (kg $ha^{-1}yr^{-1}$)		
	1990	2020	2050	1990	2020	2050
Plant species diversity	81.2	62.0	50.1	8.1	3.5	1.77
Soil and soil solution quality	54.1	35.7	23.6	5.1	1.8	0.72
Forest growth	11.0	1.3	0.13	0.44	0.05	0.01

The SMART-SUMO model chain simulated the carbon sequestration as a result of nitrogen deposition and provided the NTM model with information on pH and nitrogen availability. They predicted that a gradual decrease in nitrogen deposition from 40 to 10 kg N $ha^{-1}yr^{-1}$in the next 25 years will cause a drop in the net carbon sequestration of forest in The Netherlands of 27 %, while biodiversity remains constant in forest, but may increase in heathland and grassland.

The positive response of forest growth to N deposition, simulated by the combined VSD-EUgrow model, is an issue that has received wide attention during the last decade. Due to N limitation in most forest ecosystems (Vitousek and Howarth 1991), increased N deposition usually decreases biodiversity (Bobbink et al. 2010), while increasing net primary production (NPP), thus stimulating C sequestration in trees (LeBauer and Treseder 2008; Thomas et al. 2010; Zaehle et al. 2011). Increased productivity may also increase C sequestration in the soil due to increased soil C inputs by litterfall (Lu et al. 2011) and reduced decomposition of organic matter (Berg and Matzner 1997, Janssens et al. 2010). An important question is at which

N deposition level are adverse impacts to be expected over the course of time due to N saturation, soil acidification and nutrient imbalances. As mentioned before, N addition studies and field studies indicate threshold values of 20–50 kgN $ha^{-1}yr^{-1}$ (De Vries et al. 2014a). In this study, a critical value of 25 kg N $ha^{-1}yr^{-1}$was used, since higher values, mostly based on N addition studies neglect the possible adverse impact of chronic N deposition over longer time horizons. Integrated carbon-nitrogen models indicate that the N deposition effect on productivity tends to decrease with time, due to progressive N enrichment (Townsend et al. 1996). However, these models also show that the positive impact will remain important for decades and even centuries to come (Zaehle et al. 2011). In this context, apart from N, the availability of other nutrients such as phosphorous and base cations also plays a role (Fernández-Martínez et al. 2014), but they are hardly ever accounted for in existing global C-N models (De Vries 2014). EUgrow includes impacts of changed availabilities of calcium, magnesium and potassium in a rather simple approach, but phosphorous is not included. Considering uncertainties caused by such simplifications, model predictions of the long-term C responses of forests to N deposition are uncertain.

The combined VSD-PROPS model has the potential for evaluating the effects of changes of climate and deposition on soil chemistry and species diversity. The PROPS model described here is, however, tentative. For example, the range in abiotic conditions is not broad enough to derive reliable response curves for a substantial number of species. As a consequence, response functions are continuously increasing or decreasing with changing environmental factors. More data sets need to be included in the PROPS data base to construct better responses. Moreover, explanatory environmental variables should be broadened to include e.g. drought stress. Furthermore, in this version of PROPS we have used N deposition as a driver. It is likely however, that species occurrence probability is stronger related to N availability or N concentration (De Vries et al. 2010). Therefore it is likely that PROPS overestimates the effects of N on species diversity as future changes in N deposition are larger than in N concentration. The PROPS model is currently being revised to include pH and NO_3 concentrations of more than 600,000 relevés, based on measurements for indicator species in each relevé, and annual mean temperature and precipitation derived from the CRU meteorological data set (see also Chap. 11).

Conclusions: From this set of studies it is concluded that:

- Nitrogen saturation, in terms of elevated NO_3 concentrations, and acidification, in terms of Al and BC depletion and elevated Al concentrations, was (very) low in 1900 and increased only slightly during the period 1900–1950.The values of these soil and water quality parameters showed a strong increase between 1950 and 1980 followed by a decrease between 1980 and 2010, reflecting the strong initial increase and subsequent decrease in N and S deposition in both periods, respectively. The impact of future emission reductions on the various parameters in the period 2010–2050 is larger than the climate change impact, specifically for the maximum feasible reductions scenario.

- Unlike the adverse impact on soil and water quality, N deposition has a positive impact on carbon sequestration, and the same holds on average for climate change. For the period 1900–2000, N deposition is the dominant driver of changes in forest growth, with an estimated average tree response (C-sequestration) to N deposition of 25 kgC kgN^{-1}. Unlike in the past, climate change was estimated to be the dominant driver of forest growth changes between 2000 and 2050.
- Plant species diversity changes hardly (slight decrease in Bray-Curtis index) in scenarios assuming constant climate and small N deposition reduction (CLE), significantly at constant climate and strongly decreasing N deposition (MFR), and sharply when both climate and N deposition changed, especially in areas with pronounced temperature change.

Acknowledgements The research described in this chapter has been funded by the European Commission under the project "Effects of Climate Change on Air Pollution Impacts and Response Strategies for European Ecosystems" (Grant agreement no: 282910). This research is also part of the strategic research program "Sustainable spatial development of ecosystems, landscapes, seas and regions" funded by the Dutch Ministry of Economic Affairs and carried out by Wageningen University and Research Centre (project code KB-14-001-036).

References

Aber, J. D., Nadelhoffer, K. J., Steudler, P., & Melillo, J. M. (1989). Nitrogen saturation in northern forest ecosystems. *Bioscience, 39,* 378–386.

Aber, J. D., McDowell, W., Nadelhoffer, K., Magill, A., Berntsen, G., Kamakea, M., McNulty, S., Currie, W., Rustad, L., & Fernandez, I. (1998). Nitrogen saturation in temperate forest ecosystems: Hypotheses revisited. *Bioscience, 48,* 921–934.

Amann, M., Asman, W., Bertok, I., Cofala, J., Heyes, C., Klimont, Z., Schöpp, W., & Wagner, F. (2007). *Cost-effective emission reductions to meet the environmental targets of the thematic strategy on air pollution under greenhouse gas constraints.* (NEC Scenario Analysis Report Nr. 5). Laxenburg: International Institute for Applied Systems Analysis (IIASA).

Berg, B., & Matzner, E. (1997). Effect of N deposition on decomposition of plant litter and soil organic matter in forest systems. *Environmental Reviews, 5,* 1–25.

Bobbink, R., & Hettelingh, J.-P. (2011). Review and revision of empirical critical loads and dose-response relationships: Proceedings of an expert workshop, Noordwijkerhout, 23–25 June 2010. (Report 680359002/2011). Bilthoven: Coordination Centre for Effects, National Institute for Public Health and the Environment.

Bobbink, R., Hornung, M., & Roelofs, J. G. M. (1998). The effects of air-borne nitrogen pollutants on species diversity in natural and semi-natural European vegetation. *Journal of Ecology, 86,* 717–738.

Bobbink, R., Hicks, K., Galloway, J., Spranger, T., Alkemade, R., Ashmore, M., Bustamante, M., Cinderby, S., Davidson, E., Dentener, F., Emmett, B., Erisman, J. W., Fenn, M., Gilliam, F., Nordin, A., Pardo, L., & De Vries, W. (2010). Global assessment of nitrogen deposition effects on terrestrial plant diversity: A synthesis. *Ecological Applications, 20,* 30–59.

Camargo, J. A., & Alonso, A. (2006). Ecological and toxicological effects of inorganic nitrogen pollution in aquatic ecosystems: A global assessment. *Environment International, 32,* 831–849.

Carter, T. R. (2007). General guidelines on the use of scenario data for climate impact and adaption assessment. Version 2. Task Group on Data and Scenario Support for Impact and Climate Assessment (TGICA), Intergovernmental Panel on Climate Change.

Costanza, R., d'Arge, R., de Groot, R. S., Farber, S., Grasso, M., Hannon, B., Limburg, K., Naeem, S., O'Neill, R. V., Paruelo, J., Raskin, R. G., Sutton, P., & van den Belt, M. (1997). The value of the world's ecosystem services and natural capital. *Nature, 387,* 253–260.

Costanza, R., de Groot, R., Sutton, P., van der Ploeg, S., Anderson, S. J., Kubiszewski, I., Farber, S., & Turner, R. K. (2014). Changes in the global value of ecosystem services. *Global Environmental Change, 26,* 152–158.

De Groot, R. S., Wilson, M. A., & Boumans, R. M. J. (2002). A typology for the classification, description and valuation of ecosystem functions, goods and services. *Ecological Economics, 41,* 393–408.

De Groot, R. S., Alkemade, R., Braat, L., Hein, L., & Willemen, L. (2010). Challenges in integrating the concept of ecosystem services and values in landscape planning, management and decision making. *Ecological Complexity, 7,* 260–272.

De Vries, W. (2014). Nutrients trigger carbon storage. *Nature Climate Change, 4,* 425–426.

De Vries, W., & Posch, M. (2011). Modelling the impact of nitrogen deposition, climate change and nutrient limitations on tree carbon sequestration in Europe for the period 1900–2050. *Environmental Pollution, 159,* 2289–2299.

De Vries, W., Reinds, G. J., Posch, M., & Kämäri, J. (1994). Simulation of soil response to acidic deposition scenarios in Europe. *Water Air & Soil Pollution, 78,* 215–246.

De Vries, W., Kros, J., Reinds, G. J., Wamelink, G. W. W., Mol, J., van Dobben, H., Bobbink, R., Emmett, B., Smart, S., Evans, C., Schlutow, A., Kraft, P., Belyazid, S., Sverdrup, H. U., van Hinsberg, A., Posch, M., & Hettelingh, J.-P. (2007). *Developments in deriving critical limits and modelling critical loads of nitrogen for terrestrial ecosystems in Europe.* (Report 1382). Wageningen: Alterra Wageningen UR.

De Vries, W., Wamelink, G. W. W., van Dobben, H., Kros, J., Reinds, G. J., Mol-Dijkstra, J. P., Smart, S. M., Evans, C. D., Rowe, E. C., Belyazid, S., Sverdrup, H. U., van Hinsberg, A., Posch, M., Hettelingh, J.-P., Spranger, T., & Bobbink, R. (2010). Use of dynamic soil–vegetation models to assess impacts of nitrogen deposition on plant species composition: An overview. *Ecological Applications, 20,* 60–79.

De Vries, W., Du, E., & Butterbach-Bahl, K. (2014a). Short and long-term impacts of nitrogen deposition on carbon sequestration by forest ecosystems. *Current Opinion in Environmental Sustainability,* 9–10: 90–104.

De Vries, W., Goodale, C., Erisman, J. W., & Hettelingh, J.-P. (2014b). Impacts of nitrogen deposition on ecosystem services in interaction with other nutrients, air pollutants and climate change. In M. A. Sutton, K. E. Mason, L. J. Sheppard, H. Sverdrup, R. Haeuber, & W. K. Hicks (Eds.), *Nitrogen deposition, Critical loads and Biodiversity* (pp. 387–396). Springer. (Chap. 41).

EC. (1998). Council Directive 98/83/EC of 3 November 1998 on the quality of water intended for human consumption. Brussel: European Commision.

Ellenberg, H., Jr. (1985). Veränderungen der Flora Mitteleuropas unter dem Einfluss von Düngung und Immissionen. *Schweizerische Zeitschrift fur Forstwesn, 136,* 19–39.

Erisman, J. W., & De Vries, W. (2000). Nitrogen deposition and effects on European forests. *Environmental Reviews, 8,* 65–93.

Fernández-Martínez, M., Vicca, S., Janssens, I., Sardans, J., Luyssaert, S., Campioli, M., Chapin, F. S., III, Ciais, P., Malhi, Y., Obersteiner, M., Papale, D., Piao, S. L., Reichstein, M., Rodà, F., & Peñuelas, J. (2014). Nutrient availability as the key regulator of global forest carbon balance. *Nature Climate, Change, 4,* 471–476.

Gundersen, P. (1991). Nitrogen deposition and the forest nitrogen cycle: Role of denitrification. *Forest Ecology and Management, 44,* 15–28.

Gundersen, P., Schmidt, I. K., & Raulund-Rasmussen, K. (2006). Leaching of nitrate from temperate forests—effects of air pollution and forest management. *Environmental Reviews, 14,* 1–57.

Helliwell, D. R. (1969). Valuation of wildlife resources. *Regional Studies, 3,* 41–47.

Janssens, I. A., Dieleman, W., Luyssaert, S., Subke, J.-A., Reichstein, M., Ceulemans, R., Ciais, P., Dolman, A. J., Grace, J., Matteucci, G., Papale, D., Piao, S. L., Schulze, E.-D., Tang, J., & Law, B. E. (2010). Reduction of forest soil respiration in response to nitrogen deposition. *Nature Geoscience, 3,* 315–322.

Laane, R. W. P. M. (2005). Applying the critical load concept to the nitrogen load of the river Rhine to the Dutch coastal zone. *Estuarine, Coastal and Shelf Science, 62,* 487–493.
LeBauer, D. S., & Treseder, K. K. (2008). Nitrogen limitation of net primary productivity in terrestrial ecosystems is globally distributed. *Ecology, 89,* 371–379.
Liu, C., Kroeze, C., Hoekstra, A. Y., & Gerbens-Leenes, W. (2011). Past and future trends in grey water footprints of anthropogenic nitrogen and phosphorus inputs to major world rivers. *Ecological Indicators, 18,* 42–49.
Lu, M., Zhou, X., Luo, Y., Yang, Y., Fang, C., Chen, J., & Li, B. (2011). Minor stimulation of soil carbon storage by nitrogen addition: A meta-analysis. *Agriculture Ecosystems & Environment, 140,* 234–244.
Magill, A. H., Aber, J. D., Currie, W., Nadelhoffer, K., Martin, M., McDowell, W. H., Melillo, J. M., & Steudler, P. (2004). Ecosystem response to 15 years of chronic nitrogen additions at the Harvard Forest LTER, Massachusetts, USA. *Forest Ecology and Management, 196,* 7–28.
Millennium Ecosystem Assessment. (2005). *Ecosystems and human well-being: Synthesis.* (A Report of the Millennium Ecosystem Assessment). Washington, DC: Island Press. <http://www.millenniumassessment.org/documents/document.356.aspx.pdf>.
Mitchell, T., Carter, T. R., Jones, P. D., Hulme, M., & New, M. (2004). A comprehensive set of high-resolution grids of monthly climate for Europe and the globe: the observed record (1901–2000) and 16 scenarios (2001–2100). (Working Paper 55). Tyndall Centre.
Nakícenovíc, N., Alcamo, J., Davis, G., De Vries, B., Fenhann, J., Gaffin, S., Gregory, K., Grübler, A., Jung, T. Y., Kram, T., Emilio la Rovere, E., Michaelis, L., Mori, S., Morita, T., Pepper, W., Pitcher, H., Price, L., Riahi, K., Roehrl, A., Rogner, H.-H., Sankovski, A., Schlesinger, M. E., Shukla, P. R., Smith, S., Swart, R. J., van Rooyen, S., Victor, N., & Dadi, Z. (2001). *Special report on emission scenarios: Intergovernmental panel on climate change.* Cambridge: Cambridge University Press.
Posch, M., & Reinds, G. J. (2009). A very simple dynamic soil acidification model for scenario analyses and target load calculations. *Environmental Modelling & Software, 24,* 329–340.
Rahmstorf, S., Cazenave, A., Church, J. A., Hansen, J. E., Keeling, R. F., Parker, D. E., & Somerville, R. C. J. (2007). Recent climate observations compared to projections. *Science, 316,* 709.
Reinds, G. J., Posch, M., & Leemans, R. (2009). Modelling recovery from soil acidification in European forests under climate change. *Science of the Total Environment, 407,* 5663–5673.
Reinds, G. J., Bonten, L., Mol-Dijkstra, J. P., Wamelink, G. W. W., & Goedhart, P. (2012). Combined effects of air pollution and climate change on species diversity in Europe: First assessments with VSD+ linked to vegetation models. In M. Posch & J.-P. Hettelingh (Eds.), *CCE status report 2012* (pp. 49–61). Coordination Centre for Effects.
Schöpp, W., Posch, M., Mylona, S., & Johansson, M. (2003). Long-term development of acid deposition (1880–2030) in sensitive freshwater regions in Europe. *Hydrology and Earth System Sciences, 7,* 436–446.
Simpson, D., Fagerli, H., Jonson, J., Tsyro, S., Wind, P., & Tuovinen, J.-P. (2003). *The EMEP unified Eulerian model. Model description.* (EMEP MSC-W Report 1/2003). Oslo: The Norwegian Meteorological Institute.
Solberg, S., Dobbertin, M., Reinds, G. J., Lange, H., Andreassen, K., Fernandez, P. G., Hildingsson, A., & De Vries, W. (2009). Analyses of the impact of changes in atmospheric deposition and climate on forest growth in European monitoring plots: A stand growth approach. *Forest Ecology and Management, 258,* 1735–1750.
Strengers, B., Leemans, R., Eickhout, B., De Vries, B., & Bouwman, A. (2005). The land-use projections and resulting emissions in the IPCC SRES scenarios scenarios as simulated by the IMAGE 2.2 model. *GeoJournal, 61,* 381–393.
Tarrasón, L., Fagerli, H., Jonson, J. E., Simpson, D., Benedictow, A., Klein, H., Vestreng, V., Aas, W., & Hjellbrekke, A. G. (2007). *Transboundary acidification, eutrophication and ground level ozone in Europe in 2005.* (EMEP Report 1/2007). Oslo: Norwegian Meteorological Institute.
TEEB. (2010). The economics of ecosystems and biodiversity: Mainstreaming the economics of nature: A synthesis of the approach, conclusions and recommendations of TEEB.
Thomas, R. Q., Canham, C. D., Weathers, K. C., & Goodale, C. L. (2010). Increased tree carbon storage in response to nitrogen deposition in the US. *Nature Geoscience, 3,* 13–17.

Townsend, A. R., Braswell, B. H., Holland, E. A., & Penner, J. E. (1996). Spatial and temporal patterns in terrestrial carbon storage due to deposition of fossil fuel nitrogen. *Ecological Applications, 6,* 806–814.

Vitousek, P. M., & Howarth, R. W. (1991). Nitrogen limitation on land and in the sea: How can it occur? *Biogeochem, 13,* 87–115.

Wamelink, G. W. W., van Dobben, H. F., Mol-Dijkstra, J. P., Schouwenberg, E. P. A. G., Kros, J., De Vries, W., & Berendse, F. (2009). Effect of nitrogen deposition reduction on biodiversity and carbon sequestration. *Forest Ecology and Management, 258,* 1774–1779.

Zaehle, S., Ciais, P., Friend, A. D., & Prieur, V. (2011). Carbon benefits of anthropogenic reactive nitrogen offset by nitrous oxide emissions. *Nature Geoscience, 4,* 601–605.

Chapter 25
Effects-Based Integrated Assessment Modelling for the Support of European Air Pollution Abatement Policies

Jean-Paul Hettelingh, Maximilian Posch, Jaap Slootweg, Gert Jan Reinds, Wim de Vries, Anne-Christine Le Gall and Rob Maas

25.1 Introduction

Since its development some three decades ago, the critical load approach has been applied to provide scenario-specific forecasts of exceedances in terms of magnitudes, extent and geographical locations. Critical loads and information on exceedances and dose-response relationships have been used extensively to support protocols under the Convention on Long-range Transboundary Air Pollution (LRTAP) including the second Sulphur Protocol (Hettelingh et al. 1997; Hettelingh et al. 1995; UNECE 1994) the multi-pollutant multi-effect protocol addressing causes and effects of acidification, eutrophication and ozone (Bull et al. 2001; Hettelingh et al. 2001; Hettelingh et al. 2007a; Posch et al. 2001a; Spranger et al. 2008; UNECE 1999) and, in addition, of fine particulate matter in its revision in 2012 (Reis et al. 2012; UNECE 2012). Critical loads were also key to the scientific support of air pollution abatement strategies and directives of the European Union (EC 2001; EEA 2012) and their scientific support (Hettelingh et al. 2013).

Integrated assessment (see also Hettelingh et al. 2009) is the framework methodology with which the scientific knowledge on effects of air pollution is combined with their sources and abatement options and costs, according to the logic of Driving Forces, Pressures, State, Impact, Response (DPSIR). The Greenhouse gas and Air pollution Interactions and Synergies (GAINS; Amann et al. 2011) model is the

J.-P. Hettelingh (✉) · M. Posch · J. Slootweg
Coordination Centre for Effects (CCE), RIVM, Bilthoven, The Netherlands
e-mail: jean-paul.hettelingh@rivm.nl

G. J. Reinds · W. de Vries
Alterra Wageningen University and Research Centre, Wageningen, The Netherlands

A.-C. Le Gall
LRTAP-Task Force Modelling and Mapping, INERIS, Verneuil-en-Halatte, France

R. Maas
LRTAP-Task Force Integrated Assessment Modelling, RIVM, Bilthoven, The Netherlands

W. de Vries et al. (eds.), *Critical Loads and Dynamic Risk Assessments*,
Environmental Pollution 25, DOI 10.1007/978-94-017-9508-1_25

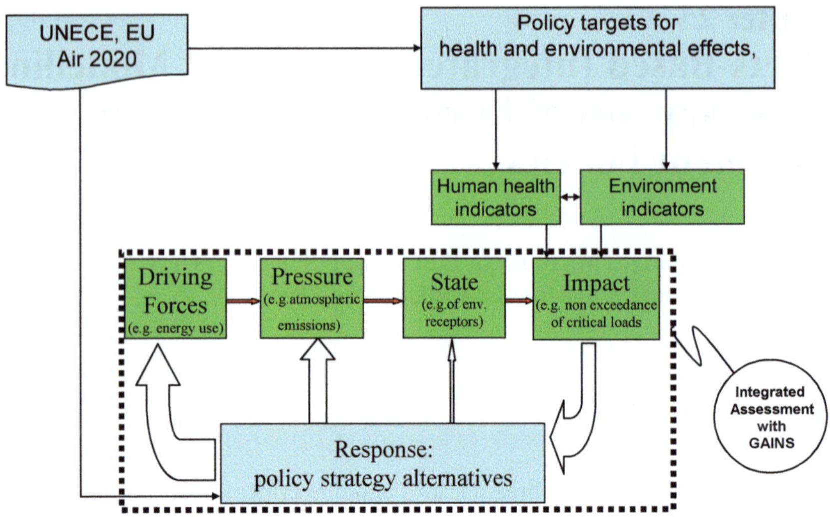

Fig. 25.1 Integrated assessment modelling (e.g. with GAINS) offers a science-policy interface (flowchart within the *dotted box*) where policy objectives for air pollution abatement strategies of the LRTAP Convention and European Union policies are modelled and iteratively evaluated in terms of causes, impacts and responses in the *Driving Forces, Pressures, State, Impact, Reponse* (DPSIR) logic

model of choice for the appraisal of emissions, emission abatement measures and their costs as well as health and ecosystem impacts in support of European air pollution policies. The logic of the GAINS model as illustrated in Fig. 25.1 consists of establishing an energy mix for each European country, from which air pollution emissions, including sulphur dioxide, oxidized and reduced nitrogen, ozone precursors and fine particulate matter, can be computed. These national emissions are then used as input to the EMEP atmospheric transport model to compute the distribution of ambient concentrations and depositions in grid cells, the size of which has decreased from 150×150 km^2 in the early days of the EMEP model to about 28×28 km^2 at present (Eliassen et al. 1982; EMEP 1998; Simpson et al. 2012). While the full EMEP model needs powerful computers to run, the GAINS model uses so-called source-receptor matrices, relating emissions in a country to depositions in all grid cells. Finally, the impact module of GAINS includes the comparison of depositions to critical loads for ecosystem areas in each grid cell, results of which are described in this chapter.

Note that measures can affect "Driving Forces", involving e.g. structural changes to the energy and transport sector, "Pressures", such as the application of end-of-pipe emission abatement techniques and the "State", involving effect-based measures such as liming of soils. The size of the arrows in Fig. 25.1 indicate the likeliness of policies, illustrating that emission abatement is dominant in policies, rather than the implementation of effect-based measures.

For the design of air pollution abatement policies it is important to know where (in Europe or in a country) adverse impacts can be expected to occur as a result of the dispersion of national emissions and resulting excessive depositions. Moreover, policy analysts also wish to know the magnitude of the exceedance, because it is assumed that an adverse effect may occur 'sooner' when the exceedance is higher.

In this chapter the effects-based support of European air pollution abatement policies since the early 1990s is described with respect to acidification and eutrophication, taking the history of excessive sulphur and nitrogen deposition since 1880 into account.

Assessments of future air pollution impacts include target years 2020, 2030 and 2100 based on future depositions that stem from two emission reduction scenarios, i.e. "Current Legislation" and "Maximum Feasible Reductions". Current legislation is based on the revision in 2012 of the Gothenburg Protocol. The assessment involves a chain of methods, i.e. using empirical critical loads (see Chap. 4), computed critical loads (see Chap. 6), dynamic modelling (see Chap. 8) and dose-response curves (see Chap. 23). Finally, an ensemble assessment of the different approaches is presented as a means to improve the robustness of the identification of European areas at risk of nitrogen effects.

25.2 Methods

The logic of the impact analysis in integrated assessment (dotted box in Fig. 25.1) is based on both modelled and empirical indicators.

25.2.1 Environmental Impact Analysis

Figure 25.2 illustrates that an indication of the risk of acidification or eutrophication to European terrestrial and aquatic ecosystems is obtained by comparing depositions to critical loads in ecosystem areas in every grid cell following two pathways, i.e. impact analysis based on modelled or empirical indicators, i.e. critical loads or plant species diversity. In the remainder of this chapter impacts are evaluated for emission scenarios that policy makers identified on the basis of their technical and economic potential. These are emissions according to the revised Gothenburg Protocol in 2020 (GP-CLE), and for 2030 either Current Legislation or Maximum Feasible Reductions (MFR2030) are applied.

The impacts on the environment are assessed by using a European database on computed critical loads (see next section) that is also embedded in the GAINS-model. The dynamic modelling part of the "impact analysis based on computed knowledge", which is not part of the GAINS model, can be either used for scenario analysis (dotted arrows in Fig. 25.2) or to simulate the delay of damage or recovery (i.e. when critical loads are exceeded or no longer exceeded, respectively). Depositions required to achieve recovery within a prescribed time horizon, so-called target

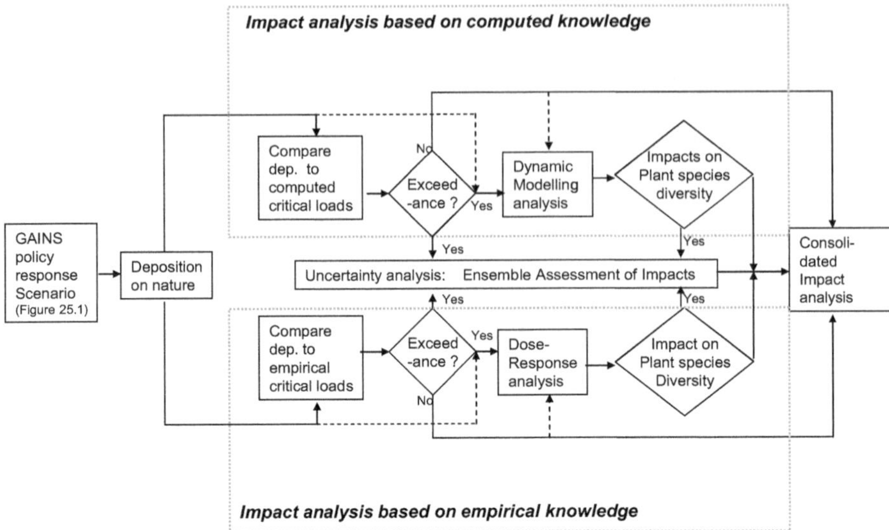

Fig. 25.2 The environmental impact analysis based on modelled and empirical indicators (adapted from Hettelingh et al. 2008). A *dotted arrow* indicates an optional pathway, e.g. dynamic modelling can also be applied without first assessing exceedance, or in the case critical loads are not exceeded (i.e. to assess recovery)

loads, are also computed with dynamic models (see Chap. 8), and below we show exceedances of such target loads on a European scale.

The "impact analysis based on empirical knowledge" (lower part of Fig. 25.2) consists of the use of empirical critical loads and dose-response relationships. While computed impact assessments are performed for both acidification and eutrophication, empirical indicators are only available for the latter. One reason is that acidification effects in terrestrial ecosystems are less visible in the field (Chap. 2), and empirical critical loads for acidification have not been developed further, once mass-balance acidity critical loads (Chap. 6) had become available. Finally, information on the robustness of the geographic location and magnitude of the risk of eutrophication is obtained through a combination of both kinds of impact analyses (middle part of Fig. 25.2).

25.2.2 *The European Critical Loads Database*

For more than two decades, critical loads of sulphur and nitrogen have been compiled by parties under the LRTAP Convention to support effects-based emission reduction agreements (Hettelingh et al. 1995, 2001, 2007a). The critical loads database of 1998 (Posch et al. 1999) was used for the support of the Gothenburg Protocol to the LRTAP Convention and the EU National Emissions Ceiling (NEC) Directive, while the 2008 database (Slootweg et al. 2008) was used for the revision of the Gothenburg Protocol, whereas the latest critical loads database (Posch et al. 2012) is used for the revision of the NEC Directive. Critical loads for countries that did not submit data, were computed using a European 'background data base' (Posch and Reinds 2005; Slootweg et al. 2011). Periodic updates of the critical load data bases are necessary for a number of reasons, e.g., to increase the resolution of mapped critical loads to match the size of smaller deposition grid cells, and to classify natural areas following the European Nature Information System (EUNIS; Davies and Moss 1999). The European background information covers forest ecosystems (EUNIS class G) and semi-natural vegetation (EUNIS classes D, E and F). The distinction between EUNIS classes leads to a broad range of critical loads of nitrogen in particular, based on a range of acceptable nitrogen concentrations in soil solution (De Vries et al. 2007). The present critical loads database includes more than 2 million receptors covering about 380 million ha of ecosystems. Consequently, every EMEP grid cell contains up to several thousand receptors, for which the (area-weighted) cumulative distribution functions of the critical loads are established. In Fig. 25.3 maps of the 5th percentile of the critical loads of acidity and nutrient N (computed and empirical) are shown.

The 5th percentile of the computed nutrient N critical load tends to be lower (lowest range 200 eq $ha^{-1}yr^{-1}$; Fig. 25.3 centre) than the empirical critical load (lowest range 200–400 eq $ha^{-1}yr^{-1}$; Fig. 25.3 right). However, the geographical distribution of sensitive areas is similar, with areas most sensitive to nitrogen located in the north, east and south of Europe, whereas areas that are sensitive to acidification (Fig. 25.3 left) are concentrated in northern Europe. Note that the sharp differences along national boundaries for computed critical loads (especially Fig. 25.3 centre), reflect the different assumptions on input parameters, e.g. the magnitude of long-term net immobilization of N.

25.2.3 Critical Load Exceedances, Dynamic Modelling and Dose-Response Relationships

If a deposition at a site is higher than the critical load, one says that the critical load is exceeded, and the difference is called exceedance. In this case the area represented by this site is said to be at risk (of acidification and/or eutrophication). To characterise exceedances within a grid cell or region, the so-called 'average accumulated exceedance' (AAE) has been introduced as the area-weighted average of exceedances accumulated over all ecosystems in the respective mapping unit (e.g. grid cell) (Posch et al. 2001b; see also Chap. 6). Note that for the revision of the Gothenburg Protocol and the EU NEC Directive, ecosystem-specific (forests and semi-natural vegetation) depositions are available, in addition to grid-average depositions (see also Hettelingh et al. 2013).

Dynamic modelling is the logical extension of steady-state critical loads in support of the effects-oriented work under the LRTAP Convention (Hettelingh and Posch 1994; Posch et al. 2003, 2008). Critical loads are based on a steady-state

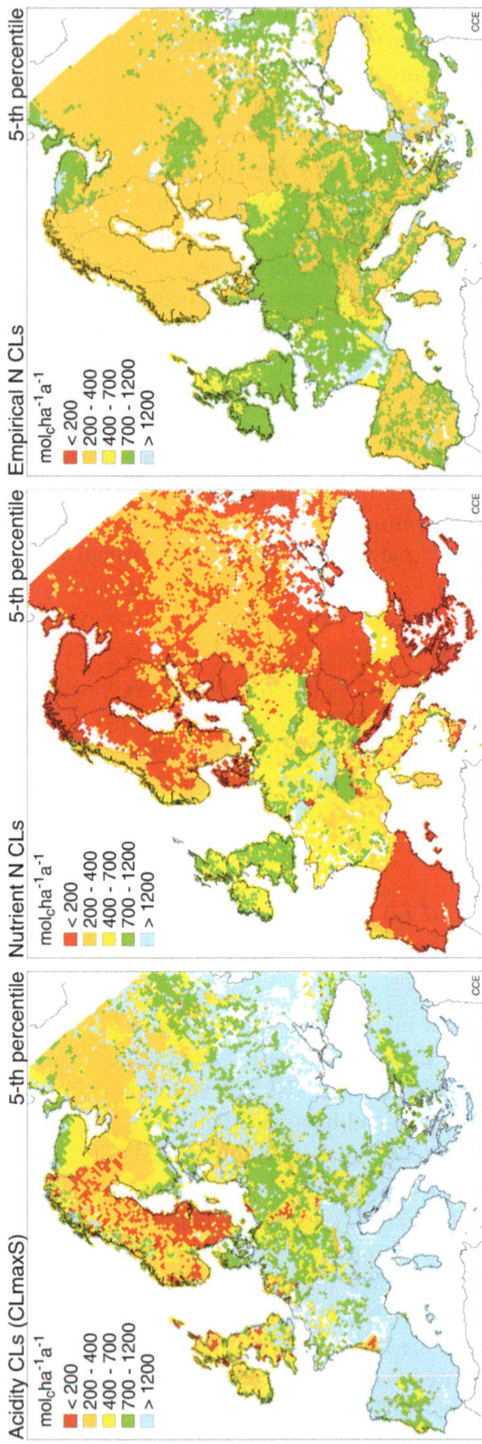

Fig. 25.3 Fifth percentile of the computed critical load of acidity (*left*), computed (*centre*) and empirical (*right*) critical load of nutrient nitrogen in each 25×25 km^2 grid cell

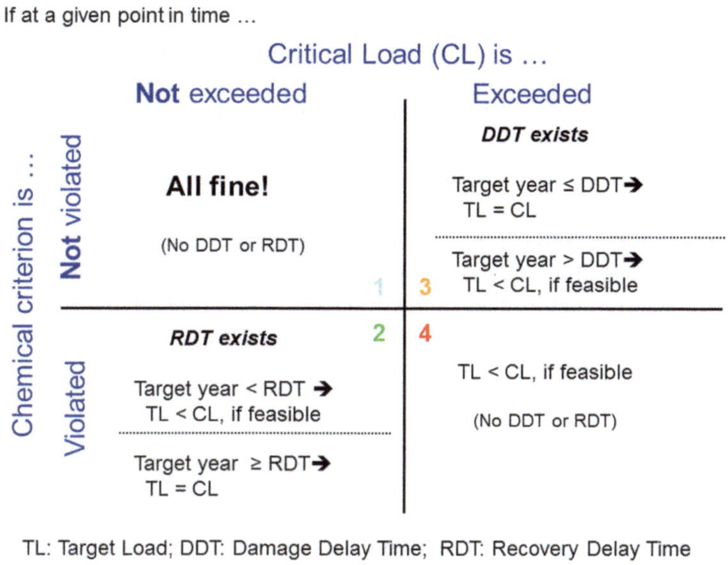

Fig. 25.4 Four combinations of critical load (non-)exceedance and criterion (non-)violation

concept, they are the constant depositions an ecosystem can tolerate in the long run, i.e. after it has equilibrated with these depositions. However, many ecosystems are not in equilibrium with present or projected depositions, since there are processes ('buffer mechanisms') at work, which delay the reaching of an equilibrium (steady state) for years, decades or even centuries. By definition, critical loads do not provide any information on these time scales. Dynamic models enable the assessment of time delays of recovery in regions where critical loads cease being exceeded and time delays of damage in regions where critical loads continue to be exceeded. Whether long-term deposition leads to recovery or to damage also depends on the value of the underlying critical limit (Fig. 25.4). Four cases can be distinguished by combining (non-)exceedance of critical loads with the (non-)violation of critical limits.

Case 1 applies to deposition that does not exceed a critical load while the critical limit is not violated. In case 2 there is no exceedance, however the critical limit is (still) violated. In case 3 the critical load is exceeded but the critical limit not (yet) violated, while finally case 4 applies to the situation where both the critical load is exceeded and the critical limit is violated. Case 4 includes ecosystems that are subject to immediate risk of damage by deposition. Case 1 implies full protection. Cases 2 and 3 include ecosystems for which recovery delay times (RDT) and damage delay times (DDT) can be identified, respectively. Obviously, addressing ecosystems for which case 2 holds are the most interesting from the point of view of establishing deposition values that target recovery. Therefore, finding such a target load to establish recovery in a selected year (target year) is an interesting element of effect-oriented integrated assessments for ecosystems. Target loads are also relevant for addressing deposition reductions in case 4. A target load is (by definition) lower than critical loads and is defined as the maximum deposition allowed to obtain (and sustain) a desired chemical state (critical limit value) in a prescribed year. In this chapter the distribution of European ecosystem areas at risk of acidification and eutrophication over cases 1–4 is assessed for two historical (1980 and 2010) and two future (2030 and 2100) years. Future assessments are based on depositions that follow from the CLE and MFR emission reduction scenarios, with depositions kept constant after 2030.

Dose-response relationships, describing the dependence of grassland species richness on N deposition and based on gradient studies, are described in Chap. 23. For their application in this chapter, the European background base is used, because it includes the necessary information of grasslands in addition to location (e.g., pH and altitudes).

25.2.4 Robustness and Ensemble Assessment of Impacts

Ensemble Assessment of Impacts (EAI), as tentatively introduced in Hettelingh et al. (2007b), is presented in this chapter to explore the robustness of exceedances on a scale that could range from "exceptionally unlikely" to "virtually certain". This, in analogy to the manner in which uncertainties have been addressed in the IPCC Fourth Assessment Report (IPCC-AR4) as summarized in IPCC (2005). The term "Ensemble Assessment" is borrowed from "Ensemble Modelling" used in different scientific disciplines to improve the accuracy of predictions by the pooling of results from different models addressing the same issue.

The main aim of the critical load approach is the best possible estimation of the geographical location of ecosystems of which the critical load is exceeded by atmospheric deposition. Therefore, when addressing ecosystem impacts, integrated assessment modellers and policy analysts are primarily interested in the likelihood of (the occurrence of) an exceedance, and its emission scenario-dependent trend. It is known that the uncertainty in exceedances depends on methods and data in the chain from emissions to depositions, and their spatial and temporal resolution. These include data and emission factors behind national emission reports, input data, meteorology and climate conditions behind atmospheric dispersion models and input data, soil-vegetation characteristics and modelling methods behind critical loads. All these are elements relevant to uncertainty analyses that have been conducted and reported under the LRTAP Convention (e.g., Hettelingh and Posch 1997; Suutari et al. 2001). An important element is that integrated assessment is concerned with scenario analysis, that is, with the change in selected indicators when comparing scenarios. Model and data uncertainty, to a certain extent, is 'cancelled out' when comparing one scenario to another. More important is the uncertainty caused by false positives; for example, not taking climate change into account. The impacts of climate change on target loads and delay times have recently been investigated in Reinds et al. (2009).

Use was made of the IPCC Guidance Notes for lead authors of the IPCC AR4 on addressing uncertainties (IPCC 2005) for the analysis of the robustness of exceedances (Hettelingh et al. 2007b). This note distinguishes:

- **Plan to treat issues of uncertainty and confidence**: the aim is to explore the robustness of concluding that a grid-cell in Europe contains ecosystems at risk, under a particular emission scenario and related depositions.
- **Review the information available**: The robustness of the occurrence of an exceedance could be increased by including more methods to compute critical loads (e.g. reverse dynamic modelling with geo-chemical and vegetation type models), or methods to assess exceedances (e.g. distinguish between special protection areas such as Natura2000 from other sensitive areas)
- **Make expert judgements:** The critical limits used in modelled critical loads have not been derived from empirical critical loads, nor has one method be validated on the basis of the other. Therefore, we assume that both methods lead to critical loads of which the distributions (in a single grid cell) are independent, and that they reflect effects that are complementary. Furthermore we assume that each of the two sets of critical loads in a single grid cell is representative for the (sensitive) ecosystems in that grid cell.
- **Use the appropriate level of precision to describe effects:** The IPCC Guidance Notes propose a hierarchy of 5 steps—with increasing specificity—by which statements for key findings can be substantiated. We use a likelihood scale that is derived from Table 3 in that document to develop statements with respect to exceedances.
- **Communicate carefully, using calibrated language:** Communication should be tailored to the interpretation of scenario-dependent exceedances. For example, areas where the critical load remains exceeded even after application of MFR2030 could, tentatively, be judged as persistently at risk. However, even the reduction of the area and magnitude of exceedances can be policy relevant.

The joint probability of an exceedance of empirical and modelled critical loads is assumed to be the product of both percentages of areas at risk, because of them being independent. The likelihood of AAE > 0 in an EMEP grid is said to be "likely", "very likely" or "virtually certain" if the square root of the product (i.e. the geometric mean) of the protection percentages based on empirical and modelled critical loads are in the ranges 0–33, 33–67 and > 67 %, respectively. The likelihood is "unlikely" if both protection percentages are equal to zero. If only one of the two protection percentages is zero the likelihood of an exceedance is said to be "as likely as not". A graphical representation of the scaling method is provided in Fig. 25.5.

25.3 Results

First, we present critical load exceedances over the whole of Europe since 1880 (25.3.1). The remainder of Sect. 25.3 follows the logic of Fig. 25.2, starting with the impact analysis based on computed knowledge, i.e. consisting of exceedances

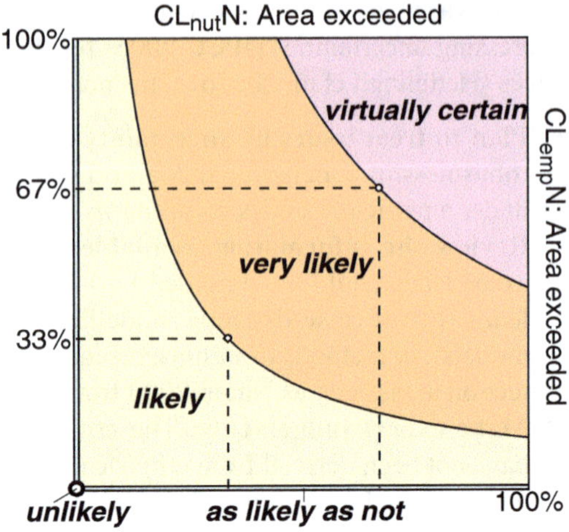

Fig. 25.5 The likelihood scale indicating the simultaneous probability of an exceedance of the empirical (*vertical* axis) and modelled (*horizontal* axis) critical loads in a grid cell

of computed critical loads and critical limits by application of steady-state and dynamic modelling in Sect. 25.3.2. This is followed by Sect. 25.3.3 describing impact analyses using empirical knowledge, i.e. empirical critical loads followed by the an application of empirical dose-response functions in terms of N input and plant species diversity. Finally, ensemble assessment (25.3.4) is presented as a means to improve the robustness of the identification of European areas at risk of nitrogen effects.

25.3.1 Critical Load Exceedances for Nitrogen and Acidity Since 1880

Figure 25.6 illustrates computed critical load exceedances for nitrogen and acidity between 1880 and 2030 based on emission trends since 1880 (Schöpp et al. 2003),

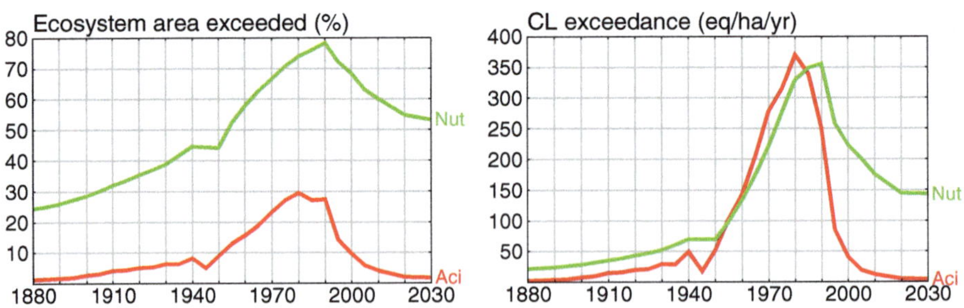

Fig. 25.6 Temporal development since 1880 of the area at risk (in %; *left*) and Average Accumulated Exceedance (*AAE*) (in eq $ha^{-1}yr^{-1}$; *right*) of critical loads of acidification (*red*) and eutrophication (*green*). Future exceedances are based on the implementation of agreed legislation under the LRTAP Convention in 2010 up to 2030

harmonised deposition patterns from different versions of the EMEP model (e.g. Hettelingh et al. 2013) and the most recent critical load database (Posch et al. 2012).

The following can be noted from Fig. 25.6. Assuming that emission reduction techniques are implemented as foreseen under Current Legislation in 2030, both the area at risk of acidification and its AAE magnitude in 2030 are similar to 1880 i.e. near 2 % exceedance only, with an AAE near 2 eq $ha^{-1}yr^{-1}$. The peaks of the acidified area and exceedance magnitude are in 1980. In contrast, eutrophication remains a serious issue, affecting about 53 % of the European ecosystem area in 2030 with an AAE of about 140 eq $ha^{-1}yr^{-1}$. However, in contrast to acidification, it appears that eutrophication already existed in 1880. The area at risk then is already about 24 %, which is likely caused by emissions of reduced nitrogen (from agricultural activities) in particular (Kopáček and Posch 2011). The peaks of eutrophication is reached a decade later than for acidification, due to the fact that policies to curb nitrogen emissions seemed less urgent (emission abatement policies focussed on SO_2 reduction) and started later.

25.3.2 Impact Analysis Based on Models

Impact analysis based on 'computed knowledge' (see Fig. 25.2) consists of the assessment of exceedances of the critical loads for acidification and eutrophication and of the application of dynamic modelling to estimate depositions required to reach recovery in a target year, e.g. 2050.

25.3.2.1 Acidification

The trend between 1980 and 2030 of European ecosystem areas where critical loads for acidification are exceeded (Fig. 25.7) illustrates the decrease of the area at risk from about 30 % (43 % in the European Union consisting of 28 Member States (EU28)) in 1980 to 2 % (4 % in the EU28) in 2020 under the revised Gothenburg Protocol including the implementation of current legislation. The implementation of maximum technically feasible reductions by 2030 would about half the average exceedance (AAE) of 2020 and result in about 1 % of the area remaining at risk of acidification.

These reductions of acidifying deposition over time has led to lower measured concentrations in soils and surface waters, and also to the recovery of sensitive species in lakes and streams (Skjelkvåle et al. 2005; Stoddard et al. 1999).

25.3.2.2 Eutrophication

The trend of areas at risk of eutrophication between 1980 and 2030 (Fig. 25.8) confirms the urgency of reducing excessive nitrogen deposition. The area at risk of eutrophication decreases from 75 % in 1980 (80 % in the EU28) to about 55 % in

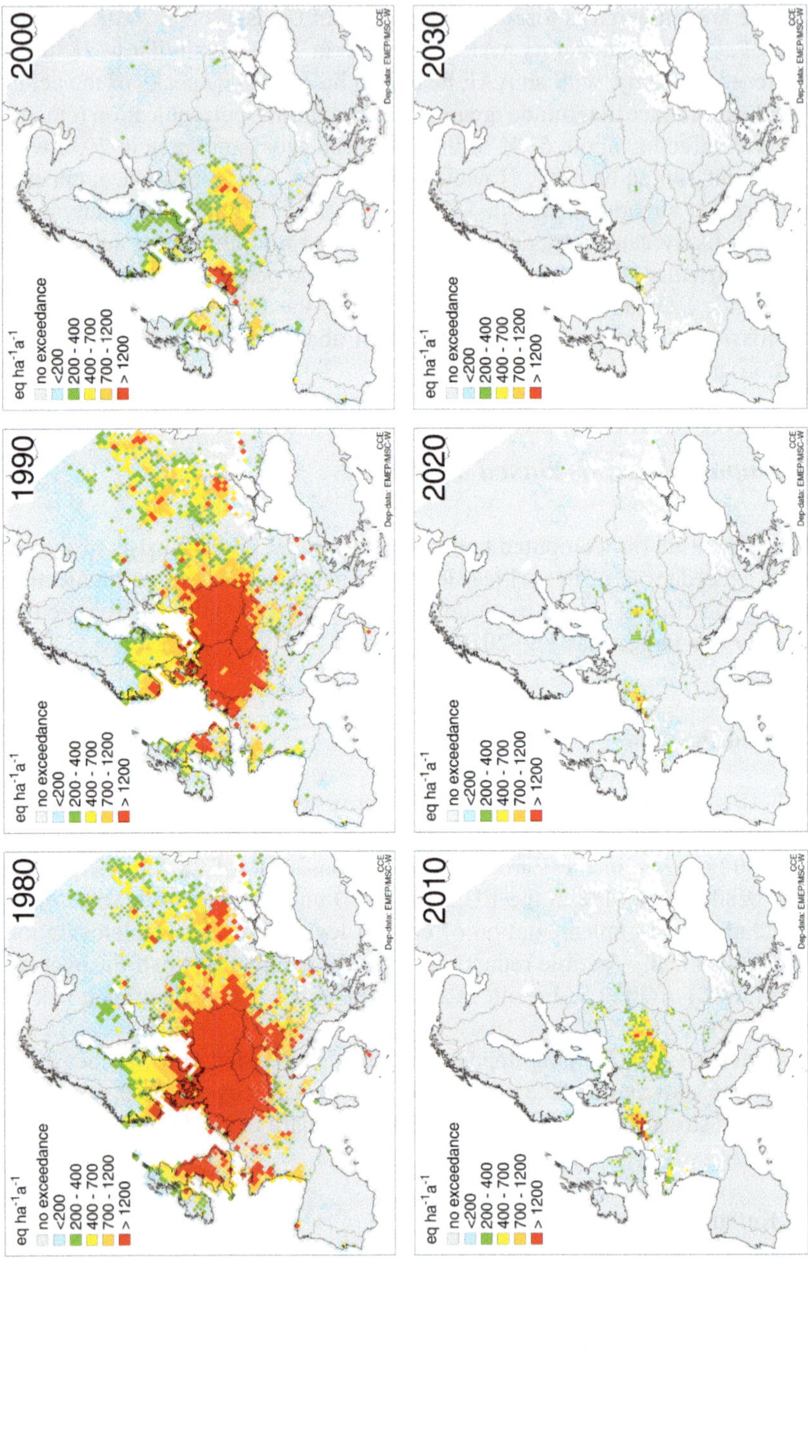

Fig. 25.7 Average Accumulated Exceedance (*AAE*) of computed critical loads for acidification in 1980 (*top left*), 1990 (*top centre*), 2000 (*top right*), 2010 (*bottom left*), 2020 under the revised Gothenburg protocol (GP-CLE scenario) emission reduction agreements (*bottom centre*), and in 2030 under Maximum Feasible Reductions (*bottom right*)

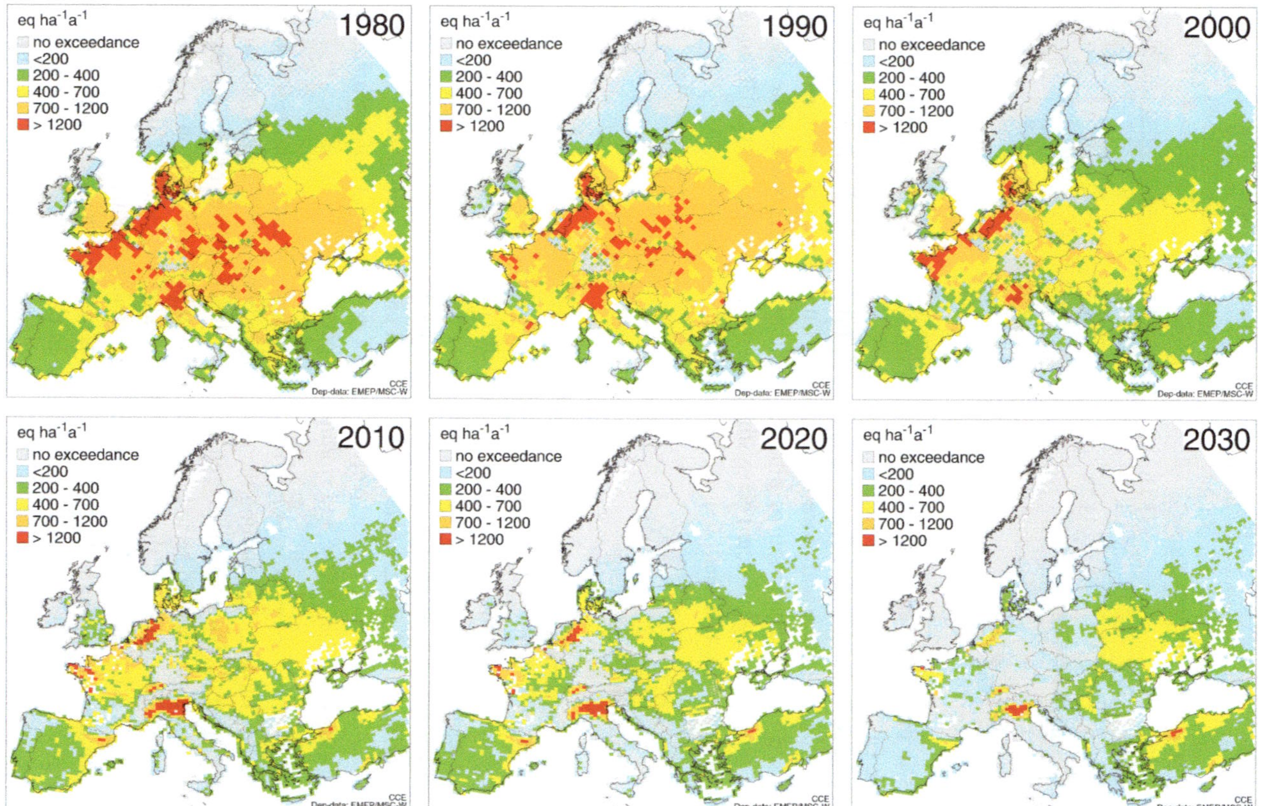

Fig. 25.8 Average Accumulated Exceedance (*AAE*) of computed critical loads for eutrophication in 1980 (*top left*), 1990 (*top centre*), 2000 (*top right*), 2010 (*bottom left*), 2020 under the revised Gothenburg protocol (GP-CLE scenario) emission reduction agreements (*bottom centre*), and in 2030 when applying Maximum Feasible Reductions (*bottom right*)

Table 25.1 Distribution of area percentages over categories 1–4 (see Fig. 25.4) covering all combinations of critical load (non-)exceedance and criterion (non-)violation for acidification

Cases (Fig. 25.4)	Historical years		CLE		MFR	
	1980	2010	2030	2100	2030	2100
1 Not-exceeded and not violated	61.63	92.81	94.48	94.55	96.58	97.26
2 Not-exceeded and violated	1.11	1.80	0.62	0.55	2.22	1.53
3 Exceeded and not violated	33.48	2.90	2.62	2.39	0.31	0.30
4 Exceeded and violated	3.79	2.48	2.28	2.51	0.89	0.90

2020 (54 % in the EU28). In 2030 a further reduction to about 49 % under MFR2030 (40 % in the EU28) could be technically feasible.

Areas at risk in 2020 with exceedances above 700 eq ha^{-1} yr^{-1} (orange and red shading) are much smaller and more scattered over Europe than in 1980. However, the risk posed by the deposition of nitrogen remains obvious since none of the European countries show protected ecosystems everywhere, not even in 2030 under Maximum Feasible Reductions.

25.3.2.3 Dynamic Modelling

Focussing on the persistent exceedances of critical loads of nutrient nitrogen even under depositions of MFR2030, the use of dynamic models becomes an interesting tool to assess whether an ecosystem is at risk in terms of its critical load exceedance as well as in terms of the violation of the chemical criterion (linking the critical load to the effect). For this type of analysis four cases have been identified in Fig. 25.4, i.e. case 1 (critical load *not exceeded* and chemical criterion *not violated*), case 2 (*not exceeded* and *violated,* respectively), case 3 (*exceeded* and *not violated,* respectively) and case 4 (*exceeded* and *violated,* respectively).

(Non-)Protection and (Non-)Recovery from Acidification: Results for acidification are presented in Table 25.1, enabling the comparison between historical and future years of the percentage of ecosystem area that meet the conditions of the four cases. Ideally, all ecosystems should be in case 1. In 1980, when acidification reached its peak (Fig. 25.6), about 62 % of the ecosystems are 'safe', i.e. case 1. This improves to about 97 % in 2030 and 2100 under Maximum Feasible Reductions. The difference between these two years of about 0.7 % is interesting, because it corresponds to the change between 2030 and 2100 of the area in case 2. Areas can become less at risk as emissions, and thus depositions, are reduced. This can imply a shift of areas from case 2, 3 and 4 to case 1. However, the shift from 4 to 1 is not likely to occur on a broad spatial and short temporal scale. For these areas, where both critical loads were exceeded and critical limits violated, emission reductions can rather cause a shift from case 4 to 2, i.e. critical loads of areas are no longer exceeded, but not necessary to a level where the critical limit is met.

Table 25.2 Distribution of area percentages over categories 1–4 (see Fig. 25.4) covering all combinations of critical load (non-)exceedance and criterion (non-)violation for eutrophication

Cases (Fig. 25.4)	Historical years		CLE		MFR	
	1980	2010	2030	2100	2030	2100
1 Not-exceeded and not violated	21.47	35.09	36.60	36.61	64.39	64.70
2 Not-exceeded and violated	0.00	0.05	0.02	0.01	0.40	0.09
3 Exceeded and not violated	12.44	8.51	8.41	8.18	5.88	5.75
4 Exceeded and violated	66.09	56.35	54.97	55.20	29.33	29.47

This can clearly be concluded from the percentage of the area in case 2, which increases from about 1.1 to 1.8 % between 1980 and 2010, while in case 4 the area percentage diminishes from about 3.8 % in 1980 to about 2.5 % in 2010. In the years following 2010 the case 4 to case 2 shift continues, and the percentage area in case 2 further increases from 1.8 % in 2010 to 2.2 % in MFR2030. The decrease of the percentage area in 2030 and 2100 in case 2 under CLE in comparison to 1980 and 2030 is due to a shift of areas from case 2 to case 1. In summary CLE tends to shift areas from case 2 to case 1, whereas MFR tends to shift areas from case 4 to case 2. Note that the percentage in MFR2030 of 2.2 is reduced by about 0.7 % in MFR2100 which corresponds to areas that move from case 2 to case 1. Inspecting the percentage of the areas in case 3 (*exceeded* and *non-violated*) the peak in 1980 (about 33 %) diminishes to about 3 % in 2010 as areas move to case 1, the percentage areas in case 3 are further reduced in 2100 to 2.4 % and 0.3 % under CLE and MFR respectively. Finally, the area percentage in case 4, where both critical loads for acidification are exceeded and the critical limits violated, is reduced from about 4 % in 1980 to about 2 % in 2100 under CLE. Under the application of MFR a further reduction of the area percentage to about 1% in 2030 would be achieved. This hardly changes in 2100; the small increase between 2030 and 2100 of 0.1 % can be due to a shift from areas in case 3 to case 4. The latter can occur when areas in case 3, i.e. exceeded, flip from non-violation to violation of the criterion, even under MFR.

(Non-)Protection and (Non-)Recovery from Eutrophication: Table 25.2 illustrates the distribution of area percentages in any of four cases between 1980 and 2100 for eutrophication. As with acidification there is a marked increase of areas that become "safe" (i.e. case 1) between 1980 (about 21 %) and 2100 (about 65 % under MFR) as a result of emission reductions. However, even under MFR in 2100, still about 35 % of the area remains at risk of eutrophication, i.e. falls in cases 2, 3 or 4. Areas in case 4, i.e. critical loads exceeded and the critical limit violated, diminishes from about 66 % in 1980 to about 55% under CLE and 29% under MFR in 2100. This implies that target loads for depositions may exist that would demand nitrogen emission reductions that are more stringent than MFR to achieve recovery as of 2100.

Finally, another way of using dynamic modelling is for the calculation of the time delay before recovery of the soil chemistry, base saturation in particular, occurs. Posch et al. (2008) have shown – using the European background database

– that a recovery of the base saturation to be "safe", i.e. close to 20 %, would take about 200 years from the year of implementation of Maximum Feasible Reductions.

25.3.3 *Impact Analysis Based on Empirical Knowledge*

Impact analysis based on empirical knowledge (see Fig. 25.2) consists of the assessment of exceedances of the empirical critical loads of nutrient nitrogen, based on Bobbink and Hettelingh (2011) and of the application of relationships between N-deposition and the change of species richness (Stevens et al. 2010) after regionalization over European grasslands (see Chap. 23).

25.3.3.1 Exceedance of Empirical Nitrogen Critical Loads

Figure 25.9 shows the areas at risk of nutrient nitrogen deposition. The exceedances of empirical critical loads turn out to be lower than those for computed critical loads (see Fig. 25.8). This is because empirical critical loads are generally higher that computed critical loads. Peaks of exceedances (red shading) are in 1980 less widespread than for computed critical loads, whereas in 2030 more areas with no exceedances (grey shading) appear.

25.3.3.2 Empirical Relation Between Nitrogen Loads and Plant Species Richness

An illustration of how the deposition of nitrogen affects plant species richness between 1980 and 2030 is shown in Fig. 25.10. In 1980 the species richness in grasslands (EUNIS code E1, E2 and E3) is lower than 70 % (red shading) in major parts of central-western Europe. Due to the emission reductions foreseen under the revised Gothenburg Protocol, species richness in this area increases to a percentage between 85 % and 90 %. Finally, under Maximum Feasible Reductions of nitrogen emissions in 2030, species richness is computed to be lower than 75 % only in the Netherlands and its bordering areas with Belgium and Germany, in western France, northern Italy and Switzerland.

25.3.4 *Ensemble Assessment*

Ensemble Assessment of Impacts (EAI) explores the robustness of exceedances on a scale that ranges from 'exceptionally unlikely' to 'virtually certain', in analogy to the manner in which uncertainties are proposed to be addressed in the IPCC Fourth Assessment Report (IPCC 2005) (see also Hettelingh et al. 2007b). The result of the application of the ensemble assessment is shown in Fig. 25.11.

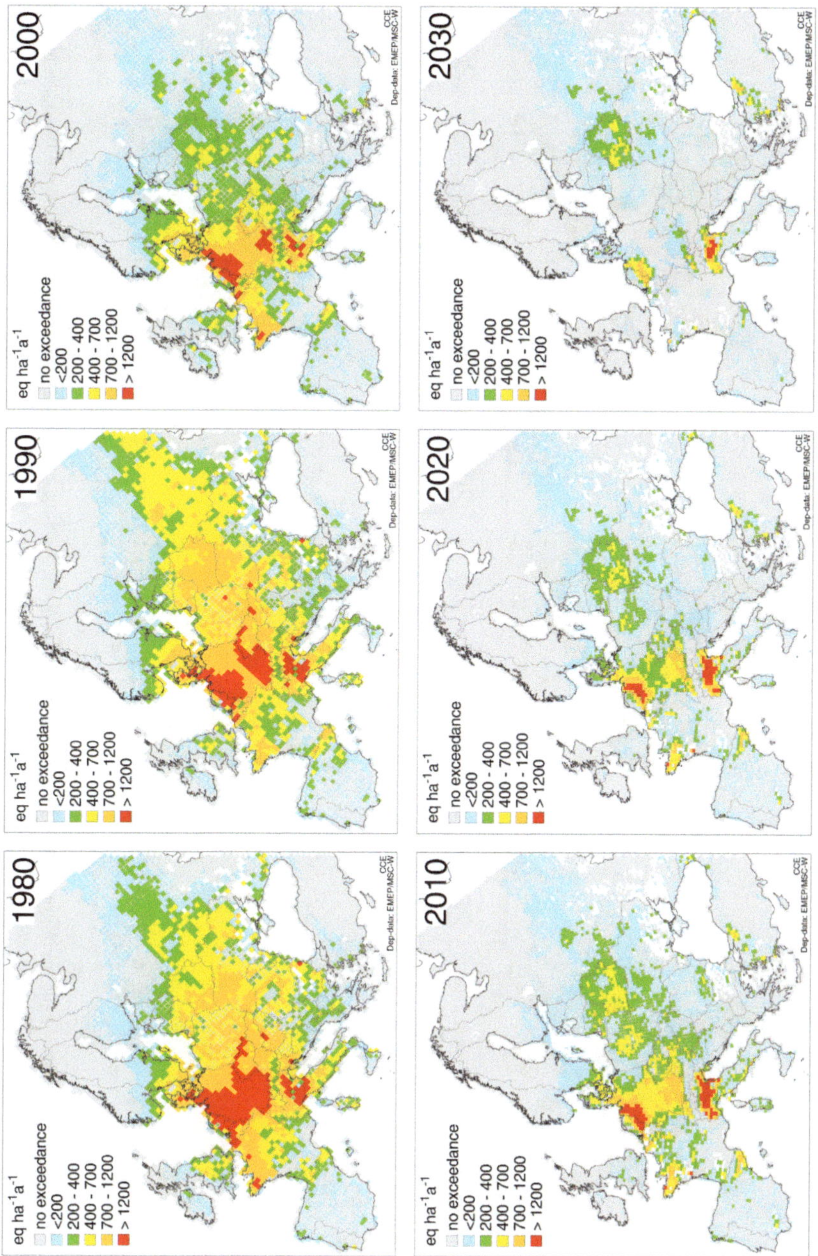

Fig. 25.9 Average Accumulated Exceedance (*AAE*) of empirical critical loads of nutrient nitrogen in 1980 (*top left*), 1990 (*top centre*), 2000 (*top right*), 2010 (*bottom left*), 2020 under the revised Gothenburg protocol (GP-CLE scenario) emission reduction agreements (*bottom centre*), and in 2030 under Maximum Feasible Reductions (*bottom right*)

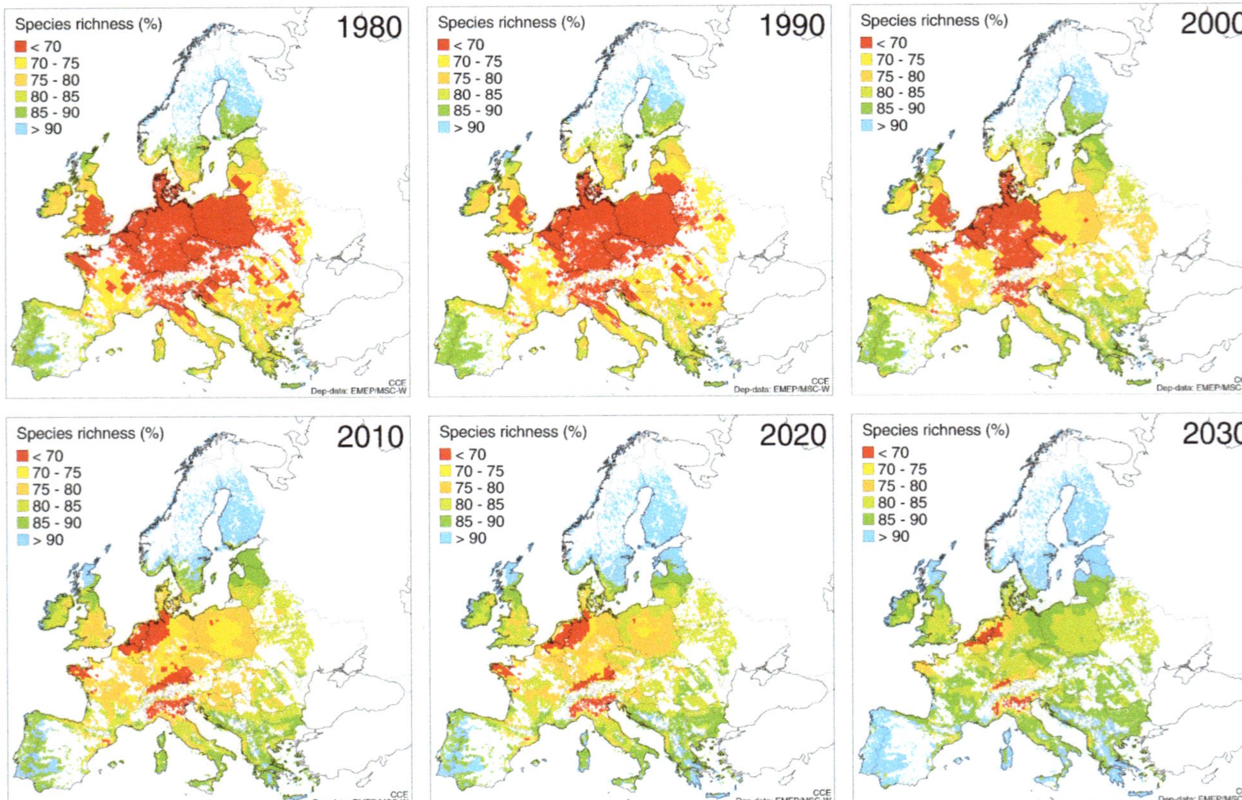

Fig. 25.10 Relative species richness (in %) in grasslands (EUNIS codes E1, E2 and E3) using a dose-response relationship derived from *gradient studies* (see Chap. 23) under N depositions in 1980 (*top left*), 1990 (*top centre*), 2000 (*top right*), 2010 (*bottom left*), 2020 under the revised Gothenburg protocol (GP-CLE) emission reduction agreements (*bottom centre*), and in 2030 under Maximum Feasible Reductions (*bottom right*)

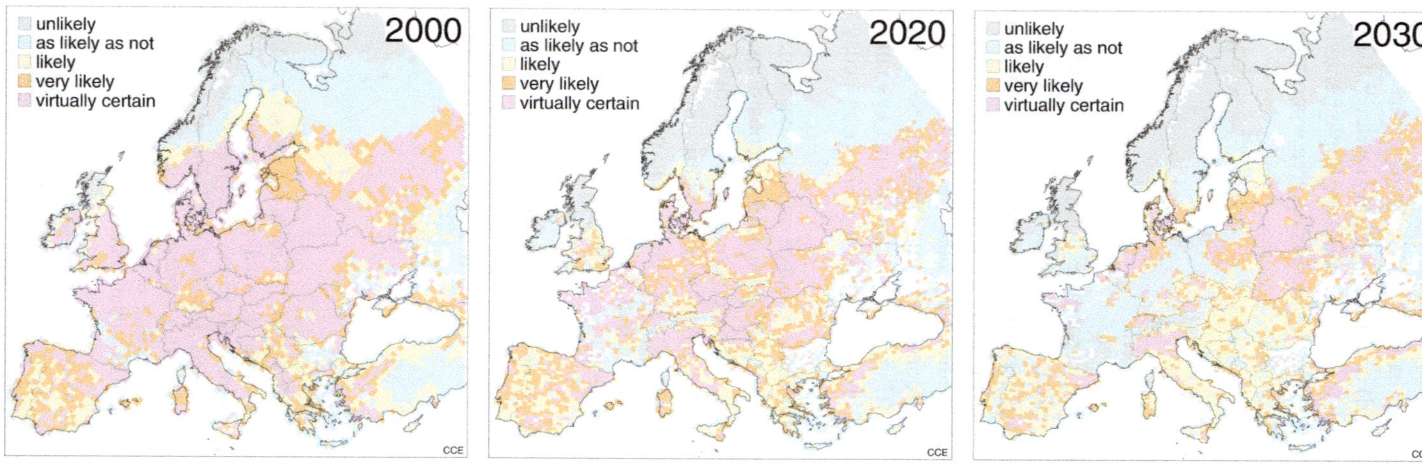

Fig. 25.11 The likelihood of an exceedance of the critical load of nutrient nitrogen in an EMEP grid cell in 2000 (*left*), in 2020 under the revised Gothenburg protocol (GP-CLE) (*centre*) and in 2030 under the implementation of Maximum Feasible Reductions (*right*)

The area where exceedance is 'virtually certain' (purple shading) covered most of Europe in 1980. This area is gradually being replaced by areas where exceedances are 'very likely' (red) or 'likely' (yellow) towards 2020 when emission reductions from the revised Gothenburg Protocol are implemented. Finally, in 2030 under Maximum Feasible Reductions an increasing area where exceedances are 'as likely as not' (light blue) can be noted.

Conclusions

Acidifying and eutrophying depositions compared to computed critical loads, show that only a minor area remains at risk of acidification in the future. However, it is shown that eutrophication persists as a significant policy issue, with 53 % of European ecosystems still at risk in 2030 under Current Legislation. Dynamic modelling is then applied to explore, the delay that can be expected before areas at risk of critical load exceedance become endangered because a given geochemical criterion violates the critical limit (Chap. 2). It turns out that even under Maximum Feasible Reductions in 2030 still about 29 % of the European ecosystem area falls in this category

Empirical critical loads that are based on field experiments addressing nitrogen sensitivity of plant species are also compared to depositions. Areas at risk of nutrient nitrogen are smaller than those based on computed critical loads, because empirical critical loads tend to be higher than computed critical loads. Subsequently, experimental findings of the response of grasslands to nitrogen deposition (Chap. 23) is tentatively applied to conclude that species richness in European acidic grasslands can increase compared to 1980 by about 10–15 % in 2020 under Current Legislation.

Finally, ensemble assessment of areas where both computed and empirical critical loads are exceeded confirms the robustness of the forecasted area at risk of eutrophication, if nitrogen emissions are not mitigated further than currently foreseen.

References

Amann, M., Bertok, I., Borken-Kleefeld, J., Cofala, J., Heyes, C., Höglund-Isaksson, L., Klimont, Z., Nguyen, B., Posch, M., Rafaj, P., Sandler, R., Schöpp, W., Wagner, F., & Winiwarter, W. (2011). Cost-effective control of air quality and greenhouse gases in Europe: Modeling and policy applications. *Environmental Modelling & Software, 26,* 1489–1501.

Bobbink, R., & Hettelingh, J.-P. (2011). *Review and revision of empirical critical loads and dose-response relationships: Proceedings of an expert workshop, Noordwijkerhout, 23-25 June 2010* (Report 680359002/2011). Bilthoven: Coordination Centre for Effects, National Institute for Public Health and the Environment.

Bull, K. R., Achermann, B., Bashkin, V., Chrast, R., Fenech, G., Forsius, M., Gregor, H.-D., Guardans, R., Haußmann, T., Hayes, F., Hettelingh, J.-P., Johannessen, T., Krzyzanowski, M., Kucera, V., Kvaeven, B., Lorenz, M., Lundin, L., Mills, G., Posch, M., Skjelkvåle, B. L., & Ulstein, M. J. (2001). Coordinated effects monitoring and modelling for developing and supporting international air pollution control agreements. *Water, Air, and Soil Pollution, 130,* 119–130.

Davies, C. E., & Moss, D. (1999). *EUNIS habitat classification*. Copenhagen: European Environment Agency.

De Vries, W., Kros, J., Reinds, G. J., Wamelink, G. W. W., Mol, J., van Dobben, H., Bobbink, R., Emmett, B., Smart, S., Evans, C., Schlutow, A., Kraft, P., Belyazid, S., Sverdrup, H. U., van Hinsberg, A., Posch, M., & Hettelingh, J.-P. (2007). *Developments in deriving critical limits and modelling critical loads of nitrogen for terrestrial ecosystems in Europe*. (Report 1382). Wageningen: Alterra Wageningen UR.

EC. (2001). *Directive 2001/81/EC of the European Parliament and of the Council of 23 October 2001 on national emission ceilings for certain atmospheric pollutants*. Brussel: European Commision.

EEA. (2012). *NEC Directive Status Report 2011*. (Technical report 6/2012). European Environment Agency.

Eliassen, A., Hov, Ø., Isaksen, I. S. A., Saltbones, J., & Stordal, F. (1982). A Lagrangian long-range transport model with atmospheric boundary layer chemistry. *Journal of Applied Meteorology, 21,* 1645–1661.

EMEP. (1998). *Transboundary acidifying air pollution in Europe. Part 1: Estimated dispersion of acidifying and eutrophying compounds and comparison with observations. Part 2: Numerical Addendum*. Oslo: EMEP/MSC-W Report 1/98 Meteorological Synthesizing Centre-West Norwegian Meteorological Institute.

Hettelingh, J.-P., & Posch, M. (1994). Critical loads and a dynamic assessment of ecosystem recovery. In J. Grasman & G. Van Straten (Eds.), *Predictability and nonlinear modelling in natural sciences and economics* (pp. 439–446). Dordrecht: Kluwer Academic.

Hettelingh, J.-P., & Posch, M. (1997). An analysis of the critical load and input data variability. In M. Posch, J.-P. Hettelingh, P. A. M. de Smet, & R. J. Downing (Eds.), *Calculation and mapping of critical thresholds in Europe. CCE Status Report 1997* (RIVM Report 259101007) (pp. 29–39). Bilthoven: Coordination Centre for Effects, National Institute for Public Health and the Environment.

Hettelingh, J. P., Posch, M., de Smet, P. A. M., & Downing, R. J. (1995). The use of critical loads in emission reduction agreements in Europe. *Water, Air, and Soil Pollution, 85,* 2381–2388.

Hettelingh, J.-P., Posch, M., & De Smet, P. A. M. (1997). The use of critical thresholds to assess natural stock at risk. *Environmental Research Forum, 7–8,* 536–544.

Hettelingh, J.-P., Posch, M., & de Smet, P. A. M. (2001). Multi-effect critical loads used in multi-pollutant reduction agreements in Europe. *Water, Air, and Soil Pollution, 130,* 1133–1138.

Hettelingh, J.-P., Posch, M., Slootweg, J., Reinds, G. J., Spranger, T., & Tarrasón, L. (2007a). Critical loads and dynamic modelling to assess European areas at risk of acidification and eutrophication. *Water, Air, & Soil Pollution: Focus, 7,* 379–384.

Hettelingh, J.-P., Posch, M., & Slootweg, J. (2007b). Tentatively exploring the likelihood of exceedances: Ensemble Assessment of Impacts (EAI). In J. Slootweg, M. Posch, & J.-P. Hettelingh (Eds.), *Critical loads of nitrogen and dynamic modelling, CCE Progress Report 2007* (pp. 53–58). MNP Report 500090001.

Hettelingh, J. P., Posch, M., & Slootweg, J. (2008). *Critical load, dynamic modelling and impact assessment in Europe. CCE status report 2008* (RIVM Report 500090003). Bilthoven: Coordination Centre for Effects, National Institute for Public Health and the Environment.

Hettelingh, J.-P., De Vries, B., & Hordijk, L. (2009). Integrated assessment. In J. Boersema & L. Reijnders (Eds.), *Principles of environmental sciences* (pp. 385–417). New York: Springer.

Hettelingh, J.-P., Posch, M., Velders, J. M., Ruyssenaars, P., Adam, M., de Leeuw, F., Lükewille, A., Maas, R., Sliggers, J., & Slootweg, J. (2013). Assessing interim objectives for acidification, eutrophication and ground-level ozone of the EU National Emissions Ceilings Directive with 2001 and 2012 knowledge. *Atmospheric Environment, 75,* 129–140.

IPCC. (2005). *Guidance notes for lead authors of the IPCC fourth assessment report on addressing uncertainties*. (http://www.ipcc-wg1.unibe.ch/publications/supportingmaterial/uncertainty-guidance-note.pdf). Accessed 26 Nov 2014.

Kopáček, J., & Posch, M. (2011). Anthropogenic nitrogen emissions during the Holocene and their possible effects on remote ecosystems. *Global Biogeochemical Cycles, 25,* GB2017.

Posch, M., & Reinds, G. J. (2005). The European background database. In M. Posch, J.-P. Hettelingh, & J. Slootweg (Eds.), *European critical loads and dynamic modelling. CCE status report 2005* (pp. 33–37). Bilthoven: RIVM Report 259101016/2005, ISBN 9069601281, Coordination Centre for Effects, National Institute for Public Health and the Environment.

Posch, M., de Smedt, P. A. M., Hettelingh, J. P., & Downing, R. J. (1999). *Calculation and mapping of critical thresholds in Europe. CCE status report 1999* (RIVM report 259101009). Bilthoven: Coordination Center for Effects, National Institute of Public Health and the Environment.

Posch, M., Hettelingh, J.-P., & Mayerhofer, P. (2001a). Past and future exceedances of nitrogen critical loads in Europe. *The Scientific World Journal, 1,* 945–952.

Posch, M., Hettelingh, J.-P., & De Smet, P. A. M. (2001b). Characterization of critical load exceedances in Europe. *Water, Air, and Soil Pollution, 130,* 1139–1144.

Posch, M., Hettelingh, J.-P., & Slootweg, J. (2003). *Manual for dynamic modelling of soil response to atmospheric deposition* (RIVM report 259101 012). Bilthoven: National Institute for Public Health and the Environment.

Posch, M., Hettelingh, J.-P., Slootweg, J., & Maas, R. (2008). Illustrative dynamic modeling applications for use in support of European air pollution abatement policies. In J. P. Hettelingh, M. Posch, & J. Slootweg (Eds.), *Critical load, dynamic modelling and impact assessment in Europe. CCE status report 2008* (pp. 51–62). Bilthoven: Coordination Centre for Effects, National Institute for Public Health and the Environment.

Posch, M., Slootweg, J., & Hettelingh, J.-P. (2012). *Modelling and Mapping of atmospherically-induced ecosystem impacts in Europe: CCE status report 2012* (RIVM Report 680359004). Bilthoven: Coordination Centre for Effects, National Institute for Public Health and the Environment.

Reinds, G. J., Posch, M., & Leemans, R. (2009). Modelling recovery from soil acidification in European forests under climate change. *Science of Total Environment, 407,* 5663–5673.

Reis, S., Grennfelt, P., Klimont, Z., Amann, M., ApSimon, H., Hettelingh, J.-P., Holland, M., Le-Gall, A.-C., Maas, R., Posch, M., Spranger, T., Sutton, M. A., & Williams, M. (2012). From acid rain to climate change. *Science, 338,* 1153–1154.

Schöpp, W., Posch, M., Mylona, S., & Johansson, M. (2003). Long-term development of acid deposition (1880–2030) in sensitive freshwater regions in Europe. *Hydrology and Earth System Sciences, 7,* 436–446.

Simpson, D., Benedictow, A., Berge, H., Bergström, R., Emberson, L. D., Fagerli, H., Flechard, C. R., Hayman, G. D., Gauss, M., Jonson, J. E., Jenkin, M. E., Nyiri, A., Richter, C., Semeena, V. S., Tsyro, S., Tuovinen, J.-P., Valdebenito, A., & Wind, P. (2012). The EMEP MSC-W chemical transport model—technical description. *Atmospheric Chemistry and Physics, 12,* 7825–7865.

Skjelkvåle, B. L., Stoddard, J. L., Jeffers, J. N. R., Tørseth, K., Høgåsen, T., Bowman, J., Mannio, J., Monteith, D., Mosello, R., Rogora, M., Rzychon, D., Veselý, J., Wieting, J., Wilander, A., & Worsztynowicz, A. (2005). Regional scale evidence for improvements in surface water chemistry 1990-2001. *Environmental Pollution, 137,* 165–176.

Slootweg, J., Posch, M., & Hettelingh, J.-P. (2008). Summary of national dataz. In J. P. Hettelingh, M. Posch, & J. Slootweg (Eds.), *Critical load, dynamic modelling and impact assessment in Europe* (CCE status report 2008) (pp. 29–50). Bilthoven: Coordination Centre for Effects, National Institute for Public Health and the Environment.

Slootweg, J., Posch, M., & Hettelingh, J.-P. (2011). Summary of national data. In M. Posch, J. Slootweg, & J. P. Hettelingh (Eds.), *Modelling critical thresholds and temporal changes of geochemistry and vegetation diversity: CCE status report 2011* (pp. 29–46). Bilthoven: Coordination Centre for Effects, National Institute for Public Health and the Environment.

Spranger, T., Hettelingh, J.-P., Slootweg, J., & Posch, M. (2008). Modelling and mapping long-term risks due to reactive nitrogen effects: An overview of LRTAP convention activities. *Environmental Pollution, 154,* 482–487.

Stevens, C. J., Duprè, C., Dorland, E., Gaudnik, C., Gowing, D. J. G., Bleeker, A., Diekmann, M., Alard, D., Bobbink, R., Fowler, D., Corcket, E., Mountford, J. O., Vandvik, V., Aarrestad, P. E.,

Muller, S., & Dise, N. B. (2010). Nitrogen deposition threatens species richness of grasslands across Europe. *Environmental Pollution, 158,* 2940–2945.

Stoddard, J. L., Jeffries, D. S., Lükewille, A., Clair, T. A., Dillon, P. J., Driscoll, C. T., Forsius, M., Johannessen, M., Kahl, J. S., Kellogg, J. H., Kemp, A., Mannio, J., Monteith, D. T., Murdoch, P. S., Patrick, S., Rebsdorf, A., Skjelkvale, B. L., Stainton, M. P., Traaen, T., van Dam, H., Webster, K. E., Wieting, J., & Wilander, A. (1999). Regional trends in aquatic recovery from acidification in North America and Europe. *Nature, 401,* 575–578.

Suutari, R., Amann, M., Cofala, J., Klimont, Z., Posch, M., & Schöpp, W. (2001). *From economic activities to ecosystem protection: An uncertainty analysis of two scenarios of the RAINS integrated assessment model.* (IIASA/CCE Report 1/2001). Laxenburg: IIASA.

UNECE. (1994). Protocol on the further reduction of sulphur emissions, done in Oslo, Norway on 14 June 1994. In United Nations (2004), *Handbook for the 1979 Convention on Long-range Transboundary Air Pollution and its protocols*. United Nations Publication. ISBN: 92-1-116895-3, www.unece.org/fileadmin/DAM/env/lrtap/BIBLE.E.pdf, pp. 101–128.

UNECE. (1999). Protocol to abate acidification, eutrophication and ground level ozone done at Gothenburg, Sweden, on 30 Nov. 1999. In United Nations (2004), *Handbook for the 1979 Convention on Long-range Transboundary Air Pollution and its protocols*. United Nations Publication. ISBN: 92-1-116895-3, www.unece.org/fileadmin/DAM/env/lrtap/BIBLE.E.pdf, pp. 213–274.

UNECE. (2012). Amendments to the protocol to abate acidification, eutrophication and ground level ozone, decisions 1–4 and consolidated text of the protocol. www.unece.org/env/lrtap/multi_h1.html.

Chapter 26
Synthesis

Jean-Paul Hettelingh, Wim de Vries and Maximilian Posch

26.1 Introduction

The emphasis of this book is on the science behind critical loads, in view of the protection of specified environmental endpoints using critical limits for defined indicators, and on methods (dynamic models) to forecast consequences of atmospheric deposition in relation to these critical loads. While there is a clear scientific interest in improving knowledge of impacts caused by deposition, as described in this book, there is also an important drive, especially in Europe, to use this knowledge to support air pollution abatement policies. Information on critical loads for a variety of ecosystems have successfully led to effect-based protocols under the LRTAP Convention (Reis et al. 2012) and directives of the European Union (Hettelingh et al. 2013). Acidification in the USA (see, e.g., NAPAP 1991) and European fish and forest dieback documented in the 1980s (see Chap. 1) were an important motivation for seeking ways to reduce pressures with the aim to prevent damage. The critical load approach became part of a logic whereby scientific uncertainty on actual effects was compensated by applying the precautionary principle. The critical load was considered a way to provide a scientific basis for substantiating the need for reductions of pressures to a level at which, according to the critical load definition, adverse effects do not occur according to present knowledge. While with a strict precautionary approach emissions could be targeted to become zero, this book elaborates how the critical load approach provides a scientific basis for (economically) more realistic targets, while also providing sufficient protection.

Following the definition of a critical load, the ecosystem composition and/or diversity ('structure') or the rate by which a deposited pollutant is leached,

J.-P. Hettelingh (✉) · M. Posch
Coordination Centre for Effects (CCE), RIVM, Bilthoven, The Netherlands
e-mail: jean-paul.hettelingh@rivm.nl

W. de Vries
Alterra Wageningen University and Research Centre, Wageningen, The Netherlands

W. de Vries et al. (eds.), *Critical Loads and Dynamic Risk Assessments*,
Environmental Pollution 25, DOI 10.1007/978-94-017-9508-1_26

decomposed or mineralized ('functioning') in that ecosystem is sustained when the critical load is not exceeded. By adversely affecting structure and function of ecosystems, an exceedance of critical loads can cause undesired changes to biological diversity, which among the many definitions that exist, is formulated by the Convention on Biological Diversity as "the variability among living organisms from all sources including, *inter alia*, terrestrial, marine and other aquatic ecosystems and the ecological complexes of which they are part; this includes diversity within species and ecosystems" (UN 1992). Adherence to critical loads not only helps to avoid damage to biodiversity and related human wellbeing, but can also help limiting air pollution levels that could affect ecosystem resilience against climate change. Critical loads are well suited to support the so-called AICHI biodiversity targets[1] set by the CBD in 2010 for its biodiversity strategy until 2020. For Europe, the EU specified six biodiversity targets for 2020[2].

The chapters in this book intend to contribute to the understanding of impacts of acidification, eutrophication and heavy metal mobilization by evaluating critical limits of geochemical indicators for endpoints to be protected, describing critical load and dynamic modelling approaches, applications on a national and European scale, and their use in integrated assessment modelling. In this chapter, we first present a synthesis of main findings and implications of the results described in this book, followed by a treatment of the uncertainties involved in critical load assessments and applications, and conclude with an outlook.

26.2 Findings

This book documents that methods to analyse impacts of acidification, eutrophication and heavy metal mobilization have evolved since the 1980s from describing geochemical processes alone, to the simulation of niches for plant species and to including interactions with climate change. Below, methodological aspects are summarized, including the combined use of empirical and model-based approaches, main results of these analyses and the relevance of results in view of impacts on ecosystems (including ecosystem services) and human health.

Model Approaches to Assess Impacts of Nitrogen, Acidity and Metal Inputs on Ecosystems:

In order to assess impacts of nitrogen, acidity and metal inputs on ecosystems, critical limits for geochemical and biodiversity indicators are elaborated that protect specific environmental endpoints (Chaps. 2 and 3). These critical limits are then used to model critical loads (Chaps. 6, 7, 15–17, 19, 21 and 25) and to evaluate risks of predicted geochemical indicator values by dynamic models (Chaps. 8–14, 18, 19, 22, 24, 25). By doing so, we assess impacts of the deposition of sulphur

[1] http://www.cbd.int/sp/targets/.

[2] COM(2011) 244 final: http://ec.europa.eu/environment/nature/biodiversity/comm2006/pdf/2020/1_EN_ACT_part1_v7[1].pdf. also see http://biodiversity.europa.eu/bise/policy/eu-biodiversity-strategy.

(acidification), nitrogen (acidification and eutrophication) and heavy metals (pollution) on specific environmental endpoints, but actual effects in terms of (e.g.) biodiversity decline or dieback of fish populations is not forecasted. Instead, those effects are assumed to be negligible below critical limits, and there is an increased risk of damage above the critical limits. An exception is the recent development of estimating future impacts of nitrogen deposition on plant species diversity by combining geochemical with plant species diversity models (Chaps. 11–14).

Both the spatial and temporal scales involved in assessing various endpoints are addressed in this book. The need for the inclusion of time delays between exposure and effects has led to the necessity of applying dynamic models. The various modelling approaches describe the mechanisms behind acidification, eutrophication and metal mobilization. While the exceedance of critical loads for acidification and eutrophication provides information on possible adverse impacts within the coming century, this is less the case regarding impacts of excessive heavy metal deposition. Due to extremely long time horizons involved before heavy metal impacts occur, it is argued that the impacts are more meaningfully addressed by using dynamic models than to compare the magnitude of heavy metal deposition relative to its critical load.

Combined Use of Empirical and Model-Based Approaches for Nitrogen Impacts:
Unlike acidification, effects of eutrophication on terrestrial systems are more visible in the field, which has led to the derivation of empirical critical loads of nitrogen, both in Europe (Chap. 4) and the USA (Chap. 5). Although based on visible effects, one of the drawbacks of empirical nitrogen critical loads is that they are based on comparatively short-term experiments, and thus provide limited knowledge of the evolution of impacts over time. Furthermore, the variation in empirical critical loads for a given ecosystem is not spatially explicit, but only included by defining it as a range. Conversely, model based approaches are based on critical limits of chosen endpoint indicators, allowing the assessment of risks of adverse effects over time and space. While the focus of policy evaluations of nitrogen impacts on ecosystems is based on modelling approaches, the robustness of results of these evaluations benefits from the combination with empirical critical loads (Chap. 10) and different measures of biodiversity (Chaps. 11, 22 and 24). The robustness of the assessment of impacts is further enhanced by using established empirical relationships between the rate of (excessive) depositions and changes to plant species richness in experiments and field surveys (Chap. 23), and this approach is thus also included in policy analyses (Chap. 25).

Exceedances of Critical Loads:
In line with significant reductions in sulphur dioxide emissions since the mid-1980s, the exceedance of critical loads of acidity and related soil acidification have markedly decreased over time, although is not reversed everywhere, in Europe and the USA. As with sulphur dioxide, emissions of nitrogen oxides and ammonia also decreased but less strongly than sulphur dioxide. Inversely, in China, anthropogenic emissions of both nitrogen oxides and ammonia increased strongly since 1980 (Liu et al. 2013). Consequently, N deposition is still excessive almost worldwide. Reductions of exceedances of critical loads of nitrogen are thus an important challenge for

the future, and recognized as such in European policies. Emissions of heavy metals also decreased in the past decades, but deposition levels of lead and mercury are still exceeding critical loads in many parts of Europe (Chap. 21).

Impacts on Ecosystem and Human Health:
The health of ecosystems (geo-chemistry, vegetation diversity/abundance, ecosystem functions) is the common deposition-based endpoint of acidification, eutrophication and metal mobilization. However, deposition effects are less evident for human health than those caused by ambient concentrations, e.g. smog episodes (see Annex). Note, however, that there is a link between assessing sources and impacts of critical load exceedances in ecosystem areas on the one hand and tackling the violation of WHO health guidelines in urban areas on the other hand. This is obvious from the fact that implementation of abatement techniques on local and distant pollution sources leads to improvement of urban air quality (diminishing excessive ambient concentrations) as well as the reduction of the exceedance of critical loads or levels in rural parts of ecosystem regions (diminishing excessive atmospheric depositions). Therefore, measures to improve urban air quality can also diminish the risk to biological diversity (see also Hettelingh et al. 2014).

Impacts on Ecosystem Services:
The WHO prepared a report as a contribution to the Millennium Ecosystem Assessment (MEA) to address the issue of ecosystems and human well-being (Corvalan et al. 2005). The report starts by stating that "ecosystems are essential to human well-being"—defined by the WHO as a state of complete physical, mental and social well-being. The MEA then introduces ecosystem services, distinguishing provisioning services, regulating services, supporting services and cultural services, all related to well-being. Several of those services are affected by acidification and eutrophication, such as the production of wood, fibers, clean water and regulation of some pests and diseases, as discussed in Chap. 24.

As noted before, also other drivers and effect-receptors are involved than those related to the context of this book, i.e. adverse effects of acidification, eutrophication and heavy metal mobilization. However, the critical load approach fits well with the MEA logic (Hettelingh et al. 2008); critical load exceedance can imply adverse effects on ecosystem services. Critical load exceedance is one of the causes of environmental change affecting human well-being. Such synergies and antagonies, both between different drivers and between their impacts, can be captured through integrated assessment modelling (Fig. 26.1), where policy principles (e.g., precaution, environmental sustainability, meeting CBD biodiversity targets) are translated into (ecosystem service) endpoints and pollutant mitigation alternatives.

26.3 Robustness of the Critical Load Approach

While above it is argued that the critical load approach meets precautionary requirements needed to support environmental sustainability, it may still leave the issue of scientific uncertainty not sufficiently addressed, thus hampering its wider

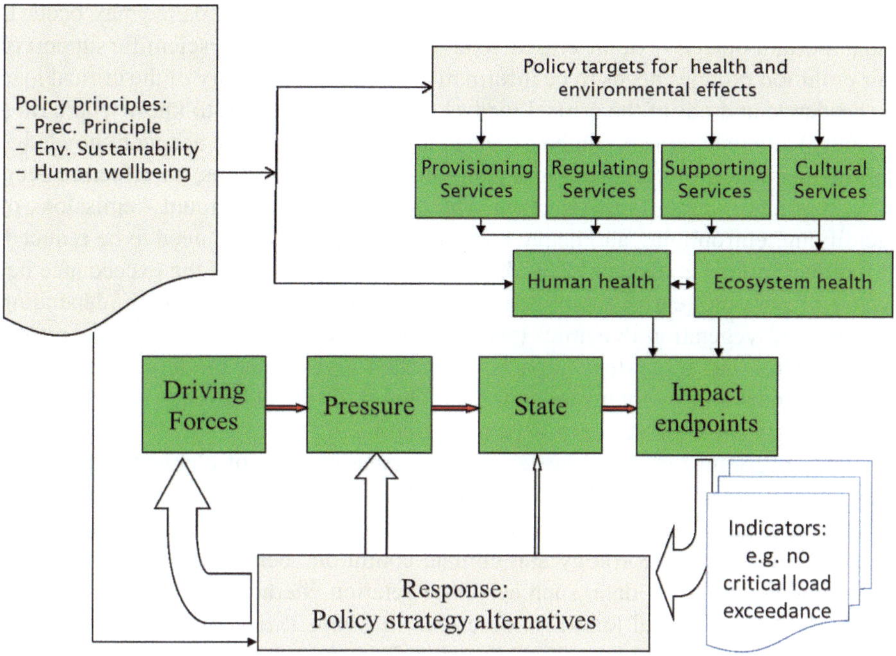

Fig. 26.1 Linkages between policy principles (e.g. precautionary principle, environmental sustainability), policy targets, ecosystem services and the iterative decision making process trading off pollution source, mitigation and impacts

acceptance. Weiss (2006)—in another context than critical loads—treats the question "what level of scientific proof is needed to justify measures to avoid or mitigate environmental threats of uncertain size and probability". Against the background of many chapters in this book, a similar question can be addressed by placing critical loads, related uncertainties and impacts of exceedances, in the context of the precautionary principle, sustainability, human well-being and its usefulness for the environmental challenges ahead.

When a critical load is exceeded, adverse effects to structure and function can occur with varying biological consequences, but not necessarily for all locations where the critical load is exceeded, nor within the same time horizon. With this ignorance in mind we can consider the critical load as a precautionary mind-set for targeting emission reductions to sufficiently limit deposition of air pollutants below 'safe' limits. The critical load approach meets a number of concepts of the precautionary principle, such as "preventive anticipation", "safe-guarding of ecological space" and "paying for past ecological debt" (O'Riordan and Cameron 1994). A critical load aims to prevent damage caused by deposition, even when scientific proof would not be conclusive, and it aims to avoid (soil) buffering mechanisms to be exhausted.

Knowing where, in which country and ecosystem, an exceedance may occur is an important objective of the critical load approach. Ultimately, scientific support of air pollution policies needs to be informative about the reliability of the critical load exceedance and not of the critical load as such. It is important to know where (in a country) adverse impacts can be expected to occur. Taking into account knowledge of the dispersion of emissions and resulting excess depositions, an effective decision requires reliable knowledge on where—and by what amount—emissions of acidifying, eutrophying and heavy metal mobilizing emissions need to be reduced. Moreover, policy analysts also wish to know the magnitude of the exceedance because of possible (early) future consequences as exceedances are high, depending on soil and vegetation dynamics (see e.g. Chaps 8, 9, 12, 18 and 19). Therefore, when addressing ecosystem impacts, integrated assessment modellers and policy analysts are primarily interested in the likelihood of (the occurrence of) an exceedance, and its emission scenario-dependent trend.

The uncertainty of exceedances depends on a number of elements—and their spatial and temporal resolution—in the chain from driving forces to impacts (Fig. 26.1). These include data and emission factors behind national emission reports, input data, meteorology and climate conditions behind atmospheric dispersion models and input data, such as soil-vegetation characteristics, and modelling methods behind critical loads. As described in Chap. 2, uncertainties also exist in using laboratory results for critical limits in field situations. This uncertainty is of particular importance when using the critical limit indicator as explanatory variable for a future status of a chosen geochemical or biological endpoint. While field experiments have established adverse effects when critical loads are exceeded (see e.g. Chaps. 2, 4, 5), more findings on critical limits under field conditions would be welcome. However, also addition experiments in the field do not lead to unique empirical critical loads (Chap. 4). The choice, out of a range of empirical critical loads, of a reliable value for specified regions (Chap. 22) can also be decisive in avoiding a false positive (predict non-exceedance when in fact there is) conclusion on exceedances. Therefore one could argue that a focus of uncertainty analysis on avoiding false positives rather than false negatives (predict exceedance when there is none) could prevent emission reductions from being too modest. Uncertainty analyses have been conducted and reported under the LRTAP Convention (e.g. Hettelingh and Posch 1997; Suutari et al. 2001).

The question is whether methods will ever be sufficiently reliable to support consensus of expected and computed impacts on soil chemistry and vegetation endpoints that could result from critical load exceedances. This is a reason why ensemble-modelling, using independent methods to explain a single phenomenon, is becoming increasingly important in dealing with environmental issues that are not easy to reverse, such as climate change and adverse effects on biodiversity. In these cases 'uncertainty' is less of an issue than 'robustness'. The latter is aimed more at avoiding false positive conclusions, than concepts surrounding uncertainty analysis and risk perception in systems in general (see, e.g., Helbing 2013) and the air pollution cause-effect system in particular.

Therefore, the combination of empirical and dynamic biogeochemical modelling may, as these methods are independent of one another, provide benefits to determine critical loads (Chap. 10). The comparison of the magnitude and location of exceedances of both empirical and modelled critical loads (Chap. 25; Hettelingh et al. 2007) can be another approach to increase the reliability of exceedance assessments. It has been based on the application of the IPCC Guidance Notes for lead authors of the IPCC 4th Assessment Report on addressing uncertainties (IPCC 2005). This has been followed up by the Guidance Note in the 5th Assessment Report (IPCC 2010) which also includes a statement "…to be aware that findings can be constructed from the perspective of minimizing false positive (Type I) or false negative (Type II) errors, with resultant trade-offs in the information emphasized." (IPCC 2010). In the context of exceedances, the 'trade-off' is that both Type I (no exceedance when there is) or Type II (exceedance when there is none) errors can lead to policy measures which are either insufficient or cost-ineffective (i.e. not necessary), respectively.

Finally it should be noted that the use of critical loads and the assessment of exceedances for policy support consists of scenario analyses in integrated assessment modelling of emission reductions (e.g. Chap. 25). Uncertainty of methods and data used in integrated assessment models, i.e. emissions, atmospheric transport and critical load exceedances hardly vary between scenarios. Therefore the comparison of exceedances (and other outputs of integrated assessment) between scenarios is more robust, and is driven by the model inputs (i.e. emission reductions and costs) of which impacts are investigated. Requirements to assess the robustness of changes in exceedances under different scenarios seems more useful to policy support than pursuing understanding of all uncertainties that are involved in the models used to assess and appraise critical load exceedances. To a certain extent a parallel can be drawn between the complexity of the models and data described in this book and the assertion by Oreskes et al. (1994) that verification and validation of numerical models that describe interacting systems is impossible. Interaction is at the basis of the air, soil, water and vegetation processes and mechanisms described in this book.

26.4 Outlook

This book aims at increasing the understanding of available methods and data that can be combined to produce robust impact assessments of atmospheric depositions in support of policies in the field of air quality. The critical load approach, in combination with dynamic modelling, provides operational methodologies and applications to assess long-term sustainability of ecosystems by defining no-effect thresholds. The assessments focus on the soil-mediated effects (acidification, eutrophication and metal accumulation) of elevated deposition of SO_2, NO_x, NH_3 and metals on the health of terrestrial and aquatic ecosystems.

In the past decade, most attention has focused on the impacts of elevated N deposition. The reason is that, in contrast to sulphur, the atmospheric deposition of

N, especially ammonia, continues to be significant and most likely remains high in the future. Increased attention on N is also warranted because of its multiple effects, not only including biodiversity impacts (acidification and eutrophication by nitrogen compounds), but also health impacts due to exposure to NO_x, ozone and fine particulate matter, and climate change impacts. An important climate change impact consists, on the one hand, in the reduction in radiative forcing by CO_2 uptake induced by atmospheric N fertilization (see Chap. 24) and, on the other hand, an increased radiative forcing by elevated ozone concentrations, mainly caused by NO_x emissions. Chapter 24 is an example of considering multiple effects, including impacts on soil quality, biodiversity and carbon sequestration by forest growth, but there are more interactions, e.g. involving soil temperature.

The interaction of climate change and air pollution, specifically N deposition and ozone exposure, is a scientific and policy issue that needs increasing attention in view of regional impacts on both biodiversity and climate. This holds for both air pollution effects on climate change (radiative forcing) and climate change effects on air pollution impacts. The relationship between sources and impacts of air pollution and climate change has been studied already in the mid-nineties (Alcamo et al. 1995; Alcamo et al. 2002; Posch et al. 1996). Those studies, which investigated air pollution effects on climate change, all focused on the beneficial effect of reducing SO_2 emissions in view of reduced acidification and the detrimental effect of an enhanced climate change due to smaller amounts of sulphate aerosols in the atmosphere. Increased attention is warranted for combined impacts of: (i) atmospheric SO_2, NO_x and NH_3 emissions on the formation of ammonium nitrate and ammonium sulphate aerosols in the atmosphere, leading to a cooling effect, (ii) NO_x emission induced ground-level O_3 formation, causing radiative forcing, since O_3 is the third most important greenhouse gas, and (iii) NO_x and NH_3 emission-induced CO_2 uptake caused by atmospheric N fertilization (e.g. Butterbach-Bahl et al. 2011; De Vries et al. 2011; Erisman et al. 2011).

Conversely, climate change also affects air pollution impacts. Since temperature and moisture affect nearly all geochemical and biochemical processes in the soil, including weathering rates, growth and N transformation rates, climate change is likely to influence ecosystem sensitivity, and thus critical loads. Posch (2002) investigated those impacts on critical loads of acidity and nutrient nitrogen derived by the SMB model. Results of that study indicated that that higher temperatures, changed precipitation patterns, and modified net primary production mainly increase critical loads, except in mountainous and arid regions. This study, however, only considered impacts on weathering rates, forest growth and precipitation surplus. There is ample evidence that climate change enhances decomposition and thereby N availability (e.g. Eliasson et al. 2005; Melillo et al. 2002; Melillo et al. 2011) and thus the impacts of N deposition on both plant species diversity and growth. Furthermore, there is an interaction between elevated CO_2 concentration and N availability with respect to growth responses. Elevated CO_2 concentrations increase N uptake (e.g. Luo et al. 2004 and elevated N availability increases the CO_2 fertilizing effects (e.g. De Graaff et al. 2006; Finzi et al. 2007).

Another important issue, also in policy terms, is the trade-off between the positive effect of N deposition on productivity and carbon sequestration and its negative effect on biodiversity. This trade-off is related to possible thresholds of nitrogen deposition at which productivity is reduced or even reversed over the course of time due to N saturation and acidification. Both experimental and field data about N retention and growth responses as well as integrated model approaches indicate that C sequestration diminishes over time at N deposition thresholds that may vary locally (see Chap. 24) and more insight in their values depending on local circumstances is a relevant topic for further investigation.

In conclusion, critical load methods and dynamic risk assessments are instrumental to increase our knowledge on impacts of the interactions between climate change and air pollution. This knowledge, to which the chapters in this book have aimed to contribute, is vital for providing effective support to future policies on air pollution, climate change and biodiversity.

References

Alcamo, J., Krol, M., & Posch, M. (1995). An integrated analysis of sulfur emissions, acid deposition and climate change. *Water Air and Soil Pollution, 85,* 1539–1550.

Alcamo, J., Mayerhofer, P., Guardans, R., van Harmelen, T., van Minnen, J., Onigkeit, J., Posch, M., & De Vries, B. (2002). An integrated assessment of regional air pollution and climate change in Europe: Findings of the AIR-CLIM Project. *Environmental Science & Policy, 5,* 257–272.

Butterbach-Bahl, K., Nemitz, E., Zaehle, S., Billen, G., Boeckx, P., Erisman, J. W., Garnier, J., Upstill-Goddard, R., Kreuzer, M., Oenema, O., Reis, S., Schaap, M., Simpson, D., De Vries, W., Winiwarter, W., & Sutton, M. A. (2011). Nitrogen as a threat to the European greenhouse balance. In M. A. Sutton, C. M. Howard, J. W. Erisman, G. Billen, A. Bleeker, P. Grennfelt, H. van Grinsven, & B. Grizzetti (Eds.), *The European nitrogen assessment* (pp. 434–462). Cambridge: Cambridge University Press, (Chapter 19).

Corvalan, C., Hales, S., & McMichael, A. (2005). *Ecosystems and human well-being: Health synthesis. Report prepared by WHO as contribution to the millennium ecosystem assessment.* Geneva: WHO.

De Graaff, M. A., van Groenigen, K. J., Six, J., Hungate, B., & van Kessel, C. (2006). Interactions between plant growth and soil nutrient cycling under elevated CO_2: A meta-analysis. *Global Change Biology, 12,* 2077–2091.

De Vries, W., Kros, J., Reinds, G. J., & Butterbach-Bahl, K. (2011). Quantifying impacts of nitrogen use in European agriculture on global warming potential. *Current Opinion in Environmental Sustainability, 3,* 291–302.

Eliasson, P., McMurtrie, R. E., Pepper, D. A., Strömgren, M., Linder, S., & Ågren, G. I. (2005). The response of heterotrophic CO_2 flux to soil warming *Global Change Biology, 11,* 167–181.

Erisman, J. W., Galloway, J., Seitzinger, S., Bleeker, A., & Butterbach-Bahl, K. (2011). Reactive nitrogen in the environment and its effect on climate change. *Current Opinion in Environmental Sustainability, 3,* 281–290.

Finzi, A. C., Norby, R. J., Calfapietra, C., Gallet-Budynek, A., Gielen, B., Holmes, W. E., Hoosbeek, M. R., Iversen, C. M., Jackson, R. B., Kubiske, M. E., Ledford, J., Liberloo, M., Oren, R., Polle, A., Pritchard, S., Zak, D. R., Schlesinger, W. H., & Ceulemans, R. (2007). Increases in nitrogen uptake rather than nitrogen-use efficiency support higher rates of temperate forest productivity under elevated CO_2. *Proceedings of the National Academy of Sciences USA, 104,* 14014–14019.

Helbing, D. (2013). Globally networked risk and how to respond. *Nature, 497,* 51–59.
Hettelingh, J.-P., & Posch, M. (1997). An analysis of the critical load and input data variability. In M. Posch, J.-P. Hettelingh, P. A. M. de Smet, & R.J. Downing (Eds.), *Calculation and mapping of critical thresholds in Europe. CCE status report 1997. RIVM report 259101007* (pp. 29–39). Bilthoven: Coordination Centre for Effects, National Institute for Public Health and the Environment.
Hettelingh, J.-P., Posch, M., & Slootweg, J. (2007). Tentatively exploring the likelihood of exceedances: Ensemble assessment of impacts (EAI). In J. Slootweg, M. Posch, & J.-P. Hettelingh (Eds.), *Critical loads of nitrogen and dynamic modelling, CCE progress report 2007* (pp. 53–58). MNP Report 500090001.
Hettelingh, J.-P., Posch, M., & Slootweg, J. (2008). Status of critical loads database and impact assessment. In J.-P., Hettelingh, M. Posch, & J. Slootweg (Eds.), *Critical load, dynamic modelling and impact assessment in Europe, CCE status report 2008* (pp. 15–28). MNP Report 500090003, Bilthoven.
Hettelingh, J.-P., Posch, M., Velders, J. M., Ruyssenaars, P., Adam, M., de Leeuw, F., Lükewille, A., Maas, R., Sliggers, J.,& Slootweg, J. (2013). Assessing interim objectives for acidification, eutrophication and ground-level ozone of the EU national emissions ceilings directive with 2001 and 2012 knowledge. *Atmospheric Environment, 75,* 129–140.
Hettelingh, J.-P., De Vries, W., Posch, M., Reinds, G. J., Slootweg, J., & Hicks, W. K. (2014). Development of the critical loads concept and current and potential applications to different regions of the world. Chapter 30. In M. A. Sutton, K. E. Mason, L. J. Sheppard, H. Sverdrup, R. Haeuber, & W. K. Hicks (Eds.), *Nitrogen deposition, critical loads and biodiversity*. Dordrecht: Springer Science+Business Media.
IPCC. (2005). Guidance notes for lead authors of the IPCC fourth assessment report on addressing uncertainties. http://www.ipcc-wg1.unibe.ch/publications/supportingmaterial/uncertainty-guidance-note.pdf.
IPCC. (2010). Guidance note for lead authors of the IPCC fifth assessment report on consistent treatment of uncertainties. http://www.ipcc.ch/pdf/supporting-material/uncertainty-guidance-note.pdf.
Liu, X., Zhang, Y., Han, W., Tang, A., Shen, J., Cui, Z., Vitousek, P., Erisman, J. W., Goulding, K., Christie, P., Fangmeier, A., & Zhang, F. (2013). Enhanced nitrogen deposition over China. *Nature, 494,* 459–462.
Luo, Y., Su, B., Currie, W. S., Dukes, J. S., Finzi, A., Hartwig, U., Hungate, B., McMurtrie, R. E., Oren, R., Parton, W. J., Pataki, D. E., Shaw, R., Zak, D. R., & Field, C. B. (2004). Progressive nitrogen limitation of ecosystem responses to rising atmospheric carbon dioxide. *Bioscience, 54,* 731–739.
Melillo, J. M., Steudler, P. A., Aber, J. D., Newkirk, K., Lux, H., Bowles, F. P., Catricala, C., Magill, A., Ahrens, T., & Morrisseau, S. (2002). Soil warming and carbon-cycle feedbacks to the climate system. *Science, 298,* 2173–2176.
Melillo, J. M., Butler, S., Johnson, J., Mohan, J., Steudler, P., Lux, H., Burrows, E., Bowles, F., Smith, R., Scott, L., Vario, C., Hill, T., Burton, A., Zhou, Y.-M., & Tang, J. (2011). Soil warming, carbon-nitrogen interactions, and forest carbon budgets. *Proceedings of the National Academy of Science USA, 108,* 9508–9512.
NAPAP. (1991). *The U.S. national acid precipitation assessment program: 1990 integrated assessment report*. Washington, DC: The NAPAP Office of the Director.
O'Riordan, T., & Cameron, J. (1994). *Interpreting the precautionary principle*. London: Earthscan.
Oreskes, N., Shrader-Frechette, K., & Belitz, K. (1994). Verification, validation, and confirmation of numerical models in earth sciences. *Science, 263,* 641–646.
Posch, M. (2002). Impacts of climate change on critical loads and their exceedances in Europe. *Environmental Science & Policy, 5,* 307–317.
Posch, M., Hettelingh, J.-P., Alcamo, J., & Krol, M. (1996). Integrated scenarios of acidification and climate change in Asia and Europe. *Global Environmental Change, 6,* 375–395.
Reis, S., Grennfelt, P., Klimont, Z., Amann, M., ApSimon, H., Hettelingh, J.-P., Holland, M., LeGall, A.-C., Maas, R., Posch, M., Spranger, T., Sutton, M. A., & Williams, M. (2012). From Acid Rain to Climate Change. *Science, 338,* 1153–1154.

Suutari, R., Amann, M., Cofala, J., Klimont, Z., Posch, M. & Schöpp, W. (2001). *From economic activities to ecosystem protection: An uncertainty analysis of two scenarios of the RAINS integrated assessment model.* (IIASA/CCE Report 1/2001). Laxenburg: IIASA.
UN. (1992). *Convention on biological diversity.* Concluded at Rio de Janeiro on 5 June 1992.
Weiss, C. (2006). Can there be science-based precaution? *Environmental Research Letters, 1,* 014003.

Annex: Direct Impacts on Ecosystems and Human Health Induced by Exposure to Ambient Concentrations of Air Pollutants and Related Critical Levels

This annex presents an overview of the direct impacts of elevated air concentrations of sulphur dioxide (SO_2), nitrogen oxides (NO_x), ammonia (NH_3) and NO_x-induced ozone (O_3) on ecosystems and human health and related critical levels. Furthermore it mentions the adverse impacts of fine particulate matter, to which those pollutants also contribute.

Direct Impacts of Air Pollutants

Ecosystem Impacts

Direct effects are caused by exposure of plants (and humans) to ambient concentrations of air pollutants. Excessive concentrations, i.e. exceeding critical levels, have been recognized as an important cause for damage, potentially in addition to critical load exceedance. Sulphur dioxide was earlier viewed as the most important phytotoxic pollutant in Europe, and until the early 1980s was the subject of most research in this field. The extensive forest damage observed in rural areas in Central Europe in the beginning of the 1980s (e.g. Lammel 1984; Schütt et al. 1983) was particularly severe in the Ore Mountains on the border between Germany, Poland and the Czech Republic (formerly called the "Black Triangle"), which were heavily exposed to SO_2. High concentrations of SO_2 can produce acute injury in the form of foliar necrosis, even after relatively short exposure. Furthermore, chronic injury may result from long-term exposure to relative low concentrations of the gas and is essentially cumulative in nature. This may lead to reduced growth and yield as well as increased senescence, often with no clear visible symptoms or with some degree of chlorosis. Forest damage was reported after periods of temperature inversion, when daily SO_2 concentrations in a range of 100–1000 $\mu g\ m^{-3}$ were measured (Materna 1974, 2002). SO_2 appeared to have been the main cause, alone and in synergy with other stressors (Materna 2002).

An important effect of nitrous gases, aerosols and dissolved compounds (NH_3, NO_2, NO, HNO_3 and NH_4^+) is their direct toxicity to the above-ground parts of

W. de Vries et al. (eds.), *Critical Loads and Dynamic Risk Assessments*, Environmental Pollution 25, DOI 10.1007/978-94-017-9508-1 plants, lichens and bryophytes. Studies on crops, saplings and native plant species in open-top chambers (OTCs) and free-air fumigation experiments have demonstrated leaf injury, changes in physiology, and growth reductions at increased concentrations of those N pollutants (e.g. Krupa 2003; Pearson and Stewart 1993; Sheppard et al. 2009). Considering the effects of NH_3, direct uptake through the foliage is the dominant pathway (Cape et al. 2009). For an overview of direct effects of atmospheric ammonia on terrestrial vegetation, see Krupa (2003) and Cape et al. (2009).

At low concentrations, NO_x may enhance the growth of agricultural crops and natural vegetation, although Adaros et al. (1991) observed significant adverse interactive effects of low levels of NO_2 and O_3 on the growth and yield of spring rape. Adverse impacts of NO_x, such as growth reduction, are furthermore known at increased concentrations for both (semi-) natural vegetation and agricultural crops (cf. Van der Eerden et al. 1998; WHO 2000), and may lead to a considerable crop yield reductions in high concentration areas (Van der Eerden and Duym 1988). Furthermore, it may result in forest damage. Natural vegetation is more sensitive to NO_x than agricultural crops. Direct toxicity impacts of NO_2 were observed in parts of Europe and North America in the 1980s, but have become rare in these regions, except in cities or in the direct neighbourhood of roads with heavy traffic. However, concentrations of nitrogen oxides and especially ammonia in air are now increasing in large areas of Asia (primarily China and India) and may lead to direct foliar impacts in these regions (Liu et al. 2011, 2013). Lichens and bryophytes are considered the most sensitive groups in the vegetation with respect to direct toxicity of NH_3 (e.g. Sheppard et al. 2009; Van Herk et al. 2003).

Negative impacts of ozone on terrestrial ecosystems vary between receptors and include growth changes, yield loss, visible injury and reduced seed production. Most studies have been carried out for agricultural (mainly wheat and potato) and horticultural crops (mainly tomato) and forest trees (mainly Norway spruce, birch and beech). Negative impacts of ozone on biomass production and physiological functions have been investigated in a number of experiments, mostly including tree seedlings in open top chambers. Many of these demonstrated relationships between O_3 exposure and reductions in both growth and leaf gas exchange (Matyssek et al. 1995; Novak et al. 2005; Sandermann et al. 1997). In Europe, ambient ozone concentrations in rural and forested areas have been shown to reach concentrations high enough to produce phytotoxic effects in native vegetation and reducing forest productivity (Matyssek and Innes 1999; Skärby et al. 1998; Taylor et al. 1994) causing yield losses of considerable economic importance (Ashmore 2005; King et al. 2005; Morgan et al. 2006).

Health Effects

Apart from ecosystem effects, NO_x also effects human health. NO_2 is an irritant gas, and if inhaled can cause severe damage to the lungs. High indoor NO_2 levels can also induce a variety of respiratory illnesses. The lethal concentration is about

200 ppm. Concentrations above 60–150 ppm can cause cough and burning sensations deep inside the lungs. Damage to the lungs can be visible after 2–24 h. These concentrations are, however, much higher than ambient levels. Continuous exposure to low concentrations of NO_2 can cause cough, headache, loss of appetite and stomach problems. Epidemiological studies have proven that children continuously exposed to NO_2 end up with an increased incidences of breathing diseases and a reduced breathing efficiency (WHO 2003).

Ozone affects human health almost exclusively through inhalation (Von Mutius 2000; WHO 2003). Adverse health impacts that can be initiated and exacerbated by O_3 exposure include coughs and asthma (reactive airways diseases, RAD), short-term reductions in lung function, and chronic respiratory disease (Von Mutius 2000; WHO 2003). Children exercising in environments high in O_3 are 40 % more likely to develop asthma (McConnell et al. 2002). A recent overview of health risks of ozone by the World Health Organisation (WHO) indicates a clear increase in mortality and respiratory morbidity rates based on epidemiological short-term studies on lung function, lung inflammation, lung permeability, respiratory symptoms, morbidity and mortality (WHO 2008).

Furthermore, sulphate, nitrate, ammonium and heavy metals form constituents of fine particulate matter (PM). Multiple studies have shown positive correlations between fine particulate air pollution and cardiovascular diseases, respiratory diseases, asthma, reduced lung function, and overall mortality (Pope et al. 2002). Ambient air pollution, as annual $PM_{2.5}$, has been estimated to account for 3.1 million deaths globally in 2010 (WHO 2013).

Critical Levels for Air Pollutants

Ecosystem Impacts

Similar to critical loads, critical levels for ambient air concentrations have been introduced for the protection of ecosystems. Critical levels refer to "the atmospheric concentrations of pollutants in the atmosphere above which adverse effects on receptors, such as human beings, plants, ecosystems or materials, may occur according to present knowledge" (Mapping Manual; www.icpmapping.org). Critical (concentration) levels of SO_2, NO_x, NH_3 and O_3 have been derived for vegetation from a compilation of literature on dose (i.e. pollutant concentration × duration of exposure) response relationships. Critical levels for these pollutants were first defined at UNECE workshops in Bad Harzburg (Germany) in the late 1980s and have been further elaborated since then (e.g. Ashmore and Wilson 1994), with a major focus on ozone. Critical levels are described in different ways for different pollutants, including mean annual or monthly/daily concentrations (SO_2, NO_x and NH_3) and as cumulative exposures or fluxes through plant stomata (O_3). A summary of critical levels is given below, in accordance with the Mapping Manual (www.icpmapping.org).

Table A1 Critical levels (μg m^{-3}) in view of vegetation impacts used for short-term and long-term exposures to SO_2, NO_x and NH_3. (Based upon Mapping Manual (www.icpmapping.org))

Exposure	Vegetation	SO_2	NO_x	NH_3
Long term (Annual mean)	Lichens and bryophytes[a]	10	–	1
	Higher plants	20–30[b]	30[c]	3[c]
Short term[c]	Higher plants[d]	–	75	23

[a] Includes ecosystems where lichens and bryophytes are key parts of the ecosystem
[b] Limit is 20 μg m^{-3} for semi-natural vegetation and forests and 30 g m^{-3} for arable crops
[c] For NO_x it includes semi-natural vegetation. For NH_3 it includes heathland, grassland and forest ground flora. A range is suggested here of 2–4 μg m^{-3}
[d] 24-h mean for NO_x and monthly mean for NH_3

Critical Levels for Sulphur Dioxide, Nitrogen Oxides and Ammonia A summary of critical levels of SO_2, NO_x and NH_3related to acute effects by short-term exposures (daily or monthly means) and to chronic effects by long-term exposures (annual means) is given in Table A1. For NO_x separate critical levels have not been set for different vegetation classes because of a lack of information. The sensitivity is, however, thought to decrease according to (semi-)natural vegetation > forests > crops. Critical levels are related to both growth stimulation in response to the fertiliser effect of nitrogen, and adverse physiological effects at toxic levels. The critical level for NO_x is based on the sum of NO and NO_2 concentrations. The knowledge to establish separate critical levels for the two gases is still lacking, even though there is evidence that NO is more phytotoxic than NO_2 (Morgan et al. 1992). Critical levels for NO_x and NH_3 are not defined for intensively managed agricultural grasslands (pastures) and arable crops, which are often sources rather than sinks of ammonia.

Critical Levels for Ozone Unlike critical levels for NO_x and NH_3, which are described in terms of mean annual, monthly or daily concentrations, critical levels for O_3 are defined as cumulative exposures or fluxes through plant stomata. Until recently, the main indicator used (e.g. in the Gothenburg Protocol) to protect crops, trees and semi-natural vegetation was the AOT40 (accumulated hourly ozone concentration over the threshold of 40 ppb), which is a concentration based critical level. More precisely, AOT40 is derived by the sum of the hourly ozone concentrations during daylight hours (defined as the time with global radiation >50 W m^{-2}) above a threshold value of 40 parts per billion (ppb) during the growing season. Based on a typical exposure duration, the AOT40 is calculated for crops over 3 months (e.g. May–July) and for forest trees over 6 months (April–September). The threshold of 40 ppb is based on a reduction in annual biomass increment. The use of daylight hours only is based on the observation that ozone uptake during night time is negligible due to the closure of stomatal pores (Ashmore and Wilson 1994). This accumulated ozone exposure over a threshold of 40 ppb, called AOT40, is expressed in ppm hours. Critical AOT40 levels, corresponding to a 5% reduction in yield or biomass, were already derived in the early 1990s and have been updated and expanded since then, as summarized in Table A2 (see also Mapping Manual; www.icpmapping.org).

Table A2 Critical levels used for concentration based exposures (AOT40) and flux based exposures (POD_1, POD_3 or POD_6) to O_3

Vegetation	Effect considered[a]	O_3 exposure term	
		AOT40 (ppm h)[b]	POD (mmol m^{-2})[c]
Crops			
Wheat	5 % yield reduction	3	POD_6 of 1
Potato	5 % yield reduction	–	POD_6 of 5[d]
Tomato	5 % yield reduction	6	POD_6 of 2
Generic crop	5 % yield reduction	–	POD_3 of 3[e]
Forests			
Coniferous trees (Norway spruce)	2 % growth reduction	5[d]	POD_1 of 8[f]
Deciduous trees (birch and beech)	4 % growth reduction	5[d]	POD_1 of 4[f]
Semi-natural vegetation			
Grasslands	10 % reduction in biomass production	5	POD_1 of 2[g]

[a] Levels aimed to protect against a 5 % reduction in yield or biomass
[b] Based on Mapping Manual (www.icpmapping.org)
[c] Based on Harmens et al. (2010)
[d] A correction ofthe value of 4, as given Harmens et al. (2010), corrected in UNECE document ECE/EB.AIR/WG.1/2010/13/Corr.1
[e] Based on a 5 % yield reduction, using results presented in Harmens et al. (2010)
[f] Using 5% (10 %) growth reduction as the effect considered leads to a POD_1 of approximately 20 (40) for Norway spruce and 5 (10) for beech/birch
[g] This refers to both productive grasslands and grassland areas of high conservation value for which *Trifolium* spp. (clover species) are used as an indicator species

Considering the above mentioned drawbacks of AOT40, the accumulated ozone flux via plant stomata, referred to as Phytotoxic Ozone Dose above a threshold of Y, POD_Y, is currently considered to provide a biologically sound method for describing observed effects. This stomatal flux-based critical level for ozone is calculated from the effects of climate (temperature, humidity, light), ozone, soil (moisture availability) and plant development (growth stage) on the extent of opening of the stomatal pores on leaf surfaces through which ozone enters the plant (e.g. Ashmore et al. 2004). In this approach, the hourly mean stomatal flux of ozone is accumulated over a stomatal flux threshold of Y nmol $m^{-2}s^{-1}$(Mapping Manual; www.icpmapping.org). The flux is based on the projected leaf area (PLA), being the total area of the sides of the leaves that are projected towards the sun. The stomatal flux-based POD (in mmol m^{-2} PLA) is further divided on the basis of stomatal flux threshold of either 1 or 6 nmol m^{-2} s^{-1} (POD_1 or POD_6). For (semi-)natural vegetation, flux-based response relationships are strongest when there is a small threshold above which flux is accumulated (POD_1), whereas a high threshold gives strongest responses for agricultural and horticultural crops (POD_6). Critical POD values thus derived are also given in Table A2. In line with earlier concepts used for crop critical

Table A3 Limit values (μg m^{-3}) of NO_2, O_3 and particulate matter (PM) as defined in (i) the European Air Quality Directive 2008/50/EC (2008) for the protection of human health and (ii) the air quality standards as recommended by the WHO

Period	Type of limit	NO_2	O_3	PM_{10}	$PM_{2.5}$
Calendar year	EU	40[a]	–	40[a]	25[a]
	WHO	40	–	20	10
Short term[b]	EU	200[a]	120[a]	50[a]	–
	WHO	200	100	50	25

[a] The EU limit values for NO_2, $O_3$and PM_{10} are legally binding since 1 January 2010; the $PM_{2.5}$ value is set as target value since 2010 and will become legally binding by 2015
[b] The values refer to an hourly average for NO_2, the highest 8 hour average during a day for O_3 and a daily average for PM_{10} and $PM_{2.5}$. Exceedances of the EU limit/target values are allowed for 18 h (NO_2), 25 days (O_3) and 35 days (PM_{10})

levels, a 5 % yield reduction was used as the loss criterion for the identification of stomatal flux-based critical levels.

Health Effects

A summary of critical levels of NO_2, O_3, PM_{10} and $PM_{2.5}$, based on the European Air Quality Directive, related to acute effects due to short-term exposure and to chronic effects due to long-term exposure (annual means) is given in Table A3. There are no critical levels for NO in terms of human health, as toxic values are far above ambient air quality levels. For NO_2, the EU limit value for the annual average concentration of 40 μg m^{-3} appears to be the most stringent, and its attainment generally implies achievement of the hourly limit value. For O_3, the reference level is limited to short-term exposure, although there is new epidemiological evidence for long-term impacts of O_3 (Jerrett et al. 2009), but this evidence is still too limited for setting a critical level. A measure to assess the risk of repeated exposure to elevated daily O_3 concentrations is SOMO35, defined as the sum of maximum daily 8-h average O_3 levels above 35 ppb (70 μg m^{-3}) during a year. It is assumed that there is no no-effect level for SOMO35, i.e. effects occur whenever SOMO35 is above zero. There are indications that health effects of PM_{10} or $PM_{2.5}$ are to be expected by exposures to levels above the WHO air quality guidelines of 20 (for PM_{10}) and 10 μg m^{-3} (for $PM_{2.5}$). The EU target values are less stringent than the WHO air quality standards (Table A3). At annual mean levels of about 33 μg m^{-3} the attainment of the daily limit value can be expected. The $PM_{2.5}$ value of 25 μg m^{-3} is a target value for 2010 and a binding limit value as of 2015. From 2020 onwards a provisional limit value of 20 μg m^{-3} is foreseen for $PM_{2.5}$.

References

Adaros, G., Weigel, H. J., & Jager, H. J. (1991). Single and interactive effects of low levels of O_3, SO_2 and NO_2 on the growth and yield of spring rape. *Environmental Pollution, 72,* 269–286.

Ashmore, M. R., & Wilson, R. B. (1994). *Critical levels of air pollutants for Europe*. Background papers prepared for ECE workshop on critical levels. Egham, UK, 23–26 March 1992 UK Department of the Environment.

Ashmore, M. R., Emberson, L., Karlsson, P. E., & Pleijel, H. (2004). New directions: A new generation of ozone critical levels for the protection of vegetation in Europe. *Atmospheric Environment, 38,* 2213–2214.

Ashmore, M. R. (2005). Assessing the future global impacts of ozone on vegetation. *Plant Cell and Environment, 28,* 949–964.

Cape, J. N., Van der Eerden, L. J., Sheppard, L. J., Leith, I. D., & Sutton, M. A. (2009). Evidence for changing the critical level for ammonia. *Environmental Pollution, 157,* 1033–1037.

Harmens, H., Mills, G., Hayes, F., Norris, D., & The Participants of the ICP Vegetation. (2010). *Air pollution and vegetation* (ICP Vegetation Annual report 2009/2010). Centre for Ecology & Hydrology, Bangor, UK.

Jerrett, M., Burnett, R. T., C. Arden Pope, I., Ito, K., Thurston, G., Krewski, D., Shi, Y., Calle, E., & Thun, M. (2009). Long-term ozone exposure and mortality. *New England Journal of Medicine, 360,* 1085–1095.

King, J. S., Kubiske, M. E., Pregitzer, K. S., Hendrey, G. R., McDonald, E. P., Giardina, C. P., Quinn, V. S., & Karnosky, D. F. (2005). Tropospheric O_3 compromises net primary production in young stands of trembling aspen, paper birch and sugar maple in response to elevated atmospheric CO_2. *New Phytologist, 168,* 623–636.

Krupa, S. V. (2003). Effects of atmospheric ammonia (NH_3) on terrestrial vegetation: A review. *Environmental Pollution, 124,* 179–221.

Lammel, R. (1984). Endgültige Ergebnisse und bundesweite Kartierung der Waldschadenserhebung 1983. *AFZ-Der Wald, 39,* 340–344.

Liu, X. J., Duan, L., Mo, J. M., Du, E., Shen, J., Lu, X., Zhang, Y., Zhou, X., He, C., & Zhang, F. S. (2011). Nitrogen deposition and its ecological impact in China: An overview. *Environmental Pollution, 159,* 2251–2264.

Liu, X., Zhang, Y., Han, W., Tang, A., Shen, J., Cui, Z., Vitousek, P., Erisman, J. W., Goulding, K., Christie, P., Fangmeier, A., & Zhang, F. (2013). Enhanced nitrogen deposition over China. *Nature, 494,* 459–462.

Materna, J. (1974). Einfluss der SO_2-Immisionen auf Fichtenpflanzen in den Wintermonaten. In *IX. Int. Tagung. Luftverureinigung und Forstwirtschaft. Marianske Lazné* (pp. 107–114).

Materna, J. (2002). Impact of air pollution on forests. In B. Lomský, J. Materna, & H. Pfanz (Eds.), *SO_2-pollution and forest decline in the ore mountains* (pp. 117–138). Jiloviste-Strnady: Ministry of Agriculture of the Chech Republic. Forestry and Game Management Research Institute.

Matyssek, R., Günthardt-Goerg, M. S., Maurer, S., & Keller, T. (1995). Nighttime exposure to ozone reduces whole-plant production in betula-pendula. *Tree Physiology, 15,* 159–165.

Matyssek, R., & Innes, J. L. (1999). Ozone—a risk factor for trees and forests in Europe? *Water Air & Soil Pollution, 116,* 199–226.

McConnell, R., Berhane, K., Gilliland, F., London, S. J., Islam, T., Gauderman, W. J., Avol, E., Margolis, H. G., & Peters, J. M. (2002). Asthma in exercising children exposed to ozone: A cohort study. *Lancet, 359,* 386–391.

Morgan, P. B., Mies, T., Bollero, G., Nelson, R. L., & Long, S. P. (2006). Season-long elevation of ozone concentration to projected 2050 levels under fully open-air conditions substantially decreases the growth and production of soybean. *New Phytologist, 170,* 333–343.

Morgan, S. M., Lee, J. A., & Ashenden, T. W. (1992). Effects of nitrogen oxides on nitrate assimilation in bryophytes. *New Phytologist, 120,* 89–97.

Novak, K., Schaub, M., Fuhrer, J., Skelly, J. M., Hug, C., Landolt, W., Bleuler, P., & Kräuchi, N. (2005). Seasonal trends in reduced leaf gas exchange and ozone-induced foliar injury in three ozone sensitive woody plant species. *Environmental Pollution, 136,* 33–45.

Pearson, J., & Stewart, G. R. (1993). The deposition of atmospheric ammonia and its effects on plants. Tansley Review 56. *New Phytologist, 125,* 283–305.

Pope, C. A. III, Burnett, R. T., Thun, M. J., Calle, E. E., Krewski, D., Ito, K., & Thurston, G. D. (2002). Lung cancer, cardiopulmonary mortality, and long-term exposure to fine particulate air pollution. *The Journal of the American Medical Association, 287,* 1132–1141.

Sandermann, H., Wellburn, A. R., & Health, R. L. (1997). Forest decline and ozone: Synopsis. In *Forest decline and ozone: A comparison of controlled chamber and field experiments*. Ecological Studies Series 127 (pp. 369–377). Berlin: Springer-Verlag.

Schütt, P., Blaschke, H., Hoque, E., Koch, W., Lang, K. J., & Schuck, H. J. (1983). Erste Ergebnisse einer botanischen Inventur des 'Fichtensterbens'. *Forstwissenschaftliches Centralblatt, 102,* 177–186.

Sheppard, L. J., Leith, I. D., Crossley, A., van Dijk, N., Cape, J. N., Fowler, D., & Sutton, M. A. (2009). Long term cumulative exposure exacerbates the effects of atmospheric ammonia on an ombrothrophic bog: Implications for critical levels. In M. A. Sutton, S. Reis, & S. M. H. Baker (Eds.), *Atmospheric ammonia—Detecting emission changes and environmental impacts*. Springer Science.

Skärby, L., Ro-Poulsen, H., Wellburn, F. A. M., & Sheppard, L. J. (1998). Impacts of ozone on forests: A European perspective. *New Phytologist, 139,* 109–122.

Taylor, G. E., Johnson, D. W., & Andersen, C. P. (1994). Air pollution and forest ecosystems: A regional to global perspective. *Ecological Applications, 4,* 662–689.

Van der Eerden, L. J., & Duym, N. (1988). An evaluation method for combined effects of SO_2 and NO_2 on vegetation. *Environmental Pollution, 53,* 468–470.

Van der Eerden, L. J. M., de Visser, P. H. B., & van Dijk, C. J. (1998). Risk of damage to crops in the direct neighbourhood of ammonia sources. *Environmental Pollution, 102,* 49–53.

Van Herk, C. M., Mathijssen-Spiekman, E. A. M., & de Zwart, D. (2003). Long distance nitrogen air pollution effects on lichens in Europe. *Lichenologist, 35,* 347–359.

Von Mutius, E. (2000). The environmental predictors of allergic disease. *Journal of Allergy and Clinical Immunology, 105,* 9–19.

WHO. (2003). *Health aspects of air pollution with particulate matter, ozone and nitrogen dioxide*. Report on a WHO Working Group; Bonn, Germany; 13–15 January 2003. WHO, Copenhagen.

WHO. (2008). *Health risks of ozone form long-range transboundary air pollution* (Report of World health organisation). Regional Office for Europe, Copenhagen, Denmark.

WHO. (2013). *Health effects of particulate matter: Policy implications for countries in eastern Europe, the Caucasus and central Asia* (Report of the World Health Organization). Denmark Regional Office for Europe, Copenhagen.

Index

W. de Vries et al. (eds.), *Critical Loads and Dynamic Risk Assessments*, Environmental Pollution 25, DOI 10.1007/978-94-017-9508-1

MIX
Papier aus verantwortungsvollen Quellen
Paper from responsible sources
FSC® C105338

If you have any concerns about our products,
you can contact us on
ProductSafety@springernature.com

In case Publisher is established outside the EU,
the EU authorized representative is:
Springer Nature Customer Service Center GmbH
Europaplatz 3, 69115 Heidelberg, Germany

Printed by Libri Plureos GmbH
in Hamburg, Germany